AF477939

The Santo 2006 expedition was organized,
with the support, among others, of

and

Santo

edited by
Philippe Bouchet, Hervé Le Guyader & Olivier Pascal

Institut de recherche
pour le développement

Publications scientifiques du Muséum national d'Histoire naturelle

Cette publication constitue le volume **70** de la collection Patrimoines Naturels

Directeur de la publication : Thomas Grenon
Directeur général du Muséum national d'Histoire naturelle

Rédacteur en chef : Jean-Philippe Siblet
Secrétaire de rédaction : Gwénaëlle Chavassieu

Photos 1ʳᵉ de couverture :

a	b	c	d	e

a: photo O. Pascal
b: photo X. Desmier
c: photo C. Rives
d: photo E. Boitier
e: photo C. Rives

Photos 4ᵉ de couverture :

f	g	h	i	j
k	l	m	n	

f : *Pandanus nogarete* (photo J.-N. Labat)
g : *Figulus foveicollis* (photo A. Tishechkin)
h : *Trapezia rufopunctata* (photo T.-Y. Chan)
i : *Bulbophyllum stenophyllum* (photo J.-N. Labat)
j : *Halimeda macroloba* (photo J.-L. Menou IRD Nouméa)
k : *Hebridea rufotibialis* (photo S. Hugel)
l : *Micromelo undatus* (photo Y. Camacho)
m : *Sturanya* cf. *albescens* (photo O. Gargominy)
n : *Emoia cyanura* (photo I. Ineich)

Design : Catherine Lasnier

ISSN 1281-6213
ISBN MNHN 978-2-85653-627-8
ISBN IRD 978-2-7099-1708-7

This volume is best cited as follows/ Cet ouvrage doit être référencé comme suit :
BOUCHET P., LE GUYADER H. & PASCAL O. (Eds) 2011. — *The Natural History of Santo.* Muséum national
d'Histoire naturelle, Paris ; IRD, Marseille ; Pro-Natura international, Paris, 572 p. (Patrimoines naturels ; 70).

Edward Natapei
Prime Minister of Vanuatu

As an island state, Vanuatu has a biodiversity that is unique in many ways, with many endemic species of fauna and flora. The Government of Vanuatu, through its Ministry of Lands and Natural Resources, which is responsible for biodiversity management, has given its full support to the Santo Global Biodiversity Survey. We, the people of Vanuatu, are grateful to the expedition team for conducting this research on an island of our country. We live in a world where development is accelerating to the detriment of the biodiversity. Vanuatu is not immune from that, and this is why, as unique as we are, we are urged to know what our biodiversity has to offer to the world.

This biodiversity expedition conducted on the largest of the Vanuatu Islands has shown us how much species bounty a single island can hold. This makes us wonder how much there is in the whole of the archipelago. Today we have a book that all Vanuatu citizens can be proud of. This book will not only make Vanuatu known to the world of science, it will also make Vanuatu known to the rest of the world. The results produced in this book will not only help the country move forward in establishing its long term planning and policies, but it will also help us to respond more clearly to international requirements and reporting as requested by the international community as a whole.

Beyond contributing greatly to knowledge of the whole of Vanuatu in terms of its biodiversity, *The Natural History of Santo* will also impact education. It will be used as a mean of thought to our present but mostly to our young generations for their learning path, and will boost their interests in taking up studies in the field of sciences.

It is with great honour and on behalf of the people of Vanuatu that I would like to thank all the Santo 2006 expedition team for producing such a state of the art book that will contribute immensely towards the development of Vanuatu as a country.

The Natural History of Santo: An Attempt to Bridge the Gap between Academic Research and Conservation and Education

When the description of the gecko *Lepidodactylus buleli* was published in October 2008, one of the journalists that reported the discovery in Vanuatu entitled his press release "Better later than never". Yet, less than two calendar years had elapsed between the collecting of suspect eggs in the forests of Penaoru, their rearing in captivity by reptile buffs, the recognition of a new species and its description by Ivan Ineich, and finally its publication in the research journal Zootaxa. To an academic research scientist, this was a remarkably swift sequence of events. To a lay person, this is an agonizingly long period of time. We live in an age of immediacy, and the journalist's "Better later than never" epitomizes the difficulty in reconciling the pace of academic research with that of environmental decision-making.

Historically, the time-proven approach to documenting biodiversity is undoubtedly that of taxonomical inventories. Taxonomists travel the world to discover species, document where they live, name them and establish their classification. To a taxonomist, "every species counts". They return to their "home" institution with specimens of taxa that have attracted their attention for one reason or another (suspected new species, rare or seldom seen species, population with unusual variation, etc.).

After two and half centuries of such exploration, taxonomists have successfully documented around 1.9 million species, and continue to describe new species at the pace of 17 000 per year. By the end of the 19th century, the big picture of biogeographical realms was already clear: the "Southern Seas" formed a single biogeographical marine province (part of the vast Indo-West Pacific region), while each island or island group was rich in terrestrial endemics.

The quest to delineate the finer details of this big picture continues to this day, and the Santo 2006 Global Biodiversity Survey was part of this long tradition. But we now live in an age of environmental anxiety, and taxonomists are not good at delivering in a timely fashion facts and data that are meaningful for nature management and conservation.

Taxonomists are obsessed with species and their names, and they consider their work "done" only when the last specimen of the last sample has been bestowed with a species identification. As a result, it takes years —many years— for taxonomists to deliver their findings: this time lag is part of what has been called the "taxonomic impediment". Ten years, even five years, is not a time frame that fares well in our age of immediacy: managers, funding bodies, decision makers, like to have "immediate" results.

These limitations of taxonomical work have paved the way for a new approach to biodiversity research and monitoring, i.e. biodiversity assessments. Conservationists need "immediate" science-based facts to inspire decisions and policy on land and sea use and management. But conservationists are daunted by the magnitude of the biodiversity they want to highlight, promote, and conserve. The "taxonomic impediment" is real. As a consequence, biodiversity assessments focus on a few selected taxa for which there is the knowledge and work force to identify them on the spot: birds and mammals, trees, reef corals and fishes, and, at best, a handful of charismatic invertebrates such as butterflies and dung beetles. The term Rapid Biological Assessment has been coined for this approach. Typically, conservationists leave the field with "data" and species lists for selected taxa, and (very) marginally with specimens. On a global scale, the Rapid Assessment approach has been successful in highlighting areas of conservation interest, in raising and disseminating environmental awareness, and in bringing together the worlds of public agencies (the World Bank, USAID, etc.), private funding (corporate and foundations) and public opinion. But Vanuatu does not have any of the charismatic vertebrates that, elsewhere, are the flagships of conservation (e.g. primates, cats, or even crocodiles), and this may be the reason why none of the major international NGOs are currently operating in the country.

The present introduction is adapted from:
BOUCHET P., LE GUYADEr H. & PASCAL O. 2009. — The SANTO 2006 Global Biodiversity Survey: an attempt to reconcile the pace of taxonomy and conservation. *Zoosystema* 31(3): 401-406.

Yet, serious science-based conservation in the South Pacific cannot afford to ignore, e.g., the snails, the weevils or the geckos, all of which have astonishingly high levels of single-island or island-group endemism, and are threatened by loss of habitat and the spread of aliens.

It is precisely this gap between academic biodiversity exploration and operational conservation that the Santo expedition has attempted to bridge. In this, *The Natural History of Santo* was inspired by Tony Whitten's *Ecology of Indonesia* series. To a large extent, it is the book we all would have liked to find at the onset of the expedition. Our ambition is that it will emulate companion volumes throughout the South Pacific, and stimulate awareness of, and interest for, the "neglected" components of biodiversity.

Philippe Bouchet, Hervé Le Guyader, Olivier Pascal

Directors, Santo 2006 Global Biodiversity Survey.

Vanuatu in the South Pacific

Benoît Antheaume

Previously known as the Anglo-French Condominium of the New Hebrides set up in 1906, Vanuatu, "Our Land Forever", is one of the newest sovereign state in the world, born in 1980. Broadly Y shaped, the Vanuatu archipelago is made of 13 principal and many small islands, extending from North to South for about 850 km in the South-West of the Pacific Ocean. Vanuatu covers a land area of 12190 km² and an Exclusive Economic zone of 680000 km². With Papua New Guinea, the Solomon Islands, New Caledonia and Fiji, Vanuatu belongs to the subregion of Oceania called Melanesia, "black islands", distinct from Polynesia, "multiple islands" and Micronesia, "very small islands", which are the three main anthropological grouping of islands in the South Pacific.

Located 800 km west of Fiji and respectively 540 and 1770 km east of New Caledonia and Australia, the archipelago of Vanuatu is made of rugged mountains, plateaus, coastal terraces and offshore coral reefs. With 35% of the land above 300 m and the highest mountain (Mt Tabwemasana, 1877 m) on Espiritu Santo, the largest island, it shows that

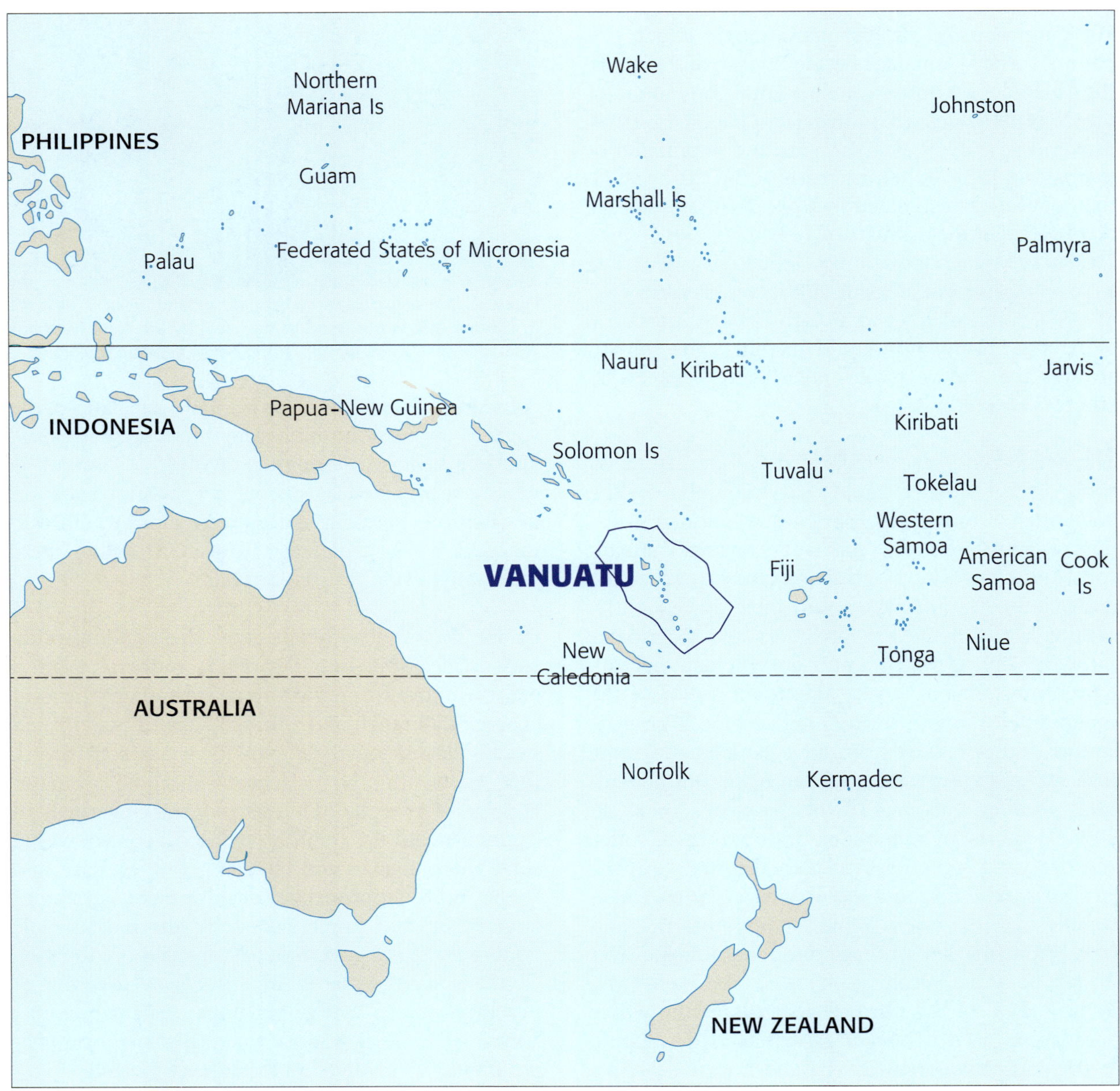

Vanuatu consists mainly of high islands, and counts no atolls except in the North (Reef Island) and in the South (Aniwa).

Active volcanoes are to be found on the islands of Tanna (Mt Yasur), Ambrym (Mt Benbow and Mt Marium) ,Matthew and Hunter, notwithstanding several submarine volcanoes. Because Vanuatu is located on the margin of the Pacific plate where it collides with the Indo-Australian plate, the country experiences significant seismic activity, with earthquakes of a magnitude up to 7.4 in 2010.

Vanuatu's climate is tropical (from wet in the North to sub-tropical in the South). There are severe contrasts between seasons: May to October being the dry season with fresh south- easterly breezes producing fine days and cool nights; November to April being the wet season, when higher temperatures, heavy rains and occasional cyclones are experienced. Average rainfall is over 2 m in Port Vila and over 3 m in Luganville!

Mainly Melanesian, the people of Vanuatu (called Ni-Vanuatu or, simply, Ni-van) speak more than 100 languages, beside Bislama, a pidgin which is a national official language beside English and French. Just before becoming Vanuatu, the population of the New Hebrides accounted for 110000 in 1979. The end-year 2009 estimate gave the population of Vanuatu as 243000, an increase of 54000 over the total of 186000 estimated in 1999, itself an increase of 44000 over the estimate of 142000 in 1989. Some 76% of the total population of Vanuatu was recorded as residing in rural areas in 2009, compared to 78.5 in 1999, 82% in 1989 and 84% in 1979. Almost 50% of all agricultural household members in Vanuatu are less than 20 years old, 44% being between 20 and 59 years old.

The GDP (Purchase Power Parity) was in 2009 about US$ 4700 per capita. The four pilars of the Vanuatuan economy are agriculture, tourism, offshore financial services, and cattle raising. Vanuatu constitution rules land rights: "All land belongs to the indigenous custom owners and their descendants", making adjustments for foreigners and expatriates through a leasing system under governement control. Agriculture remains a very important sector of the economy of Vanuatu. Household income from agriculture comes mainly from crop gardens and some cash crops. Exports are made up mainly of agricultural products (about 73% of the total exports in 2007). Exports are dominated by copra, kava, coffee and beef, and sometimes cocoa and timber. But the level of production and exports have varied considerably over the years. Value for imported rice and wheat flour in Vanuatu has more than doubled over the last ten years, meaning imported food is covering by now 27% of the nutritional needs of the whole population, even if market gardening is developing in the peri-urban areas of Vila and Luganville. As a

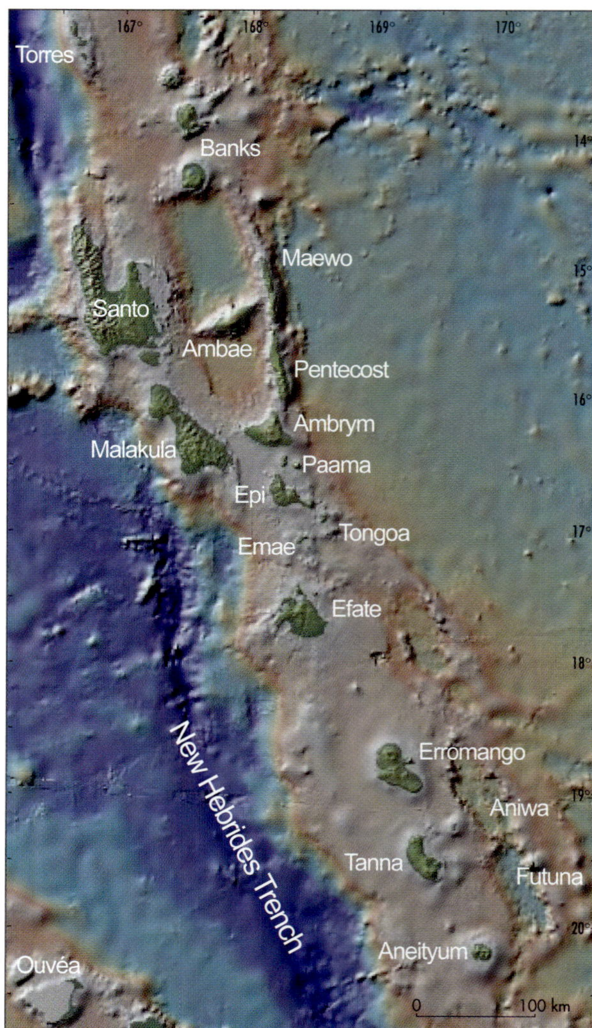

Figure 2: The archipelago of Vanuatu extends over 7 degrees of latitude and is bordered to the west by the New Hebrides Trench. Source: GeoMapApp, http://www.geomapapp.org.

consequence, external trade is totally imbalanced, the value of imports outnumbering the value of imports. Tourism brings in foreign exchange, as Vanuatu is widely recognized as one of a premium vacation destination. Foreign visitors were close to 200000 in 2008, a sharp increase from 2000, when there were only 57000 tourist arrivals.

Under the Anglo-French rule, the condominium was divided into four Districts (Southern, Central No. 1, Central No. 2, Northern) each under the joint charge of a French and a British agent, Port Vila and Luganville having their own town advisory board. Five years after New Hebrides became Vanuatu, Districts were replaced by eleven Local Governement Councils, with the ambition that local control would solve the injustices and imbalances of the past, and shape more appropriate development strategies targeted at, and implemented by, the local people under the central government's supervision. In 1994, Councils were replaced by six Local Government Provinces (Tafea, Shefa, Malampa, Penama, Sanma, Torba) with the view, once again, that the local people should be involved in local development.

The Republic of Vanuatu is a parliamentary democracy, with President, Prime Minister and Parliament. Besides national authorities, Vanuatu also has high-placed people (chiefs and leaders) at the village level, giving a sense of place to the locals as Vanuatu governement is still believing in community-based development. The constitution provides for the establishment of village courts to deal with questions of customary law. The constitution also gives clear recognition to the rights and interests of traditional landholders, who remain the principal managers and users of biological resources and systems in Vanuatu.

Vanuatu has joined several multinational institutions. Since 1980, Australia, the United Kingdom (until 2005), France, and New Zealand have provided the bulk of Vanuatu's development aid. However, more recently China has been providing increasing amounts of aid but Australia is still providing most of the external assistance.

In terms of international biodiversity networks and conventions, Vanuatu is a member of the Secretariat of the Pacific Regional Environment Programme (SPREP; http://www.sprep.org/members/map.htm), or *Programme Régional Océanien de l'Environnement* (http://www.sprep.org/Francais/PROE.htm).
Vanuatu is, since 1989, a Contracting Party to the Convention on the International Trade in Endangered Species of Wild Fauna and Flora (CITES; http://www.cites.org/). Vanuatu also ratified the Convention on Biological Diversity in 1993 (http://www.cbd.int/countries/?country=vu), under which it completed its National Biodiversity Conservation Strategy in 1999, and submitted its Third national report in 2006. Vanuatu is not yet a contracting party to the Ramsar Convention on Wetlands, nor to the Global

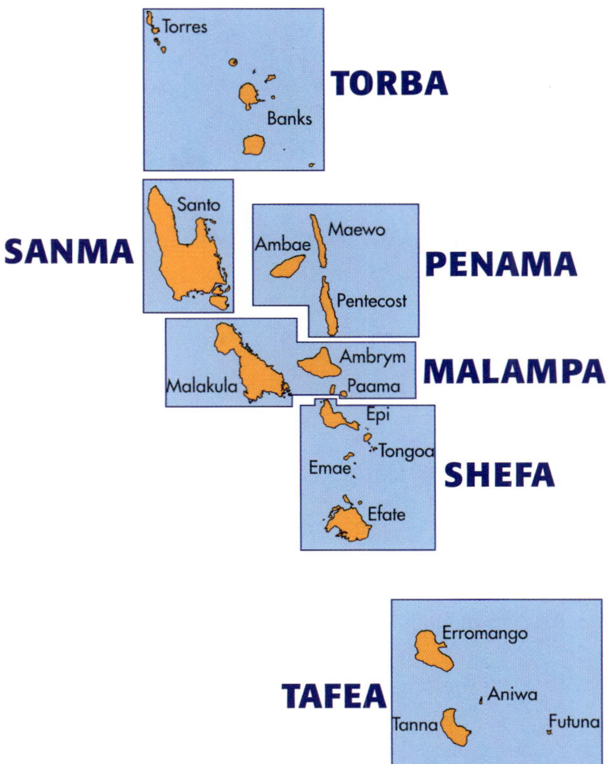

Figure 3: Administrative divisions of Vanuatu. The province of Sanma derives its name from the islands of SANto and MAlo.

Biodiversity Information Facility (GBIF).
Vanuatu, the Solomon Islands, the Bismarck and Admiralty Islands (Papua New Guinea) constitute *Conservation International*'s East Melanesian Islands Biodiversity Hotspot (Fig. 4, below). Together with Solomon's Santa Cruz Islands (Temotu Province), Vanuatu also constitutes one of *Birdlife International's* Endemic Bird Area.

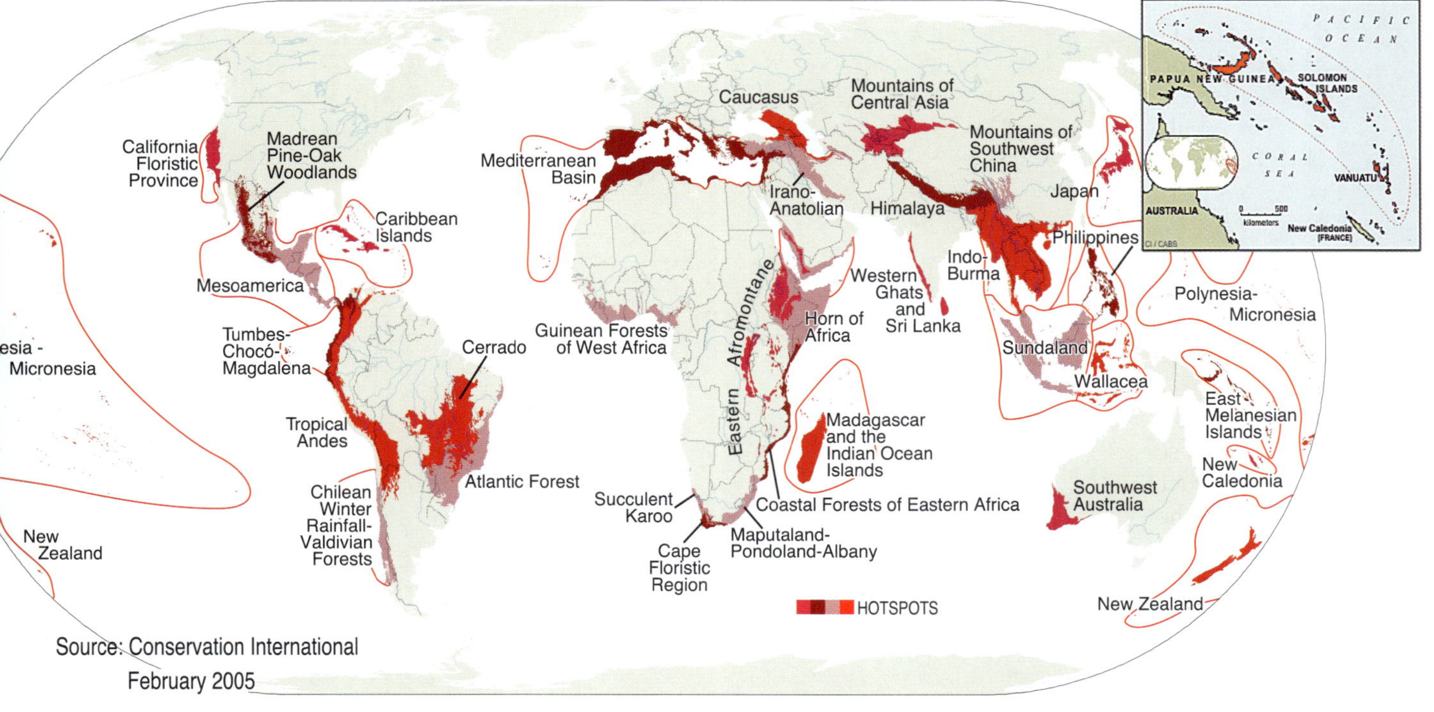

Source: Conservation International
February 2005

Espiritu Santo

in Space and Time

coordinated by Bruno Corbara

Quaternary Reefs

Guy Cabioch & Frederick W. Taylor

The Vanuatu island arc, at the convergent boundary of the Australian and Pacific plates, is characterized by some uncommon features in its central part. The central part of the Vanuatu arc is characterized by the absence of a physiographic trench where ridges and seamounts, the d'Entrecasteaux zone in particular, intersect the arc and are subducted. Moreover, the islands of Santo and Malakula occur anomalously close to the plate boundary and undergo rapid Holocene uplift rate up to 5.5 mm.yr^{-1} at south-west Santo and up to 3.4 mm.yr^{-1} in northern Malakula. Another characteristic is the presence of three parallel chains of islands:

- The western chain (Santo and Malakula);
- The central chain corresponding to active volcanoes (Aoba and Ambrym);
- The eastern chain (Maewo and Pentecost).

Eastern and southern Santo is covered by a series of raised reef terraces. The eastern part is topped by a broad reef plateau (Fig. 5). Several studies of these terraces have investigated various aspects of the island's neotectonics and paleoclimate.

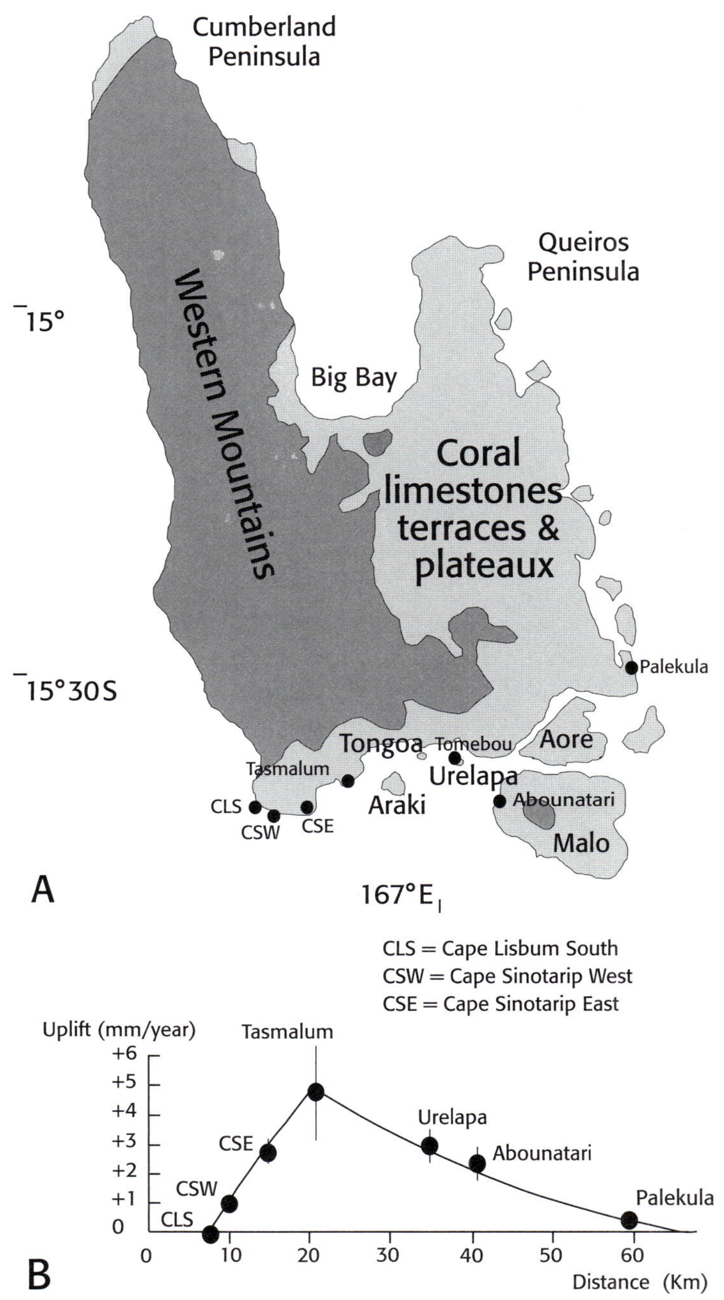

Figure 5: **A**: Location of uplifted coral reefs in Santo and location of the sites quoted in the text. **B**: Holocene uplift rates along the south coast of Santo (modified after Taylor *et al.* 1980).

NEOTECTONICS

Coral reefs can be used as recorders of neotectonic movements. (Fig. 7). Most previous studies devoted to coral reefs in Vanuatu were published in the 1970s and 1980s on this topic. In Santo, the uplifted coral reefs mainly occur in the eastern half of the island offlapping broad plateaux and in the south including the islands of Aore, Malo, Araki, Tangoa and Urelapa (Figs 5 & 8). The altitude of these terraces reaches a maximum of 784 m at Mt Tankara in the center of the island. Generally, their altitude increases to the west and reveals a tilting down to the east. The Holocene uplift rate displays a maximum of ~6-7 mm.yr^{-1} just west of Tasmalum and decreases both to the west and the east (Fig. 5).

Uranium-series and ^{14}C dating of both subsurface and surface samples provides information on the tectonic behavior recently analyzed and interpreted by Taylor and coauthors in 2005. Several sites from west to east were examined including Tasmalum, Araki Island, Urelapa Island, Tomebou Hill and Malo Island (Fig. 5).

At Tasmalum, the 6 ka (1 ka = 1 000 years) mid-Holocene reef flat is at an altitude of 35 m (Fig. 6A)

indicating an uplift of 5.5 mm.yr[-1]. Behind this reef flat, two corals from a terrace extending in altitude from 35 to 45 m are dated as 210.9 to 222.1 ka, (Fig. 6). The highest reef terraces behind Tasmalum, reaching about 400 m elevation, are thought to be in the 100-130 ka age range by correlation to dated localities on Araki Island, Malo Island, and southeastern Santo. This older reef corresponds to the interglacial sea level highstand called Marine Isotope Stage (MIS) 7c (216 ka) when

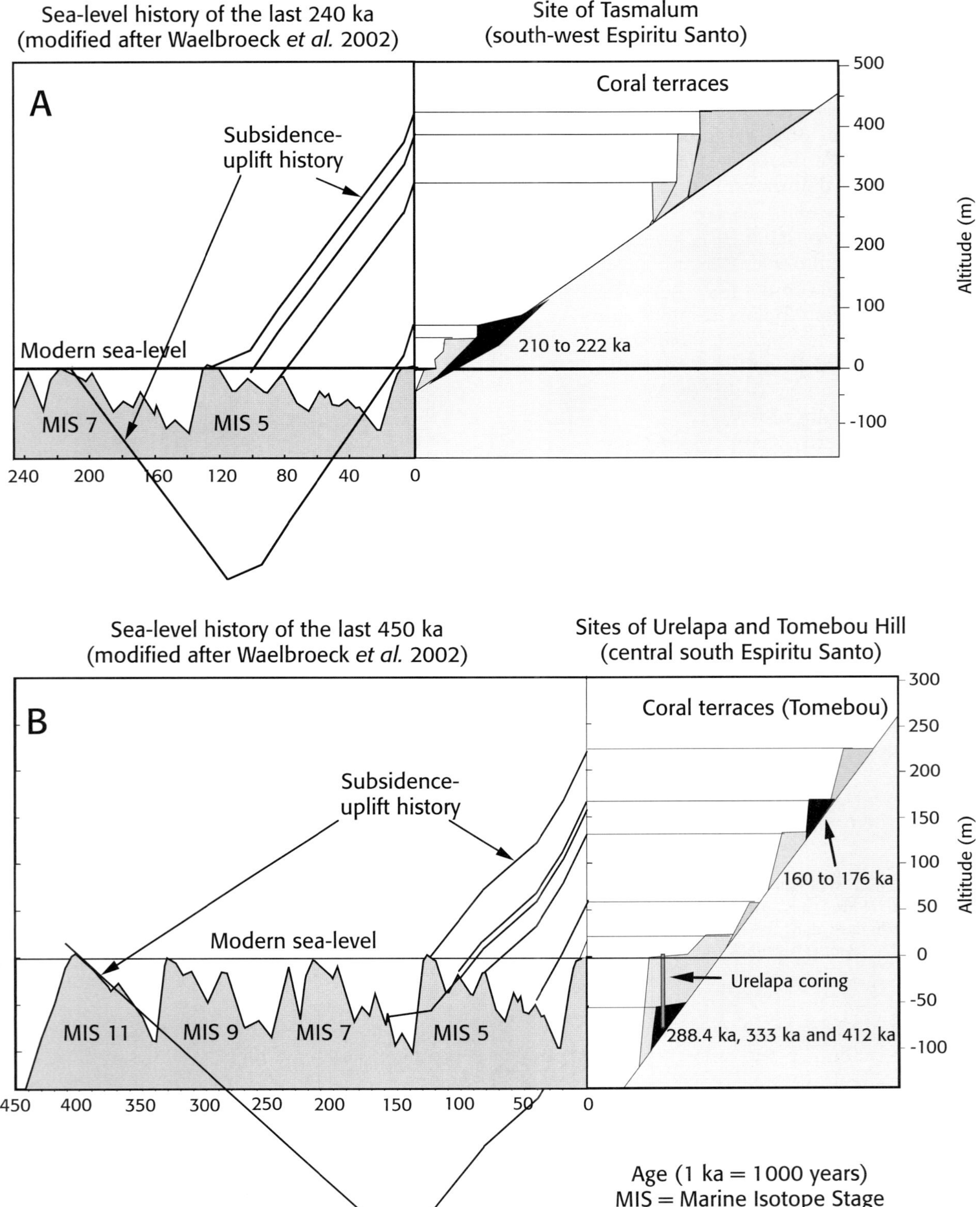

Figure 6: Subsidence-uplift cycle in Santo. **A**: Tasmalum site. **B**: sites of Urelapa Island and Tomebou Hill (figure after Taylor *et al.* 2005).

Figure 7: Porites microatoll, marker of sea-level providing evidence of uplifts.

sea-level was at an elevation similar to present. This reef grew before the younger MIS 5 reefs were uplifted to an altitude of 400 m. Taking in account that the MIS 7c and MIS 5 reefs formed at a sea-level similar to the present, the 216 ka reef must have subsided at least of 350 m by the time the MIS 5 reefs formed in order for them to now be separated vertically by 350 m. Such a succession is typical of a subsidence-uplift cycle.

Coral dating both in cores from Urelapa Island and Tomebou terraces which share a Holocene uplift rate of ~3 mm.yr^{-1}, reveals the existence of the last deglacial sequence (the last 23 ka) extending from +20 to -55 m (Fig. 6B). Below this sequence and deeper than 55 m, corals are dated of 288.4 ka, 333 ka and 412 ka at -58 m, -64 m and -75 m, respectively. The flanks and top of the 220 m flat-topped hill of Tomebou is dated of 37 and 38 ka at 41 m, 160 and 176 ka at 160 m and 149 ka at 220 m, respectively. The corals below the modern reef at Urelapa correspond to the MIS 9 to 11 (290 to 330 ka and 410 ka respectively). Assuming that these corals grew when sea level was in the 23 ka to present range of -125-0 m, then they have undergone a minimum subsidence of at least 150-300 m of these 300-400 ka reefs to explain their position below the modern reef and the occurrence of the 130 ka reef at +220 m. This

Figure 8: Coral terraces from Araki Island, South of Santo.

scenario also supports the existence of cycles of subsidence and uplift.

These data indicate that Santo rapidly subsided at least once during the past 400 ka before being uplifted during the last 100-120 ka. This Late Quaternary cycle of uplift and subsidence occurs over one or two hundred thousand years. Subsidence and uplift of this large area on a 200 ka timescale might be explained by impinging seamounts and ridges that greatly increase friction between the subducting plate causing uplift. When the impinging objects break or are subducted, subsidence of the uplifted areas may follow.

THE MODERN REEFS

Studies of modern and living coral reefs in Vanuatu remain rare. In 1974, Guilcher analyzed the morphology of the coral reefs and defined the fringing and open-sea reefs as primary types. In 1990, Veron recognized about 62 genera and 296 hermatypic coral species in Vanuatu. In 1990, Done and Navin studied habitats of the shallow water communities of many coral reefs in different islands including Santo. From their study on reef zonation, they observed marked differences in the assemblages between the oceanic exposed reefs and the sheltered reefs and they recognized four typical assemblages:

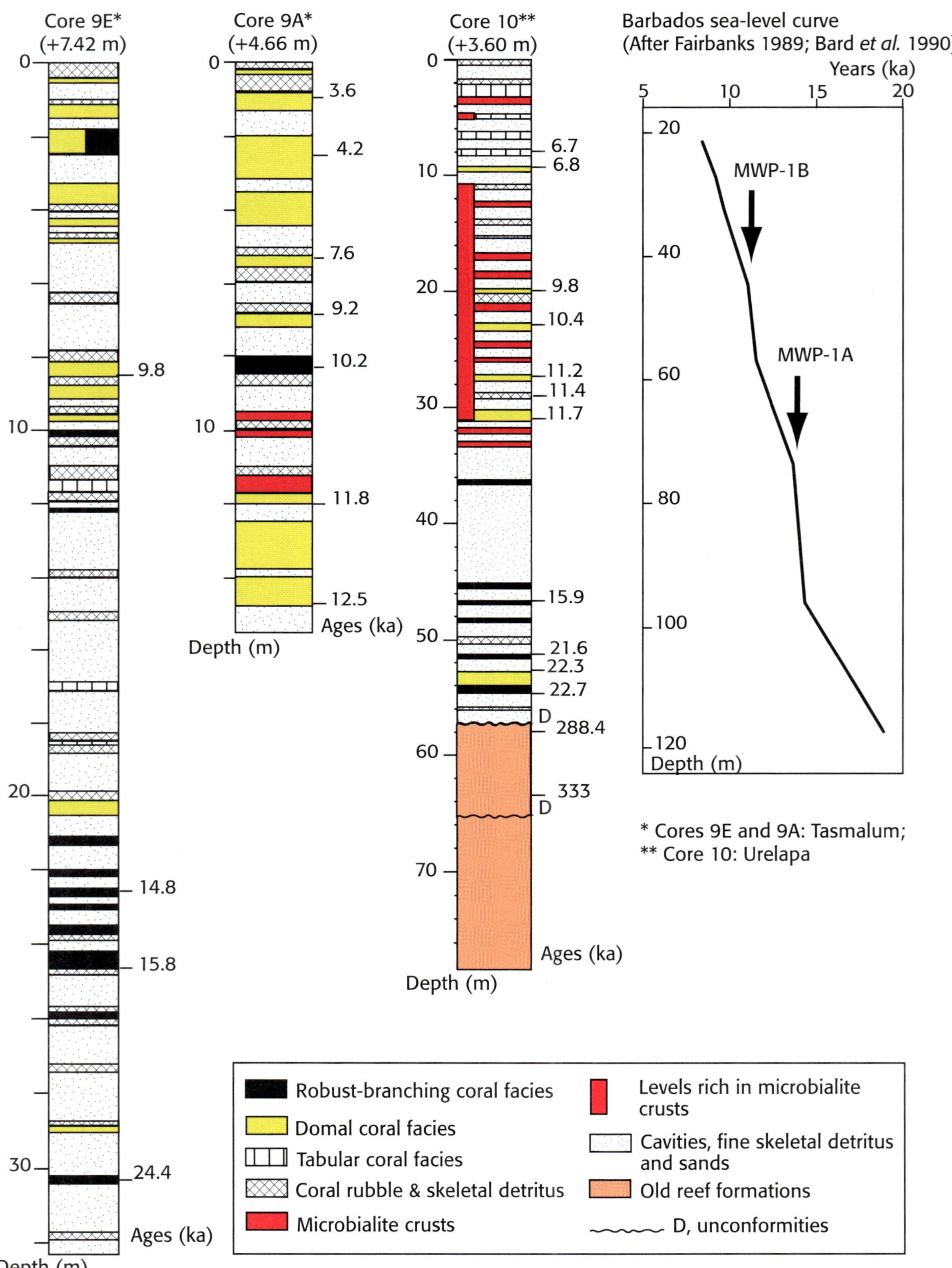

Figure 9: Biofacies of the 23 000 years old reefs from Tasmalum and Urelapa (adapted from Cabioch et al. 1998, 2003).

• In the outer reef slopes, the biological communities are characterised by abundant coralline algae and robust-branching corals (*Acropora* spp. and pocilloporids) in the reef crest zones and by massive and branching corals including *Diploastrea heliopora* and *Goniopora* spp. in the steep slopes.

• The sheltered parts of outer reefs are characterized by various species belonging to *Acropora* and *Montipora*.

• Open embayments are characterized by massive domal *Porites* spp. accompanied by *Acropora* spp.

• In the sheltered embayments, soft corals are dominant accompanied or not by domal *Porites* spp. Various branching forms of *Acropora* and *Porites* were also observed.

See also the chapter by Payri and coauthors in this volume.

REEF DEVELOPMENT DURING THE LAST DEGLACIAL SEA-LEVEL RISE (THE LAST 23 000 YEARS)

The modern coral reefs of Santo is primarily a result of their growth during the last 23 000 years. This period is marked by the sea-level rise from about -120 m to the present sea surface as a consequence of the melting of continental ice sheets in the polar zones. During this sea-level rise covering the last 23 ka, some brief periods of accelerated sea-level rise occurred (Fig. 9). These events, interpreted as accelerated ice cap melting are still debated, but are called meltwater pulses (or MWP) and observed at around 14 ka (MWP-1A) and 11.3 ka (MWP-1B). Species identifications of corals and coralline algae are used to reconstruct the succession of the past reef environments. The ecological significance of these organisms must be be inferred by comparison with their modern counterparts because their typical depth ranges vary in response to local ecological conditions particularly wave energy and irradiance.

In Santo, the growth of two modern reefs was investigated by drilling (see location in Fig. 5A). Several cores ranging in age from present to 24 ka were recovered in an oceanic exposed reef and a sheltered reef at Tasmalum and Urelapa, respectively. The Tasmalum area, in south-west Santo, underwent rapid uplift of 5.5 mm.yr^{-1} (Fig. 5B). The modern fringing coral reef is narrow and characterized by reef fronts very close to the shores. Inland a series of several narrow and broad terraces bordered by more or less steep slopes terminate at an altitude of 35 m. At this altitude this terrace is the broadest and dated at 6.6 ka by Gilpin in 1982 and 7.1 ka by Bloom and Yonekura in 1985.

Offshore from the southeastern coast of Santo, Urelapa is an islet uplifted at a rate of 3 mm.yr^{-1} (Fig. 5B). Urelapa is surrounded by a continuous narrow fringing reef characterized by steeply sloping forereefs very close to the shores.

The development of the Tasmalum reef is characterized by several stages (Fig. 9):
• A lower unit, from 24 to 15 ka, composed of fine skeletal detrital facies and robust-branching acroporids.

• A middle unit, from 15 to 10 ka, composed usually of a higher proportion of framework facies dominated by the robust branching assemblage indicating another shallowing up sequence.

• An upper unit, from 10 to 6 ka, dominated by a mixture of the robust branching assemblage with foliaceous corals typical of deeper waters. This unusual mixture of shallow and deep coral assemblages may be due to down-slope transport of corals during seismic events and, it appears therefore, that the robust-branching corals in this facies could be reworked. This may be a peculiarity of reef development on uplifting coasts. Another characteristic is the occurrence of microbialite crusts.

• From 6 ka to Present, reef growth is marked by a succession of emergence events resulting from the combination of sea-level stabilization and incremental uplift movements.

In cores drilled at Urelapa, the reef development is characterized by two stages as observed in the coralgal assemblages (Fig. 9):
• From 23 to 11.5 ka, by a branching coral facies of *Acropora* spp., various small coral buildups including favids, and encrusting coralline algae characteristic of medium to high energy conditions close to the sea surface.

• From 11.5 to 6 ka, by domal coral facies including *Porites* spp. accompanied by occasional branching coral forms, usually reflecting more sheltered habitats at 10 to 20 m depth. Occurrence of plan-laminar microbialites, especially abundant at the top of sequences, from around 12 ka to around 6 ka, provides evidence of reef growth in sheltered environments on relatively deeper slopes. The succession of two types of coralgal assemblages indicates a variation in growth mode, which can be subdivided into a keep-up growth mode from 23 to around 11.5 ka and a catch-up growth mode from around 11.5 to 6 ka. Such a change probably indicates global palaeoceanographic changes.

ROLE OF THE SUBSTRATUM IN THE INITIATION OF CORAL REEFS

At Tasmalum, reef growth initiated during the glacial period, older than 24 ka and probably as old as, at least, 30 ka, on a thick deposit of bioclastic sands and gravels rich in benthic foraminiferids, mollusc debris, and *Halimeda* segments. Then pebbles, gravels and conglomerates, encrusted by coralline algae, cap the thick sand formation.

At Urelapa, the substrate upon which the reef grew was reached only in the deepest cores at depths ranging from 60 m to 72 m and consists of recrystallized limestones rich in large benthic foraminiferids, calcareous algae and mollusks.

Such substrata both in Tasmalum and Urelapa have probably provided a favourable ground for reef initiation due to their roughness and carbonate composition which is know to be conducive to coral larval recruitment.

CONSEQUENCES OF THE RELATIVE SEA-LEVEL VARIATIONS ON THE MARINE HABITATS

In Santo, sea-level varied according to the eustatic variations and the tectonic vertical movements so that the modern marine habitats appear to be recent in terms of the geological time scale. During the last one million years, the high sea stands (interglacial periods), similar to the present sea-level, alternate with low sea stands (glacial periods) reaching more than 100 m deep at a cyclicity of 100 ka. Moreover, in combination with the vertical movements, comprising uplift and/or subsidence, the up and down movements of the coastline were amplified over time. In such conditions, the marine biodiversity probably benefited from such conditions in this island.

and Pleistocene Marine Faunas Reconsidered

Pierre Lozouet, Alan Beu, Philippe Maestrati, Rufino Pineda & Jean-Louis Reyss

The earliest evidence of rich Quaternary faunas in the southern part of Santo dates back to the report by Mawson who collected in 1905 many molluscs close to the southwest coast. The molluscs were partly identified by Hedley (12 bivalves and 28 gastropods). However, the presence of mollusc-rich Pleistocene deposits has been recorded only recently, as a result of mapping by Mallick who reported a beautifully preserved fauna from the lower Navaka River in 1971. Later the same author announced the discovery of the Kere Shellbed while Mallick & Greenbaum gave some additional information on the fauna of the Navaka River and preliminary data on the Kere Shellbed. These two exposures were assigned to the same lithostratigraphic unit, "Navaka sands".

Since their discovery in 1970, collections of fossils made by Mallick and Greenbaum were sent to the U.S. National Museum for identification. The interest of this deposit was so great that Warren Blow and Thomas Waller organised an expedition to the Kere River in 1974 and collected bulk samples. Ladd included 167 species from the Kere River in his description of marine fossil gastropods of the Western Pacific. He also described two new species of Volutidae and eleven other new species and a new genus of Nassariidae (*Bathynassa*). In total he described 29 new species from these deposits. His opinion of fossil assemblages of South Santo merits quotation: "*Santo sediments contain perhaps the richest and most diversified and certainly the best preserved fauna of fossil mollusks yet discovered in the islands of the Pacific, possibly in all of the Indo-Pacific region*".

One of the characteristics of the fauna of the Navaka and Kere Rivers is that many of the fossil species were unknown in Vanuatu waters at the time. This was not only because their habitats are not easily accessible (because of the offshore origin of this fauna) but also because before the Santo 2006 Expedition very little systematic dredging had been done around Vanuatu. However, since the publications by Ladd, several new species described from the Navaka or Kere Rivers deposits were reported living in other regions of the Pacific (e.g. *Siphonofusus walleri*). But for several other species named as fossil, no close living relative has been indicated. Another characteristic of the Santo fauna, and especially of the Kere Shellbed, is the abundance of shells considered today to be very rare and precious "*Cypraea*", such as *Perisserosa guttata*, *Lyncina porteri*, and *Austrasiatica langfordi*. But, to date, knowledge of this fauna is incomplete, based only on the museum material collected by Mallick (during 1971-75) and by Warren Blow and Thomas Waller (in 1974).

For that reason, the Santo Expedition initiated a preliminary exploration of the fossil sites. This mission by Alan Beu, Pierre Lozouet, Philippe Maestrati and Rufino Pineda visited in February 2006 (18 February-2 March) the outcrops of the Kere River, Navaka River and Funato River (Fig. 10), located previously by Rufino Pineda. During the Santo Marine Biodiversity Survey (September-October 2006) an additional sample of fossils was collected in the Wounaouss River by Pierre Lozouet and Jean-Claude Plaziat.

The material was sorted to the species level and a provisional listing of molluscs for Kere 1 and Kere 2 records several thousand specimens representing more than 450 species.

THE NAVAKA RIVER LOCALITY

The grey marl-siltstone of the Navaka River are accessible mainly on the right bank and cliffs of the river. Among a particularly rich and well-preserved fauna of various organisms, W. Blow revealed an exceptional concentration of Crustacea Decapoda, among the richest in the world. Ladd described some molluscs collected by Mallick, Blow and Ward. Nineteen new species and the new nassariid genus *Bathynassa* have been described from this locality.

Heavy rains during our visit resulted in only one field collection from this deposit (Fig. 11). High waters that covered the outcrops visited by Mallick & Greenbaum in 1975 made them inaccessible. Fortunately, the local people led us to another outcrop located in the bed of one of the tributaries of the river that was not yet flooded. In this section, molluscs and some zooxanthellate scleractinian corals, packed in irregular

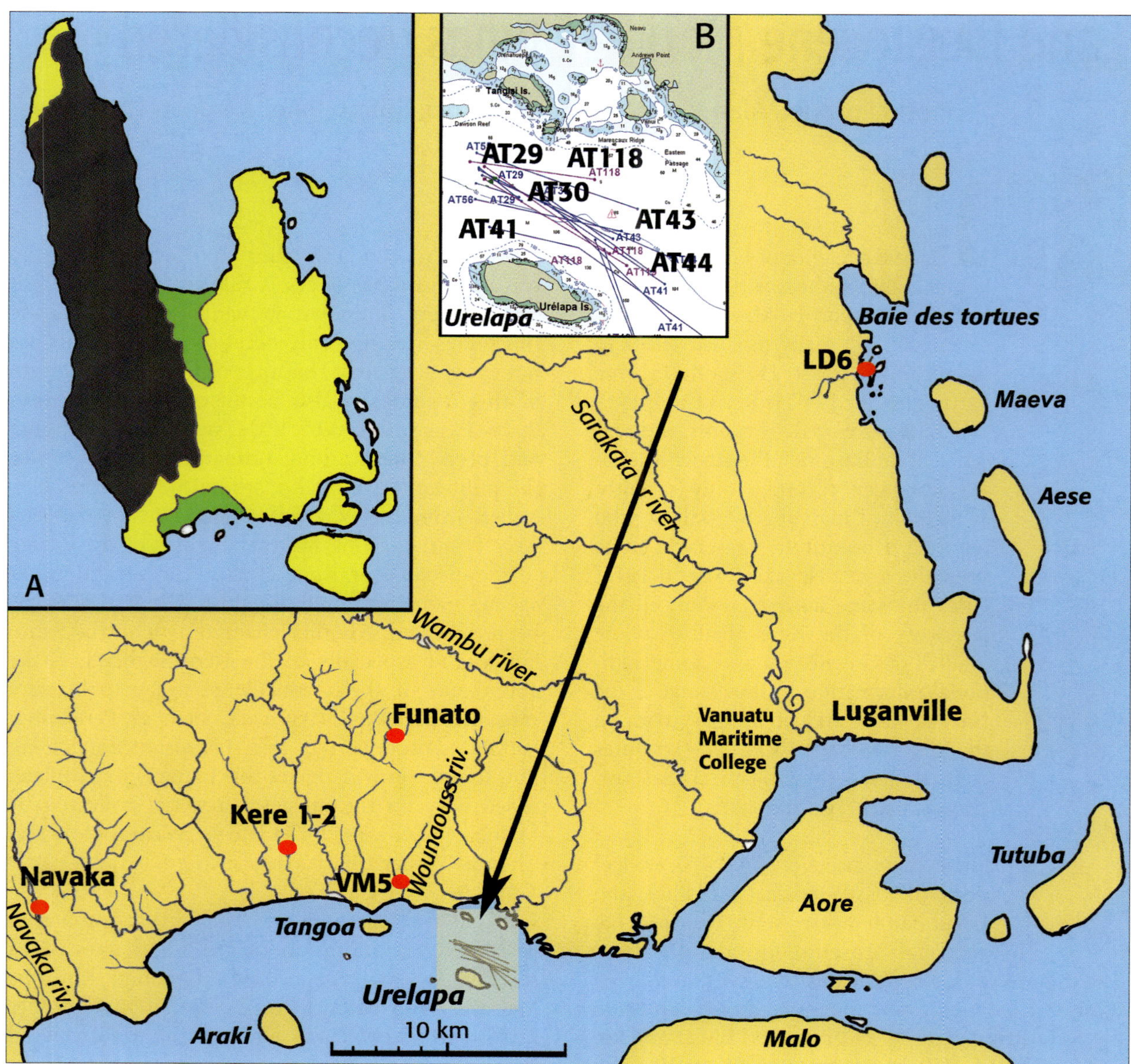

Figure 10: Location of the Recent and fossil sites. **A**: Geological map. Black area represents Miocene volcanics and volcaniclastics; green areas represent Pliocene to Pleistocene deposits; yellow areas represent Pleistocene to Recent Carbonates (modified from Meffre & Crawford 2001). **B**: Location of samples collected with the research vessel "ALIS" (dredging and trawling) with assemblages close to Kere 1 and Kere 2 faunas.

layers, were especially abundant and well preserved. The transported assemblage included notably two species of Turbinidae, six small Strombidae (including two *Canarium* sp.), various littoral Cypraeoidea, two species of Harpidae and other shallow-water molluscs. This does not correspond to the assemblage described by Mallick & Greenbaum or Ladd, which was considered to be the result of an offshore environment of deposition. Our sampling work was stopped by the rapid rise of the river following heavy rain in the mountains.

THE FUNATO RIVER (BELMOL)

The Funato River (Fig. 12) is a small tributary of the Adsone River. This creek is particularly deeply embanked. The fossiliferous site discovered by R. Pineda (January 2006) is the farthest from the coast (7.5 km) of those we collected. We sampled on only one day. The fine, grey marl-siltstone exposed along the banks and cliffs of the river includes several shellbeds.

Strombidae (*Strombus*, *Terebellum*) dominate the gastropods and Cardiidae the Bivalvia. We note the abundance of *Strombus plicatus columba* similar to the specimen collected by Ladd from the alluvium of the Sarakata River. The abundance of the cardiid *Nemocardium* cf. *bechei* is especially striking.
While this is clearly a shallow-water deposit, most of its species occur also in the fauna of the deep-water outcrops of the Kere River.

Figure 11: The Navaka River. General view of the outcrop, Alan Beu is standing in the foreground and examines a shell, behind Philippe Maestrati sorts with an orange sieve. Coord.: 5°35.290'S - 166°51.089'E.

Figure 12: The Funato River. Coord.: 15°31.365'S - 166°58.872'E.

THE KERE RIVER

In the original outcrop a shell lag facies tops a very thick marl sequence. These contrasting deposits have been attributed to distinct sedimentation phases related to changes in current dynamics. In this study, we name these units respectively "Kere 2" (the lower sequence) and "Kere 1" (the classic Kere Shellbed). The previous collectors concentrated their attention on the very rich shelly sand deposit (Kere 1) (Fig 13).

The Kere locality is situated on the low banks of the river that crosses a limestone plateau (Rabua) at about 70 m above sea-level. Kere 1 and Kere 2 are exposed on both sides of the river. Kere 1 is accessible for only 10-20 m on the east bank and 5-6 m on the west bank (Fig. 13). Kere 2 underlies Kere 1, being exposed along the river crossed by a series of waterfalls; one of them, about 200 m downstream from the Kere 1 outcrop, is 5-6 m high.

••• Kere 1 (= Kere Shellbed of Mallick and Ladd. Main fossiliferous deposit)

The Kere 1 deposit (a biocalcarenite) is attributed to the Upper Pleistocene (Fig 14). It is an unconsolidated (or poorly consolidated) calcareous silty gravel consisting of 50/70% shell fragments. Dissolution and recrystallisation have cemented parts of the deposit. This process concerns principally the exposed outcrops.

••• Kere 2

The Kere 2 deposit is a grey marl-siltstone with some fine fossiliferous layers. Our excavations were in the top of the series, between 50 and 100 m SE of Kere 1.

••• The molluscan fauna

Some taxonomic groups from Kere 1 and Kere 2 have been examined by specialists (Fig. 15). Accordingly, our samples contain 50 species of Cypraeoidea (Fig. 16), 29 species of Tonnoidea (Fig. 17), 23 species of Muricidae, 10 species of Nassariidae, at least 38 species of *Conus* and 14 species of the bivalve family Cardiidae. In all groups, similarities between the New Caledonian fauna and the southern Japan/ Philippines faunas are noted, showing an unusual mixture of faunas that now extend from Japan to New Caledonia, throughout the western Pacific. The present range of *Austrasiatica langfordi* illustrates this pattern. This species is known in southern Japan, in the Philippines, in the south of New Caledonia (Norfolk Ridge) and in Queensland. Thus, the Kere River specimens fill the gap between the northern and southern parts of its present distribution.

Kere 2 (Fig. 15) has yielded 220 species of gastropods, 90 species of bivalves and 3 scaphopods. The dominant groups of Gastropoda are Turritellidae, Strombidae, Naticidae, Nassariidae, Mitridae, Costellariidae, Turridae and Cephalaspidea. The dominant families of bivalves are the Arcidae (*Anadara*), Malleidae (*Malleus*), Tellinidae (*Apolymetis* sp.) and Corbulidae. The abundance of byssally attached *Malleus* is worthy of note. In the fine fraction the Scaphopoda Gadilidae (*Gadila* sp.) is very common.

Our collection from Kere 1 contains more than 300 species of gastropods and 80 species of bivalves. The dominant gastropod groups are Siliquariidae, Hipponicidae, Strombidae (*Tibia*), Naticidae, Ranellidae/Bursidae, Nassariidae, Buccinidae, and Conidae (*Conus*). The dominant bivalve groups are Arcidae, Cucullaeidae (*Cucullaea labiat*), Pectinidae, Spondylidae, Veneridae, Cardiidae, Chamidae, and Glossidae (*Meiocardia vulgaris*). Compared to Kere 2 we note the abundance of bivalves (Spondylidae, Chamidae) attached to hard substrates, gastropods living in sponges (*Siliquaria*) or temporarily attached (Hipponicidae). The presence of at least six

Figure 13: View of the right bank of the Kere River (Kere 1 outcrop). Coord.: 15°33.322'S - 166°57.200'E.

Figure 14: General view and detail of the shellbeds of Kere 1 (Pleistocene).

species of Ovulidae and four species of Triviidae suggests the presence of a rich cover of gorgonians and ascidians.

This fauna is also unusual for the abundance of whale barnacles (Crustacea Cirripedia: *Coronula diadema*), as mentioned previously.

Figure 15: Overview of the Kere 2 assemblage.
1: Fine fraction with sorted bivalves. 2: Fine fraction with sorted gastropods and scaphopods. 3: *Malleus* sp. [Length 50 mm]. 4-5: *Apolymetis* sp. [Length 59 mm]. 6: *Anadara* sp. [Length 40 mm]. 7-8: *Corbula* sp. [Length 19 mm]. 9: *Cymatium sinense* (Reeve, 1844) [Length 61 mm]. 10: *Ficus ficus* (Linnaeus, 1758) [Length 68 mm]. 11: *Strombus plicatus* (Röding, 1798) [Length 47 mm]. 12: *Diminovula* aff. *culmen* (Cate, 1973) [Length 20 mm]. 13: *Cymatium gutturnium* (Röding, 1798) [Length 54 mm]. 14: *Nassarius variegatus* (A. Adams, 1852) [Length 17 mm]. 15: *Metula* cf. *mitrella* (Adams & Reeve, 1850) [Length 27 mm]. 16: *Conus praecellens* A. Adams, 1854 [Length 40 mm]. (Photos P. Lozouet).

Figure 16: Kere River Cypraeoidea (Cypraeidae, Ovulidae) and Velutinoidea (Triviidae) gastropods (Identifications by L. Dolin). **1-3**: *Lyncina nigromaculata* Lorenz, 2002 [Length 33.5 mm] Kere 1. **4-6**: *Paulonaria beckii* (Gaskoin, 1836) [Length 12.8 mm] Kere 1. **7-9**: *Palmulacypraea boucheti* (Lorenz, 2002) [Length 20 mm] Kere 1. **10-12**: *Perisserosa guttata* (Gmelin, 1791) [Length 46 mm] Kere 1. **13-15**: *Hiraseadusta hirasei* Roberts, 1913 [Length 33.5 mm] juvenile specimen, Kere 1. **16-18**: *Xandarovula xanthochila* (Kuroda, 1928) [Length 14.8 mm] Kere 2. **19-21**: *Adamantia dubia* Cate, 1973 [Length 4 mm] Kere 1. **22-24**: *Carpiscula bullata* (G.B. Sowerby *in* A. Adams & Reeve, 1848) [Length 8.2 mm] Kere 2. **25-27**: *Trivellona paucicostata* (Schepman, 1909) [Length 8.5 mm] Kere 1. **28-29**: *Cleotrivia euclaensis* Cate, 1979 [Length 2.8 mm] Kere 1. (Photos P. Lozouet).

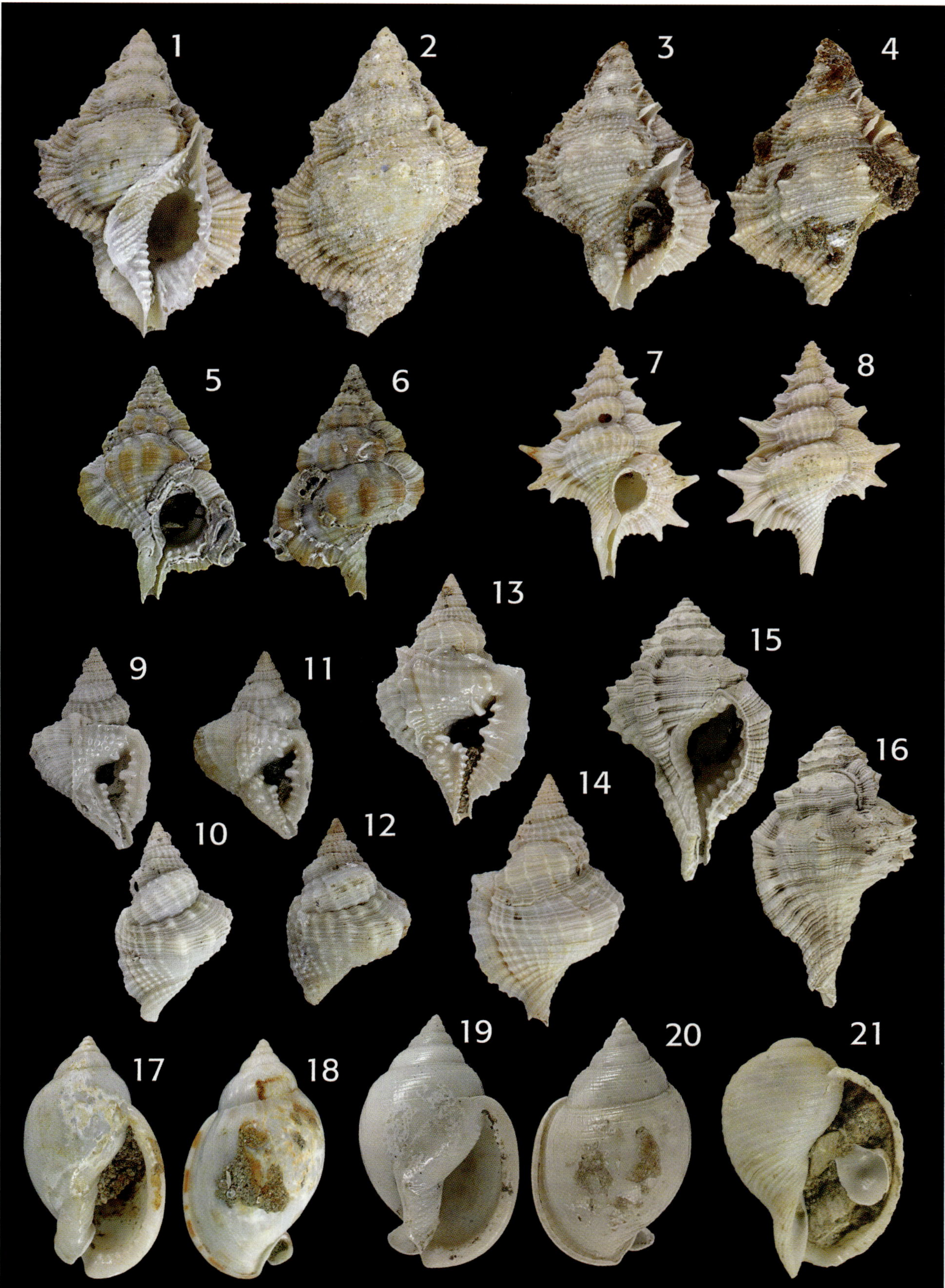

Figure 17: Kere River (outcrop 1) Tonnoidean gastropods.
Family Bursidae - **1-2**: *Bursina nobilis* (Reeve, 1844) [Length 67 mm]. **3-4**: *Bursina ignobilis* (Beu, 1987) [Length 40 mm].
Family Ranellidae - **5-6**: *Gyrineum bituberculare* (Lamarck, 1816) [Length 33.5 mm]. **7-8**: *Biplex pulchra* (G.B. Sowerby I, 1835) [Length 42 mm]. **15-16**: *Cymatium (Ranularia) dunkeri* (Lischke, 1868) [Length 63 mm].
Family Personidae - **9-10**: *Distorsio graceiellae* Parth, 1989 [Length 25 mm]. **11-12**: *Distorsio euconstricta* Beu, 1987 [Length 24 mm]. **13-14**: *Distorsio reticularis* (Linnaeus, 1758) [Length 43 mm].
Family Cassidae - **17-18**: *Casmaria ponderosa* (Gmelin, 1791) [Length 30 mm]. **19-20**: *Semicassis bulla* Habe, 1961 [Length 44 mm].
Family Tonnidae - **21**: *Tonna zonata* (Green, 1830) [Length 30 mm]. (Photos P. Lozouet).

THE WOUNAOUSS RIVER

Figure 18: The Wounaouss River, coord. 15°34.3'S 167°00.2'E. Several metres of very fossiliferous fine grey sand (Holocene) are exposed on the left bank.

A deposit of several metres of very fossiliferous fine grey sand is exposed on the left bank of the Wounaouss River (Figs 18 & 19), NW of Tangis-Tapuntaris, including layers with hermatypic corals and lenses of shells. Sieving revealed the presence of many millimetric and centimetric remains of wood. The molluscan shells are perfectly preserved.

The assemblage is dominated by a minute species of *Cerithium* sp., Strombidae (three species of *Strombus*, *Terebellum*), Nassariidae (at least five species), Costellariidae and various opisthobranchs (Acteonidae, Cephalaspidea). Mid-littoral Ellobiidae, freshwater Thiaridae and several land-snails are also present. The most abundant and characteristic species is a small strombid, *Strombus* (*Dolomena*) *minima*. The Tellinidae (*Tellina*, *Macoma*) are the dominant bivalves.

Figure 19: View of one of the outcrops and enlarged detail of concentration of shells consisting largely of Bivalvia but also gastropods and pieces of corals.

RADIOMETRIC DATING AND CORRELATION
WITH MARINE ISOTOPE STAGE OF THE QUATERNARY

The deposits of the Navaka and Kere Rivers have generally been referred to as "Pleistocene" or "Upper Pleistocene". Two radiometric ages have been suggested giving a ^{14}C age of 25 280 (+/- 460) years BP and a Uranium/Thorium age of 14 000 years for Kere 1 (Kere Shellbed of Mallick and Ladd), both dates being based on an ahermatypic coral. Ladd

retained the older date because it was admitted that the younger age was not consistent with the originality of the fauna. Three new dates provide contradictory chronological data.

Very well-preserved ahermatypic corals (*Flabellum* spp.) collected in 2006 from Kere 2 have been dated by the Uranium/Thorium method and established an age of 232 +/- 10 ka. The same species of coral from the overlying deposit of Kere 1 gave an age of 133 +/- 5.5 ka. These reliable benchmarks suggest that the Kere sequence was deposited during the warmer Marine Isotope Stages (also named OIS for "Oxygen Isotope Stage") MIS 7 (Kere 2) and MIS 5 (Kere 1). Thus the change between the fossil assemblages of Kere 1 and Kere 2 apparently reflects a major sedimentary gap corresponding to glacial stage MIS 6.

A hermatypic coral (*Montipora*) from the deposit of the Wounaouss River (VM5) has been placed in the early Holocene according to its Uranium/Thorium date of 9 200 +/- 0.27 yr.

These results demonstrate that the raised outcrops were deposited during high sea-level (interglacial) stages. Despite the tectonic instability of this area, the glacial periods are not represented by Lowstand Systems Tracts (LST) with fossiliferous deposits. It is well known that in shallow-water deposits on gently uplifting shorelines, low sea-level stands are frequently represented by hiatuses, as the sea left the deposition site altogether during glacial periods because of the sea-level difference of 100-130 m between glacial and interglacial periods. However, considering the high tectonic activity of this region and the relatively deep deposition site (100-120 m) of Kere 1, more complete deposition during the glacial periods might have been expected in southern Santo.

COMPARISON WITH RECENT ASSEMBLAGES

During the Santo 2006 Expedition, more than 340 bottom samples (dredging and trawling) were collected between 16 and 1 285 m with the research vessel *Alis* (208 samples between 16 and 350 m). Forty three additional dredgings were acquired by the *Evolan* and utility boat of *Alis* between 1.5 and 31 m. These samples (see chapter by Bouchet and coauthors, this volume) provide an inestimable source of data and the means of interpreting the past environment of the fossil assemblages. The molluscs collected were sorted and stored at the family level and we have examined all the samples during the drying phase or just after sampling.

Six stations (AT29, AT30, AT41, AT43, AT44 and AT118) have yielded assemblages that fit very well with the Kere 1 and Kere 2 faunas. These assemblages include the same species of *Tibia*, Ranellidae, Nassariidae, Mitridae, and Conidae, and the same Tellinidae, in particular a poorly known species of *Apolymetis*. All these stations are grouped in the same area at depths between 71 and 122 m, the average depth being 94 m (Fig. 10B). The assemblages of stations AT30 and AT29 are very similar to the Kere 2 fossil assemblage, while some others are more similar to Kere 1. We may also deduce that local factors such as hydrodynamic conditions would have introduced appreciable variations in molluscan communities.

The modern molluscan assemblage (Station LD6: 2-5 m) that most closely resembles the Wounaouss River fossil site was collected from shallow water around Malparavu Island (= Oyster Island), where *Strombus* (*Dolomena*) *minima* is especially common. The two assemblages also have several Nassariidae and Costellariidae in common, while the dominant bivalves are the Veneridae in the modern assemblage and the Tellinidae in the fossil one.

CONCLUSION

This preliminary study shows that the richest Quaternary outcrops range from Middle Pleistocene to Holocene in age and were deposited during marine glacio-eustatic transgressions. The Holocene Wounaouss River deposit consists of several meters of shallow-water fossiliferous sands including macrofossil-rich beds (shellbeds, coral beds). This assemblage bears similarities with a faunule collected alive at 5 m. On the other hand, the Pleistocene assemblages of the Kere River are consistent with an offshore habitat between 70 and 120 m depth. Kere 1 is attributed to the MIS 5 Last Interglacial High Stand and Kere 2 to the preceding MIS 7 interglacial. The deposits of Navaka River and Funato River are not dated but are also probably Middle Pleistocene.

Only further intensive field work, collecting, surveying and dating the deposits through well-preserved fossils, is likely to contribute to our understanding of the exceptionally rich fossil deposits of Santo, in parallel with the systematic inventory of the modern fauna. Because Vanuatu (including Santo) is located in an uplifting island arc, it was subject to major, random tectonic movements that resulted in a complex tectono-sedimentary history (see chapter by Cabioch & Taylor, this volume). We shall conclude by stating the need for a program involving the collaboration of various disciplines. It is a vast but realistic project consistent with the multidisciplinary ambition of Santo 2006.

# of Santo and of the Sanma Province

Patricia Siméoni

Sanma province (Fig. 20) is located in the north-central part of the archipelago. Santo and its small neighboring islands including Malo, Aoré, Tutouba, Araki, Tangoa, Aïs, Mafea, Litaro, Elephant, Dauphin and Sakao are located between latitudes 14°39' and 15°46'S, and between longitudes 166°32' and 167°18'E and cover an area of almost 4250.8 km².

Sanma's provincial capital is Luganville: the second largest city of the country. The province is divided into nine districts: seven on Santo and two on Malo-Aoré (Fig. 20).

Flag of Sanma province

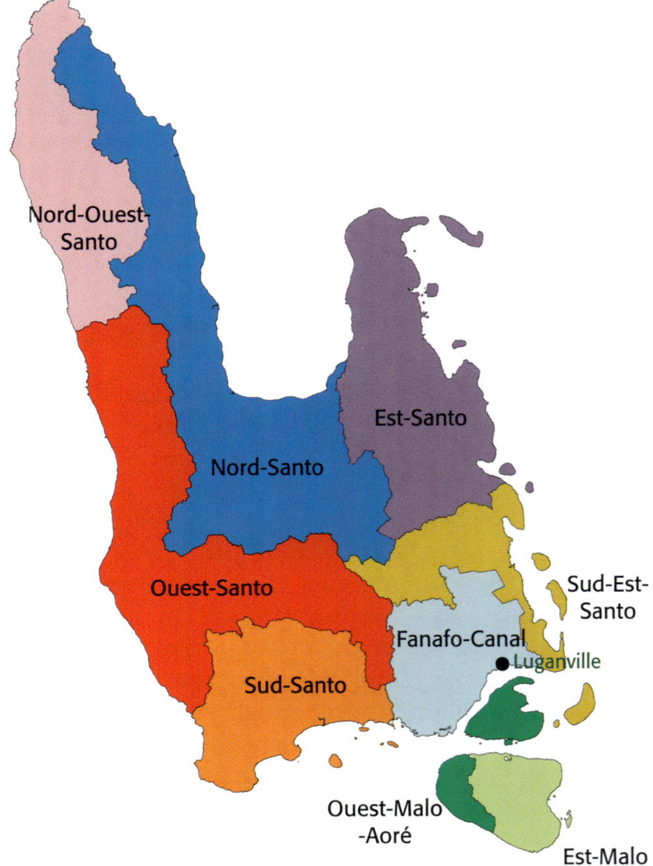

Figure 20: Administrative districts of Sanma province.
© Géo-consulte

Santo, Malo and the islets in figures

1 large island and 45 smaller islands and islets

Total area: 4 250.8 km² including 3 959 km² for Santo and 182.7 km² for Malo

Number of inhabited small islands and islets other than Santo and Malo: 9

Area (inhabited small islands and islets): 88.8 km²

Total population (2009): 47 872

Population on small islands and islets (2009): 1 981

Urban population (2009): 13 484

Average population density: 11.4 inhabitants/km²

Density (small islands and islets in 2009): 22.3 inhabitants/km²

Other uninhabited islands and islets: about 30

Area of the uninhabited islands: around 21 km²

Number of vernacular languages: 28

The 2009 census gave a population of 47 872 inhabitants, of which 48.3% were women and 51.7% were men. The average annual demographic growth is 3.2% for the rural areas of the province and 2.4% for Luganville. The town, which is the second urban center of the country, has a population of 13 484 inhabitants, or 28.2% of the province population. This ratio is slightly lower than in 1999 when urban population represented 30% of the province population. The population growth rate of this province is one of the highest in the country;

This article is the revised English version of Patricia Siméoni's Atlas du Vanouatou (Éditions Géo-Consulte, geo-consulte@vanuatu.com.vu). All the maps in this article originate from the Atlas; toponyms on the maps follow the French spelling.
Translation: Béatrice Marx

this growth is mostly because of a migration movement between islands, both in rural and urban areas. The population is mostly distributed on the southern and eastern coasts, but new villages have been created inland along newly built road sections.

GEOGRAPHIC AND GEOMORPHOLOGIC PRESENTATION

Santo Island covers an area of about 3 959 km²; the four other biggest islands in terms of area are Malo (182.7 km²), Aoré (58.2 km²), Toutouba (13 km²) and Sakao (11.5 km²). They form the province of Sanma (province of the islands of Santo and Malo).

Santo Island is the largest and the highest island of the country. It is composed of two very different geomorphological features. To the west is a mountain range (Figs 21 & 22) with a north-south crest that reaches an altitude of more than 1 000 m and includes the two highest peaks of the archipelago: Mt Tabwemasana (1 879 m) and Mt Santo (1 704 m). Its landscape is very dramatic with amphitheatre-like valleys. The eastern part of the island is composed of coral reef plateaux and uplifted sedimentary terraces which are divided by rare deep valleys (Sarakata) and serviced by a mainly underground hydrographic network. These plateaux do not exceed 300 m in altitude, except for a few peaks raised in a sugar-cone shape (Mt Tankara, 784 m and Tiouri, 583 m). The eastern coastal terrace, recently submerged, is composed of fringing reef plateaux and coral reef beaches.

Figure 21: The summit of Santo: Mt Tabwemasana.

There is a clear assymetry between the mountainous West and the dissected plateaux of the East. In between is a huge graben[1] where the Jordan plain and Big Bay are located.

1 - depressed block of land bordered by two horsts (synonym: tectonic rift).

Figure 22: The roof of Santo: view of the mountain range of West Santo.

© Géo-consulte

TOPONYMY AND HISTORY

• • • Santo

The name of the big island comes from Quiros, the explorer who, on May 1, 1606, dropped anchor in Big Bay which he named "Bay of Saint-Philip and Saint-Jack" after the saints traditionally celebrated on that day. After three days exploring the coastline, Quiros found a suitable spot to moor in the south-eastern part of the bay. Quiros would call the location, El Puerto de Vera Cruz, just located at the mouth of a small clearwater river which was named El Salvador. This anchorage, nowadays named Table Anchorage, near the current village of Matantas at the mouth of the river of the same name is still the only mooring location of the bay.

The bay was so huge that Quiros believed that he had discovered the great southern continent, and, on May 14, 1606, he officially claimed those lands for Pope Clemente VIII and the King of Spain, Philip III. This land was named Terra Austrialis del Espiritu Santo (Austrialis, not Australis, because of the Austrian heritage of the Spanish king of that time). He established the first Christian colony on the island and called it "The New Jerusalem"; he named the river that flows into the bay the "Jordan".

The people of Santo traditionally called the island with vernacular names from their respective languages: *Inéréï*, *Ta Marina*, etc. The existence of twenty-eight native languages makes this island one of the most culturally diverse regions of the archipelago. The name given to the island by Quiros, probably the most neutral one in terms of traditional cultures of the island was kept. The name was later shortened to Espiritu Santo and, eventually, evolved into its modern name of Santo. The islets located on the eastern and southern fringes of Santo mostly kept their vernacular names except for a few. Malo Island was named "Saint Barthélémy" by Cook in August 1774, but took its vernacular name back a century later.

What distinguishes Santo from the rest of the country is the survival of its traditional pottery art in Olpoï, Nokoukou and Wousi on the western coast of the island, a tradition that dates back more than 3 000 years. Indeed, archeologists have unearthed the oldest Lapita[2] pottery of the archipelago —dated between 3 100 and 3 000 BP— on the islands of Aoré and Malo. The traditional political system on the big island and the neighboring islets was the graded society system. Chiefs progressively acquired their status based on the trust granted to them by the population, who organized ritual grading ceremonies in their honor and offered them high ritual value tusker

2 - vernacular name to designate a style of pottery first discovered on the eponym site of Lapita in New Caledonia. Lapita is also the name of the first people who colonized Oceania 3000 years ago, who are archeologically identified by the pottery bearing the same name.

pigs. Each local community created alliances with other groups and maintained exchange networks that extended sometimes very far. It seems that two major groups coexisted on Santo. The first, located on the Cumberland Peninsula, maintained a hierarchical grading system of the *Suque* kind, originally from the Banks Islands which was closer to Santo. This alliance was characterized by the use of shell money as a medium of exchange. The second group located in the south and the east of the island was governed under the *Sumbwe*, or *Hungwe*, political system and used the tusker pig as medium of exchange. They maintained favorable relationships with the islands of Ambaé and Mallicolo. At the beginning of the 20th century, the German anthropologist Felix Speiser identified five major ethnic settlement areas: the Sakaos in the east of the island, the population of the center of the island, the people from the Big Bay area, the highlanders from the western side of the island, and, finally, the people from southern Santo and Malo. The small body-size of the people from the mountains of Santo led to the mistaken belief that a group of pygmies lived there. These people proved to be a Melanesian people of short stature. This population classification was used for a long time as a way of distinguishing more or less homogeneous cultural areas on the island; it was also used to set up the first electoral precincts in the first decades that followed the Independence.

When missionaries arrived in the 19th century, the population from the eastern side of the island left the interior bush, where they used to live in large families scattered among small hamlets, and settled along the coast in villages that were created around the missions. Instead of dividing the island between both groups, Catholic and Protestant missions co-existed in each region. This created tense relationships among the members of the different congregations which sometimes shared the same language and strong family ties; notably, this occurred between Port-Olry and Hog Harbour to the east, Tolomako and Tourébiou in Big Bay, and, to a lesser extent, in the south and the southwest of the island. Generally speaking, the inhabitants of the central part of Santo have remained isolated and have been less subjected to the influence of the outside world than the coastal populations. Few in numbers, they are spread geographically over a large area. In the mid 1960's, these populations became involved in the *Nagriamel* Movement. This neotraditional movement, the forerunner of sovereignty and land claims (1965), was created by the charismatic leader Jimmy Stevens, whose headquarters were located in Fanafo (southeast Santo). Stevens emphasized the importance of "custom" as a basis for the Melanesian society. He was supported in

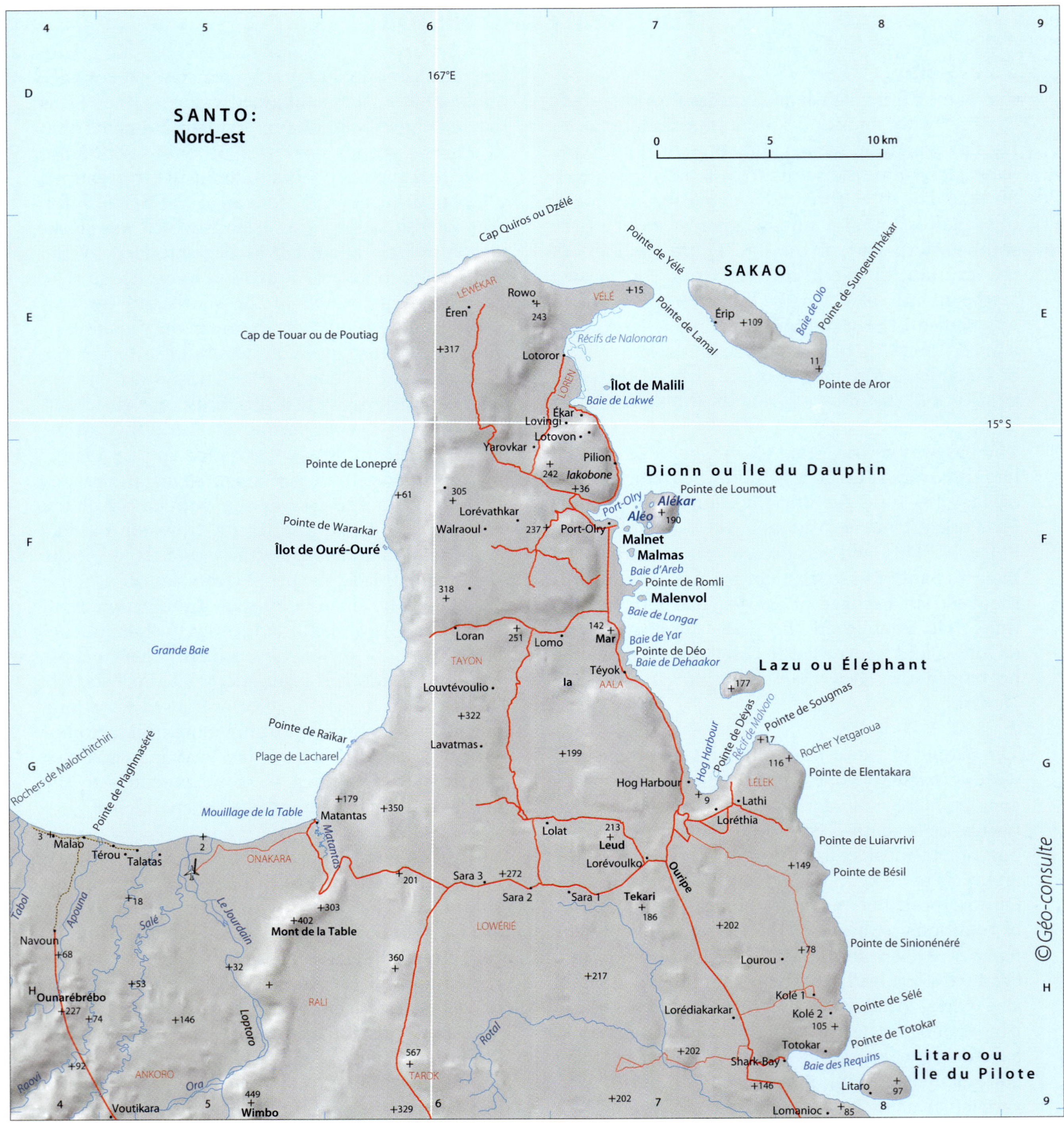

his claim by like-minded bush dwellers who held strong distrust towards the outside world and politicians.

••• Malo

One of the oldest Lapita archeological sites of the archipelago was discovered on Malo Island. The island's ancient people are divided into two linguistic groups: the Tinjivo, to the east, and the Aouta, to the west. Traditionally, Malo shared important relationships with some areas of the north of Mallicolo, Ambaé, Maéwo and Pentecôte, which undoubtedly influenced its cultural evolution. Though these traditional ties have not been maintained, the population still acknowledges this heritage, which allows for an open-minded spirit and the feeling of belonging to the large community of the northern islands.

The arrival of Europeans in the 19th century brought substantial changes to Malo. Its proximity to Tangoa, a small island located south of Santo which became the hub of the Presbyterian Church for the archipelago, exposed Malo to its missionary influence. With the exception of an attempt led by the Catholic missionaries to settle in the east, the Presbyterian Church remained the main Christian mission in Malo until the arrival of several new religious congregations in the second half of the 20th century.

3 - The Société Française des Nouvelles-Hébrides (SFNH) was created in 1886 to replace the CCNH, la Compagnie Calédonienne des Nouvelles-Hébrides which, under the direction of John Higginson, vowed to buy the archipelago from the Melanesians and the English colonizers who found themselves in dire straits. In 1886, SFNH became a shareholding company supported by the French government because of the political interest it represented. At that time, the company claimed virtual ownership of 780000 hectares which included the best land in terms of quality and location in the archipelago. The company attracted French colonizers from New Caledonia by offering plots of land of 25 to 50 hectares to newcomers on the condition that they would develop it.

The role that Malo played in the economic development of the New Hebrides was a major component of its own development. In the 19th century, the Europeans had taken possession of most of the east end of the island where they set up coconut plantations for the production of copra. The planters were French and settled on lands bought by the SFNH[3]. Local populations started to express their strong opposition to the presence of Europeans after World War II, and planters were not allowed to expand their plantations beyond the eastern flat coastal plain of the island. Land became a major topic during the elections that took place before the Independence.

MODERN ORGANIZATION

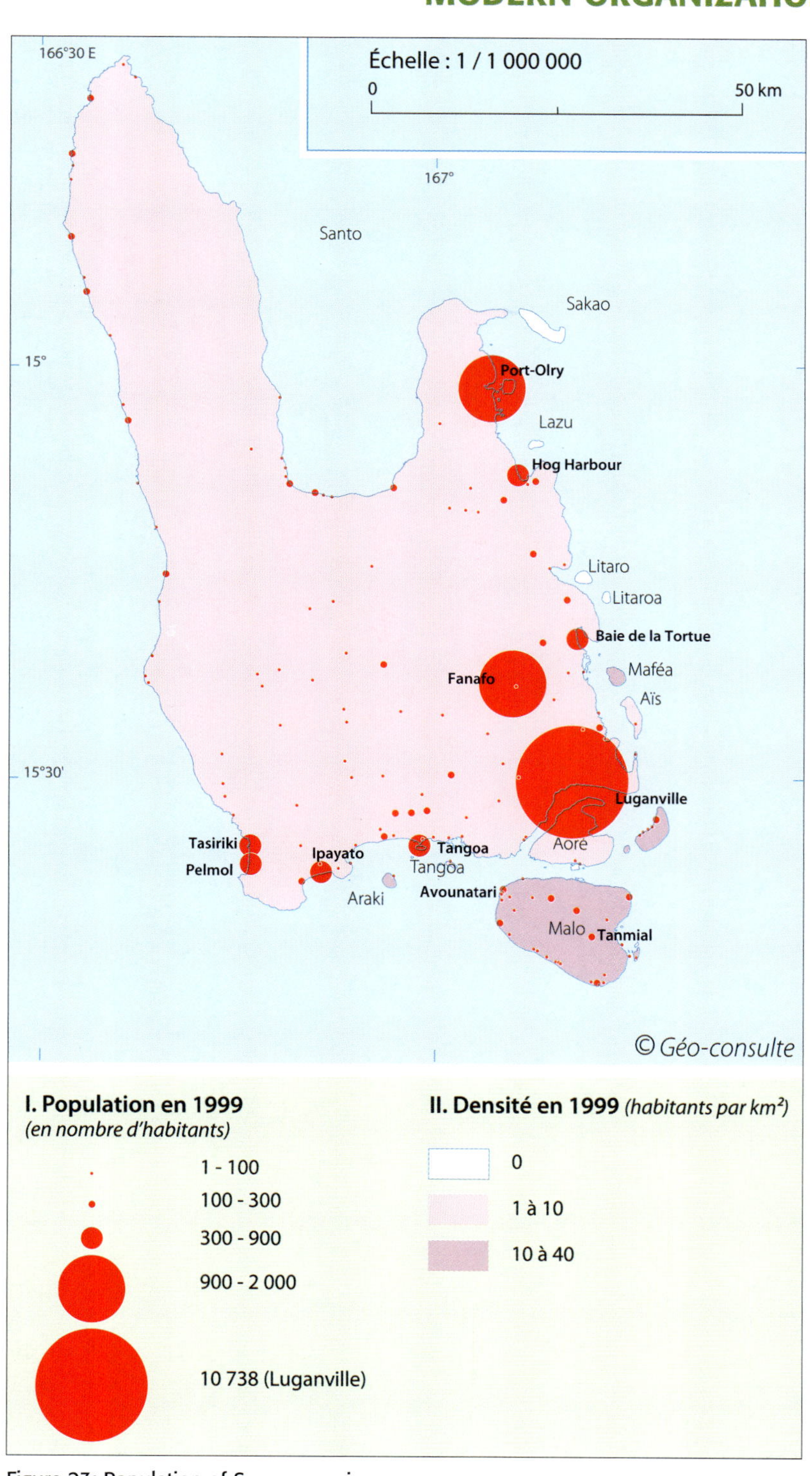

Figure 23: Population of Sanma province.

Santo is the biggest island of the archipelago and, together with Malo and its neighboring islets, the second most populated province of the archipelago with a 2009 population of 47872 (which includes Luganville, the second largest urban center of the country, with 13484 inhabitants) (Fig. 23). The island is divided into seven districts: Northern, Eastern, Western, Northwestern, Southwestern and Canal/Fanafo. The rural population is estimated at 34388 inhabitants. Rural islanders live mostly along the coasts except on the eastern and southern fringes of the island where settlements are found as far inland as 15 km (Figs 24 & 25). On the western part of the island, the population lives in small isolated villages on the coastal fringe of the volcanic Cape Cumberland. Inland, especially to the south, was far more populated before the arrival of the missionaries and farmers who attracted Santo's natives to the coasts of an area extending from Port-Olry to Tasmalum, through Hog Harbour and Tangoa. The extensive toponymy between Santo Peak and the Butmas plateau confirms the existence of these ancient settlements. The eastern coast remains the most populated area, mostly because of Luganville, but also because of two historical factors: the concentration of old European plantations and the geographical location of the Nagriamel settlements in the Fanafo area. Finally, the islets located along the main island on the east and southeast sides account for a population of 1652 inhabitants (1999). Malo is the thirteenth island of the archipelago in size, just after Maéwo, and the ninth in population. Its area (180 km²) and its population

166°30'E
15°30 S
Naoak
Ourouna
685
Rochers de Kariaghaghara
Tasmaté
178
+57
+605
Élia 2
Élia 1
252
Wounaro
Liaponi
Pararata
Kéréboa
Kéréboa
Manoura
Poua
Manii
Péaroé
Pointe de Tsarova
Noïwousi
Wousi
28
Tsaroï
Linsoubé
Poulouai
914
Lindouri
Nono
Vakola
Pointe de Vakola
Vakol
74
Navoutlapa
12
Saouri
+141
Nasouravé ou Pérépépéo
+236
918
896
Pareo
Mataaipi
58
Papaïsalé
Patoui
Batounasébou
356
Kéréway
Lowoumaliou
575
Soulémaori
Toramaori
Pointe de Nambouto
Laarou
261
Lalaolo
Pointe de Tsinionibavana
Pointe de Tsaramélé
Kérévinoumpou
Rocher Oukouan
+202
Pointe de Tsinioni
Tassiriki
Waïnayoa
263
Rocher Épénévé
Pelmol
311
Bouvo
Cap de Mataabé ou Cap de Lisburne
Lovénoué
Tïra
164
Tanbouipil
Baie de Tchanapapa
VOULQUA
Nub Hill
239
Pointe de Tsinionivounouna
+875
Poulouma
234
Ounay
53
Tabol
Apouna
Salé
+18
Léletstivi
886
+178
+290
Navoun
+68
+53
Woké
+901
+190
Tabol
+474
Ounarébrébo
+227
+74
+146
+372
+868
PWELINOKÉVOKE
+193
Roovi
+92
92
Ankoro
+1 182
Apouna
747
+338
Ora
+478
Voutikara
121
Loptchégolono
+202
Batnarouta ou
Térémaratsa
609
+1 034
1 093
Pouria
Voutioro
Toupélel
762
148
Bengié
Lébot
52
Wimbos
1 578
394
MAMARIN
1 059
+1 621
+850
Vololo
+212
+852
Tavel Ndanga
+203
Namatama
WOUNAMOUTSARSARI
Nokowoula
1 132
Kotamtam
+1 747
Nendélié
Ora
Tombet
Vittgo
1 879
Linsoubé
Tabwémassana
+1 043
Ora
Vataraïndin
Vatroto
+1 018
+721
Pounoro
1 742
Taoaloalo
1 541
Paka
Saktouy
Labémaroua
1 044
Tabwimol
Vatépousa
Tambouri
+376
1 084
Batounasébou
Ousièv
Toravongou
Louari
896
322
Balisololo
948
+1 459
Ora
+1 041
+371
Navoutiloui
Morgrif
1 171
+1 484
Latavoa
Batounlamévous
1 251
Fortsenal
785
Lotipa
Naenpoé
+1 281
Vatinosousou
1 059
Matsakaraï
Vouimélé
1 685
Bilibil
Paramavok
Mavounou
TCHIMAKA
437
Tataïkalo
Napona
+1 704
+1 438
Morioul
Papaïsalé
Pic Santo (Laïriri)
Arouvouai
530
Awi
Osovi
Lalaolo
Outalapa
Weylang
1 321
Lawéré
+348
NALEGO
Lalaolo
NAKILO
Morokari
Tsaraépaé
MALAGUILAN
PAYESOULÉ
502
373
238
Kéréi
Batounilovoutsavira
632
26
390
Mataïpévou
Patouatawa
404
Mont Ablot
497
Souléépé
253
Bayalo
Tsarati
83
+371
76
Viasé
Lasoulésoulé
Pouama
Novaka
Ouréouré
Lalouvoukaka
+21
Tasmaloum
Malovira
Vimala
41
Ipayato
+65
Pointe de Powell
Pointe de Mamavoun
Valaé
Tanovousivousi
Taviako
Cap de Sinotarip
SANTO:
Sud-ouest
0   5   10 km
178

SANTO: Sud-est
Mont de la Table
+303
+402
+332
RALI
+32
Loptoro
Wimbo
449
+
+238
Pont de Lapé
NAVARAS
Lapé
567
TAROK
+329
SIKETKET
+399
+482
TANAKAR LATI
735
+387
MANIOK
+151
360
+
217
Rotal
LOWÉRIÉ
Sara 2
Sara 1
Tekari
186
202
Lourou
+78
Pointe de Sinionénéré
Lorédiakarkar
Kolé 1
Kolé 2
105
Totokar
Pointe de Sélé
Pointe de Totokar
+202
Shark-Bay
146
Baie des Requins
Litaro
Lomanioc
+85
Litaro ou Île du Pilote
97
Natawa
+139
Litaroa
68
Litaroa ou Île du Pilotin
Mont Tankara
+784
Tanakar
+98
Toro
473
Patounbanga
Wimbou
+330
Morouas
643
Tanmet
+546
426
Maloutfol
Patounvava
+572
Péréivaé
+464
Patéliou
524
Souléparav
Patoumbéï
+192
+21
Namafoun
TANORIK
Fimélé
Wailapa
Namorou
Wailapa
Namalo
Parisa
Maloka
Tangoa
Élia
Tangoa
Toubana
Oulilapa
Araki
227
Narafoua
ARAKI
167°E
Boutmas
261
602
377
583
Tiouri
+607
223
Lotounaï
+606
Laonasmata
418
185
173
Nambél
Flror
+517
Tafakar
+239
165
+
Tambotalo
Belnatsa
Bélérou
Belembout
355
Laboué
Wamb
42
Mavoun Lif
+202
+181
Sarété
Narango
241
Tomébou
Nasouloun
+198
84
Artatcha
Néafou
Navota
Tangis
Vénoui
Tangis
Rose Point
Pointe de Voutia
Ourélapa
Pointe de Djénaro
23
Lagon de Belmoul
Belmoul
Ratard
Naonébane
Baie de Belmoul
Pointe de Brigstocke
Pointe de Ratard
Pointe de Richard
Wékéssa
Ratoua
Canal
Malo-Pass
Béaou
Avounatari
Natanowaha
Narango
Vatouemba
Aravita
+326
102
Pic Malo
Avounarani
Béloumbévou
Ambilahé
+72
Nanoukou
Récif de Ourénimbavi
Avorani
Amahiroua
Ativousa
Adolala
Arémanapoï
Alowarou
Antaripoï
Avounarara
Amalo
Kona
Baie de la Tortue
Palon
128
Péren
Fanafo
+334
+461
+610
377
Woundrpoutsour
Tanmet
Tavoa
+606
+197
+135
+170
Matéwoulou
Mon Exil
+161
Sarami
17
Saraotou
Valétérourou
+104
Sarakata
15
78
Le Canal
6
La Roseraie
16
147
Aouta
86
54
Port-Latour
13
Ratoua
AORÉ
+99
Port Bénier
+5
Bonassaria
Pointe de Chapuis
Baie de la Délimitation
Pointe de Naourata
69
+5
Pointe de
Naourata
Souchounlagré
Étang de Nandioutou
+81
Naviarou
Avounaléléo
Amambwélao
Naviova
+62
Tanmial
+96
127
275
MALO
10
Samarandas
+82
Asanawari
Vapépoï
+13
Baie de Bougainville
Pointe de Sanif
Pointe de Sanif
Île de la Tortue
Passage du Vaucluse
Mounparap
Malon
Maféa
48
Amséran
Passe de l'Undine
Pointe de Semtarbour
19
Aïs
Aésé
Pointe de Ouren
Cap de Ondine
49
Passe du Diamont
Palikoulo
Baie de Palikoulo
2
Pointe du Scorf
Pointe des Millions de Dollars
Pointe de Narouroundo
Passe du Scorf
Baie de Natoubou
Vounamalaé
Vounasori
Pouéli
Anoussa
Taboumbatou
+36
Toutouba
+5
Abokissa
Pointe de Sabato
Passe du Fabert
Passe de la Dives
Malobolébolé
Baie de Anaoné Pwaravou
Malotina
+2
Malokilikili
Malokilikili
CTRAV ex-IRHO
71
Sourunda
Luganville
Chapuis
Pékoa
18
10
Port-Santo
Canal du Segond
Matafou
Pointe de Taoundou
19
Baie de la Tortue
Pointe le Hénaff
Passe du Tambo
2+
Rotal
Sarakata
Détroit de Bougainville
Pont de Lapé
0   5   10 km
15°30 S
6   7   8   9
© Géo-consulte

(3 532 inhabitants in 1999) are the features that give Malo its special status. Malo is divided into two districts: Western and Eastern Malo.

• • • Infrastructures

Long before its transformation by the American soldiers in 1942, Luganville (Fig. 26) was an important trade center for the northern islands. After the Second World War and the removal of their army, the Americans left an improved infrastructure of ports, roads, landing strips, buildings, and equipment that enabled the development of Luganville's economy; this growth was favorable both to the northern and to the central-northern islands of the archipelago. Luganville enjoys a naturally favorable environment: an exceptional port location, hydroelectric energy produced by the Sarakata River, a mostly rural and diverse backcountry, and many islets that represent ideal locations for touristic development.

Though Santo has the longest road network of the archipelago, it is still commensurate to the size of the island. The main constraint to economic development remains the lack of infrastructure. The roads of Luganville are the only ones to be paved with tarmac, all other road networks are unsealed crushed coral roads. This road network is important around the urban center and in the neighboring districts of Canal-Fanafo, Southeast Santo, and East-Santo. A coral road runs along the eastern coast from Luganville to North Port-Olry and a second road runs inland near Fanafo, Butmas, and goes up to Matantas. South-Santo has a coastal road but the island lacks secondary roads to link the coast to the

Figure 24: Schoolchildren of a rural village in South Santo.

Figure 25: Village scene on the south coast near the Kere River.

Figure 26: Luganville's main street.

inland. Because of the difficult construction conditions of such roads, the districts of North-Santo, Northwest-Santo, and West-Santo have very few of them. Thus many villages are connected by irregular boat service. On Malo, the trail around the island was never completed. Santo Island has an international airport, Pekoa, located to the east of Luganville, as well as a landing strip in Lajmoli in the Northwest Santo district. There are several other airfields dating back to the Second World War at Big Bay, Palikoulo and Roseraie but they have fallen into disuse. There is no air connection available with Malo; the only way to get there is to take a boat from Avounatari to Pointe Ratard, Malo's wharf.

••• Resources and economic activities

Arable land in Santo, Malo, and their islets is estimated at approximately 1 831 km², which represents a little bit more than 43% of the total area. Food-producing agriculture is not only for subsistence, it also benefits from an urban market (Fig. 27) and exportation. Food crops are diversified, with products such as taros (*Colocasia* and *Xanthosoma*), cassava roots, and sweet potatoes. *Colocasia* is the dominating crop in the center and the south, thanks to an irrigation system of taro terraces (Fig. 28), and yam is the most important crop of the east coast and Malo. Malo is the producer of a well-known yam in the markets of Luganville and Port-Vila, named "*Marou*" (*Dioscorea transversa*). In the highlands, the area of food-producing agriculture can reach an altitude of 1 000 m. The production of kava (*Piper methysticum*) for home consumption and marketing has increased considerably over the past decade in Santo Island where it has become the main crop of the east-central area of the island between the regions of Belerou, Fanafo and Ankoro, and in the Navaka valley of the southwest where it allows food-producing agriculture in areas otherwise difficult

Figure 27: The urban market of Luganville.

Figure 28: Taro plantation.

to grow because of their inhospitable topography. Coconut farms are widely developed on the southern

Figure 29: Cattle and coconut plantation: the symbol of Santo farms.

Luganville

The Canal du Segond, the marine channel that separates the southeastern coast of Santo Island from the island of Aoré, is a natural port whose exceptional location attracted the first planters as early as the end of the 19th century. The Canal du Segond is also the name of the first village that was created on the shore of Santo. Respectively in 1882 and 1909, the CCNH (see SFNH) and Catholic missionaries settled on the right shore of the Sarakata River. The village was later called Lugan, after the captain of the ship from the *Messageries Maritimes* that brought supplies quarterly to the local colonizers. Eventually, the town became known as Luganville. European settlement was limited to trading companies (the SFNH, Barrau, de Béchade, Burns Philp), to planters who, for the most part, came from France (about 150 people in 1910), to the mission, and to the Condominium offices. On the other side of the Sarakata, the large swamp that belonged to the SFNH was of no interest to anyone until the American army arrived in 1942 and decided to fill it in and clean it up to enable the installation of their main military base of the South Pacific. The American army transformed the village into a town and brought to it, as in Port-Vila, the infrastructure that was missing (three landing strips and a fourth one in Matewoulou, roads, bridges, piers, warehouses, buildings, etc.).

Luganville competed for a long time with Port-Vila for the first rank in the urban landscape. It was considered the capital of the northern portion of the archipelago. Until 1980, both towns benefited from the same favorable development opportunities, and the insular migrants, attracted by the short term work opportunities that the towns offered, tended to leave for the closest urban center from their island. Though both towns grew at the same rate during the years of 1967-1979, in the decade that followed, Luganville grew approximately half as fast as Port-Vila (2.9 % vs. 5.8 %). This was partly because of the tense political context and the desire of the Nagriamel supporters to withdraw; this atmosphere led many European and Asian investors to leave Luganville by the end of the 1970's. The Independence and the rebellion that followed stopped the growth rate of the second town of the archipelago. The eviction of the French colonizers, who were suspected of participating in the rebellion, put an end to the French cultural influence that prevailed in the northern town. The demographic expansion of Luganville was far less dramatic than in Port-Vila (with a migratory growth of 7 % for Luganville and 61 % for Port-Vila between 1979 and 1989). In the 1990's, migrants began to settle in Luganville and, along with the natural demographic growth, contributed to an increase in Luganville's urban population. Since the last decade, the arrival of foreign investors, tourism, and construction have given a strong boost to Luganville's expansion.

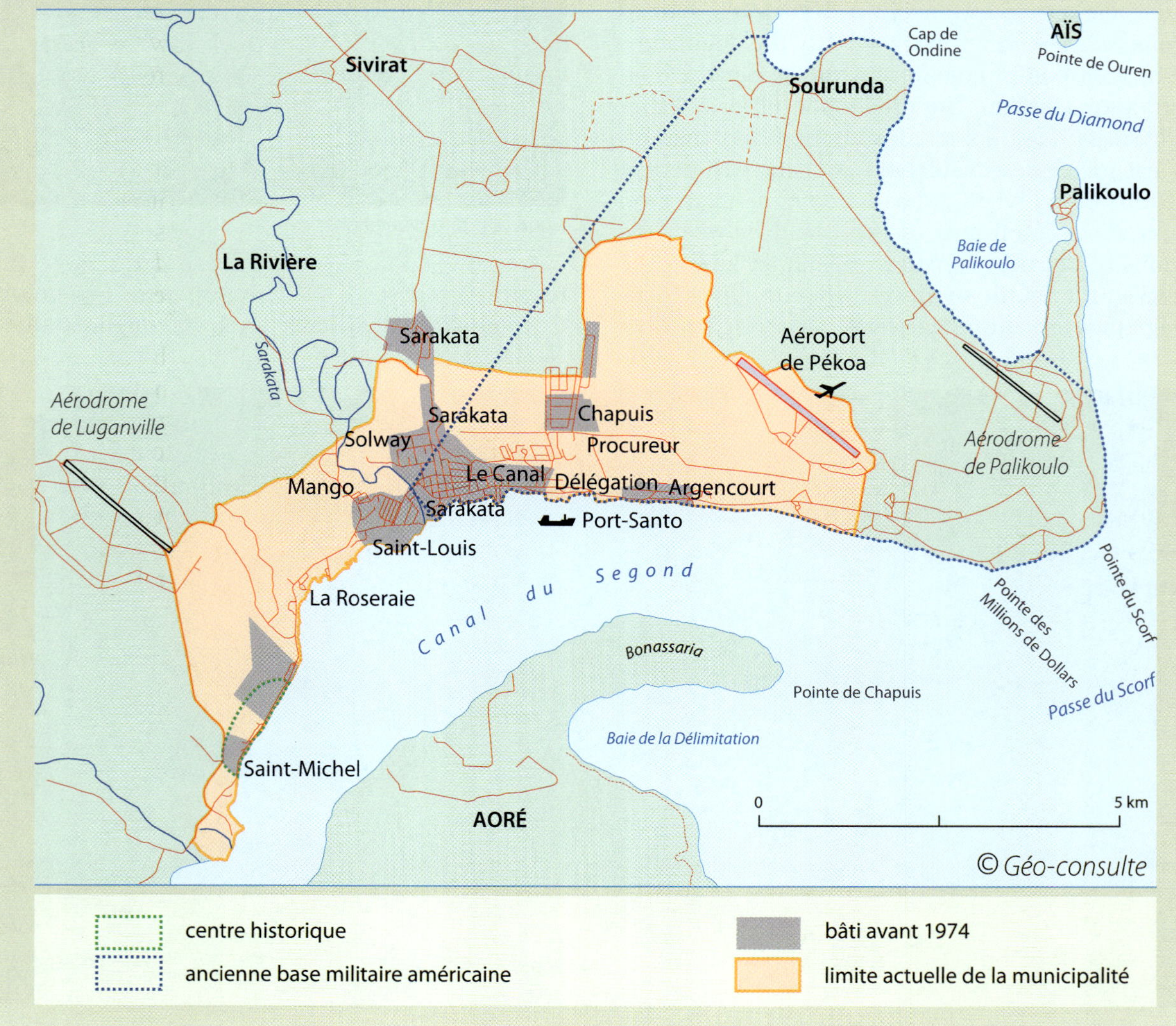

and eastern coasts where they cover an area of about 1 370 hectares. Cacao farms account for about 370 hectares, mainly in the Western Malo district. Converted pastures or woody fallows cover about 1 350 hectares, and cattle grazing under coconuts about 1 050 hectares (Fig. 29). Some of the remaining colonial plantations have been abandoned and subsequently transformed into pastures. Thus, more than 128 km² are dedicated to agriculture; this represents 3 % of the total area of the islands and 7 % of the arable land. There is still a considerable amount of land that has the potential to be developed.

Santo's economy developed mainly around export products: mostly copra (dried kernel of the coconut), cattle, and, to a lesser extent, cocoa. The first copra export dates back to the early 1870's. Local rural labor was scarce in the European plantations located in the southeast of the island, thus thousands of Vietnamese workers were hired in the 1920's. In the 1930's, the production of coffee and cocoa developed but the lack of a labor force favored the development of cattle. Cattle was first introduced to supply milk to Europeans, but it soon became appreciated for the help it provided in the maintenance of the coconut farms, controlling grass growing, and for the meat it provided to workers. Coffee is no longer produced today, but copra and cocoa remain important crops. Small plantation owners produce 80 % of the exported copra and cocoa. Furthermore, small livestock farmers supply 30 % of the meat to Santo's slaughterhouse. Smaller livestock farming of pigs and poultry is well developed but for self-subsistence purposes only.

Santo Island has benefited from major agricultural development projects such as the development of hybrid coconuts, poultry and pig farming, the development of cattle farming in southern Santo, as well as the creation of two research centers on oil and oil-yielding plants. Yet it seems that the results of this research do not impress decision makers in their choice of investment, as demonstrated by the very controversial project of establishing 10 000 hectares of oil palm (*Elaeis guineensis*) plantations even though such culture has proven to be intractable, or the various attempts at introducing rice culture, which have all failed.

Three-quarters of the island are unoccupied and covered with a dense vegetation cover. This evergreen forest is a dense tropical rainforest, typical from the equatorial climate, yet lower in height compared to the height of the Amazon or the Congo rainforests (see the part "Vegetation and flora" and the chapters "Deforestation of Santo and logging operations" and "Conservation efforts in Santo" in this part):

The fishing industry expanded tremendously with the creation of the South Pacific Fishing Company (SPFC) in Palikoulo in the 1950's. The company worked as a recruiting agent of fishermen from Vanuatu to work on foreign fishing boats. The company stopped its activity in the 1980's. The current fishing industry operates mostly through independent fishermen who provide the local market with fresh fish.

There are no viable mineral deposits. Copper mining in Navaka was disappointing.

The private business outside of Luganville is mainly comprised of the grocery trade and kava bars. In terms of transportation, there are bush taxis and water taxis. Other trades include gas stations, inns, bakeries, and sawmills. The fish and meat canneries were abandoned in the 1980's because of a lack of supply and the same happened with the island's oil mills in 2006. There are no more industries on the island.

Tourism is slowly developing. Santo's most touristic destination is its urban area. Santo is known for its diving opportunities; there is a specific draw to the American ship *President Coolidge* and what is left of the American vehicles and heavy equipment that were sunk near Luganville during WWII. The main constraints that hinder the development of tourism are the lack of coordination (tourism office) and the difficulty of gaining access to touristic sites (roads, landing strips, cost of transportation). Yet there is a huge potential for touristic development with its quality beaches, historic sites including WWII artifacts, blue holes and waterfalls, the highest peaks of the archipelago, and unique plants and animals.

Hydrology and Fluvial Geomorphology

James P. Terry

DRAINAGE PATTERNS

Although Santo is the largest island in Vanuatu, covering an area of 3959 km², its unusual shape with two northward extending peninsulas means that most rivers have small catchments, usually less than 100 km² in area (Fig. 30). Two notable exceptions are the dominant Jourdain River, the largest drainage system on Santo and in the Vanuatu archipelago as a whole (Table 1), which drains most of the central region of Santo, flowing northwards into Big Bay, and the Sarakata River that is used for hydro-power generation and drains south east into Segond Channel at Luganville town.

The geomorphology of the western portion of the island, including the north-west peninsula, is dominated by a chain of Miocene-age volcanic mountains, of volcaniclastic geology, aligned in a SSE to NNW orientation, thereby forming an elongated highland spine. The rivers draining the rugged western peninsula are controlled by the linear arrangement of these volcanoes and therefore flow either west to the west coast or east into Big Bay. The catchments are thickly covered with native rainforest. The tallest peak in the chain is Mt Tabwemasana (15°21'S, 166°45'E) which climbs to 1879 m in Santo's south-west, but several other peaks along the spine also attain elevations above 1500 m. In the south-west of Santo, a radial pattern of drainage is identified on the flanks of the Tabwemasana volcano complex. Within individual stream catchments, however, simple dendritic drainage patterns predominate.

The narrow stream headwater courses in western Santo are typical of those on Pacific volcanic islands with mountainous interiors. Upper tributaries are very steep with bouldery channels. Stream channels are sinuous and coarse streambed deposits within them are arranged in step-and-pool sequences. Neighbouring catchments are separated by narrow, serrated interfluves, where slope angles frequently approach 40° or more. The majority of highland stream channels in western Santo are comprised of large, rounded, volcanic boulders, derived from underlying volcaniclastics, breccias and sandstones through *in situ* weathering and exhumation. Short sections of bedrock channel and waterfalls occur

Table 1: Largest river basins on selected islands in the tropical South Pacific.

Island	Country	Largest river	Approximate basin area
Viti Levu	Fiji	Rewa	2918 km²
Malaita	Solomon Islands	Wairaha	486 km²
Grande Terre	New Caledonia	Yate	437 km²
Guadalcanal	Solomon Islands	Lungga	394 km²
Santo	Vanuatu	Jourdain	369 km²
Vanua Levu	Fiji	Dreketi	317 km²
Efate	Vanuatu	Teouma	91 km²
Savai'i	Samoa	Sili	51 km²
Upolu	Samoa	Vaisigano	33 km²

frequently, where numerous small dykes and other resistant intrusions have been exposed by erosion. Stream banks are steep, composed of soil, regolith and boulders and obscured by dense rainforest. Well-rounded gravels of cobble and pebble-sized material is extremely abundant as stream bedload. The high annual rainfall, the precipitous slopes that encourage runoff, and the steep long profiles of stream channels, are all factors promoting a highly energetic fluvial environment and associated rapid rates of sediment transport, clast attrition and bedload rounding in the watercourses of western Santo.

In their lowland reaches, most of the rivers and larger streams of western and central Santo have braided rather than meandering channel patterns, and have formed braidplains and deltas of shifting, interconnecting channels that are separated by braid islands, comprising enormous quantities of coarse gravels of various volcanic lithologies. This interesting fluvial geomorphology is considered separately in "Braiding in the River Jourdain".

The relatively larger river systems of south-eastern Santo, such as the Wambu and Sarakata Rivers that have their estuaries near Luganville (Fig. 31), are allogenic rivers with headwaters rising on volcanic geology, but which then flow across limestone terrain of conical hills lower than 1000 m, descending to rolling coastal hinterlands 0-300 m in elevation. The south-easterly trend in drainage pattern is controlled by prominent tectonic lineaments and faults. Incised, gorge-like sections are common,

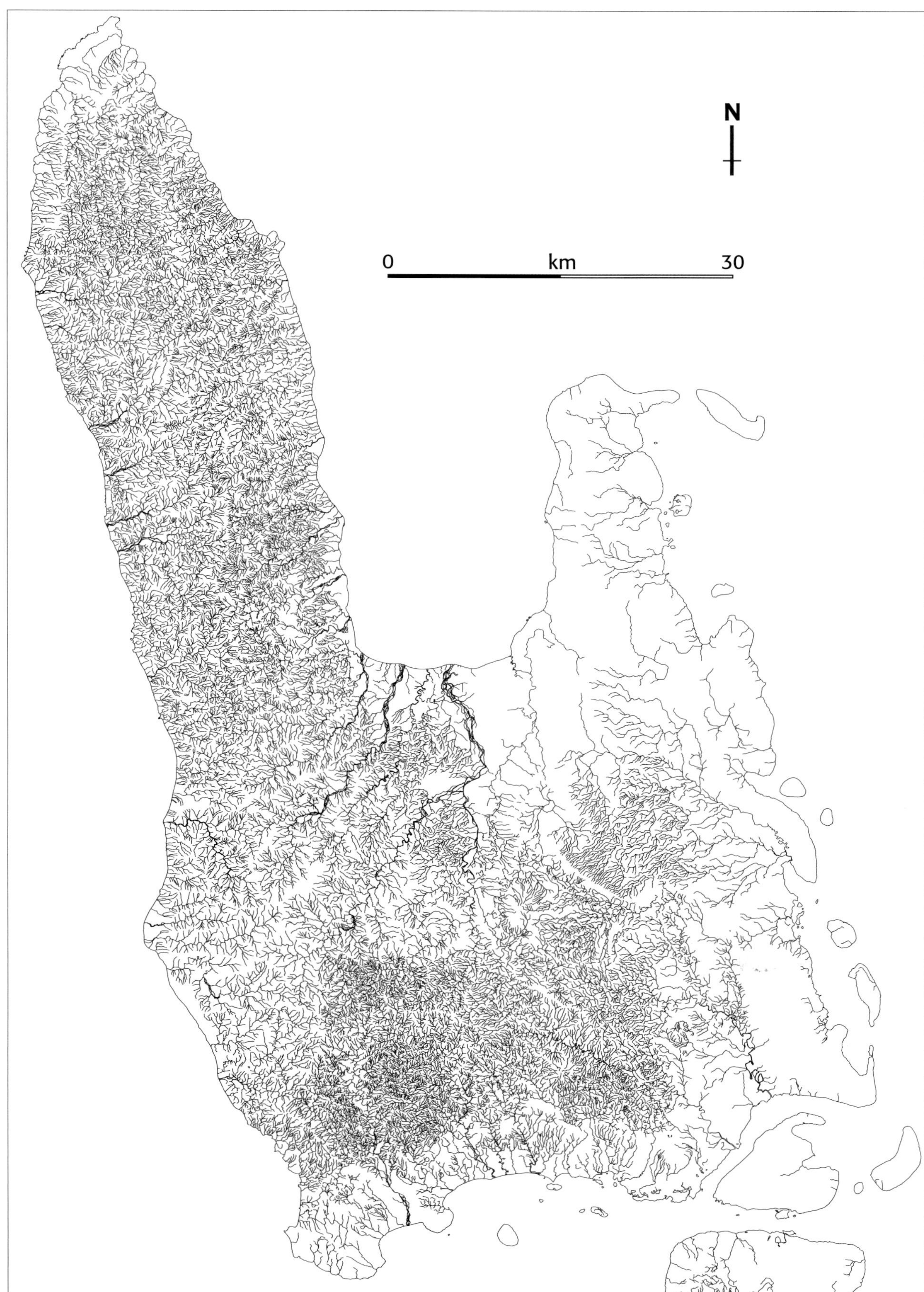

Figure 30: Drainage network on Santo. Base map courtesy of Vanuatu Lands Department.

formed by series of collapsed dolines coalescing with one another along a lineament. Rivers draining the south of the island have more extensive lowland valley sections than streams draining western Santo, with meandering alluvial channels that have built floodplains and terraces comprising fine sediments (Fig. 32). These flat areas are used for settlement, agriculture, coconut plantations and cattle pasture.

Figure 31: **A**: Source of the Sarakata River flowing out of Millennium Cave. **B**: Exit of the Sarakata River into Segond Channel at Luganville.

In the eastern portion of Santo and on the north-east peninsula, surface drainage is either less well-developed, intermittent in nature, or in some areas entirely absent. Drainage density is notably lower than on the rest of the island. These characteristics are associated with the relatively permeable basement geology of uplifted Quaternary and Holocene coral reef limestones in eastern Santo, occurring in the form of denuded plateaus and adjacent "staircases" of emerged marine terraces. Essentially this is a region that has been karstified to varying degrees under the humid tropical regime, and subterranean drainage has developed. The overall 3-5° eastward tilt of the plateaus and terraces means that extant surface drainage has formed parallel drainage patterns, generally flowing through cone karst and nascent cone karst topography to the east coast. Several of the short coastal streams in the East begin at flooded dolines (karstic depressions), either formed by solution or collapse, locally called "blue holes", which are supplied by springs of resurgent groundwaters (Fig. 33).

Figure 32: 1986 aerial photograph of the lower Sarakata River meandering across its adjacent floodplains, immediately north west of Luganville town. Courtesy of the Vanuatu Lands Department.

Figure 33: Blue holes – flooded dolines that are the source of several small streams in the coastal hinterland of karstified eastern Santo.

HYDROLOGY

Little continuous discharge monitoring has been carried out in the rivers of Santo Island. Occasional stage recording and discharge measurements have been undertaken for the major Sarakata and Jourdain systems. Calculating flow rates in the lower Jourdain is extremely difficult because this river has a complex braided network of multiple, shifting channels. A meaningful stage-discharge relationship for the sum of all channels within a valley-bottom cross-section is therefore almost impossible to establish. In October 1981 the Hydrology Section of ORSTOM (*Office de la recherche scientifique et technique outre-mer, Centre de Noumea*; now IRD) embarked on a campaign of flow monitoring near the estuary of the Jourdain. An OTT R20 type stage recorder flow was installed near the river mouth, but the entire station was promptly washed away two months later by the flood produced on 23rd December 1981 by Cyclone Gyan. Replacement equipment installed at another site farther upstream the following year, 7 km from the estuary, was in operation from March 1982 until the end of 1983. Topographic surveys across the two main river channels, in conjunction with velocity measurements and stage readings, allowed daily flow data to be recorded for this period. Daily average discharges of baseflows varied from a minimum 11.6 m³/s in September 1983, which was the middle of the dry season in an exceptionally dry year, up to 82.3 m³/s in the wet season month of February 1982.

For the Sarakata River, ORSTOM also measured discharge for a similar period at a gauging station installed approximately 10 km above the estuary. The site was established above two major tributary confluences, and so measured discharge for only a portion of the Sarakata drainage network, an area of roughly 90 km². The lowest mean daily flows for 1981, 1982 and 1983, recorded in September or October of these three years, were 3.62, 3.89 and 2.72 m³/s respectively. Manual gauging allowed the establishment of satisfactory stage-discharge relationships, and extrapolation of the rating curves meant that peak flood discharges for two cyclones that occurred in this period could be estimated, based on maximum flood heights. Maximum flows were approximately 280 m³/s during Cyclone Gyan on 24 December 1981 and 410 m³/s during Cyclone Joti on 4 November 1982. These are enormous discharges considering the modest size of the monitored basin.

Recent work on reliable long-term flow data for the Tontouta River, a major system in the south of mountainous Grande Terre Island in neighbouring New Caledonia, showed that 65% of the largest historical flows were caused by tropical cyclones, and 75% of cyclone-induced floods were overbank events. A similar situation probably exists for the Sarakata River on Santo, so better understanding of the hydrology of cyclone-generated floods remains a priority for Vanuatu, as well as for other Pacific Island states.

BRAIDING IN THE JOURDAIN RIVER

Figure 34: **A**: Braidplains of the lower Navaka River. **B**: Jourdain River.

The Jourdain River is the largest drainage system in Vanuatu, and dominates the fluvial geomorphology of central Santo (Fig. 34). The river flows generally northwards, draining into Big Bay through the Vatthe Conservation Area. Vatthe is the first designated national forest park in Vanuatu, and protects 2 276 hectares of rich and diverse lowland alluvial rainforest. The Jourdain rises from over 1 800 m on the flanks of the Mt Tabwemasana volcanic peak. The drainage basin covers an area of about 370 km², occupying some 11% of the island. This river, and many of the other bigger streams of central and western Santo that drain across volcanic rock types, such as the Apuna (central), Navaka (south-west) and Vakola (west) Rivers, have formed gravelly braidplains in their lower reaches.

Braided rivers consist of numerous wide, shallow and fast-flowing alluvial channels that subdivide and rejoin repeatedly around bars and islands, forming an intertwining structure. Although braided systems are common in both pro-glacial and semi-arid environments, where vegetation cover is lacking and coarse bedload sediment is plentiful, most large rivers in the tropical South Pacific Islands normally develop a meandering pattern in their downstream alluvial reaches. Thus, an explanation is needed for the origin and evolution of the braided channel patterns on Santo.

The interaction of several controls is probably the best explanation for the active braiding in the Jourdain and other rivers of western Santo. First, a steepening angle of plate subduction at the Vanuatu Trench during the Pliocene to Holocene period (5-0 ma) has been associated with rapid vertical tectonic uplift of Santo Island, at rates up to 7-8 mm/yr. This fast rate of uplift provides the Jourdain River excess gradient, such that the long profile is steep (Fig. 35), which promotes active fluvial downcutting. Second, regular earthquakes associated with local tectonic activity have triggered many landslides, several over 5 km² in size, in the precipitous terrain of the highlands. These slope failures provide sources of fresh sediments into the drainage channels. Third, the regular passage of tropical cyclones produces large and very powerful river floods, which contribute to high rates of bedload transport. Fourth, sequences of Quaternary marine gravels underlie much of the lower Jourdain basin, the exposure and reworking of which has provided abundant coarse gravels to form the channel bars and braid islands in between the shifting channels (Fig. 36).

On top of many of the larger gravel islands of the Jourdain River braidplain, which have remained stable long enough for vegetation colonisation, sequences of fine-grained sediments have been deposited (Fig. 37). Analysis of the caesium-137 (^{137}Cs) content down through the profile of these fine deposits, as determined from laboratory measurements of the gamma-ray spectra of sediment samples, allows an estimation of the rate of vertical sediment accretion. The estimated rate from a single core of alluvium, collected from the bank of the Jourdain in 2004 (Fig. 37), is at least 4.8 cm/yr over the last four decades. It is found by comparison that this sediment accumulation rate is high compared to that measured in other tropical Pacific Island river basins. This indicates both high erosion rates in the upper Jourdain and effective sediment transport in channels, and suggests that rapid tectonic uplift and the effects of cyclone-induced flows on Santo are probably important geomorphic controls on braidplain depositional processes and long-term evolution.

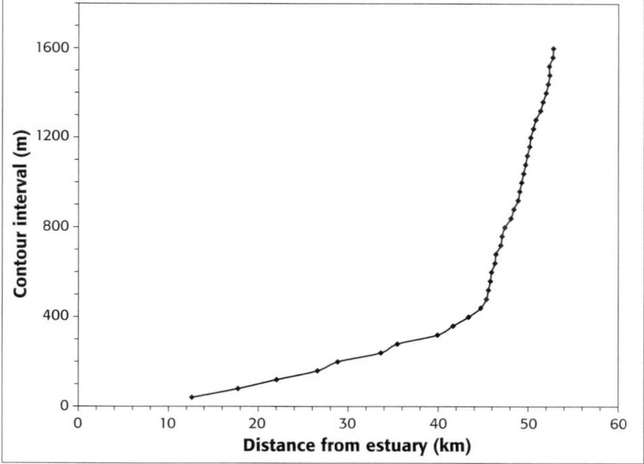

Figure 35: Long profile of the Jourdain River.

Figure 36: Gravelly bars and braid islands, some bare, others vegetated, and multiple interconnecting channels of the lower Jourdain River. The raised limestone plateau of eastern Santo is seen in the background.

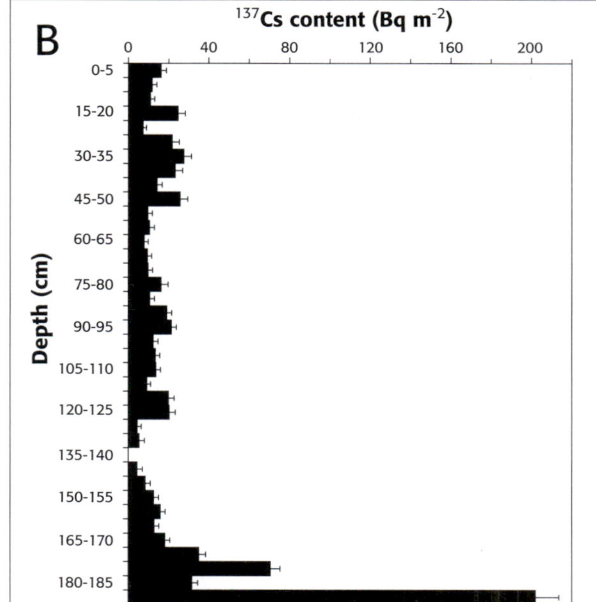

Figure 37: **A**: Sampling a core of fine alluvium on the eastern bank of the main channel of the Jourdain River. **B**: Resulting vertical profile of ^{132}Cs activity in the sediments.

The climate of Santo

James P. Terry

Santo Island lies at 15°S, 167°E in the western South Pacific Ocean. By virtue of its tropical location, surrounded by the warm waters of the South Equatorial Current, the climate of the island is described as 'tropical maritime', or *Af* (Tropical Wet) according to the Köppen system of climate classification. Although the climate is warm and wet all year round, there is a seasonal pattern to precipitation, temperature and other climatic variables. The hot wet season lasts six months during the southern hemisphere summer from November to April, and a drier, cooler season is experienced during the southern hemisphere winter from May to October.

The main regional influence on the climate of Santo is the South Pacific Convergence Zone (SPCZ). This is a broad zone of convergence, low pressure, convectional activity, cloudiness and precipitation (Fig. 38). The SPCZ is a branch of the ITCZ and has an approximately diagonal orientation from north west to south east over the south west Pacific (Fig. 39). It stretches from the 'warm pool' (of tropical ocean water) centred in the far west South Pacific near New Guinea, and extends as a band of cloud and rain across Solomon Islands, Vanuatu, to Fiji, the Cook Islands and often as far as French Polynesia. The convergence zone forms a boundary between the south east trade winds and the region of divergent easterly winds farther to the east that are produced by the semi-permanent anticyclone located in the eastern South Pacific. The SPCZ does not remain static but shows seasonal migration, lying generally to the north of its mean position in midwinter (July) and generally south of its mean position in mid-summer (January). As a general rule, it also tends to be better defined, and as such is characterised by stronger convergence, during the summer. The moisture advection associated with the convergence, uplift and condensation in the low-pressure belt allows stratiform and cumulus clouds to form, giving showery weather. Well-defined cells of low pressure often form embedded within the convergence zone and may develop into tropical depressions, with thick towers of cumulonimbus cloud bringing thunderstorms and intense downpours. The seasonal north-to-south shifting and alternate weak and strong activity of the SPCZ is reflected in the seasonal pattern of Santo's annual rainfall (see below).

The trade winds are another important regional climatic influence on Santo (see Large scale climatic and oceanic conditions around Santo). Vanuatu lies in that part of the western tropical Pacific which benefits from the south east trades, and so the predominant wind pattern across Santo in any season is trade wind flow from the east to southeast, averaging around 5 knots. The wind flow during the summer hot season is generally lighter and more variable, whereas during the cooler winter season, trade wind flows become more persistent and freshen to an average 10 knots. Because the south east trade winds blow uninterrupted across vast stretches of open ocean, they hold large amounts of moisture derived by evaporation at the sea surface. Santo's western mountain range interacts with the

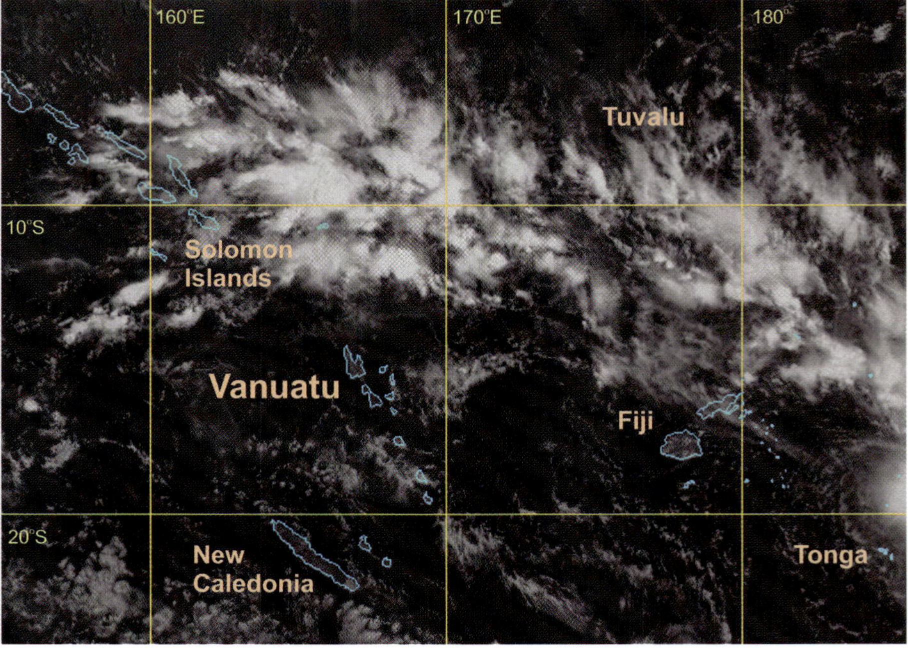

Figure 38: Satellite image on 7 March 2000, showing cloud formation organised along the South Pacific Convergence Zone. The SPCZ lay to the north of Vanuatu, so Santo experienced generally fine weather. Base image courtesy of the Japan Meteorological Agency.

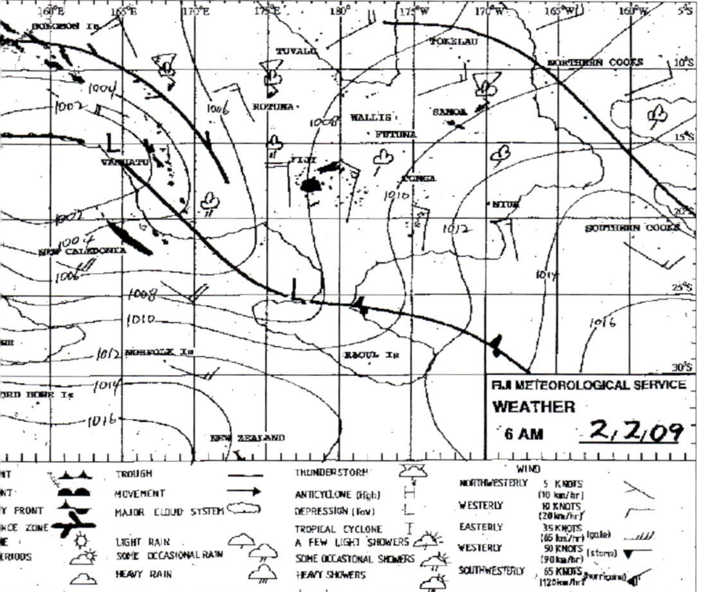

moist south east wind flow. This gives orographic rainfall enhancement in the highlands, but also produces a rain shadow effect across the island, such that the south and east windward side of the mountains receive more moisture than the leeward west coast. Episodes of the positive phase of the El Niño Southern Oscillation (ENSO) can cause the south east trade winds to die down or reverse in direction.

Figure 39: Synoptic weather chart for the western South Pacific on the morning of 2nd February 2009 (5 a.m. Vanuatu Standard Time). The centre of the South Pacific Convergence Zone is drawn through Solomon Islands, passing to the north and east of the Vanuatu group. Note the large low pressure system forming to the west of Santo which extends a trough to the central and southern islands of Vanuatu. The low gradually deepened and moved south, and brought thunderstorms and heavy rain for several days to Santo. Courtesy of the Fiji Meteorological Service.

AVERAGE CLIMATIC CONDITIONS

Climate data for Santo are available for one long-term weather station, located by the coast at Pekoa International Airport (15° 31'S, 167° 13'E), so detailed climatic description must necessarily be limited mostly to this station. The airport is situated outside Luganville in the south east corner of the island. The weather station lies at an elevation of 45 m a.s.l. since the airport is built on an uplifted limestone terrace. Reliable data records exist for temperature, rainfall and humidity back to the early 1970s and for wind measurements to the mid-1980s. Figure 40 summarises the long-term monthly averages at Pekoa station. Precipitation totals 2 252 mm per annum, of which 66 % is distributed in the wet season and 34 % in the dry season. Rainfall in the highlands of the western peninsula, and on the windward flanks of Mt Tabwemasana which rises to 1 879 m in the south west of the island, is expected to be much greater due to orographic effects. Elsewhere, mainly in the basins of the Sarakata and Jourdain Rivers, some other precipitation measurements are also available owing to hydrological research undertaken by ORSTOM (now IRD) from 1981 to 1984. Eight raingauges were installed in coastal and inland locations additional to Pekoa, for which average yearly rainfall was estimated (Table 2). It is seen that there is good correlation between elevation and annual precipitation (Fig. 41).

February is the wettest month on Santo, on average receiving 299 mm of rainfall at Pekoa, whereas August is the driest month, receiving 83 mm. Temperature is continually warm owing to the low latitude and surrounding warm ocean, and varies little throughout the year; there is just a 2.5 degree mean annual temperature range between August

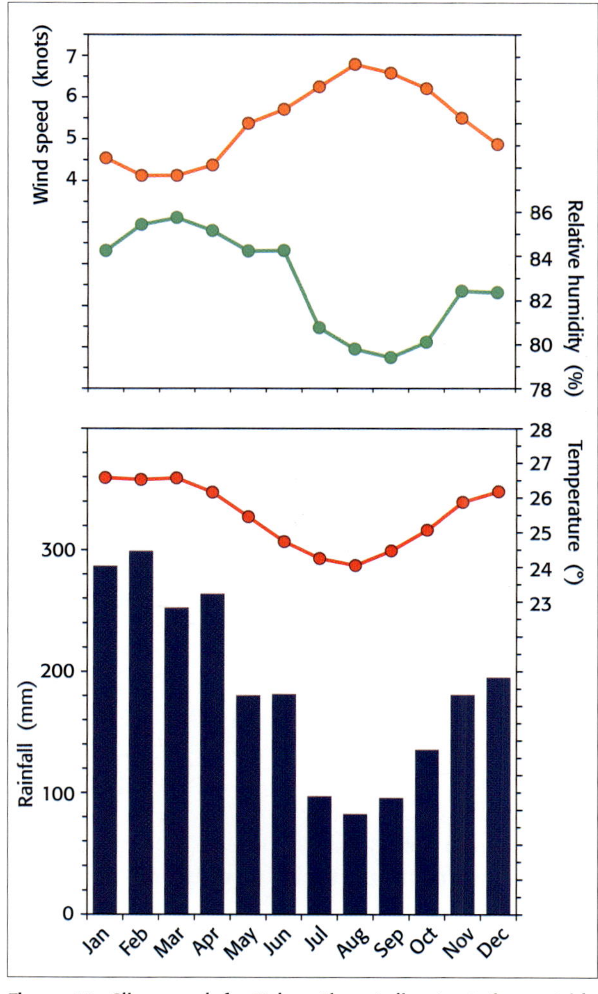

Figure 40: Climograph for Pekoa Airport climate station outside Luganville (15° 31'S, 167° 13'E; 45 m a.s.l.). Means are based on long-term records: rainfall 1974-2007, temperature 1973-2007, humidity 1973-2005 and wind speed 1985-2008. The climate has two seasons, a hot wet season from November to April and a cooler 'dry' season from May to October. Data kindly supplied by Vanuatu Meteorological Services.

Table 2: Average annual rainfall at nine sites on Santo. Values are based on 1974-2007 data for Pekoa and estimated from 1981-84 data for all other sites. Data provided by Vanuatu Meteorological Service and from ORSTOM (1985).

Site	Elevation (m)	Annual rainfall (mm)
Pekoa	45	2252
Saraoutou	15	2620
Fanafo	160	4000
Butmas	440	5120
Peren	190	4300
Tankara	606	5550
Matantas	10	3010
Nassara	260	3560
Bakakara	762	5480

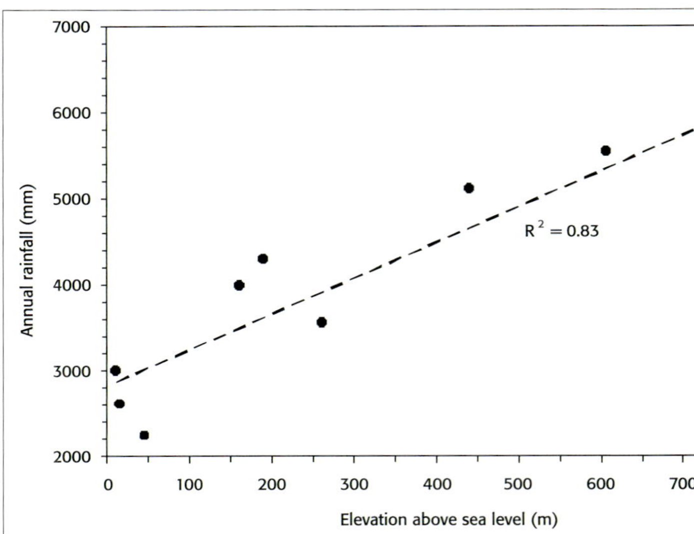

Figure 41: Linear regression between average annual precipitation and elevation at nine sites on Santo (sites are listed in Table 2).

(24.1 °C) and January (26.6 °C). Diurnal temperature range is around 6 degrees. Mountain temperatures are appreciably cooler than at sea level, especially at night. Relative humidity likewise displays little average variation between 79.4% in September and 85.8% in March. The combination of lower rainfall, slightly decreased temperatures and atmospheric humidity, and fresher trade winds during the dry season, make conditions feel more comfortable at this time of year (Fig. 42).

Figure 42: Clouds form above the peaks of western Santo's volcanic ranges, whilst the coastal lowlands of the Big Bay area enjoy morning sunshine on 6th August 2006.

CLIMATIC EXTREMES

Extremes in climate have been recorded that deviate far from the mean conditions described above. Drought effects can be experienced owing to moisture shortage in the dry season of drier-than-average years. The driest month on record is August 1979, which received only 1.4 mm of precipitation, closely followed by September 1983 when only 2.8 mm of rain fell. Often, prolonged rain failure resulting in exceptionally dry years is linked to positive phases of the El Niño Southern Oscillation (El Niño episodes), when rainfall across Vanuatu drops below average. One index of the strength of ENSO is the Southern Oscillation Index (SOI), which is a measure of the mean monthly atmospheric pressure difference between Tahiti (French Polynesia) in the central South Pacific and Darwin, Australia. Monthly values of the SOI show no correlation with monthly rainfall totals in Santo because of the seasonal nature of Vanuatu's rain distribution. However, figure 43 illustrates how the pattern in average SOI values (12-month running means)

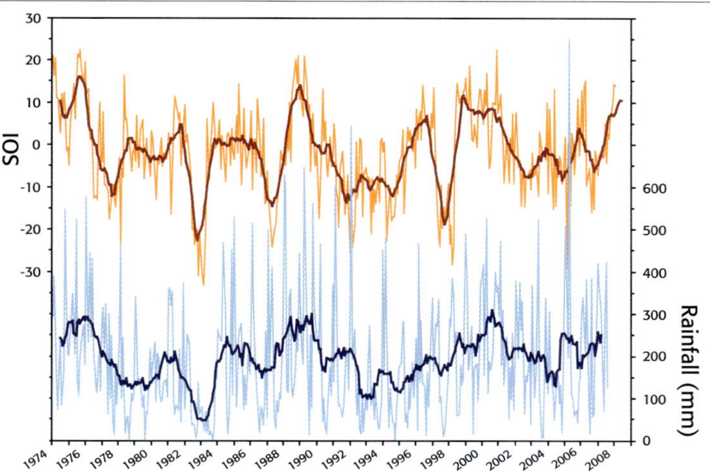

Figure 43: Time series from 1974 to 2008 of monthly values of the Southern Oscillation Index (SOI) and monthly rainfall totals for Pekoa climate station. To remove seasonality effects, 12-month central moving averages are also drawn (bold brown and blue lines). These average plots show some similarity in pattern and some (although not all) periods of relatively low and high rainfall, e.g. the 1982/3 (dry) and 1988/89 (wet) periods, correspond to episodes of strong ENSO activity. SOI data provided by the Australian Bureau of Meteorology (BoM, 2009).

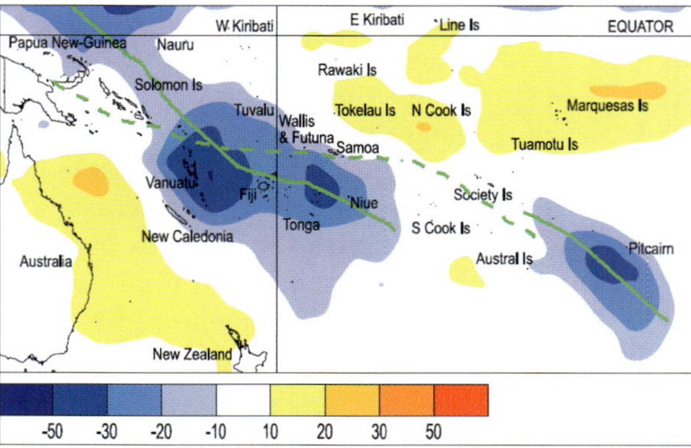

Figure 44: Outgoing Long-wave Radiation (OLR) anomalies, in Wm[-2]. Blue shading equals high rainfall and yellow equals low rainfall. The April 2005 position of the SPCZ, as identified from total rainfall, is indicated by the solid green line, compared to the average SPCZ position shown by the dashed green line. Image courtesy of NIWA (2005).

bears good relationship with equivalent averages (that remove seasonality influences) for precipitation patterns on Santo. The 1982/83 El Niño episode resulted in the two driest years on record: 1075.0 mm (1982) and 685.3 mm (1983), which are only 48% and 30% respectively of the mean annual precipitation.

April 2005 is the wettest month on record. In this month, the South Pacific Convergence Zone was strong and active, extending from the region north of New Guinea southeast towards Fiji (Fig. 44). Precipitation was at least 125% of average over much of Micronesia, Vanuatu, and parts of New Caledonia, Fiji, Tonga and Niue, with some locations measuring more than 300% of normal rainfall. As a result of this convergence activity, weather was very wet across Vanuatu generally. Pekoa measured 950 mm for April 2005, with five days in the month receiving daily rainfalls exceeding 100 mm.

TROPICAL CYCLONES

The most extreme meteorological conditions on Santo occur during tropical cyclones. These rotating migratory storms normally bring furious winds, intense rainfall and powerful waves. The most intense storm so far this decade to afflict Santo was Cyclone Ivy in February 2004, which attained category-4 (hurricane intensity) strength. On 26th February (local time) Ivy tracked immediately east of Santo as a mature system with a well-developed, symmetrical eye and maximum sustained winds[4] estimated up to 90 knots near the centre (Fig. 45). More recently, Cyclone Funa passed directly over Santo on 17th January 2008 at category-2 (storm intensity) (Fig. 46), with a minimum central pressure of 970 mb and sustained winds[5] of 55 knots.

4 - Estimated by the Joint Typhoon Warning Center, Hawaii, using 1-minute averages.
5 - Estimated by Vanuatu Meteorological Services, using 10-minute averages.

Vanuatu waters are a common route for cyclone paths, compared to areas farther east in the western tropical South Pacific. From 1970 to 2006, 84 cyclones affected parts of Vanuatu with strong winds, giving an average frequency of about 23 per decade. Peripheral cloud bands associated with other cyclones that did not pass close to Vanuatu may also have delivered heavy rains to Santo. Cyclone

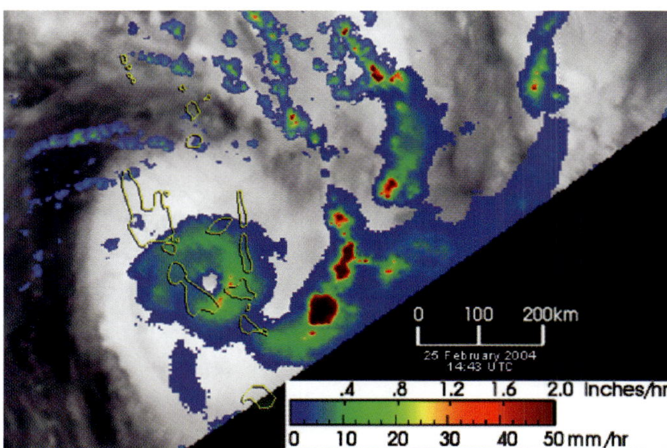

Figure 45: Satellite image of Cyclone Ivy captured by the Tropical Rainfall Measuring Mission (TRMM) at 14:43 UTC on 25/02/04, making landfall on the eastern coast of Malakula Island. In spite of the strong winds experienced, the cyclone was a relatively 'dry' storm for Santo; the eye is surrounded by mainly moderate rain rates (green areas) with the most intense rates (dark reds) present in rain bands off to the east. Image produced by H. Pierce (NASA-GSFC).
TRMM is a joint mission between NASA and the Japanese Space Agency.

tracks between 1990 and 1999 are illustrated in figure 47. This shows that most storms affecting Santo do not develop nearby, but instead form at lower latitudes (farther north) and approach the island either directly, or sometimes following convoluted tracks, before passing away to the south. Individual cyclones are often known for their erratic movement and therefore cause problems for forecasters to predict, but in general terms they tend to maintain southerly or south easterly movement until south of 20°S, when they steer along parabolic curves more

towards to the east, before entering and dissipating in extra-tropical waters beyond 25°S. Cyclones can affect Santo in any of the hot season months from November to April, but the greatest frequency is in January and February.

Several of the highest daily rainfalls on Santo were produced when cyclones followed unusually complex or looping tracks through Vanuatu, thus extending the time for moisture delivery, for example cyclones Bola in 1988 and Esau in 1992. Table 3 lists maximum daily rainfalls generated by cyclones in recent decades, above an arbitrary threshold of 50 mm; the highest recorded 1-day total was 266.6 mm on 24th January 1994 during the passage of Cyclone Sarah.

Table 3: Maximum daily rainfalls greater than 50 mm, produced by tropical cyclones from 1981-2008, at Pekoa climate station. Data provided by the Vanuatu Meteorological Services.

Cyclone Name	Maximum 1-day rainfall (mm)	Date of occurrence		
Cliff	92.4	15th	Feb	1981
Gyan	112.8	23rd	Dec	1981
Joti	57,2	4th	Nov	1982
Beti	75.4	6th	Feb	1984
Eric	100.8	15th	Jan.	1985
Nigel	246.4	18th	Jan.	1985
Hina	52.8	14th	Mar	1985
Keli	122.0	13th	Feb	1986
Patsy	112.5	15th	Dec	1986
Uma	131.3	5th	Feb	1987
Anne	247.6	11th	Jan	1988
Bola	262.2	29th	Feb	1988
Eseta	87.1	18th	Dec	1988
Lili	171.5	8th	Apr	1989
Betsy	77.6	9th	Jan	1992
Daman	117.5	15th	Feb	1992
Esau	219.3	25th	Feb	1992
Prema	57.7	29th	Mar	1993
Sarah	266.6	24th	Jan	1994
Tomas	127.5	25th	Mar	1994
Fergus	65.0	26th	Dec	1996
Susan	78.8	6th	Jan	1998
Yali	98.0	20th	Mar.	1998
Zuman	223.7	1st	Apr	1998
Dani	148.0	19th	Jan	1999
Ella	107.7	5th	Feb	1999
Iris	115.0	10th	Jan	2000
Paula	119.6	27th	Feb	2001
Sose	129.6	7th	Apr	2001
Gina	115.0	9th	Jun	2003
Ivy	74.0	25th	Feb	2004
Kerry	89.0	6th	Jan.	2005
Funa	87.3	18th	Jan.	2008

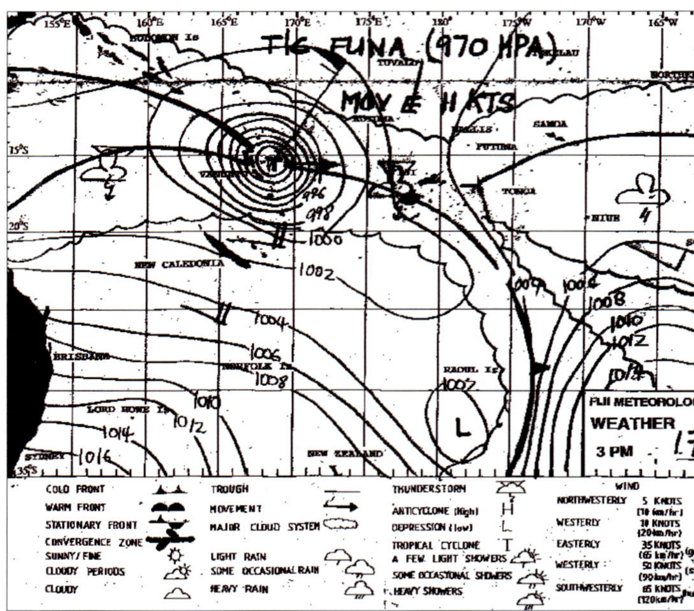

Figure 46: Synoptic map of the south west Pacific on the afternoon of 17th January 2008. Cyclone Funa has tracked across Santo from west to east within the previous hour, bringing peak sustained winds of 55 kts. Maximum daily rainfall of 87.3 mm was recorded at Pekoa station at 9 a.m. on 18th January for the previous 24 hours. Map courtesy of Fiji Meteorological Service.

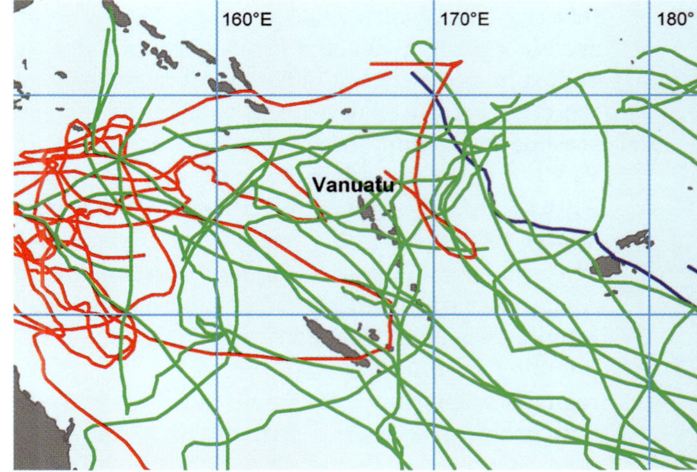

Figure 47: Tracks of tropical cyclones in the decade 1990-99 affecting Vanuatu (centre of map) and surrounding areas of the south west Pacific. Line colours indicate the degree of track complexity: **blue** - relatively straight; **green** - moderately curvy, **red** - sinuous. Image courtesy of Nick Rollings, formerly GIS Unit, University of the South Pacific.

Climatic and Oceanic Conditions around Santo

Christophe Maes & David Varillon

The Pacific Ocean is the largest of the oceans. It is divided by an inter-oceanic ridge system close to its eastern boundary, producing sub-basins in the central and western Pacific Ocean similar in size to the Atlantic and Indian Oceans. In the southwest Pacific, New Zealand and the various Melanesian Islands and Archipelagos provide natural boundaries for the adjacent Tasman and Coral Seas. The Vanuatu archipelago, including Santo, marks the main entrance of the Coral Sea that is bounded, respectively, northward and southward by the Solomon Sea and the New Caledonia basin. Over such a vast area, the climate of the region is mainly controlled by its oceanic context and by large-scale atmospheric circulation features. The latter include the trade winds, the Hadley and the Walker circulations, the tropical convergence zones, the subtropical high-pressure belt and the zonal westerlies to the south. The InterTropical Convergence Zone (ITCZ) lies just north of the Equator and the South Pacific Convergence Zone (SPCZ), extends from near the Solomon Islands to Samoa and beyond. Convergence zones are regions of lower pressure where converging, rising air produces clouds and rainfall. The southeastern Trades associated with the SPCZ are weaker than their northeastern counterparts, but they are extremely steady such that completely calm conditions under the SPCZ are encountered not more than 30% of the time over the course of a year and the region is one of the most persistently cloudy regions on earth. At seasonal and biannual timescales the southwest Pacific region is also under the influence of the Australian summer low pressure that produces a monsoonal wind pattern in the northern Coral Sea and across the Vanuatu archipelago. The Australian monsoon circulation results mainly from feedbacks in the seasonal cycle of the atmosphere-ocean interaction in the warm water pool region (a vast area of the western tropical Pacific with surface water warmer than 28°C). The warm pool and its local variability also influence the generation and propagation of tropical cyclones, which occur in the region during the October-May period. Because of the significance of these major circulation features, long-term climate trends are largely determined by corresponding long-term trends in the strength and position of the SPCZ, the Australian summer monsoon and the trade winds.

From an oceanographic point of view, the geographical position of Santo places it within a very interesting transition zone between the largely zonal equatorial circulation and the subtropical gyre circulation to the south. In the gyre, the circulation of the upper ocean layers, typically throughout the upper one or two kilometres of the water column, results directly from the wind-driven transports in the Ekman layer and from the geostrophic flow produced by the pressure gradient forces down to bottom. Under the trade winds the major westward components of the Pacific equatorial current system are the North Equatorial Current and the South Equatorial Current (SEC). These currents respond quickly to variations in the wind fields and they are therefore strongest during the winter of their respective hemispheres when the trades reach their greatest strength. The subtropical part of the SEC (near 15° S) blends with the northern limb of the subtropical gyre circulation of the South Pacific that transports water toward the equator at the depth of the main thermocline. The mass temperature and salinity characteristics of that water were formed far away to the southeast when that water was at the surface and interacted with the atmosphere. By transporting water masses over such long distances toward key regions such as the equatorial band where climate events as El Niño occur, the various pathways of this ocean circulation are of primary importance for understanding the role of the ocean in the global climate system. A more precise understanding of the role of the global thermohaline circulation and of the confluence of many different water masses appears crucial nowadays. Consequently, the western equatorial Pacific Ocean, as a crossroads for thermocline and intermediate waters formed at higher latitude, has gained renewed attention in recent years. However, the region presents a serious challenge to our descriptive abilities because of strong temporal variability, complicated topography and non-linear dynamics. Direct observations at sea remain relatively sparse and are often confined near the main routes of commercial vessels. In addition, it had been generally assumed that the circulation in the subtropics of the southern hemisphere was weaker than in its northern counterpart. This perception was in part confirmed by large-scale analyses of hydrographic measurements that indicated a broad westward flow at a depth of a few hundred meters. This point of view began to break down only in early 2000 when several studies reported the existence of a series of zonal jets resulting from the splitting of the SEC flow by the topography associated with the main reefs and archipelagos of the southwest Pacific Ocean. Recently, direct observations from an underwater autonomous glider have offered the first description of the zonal jets entering the Coral Sea through the gap between New Caledonia and the Solomon Islands. The main patterns of the circulation across the Vanuatu archipelago thus rely on the characteristics of the so-called North Vanuatu Jet (NVJ), a 300 km wide westward current that is associated with the slope of the main thermocline between 12°S and 16°S. When entering into the Coral Sea the NVJ is choked by the topography between the Banks Islands and the northern tip of Santo resulting in a separation of the water originating from the Fiji archipelago. The relatively deep extension of the NVJ along the water column, as revealed by autonomous deep floats (at ~1000 m depth), also represents a surprising result that is not completely understood. All these recent results need further study to understand the relationship of the jets with smaller scales and coastal dynamics. However, to our knowledge, no existing investigations have focused on the circulation patterns in the immediate vicinity around Santo.

CLIMATIC VARIABILITY

••• Seasonal and interannual variations

The most important seasonal variability in the dynamics of the southwest Pacific Ocean is linked to the displacement of the SPCZ which shows more movement in austral summer than at other times of the year. Such variability is strongly coupled to the very warm sea surface temperatures (SSTs) above 28°C of the tropical warm pool. The variations observed at Santo reflect primarily its position on the southwest fringe of the warm pool as shown in figure 48. Because of the ongoing MOTEVAS program which is dedicated to geodesic studies, observations of in situ temperature and salinity at two sites on the western side of Santo, namely the Sabine and Wusi banks (near 16°S-166°20'E and 15°20'S-166°30'E, respectively), are available since the end of 1999 (Fig. 49). Thermal and conductivity sensors are installed there with a tidegauge anchored at a depth of 15 m and so, these measurements could be thus viewed as representative of the surface. These timeseries illustrate the dominance of seasonal to interannual variations in the climatic conditions near Santo. Whereas the temperature varies seasonally with a typical range of 4-5°C the salinity appears more representative of the interannual conditions. Using a 25-year timeseries of SST near the Sabine bank a spectral analysis reveals an interannual band between two and seven years with a maximum peak around 5.5 years. In addition, the variations in salinity measured from thermosalinographs installed on commercial vessels also confirm that a difference larger than 0.2 exists between periods of La Niña (positive SOI) and El Niño (negative SOI) conditions. Even if local rainfall plays an unquestionable role in the salinity variations, these differences indicate different origins of the water masses around Santo depending on the interannual state of the equatorial Pacific Ocean. Another interesting point regards the quite permanent difference in temperature as large as 2°C between Sabine and Wusi banks during the austral winter season. The difference may result from the effect of wind-driven coastal upwelling that

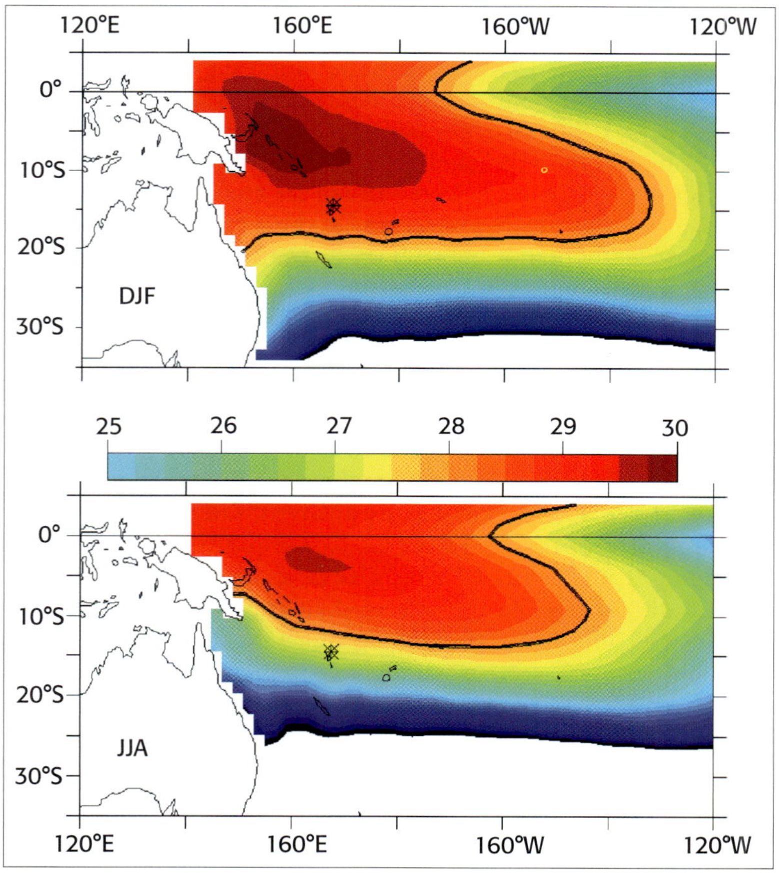

Figure 48: Mean Sea Surface Temperature (in °C) of the southwest Pacific at the seasonal timescales (**DJF** indicates December-January-February and **JJA** indicates June-July-August). The dark line represents the 28°C isotherm and subtropical values below 22°C are not represented. The black star recalls the position of Santo.

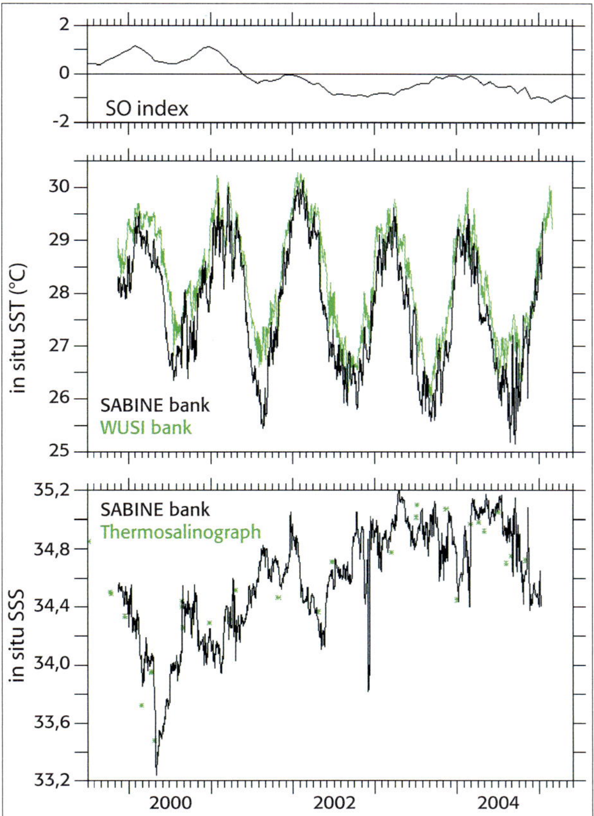

Figure 49: Timeseries of the Southern Oscillation Index, of in situ **SST** at Sabine and Wusi banks and of the sea surface salinity (**SSS**) at Sabine bank for the period 2000-2004. The thermal and conductivity sensors are located at 15 m depth. On the lower panel the star symbol represents the salinity measurements collected on commercial vessels in the context of the ORE-SSS (www.legos.obs-mip.fr/observations/sss).

appears during the strongest season for the trades. These upwelling events may share some similar origins and dynamical processes with the variability observed during strong cooling events off the western barrier reef of New Caledonia.

••• Longer term changes and trends

Connections between the tropical and subtropical oceans through the wind-driven meridional overturning ocean circulation are believed to be of primary importance for decadal and longer temperature fluctuations. This phenomenon is referred as the Pacific Decadal Oscillation in the North Pacific whereas, in the South Pacific, it is also known as the Interdecadal Pacific Oscillation (IPO). This variability is characterized by low frequency fluctuations with ~15 to ~30-year time-scales. During the 20th century three phases of the IPO have been identified:
- A positive phase (1922-1944);
- A negative phase (1946-1977);
- Another positive phase (1978-1998).

Spatial patterns in these decadal trends are strongly affected by the SPCZ, especially for the changes during the mid-70s. The shifts in the position of the SPCZ are apparently related to El Niño Southern Oscillation variability on interannual time-scales and to the IPO variability on decadal time-scales. The variations at the two time-scales appear to be of similar magnitude and they are independent. However, the physical processes involved in these different fluctuations are still the topic of an open debate that depends in part on the tropical or extratropical origin of the particular phenomenon. Among the different theories, a particular emphasis has been placed on the importance of the South Pacific in sustaining tropical decadal variability through remote influence of the atmospheric circulation.

Superimposed on the decadal variability that may be inferred from modern observations is an acceleration of the warming trend over the last 50 years that is due to the increase of anthropogenic gases in Earth's atmosphere. An increase is observed not only in the ocean heat content of the upper layers but also in the deeper layers of the oceans. These climatic changes and their future projections over the next 50 years are very important consequences for coral reefs. Although it may be tempting to link this warming to the enhanced greenhouse effect, the response of the entire Pacific to El Niño- or La Niña-like conditions remains still uncertain. Coupled numerical models as well as historical reconstructions based on sparse observations such as those most often used for the SST field have their own flaws and caution is required in using them as evidence of the present climate variability. Similar conclusions have been drawn from the various paleoclimate proxies that describe the variability during the last millennia. A great advantage of these last data is that they facilitate separating the natural from the anthropogenic effects. Coral proxies in the South Pacific indicate however that expansion of the SPCZ would imply a gradual change in the South Pacific to more La Niña-like conditions in the long term mean. Such variations in the position/displacement on decadal time scales of the SPCZ influence not only mean precipitation, but also daily rainfall and temperature extremes as observed on islands.

Another very important climate change for island communities of the South Pacific concerns the rise of the mean sea level. Sea level tendencies suffer however from the same uncertainties as the surface temperature variations with regard to the possible influence of decadal fluctuations. The large variability in climatic signals and the shortness of many of the individual records contribute to uncertainty of historical rates of sea-level rise. For 1993-2001, all the data available exhibit large rates of sea-level rise, approaching 30 mm/yr, over the western Pacific Ocean. If there is some evidence that the sea level rise observed over the last decade is largely due to thermal expansion, present estimates are still sufficiently uncertain to exclude other contributions. For instance, a clear freshening trend of the order of 0.1-0.3 per 30-yr, together with an extension of the low-salinity water at the surface, has been reported from a recent analysis of sea surface salinity measurements in the SPCZ region.

REGIONAL OCEANIC CIRCULATION

... Water masses

The SST and SSS features result principally from direct air-sea interactions and the action of the winds driving the ocean circulation of the upper layers. Heat and freshwater transfer between the atmosphere and the ocean penetrates into deeper layers through the surface mixed layer. In the tropics, winter cooling is not strong enough to destroy the seasonal stratification, and a permanent thermocline is maintained throughout the year. It is a transition zone from the warm waters of the surface layer to the cold waters of great oceanic depths. The processes that formed the permanent thermocline and fixed its properties are governed by a combination of water mass formation and vertical pumping in response to the wind. The water masses of the thermocline are injected in the subtropical convergence zone at intermediate depths through a subduction mechanism. Because their properties are fixed at the surface these water masses keep characteristics that can be traced in temperature-salinity relationships along isopycnal surfaces. Six thermocline water masses can be distinguished in the Pacific Ocean, and the most saline of them is observed in the western and southern region. Below the thermocline depth the water column is mainly filled with intermediate water characterized by a minimum in salinity near the 800-1 000 m depth. All the above features are illustrated by the vertical profiles of temperature and salinity (Fig. 50), observed on the western side of Santo and recorded by autonomous floats of the international ARGO program. The bottom of the mixed layer deepens to 120 m depth during the boreal summer (linked to the colder temperature as shown in Fig. 49) and is shallower than 20-40 m during boreal winter. The permanent thermocline corresponds to waters with temperature between 10 and 20 °C associated with the maximum salinity near 200 m depth. The intermediate waters fill the lower part of the column, down to 2 000 m depth and deeper, and they could be identified by a minimum in salinity. The rest of the ocean depths is filled with abyssal water masses those the origins are linked to the Antarctic bottom water. Such waters are renewed very slowly and their westward spreading is strongly influenced by the bottom topography.

... Currents around Santo

The most prominent feature of the ocean circulation in the South Pacific is the subtropical gyre, consisting of the South Equatorial Current (SEC) at around 15°S, the East Australian Current, and the eastward return current and the Peru/Chile current in the eastern Pacific Ocean (Fig. 51A). However, the traditional and climatological view of the SEC as a broad and weak westward flow is misleading. Because of the presence of shallow and complex topography associated with islands and reefs the SEC, the inflow toward the southwest Pacific is broken into several narrow zonal jets at the southern and northern tips of the larger islands such as Fiji, New Caledonia and Vanuatu. A more careful consideration of the influence of the topography in updated analyses based on historical hydrographic data sets has led to the recognition of these zonal structures in the ocean circulation of the southwest Pacific. However the amplitude and the properties of these different pathways for the SEC inflow remains poorly documented. These jets have been directly observed with recently developed observational tools such as sea gliders and their dynamics

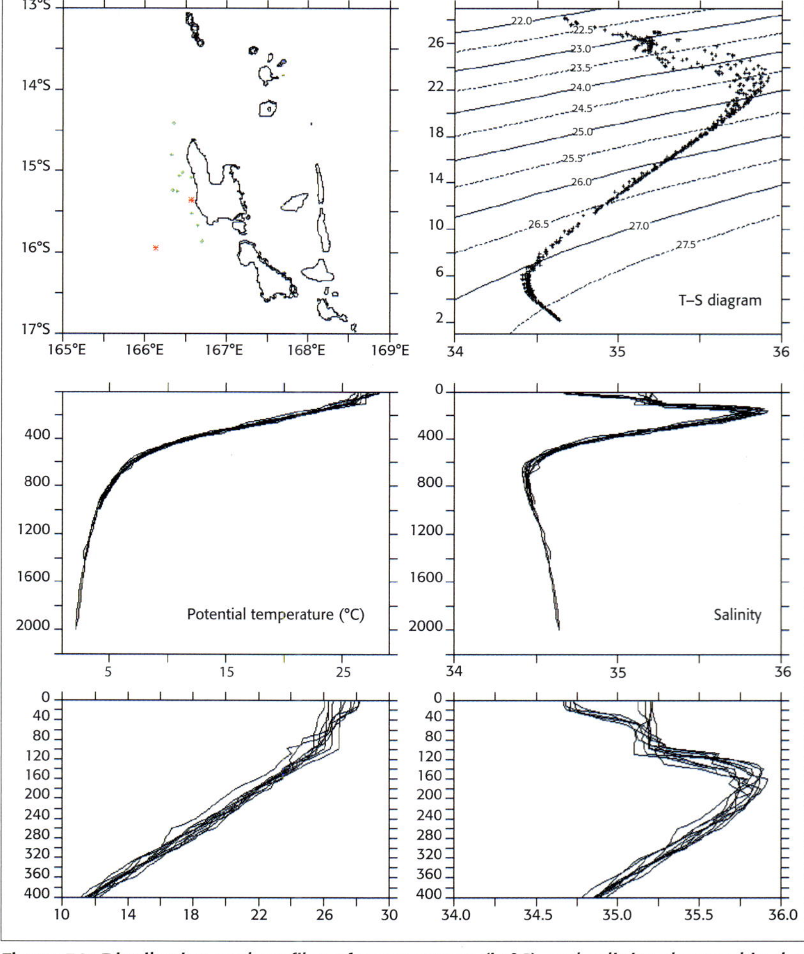

Figure 50: Distribution and profiles of temperature (in °C) and salinity observed in the southwest region of Santo (green symbols). On the geographical map the red stars indicate the position of the sensors at Sabine and Wusi banks (see figure 49). On the T-S diagram the continuous and dashed lines indicate the corresponding isopycnal values (kg/m³). The lower figures zoom on the upper layers as shown in the middle.

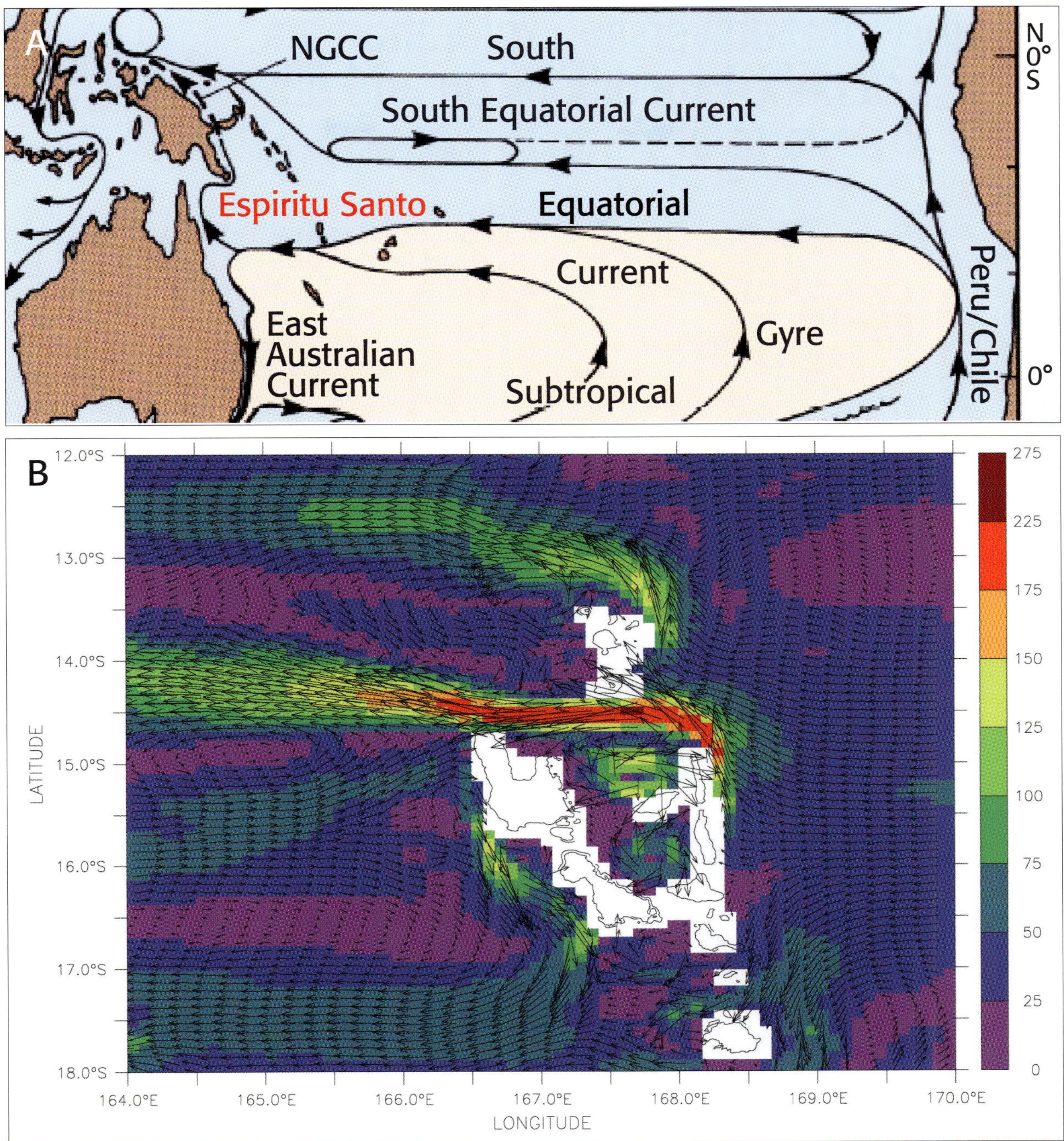

Figure 51: **A**: General sketch of the South Pacific Ocean circulation (from Tomczak & Godfrey, 1994). **B**: Integrated mean circulation (units in m²/s) in the top 1 000 m depth.

have started to be properly simulated with high-resolution numerical models. The mean integrated circulation of the top 1 000 m around Santo from one of these models is displayed in figure 51B. On the eastward side of the Vanuatu archipelago the westward flow is associated with the subtropical limb of the SEC that has been deflected by the Fiji Islands. This flow is itself deflected by the presence of several islands before entering into the Coral Sea. The circulation on the westward side of the islands is also much more complicated with the presence of several counter-currents in the lee of the Banks Islands and Santo. Most of the flow that is choked off by these islands is observed at the northern tip of Santo and is slightly deflected equatorward in the Coral Sea. Northward of this position a sec-

ond branch of the so-called North Vanuatu Jet is also deflected equatorward by the Banks Islands. These two components are nearly zonal and flow westward after the bifurcation toward the coast of Australia. On the westward side of Santo the eastward counter-current bifurcates near the coast and most of the flow is deflected toward the southeast before joining the westward flow between 17°S and 18°S that represents one of the sources of the North Caledonian Jet. The variability of these jets is still largely unknown but they must vary in phase with the interannual response of the trade winds and of ENSO. In the intermediate layers and deeper, the circulation is globally westward with typical speed amplitude of several mm/s. For such circulation the influence of the bottom topography is essential.

History of Biodiversity Exploration and Scientific Expeditions on and off the Island of Santo

Bruno Corbara & Bertrand Richer de Forges

The aim of these few lines is not to draw up an exhaustive inventory of all of the data that the botanists, zoologists, ecologists and marine biologists who have conducted research on and off of the island of Espiritu Santo have gathered since the first Europeans landed on the island. Because of the lack of easily accessible information —and despite the relative low level of interest shown by biologists towards Vanuatu during the last century compared to other parts of the world— such an aim would be out of reach for the non-historians that we are. More reasonably, we shall focus below on the data collected through the collective effort that has preceded the Santo 2006 expedition, and that has significantly increased our knowledge of the terrestrial flora and fauna of Vanuatu.

Of course, in a certain manner, the first biodiversity explorers on Santo were —as far as we know— the first Lapita settlers more than 3 000 years ago. Also, today, the vernacular botanical and zoological knowledge that the Ni-Vanuatu villagers possess illustrates the richness of the human perception of diversity; this knowledge can be linked to the potential use of the plants and animals concerned.

No doubt the first European explorers were also interested by the (for them) exotic flora and fauna of the island, and by the local domesticated plants and animals; however, for obvious geopolitical reasons, their first scientific aim was geographic (cartographic). In 1606, during Quiros' expedition, Torres, the captain and pilot of the second ship (the "*Almiranta*" San Pedro) was also a cartographer. He and Prado y Tovar drew the first representations of the coast of La Austrialia del Espiritu Santo —as Quiros had named the island. One of Tovar's maps, which indicates the ships' mooring sites inside of the present-day Big Bay, seems to reflect the strong impression the mountains and forests of the island made on the Spaniards: trees were drawn as if they were many hundreds of feet high on top of the mountains. The Spaniards, who spent 36 days on Santo made more or less reliable observations about the local flora and fauna. "We have seen on Santo", Quiros writes, "a tree with such a large trunk that 15 to 20 men could not make a circle around it with their arms."

In August 1774, James Cook reached the newly-named New Hebrides aboard the vessel, *The Resolution*. On board were Johann Forster and his son George, who gathered the first substantial botanical collections on the archipelago. *The Resolution*, however, only made a short landing on Santo, northeast of the Cumberland Peninsula. A few years before, the French government had sent Louis Antoine de Bougainville with two vessels, *La Boudeuse* and *L'Étoile*, to the South Pacific. The French botanist Philibert Commerson participated in this expedition (1766-1769) that also visited the New Hebrides.

Others scientists followed, but they seldom penetrated into Santo's interior. It was only at the very beginning of the 20th century (August, 1901) that one of them, the botanist Ollivier, crossed the southern part of the island. The trip took him six days, from the Segond Channel to Big Bay; he needed the help of a team of thirty porters to cross a country which was, at that time, considered a very inhospitable area.

THE WHITNEY SOUTH SEA EXPEDITION (1920-1932)

Under the leadership of the ornithologist Rollo Beck from San Jose, California, the "Whitney South Sea Expedition" was organized primarily to collect ornithological and other zoological specimens for the American Museum of Natural History (AMNH) in New York City (Fig. 52). The expedition was financed by a fund provided to the Museum by the philanthropist Harry Payne Whitney, a rich, thoroughbred horse breeder. During over more than 12 years, a team of scientists and collectors, travelling on board a Tahitian sailing ship named *France*, visited hundreds of islands in the South Pacific Region. They collected more than 40 000 bird skins, as well as many other animals (reptiles) and plant specimens. Beck's wife, Ida, who participated in the expedition and had a personal interest in Oceanic cultures, oversaw the extensive collection of anthropological artifacts and photographs which, for the most part, are now the property of the Department of Anthropology at The California Academy of Sciences.

Figure 52: The logo of the Whitney South Sea Expedition.

The Whitney South Sea Expedition visited Santo in 1926. In 1929, Ernst Mayr—who later became famous as an ornithologist and evolutionist— succeeded Beck as the leader of the project. As the associate curator and then the curator of the Whitney-Rothschild Collection of bird specimens in the AMNH, Mayr collected bird specimens in the Solomon Islands and studied specimens from Santo. The study of these collections enabled him to publish a series of papers in American Museum Novitates —the *AMNH journal*— under the generic title "*Birds collected during the Whitney South Sea Expedition*". They provided the basis for Mayr's noteworthy ornithological field guide published in 1945, *Birds of the Southwest Pacific*, and for other, more significant theoretical writings on ecology and evolution. The success of the expedition also induced its patron to offer US$ 750000 to the City of New York for the construction of the building known as the Whitney wing (built in 1936), one entire floor of which is dedicated to presenting the birdlife of the Pacific region.

THE OXFORD UNIVERSITY EXPEDITION TO THE NEW HEBRIDES (1933-1934)

On a smaller scale, but specifically concerning Santo Island, the "Oxford University Expedition to the New Hebrides" directed by the zoologist John Baker was undertaken under the auspices of the Oxford University Exploration Club. Baker had already visited Santo (mostly the northeastern peninsula) in 1922-23 and in 1927 with biological and geographical objectives; during these expeditions he climbed Mt Tabwemasana, which allowed him to draw up a precise map of the area that was published in the *The Geographical Journal* in 1929.

In an article published in the same journal in 1935, Baker reminds us, based on his own experience, that "it is not always easy to collect money for scientific expeditions". The organizers of Santo 2006 would undoubtedly agree with him! And, as Baker notes in his 1935 report: "our famous forerunner, Quiros, had sent in no fewer than 51 'memorials' asking for financial support for his proposed second expedition to these islands". Eventually, the Oxford University Expedition —which was also sponsored by the Royal Society, the University of Oxford, New College and the Percy Sladen Trust— could take place in Santo starting from September 1933. The small team comprised six people including Baker, his sister (a botanist) and his wife (a photographer). John Baker stresses the "two very definite main objects (of the expedition); namely (1) to study the breeding seasons of animals in a climate which scarcely varies during the course of the year, and (2) to discover which is the highest mountain in the New Hebrides, to climb it, and to survey the whole surrounding district". His results were also published in different issues of the *The Geographical Journal*. The detailed study of the breeding season of animals led to the dissection of over 3000 bats, birds and lizards. During the different outings conducted on the most mountainous part of the island, Baker and his colleagues collected numerous species of animals including invertebrates, despite the fact that "the heavy rain made the collection of insects very difficult"; they also gathered "plants in flower, including several species of Orchids". Climbing this mountain again, Baker could conclude by means of barometric readings that "there is no doubt that Tabwemasana is the highest mountain in the New Hebrides".

THE ROYAL SOCIETY-PERCY SLADEN EXPEDITION TO THE NEW HEBRIDES (1971)

In 1968, at a meeting of the Royal Society of London concerning the results of an expedition held in the Solomon Islands in 1965, the discussion focused on the biogeographical relationships of the Solomon flora and fauna with those of adjacent island groups, and consequently the organisation of an international expedition in the New Hebrides progressively took shape. Mainly funded by the Royal Society and the Percy Sladen Trust, this expedition was led by K.E. Lee from the Australian Commonwealth Scientific and Research Organisation (CSIRO), and involved 21 researchers from seven countries. Members of the expedition began arriving in the New Hebrides on June

20th, 1971, and all of the field work was completed by October 24th. Six of the main islands of the archipelago, including Santo, were explored. Participants were specialists with different backgrounds in diverse areas of terrestrial biodiversity which led to an extensive study of, inter alia, the flora and phytogeography, vertebrates, and invertebrates (e.g. earthworms, insects). The quantity of samples collected had no previous equivalent for the area concerned, with, for example, more than 15 000 samples of insects obtained uniquely through hand collection and by sweeping the foliage of low vegetation.

In his final paper published in the special issue of *The Philosophical Transactions of The Royal Society B*, which compiled a dozen articles presenting the results of the expedition, Lee writes that "before the Expedition the flora and fauna of the islands were little known and that a great deal of knowledge has resulted from the Expedition's work". Two lines later he notes that "there was much we could not cover in our four months' work and there is wide scope for further exploration of the islands' biota". No doubt that both phrases remain relevant for the Santo 2006 expedition despite its unequalled, large-scale dimension.

THE TSUKUBA BOTANICAL GARDEN, NATIONAL SCIENCE MUSEUM EXPEDITIONS (1996-2001)

In 1996, 1997, 2000 and 2001, the Tsukuba Botanical Garden (TBG) at the National Science Museum in Japan sent four successive botanical expeditions to Vanuatu (mostly to Santo Island), under the direction of Tsukasa Iwashina. Six to nine botanists were involved each time, including Ni-Vanuatu participants from the Vanuatu Environment Unit and Department of Forestry; among them was Sam Chanel, who actively participated in Santo 2006. The scope of the TBG expeditions included fern and fern allies, gymnosperms and angiosperms (Fig. 53). The results were published under the generic title "Contributions to the Flora of Vanuatu", in special issues of the Annals of the Tsukuba Botanical Garden, and are available through the website (http://ci.nii.ac.jp/vol_issue/nels/AN10009042_en.html).

Figure 53: The understory of the forest in Santo, Cumberland Peninsula, Saratsi Range, 600 m a.s.l.

DEEP SEA MARINE BIODIVERSITY DISCOVERY IN SANTO

Since the discovery of the Vanuatu archipelago by Quiros in 1606, the marine fauna has been known solely from specimens gathered by naturalists along the shore (Fig. 54). The deep sea fauna of the Indo-West Pacific was discovered during the so called "Great Expeditions".

• The 1874 *Challenger* around-the-world expedition sampled fauna from the western Pacific. This great expedition, however, only sampled a small area of the Coral Sea; after some stopovers in the Fiji Islands, they sailed towards the Torres Strait passing by the New Hebrides archipelago (Vanuatu's former name given by James Cook), without conducting any sampling.

• In 1928, *The Dana* expedition sampled in the Fiji Islands as well as at a few points in southern New Caledonia.

• In 1951, *The Galathea* expedition sampled in the Solomon Islands archipelago.

Figure 54: The seashore; South of Santo.

Hence, the first deep-sea expedition to sample on and around the Vanuatu Islands was the MUSORSTOM 8 cruise onboard the *Alis*, a small French oceanographic vessel (Fig. 55). The deep sea fauna was sampled along the slopes of the main islands of Vanuatu between 100 and 1500 m deep. While the species composition of the southern islands shows some similarities with New Caledonian deep-sea fauna, the islands north of Efate (17° 30' S) shelter different species. On the deep sea floor, the amount of food available is very limited, and all of the organic substrates sinking from the surface are integrated into the food web. By far, wood makes up the main organic substrate present in the deep sea. To the north of Santo, Big Bay — which is a very large and deep bay— was also sampled. The fauna associated with the sunken wood found there was so different from that found elsewhere that a new scientific program on sunken wood was established. As a result, several other deep sea expeditions were organized to further investigate the deep-sea fauna around Vanuatu and especially the north of Santo Island. Multibeam mapping completed during the cruises BOA 0, BOA 1 and SANTOBOA revealed the very strange seabed morphology of Big Bay: two long, underwater rivers carve through the sea floor (Fig. 56). The beds of these rivers are filled with sunken wood. Several scientific studies in the areas of zoology, genetics and biogeography have

Figure 55: The oceanographic vessel *Alis* belonging to IRD: a 28 m long scientific trawler on which the sampling of the slopes of Santo have been done.

been conducted on the deep sea fauna of Vanuatu. A large list of deep sea species from Vanuatu is already available through the Ocean Biogeographic Information System (OBIS) website (http://www. iobis.org/).

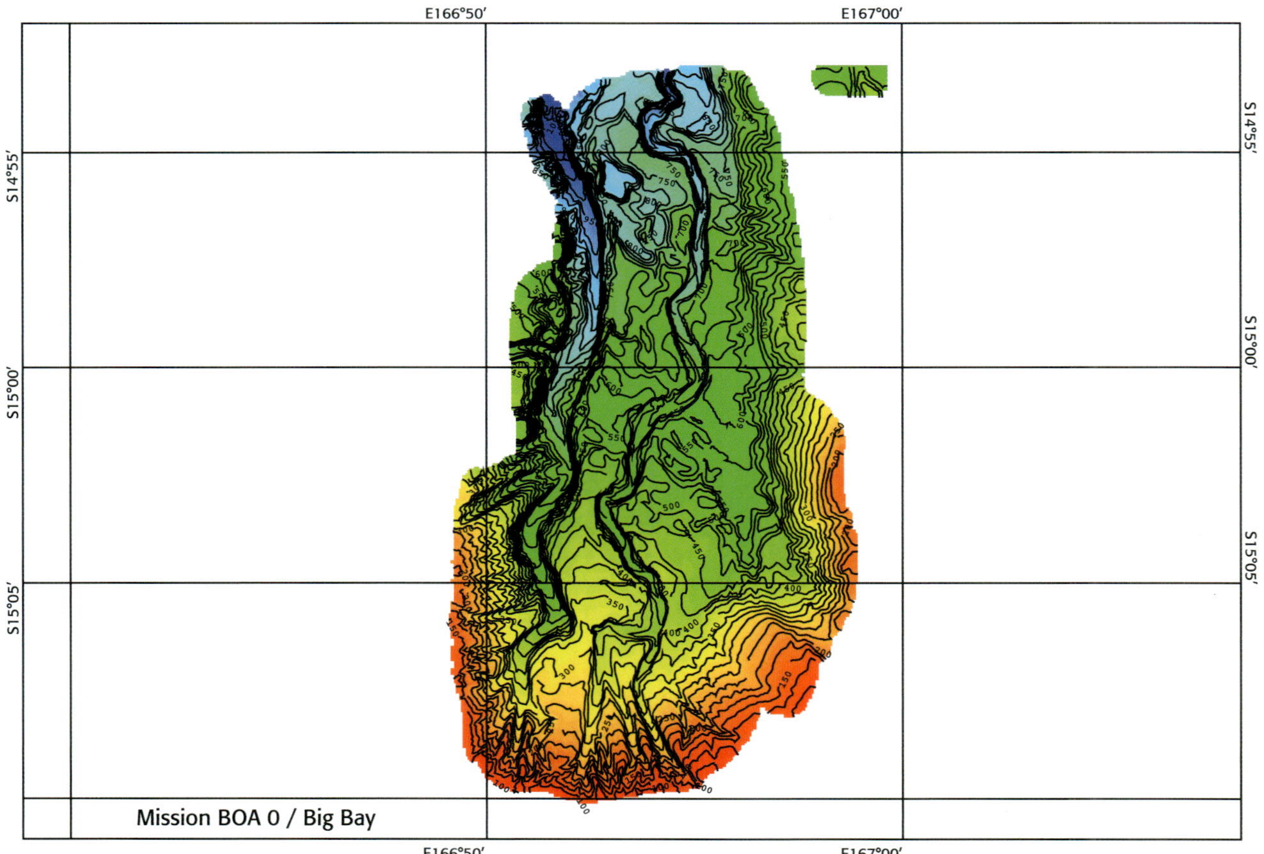

Figure 56: Mapping of the bottom with the multibeam echosondeur revealed in Big Bay a complexe network of "fossil" rivers well inscribed inside the sediments to more than 1000 m deep. In these river beds were collected plenty of sunken wood with their associated fauna. The colors are describing the depth in meters, red for the shallowest and blue for the deepest.

FAMOUS 20th CENTURY SCIENTISTS IN SANTO

Some famous 20th century biologists spent a few days on Santo or worked on samples collected from the island. Here we shall discuss the work of only three of them who, not surprisingly, are well-known for their contributions to a better understanding of island biogeography.

As we saw above, the great biologist Ernst Mayr analyzed the avian distributions of birds on 31 islands in the South Pacific, including Vanuatu, largely on the basis of collections made during the Whitney South Sea Expedition. It is generally accepted that these results have had a great impact on the development of Mayr's evolutionary theories.

In January 1955, Edward Osborne Wilson, the well known ant specialist and Robert MacArthur, the great theoretician on island ecology and coauthor of the famous book *The Theory of Island Biogeography*, spent a few days on Santo where he collected samples of his favorite insects in the immediate vicinity of Luganville.

Jared Diamond, famous for his recent, best-selling books, including *Collapse: How Societies Choose to Fail or Survive* also made some brief visits to the New Hebrides. His studies on the "distributional ecology of breeding land and fresh-water birds in light of immigration-extinction equilibria" are largely based on his own collections and mostly on data gathered in 1971 during the Royal Society-Percy Sladen Expedition to the New Hebrides.

CONCLUSION

The Santo 2006 Global Biodiversity Survey is by far the largest expedition ever to take place on Vanuatu; moreover, it is probably one of the largest ever organized on Earth since the beginning of scientific exploration. Based on the observations and collections made during this expedition, many scientific peer-reviewed articles will undoubtedly be produced; some have already been published, others are in preparation. Due to the scale of the expedition, the diversity of the disciplines involved and the unavoidable asynchronic rhythms in the processing of the collected samples and data, it is totally unrealistic to conceive of a coordinated publication of the results. Some efforts have been made towards this end: for instance, a special issue of the journal *Zoosystema* published in 2009 by the Muséum national d'Histoire naturelle and edited by one of us includes 17 contributions describing new zoological taxa collected from the bottom of the sea to montane forests. Undoubtedly, the present book will remain the best account of the expedition for generations to come —and maybe, by that time, among the young participants in Santo 2006, some will have become as famous as Ernst Mayr or Edward Wilson are for today's biologists, partly due to discoveries made on and off Espiritu Santo Island.

on Santo and Logging Operations

Rufino Pineda

When preparing the Santo 2006 Biodiversity project, project organizers involved in Forest component of the expedition were puzzled. All roads on the eastern part of Santo lead to plantations! Virtually no forest in sight! Indeed, forests on Santo, as for most if not all islands of this size, have been severely impacted by human activities.

The first significant step in the deforestation process occurred with the creation by settlers of large coconut plantations, meanwhile traditional gardening with long fallow cycle had permitted somewhat to maintain a good vegetal cover although of secondary forest.

The second and more intensive phase of deforestation is associated to the arrival of two hundred thousand soldiers during the WWII. Thousands of hectares of bushes were bulldozed to plant food crops of all kind to feed the troop. At that time one vine *Meremia peltata* commonly named in Bislama "*big lif*" (and since then American vine), was widely used for camouflage in particular of ammunition sheds.

Some claim that it was introduced by the Americans, but it is doubtful as for instance "*Tabwa Tabwa*" is its name in one of the local languages and we find it growing in places as remote as taro gardens of Penaoru away from any WWII influence. Spreading of *Meremia peltata* has occurred in the last decade, it seems to be a response to the intensive logging operations associated to the Malaysian owned logging company (Table 4).

Out of the 26 licenced mobile sawmill companies in Santo, six only appear to be still operating.

As the Department of Forestry in Santo has burnt twice, statistical data are inexistent. It is thanks to Jude Tabi Forestry Officer supervising logging operation that we can get the data contained in this paper.

The maximum logging happened in 1996 with

Table 4: Large scale logging companies operating in Santo.

Company name	Operational period	Licensed quota	Contracted areas
Sud de Pac	1974 - 1980	2 000 m³/yr	Teproma, Suranda, Fanafo areas and Turtle Bay
Island Development	1980 - 1988	2 000 m³/yr	Same as above
Santo Veneers & Timbers (Wong se sing)	1988 - 1994	5 000 m³/yr	As above including Sharkbay, Hog Harbour, Port Olry and Sara
Santo Veneers & Timbers (Malaysian)	1994 - 2002	50 000 m³/yr	All of the above including Butmas, Siketket, Malel & Lape
Melcoffee sawmills	1991	5 000 m³ export of round logs	Matevulu, Turtle bay and Suranda
Melcoffee & Hog harbor / Kole ltd.	1988	10 000 m³ export of round logs	Port Olry & Sara areas
Novat	1983 – 1987	2 000 m³/yr	Malakula & Santo
Resources development	1987 – 2007	15 000 m³/yr	Sara & Port Olry
Melcoffee sawmills	1983 - 2007	15 000 m³/yr	East & Sth. Santo
Forest Products & Plywood ltd.	1983 - 1995	5 000 m³/yr	Loran, Sarabo, Palon & Vanafo areas
Tri wood industries	1995 - 1999	2 500 m³/yr	Wunpuko, West Santo
Frank Gallo	1991 - 1999	5 000 m³/yr	West coast Santo
Santako	2000 - 2007	5 000 m³	Kohu West coast Santo
Veneer Logging	2005 - 2007	500 m³	Matevulu, Butmas

33 000 m³ logged timber registered in Santo (nearly half of the 68 000 m³/yr calculated as the sustainable harvest quota for whole Vanuatu). For the last twenty years 15 000 m³/yr of timber harvested is, according to Jude Tabi, a reasonably accurate estimate.

20 000 ha of forest have been licensed for logging but it doesn't include the forest cleared for cattle grazing, which could double this figure. Currently Melcoffee Sawmills remains the main logging company operating in Santo with Veneer logging as second, followed by an handful of mobile saw-mill owned by Ni-Vanuatu.

The following map (Fig. 57) shows the distribution of logging licenses recently attributed to large and small sawmillers in Santo.

Recent reduction of import taxes on NZ treated pine timber have dropped logging activity in Santo, which might have a beneficiary effect on forest regeneration.

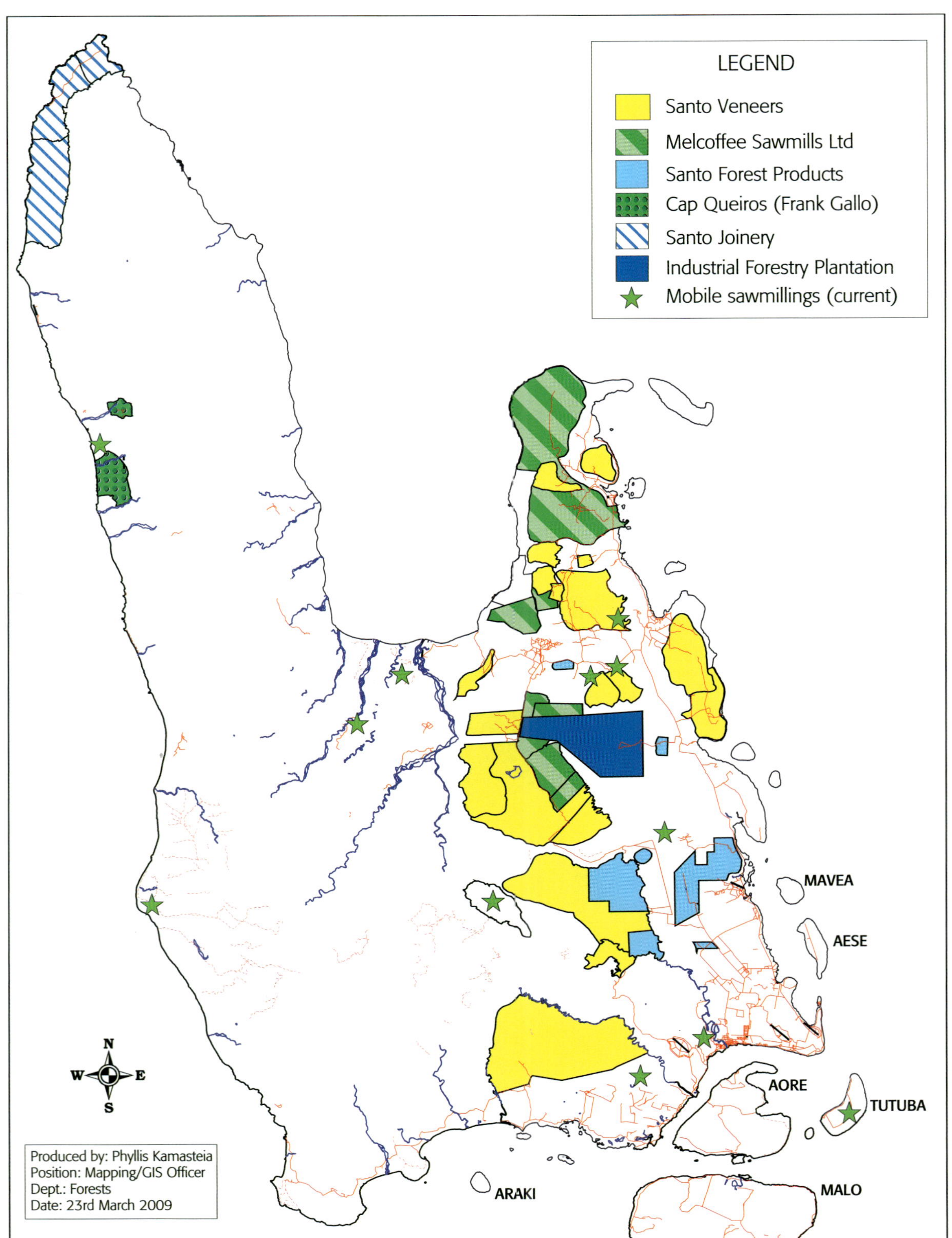

Figure 57: Logging concessions and sawmills on Santo.

# of WWII on Infrastructures and Landscape

Laurent Palka & Rufino Pineda

Shortly after the bombing of Pearl Harbour the 7th December 1941, the Japanese rapidly invaded the nearby Solomon Islands in April of the following year. The fear of invasion and occupation haunted the people of Vanuatu. One morning in May, they woke up to discover the Port-Vila Bay filled with warships. They were so terrified that a number of them ran into the hills in the belief that Japanese had arrived. Later, they learned that it was a stealthy arrival of the allied forces. Indeed, Vanuatu had been chosen as a strategic place by the US forces to stop the Japan's conquest in Pacific. The island was especially important as a base during the campaign of Guadalcanal.

TERRESTRIAL VESTIGES OF WWII

Santo was home to 100 000 permanent troops, in addition to the 60 000 of Ni-Vanuatu residents. An extremely large military base named Buttons was then constructed and numerous new infrastructures were put in place. The combat engineers built more than 50 km of road, bridges, and airstrips. Three of these strips were established around Luganville, one at Ouchard South Santo, one at Matevulu and one at Matantas. An emergency airstrip was also constructed at the northern top of the Cumberland peninsula. The old town located on the right bank of the Sarakata River was extended on the left bank toward Palekula point. One hundred hectares of land filled with coral were gained over the mangrove, and half-moon shape military sheds grew like mushrooms. Many places in the bush were cleared for dispersing munitions stocks and supplies as protection against bombing, although Santo was shelled just once with limited damages. Today, only some impacts of shells fired by the US artillery during training can be seen on the East coast of Cape Queiros. Remnants of American planes that crashed during training occur on the right side of the Jourdain River and at the end of the Matevulu airstrip. A handful of half-moon shaped sheds remain in Luganville town. Others on the outskirts leading to Fanafo are half buried and covered with grass. Perforated steel matting material, that was extensively used for rapid construction of temporary runways and landing strips, is ubiquitous and still used for fences, roads and building of houses.

THE SS PRESIDENT COOLIDGE

Figure 58: Sinking of the Coolidge between Santo and Aore Island.

After the establishment of American troops, a large harbour was built and subsequently secured by mines blocking the entrance of the Segond Channel between Santo and Aore Island. On October 1942, the President Coolidge (Fig. 58), a luxury cruise liner converted to carrying troops, attempted to enter the harbour with over 5 000 men on board. Suddenly, two friendly mines hit successively the engine room and near the stern. The Coolidge quickly capsized and sank. Nearly all the men were saved, only a mechanic and the captain were reported missing. When WWII ended, salvage operations enabled the recovery of a substantial amount of equipments from the wreck. The recovery process was ended in November 1983, and ever since the area including the President Coolidge was declared as a marine reserve by the law of Vanuatu. Today, her bow stays at a depth of about 20 m and her stern at over 70 metres. The wreck is covered with coral and homes barracudas, sharks and others reef fishes as permanent residents. For a long while, a 200 kg grouper called Boris, that is now gone, lived in the ship and was hand fed by divers. The local fauna seems to contain also the dugong, an endangered marine mammal. Submarine landscapes were also affected by large nets placed during the war in the Segond Channel to prevent the entry of Japanese submarines. The nets collapsed over time making way for corals and gorgonians.

MILLION DOLLAR POINT

After 1945, the US army sold the enormous quantity of surplus and Santo became an immense scrap and metal auction. Buyers came even from New Caledonia, New Zealand and Australia. Hurrying to leave Vanuatu, US army decided to throw what they hadn't sold into the water. Hundreds of vehicles as well as tonnes of equipments and supplies were carelessly thrown aside. They built a ramp running into the sea and every day, the boys drove trucks, jeeps, tanks, and more into the channel. By night, the people snuck to the site in search of useful goods, for instance building materials or dishes that could be transported into the villages of the bush. Since then, the locals call this place Million Dollar Point (Fig. 59) thereby alluding to the quantity of money wasted. Today, the remnants of this unfortunate episode can still be seen. They form an incredible trash heap just off the coast of Santo from the surface down to 40 metres of depth. At low tide, coral-encrusted metal objects sprinkle the beach.

Figure 59: Million dollar point located on Segond Channel.

Efforts in Santo

Samson Vilvil-Fare

Before European came to Vanuatu village chiefs and landholder families managed resource use on behalf of their communities in accordance with locally defined custom law. As elsewhere, colonization and 30 years of independence have overseen many changes. While custom still defines people's relationship with their land, there is considerable local variation in degree to which custom controls on resource use are practiced, enforced and respected. There have also been many changes in the ways natural resources are used. There are significant pressures from other countries with influence in the region, from international organizations, from national leaders and from young educated people to continue to change people's relationship with the environment. There is international pressure to participate in conservation programmes and conform to environmental standards promoted by flora such as United Nations Environment Programme (UNEP) and recently United Nations Development Programme (UNDP).

••• Protected areas in Vanuatu

At present there are three recognized Protected Areas (Pas) in Vanuatu:

- Million Dollar Point and Coolidge Marine Sanctuary;
- Vatthe Conservation Area located on the island of Santo;
- Matnakara Park on Efate Island.

In addition there are numerous informal PAs, commonly known as "tabu" areas, established and managed at a local level to conserve or improve access to locally important natural resources. The extent to which these contribute to meeting national biodiversity conservation goals is not quantified. A survey of 21 coastal villages in 2001 showed an average of 4.1 marine resource management measures per village and revealed a high level of motivation to manage marine resources at a local level.

••• Existing Conservation Areas in Santo

••• Vanuatu Protected Areas Initiative (VPAI)

In 1993, the Melanesian Forest Expedition (MFE '93) visited Vanuatu to look at the status of bird populations. Whilst on some of the islands the expedition was approached by a number of locals who expressed concerns about the future of their forest. As a result of this the Vanuatu Protected Areas Initiative (VPAI) that is base at Khole village, East Santo was formed and a return trip to the islands was organized in order to assist those that had requested help. Out of the six sites, Loru was chosen as Chief Kaleb had the vision to put a "Tabu" on an area of his land in order to prevent the removal of flora and fauna. Chief Kaleb and his family also understood the concept of a disappearing forest and had begun to inform others about the detrimental effects of removing their livelihood. Two other sites which were visited were Duck Lake and Tebi Beach.

Loru is a family owned Protected Area and is managed by an environment management committee from Khole village (East Santo). It is 220 ha in size with conservation objectives to maintain and protect coconut crabs; the endangered and largest land crab found in the world and protects the forests and birds.

Despite its small size, 23 of the 64 breeding land and freshwater birds of Vanuatu were sighted in Loru and 10 of the 23 restricted-range species, including two vulnerable, one near-threatened and five species endemic to the archipelago.

••• Vatthe Conservation Area

South Pacific Regional Environment Programme (SPREP) through the South Pacific Biodiversity Conservation Programme (SPBCP) initiated a more formal conservation programme with Vatthe Conservation Area in 1993. This initiative came about following a biodiversity research mission conducted by Royal Birds and Forest of New Zealand. Santo is the largest and oldest island of Vanuatu archipelago that has more diverse and rich biodiversity compared to the other islands. After the mission conducted by the Royal Bird and Forest New Zealand with the help of the then Vanuatu Environment Unit (now Department of Environmental Protection and Conservation) the Vatthe Area was proposed for Conservation through the SPBCP programme. The Vatthe forest is mostly secondary and not as well preserved than previously thought. But the last patches certainly deserve protection being the last witnesses of lowland rainforest on alluvial plains in Vanuatu and is the only area of such forest to extend to the coast. The forest harbours a rich flora and fauna, including native land and freshwater birds in Vanuatu and five endemic bird species. Recent freshwater research at Jordan and Matantas River lying inside the Conservation Area recorded three endemic freshwater fish. It is one of the two natural sites that are on the Vanuatu tentative list for the World Heritage Site.

Vathe conservation area is managed primarily by locals from the villages chosen by the community and their respective chiefs. It has a management plan that guides the communities to conserve and carry out management activities within the conservation area. In 1998, there was an ecotourism which set up to help the community bringing a bit of income into the community and to help sustain the local workers and their family.

In 2002, the environmental Act of No.12 2002 came into effect and required that all National Parks or Conservation areas around Vanuatu be registered. The local counterpart with the help of a Peace Corp Volunteer took up the work, after her training to register the park and it became officially a National Park.

••• Penaoru Conservation Area

Penaoru conservation area is a community initiative that began in year 2003. It covers an area of 1207 ha. The area is entirely managed by local community through an environment management committee. It has a management plan that guides the community to conserve and implement activities to contribute to its conservation objectives.

The area was initially conserved to maintain the number of wild resources in the upper courses of Penaoru River for the community's livelihood needs. During the Santo 2006 global expedition the area became known to the Department of Environmental Protection and Conservation

Figure 60: Vatthe Conservation Area facilities run by the local communities.

(DEPC) through its participation in the expedition. The Conservation Area then got the attention of the Landholders Conservation Initiatives Project under the DEPC and which then worked with the communities from April 2007. The project helped strengthened the conservation area activities and helped developed its management plan with the Penaoru Community. Penaoru Conservation Area is one of the conservation sites to be registered in year 2010 under the Environment Management and Conservation Act of No.12 2002.

During the 90's until today, conservation efforts have been put in place in order to protect the fauna and the flora of the islands of Vanuatu. There is a trend going from Local Initiatives to a more National Initiatives on Protected Areas. This can be regarded as the fruits of the tremendous efforts carried out by the local community, together with local and regional Non-Governmental Organizations and other stakeholders. Vanuatu through its Department of Environmental Protection and Conservation is working with other stakeholders for the betterment of its biodiversity. The efforts shown in creating more Protected Areas can be seen as a step forward to protect and to safeguard the fauna and flora of the Island of Santo as a whole.

Vegetation and Flora

coordinated by Jérôme Munzinger & Porter P. Lowry II

by the Santo 2006 Botany Team

Porter P. Lowry II & Jérôme Munzinger

European botanical exploration of Vanuatu began in 1774, when the Forsters (father and son) and W. Anderson visited the islands of Tanna and Malekula during James Cook's second voyage. Significant collecting expeditions, including to Santo, were conducted by F.A. Campbell in 1872-1873 and by a team led by J.R. Baker in 1933. More recently, scientists from several institutes participated in the 1971 Royal Society-Percy Sladen Expedition; botanists from IRD (formerly ORSTOM) in Nouméa, New Caledonia, visited Santo between 1965 and 1985; and a team from Tsukuba Botanical Garden collected there in 1996-97.

The Santo 2006 Botany team included 16 members from eight institutions in the Pacific, Europe and North America — the largest ever to work in Vanuatu. They focused on two complementary themes: Mountains & Rivers, to document the native flora of largely undisturbed habitats, and Fallows & Aliens, to record plants in human-modified environments. Most field work was conducted in a few key areas, including the Cumberland Peninsula (especially the Penaoru valley), Mt Tabwemasana, the area around Butmas, and the Vatthe Conservation Area near Matantas.

During the three months when team members were on Santo, they made a total of 1950 fertile collections, almost always in multiple sets. In addition, 943 trees were marked and identified in a series of permanent plots established in the Penaoru area. One duplicate of each fertile collection was deposited at the Vanuatu national herbarium in Port Vila (PVNH) and additional sets were sent to the Muséum in Paris (P) and the herbarium at the IRD Center in Nouméa (NOU); duplicates were sent to specialists working at other institutions.

Members of the Santo 2006 Botany Team made preliminary identifications as material was collected. Most specimens were dried and processed at the Penaoru base camp (Fig. 61), although some were preserved temporarily in alcohol and dried later, especially material collected during the difficult Mt Tabwemasana ascension. Following the expedition, field notes were entered into a database. Most angiosperms collections were identified by M. Tuiwawa, G. McPherson and J. Munzinger in Nouméa in February-March 2007, and by G. McPherson in Paris in June 2007. Ferns were identified by G. Rouhan (April and August 2007) and orchids by M. Pignal (2006-2008), both working in Paris. In parallel, specialists with knowledge on certain groups provided identifications.

Much of the identification work has now been completed, although a few problematical groups still require study. The team documented *c.* 650 species on Santo, belonging to 366 genera and 140 families of flowering plants, gymnosperms, ferns and bryophytes. An estimated 20 species collected during Santo 2006 are new to science and are being named by members of the team.

This exemplary effort has made a significant contribution to our knowledge of the flora of Vanuatu and toward documenting the plants of Santo, the archipelago's largest, highest, and botanically most diverse island. Establishing permanent plots also enabled the first-ever characterization of the structure and composition Santo's humid forests. Our work, summarized in the following pages, has helped fill a major gap in our botanical understanding of the southwest Pacific, and gave the team members an unparalleled opportunity to discover and share some of Santo's most exciting botanical treasures.

Figure 61: The Santo botany team at the Penaoru base camp, processing plant specimens after a long day in the field.

Types of Vegetation Occurring on Santo

Jérôme Munzinger, Porter P. Lowry II & Jean-Noël Labat

The Santo 2006 expedition was designed to carry out detailed exploration of the botanical diversity present on the island. A wide diversity of vegetation types were therefore studied, covering the full range extending from what can be regarded as "extremes" on a scale from natural, nearly undisturbed areas to those that have been profoundly modified by man. Large areas have been transformed by humans —partially or completely— through clearing, fire, and other means, in an effort to meet basic needs for food, shelter, fiber, grazing land for livestock, etc., although such habitats exist only because they are created and maintained by man or by domesticated animals. At the other extreme, Santo's vegetation includes nearly pristine formations that result from the natural processes of evolution and succession and are self-maintaining, provided they are not subject to excessive human disturbance. These natural vegetation types include humid forests, which

Table 5: Vegetation types in Vanuatu proposed by Mueller-Dombois and Fosberg.

1. Lowland rain forest

a. High-stature forests on old volcanic ash

b. Medium-stature forest heavily covered with lianas

c. Complex forest scrub densely covered with lianas

d. Alluvial and floodplain forests

e. *Agathis-Calophyllum* forest

f. Mixed-species forests without gymnosperms and *Calophyllum*

2. Montane cloud forest and related vegetation

3. Seasonal forest, scrub and grassland

a. Semi-deciduous transitions forests

b. *Acacia spirorbis* forest

c. *Leucaena* thicket, savanna and grassland

4. Vegetation on new volcanic surfaces

5. Coastal vegetation (including mangroves)

6. Secondary and cultivated woody vegetation

Figure 62: View up the Penaoru Valley toward the center of the Cumberland Peninsula, with rainforest in the foreground and montane cloud forest at the highest elevations.

were the main focus of the research carried out during Santo 2006. Some areas of humid forest were only visited briefly by botanists passing through various parts of the island, but a large area on the west coast of the Cumberland Peninsula, especially in the Penaoru Valley (Figs 62 & 63), was the subject of detailed study and will be the main focus of the present chapter. Additional field work was conducted by a team on the western slope of Mt Tabwemasana (1879 m), the highest point on Santo and indeed in the entire Vanuatu archipelago.

The vegetation of Vanuatu can be divided into six main categories (Table 5):
- Lowland rain forest;
- Montane cloud forest and related vegetation;
- Seasonal forest, scrub and grassland;
- Vegetation on new volcanic surfaces;
- Coastal vegetation, including mangroves;
- Secondary and cultivated woody vegetation.

We will limit our discussion to the vegetation types that were examined during the Santo 2006 expedition (indicated in bold in Table 5).

Phytogeographic Relationships

Gordon McPherson

A visitor to the forests of Santo familiar with New Caledonia finds himself surrounded by a strange mixture of plant species, many of them completely unfamiliar but others that are only slightly different from their close relatives found to the south. The abundance of species and individuals in such Asian genera as *Myristica* (Myristicaceae), *Pterocarpus* (Fabaceae), *Dracontomelon* (Anacardiaceae), *Dendrocnide* (Urticaceae), *Pangium* (Acariaceae), *Medinilla* (Melastomataceae), *Hydnophytum* and *Timonius* (both Rubiaceae), *Vavaea* (Meliaceae) and *Cyrtandra* (Gesneriaceae) will convince the visitor that the influence of the Southeast Asian flora is much stronger on Santo than it is on the geographically more isolated and geologically much older New Caledonia. At the same time, he or she might be surprised to discover that some groups regarded as characteristic of New Caledonia are represented in Santo by distinct species. These include such genera as *Agathis*, *Ascarina*, *Hedycarya* (Monimiaceae), *Trimenia* (Trimeniaceae), *Balanops* (Balanopaceae), *Geissois* (see "Focus on *Geissois* (Cunoniaceae)") and *Spiraeanthemum* (both Cunoniaceae), *Meryta* and *Schefflera* (both Araliaceae – see "Focus on Araliaceae").

Travelers coming from lands to the north or west of Santo or from most of the islands of the Pacific will enjoy seeing other familiar genera. These include *Calophyllum* and *Garcinia* (both Clusiaceae), *Castanospermum* (Fabaceae), *Dacrycarpus* (Podocarpaceae), *Dysoxylum* (Meliaceae), *Elaeocarpus* (Elaeocarpaceae), *Freycinetia* (Pandanaceae), *Joinvillea* (Joinvilleaceae), *Metrosideros* and *Syzygium* (both Myrtaceae), *Morinda* (Rubiaceae), *Neuburgia* (Loganiaceae), *Pittosporum* (Pittosporaceae), *Rapanea* and *Tapeinosperma* (both Myrsinaceae) and *Semecarpus* (Anacardiaceae), although in most cases the species on Santo will be unfamiliar. Figs (members of the genus *Ficus*, Moraceae) and orchids (Orchidaceae) abound here, as in most tropical forests.

Figure 63: View down the Penaoru Valley, with rainforest in the foreground and seasonal forest at lower elevations.

MEDIUM-STATURE FOREST

Figure 64: Understorey of medium-stature forest above Penaoru base camp, showing numerous buttresses on large tree trunks.

Medium-stature forest (Fig. 64), which is characteristic of low elevation sites, is frequently degraded as a result of its proximity to inhabited areas. Many of the trees are tall (Fig. 65) and especially at lower elevations some deciduous species can be found such as *Antiaris toxicaria* (Moraceae) and *Pterocarpus indicus* (Fabaceae). The most characteristic element is *Castanospermum australe* (Fabaceae), known locally as *puilapuila*, and easily recognized by its majestic habit, straight bole and especially its beautiful flowers, which vary from orange to red depending on their maturity (Fig. 66). This is one of the largest trees in medium-stature forest, and it stands out when in flower because the ground underneath is carpeted with flowers and the emergent canopy above is multicolored. This type of forest is also distinctive in having a well developed layer of small trees 6 to 15 m tall. Charateristic species include *Dendrocnide latifolia* and *Pipterus argenteus* (Urticaceae) with their small flowers. Various species of fig also occur in this type of forest, such as *Ficus wassa* (Moraceae), a cauliflorous tree that bears numerous clusters of fruits on the trunk (Fig. 67) and produces abundant white latex, along with more discrete members of the genus such as *F. kajewskii. Gardenia tannaensis* (Rubiaceae), locally named "*uka*", is usually 5 to 10 m in height but can occasionally reach 20 m, has large white flowers (Fig. 68) with a strong, pleasant odor and is thus often planted around villages and in fields.

On ridges at about 500 m elevation, the lower stature forest is dominated by several species, including *Meryta neoëbudica* (Araliaceae) (Fig. 69; see "Focus on Araliaceae"), *Myristica inutilis* (Myristicaceae) and *Ficus septica* (Fig. 70), all of which are often covered with climbing plants belonging to several species of *Freycinetia* (Pandanaceae), whose large strongly scented flowers are eaten by bats (see "Focus on Pandans").

Figure 65: Jean-Noël Labat examining the canopy in medium-stature forest.

Figure 66: *Castanospermum australe*, one of the largest trees in lowland forest, is easy to spot when it flowers because the ground is carpeted with its red-orange petals.

Figure 68: Flower of *Gardenia tannaensis*, a sweet-smelling member of the coffee family.

Figure 69: A female individual of *Meryta neoëbudica* with young multiple fruits above a whorl of large leaves.

Figure 67: *Ficus wassa*, a member of the fig genus whose fruits are borne directly on the trunk.

Figure 70: A branched infructescence of *Ficus septica*, another fig species.

KAURI-TAMANU FOREST (= *AGATHIS-CALOPHYLLUM* FOREST)

A remarkable type of forest occurs on the west coast of the Cumberland Peninsula at altitudes above about 600 m, with trees so large that a person's arms can not reach around the trunk. These majestic giants no doubt live for many years: in 1938, Guillaumin reported to have seen an individual with a circumference of 10 m (i.e. a diameter of about 3.15 m). The largest tree observed during the expedition had a diameter at breast height of 2.4 m (a circumference of about 7 m). The most impressive of these is the Kauri (*Agathis macrophylla*), easily recognized by its distinctive cones (Fig. 71) and by its bark which forms plates (Fig. 74). Also, resin can often be seen oozing out of the trunk, which hunters use to start fires. Kauris belong to the family Araucariaceae, which also includes the columnar pine (*Araucaria columnaris*), endemic to New Caledonia but sometimes found planted as an ornamental tree in Vanuatu. Kauris not only dominate the forest by the diameter of their trunk, but also by their height, emerging far above the canopy (Fig. 74). Species of *Agathis* produce a high quality wood, which has led to heavy exploitation in most places, including Vanuatu, where it is still being harvested today. The forests where members of the Santo 2006 team conducted field work, however, have thus far escaped this fate, making the Penaoru area one of the few remaining anywhere in the Pacific where true Kauri forest still exists.

The second species characteristic of the Kauri-Tamanu forest is *Calophyllum neoëbudicum* (Fig. 72), locally known as "*Tamanu*" ("of the bush"). It too can easily be recognized by its distinctive bark and also by the yellow-orange latex that exudes when the plant is cut or bruised (a characteristic of many members of the family Clusiaceae) and its small leaves with closely spaced parallel venation. Clusiaceae are abundant in the Cumberland forest, with a second genus, *Garcinia*, represented by three species, *G. vitiensis*, *G. pseudoguttifera*, and a third as-yet unidentified species, locally known as "*Malkevic*", which is by far the most common of the three. Unlike *Calophyllum*, the species of *Garcinia* do not have closely spaced parallel secondary veins in their leaves, and their fruit is fleshy, with several seeds, the best known being the mangosteen (*G. mangostana*).

Another species deserves to be mentioned as well, *Astronidium novaeëbudaense* (Fig. 73), a shrub or small tree belonging to the family Melastomataceae, which was the most frequently encountered of all in our studies. We recorded individuals from between 600 m and 1 200 m altitude, indicating that this plant is capable of growing in a fairly wide range of forest types, from Lowland rain forest to Montane forest. The Melastomataceae are a large family occurring throughout the tropics.

Figure 71: Cone of a giant Kauri (*Agathis macrophylla*), a dominant tree in forest above 600 m elevation.

Figure 72: A ripe fruit of *Calophyllum neoëbudicum*, common tree in mid-elevation forest.

Figure 73: *Astronidium novaeëbudaense*, bearing young green fruits and showing the characteristic venation of the family Melastomataceae.

They can almost always be recognized by the very distinctive venation in their leaves, in which the tertiary veins are parallel to one another and oriente at right angles to the secondary veins, giving them an appearance that resembles a ladder, as clearly observed on *Astronidium novaeëbudaense* (Fig. 73). Some members of the family are also epiphytes, such as *Medinilla cauliflora* and *M. heteromorphophylla* (Fig. 77). Two tree species in the olive family (Oleaceae) should also be noted,

Figure 74: The huge trunk of a giant Kauri (*Agathis macrophylla*) with its characteristic bark forming large plates.

Chionanthus brachystachys, observed at 600 m and 900 m, and *Ligustrum neoebudicum* (Fig. 78), reaching up to 1 200 m elevation. Oleaceae have opposite leaves (or sometimes sub-oppposite at the top of branches) and are easy to recognize when flowers are present as they have four sepals, four petals and only two stamens. We also encountered *Hedycarya dorstenioides*, a member of the ancient, primitive family Monimiaceae, which occurs sporadically at 600 m and more commonly at 1 200 m,

How Old are the Kauri (*Agathis microphylla*) Trees?

Jonathan Palmer

One of the most common questions to ask when standing in a forest is how old are the trees? After climbing up past 600 m above sea-level and first encountering the large kauris at Santo, you have a feeling of awe at their majestic size and sense that they must have been standing there for a very long period of time. But exactly how long? This was the question I wanted to try to answer.

I would describe the trees as being wide but with short trunks (boles) holding up massive, open, spreading crowns. Buttressing is absent but there are large rounded roots spreading over a considerable distance from the trunk. The open architecture of their crowns is no doubt a means of withstanding passing cyclones. Typically kauris appear most abundant along ridges, where they sometimes form groves of mixed sized trees.

The main method used to obtain the age of a tree is simply to count the rings. This is fine for temperate locations but this approach can be difficult in tropical climates. The time constraints prevented us from using other options such as placing markers into the wood (pinning method) or bands around the tree and periodically returning to measure the change in circumference (dendrometer bands). Our approach was to obtain small cores from a selection of 15 trees using manual increment corers (Fig. 75). These have been widely used throughout the world and found not to cause any harm to trees. The cores extracted from the trees were stored in straws and transported back to the laboratory where they were dried, glued onto mounts and then sanded to produce a highly polished surface. They were then looked at under a binocular microscope and the number of visible rings counted. Normally the growth rings are measured and their pattern is compared to other cores from the same tree and then to other trees. If patterns match between different trees they are said to "crossdate" and in many situations this has led to the successful reconstruction of past climatic conditions.

The Santo kauri cores were found to have ring boundaries that varied greatly from distinct to diffuse and impossible to determine. In some situations there were no "visible" rings over many centimetres of growth. Although the visible rings were counted, we knew the age estimate would not be accurate. Even so, we then checked if there was a general systematic relationship between the number of visible rings and the diameter of the trees (Fig. 76). There is no relationship. Many other studies have shown the same result: tree size ≠ age. This means for a tree found of a certain size, there is a large range of possible ages or visible rings. To check this we submitted two samples for radiocarbon dating. They were taken from inner-most ends of cores, the part closest to the centre (i.e. oldest) portion of two different trees. The results (Table 6) showed us two things. Firstly, that both samples were too young for accurate radiocarbon dating. The second thing was that although the tree Santo-15 was twice the diameter of Santo-14 (208 cm vs 102 cm respectively) there was not the same magnitude of difference in their calibrated radiocarbon ages; the trees were in fact fairly similar in probable age (note: the dates in Table 6 show the reverse trend if anything). Again, this supports the earlier result of tree size not being directly correlated to age.

Figure 75: The extraction of small samples of wood being taken from a kauri tree using an increment corer. The samples are used to try to count the number of rings formed so that the age of the tree might be estimated.

Table 6: Radiocarbon dating results from two *Agathis microphylla* wood samples. (Waikato Radiocarbon Dating Laboratory, New Zealand).

Tree label	Trunk diameter (cm)	Radiocarbon date (years BP: before 1950)	Calibrated dates (95 % probability) AD (Christian calendar)
Santo 14.1	102	241 ± 30	AD 1630-1690 AD 1730-1810
Santo 15.1	208	149 ± 30	AD 1660-1890

What can we conclude? Firstly, the kauri tree-rings are not clearly defined enough to enable accurate counting and there was no consistent pattern found either within or between trees. This is despite trying to sample trees on exposed ridgelines and also towards the upper end of their elevational range (i.e. where growth might be more well defined by climatic conditions such as seasonal drought and exposure). Secondly, based on the radiocarbon dates, the trees are not immensely old. Unlike *Agathis ovata* in New Caledonia, the trees on Santo are unlikely to be much more than 350 years old. Thirdly, there seems to be a great range of possible sizes for any given age, a result that is consistent with many other studies.

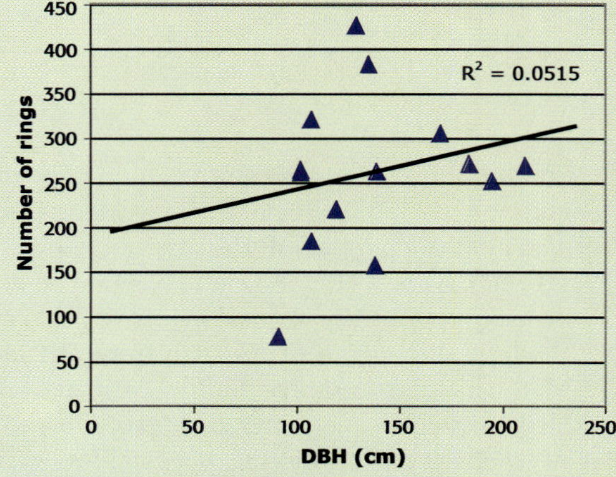

Figure 76: Plot of the number of visible rings against the diameter at breast height (DBH) of the Santo kauri trees. The regression line shows there is no significant relationship.

Figure 77: Flowers of the epiphyte *Medinilla heteromorphophylla*, which grows on trunks of large trees.

Figure 78: *Ligustrum neoebudicum,* a member of the olive family whose distinctive flowers have four white petals (which are quickly shed), four persistent sepals, and only two yellow stamens.

Figure 79: Fruits of *Hedycarya dorstenioides*, a member of the ancient, primitive family Monimiaceae.

Figure 80: Young flowers of *Cyrtandra efatensis* in humid forest.

Figure 81: Flowers of *Tapeinosperma netor*, a member of a complex genus.

Figure 82: *Psychotria milnei*, a member of the coffee family, is easy to recognize by it showy white calyx that persists long after the petals fall off.

and whose flowers have separate carpels each of which develops into a bright red drupe (Fig. 79).

Some species are very abundant in the shrubby component of this type of forest, such as *Cyrtandra efatensis*, a member of the African violet family (Gesneriaceae) (Fig. 80) and several species of *Tapeinosperma* (Myrsinaceae) (Fig. 81) that are particularly difficult to distinguish from one another. Members of the coffee family (Rubiaceae) were also frequently encountered, notably in the genus *Psychotria*, including the spectacular species *P. milnei*, which has an enlarged, showy calyx (Fig. 82). We also encountered herbaceous plants, as for example *Begonia vitiensis*, with distinctive white flowers in the understorey (Fig. 83).

Figure 83: *Begonia vitiensis*, a native relative of the commonly cultivated begonias grown throughout the world.

MONTANE CLOUD FOREST

Figure 84: Mist covered canopy (above) and understorey (below) of montane could forest at about 1 500 m above Penaoru valley.

The montane cloud forest (Fig. 84) is typically dominated by species belonging to several genera, including *Metrosideros* and *Syzygium* (both members of the family Myrtaceae), *Weinmannia* (Fig. 85) and *Geissois* (see "Focus on *Geissois* (Cunoniaceae)") (both Cunoniaceae), *Quintinia* (Quintiniaceae) and *Ascarina* (Chloranthaceae). Species of *Metrosideros* are easy to recognize by their abundant adventitious roots, which reach from the branches down to the ground: the longest seen during the expedition reached 1.9 m in length. This plant has small leaves (2.5-8.5 × 0.8-6 cm) that are opposite and a dry fruit, which distinguishes it from species in two other genera of Myrtaceae that occur in the forests on Santo, *Syzygium* and *Eugenia*, whose fruits are fleshy. The flowers of *Metrosideros collinum* are variable in color, ranging from red or pink to yellow (Fig. 86). In humid summit areas, the trees are smaller and covered with epiphytes, mostly filmy ferns and liverworts, but also epiphytic shrubs such as *Vaccinium* (Ericaceae) (Fig. 87), which is easily recognized by its small, pink, urn-shaped flowers.

In some places the montane cloud forest can take a distinctive appearance when tree ferns and/or palms are present. In the Penaoru Valley, at about 1 200 m elevation, we visited an area of forest in which a single regionally endemic palm was frequently seen, *Cyphosperma voutmelense* (Fig. 88). The species name refers to the place where it was first collected, Voutmélé Peak. This palm alone represented fully 14% of the trees with a DBH (diameter at breast height) of 5 cm or more, but the presence of several tree ferns, belonging to three different genera, *Cyathea*, *Dicksonia* and *Leptopteris*, was even more striking as they accounted for 20% of the individuals with a DBH ≥ 5 cm. The showy white flowers of *Fagraea ceilanica* (Gentianaceae) were also very distinctive (Fig. 89), although only a few individuals were observed.

Figure 85: Flowers of *Weinmannia denhamii*, a typical dominant in montane cloud forest on Santo.

Figure 86: *Metrosideros collinum* in its yellow flowered form.

Figure 87: A flower of the epiphytic shrub *Vaccinium whiteanum*, which grows high in the trunks of the largest trees.

Figure 88: Flowers and fruits of the rare endemic palm *Cyphosperma voutmelense.*

Figure 89: Large, showy flowers of *Fagraea ceilanica.*

SEASONAL FOREST, SCRUB AND GRASSLAND

Figure 90: Gaiac (*Acacia spirorbis*) is characteristic of a "gaiac forest".

The seasonal forest, scrub and grassland can also be divided into three variants (Table 5), one of which, *Acacia spirorbis* (Fabaceae) forest (Figs 90 & 91), is locally referred to as "gaiac forest". Another characteristic species of this vegetation type is *Kleinhovia hospita* (Sterculiaceae) (Fig. 92), locally known as Matalo and recognizable by its pink flowers and light, ribbed fruits. This formation can contain sandalwood, and in drier locations it may be dominated by introduced shrubs such as *Leucaena leucocephala* and *Acacia farnesiana* (both Fabaceae) and *Psidium guajava* (Myrtaceae). Sandalwood (*Santalum austrocaledonicum*; Santalaceae) is an important element of the flora from a historical point of view, as this shrub or small tree was exploited for centuries for its prized, long-lasting essential oils that are used for perfumes, incense and in aromatherapy. Sandalwood trees can be distinguished by their opposite leaves that are slightly bluish below, their cream-colored flowers with four petals, and their angular fruit.

Figure 91: *Acacia spirorbis*. Its fruits open to reveal black seeds and their brightly colored arils that attract dispersers, primarily birds.

Figure 92: *Kleinhovia hospita* is another characteristic species in low elevation gaiac forest.

COASTAL VEGETATION

Figure 93: Coastal vegetation along the beach, Cumberland cape.

The littoral forest (Fig. 93), with characteristic species such as *Pandanus tectorius* (Pandanaceae) (Fig. 94) and *Casuarina equisetifolia* (Casuarinaceae) (Fig. 95), is common in coastal areas on Santo. These two species are quite easy to recognize; *Pandanus tectorius* has bluish leaves and a typical round, pineapple-like "fruit"[6] and *Casuarina* resembles a conifer, with long drooping branches and small, odd-looking bumpy- spiny multiple fruit that open by many small pores. *Callophyllum inophyllum* (Clusiaceae), which occurs in coastal areas (Tamanu of the sea), is also easy to distinguish as it bears the same characteristics as the Tamanu of the bush (see above), but with larger leaves and round fruits like golf balls scattered on the beach.

6 - The "fruit" of members of the Pandanaceae family is technically not really a fruit, but rather a compound fruit, developing from an inflorescence made up of many individual flowers instead of just a single flower.

Figure 94: *Pandanus tectorius*, a widespread coastal tree in the Pacific, with its typical "fruits".

Figure 95: *Casuarina equisetifolia*, a common species in littoral forest along the coast, has leaves and multiple fruits that resemble those of a conifer, but it is in fact a flowering plant.

# of Santo

SOME NEW, CHARACTERISTIC OR REMARKABLE SPECIES

Gordon McPherson & Jérôme Munzinger

The recent botanical inventory work done on Santo has brought to light a number of previously undiscovered taxa. These include two new species of *Schefflera* (see "Focus on Araliaceae") and probable novelties in the following genera: *Alangium* (Alangiaceae), *Alphitonia* (Rhamnaceae), *Citronella* (Cardiopteridaceae), *Cyrtandra* (Gesneriaceae), *Elaeocarpus* (Elaeocarpaceae), *Eugenia* (Myrtaceae), *Ficus* (Moraceae), *Freycinetia* (Pandanaceae), *Ilex* (Aquifoliaceae), *Parsonsia* (Apocynaceae), *Sciaphila* (Triuridaceae), *Semecarpus* (Anacardiaceae), *Tapeinosperma* (Myrsinaceae), *Terminalia* (Combretaceae), and in three genera of Rubiaceae (*Guettardella*, *Ixora* and *Psychotria*), all of which are now in various stages of closer study or preparation for publication. Some of these potential novelties are discussed in more detail below.

••• *Alangium*

Prior to the Santo 2006 expedition, one member of this genus, *A. vitiense*, had been reported from Vanuatu by Guillaumin, although Smith, in his *Flora Vitiensis Nova*, later indicated that this species was restricted to Fiji and that Guillaumin's identification was incorrect. The original description of *A. vitiense* indeed refers to a plant that differs from the material collected on Santo (Fig. 96), whose calyx has long teeth and whose leaves have evident domatia and a very acuminate apex. Moreover, the species of *Alangium* occurring in Vanuatu does not match anything from New Caledonia, and thus appears to be new.

••• *Cyrtandra*

We identified several species of this genus on Santo: *C. efatensis*, *C. vesiculata*, *C. neohebridensis* and *C. schizocalyx*. Several specimens couldn't be related to any of these species (Fig. 97), so we suspect novelties in the genus.

••• *Elaeocarpus*

Four species were observed during the mission, *E. floridanus*, *E. hortensis*, *E. hebridarum* (this latter considered by some authors as conspecific with *E. angustifolius*), and an unidentified taxon (Fig. 98), which might be new.

Figure 97: Flowers of a *Cyrtandra* species, which is suspected as being undescribed.

Figure 96: Flowers of a probably new species of *Alangium*.

Figure 98: A species of *Elaeocarpus* in fruit, which might represent a novelty.

••• *Gmelina*

A tree belonging to this genus was collected in the Penaoru valley, representing the first record ever from Vanuatu. The plant on Santo appears to be close to *Gmelina vitiensis* from Fiji, but it is not identical, differing in several characters, which suggests that it may be a new species.

••• *Parsonsia*

Several lianescent genera of Apocynaceae were collected, including *Alyxia* and *Hoya*, and the specimens are thought to be closely comparable to already described species. However, a liana collected at 1 200 m (Fig. 99) in dense primary forest was with difficulty assigned to *Parsonsia* cf. *laevis*, and this plant might be new and in need of description.

••• *Sciaphila*

This genus in the family Triuridaceae was previously unknown from Santo, although one species, *S. aneityensis*, had been recorded elsewhere in the archipelago. Material of two distinct species was collected during the Santo 2006 expedition, one of which appears to be new to science.

Figure 99: A liana belonging to family Apocynaceae, initially identified as *Parsonsia* cf. *laevis* with doubt, might be a novelty.

FOCUS ON ARALIACEAE: SEVERAL GENERA EXEMPLIFY SANTO'S MELANESIAN BIOGEOGRAPHIC RELATIONS

Porter P. Lowry II & Gregory M. Plunkett

The family Araliaceae and the closely related family Myodocarpaceae are among the few groups to have been studied in detail for Vanuatu within the last several decades. Araliaceae in particular are well represented on Santo, and field work conducted by the Santo 2006 team offered new insights into the diversity and biogeography of several genera.

Delarbrea paradoxa, the sole member of Myodocarpaceae in Vanuatu, ranges from Norfolk Island and New Caledonia in the south through to the Solomon Islands in the north and then farther west across part of Indonesia. This species was collected at several localities in intact forest on Santo, and the material gathered will be used for phylogenetic studies to test whether the populations in Vanuatu were derived from ancestors on New Caledonia, reaching the archipelago by long-distance dispersal by birds, as has been hypothesized. The expedition team also made several collections of *Meryta neoëbudica* (Fig. 100), a species of Araliaceae endemic to Vanuatu and recorded from throughout the archipelago, which is most closely related to an endemic New Caledonia species, *M. denhamii*, sharing distinctive golf-ball-sized multiple fruits.

Figure 100: Multiple fruits of the Vanuatu endemic *Meryta neoëbudica*. Each of these globose structures is formed by the fusion of several separate fruits.

Figure 101: Vanuatu is home to two members of *Polyscias* subgenus *Tieghemopanax*, *P. schmidii* (left), endemic to Vanuatu, was recorded for the first time on Santo in 2006; *P. cissodendron* (right) ranges from the Santa Cruz Islands just north of Vanuatu through New Caledonia south to Lord Howe Island.

Two species belonging to the New Caledonia-centered group *Polyscias* subgenus *Tieghemopanax* were also found on Santo, *P. schmidii* and *P. cissodendron*, again showing the strong biogeographic affinities between Vanuatu and its closest neighbor. *Polyscias schmidii* (Fig. 101), described less than 20 years ago from a single collection made on Erromango, *c.* 400 km to the SSE, and later recorded on Tanna as well, was found in seasonal forest at low elevations near the coast, not far from the village of Penaoru. The material collected during the Santo 2006 expedition will now make it possible to determine which of several species on New Caledonia is its closest relative. The second species, *P. cissodendron* (Fig. 101), found at both mid- and high-elevation sites on Santo, has been recorded on several islands in Vanuatu as well as on New Caledonia and Lord Howe Island farther to the south, making it the only member of the *Tieghemopanax* group that is not endemic to a single archipelago (two species each are restricted to Fiji and Australia, and all 22 of the remaining species are found only on New Caledonia).

Schefflera cabalionii (Fig. 102), endemic to Santo, belongs to the Plerandra group, which is centered on Fiji (eight species, all endemic), with one species restricted to the Solomon Islands and another found there and in New Guinea. Members of this group have flowers that produce abundant nectar and bear up to several hundred stamens, both adaptations to pollination by birds and/or bats. Unlike most other Araliaceae in Vanuatu, the ancestor of *Schefflera cabalionii* may have come from Fiji rather than New Caledonia.

While on Santo, the field team found several large trees belonging to a species in another group of *Schefflera*, the Gabriellae group, otherwise represented by two species in New Caledonia, one in Fiji, and one in Vanuatu. The collections made by the Santo 2006 team (Fig. 103) were initially thought to belong to *S. vanuatua*, a species previously known only from a few collection made on Aneityum, the southern-most island in the archipelago situated about 600 km away. However, field observations made on Aneityum just after the end of the Santo 2006 expedition showed that the trees on that island are not as tall and have smaller fruits and much wider leaflets. Subsequent studies using DNA sequence data confirmed that the trees on Santo represent a different species, perhaps more closely related to the Fijian species than to *S. vanuatua*. This new species is now being described. A second novelty in *Schefflera*, also endemic to Santo, was discovered on Mount Tabwemasana and belongs to the Dizygotheca group, which has 16 species in New Caledonia and two others in Vanuatu restricted to islands farther to the south.

Figure 102: The flowers of *Schefflera cabalionii* (above), a species endemic to high elevation forest on Santo, produce abundant nectar and are pollinated by birds. The fruits (below) are located below the leaves and may be bird-dispersed.

Figure 103: This new species, belonging to the Gabriellae group in the genus *Schefflera*, was discovered during the Santo expedition and appears to occur only on the Cumberland Peninsula.

FOCUS ON *GEISSOIS* (CUNONIACEAE): ANOTHER EXAMPLE OF THE MELANESIAN CONNECTION

Yohan Pillon

Figure 104: *Geissois denhamii* at Penaoru, Santo.

The genus *Geissois*, taken in the strict sense (i.e. excluding the Australian species), a member of the family Cunoniaceae, is a group of trees and shrubs endemic to the islands of the Pacific. It is easily distinguished by its opposite palmately compound leaves and its bright red flowers arranged in bottle-brush like inflorescences that attract nectarivorous birds (Fig. 104). Besides their ornamental potential, species of *Geissois* provide valuable timber and several species possess antibiotic and anti-oxidant properties, a feature shared by many Cunoniaceae.

The genus encompasses 19 species, 13 of which are endemic to New Caledonia, four to Fiji, one to Temotu Province (Vanikoro) in the Solomon Islands, and one to Vanuatu (Fig. 105). *Geissois denhamii*, the Vanuatu species, ranges from Vanua-Lava in the North to Aneytium in the South and is present on most of the larger islands, including Santo. It is a small to large tree found in primary rainforest or sometimes in secondary vegetation (white grass) at medium to high elevation, including the summit of Mount Tabwemasana.

The Santo 2006 expedition provided an opportunity to collect material for a phylogenetic study of Cunoniaceae largely focused on the taxa occurring in New Caledonia. Two nuclear genes were sequenced for all members of *Geissois* from New Caledonia and Vanuatu, and the results were used to reconstruct the relationships among the species. The phylogenetic tree (Fig. 106) indicates a close relationship between *G. denhamii* and the New Caledonia species, with *G. denhamii* nested within the New Caledonia group, suggesting that its ancestor originated in New Caledonia. Although the Fijian and Solomon Island species

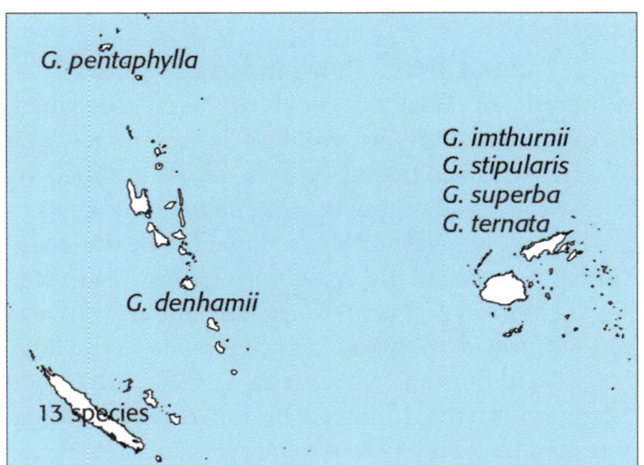

Figure 105: Distribution of the species of *Geissois* within Melanesia.

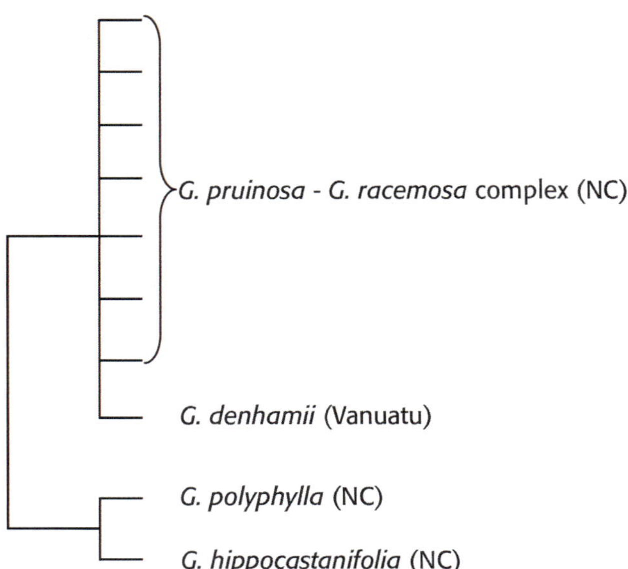

Figure 106: Phylogeny of the genus *Geissois* according to molecular data.

have not yet been included in phylogenetic studies, a similar position can be expected for them.

New Caledonia is home to the largest number of species and the greatest morphological diversity within *Geissois*, and it seems plausible that this old archipelago was the cradle of the genus. As the islands of Vanuatu emerged from the sea through the process of volcanism, the genus no doubt reached them through long-distance dispersal, which was facilitated by small winged seeds that can easily be carried by the wind. Thus *Geissois* is similar to other genera such as *Megastylis* (Orchidaceae), *Oxera* (Lamiaceae) and the genera of Araliaceae mentioned above, all of which likewise had members that dispersed from New Caledonia (or in one case perhaps Fiji) to Vanuatu, demonstrating the biogeographic affinities between these neighboring Melanesian archipelagos.

FOCUS ON PANDANS

Thomas Haevermans

This chapter deals with the screwpine family, Pandanaceae, a member of the monocotyledons, a major flowering plant group that also includes famous plants such as the palms (Arecaceae), orchids (Orchidaceae) and grasses (Poaceae). Species of Pandanaceae are commonly seen in the tropics, and *Pandanus* is widely grown as an ornamental. Some of these plants are used for a variety of purposes, especially as a source of food and as a condiment (for example, leaf extracts from *Pandanus amaryllifolius* add a distinctive aroma and green color to dishes, especially in southeast Asian cuisine). The edible fruit of some species are also used to make a unique red sauce, especially in Oceania.

Pandanaceae can be recognized by their long, green, entire, generally spiny-toothed leaves that are usually spirally arranged at the branch tips, and by the characteristic stilt roots produced at the base of the stem. Species of Pandanaceae are dioecious, with male and female flowers borne on different plants. The female flowers often resemble a pineapple and develop into distinctive multiple fruits (Fig. 107) made of tightly arranged phalanges (containing the seeds) also called "keys", because when one is removed the others come apart easily (Fig. 108). The male flowers are especially ephemeral and are therefore under-represented in herbarium specimens. Some species contain cultivars with variegated leaves while others show various degrees of glaucousness.

••• Taxonomy of Pandanaceae

Hundreds of Pandanaceae have been described, many of which are not valid or do not correspond to currently recognized species, as exemplified by the 3-page list of synonyms for *Pandanus tectorius* in the Pacific, which includes a name for nearly every known variant of the fruit. The family comprises four genera, *Freycinetia*, *Martellidendron*, *Pandanus* and *Sararanga*. Among these, *Pandanus* is the largest, with about 600 recognized species and a huge list of synonyms, followed by *Freycinetia* with ca. 300 species. The two remaining genera are less speciose: *Martellidendron* has seven species and *Sararanga* two.

Figures 107: Multiple fruits of two species of *Freycinetia*, resembling a pineapple.

Pandanaceae are restricted to the Old World, and *Pandanus* is the most widely distributed genus, extending from West Africa to Oceania. *Freycinetia* ranges from Southeast Asia to Oceania, *Martellidendron* occurs only in the western Indian Ocean Islands (Madagascar and the Seychelles), and *Sararanga* is restricted to the Philippines and New Guinea.

In Vanuatu, seven species of *Freycinetia* have been recorded, none of which is endemic. By contrast, 14 of the 19 species of *Pandanus* currently listed from Vanuatu are known only from the archipelago (Table 7), although in reality there may be fewer species as the genus has a long history of over-description, a phenomenon that has been less pronounced in *Freycinetia*. On Santo we collected one endemic taxon, *Pandanus nogarete* (Fig. 108), which was previously only known from Vanua Lava Island farther to the north.

••• Major ecological and morphological characteristics of Pandanaceae

Species of *Freycinetia* are lianas and therefore require support to grow. They often "crawl" on tree trunks, reaching toward the canopy to flower and attract pollinators with their very showy bracts

Figure 108: Fruits of *Pandanus nogarete* break into individual phalanges showing a distinctive red color that is characteristic of this species and its relatives.

Table 7: List of Vanuatu Pandanaceae (adapted from the World Checklist of Monocotyledons, 2008), taxa restricted to Vanuatu preceded by an asterisk *.

Freycinetia (7 species)

Freycinetia arborea Gaudich., 1829

Freycinetia demissa Benn. in J.J.Bennett & R.Brown, 1838

Freycinetia microdonta Martelli, 1910

Freycinetia monticola Rendle, 1921

Freycinetia reineckei Warb. ex Reinecke, 1898

Freycinetia schlechteri Warb., 1906

Freycinetia sulcata Warb., 1906

Pandanus (19 species)

**Pandanus arrectialatus* H.St.John, 1989

Pandanus cominsii Hemsl. var. *cominsii*, 1900

**Pandanus cuneiformis* H.St.John, 1989

Pandanus dubius Spreng. var. *dubius*, 1826

**Pandanus efateensis* H.St.John, 1989

**Pandanus halleorum* B.C.Stone, 1976

**Pandanus mauricei* H.St.John, 1989

**Pandanus minimus* H.St.John, 1989

**Pandanus multidentatus* H.St.John, 1989

**Pandanus ngunaensis* H.St.John, 1989

**Pandanus nogarete* H.St.John, 1989

**Pandanus onesuaensis* H.St.John, 1989

**Pandanus parou* H.St.John, 1989

**Pandanus pweleensis* H.St.John, 1989

**Pandanus pyramidos* H.St.John, 1989

**Pandanus quinarius* H.St.John, 1989

**Pandanus radifer* H.St.John, 1989

Pandanus tectorius Parkinson ex Du Roi, 1774

Pandanus whitmeeanus Martelli, 1905

(modified, brightly colored leaves) containing numerous flowers borne on fleshy edible parts (Fig. 109) that attract bats and birds, which are the main pollinators. *Freycinetia*, along with *Pandanus* and *Martellidendron*, lacks a perianth (petals and sepals), although these structures are found in Sararanga. Species of *Pandanus* are free-standing trees, often with lateral stilt roots, and thus do not require any physical support. Some members of the genus are very large, and the species exhibit a variety of growth forms and architectural models.

In Vanuatu, as elsewhere, Pandanaceae occur primarily in forests and in coastal areas, although *Pandanus* contains some specialized endemics that are adapted to swampy habitats or are restricted to dry, rocky areas (such as limestone formations in Southeast Asia and Madagascar).

••• Uses of Pandans

Hundreds of varieties of *Pandanus tectorius* are grown in the Pacific, and in Vanuatu this species is know as *pandanas* in Bislama. Some varieties are selected for their fruits and others for their leaves. The fruits (keys) of edible varieties are consumed in various ways. For example, they are used to make drinks, a dried paste and flour, and the seeds of some species are also eaten. The leaves are used to make high quality baskets, mats and hats, and are also used for roof thatching. Moreover, the stems are used for house construction and for craft making, and the roots for their medicinal properties.

Figure 109: Inflorescence of *Freycinetia sp.*, with its distinctive showy, fleshy bracts that are often eaten by pollinators.

FOCUS ON ORCHIDS

Marc Pignal

Orchids probably arrived on Santo by wind dispersal. Of the islands that make up Vanuatu, Santo is the one that hosts the most species of this emblematic and well known plant family. The island's broken relief and the diversity of its habitats have made it possible for many species to grow there and for groups to diversify.

Orchidaceae surprise botanists every day, and there is probably no family that is more studied. They occupy the first place among flowering plants in terms of the number of species, with an estimated 20 000 worldwide. Globally, orchids have colonized almost all milieus, with the exception of two extremes: aquatic environments and deserts. Elsewhere, however, from cold temperate zones to hot and wet climates, one can find members of this family everywhere, but it is in the intertropical zones that the largest number of species are found. Orchids grow on many substrates: shallow and deep soils, bare rocks, and even other plants, on which they occur as epiphytes, one of the most common habits in the family.

Numerous adaptations can be observed among orchids. Relations with the insect world are innumerable. Charles Darwin was the first to study the patterns of pollination in this extremely diverse group, and he even postulated the existence in Madagascar of a butterfly whose proboscis exceeded 30 cm in length based on the existence of a spur that long in the orchid *Angraecum sesquipedale*. The insect, a sphinx named *Xantopan morganii praedicta*, was indeed discovered, more than four decades later.

Orchids aroused passion and speculative fever in the nineteenth century. Even today, facts such as "the illegal trade of fauna and flora generates the third highest fraudulent sales after drugs and weapons, and often uses the same networks" can be read on the website of the French Ministry of Environment. Among plants, Orchidaceae along with Cactaceae are subject to the highest levels of illegal commerce. Their trade is highly regulated: beyond regional protection, all orchid genera are listed on Appendix 1 or 2 of the CITES Convention. And among plant lovers, the ironic term "orchidomania" has often been employed.

• • • Affinities of the Vanuatu orchid flora

According to Lewis and Cribb, the orchid flora of the Vanuatu archipelago has a strong influence from the Indo-Malaysian region and to a lesser extent from New Caledonia. The absence of absolute isolation and the relative youth of these islands, which emerged about 1.5 million years ago, make the Vanuatu flora relatively poor, although there is still some endemism. The archipelago has never been connected with the Australian mainland, so it has been derived entirely from trans-oceanic elements. The small size of orchid seeds is particularly favorable for dispersal by wind.

• • • The floristic affinities of orchids on Santo Island

Santo is the largest island of the Vanuatu archipelago. It has the widest variety of geological zones, including coastal plains in the north and south, a mountainous massif dating from the Oligocene

Table 8: Comparison of Orchidaceae species on Santo with those occurring in the other islands of the Vanuatu archipelago.

	Species in common with Santo	Total number of species	% of species shared with Santo	Species/km²
Vanuatu (all islands)		161		
Santo	98	98	100	0.03
Erromango	46	70	66	0.06
Anatom	36	62	59	0.39
Efate	40	57	70	0.07
Pentecost	35	48	73	0.15
Vanua Lava	26	45	58	0.56
Ambae	34	43	79	0.10
Tanna	22	31	71	0.08
Malekula	17	22	77	0.02
Epi	7	8	87	0.03
Ambrym	5	8	62	0.02
Maewo	3	6	50	0.03

Figure 110: *Dendrobium polysema*, an epiphytic orchid growing on a tree trunk.

in the west, and plateaus of Quaternary age in the east. Similarly, Santo has the largest number of orchids in the archipelago: 98 of the 161 species (61%) recorded in Vanuatu are found on the island. However, within the archipelago, Santo's orchid flora is among the lowest in terms of species per unit area.

• • • Morphology

••• Monopodial and sympodial orchids
Orchids can be classified according to their mode of vegetative growth. Monopodial groups comprise those whose stem grows indefinitely and whose inflorescences are born in a lateral position (such as for example the genera *Saccolabium* and *Gunnarella*). Sympodial groups have stems that terminate their growth in an inflorescence (such as *Dendrobium* and *Bulbophyllum*), with subsequent growth taking place from buds located lower down on the stem.

••• Terrestrial, lithophytic and epiphytic orchids
Orchids that develop on the ground (referred to as terrestrial) are relatively infrequent, whereas those that use another substrate for support (such as rocks for lithophytes, or another plant for epiphytes) are more numerous (Fig. 110). The roots of these orchids are adapted so that they can attach to their support. Their exposure to the free air has resulted in another adaptation, indicated below.

••• Floral morphology
Orchids have type III flowers, meaning that they have a regular plan based on three sepals, three petals, two whorls of three stamens, and an ovary with three locules. This basic plan is, however, often deformed. In many genera, the ovary undergoes a torsion such that the upper parts of the flower are positioned below. The three sepals are often unequal in size and the dorsal one (i.e. the one in the upper position) sometimes differs considerably from the two occupying the lateral positions. The situation is reversed in the petals: it is the basal petal, which forms the lip, that differs from the others. This lip plays a major role in the attraction of pollinating insects. Lastly, the reproductive parts are amalgamated or united to form a common structure known as the gynostem (or column) in which the stigma forms a cavity, and only some of the stamens develop, while the others may form structures that prevent self pollination. Among the orchids on Santo, only a single stamen usually develops and the anthers form masses of pollen that can stick to pollinating insects.

••• Seed structure
The seeds of orchids, which are very numerous (more than one million have been counted in a single fruit of some species), are among the smallest in the plant world, and are also the simplest morphologically, comprising a solid mass of undifferentiated cells surrounded by a dead, unicellular, generally very thin

and absorbent tegument. With the exception of some type of seeds that exhibit ancestral characters within the family, most orchid seeds have no cotyledon and very few if any reserves. The undifferentiated embryo is thus dependent on its immediate environment for development upon germination.

A century ago Bernard highlighted the complex relations that exist between the seeds of an orchid and an endomycorhiza, *Rhizoctonia*. As shown by this author, symbiosis between the two organisms is a permanent struggle that must lead to the formation of a seedling. It is clear that the fungus provides its host with the nutrients necessary for its development and cellular multiplication. The Orchid embryo is not, however, as undifferentiated as its simple morphology might suggest. Specialized cells are present, which Bernard called phagocytes, whose function is to digest the fungus filaments as they develop, and without them the embryo would be quickly infested. This is where the specificity of the orchid-fungus relationship intervenes: if the phagocytes are too effective, they will completely block development of the fungus and thus the growth of the orchid embryo. Symbiosis between these organisms thus takes the form of a carefully balanced competition between them. But this relationship is still more complex: Bernard also showed that an attenuated strain of fungus loses the capacity to bring about germination in the orchid.

... Diversity in the genera *Dendrobium* and *Bulbophyllum*

... *Dendrobium*
The large genus *Dendrobium*, which comprises more than 1 000 tropical and subtropical species, is represented in Vanuatu by 28 species of which 19 grow on Santo. These orchids are very variable vegetatively, leading some botanists to recognize many additional taxa within the genus, a matter of debate among specialists. All species of *Dendrobium* share several floral features, and the search for similarities and differences within this group is fascinating. Its evolution and the multiple adaptations of its members are vast subjects for study.

What are the common aspects shared by all the members of this genus? As in the majority of plant groups, key features are related to the reproductive system, i.e. the flower. Like all orchids, *Dendrobium* species have a lip —a specialized petal adapted to a mechanism of pollination involving insects— and a gynostem or column. Generally, in orchids the column is positioned directly in the center of the flower, whereas the lip is in the same whorl as the two other petals. In this respect, *Dendrobium* flowers are even more asymmetrical than those of other orchids: the column is attached in its middle to the inferior ovary, the lower part forming a mentum (a chin), which together with the lip below it, forms a

kind of more or less articulated clip, shaped almost like a tube in some species. This asymmetry is variably marked depending on the species, and the range of variation based on this simple floral model almost exceeds the imagination.

The inflorescences are also very diversified in *Dendrobium*. They are short and few-flowered in *D. delicatulum*, which compensates for its low number of flowers by their bright coloring. In *D. macranthum*, the inflorescences are long and full of flowers, which can be observed on plants growing on fallen tree trunks along the coast, or more frequently in *D. conanthum*, which occurs inland. Some other species with brightly colored flowers, such as *D. rarum*, bloom once the stem has lost its leaves, whereas those of others such as *D. involutum* are scattered among the foliage (Fig. 111).

Botany students are taught that buds (with the exception of the terminal bud) are born in the axil of a leaf. In *Dendrobium*, this is generally the case, but in certain species, a curious adaptation has led to inflorescence buds being positioned opposite a leaf. This is the case for example in all members of section *Grastidium*. It reality, this is an artifact: the base of each leaf comprises a sheath that sometimes surrounds the stem for 1-2 cm. The inflorescence bud is borne in a leaf axil, but it pierces the sheath and the inflorescence thus appears to be positioned opposite the leaf immediately above.

It would be nearly impossible to identify vegetative characters shared by all members of *Dendrobium*. Certain groups have developed robust pseudobulbs, as can be seen in *D. macrophyllum*, whereas others have thin stems that look like grasses, as in subgenus *Grastidium*. The leaves of various taxa can be thin or fleshy, and the roots, which can sometimes be fine or sometimes thick, are fixed to the ground, stones or branches.

... *Bulbophyllum*
A similar pattern can be seen in the second largest genus on Vanuatu, *Bulbophyllum*, which comprises more than 1 000 species in tropical and subtropical areas and is represented by 17 species in the archipelago, including 12 on Santo.

The generic name is derived from a uniform feature of this group: all species have a pseudobulb surmounted by one or two thick leaves, as in *Bulbophyllum stenophyllum* (Fig. 112). The genus is a record-holder: is has the greatest variation in size of any orchid genus. The New Caledonian species *B. keekee* has more than twenty pseudobulbes of a few millimeters each and grows on a small square of bark, whereas *B. longiscapum* (Fig. 113) from the islands of the SW Pacific, including Vanuatu, has leaves from 25 to 30 cm in length and rhizomes that can reach 70 cm long.

Figure 111: *Dendrobium involutum*, whose paired flowers are located among the leaves.

Figure 112: *Bulbophyllum stenophyllum*.

The flowers in *Bulbophyllum* have three triangular sepals of various length and two petals that are sometimes reduced to fimbriate strips. In *B. streptosepalum*, the dorsal sepal is helmet-shaped, protecting the column, and does not exceed 1 cm long. The two dorsal sepals on the other hand reach 6 cm in length and are rolled up and twisted. *Bulbophyllum* flowers have a mentum, as in *Dendrobium*, but it is generally very short. The lip has a remarkable and constant adaptation: its apical part, which is coarsely triangular, is attached to the base by a small, flexible ligament and can thus oscillate with the slightest breath of wind or under the weight of a tiny insect, which no doubt plays a very important role in pollination. *Bulbophyllum* flowers can emit a range of odors depending on the species, from pleasant perfumes to more disagreeable odors, at least as perceived by humans.

All inflorescences in *Bulbophyllum* are lateral and grow from the basal rhizome, either at the base or between two pseudobulbs. Another orchid, *Mediocalcar paradoxum* (Fig. 114), is closely related to *Bulbophyllum* and was abundantly collected by the botanists during the Santo 2006 expedition. It differs, however, by its sepals that are united throughout most of their length and thus form a small bell, and by the position of the inflorescence, which is located at the end of the terminal axis.

Figure 113: *Bulbophyllum longiscapum*.

••• Atypical among the typical

Some morphological adaptations can be astonishing. In most plants, photosynthesis takes place in the leaves or in specialized adaptations of the stem called cladodes. In epiphytic orchids, however, the roots also have chlorophyll, especially toward their tips, and this fixes a portion of the atmospheric carbon used by the plant. This astonishing adaptation makes it possible for many orchid species to reduce their total leaf surface without sacrificing their ability to photosynthesize.

••• *Microtatorchis* and *Taeniophyllum*, from reduced leaves to their entire loss

Microtatorchis (with eight species including two on Santo) and *Taeniophyllum* (100 species, also two on Santo) grow as plates on the trunks and main branches of trees. It is probable that additional species in these genera will be discovered because they are extremely small and are often hidden among mosses or by the relief of a tree's bark.

These two genera exhibit an even more distinctive evolutionary novelty: the entire loss of leaves. In *Microtatorchis* only young individuals have leaves and as the plant matures new leaves are increasingly small, ultimately being replaced by minute scales. As for *Taeniophyllum*, their stem, which is only a few millimeters long, bears its leaves in a rosette of scales that dry up shortly after developing. The loss of leaves in both *Microtatorchis* and *Taeniophyllum* means that the roots not only absorb water and minerals, they also serve as the site where photosynthesis takes place.

••• Orchids without chlorophyll

Other terrestrial orchids have an even more radical metabolic adaptation: not only do their leaves remain in a vestigial state, but evolution has led to the total a loss of chlorophyll. Thus, like animals, they exhibit carbon heterotrophism, i.e. the incapacity to synthesize sugar from atmospheric carbon dioxide, and must therefore draw the organic compounds they need from the ground around them, a phenomenon known as saprophytism. Several species, including both tropical and temperate taxa, share this distinct feature. The mechanism of the germination of seeds has been observed in a temperate saprophytic species, *Neottia nidus-avis*. The achlorophyllous orchids occurring in Vanuatu, which belong to several genera, are as follows: *Gastrodia cunnighamii*, *Dipodium punctatum*, *Didymoplexis micradenia* and *Epipogium roseum*.

The developmental cycle of *Epipogium roseum*, studied by Docters van Leeuwen in 1937, is among the fastest of the family: the inflorescence emerges very quickly, fertilization takes place in the floral bud, and the seeds, each comprising only a few cells, develop in just a few days.

Figure 114: *Mediocalcacar paradoxum*.

FOCUS ON PALMS

Jean-Michel Dupuyoo

Palms are distinctive plants that are found in practically all tropical and subtropical areas, whether in the wild or in cultivation. The family includes a total of nearly 2800 species, many of which are endemic to small areas. Vanuatu, which has 18 species belonging to 13 genera, is one of the most important centers of palm diversity in the Pacific. Members of the family are present on all of the islands making up the archipelago and in all types of habitat, comprising a dominant element of the landscape. Palms are also used traditionally in many ways by Vanuatu's inhabitants, including for construction, as a source of food, and for sacred and religious purposes.

Figure 116: *Licuala grandis* growing in its native habitat.

••• Palms of forest habitats

••• *Cyphosperma voutmelense*

This palm is endemic to the mountainous chain in western Santo, occurring primarily in the undergrowth of humid forests, where it grows in small colonies, generally above 900 m elevation, and is often abundant (see "Principal Types of Vegetation Occuring on Santo"). A few populations are found in open areas, where the plants exhibit a more prostrate habit. Individuals of *C. voutmelense* are relatively small, reaching a maximum of about 6 m in height (Fig. 115). The leaves are pinnately compound and measure between 1 and 1.5 m in length. The fruits, which are about 1 cm long and olive-shaped, are reputed in northwestern Santo for their medicinal properties, in particular as a treatment for certain types of fever.

••• *Licuala grandis*

This species is native to the Santa Cruz archipelago and San Cristobal in the Solomon Islands as well as to the islands in the northern half of Vanuatu, including Santo. Its leaves are palmately compound and the plants rarely exceed 3 m in height (Fig. 116). The fruits are round and measure about 2 cm in diameter. *Licuala grandis* occurs in shady areas in low elevation forest and is rather uncommon, growing in sparse colonies. On Santo, the leaves of this palm are often used as an umbrella by women and for roofing on traditional houses. Of all Vanuatu's palms, this is the most common worldwide: *L. grandis* is widely cultivated in Europe

Figures 115: *Cyphosperma voutmelense* growing in perhumid forest in northwestern Santo.

because, in addition to its ornamental qualities, it can easily be grown as an interior plant.

Veitchia winin and *Veitchia arecina*
The genus *Veitchia* is restricted to Melanesia (Vanuatu, Fiji and Tonga) and is represented by four endemic species in Vanuatu, two of which (*V. arecina* and *V. winin*) occur on Santo. The first of these is widely distributed, growing in colonies and ranging from the Torres Islands in the north to Tanna in the south. On Santo, this species is primarily limited to the southeastern part of the island and on Malo. *Veitchia winin* occurs on Malakula, Pentecost and in western Santo. The trunks of this large palm, which are about 20 cm thick, often reach to over 20 m in height, and the tallest individuals are emergent, dominating the forest canopy. The "almond" (endosperm) is often extracted from the seeds of these palms while still soft, but once the seeds reach maturity and turn red they can no longer be eaten. The trunks are used for traditional construction, especially in western Santo.

Calamus vitiensis
This lianescent (climbing) rattan species has a trunk that can reach 30 m in length, and occurs on practically all of the islands comprising Vanuatu and is also found in Fiji. The leaves are pinnate and at their tip they bear a flagellum armed with small barbs, which enables the plant to remain firmly attached to the surrounding vegetation and to reach into the canopy. The fruits are 2 cm in size and covered with scales. The inhabitants of Vanuatu employ the stems of this palm as an especially strong cord or rope used in both traditional construction and artisanal works.

Cultivated palms

Cocos nucifera
The coconut probably originated in coastal areas in the islands of the Pacific and research has shown that it was already present in Vanuatu well before the first humans arrived. This salt-tolerant palm occurs primarily in the littoral zone, and while its floating fruits can be dispersed over great distances by ocean currents, man has long been responsible for its presence throughout the world.

Coconut trees can grow to 20 or 30 m tall. The trunk is smooth and can reach about 30 or 40 cm in diameter near the ground, above the swollen base. Its leaves are pinnate and measure from 4 to 6 m long.

Coconuts are often grown in gardens, around habitations, and in mono-specific plantations that also serve for raising cattle. Coconuts are one of the most important agricultural species in Vanuatu, and the sale of copra provides the sole source of income for many families. Above all, however, this is a multi-purpose species. The trunk is used in construction, the leaves are employed in basket weaving and as roofing, the terminal bud is eaten, the meat of the fruit is used in the preparation of sauces, and the milk makes a refreshing drink. Moreover, several parts of the plant, such as the bark and roots, have medicinal uses.

Carpoxylon macrospermum
This palm can become very large, with the trunk reaching 20 m in height and 30 cm in diameter at breast height (DBH). The leaves are pinnate and arching, and the fruits are red at maturity, measuring 6 cm long and 4 cm wide.

Carpoxylon is the only monospecific genus in Vanuatu (i.e. with just one species), and because it is highly threatened, it has become one of the symbols for preserving the country's flora. At one point it was only known from a few individuals in the south of the archipelago, but is now widely multiplied and distributed. On Santo, *C. macrospermum* can be seen growing in cultivation around houses and in gardens.

The young fruits of this palm are edible, although when mature the endosperm is no longer palatable. The leaves are made into brooms and the trunks can yield high quality wood. However, on Santo *C. macrospermum* is mostly grown for its ornamental qualities and for its value as a rare plant.

Metroxylon warburgii
The "sago palm" or "sagoutier" is a massive, fast-growing plant that can reach 5 to 20 m in height at maturity (Fig. 117). Its trunk is 30 to 40 cm in diameter and contains a large quantity of starch. The long pinnate leaves often exceed 3 m in length and the long leaflets have valuable properties that are utilized locally. *Metroxylon warburgii* is hapaxanthic (meaning that each stem dies after flowering) and the inflorescences are borne above the leaves. Unlike *M. sagu*, a closely related species, individuals of *M. warburgii* do not re-sprout at the base, making vegetative propagation impossible. The fruits, covered with scales, are large and pear-shaped.

Metroxylon warburgii occurs on most of Vanuatu's islands and can also be found in the Solomons and on Rotuma in northern Fiji, although the precise origin of this species is not known. Sago palms are usually found as cultivated plants in gardens and around houses, and populations that appear to be wild often in fact represent old plantations that have been abandoned. It is thought that humans brought *M. warburgii* with them when they arrived and that it thus actually represents a crop plant.

This palm is now used locally for its foliage. The leaflets serve as roofing on traditional dwellings in northern Vanuatu and when properly treated are rot-resistant. Roof "tiles" are sometimes sold commercially, and the southern islands import large quantities from the north, in particular from Santo. The leaflets of the varieties that grow in the southern islands are of lower quality and are thus less widely grown.

Figure 117: *Metroxylon warburgii* growing in southern Santo.

Several species of *Metroxylon* are currently exploited for their starch. The best known, *M. sagu*, is an important crop, especially in New Guinea and Malaysia. *Metroxylon warburgii* is also exploited in Vanuatu, although its use has decreased and is now largely restricted to more isolated parts of the archipelago. Extraction of sago was observed in 2003 in southern Santo using a technique similar to that employed elsewhere in the region. Plants are felled just as flowering is about to take place, which corresponds to the stage when the trunk contains its highest concentration of starch. The trunk is cut into several segments, and in order to release the starch from the cells in which it is contained, the core is reduced to a fine powder, a process that requires extensive pounding with a hammer that resembles those used by the local populations of New Guinea and Sulawesi. The powder is then rinsed to extract the starch, which is suspended in the water while the fibers and other residue are eliminated by filtration (Fig. 118). The extract is then left to decant in a small tub, and once dried, the starch can be stored for several months.

Metroxylon also serves as a source of salt for people living in the interior of Santo. In the past, access to the sea — even though just a few kilometers away — was limited because of conflicts with neighboring groups, which isolated those living inland. The salt was extracted by dissolving the ash of burned leaf blades and petioles, and the resulting salty water was thus used to prepare sauces, a practice that has now become very rare (Fig. 119).

The flora of Vanuatu contains a second species of Sago palm. *Metroxylon salomonense* is native to the northern part of the archipelago (the Torres and Banks Islands) and also the Solomon Islands. This large palm is sometimes cultivated on Santo, but is not used except as an ornamental. While its leaflets are larger in size that those of *M. warburgii*, they do not have the same physical qualities and as a consequence this species is rarely cultivated.

Figure 118: Extraction of sago from *Metroxylon warburgii* by locals in southern Santo.

Figure 119: Extraction of salt from the ashes of *Metroxylon warburgii* (southern Santo).

FOCUS ON FERNS

Germinal Rouhan

Pteridophytes (ferns) are one of several groups of plants that do not display flowers. They are found in nearly every type of vegetation on Santo, from the shorelines to the highest summits, including humid forests, coconut plantations, river banks and dry areas. Indeed, members of this group occur throughout the world, extending from cold regions to the tropics. They can be terrestrial, epiphytic (growing on other plants), lithophytic (growing on rocks) or climbing; some are minute with leaves that are one cell thick whereas others form large trees more than 20 m tall. In addition to their importance in several ecosystems, ferns are particularly informative for understanding the floristic affinities between Santo and other Pacific islands, and more generally for studying the origin and biogeography of insular floras. Ferns also appear to play an important role in the culture of Vanuatu and are used in numerous ways.

Although ferns have no flowers, they are not bryophytes (see "Focus on Bryophytes"), which lack internal vascular (i.e. water-conducting) tissue. Ferns and what are generally referred to as "allied plants", have similarities in their life cycles and reproduce by dispersing spores. These features, in addition to the absence of flowers and seeds, distinguish them from seed plants (spermatophytes), which include the well-known flowering plants (angiosperms) and gymnosperms (including kauris; *Agathis* spp.). Ferns and "allied plants" are most frequently referred to as pteridophytes, but they actually form two distinct natural groups: monilophytes and lycophytes (Fig. 120), which diverged from their most recent common ancestor about 400 millions years ago.

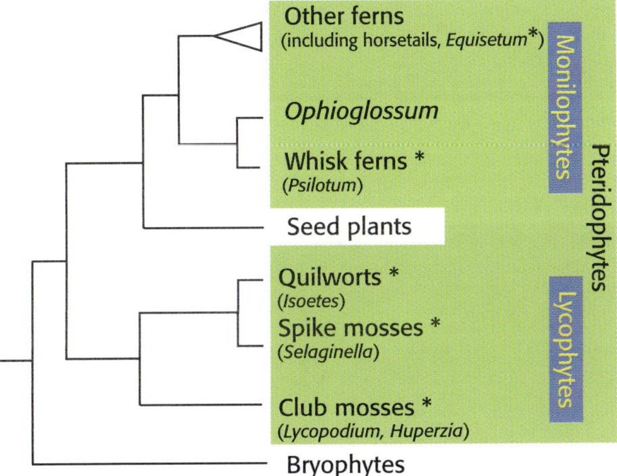

Figure 120: Evolutionary relationships between pteridophytes, shown in green, seed plants (spermatophytes), and bryophytes (mosses and liverworts), with relative placement of the two natural groups that comprise the pteridophytes: monilophytes and lycophytes; the groups often referred to as "allied plants" or "fern allies" are indicated with an asterisk (*).

• • • **A great diversity of ferns can be observed by the person who knows how to recognize them…**

All lineages of monilophytes and lycophytes are represented on Santo (Fig. 121), with just one exception, *Isoetes*. The morphological diversity exhibited by these groups on Santo is much greater than is commonly assumed by most people: ferns are not only the familiar green plants with finely dissected leaves but include many other, often surprising forms. Unfortunately, it is not possible to point to a single unique morphological character that distinguishes a plant as a fern or "allied plant". However, in most cases ferns can be distinguished by the presence of:

- croziers (also know as "fiddleheads"), which are the young leaves spirally coiled in bud (Fig. 122);
- scales (multicellular and plane epidermal outgrowths) that are more or less abundant and scattered on the leaves, petioles, and rhizomes (Fig. 123);
- rhizomes (creeping to more or less erect stems);
- sori (groups of spore-producing structures known as sporangia) of different shapes and sizes borne under the leaves, conferring patterns that are often characteristic of genera (Fig. 124).

• • • Taxonomic diversity of pteridophytes on Santo

Very few botanists have collected ferns in the northern islands of Vanuatu, including Santo. Among the first to have sampled ferns are the Reverend F.A. Campbell (1872-1873), S.F. Kajewski from the Oxford Expedition (1928-1929), and T.C. Chambers in 1963. The most intensive collecting work has been conducted more recently, in particular by the British Royal Society Expedition (1971), several Japanese expeditions (the Nippon Fernist Club of Japan in 1995; the Tsukuba Botanical Garden, National Science Museum, in 1996, 1997, 2000, and 2001), and by members of the Santo 2006 expedition (Table 9).

Table 9: Main inventories of the pteridophytes of Vanuatu. The number of endemic species is indicated in brackets.

Year of the expedition (Reference of the publication)	Number of species collected	
	Vanuatu	Santo
1971 (Braithwaite 1975)	250 (11)	167 (5)
1996-1997 (Matsumoto *et al.* 1998)		212
2000-2001 (Matsumoto *et al.* 2002)		150
2006 (Rouhan here presented)		147

Figure 121: Some species representing the main lineages of monilophytes and lycophytes present on Santo. **A**: *Lycopodium volubile*. **B**: *Huperzia* sp. **C**: *Prosaptia contigua*. **D**: *Selaginella distans*. **E**: *Psilotum complanatum*. **F**: *Cyathea truncata*. **G**: *Schizaea dichotoma*. **H**: *Blechnum pilosum*. Equisetaceae and Ophioglossaceae not shown here. *Isoetes* not recorded from Santo.

Figure 122: Crozier (young spirally coiled leaf) of *Gleichenia* aff. *milnei*.

Figure 123: Apex of the tree fern *Cyathea truncata* displaying a dense coating of hairy red-brown scales.

Figure 124: Sori of some ferns from Santo. **A**: *Selliguea* aff. *feeoides*. **B**: *Lygodium reticulatum*. **C**: *Cyathea lunulata*. **D**: *Pteris pacifica*. **E**: *Davallia repens*.

The most diversified fern groups on Santo are the genera and families that are the most diversified worldwide, such as Hymenophyllaceae (28 species), *Asplenium* (~15 species, Aspleniaceae), *Huperzia* (~12 species, Lycopodiaceae), *Cyathea* (~8 species, Cyatheaceae), *Diplazium* (8 species, Woodsiaceae), *Pteris* (~8 species, Pteridaceae), *Adiantum* (5 species, Pteridaceae), *Blechnum* (~6 species, Blechnaceae), *Lindsaea* (7 species, Lindsaeaceae), *Selaginella* (6 species, Selaginellaceae).

Observations on the ecology of ferns on Santo

Although more extensive inventories and ecological monitoring plots would be necessary to provide precise

Figure 125: Large ferns of the edges of rivers (e.g. *Hypolepis elegans*).

Figure 127: Montane cloud forest at 1450 m (Cumberland Peninsula) with numerous epiphytic ferns and tree ferns.

Figure 126: The dense vegetation formed by a creeping member of the family Gleicheniaceae (*Gleichenia* sp.) on open ridges in the western part of Santo's central mountain chain.

Figure 128: The hepatic-like filmy fern *Didymoglossum tahitense* (Hymenophyllaceae) with appressed fronds on a tree trunk (Cumberland Peninsula).

information on the distribution patterns of ferns on Santo, the botanical work conducted to date offers some insight into ecological features of the ferns growing on the island. The eastern part of the island is mainly made up of low altitude humid forests and highly secondary areas, all of which appear to be particularly poor in pteridophyte diversity, although ferns can be

locally abundant in some places. This is especially true along the edges of roads and tracks and along rivers (Fig. 125), where some of the largest ferns on Santo grow (with leaves that can reach 5 m long, e.g. in species such as *Cyclosorus costatus*, *C. magnificus*, *C. rubrinervis*, *Hypolepis elegans* and *Pteris tripartita*).

Ferns are more abundant and diverse in the western part of Santo, where more than 80 % of the species diversity probably occurs. The open ridges of the western part of the mountain chain are frequently colonized by climbing ferns such as *Gleichenia* sp., forming dense and tangled vegetation (Fig. 126). However, most fern diversity occurs in humid forest. Tree ferns belonging to the genera *Cyathea* and *Dicksonia* (with eight and one species, respectively) are well represented and diverse in forest occurring on slopes. Their abundance is higher above 600 m in montane cloud forest (Fig. 127), characterized by almost permanent atmospheric humidity (from clouds, fog, and mist) that is captured by the vegetation, adding significantly to precipitation from rainfall. Such ecological conditions are particularly favourable for epiphytic fern species, which are frequently seen in the forests of the Cumberland Peninsula and on the island's highest summits of Mount Tabwemasana and Pic Santo. Some epiphytic ferns have evolved remarkable hygrophilous strategies. On Santo, *Didymoglossum tahitense* provides a striking illustration of such adaptations. This species, a member of the most diversified fern family on Santo (Hymenophyllaceae, the "filmy fern" family, with 28 species), shows reduced, thin and rounded leaf blades that are appressed to the tree trunks, allowing them to take advantage of surface water, giving this species an unusual "liverwort-like" aspect (Fig. 128).

••• Biogeographic affinities of the Pteridophytes from Santo

More than 75% of the fern genera found on Santo have a wide geographic distribution (from paleotropical to cosmopolitan) and there is no endemism on the island at this taxonomic level. A lack of generic endemism is not unexpected for an oceanic island such as Santo. Indeed, distributions of ferns on islands suggest that they can disperse over thousands of kilometers, and it is often assumed that the distribution ranges of fern taxa are a result of dispersal rather than vicariance, reflecting the ability of pteridophytes to disperse easily via their dust-like spores. These minute unicellular propagules (< 100 μm) are produced in huge numbers (one plant of some species can generate an average of 6×10^7 spores annually) that can travel high in the atmosphere over great distances. Experiments have shown that spores are still able to germinate and develop into apparently normal prothalli after travelling above 10 000 m for several hours. A single spore arriving on Santo would thus be capable of germinating and developing into a bisexual gametophyte, which could self-fertilize and to give rise to a new spore-bearing plant.

Although geographically closer to New Caledonia and the Solomom Islands that to Fiji, Santo shows its closest pteridophytic floristic affinities with this archipelago at both the the generic and specific levels. New Caledonia and Vanuatu share about 100 species, but less than 10 of these are restricted to these two areas, fewer than five of which occur on Santo. It is amazing that New Caledonia and Vanuatu, separated by less than 350 km at their closest points, have such distinct fern floras, a situation that may be explained by geological and climatic factors. For example, numerous New Caledonian endemic terrestrial ferns grow on serpentine soils, which are essentially absent on Santo and more generally in Vanuatu.

At the species level, detailed taxonomic studies may reveal up to as many as five endemic taxa (which would still be less than 2% of the total Santo fern flora), all apparently most closely related to relatives growing on Fiji, the Solomons or New Caledonia. By comparison, more than 40% of the fern species on the nearest archipelago, New Caledonia, may occur nowhere else on Earth. Vanuatu may well have the lowest level of species endemism among ferns of any of the major archipelagos in the Pacific (Fijî, French

Figure 129: Statue carved into stylized image of figure, used in traditional rituals (from the Banks Islands).

Figure 130: Lycophyte used as hair ornaments: fertile portion of *Huperzia* sp.

Polynesia, New Caledonia, New Zealand, Samoa, Solomon Islands), suggesting that its pteridophyte flora is relatively recent and has not been isolated long enough to give rise to new species.

••• Uses of ferns in Santo and Vanuatu

Traveling in Santo, from Luganville to the Mount Tabwemasana massif, through villages and forests, it quickly becomes obvious that ferns are widely used in Vanuatu, perhaps like nowhere else in the world. Members of the tree fern genera *Cyathea* and *Dicksonia* are probably the most used. Several species can reach more than 10 m in height and have leaves (fronds) more than 5 m long. They are mainly exploited for their trunks, which are mechanically very resistant and long-lasting (decomposing very slowly). In particular, Ni-Vanuatu take advantage of the toughness of the root mantles forming the base of the trunk for flowerpots and for carving statues of stylized images or figures that are used in traditional rituals (Fig. 129). Whether carved or not, however, all species of tree ferns are protected under CITES (the convention regulating international trade in endangered species of plants and animals) so they may not be exported legally except for scientific purposes. Some other uses of ferns include as hair ornaments (any kind of fern is used, but fertile parts of *Huperzia* are the most frequently seen; Fig. 130) and for food (the young leaves of genera *Diplazium*, *Pteris* and *Cyathea* are sometimes eaten in soup).

FOCUS ON BRYOPHYTES

Elizabeth A. Brown

Mosses, liverworts and hornworts are members of plant groups that were amongst the earliest land plants to diverge some 400 million years ago. They are often described as "lower" plants because they lack roots, flowers and a sophisticated vascular system. Bryophytes require water to complete their life cycle; like ferns, they produce spores instead of seeds and water is essential for transfer of the motile gametes and fertilisation. However, recent research suggests transfer may also be accomplished by small arthropods. In spite of these limitations bryophytes have developed a range of strategies for survival and are found in virtually every habitat, including some of extreme periodic dryness. As they dry out, the plants suspend metabolic activity and only recommence functioning as they absorb water (a feature also found in some flowering plants and ferns termed "resurrection" plants). In bryophytes the re-wetting and return to activity happens very quickly as the leaves are mostly only one cell thick.

The bryophyte flora of islands is often regarded as rather depauperate. The only way for bryophytes to spread is as small pieces of plant or as spores, and many species do not produce spores regularly or even at all. There are some obvious exceptions; New Caledonia and New Zealand have very rich, interesting and diverse bryofloras but Vanuatu has a flora that is less diverse and endemism is fairly low, with many of the species being widespread. However, when size of the landmass is considered, the numbers become far more comparable. For example, Fiji has approximately 211 liverworts whereas Vanuatu has only 145, but when land area is compared there are 11.2 and and 11.7 species per 1000 km², respectively. This contrasts with the flora of New Caledonia (436 species and 23.3 species/1000 km²) and New Zealand (678 species and 2.3 species/1000 km²).

The Society Islands (including Tahiti), which have been botanically explored extensively, have 186 species of liverworts (110.7 species/1000 km²).

Research on the mosses and liverworts of Vanuatu has occurred sporadically over the last 150 years. Much of the literature has been published in scientific journals and some of these are not widely available, even in botanical libraries. The primary descriptions of many species are based on the features of specimens occuring in other countries. Frequently, the presence of the plant in Vanuatu is merely referred to in a brief note on the plant's distribution.

Vanuatu has hosted few visits by bryologists and many areas have not been extensively surveyed. The Reverend William Gunn, Medical Missionary of the United Free Church of Scotland, combined his work with botanical interests. He was based on the island of Futuna from 1883-1913 but appears to have collected only in the years 1910-1913 (with a significant percentage of collections being made on Aneityum in 1913). Many of his collections were used to describe new species and the types are held at the herbaria in Helsinki and Sydney.

Further studies on the mosses of Vanuatu were reported by Thériot in 1938, Dixon in 1948 and Tixier in 1972, 1973 and 1975. A catalogue of mosses of Melanesia was published by Schultze-Motel in 1973. As part of his work on the Pacific, H.A. Miller collected in Vanuatu and drew together all the literature. He published, in 1978 and 1983, lists of the mosses and liverworts occurring in Oceania, citing approximately 203 mosses and 142 liverworts for Vanuatu.

Dale Vitt, a Canadian bryologist, collected on Mt Tabwemasana in August 1985, publishing on elevational and latitudinal gradients of mosses on South Pacific Islands. He concluded that there was a general increase in species number with elevation (on individual islands); most species are rare on a given island, despite few of them having narrow geographic ranges. He also showed that the number of Isobryalean (moss) species increases dramatically in the tropics (and in high elevation subtropical forest) and moss floras of tropical Pacific Islands have a greater number of derived species.

Japanese bryologists have worked extensively in Vanuatu, particularly Dr. Masanobu Higuchi, who has visited the islands a number of times, including as part of the large joint expeditions of Tsukuba Botanical Garden and National Science Museum, Tokyo in 2000 and 2001. In 1996, Higuchi produced a checklist for the island moss flora, recording 236 species from 84 genera. His visits to the islands of Espiritu Santo, Efate, Tanna and Anatom (Aneityum) have resulted in a number of papers that extend the knowledge of the Vanuatuan bryoflora.

In October 1998 Heinar Streimann, then curator at the Australian National Botanic Gardens, and avid cryptogam collector (mostly mosses and lichens), visited the islands of Efate and Espiritu Santo. He died in 2001, shortly before the publication of what probably would have been the first of a series of papers extending the moss flora of Vanuatu. The manuscript includes ten new records for Vanuatu and the first report of 17 species not previously known on Efate or Santo. The following comment is made: "The reported moss species appear to have a wide Malesian – Pacific range... more mosses within this geographical range will no doubt be found". Many species occur in neighbouring countries and their absence is likely to reflect lack of collecting rather than any other factor, e.g. *Pelekium bonianum*, which is found in the Solomon Islands, New Caledonia and Samoa, is likely to occur also in Vanuatu.

The Santo 2006 expedition provided an opportunity to work in an area of Espiritu Santo that appears not to have been previously surveyed and is relatively undisturbed. Occasionally we saw or heard cattle crashing through the forest ahead of us but mostly the only evidence was *les bouses fraiches* or the cattle ticks – delightful creatures the size of a finger nail with mouth parts designed to penetrate cattle hides. The lowland and coastal forests of Penaoru can not be described as bryophytically rich, with just a handful of common species sparsely scattered through them. This situation persists through the first few hundred metres of topography; the low light levels, combined with the periodic dryness are not friendly to many bryophyte species. Although this area is dry and apparently devoid of bryophytes, closer examination (of 50 m² quadrats) at intervals of 150 m

altitude revealed at least 11 mosses and eight liverworts per quadrat (3/4 of the latter being members of the family Lejeuneaceae). Species such as the hardy and widespread *Leucophanes octoblepharoides* occur sporadically in small populations (**e.g.** this species grows in crevices at the base of tree trunks). Other hardy moss taxa include *Rhynchostegium* and *Isopterygium* species found growing on small boulders whilst a number of species in the liverwort genus *Acrolejeunea* also occur occasionally on boulders. Much as I enjoyed the cooking of our wonderful chef, I was eager to join the camp at 900 m elevation. Unfortunately, even at this altitude the range of bryophytes on the ridges we traversed was fairly limited, e.g. Lejeuneaceae, a family normally well represented in tropical regions, did not occur in great profusion (although they represent a large proportion of the species observed).

Water for the camp had to be carried up from a small stream about 100 m below. This deeply dissected and humid area proved to be a richer place to work. Many of the twigs on the trees are festooned with large pendent species of *Frullania* (Fig. 131), *Porella* and *Radula* (all liverworts) with what is probably a different suite of species in the upper canopy of rotten branches and twigs (including the moss *Chaetomitrium* found on fallen twigs). The streamside rocks are host to *Dumortiera hirsuta*, a large thalloid liverwort found throughout tropical areas and *Marchantia emarginata* subsp. *lecordiana*, a plant also recorded from New Caledonia. In the genus *Marchantia* the antheridia (sperm) and archegonia (eggs) are borne on modified areas of the thallus that stand erect, looking like miniature umbrellas (Fig. 132). Elsewhere on the slopes near the stream a thick profusion of phanerogams crowded out the bryophytes at ground level. One of the few exceptions

Figure 131: Pendent branches of a species of the liverwort *Frullania* that clothe low branches in humid valley forest.

Figure 132: *Marchantia emarginata* subsp. *lecordiana*, with antheridiophores (which produce "sperm") and archegoniophores (which produce "eggs"), from a streamside bank.

Figure 133: *Telaranea chaetocarpa* growing on the side of a rotten log.

was the resilient little moss *Isopterygiopsis*, which uses clumps of rhizoids along its creeping stems to anchor itself to the stems and leaves of the *Elatostema*.

Vitt's finding that most species are rare is brought strongly to mind by the occurrence of *Telaranea chaetocarpa* adjacent to one of the quadrats. Growing in profusion on the side of a rotten log (Fig. 133), it was not observed anywhere else on the trip. A member of the family Lepidoziaceae, one of my research interests, I would like to believe that I would have seen it if it were common; I certainly looked for it! Similarly, a few strands of *Kurzia bisetula* on the base of a tree fern (*Cyathea*) at the campsite alerted me to look carefully at every tree fern I saw, all without success.

Unlike the people studying insects, I found that every metre higher that I climbed, the bryoflora became more interesting and profuse (and yes, there are more Lepidoziaceae at the higher elevations). There is a significant change in the species composition at elevations above the level where the clouds envelope the mountains most evenings. Typically called moss forest, the species that dominate on Vanuatu (as well as in New Caledonia, Australia and New Zealand) are usually liverworts. Few people make the distinction but mosses and liverworts have been separate lineages (in spite of their superficial similarities) since before dinosaurs roamed the earth and flowering plants evolved as a separate group. At these elevations the forest is of lower stature and the higher light levels combined with the constant presence of moisture make it a very suitable habitat for bryophytes. If climate change advances as predicted, as many scientists believe it will, the cloud cover will rest on the mountains at increasing elevations and the area inhabited by these forests will become progressively smaller "islands", ultimately disappearing as the cloud fails to settle.

The track to the quadrat at 1 200 m and beyond was slippery in many places, but that moisture ensures the presence of species such as *Treubia insignis*, one of a group of closely related taxa found in New Caledonia, Papua New Guinea, Indonesia and the Philippines. Slime papillae on the surface produce quantities of mucus-like jelly that make handling this beautifully lobed and oil-cell speckled liverwort an interesting experience. Members of the genus *Haplomitrium* rarely occur in profusion and I was delighted to see a population of *H. blumii* growing on a very rotten log. Both species have only recently been reported for Vanuatu.

The highest point in our research area (1 300 or 1 400 m, depending on whose altimeter one trusts) was the richest site for bryophytes. Numerous species of *Riccardia* were encountered, considerably more species than the two currently recorded for Vanuatu. More research is needed but one of the species may be *R. eriocaula*, a species that occurs in New Zealand and southern Australia. I sighted just one clump of *Bazzania erosa* (Lepidoziaceae) growing as an epiphyte high on the trunk of a *Cordyline*, in a ball of moss. This is an unusual site for a *Bazzania* species, which more normally form dense clumps on rotten logs (and also a challenge to a short but very determined botanist in a forest full of rotten wood). This species is widespread throughout the Pacific and Melanesia but has not been reported from Vanuatu.

As the highest mountain in Vanuatu, Mt Tabwemasana, is a magnet for botanists, but the undisturbed upper slopes of Penaoru also proved to be a rich and exciting place for bryophytes. Further exploration, as well as research on the species collected during this expedition, will expand our knowledge of the bryoflora of Vanuatu.

the Forgotten Kingdom

Fungi

Bart Buyck

Mushrooms are often looked upon as some kind of odd vegetable but thanks to technological progress it has now been clearly established that they belong to a very diverse group of organisms we call "fungi", which are neither plants nor animals. As a matter of fact, the fungi represent a separate kingdom of living organisms whose affinities are more with the animal world than with plants. Fungi are not studied by botanists, but by mycologists. Unfortunately, mycologists are much harder to find than the fungi they study and we still know very little about the fungi on our planet. As past inventories focused principally on plant and animal biodiversity, we still know surprisingly little about fungal diversity. Scientists have so far named less than 5 % of the estimated minimum of 1.5 million species. It is, therefore, not surprising that we knew hardly anything about the fungi that are present on the island of Santo at the onset of this large inventory project. We have found no published records of past inventories of the fungi of Santo.

FUNGI OF SANTO: A DECOMPOSER COMMUNITY

Fungi are generally depicted as organisms that thrive on the decomposition of dead organic material and wastes. Many fungal groups, however, are entirely composed of symbionts or parasites that are highly dependant for their development on specific groups of living plants or animals. The latter is also the case for many —probably even most— of the widely known, larger forest mushrooms that appeal so much to the general public. Both these typical forest mushrooms and their host trees can not survive without forming root symbiotic structures (called "ectomycorrhiza") in the soil. Examples of such symbiotic forest mushrooms include not only such reputed edibles as morels, truffles, chanterelles and boletes but also comprise several thousand species belonging to more common and very diverse mushroom genera such as *Amanita*, *Russula*, *Lactarius*, *Cortinarius*, *Tricholoma*, *Hebeloma*, *Inocybe*, etc.

By just looking at what kind of animals and plants inhabit Santo, we may already have a hint of what groups of fungi to expect on the island. Indeed, one of the first things that may strike you when looking at the list of the forest trees that are present there, is the absence of those mushroom-associated tree families that dominate forest types elsewhere in the world: Santo lacks not only Fagaceae (oak, beech, chestnut, etc.), Betulaceae (birch, poplar, alder, etc.) and Pinaceae (pine, spruce or fir, etc.) that are so common in the northern hemisphere, but the island also has none of the locally dominant symbiotic trees from the southern hemisphere. Santo lacks for example eucalypts and some mushroom-associated Myrtaceae, common in Australia, nor is the island home to tree species of the predominantly Asian Dipterocarps or to some of Africa's widely dispersed legumes (*Isoberlinia*, *Brachystegia*, *Julbernardia*, etc.) in the family Caesalpiniaceae. It is therefore not surprising that we did not encounter any of the above-mentioned, typical symbiotic forest mushrooms on Santo, although some fungal species may perhaps be associated on the island with the roots of the amphipacific legume tree *Intsia bijuga*, which was common on Santo but is under strong pressure from deforestation, or with the roots of coastal *Pisonia* (Nyctaginaceae).

The absence of typical forest mushrooms does not by any means imply that you will encounter very few fungi when exploring Santo. Of course, the climatic conditions have to allow for their growth! Indeed, fungi need moisture for fruiting, and at least the soft fleshy mushrooms will only be present for a very limited

Figure 134: Polypores are very common and diverse in Santo's forests.

Figure 135: This still undescribed *Polyporus* was found on dead tree trunks and was said to be edible and locally well-liked.

Figure 136: It is merely a matter of one or two hours for the delicate *Dictyophora* species too change from the egg-stage to their fully expanded mature form.

time of the year when rains are abundant. Mushroom hunting will therefore be more rewarding on the much wetter eastern part of Santo. The much tougher polypores are considerably less dependant on precipitation and may remain attached to logs and trunks of fallen or standing trees for many years and continue their growth with the arrival of every new rainy season. These large, tough polypores are the most prominent

Figure 137: *Cookeina speciosa*, by far the most common species of this genus, comes in various colours, from white over yellow to pinkish red.

Figure 138: *Cookeina tricholoma* is a close relative of *C. speciosa* and easily recognized by the long hairs that cover its outer surface.

representatives of the decomposer community that is so typical of Santo's forests. One of the best places we visited was without any doubt the coastal dense forest reserve at Matantas. An overwhelming diversity of polypores in various colors and sizes was growing on almost every fallen log and branch on the ground (Fig. 134). Naturalists who have already looked for fungi in other tropical forests will easily recognize some common pantropical polypore genera such as *Cymatoderma*, *Microporus*, *Pycnoporus* and *Polyporus*. We particularly found the genus *Polyporus* itself to be strikingly diverse on Santo and we recorded close to fifteen different species during our short visit. One of these —an as yet undescribed species (Fig. 135)— is even locally consumed and bears a distinct resemblance to the European *Polyporus squamosus*.

Most of the decomposer fungi of Santo are smaller, ephemeral, fleshy species that remain present for only a few hours or perhaps a few days, the time needed to produce and disperse their spores. Examples among some of the more common, gilled mushroom genera include wood decomposers such as *Pleurotus*, *Lentinus*, *Schizophyllum* or the very beautiful yellow-orange *Cyptotrama asprata*. Around the completely

Figure 139: This strange, water-filled ascomycete is *Galiella celebica*, a pantropical species.

decomposed and rotten remains of logs and truncs, species of *Pluteus* and *Psathyrella* form sometimes vast colonies of many dozens of fruit bodies. Such very nutrient-rich sites also favour the proliferation of spectacular *Dictyophora* species (Fig. 136). Also many ascomycetes decompose wood but most species are very small. Among some of the larger ascomycetes, we can cite the many bright yellow, pink, red or purple cups of the various species of the genus *Cookeina* (Figs 137 & 138) that sprout like colourful flowers from the smaller twigs and branches that have fallen on the forest floor, whereas another wood decomposing ascomycete, *Galiella celebica* (Fig. 139), keeps it

own water reserve safely stocked inside the cavity of its gelatinous fruitbody.

Among the leaf litter on the forest floor, many short-lived species of *Coprinus*, *Lepiota*, *Marasmius* and *Mycena* specialise in the decomposition of leaves and other parts of particular plant species.

Outside the forest, in the pastures around the villages, slender species of *Coprinus*, *Psilocybe*, *Panaeolus* and other delicate mushrooms pop up everywhere after heavy rain, especially on or near cowdung or excrement left by horses, pigs and other animals.

FUNGI THAT LIGHT UP IN THE DARK!

When night falls, the dancing lights of fire flies are a familiar sight for many travellers, but mushrooms that light up in the dark remain a surprising experience. Going out for a night walk in the forest can definitely be recommended: bioluminescent fungi are common on Santo and well known by the local population, who use them as a light source.

Bioluminescent fungi are not so rare as one might think. They occur all over the globe and have been known since ancient times. They produce light continuously but the human eye needs to wait for darkness before it can detect the light emission. In a very recent overview of bioluminescence in fungi, these organisms are placed in three different phylogenetic groups: the *Omphalotus* lineage, the *Armillaria* lineage and the Mycenoid lineage. All three comprise exclusively white spored basidiomycetes. The first two lineages are also common in France for example. The Jack-O-Lantern (*Omphalotus olearius*, "le pleurote de l'olivier" in French) is a good example of a medium to large-sized Mediterranean mushroom of the first lineage. The *Armillaria* lineage comprises some extremely common mushroom species that are

responsible for the "glowing wood" phenomenon —or "foxfire" as it has been known since ancient time— because it is not the mushroom itself, but the mycelium and mycelial cords inside the wood that emit light. On Santo, we observed only bioluminescent species of the Mycenoid lineage (Fig. 140). These are predominantly much smaller fungi, in this case species of *Mycena* and *Filoboletus*. Apart from the proven fact that the light attracts insects that may (perhaps) contribute to spore dispersal (in taxa whose fruit bodies are light emitting), there is no consensus about the "why" of this phenomenon. Whether it favours detoxification of some of the metabolites resulting from the wood degradation or is simply a by-product with no particular selective advantage, we do not know.

The exploration of Santo opened a window on the incredibly diverse but still barely understood world of the fungi… and even though we will probably never be able to document the totality of the mycota of the quickly changing habitats of Santo, we can still profit from their ephemeral presence and enjoy their breath-taking beauty and ingenuity.

Figure 140: Bioluminiscent species of *Mycena* and *Filoboletus* are quite common in the wetter parts of Santo's forests. Here the same specimens of a *Mycena* photographed during daylight and at night.

Fauna

Terrestrial

coordinated by Bruno Corbara

Biodiversity Along an Altitudinal Gradient

Bruno Corbara on behalf of the IBISCA network

Identifying significant patterns in plant and animal diversity and building hypotheses about how these patterns have been generated is a key challenge for forest ecologists. Not only are such results of fundamental importance to the understanding of ecosystem dynamics, but such understanding is probably the only viable path to developing management tools reconciling sustainability and human activity.

Unfortunately, the magnitude and determinants of biodiversity are far from being even reasonably well understood; for instance, knowledge of local food webs is still rudimentary —which is not surprising when we consider the number of species that still need to be described. By far the greatest fraction of terrestrial animal diversity is made up of tropical arthropods, in general, and insects, in particular. Insomuch as a tradition of the contemporary study of arthropod biodiversity exists at all, it has been established that wresting high quality information from tropical and subtropical forests is a slow and painstaking undertaking, involving decades of careful study at best if carried out through traditional, single-research group methodologies. It took two weeks of field work, for example, for the Australian ecologist Nigel Stork to sample 15 Bornean trees (using the canopy knockdown technique, i.e. diffusing insecticide up into the foliage with a fogging machine), but more than a decade for him and his numerous collaborators to sort the collected material to morphospecies (unnamed species diagnosed by standard taxonomic procedures), not including mites. It was this apparent dilemma —the urgent need for high quality results contrasted with the essentially long-term nature of obtaining such results— that led to the development of the IBISCA "model".

••• The IBISCA programme

IBISCA —for Investigating the BIodiversity of Soil and CAnopy— is an international research programme whose main aim is to study the spatial (i.e. horizontal, vertical and altitudinal) and temporal distribution of the organisms which constitute the major part of forest biodiversity: arthropods. Interactions with plants (mostly trees) and selected other organisms are also studied in this context. The IBISCA approach is based on highly integrative research projects and the use of state-of-the art canopy access techniques. Field projects are conducted worldwide in tropical, subtropical and temperate forests. Thus far projects have been carried out in San Lorenzo (Panama, 2003-2004), Queensland (Australia, 2006-2008), Auvergne (France, 2008-2009) and, as far as what concerns us here, Santo, Vanuatu, in 2006. IBISCA was conceived by Yves Basset (Smithsonian Tropical Research Institute, Panama), Bruno Corbara (Université Blaise Pascal, Clermont-Ferrand) and Héctor Barrios (University of Panama) in response to the lack of large datasets on the diversity and distribution of arthropods over multiple scales (i.e. horizontal, vertical and temporal) in tropical rainforests. Such data are fundamental to understanding the structure of arthropod communities from the ground to the canopy, and to test hypotheses about the origin and persistence of this biodiversity.

IBISCA projects are structured around various core scientific questions, namely: quantifying the horizontal and vertical species turnover in a lowland tropical rainforest in Panama; quantifying altitudinal species turnover and its implications in the context of climatic changes in a continental (IBISCA-Queensland led by Roger Kitching, Griffith University) or insular (IBISCA-Santo led by Bruno Corbara) location; and quantifying species turnover among deciduous forest types in a temperate region (IBISCA-Auvergne led by BC). The IBISCA protocol is multi-faceted, as shown below.

••• **It is multi-scale** allowing for a comparison of the diversity and abundance of organisms studied between sites, strata (i.e. ground, understorey, canopy) and seasons. This is of particular importance since the local species richness (alpha-diversity) and the species turnover between sites (horizontal beta-diversity) may differ between the ground and the canopy. Consequently the species turnover from the ground to the canopy (vertical beta-diversity) may differ from the local to the regional scale. Without a three-dimensional approach it is difficult to assess whether the diversity is highest on the ground or in the canopy. For landscapes, a stratified sampling is conducted including the relevant habitats, strata and microhabitats. In addition, temporal replicates are carried out whenever possible since the distribution and abundance of organisms may vary seasonally, according, for example, to their life-stage. This seasonality affects the ease of capture and the identification of the specimens (immature arthropods are often very difficult to identify). Temporal replicates yield a representative picture of the functioning and dynamics of arthropod assemblages.

••• **It is multi-taxonomic** since no single arthropod taxa is representative of the pattern of distribution and abundance of the others, as each has its own ecological requirements. Because an all-taxa approach is extraordinarily costly in terms of time and resources, the IBISCA Programme has adopted a functional approach including taxa representative of various life-history strategies (e.g. bees as pollinators, termites as scavengers, hemiptera as sapsuckers). Depending on the project, non-arthropod organisms interacting with arthropods (e.g. hostplants or vertebrate predators) are also included.

••• **It is multi-methodological** since the appropriate collection method may differ based on focal taxa and because the use of numerous collection methods compensates to some extent the inevitable bias of each individual method in terms of representativeness of the real abundance of each taxa in the community. In addition, it may help in developing new and efficient protocols for studying particular arthropod assemblages. A special emphasis is placed on methods such as sticky traps, flight-intercept traps, Malaise traps, light traps that allow direct comparisons between forest strata to be made.

A characteristic of the IBISCA modus operandi is that the data collection, processing and analyses from all projects are integrated into a collective database.
During field work, project participants focus their sampling effort on the same set of plots. A typical IBISCA plot measures 20 x 20 m. Within the plot, a botanical team identifies and tags all of the trees (Fig. 141). The botanical teams are then followed by entomological teams as the scientific protocols become increasingly disruptive (e.g. when included

Figure 142: Installing a Malaise trap onto the canopy.

Figure 143: Collecting in the canopy.

insecticide fogging comes last). Tree climbers assist scientists in setting up the collection devices in the canopy (Figs 142 & 143). All participants use pre-printed labels with unique sample codes that are the only valid codes for the collective database (Fig. 144). Whenever possible, samples — especially those obtained from massive collection methods — are pre-processed on site with the help of assistants supervised by professional taxonomists involved in the fieldwork. These assistants are either students, volunteers or skilled amateur naturalists. This centralized processing of the material is ideal to be able to keep track of the specimens from collection

Figure 141: Tagging the trees.

Figure 144: Extracted Winkler samples stored in 70° ethanol with their pre-printed label with sample code.

to identification. The collective database allows specimens to be linked to samples, and, thus, for global ecological analyses to be conducted. For each taxonomic group, a workgroup leader is designated. He/She is the contact person for the project co-ordinator(s) and database manager(s). He/She co-ordinates the work of expert taxonomists and is responsible for the homogeneity (e.g. in terms of systematics) and feedback of data. Besides taxonomic workgroups, other workgroups are dedicated to special techniques (e.g. canopy access) or data analyses (e.g. multivariate statistics). In terms of canopy access, the assistance of professional tree climbers is a constant in all IBISCA projects and these projects are often used as a proving ground for innovative technologies; for instance, the prototype of a new hot air/helium balloon, the Canopy-Glider (Fig. 145), was tested in Santo for the first time.

••• IBISCA-Santo

IBISCA-Santo took place in November 2006 as a sub-programme of the "Forest, Mountains and Rivers" component of the Santo 2006 Global Biodiversity Survey. IBISCA-Santo was planned in this context, which obviously explains why site selection for IBISCA-Santo was not ideal in comparison with IBISCA-Queensland in terms of the accessibility and proximity of laboratory facilities that would have allowed temporal replications and long-term studies to be conducted. Indeed, it is highly improbable that such a large scale project will be organized again in the coming years, although strictly focused investigations (such as rapid monitoring) may be planned in the near future, as the botanical plots will be available for years.

The field sampling component of IBISCA-Santo took place in the Saratsi Range above the village of Penaoru on the west coast of the Cumberland Peninsula of Santo Island, Vanuatu.

The location was selected during two preliminary exploratory trips. Due to historical anthropic pressure, forests with a high degree of naturalness are scarce in Santo and limited to elevated areas

IBISCA-SANTO represented the first attempt to use the Canopy-Glider for sampling in the upper-canopy. The Canopy-Glider is an inflatable flying craft conceived as an observation and sampling tool for scientific and conservational purposes. The Canopy-Glider, designed to carry three persons (two scientists and the pilot), uses a "Rozière" structure (the combination of a hot air dirigible and a helium balloon) and is propelled by a helix engine. It has been designed to be highly efficient for brief flights with short intervening stops. This is particularly appropriate for the collection of botanical and zoological samples in places that are otherwise difficult to reach. Two tropical cyclones, "Xavier" (also refered to as 01P; from Oct. 23rd to Oct.25th) and "Yani" (02P; from Nov.23rd. to Nov.25th), intensely affected the surroundings of the Vanuatu archipelago during IBISCA-Santo field work and the Canopy-Glider could only fly on the final days. The Canopy-Glider was developed by Pro-Natura international and designed by Dany Cleyet-Marrel.

Figure 145: The Canopy-Glider.

(generally over 1 000 m a.s.l.), and, due to relatively recent logging, to places difficult to reach by car or by boat, such as the remote parts of the Cumberland Peninsula in the northwestern part of the island. A two-hour aerial survey of the island in November 2005 allowed the organizers to pinpoint what appeared to be —based on the aspect of the canopy— the better preserved forest stands on the Peninsula and to consider the Penaoru River Valley as the best way to reach this area from the eastern coast. Moreover, as mentioned above and unsurprisingly, this area has seldom (possibly never) been visited by scientists in the past. In March 2006, the terrestrial exploration of different valleys of the Cumberland Peninsula confirmed the relevance of this choice.

IBISCA-Santo has examined the way in which the diversity of arthropods and plants (from mosses to angiosperms) varies along an altitudinal gradient from 100 to 1 200 m above sea level. This transect represents a transition from the narrow, coastal strip —which has experienced substantial human impact through shifting agriculture— up to altitudes with only occasional and patchy disturbance (*c.* 300-500 m a.s.l.) to higher elevations of more or less pristine rainforest (500-1 200 m). The participants in the IBISCA-Santo study essentially comprised two complimentary teams: those involved in the botanical survey and those concerned with entomological endeavours.

••• Field work

Field work took place from mid-October to early December 2006. Individual participants were in the field for periods varying from three to six weeks. Two camps were set up. The base-camp was about 4 km upstream from the village of Penaoru on the Penaoru River at about 50 m altitude. A subsidiary camp was located about 5 km to the west along the Saratsi Range at 960 m altitude. A total of 12 plots were established along an altitudinal gradient at 100 m, 300 m, 600 m, 900 m and 1 200 m a.s.l., each comprising an area of 400 m² (20 x 20 m), for a total of area of 4 800 m².

Within each plot, all trees with a diameter at breast height (DBH) equal or more than 5 cm were numbered and permanently labeled, and a sample was taken of each individual. A total of 947 trees were recorded within the 12 plots. In addition to providing a basis for characterizing the forests on Santo, the plots also served as the inventory sites for the IBISCA study, ensuring a baseline of floristic and structural data to be used by the entomologists.

The IBISCA entomological team carried out complimentary sub-projects using the following techniques: light trapping; collecting endophytic organisms from the ground to the canopy; beating understorey foliage; yellow pan trapping; Malaise trapping (Fig. 142); ground flight-interception trapping (Fig. 146); and bark spraying. The "Ants and termites" programme used techniques including Winkler sifting (Figs 144 & 147), pitfall trapping, beating foliage and soil sampling.

••• Expectations

We envisage that what will be produced by many of the sub-projects will take the form of specialised ecological, faunistic and taxonomic papers. The first of these, based generally on multivariate analysis, does not require that all of the samples be identified (the Kitching paper on moths published in this issue provides an example of such results). Faunistic studies of particular interest to biogeographers, however, do require greater levels of identification than

Figure 146: Collecting flying insects with a flight-interception trap.

Figure 147: Extracting arthropods from the Winkler bags.

ecological studies. Finally, the specimens collected through IBISCA-Santo have already allowed new species and genera to be identified and described in taxonomic papers.

As mentioned above, one of the great strengths of our multi-dimensional approach versus a common experimental design is that results from individual sub-projects can be placed side-by-side and legitimately compared without any of the caveats usually necessary in such comparative analyses.

From such comparisons we will be able to identify particular taxa (i.e. species, genera, families, even orders) which show the most acute changes from altitude to altitude. It is these taxa that we can propose to managers and monitors as the most appropriate for detecting changes in climate —either from place to place or over long time frames. The coinciding of this project with the IBISCA-Queensland Project, which is based on an essentially similar experimental design, is that such comparisons can be extended so that our Vanuatu results can be evaluated on a global basis.

# of Santo

FOCUS ON ORTHOPTERA

Laure Desutter-Grandcolas, Sylvain Hugel & Tony Robillard

Orthoptera are world-wide distributed insects easily recognized by their enlarged hindlegs, adapted to jump, and by their forewings, which cover the dorsal and the lateral sides of the body. They include two infraorders, Caelifera (short-horned Orthoptera) and Ensifera (long-horned Orthoptera), subdivided into several major taxonomic groups.

Orthoptera occur in both natural and man-disturbed environments, sometimes at high density. They are diversified in forested and opened biotas, although species may have highly specialized habitats and/or ways of life. Some species are plant-specific, for food and/or oviposition, but many species depend more on vegetation structure, which defines microclimatic conditions and the availability of microhabitats, than on vegetation identity. Many species are either phytophagous, or omnivorous; some are carnivorous and feed on other insects.

One of the main biological characteristics of Orthoptera is their use of acoustic signals for communication purpose, especially for reproduction and defence. Sounds are produced by rubbing diverse morphological structures against one another, i.e. by stridulation. In Caelifera, sounds are produced mainly by rubbing hindlegs against the abdomen or a forewing vein. In Ensifera, stridulation modalities are very diverse, but the most distinctive one implies the forewings, which are raised above the body and rubbed together; specific parts of the forewings act as resonators, which greatly enhances the power of the emitted sounds. Some species sing during the day, but most sing at night, together with frogs. Each locality then resounds with a mix of all emitted sounds, which is determined by the species composition of the acoustic community. This assemblage of acoustic signals can be used to evaluate the diversity and density of co-occurring acoustic species. Such a biodiversity estimate is of course biased, as mute species cannot be revealed that way, and acoustic species do not sing all at the same time. It may anyhow be useful to survey already inventoried habitats or the density of particular species (Fig. 148).

In Santo, Orthoptera are represented by more than 55 species, belonging mainly to Ensifera. Some groups are well-diversified, with many species (some still under study), but others are "missing" on the island, compared to nearby islands or territories. For Ensifera for example, and compared to New Caledonia, Santo seems to lack Anostostomatidae (wetas) and Pseudophyllinae (true katydids). In crickets, no *Oecanthus* has been found, although the Neocaledonian species, *Oecanthus rufescens*, is widely distributed in the Southwest Pacific; there is also no *Parendacustes*, while one species has been described from the Loyalty Islands and although this genus is well-diversified in the Indo-Malaysian region. Conversely, some taxa are present or diversified in Santo, while they are lacking or poorly represented in New Caledonia; this is the case of Pentacentrinae and Trigonidiinae (Trigonidiidae). The taxa diversified in both areas, i.e. Nemobiinae (Trigonidiidae), Podoscirtinae (Podoscirtidae), Eneopterinae (Eneopteridae) and Phalangopsidae, mostly belong to different genera.

Three main situations describe the orthopteran fauna of Santo in term of species distribution, as examplified by the species of the cricket genus *Cardiodactylus*:
- some species are distributed throughout the island;
- others are restricted to the coastal margin of the island, including widely distributed species in the Southwest Pacific region, or endemic species very similar to species found in the surrounding territories; and finally,
- some species are clearly restricted to the most inner parts of Santo.

••• Caelifera (Grasshoppers, Locusts, and Pigmy Grasshoppers)

Caelifera are poorly diversified in Santo, with less than 10 species occurring on the island. All known caelifera species in Santo are mute, i.e. they are not able to produce stridulation. Most of them are widely distributed in the Southwest Pacific; they occur in man-made biota, such as pastures, gardens and roadsides with tall grasses (*Austracris guttulosa*; *Oxya japonica*, the Rice Grasshopper), and areas with more scarce vegetation (*Aiolopus thalassinus dubius*; *Heteropternis* sp.). On the ground of wooded areas some Tetrigoidea (*Eurymorphopus* sp., Pigmy Grasshoppers) are occurring.

The most remarkable caeliferan species in Santo is *Hebridea rufotibialis*, a brachypterous grasshopper

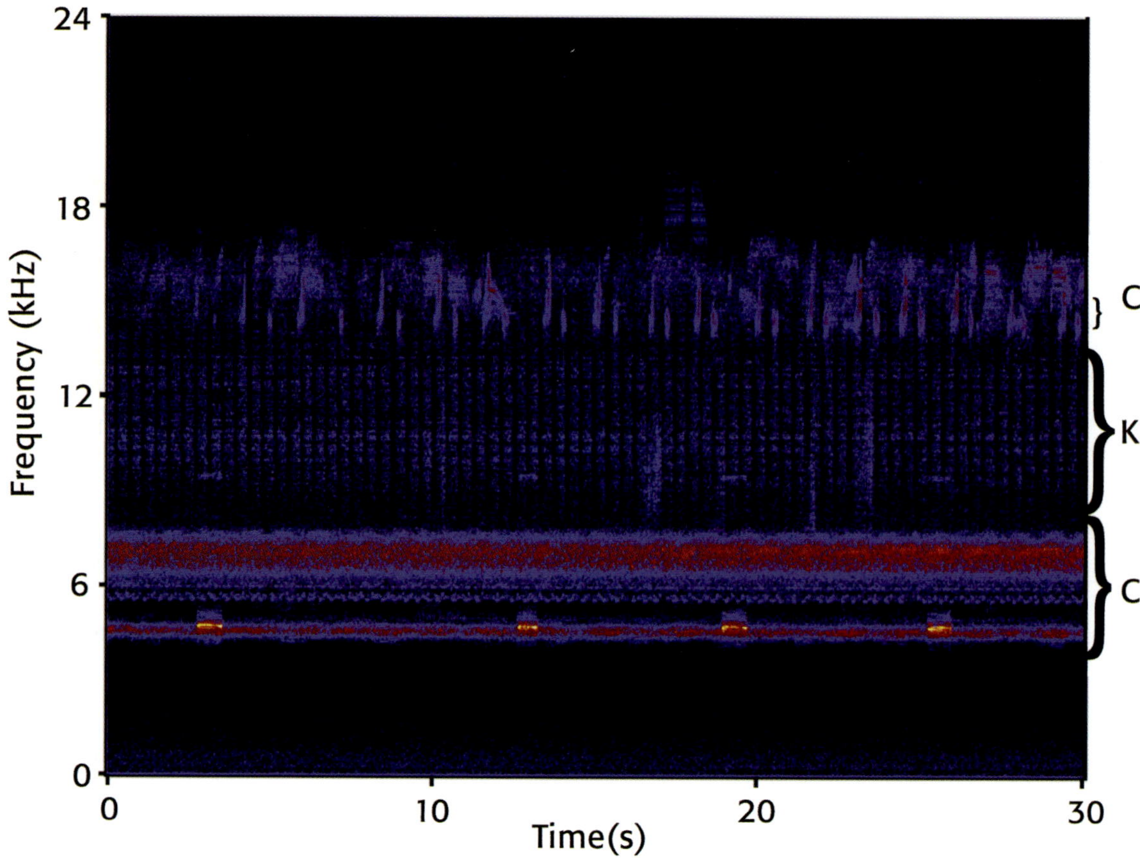

Figure 148: Sonogram of nocturnal recording at Peavot, showing the distribution of species calls in time (horizontal scale) and frequency (vertical scale). The frequencies ranging from 5 to 8 kHz are predominantly filled by crickets (**C**). Podoscirtinae and Phaloriinae are contributing to the low frequencies within this range (3, 13, 19 and 26 s), and Trigonidiinae to the high frequencies background noise within this range (and to a small extend Eneopterinae, Nemobiinae and Mogoplistidae). The frequencies ranging from 8 to 14 kHz are mostly filled by katydids (**K**). *Salomona* song corresponds to the long trill throughout the time window, whereas *Ityocephala francoisi* song corresponds to short calls at 17 and 23 s. Interestingly, the high-audio range is filled by cricket (**C**) species (*Cardiodactylus novaeguineae*), contrasting with the usual situation in most of the world areas where cricket songs are occupying lower frequencies range than katydid songs.

endemic to the island (Fig. 149). This grasshopper feeds mostly on screw palm (*Pandanus* sp.), but also on fern and *Dracaena* (Ruscaceae). This species is found on both preserved and moderately gardened wooded areas.

••• Ensifera Gryllidea

More than 30 cricket species have been found in Santo. Some are widely distributed in the Southwest Pacific, or are pantropical; they occur in man-made biota, such as pastures (Gryllidae: Gryllinae: *Teleogryllus oceanicus*) or villages (Gryllidae: Gryllinae: *Gryllodes sigillatus*), or in supralittoral vegetation near shorelines (Eneopteridae: Eneopterinae: *Cardiodactylus novaeguineae*). The large majority of species are presently known from Santo only. They live in natural environments and are found especially in forests or in extreme biotas, i.e. beaches (Mogoplistidae sp; Trigonidiidae, Nemobiinae: *Thetella* sp) or caves (Phalangopsidae; Trigonidiidae, Nemobiinae: *Cophonemobius faustini*).

The most remarkable biological feature of Santo crickets is the acoustic behavior of Eneopterinae species (three *Cardiodactylus* and two *Lebinthus* species,

three of which were discovered in 2006 and described in 2009 by T. Robillard, *C. tankara*, *L. santoensis* and *L. nattawa*). These species sing at very high frequencies, near or beyond the limit of human audible spectrum (Fig. 148).

••• Gryllidea of opened areas

Man-made opened areas, such as lawn or low pasture, are colonized by species largely distributed through the Southwest Pacific, especially *T. oceanicus*.

Figure 149: *Hebridea rufotibialis* Willemse, 1926, a brachypterous grasshopper endemic to Santo.

Along roadsides, where grass is taller, *Anaxipha fuscocinctum*, a flashy green Trigonidiinae (Trigonidiidae) with red spots (Fig. 150), can be found, sometimes at high density; this species sings and reproduces during the day. Another trigoniidiine, with yellowish brown legs and shiny dark head, body and forewings, has been found both in opened areas along the roads and in forest openings invaded with creeper. In slightly more covered places, such as along coastal tracks, an euscyrtine species (*Euscyrtus* sp.) may also occur; devoid of a stridulum, this mute cricket species is characterized by its "*Conocephalus*-like" appearance, a resemblance due to morphological features (body thin and elongated, wings short, applicated on the body, ovipositor short, flattened and incurved), coloration (yellowish with longitudinal brown lines), and behavior: individuals of *Euscyrtus* sp. actually get stretched along grass stems when disturbed, often moving back.

Mole crickets (Gryllotalpidae) are also present in Santo and their soft calls can be heard at night, for example in the viciniy of Butmas.

••• Gryllidea of wooded areas

Crickets are diversified in all the ecological compartments of tropical forests. The ecological guilds for habitats defined by L. Desutter-Grandcolas in 1997 for Newcaledonian crickets are of a general significance for the group and exist also in Vanuatu.

In Santo, the leaf litter is occupied mostly by Nemobiinae crickets, which represent the bulk of cricket specimens in forested areas. These are small, black to brown species, which sing mostly during day time. Nemobiinae spend their whole life time in the leaf litter, while other species, such as Eneopterinae, occur in leaf litter only as early juveniles. At least four Nemobiinae species have been found in all sampled forests (genera *Dictyonemobius* and *Pteronemobius*). Several species of Gryllidae have been found but sporadically in the leaf litter, represented in our sampling by at least two species of Cephalogryllini, a group not known from the area yet, although well-diversified in Australia and New Caledonia. Finally, Trigonidiinae (*Amusurgus* spp.) and Pentacentrinae (*Pentacentrus* sp) hide during the day in the leaf litter, from which they fly rapidly when disturbed; they could be active on low plants at night.

Forest vegetation is inhabited mostly by Eneopterinae, Podoscirtinae, Phaloriinae, Trigonidiinae, the genus *Pseudotrigonidium*, and to a less extend by Mogoplistidae and Itarinae. Eneopterinae crickets (*Cardiodactylus enkraussi* distributed in the whole island (Fig. 151), *C. novaeguineae* in coastal regions; *C. tankara* in the inner lands) are the most abundant cricket species on bushes and low trees, especially near edges. They

Figure 150: *Anaxipha fuscocinctum* (Chopard, 1925), male, a diurnal trigonidiine cricket living in opened areas.

are extremely abundant and feed on many local plants, from fern leaves to the invasive creeper; they generally live from 50 cm to 2 m above the ground, on variable layers of vegetation. *C. enkraussi* lives deep inside the vegetation, hidden from the edges, from where it emits a very discrete calling song made of two distinct parts. It occurs in sympatry with either *C. tankara*, or *C. novaeguineae* which are clearly more conspicuous, as they live on top of bushes or on the edges of small trees, and emit a louder calling song. Podoscirtinae (*Aphonoides* ssp) and Phaloriinae (*Phaloria chopardi*, *P. walterlinii*, *P. faponensis* described in 2009 by L. Desutter-Grandcolas) are also found on forest bushes occasionally. They probably live higher in canopy trees and can be more easily observed

Figure 151: *Cardiodactylus enkraussi* Otte, 2007, male, one of the most abundant cricket species in Santo.

on edge trees. Some of them sing quite loudly and can be heard in many places. *Pseudotrigonidium* species are elegant and shiny crickets, which are more often found at night standing on the leaves of bushes or low trees at human height, or running on lianas or twigs. They most often hide during the day below loose bark or below bush leaves. Both *Phaloria* and *Pseudotrigonidium* crickets also hide in the aerial litter of tree ferns or *Pandanus* trees during the day. Other species observed on under-storey plants and low trees at night include on one hand Trigonidiinae, such as a large *Anaxipha* species (pale green when alive, yellow after death), and several *Parametioche* species (trigonidiines with no hindwings, corneous forewings devoid of a stridulum and long, yellowish legs), and on the other hand scaly crickets (Mogoplistidae: genus *Arachnocephalus*).

Finally, tree trunks are the place where long-legged crickets (Phalangopsidae) are active at night. These crickets hide during the day in bur-rows, or in cavities under logs or dead wood; they have not been observed under loose barks or in tree cavities above the ground. These are relatively small species devoid of an acoustic stridulum; they are extremely discrete and explore their envi-ronment with their long and flexible antennae. They can be relatively abundant locally, as in the forests located on sandy soil backwards the shore-lines, but seem absent in the forests located inside the island.

••• Gryllidea of specific habitats

• Shorelines
Two different cricket communities are found in specific shoreline habitats.

In the mediolittoral zone, crickets of the genus *Thetella* (Trigonidiidae: Nemobiinae) occur either in the cavities present in rocks of alveolate calcare-ous, or on dark pebble beach (Fig. 152). In the first case, these rapid crickets are seen running on the rocks or on the sand nearby to escape the waves. Males are heard singing from the small cavities. On dark pebble beach, *Thetella* live in sympatry with a blackish scaly crickets (Mogoplistidae), together with several predators (spiders, small lizards). High waves during stormy weather seem to depauperate these pebble beach from crickets, the areas being recolonized after several days only. *Thetella* have been found also on alveolate calcareous located on riverbanks close to the coast.

In the supralittoral zone, several crickets can be observed on small trees (*Aphonoides* ssp; *Arachnocephalus* sp.), while low plants are colo-nized by *C. novaeguineae*. The leaf litter is occupied by *Dictyonemobius* and *Lebinthus* species. *Lebinthus* is found sometimes in the leaf litter of the forests

located behind the coast, or near river sides, but it never occurs far from the coast.

• Caves
Only long-legged crickets and *Cophonemobius* nemobiines are found inside caves, especially those which are inhabited by bats. The presence of *Cophonemobius* in Santo caves was documented by L. Desutter-Grandcolas in 2009. *Cophonemobius* are small, fragile insects, found mostly on cave walls, but they run and hide in small cavities and cracks when disturbed. At least four species of long-legged crickets occur in Santo caves, belonging to three different genera, one of which probably new to science.

••• Ensifera Tettigoniidea (Katydids and Raspy Crickets)

Fifteen Tettigoniidea species are known to occur in Santo as documented by S. Hugel in 2009. Most of these species can be observed in secondary habitats such as gardened or logged areas, and virtually no species are strictly restricted to unmod-ified, native biota. This might be due to the weak dependence of Tettigoniidea on specific plant species, these insects depending usually rather on the vegetation structure and climatic param-eters. Tettigoniidea are occurring in most of Santo habitats, except caves and shorelines. One third of the species are considered as endemic to Santo and/or indigenous to Vanuatu, the other two thirds are more widespread in the Indo-Australian area.

••• Tettigoniidea of opened areas
As in similar earth biota, Santo opened areas covered with grasses, such as roadsides, pasturelands, and gardens are inhabited by at least three *Conocephalus* species (meadow katydids:

Figure 152: *Thetella* sp., female, a nemobiine cricket inhabiting coastal areas such as alveolate limestones and beaches.

C. upoluensis, *C. semivittatus vittatus* and *C. oceanicus*) and also by *Pseudorhynchus lessoni* (snout-nosed katydid). These species are sometimes occurring in a relatively high density. On tall grasses of such areas, *Phaneroptera gracilis*, a nocturnal species, can also sometimes be observed. *Conocephalus* males call during both day and night whereas *Phaneroptera* are rather nocturnal; their calling song is relatively inconspicuous in the audible range. Contrasting with *Conocephalus* and *Phaneroptera* subtle song, the loud and strident buzzing call of *P. lessoni* emitted during the firsts hours of the night is easily audible.

••• Tettigoniidea of wooded areas

The apparent density of Tettigoniidea is relatively low in both preserved and secondary wooded areas, and in such biota these insects are hardly observed during day hours as all species are mostly nocturnal.

On trees and bushes foliage near the shore (*Terminalia*, *Barringtonia*, *Hibiscus*, etc.), the spiny predatory katydid *Phisis holdhausi* can be observed. This species is widespread across the Polynesian area, its range extending from New Guinea to Gambier. Another Phisidini species, *P. pseudopallida*, occurs in other secondary habitats. In wooded areas of low altitude, the large *Xanthogryllacris punctipennis* (Fig. 153) and another Gryllacrididae *Amphibologryllacris macrocera* can be observed rushing on branches and foliage while exploring their surroundings with their tremendously long antennas. *Xanthogryllacris punctipennis* is a widespread species with numerous subspecies living in Australia, Solomon Islands and New Guinea.

In such low altitude areas, several leaf mimicking Phaneropterinae and *Salomona redtenbacheri* are also occurring. *Salomona* are relatively large katydids with remarkably strong and harmful mandibles (Fig. 154). Males are producing the only conspicuous Tettigoniidea song audible in wooded areas; their continuous high-pitched song is usually emitted near the top of trees.

In both preserved and secondary wooded habitats above 30 m, together with *Salomona redtenbacheri*, *Furnia insularis* (Phaneropterinae) and *Amphibologryllacris macrocera*, the large endemic leaf mimicking Mecopodinae *Ityocephala francoisi* is found (Fig. 155). Two endemic Gryllacrididae species have only been found above 600 m: *Amphibologryllacris butmasi* and *Psilogryllacris tchancha*.

Beside undergrowth Gryllacrididae, all the other species are usually occurring near the canopy, and only few specimens (laying females, moulting specimens, etc.) can be observed from the ground.

Figure 153: *Xanthogryllacris punctipennis* (Walker, 1869) (Gryllacrididae) living in wooded lowlands.

Figure 154: *Salomona redtenbacheri* Brongniart, 1897, a large katydid with strong harmful mandibles.

Figure 155: *Ityocephala francoisi* Bolivar, 1903 (Mecopodinae), female, the large endemic leaf mimicking katydid.

TERMITES IN SANTO: LESSONS FROM A SURVEY IN THE PENAORU AREA

Yves Roisin, Bruno Corbara, Thibaut Delsinne, Jérôme Orivel & Maurice Leponce

Termites (Isoptera) are social insects. Their societies rely upon a complex system of castes, i.e. categories of morphologically distinct individuals that perform specific tasks. Among the most ubiquitous castes are the reproductives (queen and king), soldiers and workers. The reproductives are the only ones to reproduce. Soldiers are specialized for defence, whereas workers perform most of the colony's chores, such as digging, construction, food gathering and distribution, and care of the young.

The most basal termite families, such as the Kalotermitidae, live within pieces of dead wood on which they feed, but the Termitidae, often called "higher termites", have diversified to exploit a wide variety of food items such as decayed wood, grasses, lichens, leaf litter or humus. In most tropical rain forests, termites are major decomposers. Lowland continental termite assemblages are the most diversified with respect to their nesting ecology and food sources. Hypogeous nesters and soil feeders are usually poor migrants, whose species richness quickly dwindles as soon as overseas dispersal is required. In a recent revision of the Termitinae of New Guinea, Bourguignon, Leponce and Roisin showed that only a few species in this sub-family may qualify as soil or soil-wood interface

feeders, of which only one reaches the Solomons. Further east, island termite faunas comprise exclusively wood feeding species. Although the effect of altitude on termite assemblages has seldom been quantified, it is obvious for field termitologists that termite abundance and diversity decline from lowland to montane forests. Finally, within a single type of forest, termites may display a vertical stratification between ground-level and canopy strata as we have shown in a previous IBISCA programme held in a lowland forest of Panama.

The termite fauna of Vanuatu is mostly known from casual records. In 1915, Holmgren and Holmgren described *Procryptotermes speiseri* and *Nasutitermes novarumhebridarum*, as well as six new species from New Caledonia and the Loyalty Islands. The Royal Society – Percy Sladen expedition collected only six genera and 10 species from Vanuatu in 1971. Of these, two species of *Neotermes* and two *Nasutitermes* were found on Santo, bringing the reported richness of this island to six species in three genera. Subsequent collecting and identification work was carried out by John C. Buckerfield, from CSIRO, whose unpublished records mention 24 species in nine genera from Vanuatu, of which 16 species in seven genera would be present on Santo.

Table 10: Termite species collected near Penaoru during the Santo 2006 expedition. Number of occurrences in systematic ground-level samples (100-900 m) and in canopy samples (C100-C300 m).

Family / Species	100 m	300 m	600 m	900 m	C100 m	C300 m
Kalotermitidae						
Cryptotermes tropicalis Gay & Watson	2					
Cryptotermes albipes (N. & K. Holmgren)		1				1
Procryptotermes speiseri (N. & K. Holmgren)		3			1	5
Incisitermes sp.	3				2	
Neotermes sp.	2		2	5		
Glyptotermes sp.A	1	1				
Glyptotermes sp.B			1	2		
Rhinotermitidae						
Prorhinotermes inopinatus Silvestri		1				
Termitidae						
Nasutitermes novarumhebridarum (N. & K. Holmgren)	14	4	5	1	2	5
Diwaitermes kanehirae (Oshima)	6	9	5	5		
Total occurrences: 89	**28**	**19**	**13**	**13**	**5**	**11**
Total species: 10	**6**	**6**	**4**	**4**	**3**	**3**

Santo 2006: study sites and methods

During Santo 2006, termite collecting took place on the Penaoru site, in November 2006 in the context of the IBISCA-Santo programme. Systematic sampling was carried out on four plots, situated at altitudes of 100 m (plot 100B), 300 m (300A), 600 m (600B) and 900 m (900B) asl. On each plot, 5 m^2-quadrats were positioned every 10 m along a 60 m x 60 m grid (49 quadrats per plot). Each quadrat was searched for one-half man-hour, according to standard termite sampling protocols proposed by Jones and Eggleton and modified by Roisin and Leponce. At sites 100B and 300A, dead wood taken down from the canopy with the help of professional climbers was also examined for termite presence. Casual collecting at all sites as well as in the degraded vegetation situated downhill from the camp along the path to the coast provided additional material.

Termite diversity and distribution

Ten species were collected, representing eight genera and three families (Table 10).

Termitidae

Two species of Termitidae are common at all elevations. Both belong to the subfamily Nasutitermitinae, characterized by soldiers possessing a pear-shaped head with a long frontal tube, through which they can squirt a glue on intruders (Fig. 156). *Nasutitermes novarumhebridarum* was found both in the canopy and at ground level. This species builds carton nests made of partly digested wood. One such nest was found at ground level between the roots of a big tree at the 300 m site, but this species is also able to build nests higher in trees as already observed in Papua New Guinea. The small *Diwaitermes kanehirae* was limited to dead wood on the ground, where workers often formed dense feeding aggregations. Both species are widespread in the southwest Pacific region.

Rhinotermitidae

This family was represented by a single record of *Prorhinotermes inopinatus*. This genus is characterized by numerous soldiers with long, falcate mandibles and a defensive frontal gland, which opens through a small pore on the upper side of the head capsule. *Prorhinotermes* species are famous for their ability to colonize islands, presumably through colony fragments in drift wood. Substitute reproductives easily differentiate in isolated colony fragments, while soldiers and immatures are able to explore their environment to search for favourable sites for colony development.

Kalotermitidae

Known as drywood termites, the Kalotermitidae form small colonies confined to single pieces of wood on which they feed. Soldiers do not possess a frontal gland, but defend their colony by blocking passages in the wood with their head, which is often heavily thickened and endowed with powerful, crushing mandibles (Fig. 157). When their home piece of wood approaches exhaustion, most of the individuals will proceed to the winged, alate stage and attempt to found a new colony in another site. This is the most species-rich family

Figure 156: Soldiers (with pointed head capsule) and workers of *Nasutitermes novarumhebridarum*. Tunnel made of wood carton was opened to show walking termites.

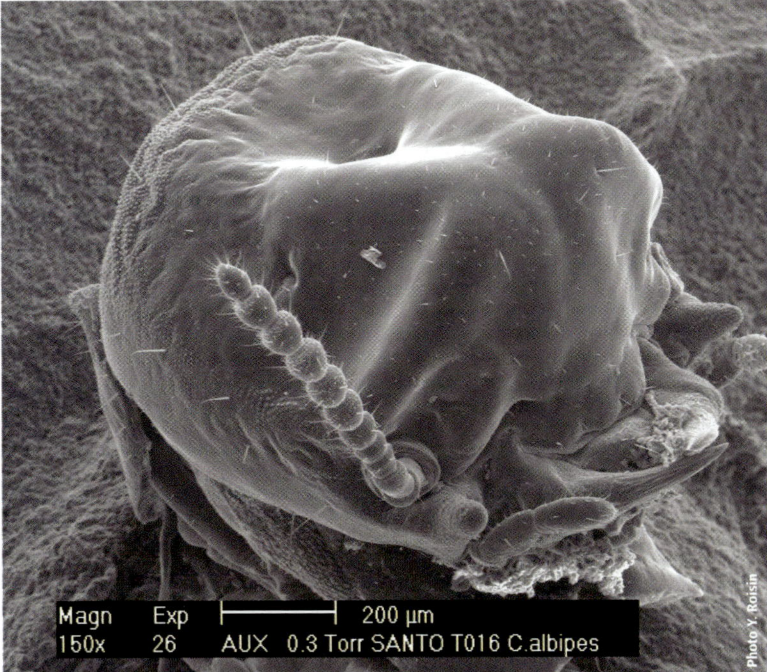

Figure 157: Scanning electron micrograph of head of *Cryptotermes albipes* soldier. Note plug-shaped head capsule and short, powerful mandibles.

on south Pacific Islands. Species identification is however made difficult by intraspecific variability, since soldier size and shape may vary not only according to their genetic constitution, but also with colony age and condition. The same two species, *Neotermes* sp. (Fig. 158) and *Glyptotermes* sp.B were found at 600 m and 900 m. *Neotermes* sp. was casually collected at 1 200 m. At lower altitudes, the fauna appears more diversified. Dead branches or stubs high in trees can be home to whole colonies, allowing this family to be well represented in the forest canopy. *Procryptotermes speiseri* was especially common in the canopy at 300 m.

••• Conclusions

As is the rule for southwest Pacific islands, the termite fauna of Santo is exclusively composed of wood feeders. Endemism of termites on Santo is probably low. Both Termitidae species and *Prorhinotermes inopinatus* are widespread in the southwest Pacific region. The geographic distribution of the two *Cryptotermes* species is probably extensive as well: *C. tropicalis* was described from Queensland, Australia, and *C. albipes* from the Loyalty Islands. This latter species also occurs in New Guinea (YR, unpublished). However, the identification of the other Kalotermitidae is still in progress. The lack of clear correspondence between our collections and published species descriptions suggests that some species might ultimately prove to be new, and perhaps endemic.

Figure 158: Part of an incipient colony of *Neotermes* sp., showing the royal pair (reddish brown), a soldier (left, with heavily sclerotized head and long mandibles), two larvae (top, white), and large working immatures (right, with pale head and reddish digestive tube).

FOCUS ON BEES AND WASPS

Claire Villemant

Wasps and bees are, together with ants, members of the Hymenoptera order that contains more than 125 000 named species. Even the most conservative approximations on the order diversity estimate between 600 000 and 1 200 000 species. Although their social and painful stinging representatives (ants, domestic bees and social wasps) are perhaps the most familiar of all insects, the other Hymenoptera are generally overlooked because of their small size or discreet solitary habits. With species ranging from minute parasitoid wasps developing within eggs of other arthropods to large digger wasps hunting grasshoppers, this order forms indisputably an important component of the terrestrial arthropod biodiversity of Santo. As plant pollinators or arthropod prey eaters, bees and wasps are also essential in sustaining biodiversity of other living organisms. While bees play a major role in fertilisation of flowering plants, wasps are critically important in natural biocontrol as almost every insect pest has at least one wasp species that lives on it. However, whereas "bees" represent a natural group of social and solitary species developing on a diet of pollen and nectar, "wasps" includes a large taxonomic diversity of insects that are recognized as being neither a bee nor an ant. In Santo, wasps can be globally categorized as true wasps (paper and potter wasps), sphecoid wasps (the closest relatives of bees), spider wasps and a large and diverse group of parasitic wasps that includes thirty Hymenoptera families. Bees, on the contrary, are almost exclusively represented by solitary species in Vanuatu, the single social bee present in the archipelago being the introduced European honeybee.

••• History of researches on Vanuatu bees and wasps

The first Hymenoptera recorded from Vanuatu were two Sphecoid wasps (*Sphex antennatus* and *Pison insulare* Fig. 161C) described by Smith between 1856 and 1869 and an endemic leafcutter bee (*Megachile similis*) described by the same author in 1879. A few other species of bees, digger or parasitoid wasps were recorded later from this archipelago until the arrival of Evelyn Cheesman, a female entomologist and explorer associated with the British Natural History Museum, who spent twelve years in naturalist expeditions throughout Pacific Islands. She explored alone Vanuatu from 1929 to 1931 but stayed only six weeks in the south of Santo. She brought from her expedition in Vanuatu a number of insect specimens that remained housed in the British Museum. From 1936 to 1948, she published (alone or in collaboration with R. Perkins) several papers on Vanuatu's Hymenoptera (mainly ichneumonid parasitoids, digger-wasps and bees) and described 30 new species or varieties, among which seven were collected in Santo. Since

their description, 12 of these new taxa were never recorded outside of this archipelago. Later, new species were described by various authors mainly from the material collected by Cheesman and four of them were dedicated to this intrepid explorer. Before Santo 2006, we numbered a hundred of species recorded from Vanuatu (in 40 papers) but only 18 from Santo (in 11 papers).

••• Bees and wasps sampled during the Santo 2006 expedition

The Hymenoptera mentioned here were collected in November 2006 mainly in the Penaoru area. Most of them were caught in Malaise traps (14 placed on the ground and four in the canopy) (Fig. 159) and yellow pan traps arranged along the IBISCA sampling device, but flying insects were also caught by sweep netting in the vicinity of the sampling sites. Before starting the Penaoru expedition, a few other specimens were collected by sweep netting in the south of the island, around Luganville. In the Penaoru area, the most species-rich zone was located at about 600 m a.s.l., in a moist lowland forest with kaori and tamanou trees.

Except for families represented by very few species or those already well studied in Vanuatu, most of the Hymenoptera collected in Santo were identified to the

Figure 159: Two Malaise traps placed in the canopy and on the ground near the Penaoru camp (plot at 100 m a.s.l.).

Table 11: Hymenoptera current known diversity (ants excluded) in the whole Vanuatu archipelago and in Santo Island. Families (**fam.**), subfamilies (**subfam.**), genera (**gen.**) and species (**spp.**) numbers after and before (∗) Santo 2006, with number of still unidentified species (**unid.**) and those already recognized as new to science (**new**).

	VANUATU				SANTO			Identification Contributors
Superfamilies	fam.	subfam.	gen.	spp. (∗)	spp. (∗)	unid.	new	
Stephanoidea	1	1	1	3 (2)	1 (0)	1	–	CV
Evanoidea	2	2	3	3 (3)	2 (0)	0	0	CV
Ceraphronoidea	1	–	-	6 (0)	6 (0)	6	–	CV
Diaprioidea	1	3	-	22 (0)	22 (0)	22	–	CV
Platygastroidea	1	4	19	27 (8)	19 (1)	19	–	J. Norman, L. Musetti
Cynipoidea	1	1	6	10 (0)	10 (0)	7	1	M. Forshage
Chalcidoidea	14	28	64	101 (13)	90 (0)	88	2	G. Delvare
Ichneumonoidea	2	27	72	130 (54)	111 (10)	90	1	CV, J. Dubois (Orthocentrines)
Chrysidoidea	3	8	10	15 (3)	15 (1)	8	2	CV, M. Olmi (Dryinids)
Vespoidea (ants excepted)	4	7	10	16 (7)	13 (2)	3	3	CV, R. Wahis, F. Durand (spider wasps), D. Brothers (M
Apoidea	6	8	14	33 (24)	21 (6)	4	2	CV, A. Pauly (bees)
TOTAL	**36**	**89**	**199**	**366 (114)**	**310 (20)**	**248**	**11**	

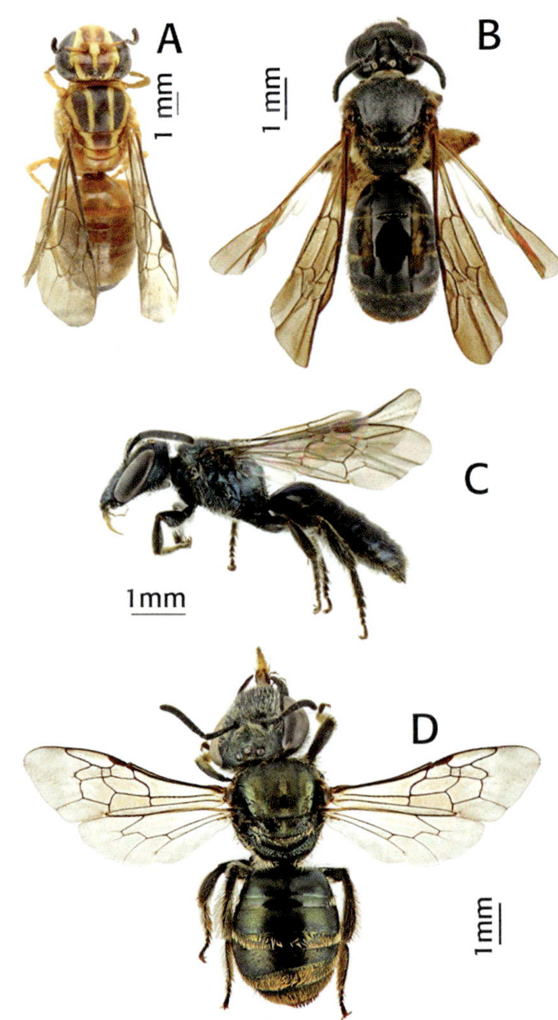

Figure 160: Bees. **A**: *Palaeorhyza maluae*. **B**: *Lasioglossum vanuatu*. **C**: *Homalictus tannaensis*. **D**: *Austronomia willeyi*. (Photos A. Touret-Alby).

genus or morpho-species level by different specialists (Table 11). Ants excluded, sorting the specimens led to the recognition of at least 310 species in Santo and thus increased threefold the known Hymenoptera biodiversity of the Vanuatu archipelago (369 species as a whole). Among the 290 species new to the Santo fauna, at least 11 have by now been recognized as new to Science and four (two spider wasps, one bee, one dryinid parasitoid) are already described. Further studies will surely greatly increase this number.

••• Bees

Bees form a natural group of species that develop on a diet of pollen and nectar and generally possess a hairy body and specialized brushes used to collect the pollen. Only three species were collected by Cheesman in Santo. Except for *Apis mellifera*, the European honeybee introduced by man all around the world, the other 12 species (one Colletidae, seven Halictidae and four Megachilidae) recorded from Santo by Alain Pauly are solitary bees, of which each female builds its own nest and supplies by-itself its progeny. The colletid *Palaeorhyza maluae* (Fig. 160A) described by E. Cheesman is a small orange bee known from Vanuatu and Solomon Islands that excavates its nest from soft wood. Halictid bees generally dig their nest into the ground; they are mainly represented by small bees of the genus *Homalictus* that are easily recognizable by their blue green metallic colour. This genus, whose representatives might have been spread from one island to another by wind, is highly diversified in Australia and Pacific Islands and comprises at least four endemic species in

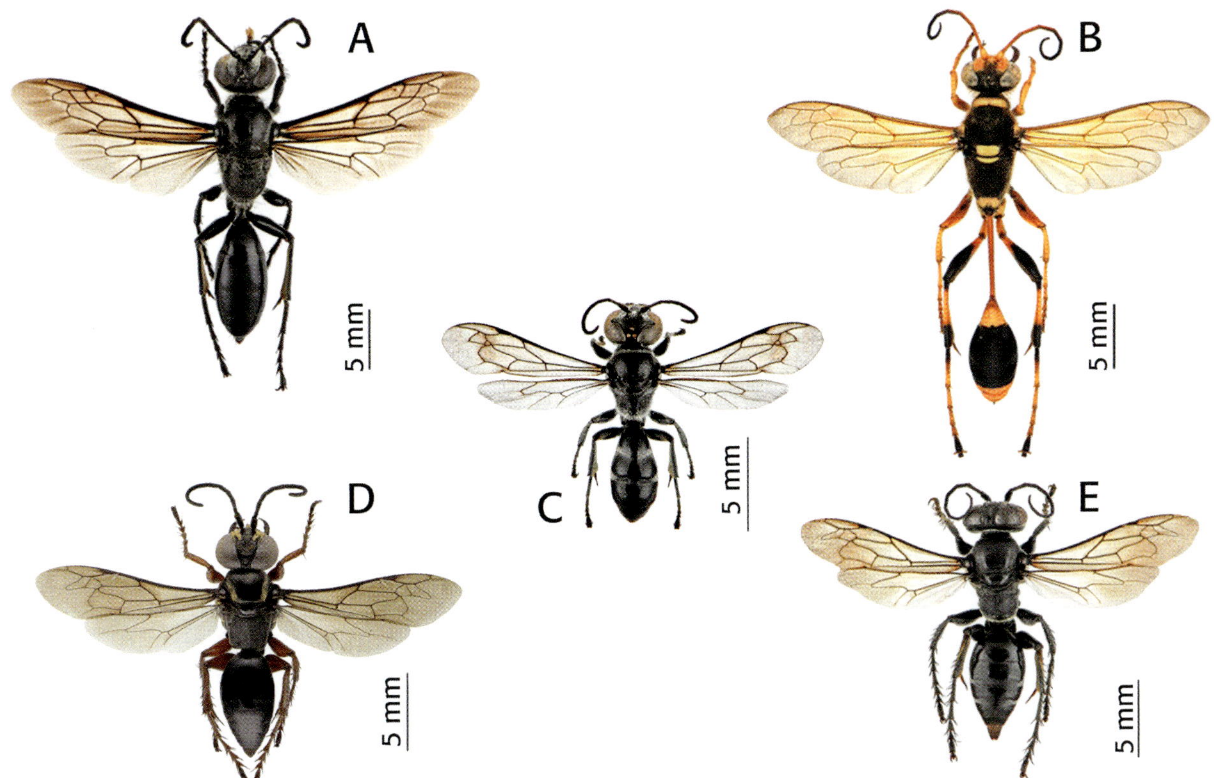

Figure 161: Sphecoid wasps. **A**: *Sphex* sp. **B**: *Sceliphron laetum*. **C**: *Pison insulare*. **D**: *Liris pacificatrix*. **E**: *Liris samoensis*. (Photos A. Touret-Alby).

Vanuatu. Among them, *H. tannaensis* (Fig. 160C) that nests in bamboos was seen gathering pollen on a variety of flowers around the Penaoru village. Two other halictids, *Austronomia* sp. and *Lasioglossum vanuatu* (Fig. 160B) were recognized as new for science. *L. vanuatu*, hitherto known only in Santo, is by far the most abundant bee species in the low and middle mountains that bordered the Penaoru village. Another Halicitid from New-Guinea, *Austronomia willeyi* (Fig. 160D), was also collected for the first time in Vanuatu. The last well represented bee family is the Megachilidae that carry the pollen on the ventral surface of their abdomen, rather than on hind legs as in other bees. *Lithurgus scabrosus* and ten species or subspecies of the genus *Megachile* are quoted from the Vanuatu archipelago. They are leafcutter bees that build the cells of their nests from leave fragments. Since they generally nest in dead wood, these species are easily transported across the sea along trade channels. *L. scabrosus* (still unrecorded in Santo) is thus widespread in Vanuatu and Pacific Islands as well as from India and Java to New-Guinea. Therefore, except *M. rambutwan* known from Vanuatu and New Caledonia, the three other *Megachile* species collected in Penaoru need further studies to determine their geographic origin. As a whole, Santo 2006 expedition led the Vanuatu's bee fauna to increase from 12 to 20 species.

• • • Sphecoid wasps

Sphecoid wasps (Sphecids and Crabronids) belong, as bees, to the Apoidea superfamily. These solitary hunting wasps feed their progeny with various preys, each species preying on a particular group of arthropods (insects or spiders). The female hunts to provision its nest, lays an egg in each cell and seals the entrance, leaving its progeny develop on paralyzed preys. Sphecidae are thread-waisted wasps that dig their nest into soil (digger wasps) or build it out of mud (mud daubers). Four species are known from Vanuatu and two were recorded for the first time in 2006. The large *Sphex* sp. (Fig. 161A) is a digger wasp that preys on big leafhoppers. *Sceliphron laetum* (Fig. 161B), a black brown and yellow mud dauber, builds its cylindrical nest cells on the surface of various substrates and provisions them with paralyzed orb-weaving spiders. Each mud cell contains one egg and is provided with several prey items. In the vicinity of the Penaoru River, where mud nests of *S. laetum* were numerous under sloping rocks, the cells left by emerged adults of the wasp sometimes sheltered gecko eggs or were reused by a potter wasp (see below). *S. laetum*, the Common Pacific Mud Dauber, occurs also in Australia, New Guinea, New Caledonia and various other Pacific Islands; it has also been found in Indonesia. It could have been introduced in Vanuatu through international trade, as have been several other *Sceliphron* species throughout the world.

Crabonidae, the second family of sphecoid wasps present in Santo, include smaller hunting wasps (five genera, 11 species known from Vanuatu). Among the four species recorded in the past from Santo, only *Pison insulare* (Fig. 161C) was collected in Penaoru and Luganville in 2006. This very common mud dauber preys on spiders. As for

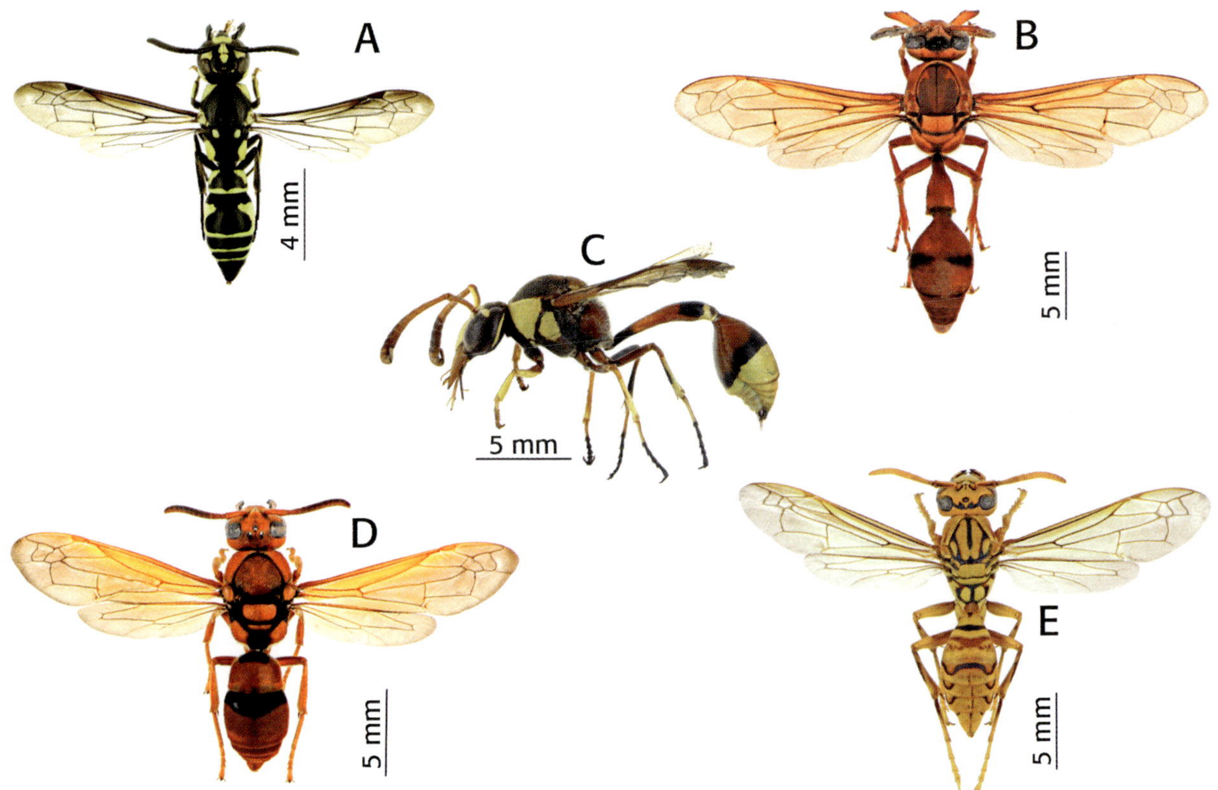

Figure 162: True wasps. **A**: *Parodynerus cheesmani*. **B**: *Delta xanthurum*. **C**: *Delta campaniforme*. **D**: *Anterhynchium alecto*. **E**: *Polistes olivaceus*. (Photos A. Touret-Alby).

S. laetum, its nesting behaviour has favoured the introduction of this species, formerly considered as endemic to Vanuatu, into other Pacific Islands as Hawaii and Tahiti. Two specimens of another undidentified *Pison* were also collected in Penaoru and Luganville. The other Crabronid wasps of Santo dig their nest into soil. *Liris* filled them with crickets; among the five *Liris* species known from Vanuatu, three were for the first time caught in Santo, of which *L. pacificatrix* et *L. samoensis* (Figs 161D & 161E). *Tachysphex* sp. is also new to Vanuatu; this unidentified species, collected in Penaoru, probably preys on grasshoppers or cockroaches. Finally, *Psen cheesmanae*, a predator of leafhoppers endemic to Santo, was never collected during the expedition. Thus, five of the 11 identified Sphecoid wasps of Santo could still be considered as endemic to the Vanuatu archipelago while two others could have been spread by man throughout a great part of the Pacific region.

● ● ● True wasps

True wasps of the Vespidae family include either social or solitary species whose wings are folded longitudinally at rest. All the social wasps share some habits: colonies are established annually and completely die after leaving fertilized queens scattering around to attempt new colonies; it is the sterile female workers that fashion the nest with paper produced from masticated wood pulp and feed the brood with insect preys. The only social wasps living in Vanuatu are Polistinae (paper wasps). The solitary true wasps are potter wasps (Eumeninae) that

built their nest out of mud and provision them with insect prey. Curiously, among the two Polistinae and four Eumeninae already known from Vanuatu, only one had been recorded from the Santo Island. Potter-wasps collected in Penaoru belong mainly to species endemic to Vanuatu (*Parodynerus cheesmani* Fig. 162A) or Vanuatu, New Caledonia and Loyalty archipelagos (*Parodynerus quodi*, *Delta xanthurum* Fig 162B and *Anterhynchium alecto* Fig. 162D). Giordani Soika, who described several varieties of these species in Vanuatu, noted that in each species, the darker varieties occur in the south of the archipelago while the lighter are present in the northern islands. In fact, the varieties found in Santo are those this author described from Malekula, one of the Vanuatu Islands closest to Santo. Two other species are new to Vanuatu: *Delta campaniforme* (Fig. 162C), also nowadays widespread in Asia, is endemic to the Pacific while *Pachodynerus nasidens* is a neotropical species that was introduced in Pacific Islands such as Hawaii and Tahiti. All the potter wasps of Santo prey on caterpillars. *Anterhynchium alecto* was seen one time in Penaoru reusing the empty cells of *Sceliphron laetum* by provisioning it with paralyzed caterpillars before laying a egg and sealing it.

It is noteworthy that *Polistes olivaceus* (Fig. 162E), a yellow paper wasp widespread in Santo, as in the other islands of the archipelago, was never recorded from Vanuatu in the literature. This invasive species however occurs in Vanuatu since at least 1935 as testified by the specimens housed

Figure 163: A *Polistes olivaceus* worker foraging in herbs near the Penaoru camp.

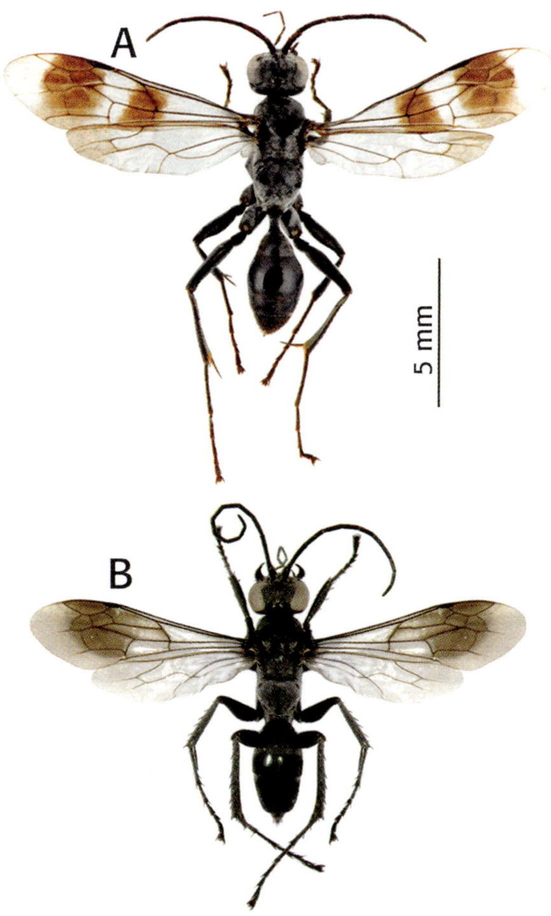

Figure 164: Spiders wasps. **A**: *Melanagenia penaoru*. **B**: *Anoplius santo*. (Photos A. Touret-Alby).

in the MNHN collection that were collected by Risbec, a French hymenopterist who frequently visited Vanuatu while working in New Caledonia. The yellow paper wasp is an oriental species which is nowadays widespread throughout the Pacific region. It is not averse to nesting near or in human habitation and builds a round, flat nest lacking an envelope, which is usually suspended from a solid sheltered surface or hidden in dense foliage. The colony frequently attains the range of 60 wasps and sometimes exceeds 100. While it tolerates activity near the nest, the workers that fiercely attack on provocation are renowned in Santo for their quite painful sting. In Penaoru, *P. olivaceus* workers were observed in herbs and bushes thoroughly checking for caterpillars and other insects on the underside of the leaves (Fig. 163). They were also seen from the Canopy Glider — the inflatable flying vehicle that was tested during the IBISCA programme — hunting insects in the crown of the high trees.

••• Spiders wasps

Pompilidae are long legged solitary species that hunt spiders to feed their progeny. A single spider is used as a host for feeding a larva. The female uses its venomous sting to paralyze the prey and then leaves it in its burrow or drags it to a place where it builds a nest. A single egg is laid on the abdomen of the spider, and the nest or burrow is closed. After consuming the host, the larva spins a silk cocoon and pupates to late emerge as an adult.

Despite the numerous Hymenoptera caught by E. Cheesman during her expedition across Vanuatu, and the diversity of the spider fauna no Pompilidae had been reported from this archipelago since R. Wahis identified *Anoplius opulentus* from

specimens of the MNHN collection that were collected in Vate, Malo and Ambrym Islands in the early 20th century. *A. opulentus* was caught in 2006 for the first time in Santo, near Luganville, on a sandy coastal road located in the vicinity of a former U.S. military base. This observation could confirm the hypothesis that the species, which is widespread throughout the Pacific Islands, could have been introduced through army transport during the Second World War. The two other endemic pompilid collected in Penaoru, *Melanagenia penaoru* and *Anoplius santo* (Figs 164A & 164B) were described in 2009. As quoted by R. Wahis, the new genus *Melanagenia* is represented in New Caledonia (one species) but mostly in New Guinea (more than 20 new species). The two new species were only collected between 600 and 900 m a.s.l., in primary forests of low and middle mountains of the Penaoru region. *Anoplius santo* was seen in clearings, standing on the sunny leaves of low branches, and females were observed landing on puddles of water, where they move by propelling with their wings as do other species of *Anoplius*.

••• Parasitoid wasps

The parasitoid wasps lay their eggs in or on various insects, each species being generally associated to a particular group of hosts. Parasitoid larvae develop by consuming their hosts, invariably killing them at least at the end of their development. In species (Cuckoo wasps, Dryinid and Bethylid wasps, Velvet ants and Scoliid wasps) belonging to Aculeata as bees, hunting and true wasps, the females have a sting with which they paralyse the host before laying a egg on its skin. In the others species (formerly named Parasitica), the females possess a longer or shorter ovipositor to lay eggs onto or more generally inside the body of their host. Parasitoid species have an enormous impact on ecosystems in sustaining their natural equilibrium, notably by keeping insect pest populations below outbreak levels. They represent in Santo 97 % of the Hymenoptera fauna.

••• Aculeata parasitoid wasps

Few Aculeates parasitoid wasps are known from Vanuatu. Among them, Vespoid wasps are represented by one species of Velvet ants (Mutillidae), which was recorded for the first time in 2006, and four species of Scoliid wasps (Scoliidae) of which only one was known in Vanuatu before Santo 2006. Specimens of three families of Chrysidoid wasps (Dryinidae, Chrysididae and Bethylidae) were also collected in 2006. Among them, three species of Dryinids had been already reported from Vanuatu and only one from Santo.

• Scoliid wasps

The few species collected in Penaoru (one *Campsomeris* and two *Scolia*) are still unidentified. No *Scolia* species had been reported yet from Vanuatu and the *Campsomeris* sp. (Figs 165A & 165B) collected in Santo differs from *C. novocaledonica*, a New Caledonian species whose presence in Vanuatu was quoted in the literature but needs to be corroborated. Scoliid wasps are black insects, often marked with yellow or orange, that parasite below ground beetle larvae. The female wasp burrows the soil in search for a grub that it paralyzes by stinging before laying an egg on its skin. The wasp larva remains in the motionless host's burrow and feeds until killing it before pupating. Scoliid wasps are, therefore, very important natural agents in the control of grubs.

• Velvet ants

Mutillidae are small to medium-sized wasps whose adults are generally heavily sclerotised and densely hairy. Males are fully winged while females are wingless and superficially resemble ants, hence the common name of the family. They search on the ground or tree trunks for nests of the bees, true or digger wasps, in which their larvae develop as ectoparasitoids fixed on the host larvae. The numerous males specimens collected in 2006 in low and middle mountain forests of the Penaoru region were studied by Denis Brothers. They belong to *Ephutomorpha* (Fig. 165C), a genus including about 200 species in Australia but considered as a complex of several undescribed genera. *Ephutomorpha* sp. of Santo is a black blue metallic species with two rows of white airs on the

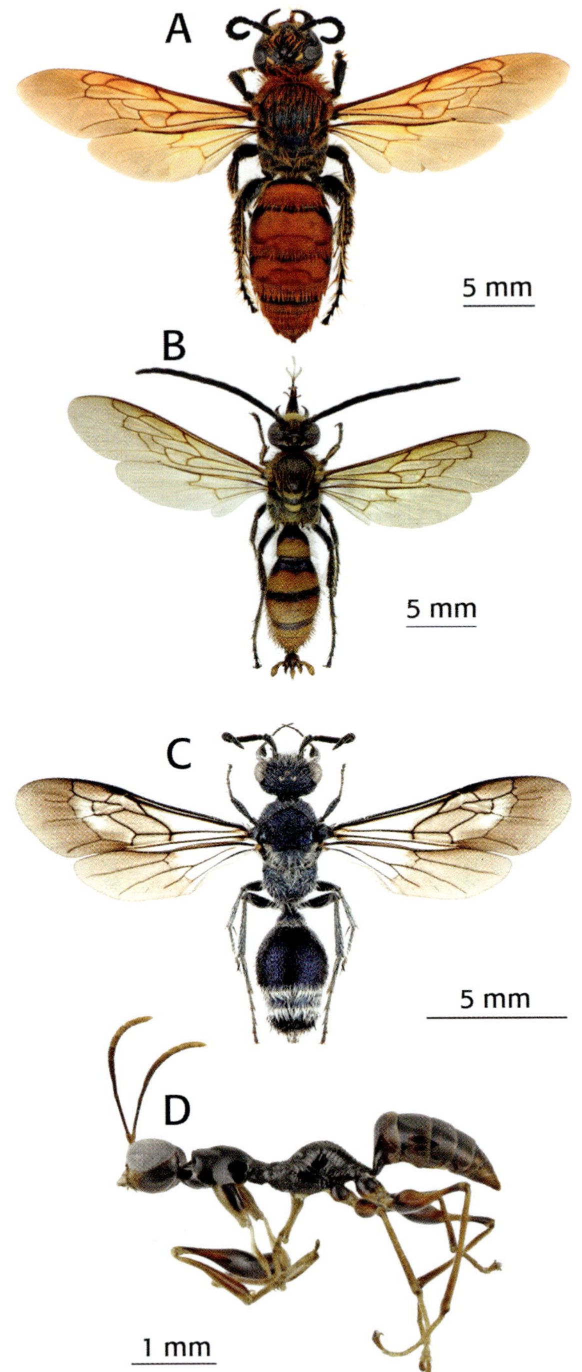

Figure 165: Aculeata parasitoid wasps. **A**: *Campsomeris* sp. female. **B**: *Campsomeris* sp. male. **C**: *Ephutomorpha* sp. male. **D**: *Gonatopus primitivus* female. (Photos A. Touret-Alby).

abdomen and wings darkened at apex. As D. Brothers said, it is probably new for science.

• Dryinid wasps

The Dryinidae are highly specialized ectoparasitoids of plant- and leafhoppers (Hemiptera). Males are usually fully winged while females are often wingless and closely resemble ants. Females generally possess modified clawed front legs with which they hold their hosts to temporarily paralyze them before injecting an egg in their body. The larva spends its early stages as endoparasitoid, but later it protrudes from the host's abdomen and continues to feed on it, protected by a sac-like "case" made from previous larval skins (left after moulting). Of the six Dryinid species (four genera) recorded from Vanuatu, only half were known before Santo 2006. Five of these species have been identified by Maximo Olmi from the material collected in the Penaoru region; among them, two: *Anteon molisae* (new for Science) and *Gonatopus primitivus* (Fig. 165D) are to date endemic to this archipelago, while the others have a wider distribution including New Guinea, Queensland, and various Pacific Islands. Their hosts are unknown except for *Taumatodryinus koebelei* that parasitizes planthoppers of the genus *Siphanta* (Flatidae) in Australia.

• Cuckoo wasps
Chrysididae are often brilliant hymenoptera with metallic blue, green, or reddish colours, hence their Latin name. They are commonly named Cuckoo wasps because a number of them develop as cleptoparasites feeding on the food stored by their hosts for their own progeny (pollen for bees, insects for wasp hosts). Other Chrysidid species however are true parasitoids developing by feeding on various insects hosts. The only Chrysidid collected in 2006 in Santo was the first recorded from Vanuatu. New to Science, it belongs to the subfamily Amiseginae and the genus *Kryptosega* that was hitherto only known from two species collected in New Guinea. Amiseginae are small cryptic wasps found in low vegetation and leaflitter that parasitize eggs of stick insects (Phasmatidae).

• Bethylid wasps
Bethylidae are small and often dark coloured wasps that have an elongate flat head. The females of many species are wingless and ant-like. They parasitize mainly Lepidoptera and Coleoptera larvae that usually live in cryptic niches, such as within rolled leaves, under bark, into rotten wood, in the ground or leaf litter. The habits of the family are very diverse, but in most cases the female drags the prey larva to a sheltered location and lays several eggs on its skin. Its larvae feed as external parasitoids and, at the end of their development, spin a silken cocoon on or near the killed host. At least eight different species of Bethylids were caught with Malaise traps in the Penaoru region. Several others have undoubtedly been collected with the other sampling devices utilized during the expedition (Berlese traps, Winkler traps…), but these specimens have still to be sorted and identified by a specialist.

••• **Parasitica parasitoid wasps**
Wasps of the Parasitica group are in Vanuatu much more numerous and diverse than Aculeata parasitoids, as is the case all over the world. More than 254 Parasitica species were already numbered in Santo, although several groups need further investigations before one may have a more clear-cut idea of their diversity. Among them, two superfamilies represented the major diversity of the group with at least 101 Chalcidoidea and 130 Ichneumomoidea species (56 Ichneumonidae and 74 Braconidae) to date recorded from Vanuatu, thus almost tripling the diversity of these taxa for the whole archipelago. The knowledge acquired on Santo's Parasitica fauna is even more significant since only a dozen species of these insects had previously been reported from this only one island.

• Ichneumonid wasps
Ichneumonidae in Santo range from very small (3 mm) to large (25 mm) wasps. Before 2006, only five species (five genera, three subfamilies) had been recorded from Santo. Most of them were reported by E. Cheesman in the 1930s. The inventory made in Penaoru identified 36 species (26 genera, 12 subfamilies) among which 30 were new to Santo. With the two Ophion wasps quoted by E. Cheesman but not recovered in 2006, 38 icheumonids are now known from this island. Species parasitizing Lepidoptera larvae and pupae are largely the most abundant and diverse are their hosts. They belong to the subfamilies Campopleginae (eight spp.), Ophioninae (seven spp.), Pimplinae (five spp.), Ichneumoninae (three spp.), Cryptinae (three spp., among which *Necolio* sp., Fig. 166A), Tryphoninae (two spp.), Anomaloninae (two spp.), Labeninae (*Certonotus mogimbensis*) and Metopiinae (*Hypsicera* sp.). Among Campopleginae (five genera), the genus *Eriborus* (four spp. in Santo) was previously known from Vanuatu, but not yet from Santo, by only one species: *E. iavilai* (Fig. 166B). Within Ophioninae, the genus *Enicospilus* (four spp. in Penaoru) is the most commonly collected by sweep netting throughout the archipelago, where 12 species as a whole have been reported. Two other Ophionin genera are present in Santo, *Leptophion* and *Xylophion*, the last one being new to Vanuatu. Ophion waps, which are yellow brown short-tailed ichneumons with a long compressed abdomen and large ocelli, are also routinely attracted to light, as *Netelia* (Fig. 166C) and *Toxochiloides* species (Tryphoninae) that are frequently confused with Ophion wasps. All these large wasps hunt in bushes for their host caterpillar. Females lay one egg inside (Ophions) or onto (*Netelia*) the host which only dies during pupal stage while wasp larva remains to pupate itself.

Pimpline wasps of Santo are mainly represented by stout brown orange (*Echthromorpha* (Fig. 166D); two spp.) or smaller yellow (*Xanthopimpla*; two spp.) insects, with a rather long ovipositor and a black spotted abdomen. Females generally lay their eggs in Lepidoptera pupae often hidden in a plant gallery or a silken cocoon. Finally, the most common species in Santo is the Banded Caterpillar Parasite Wasp, *Ichneumon promissorius* (Fig. 166E) (Ichneumoninae), a white spotted black and red Ichneumon considered as native to Australia while recorded throughout the south Pacific region. It mainly attacks Noctuid moths,

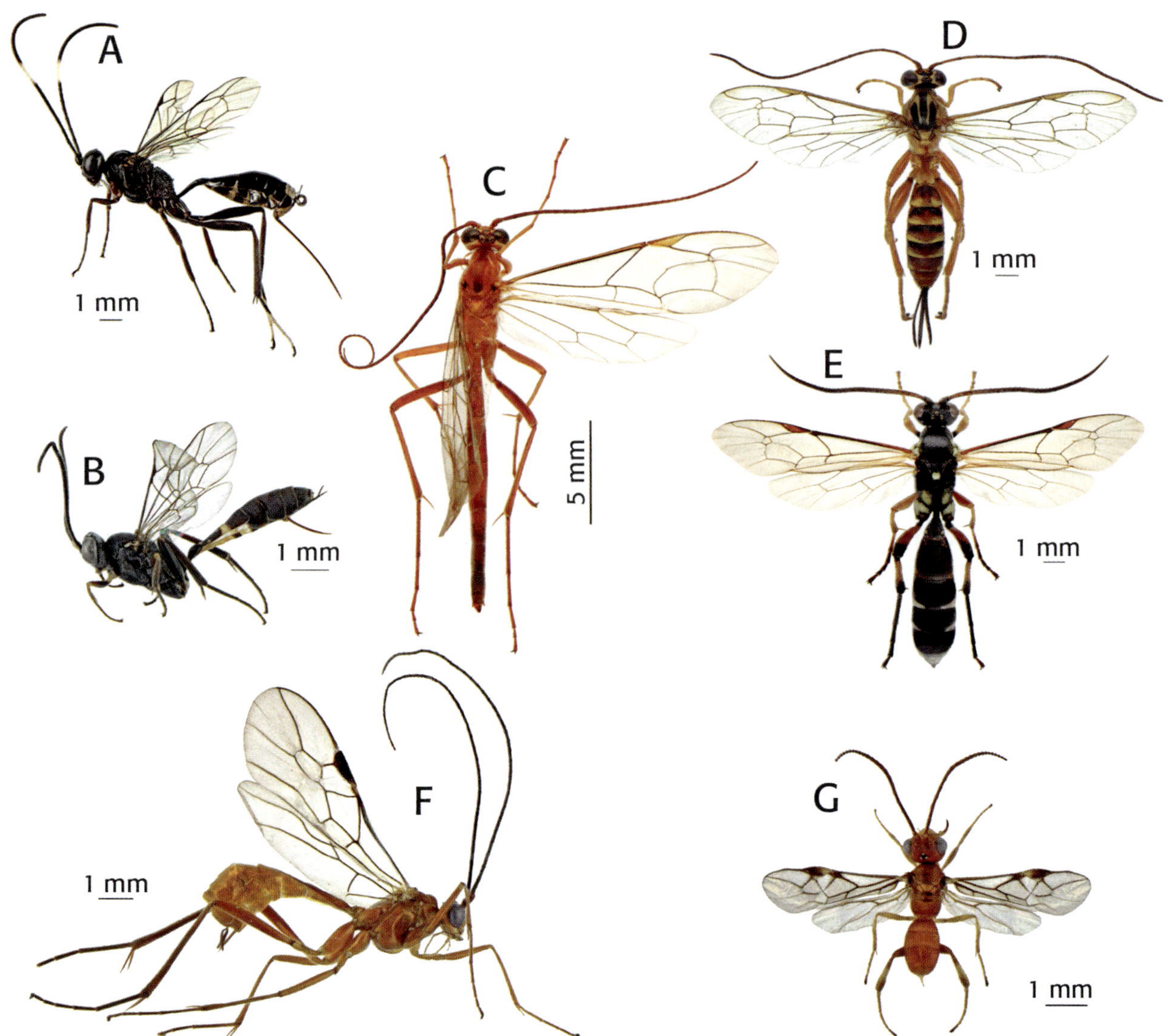

Figure 166: Parasitica parasitoid wasps. Ichneumonidae. **A**: *Necolio* sp. **B**: *Eriborus iavilai*. **C**: *Netelia samoana*. **D**: *Echthromorpha agrestoria juncta*. **E**: *Ichneumon promissorius*. Braconidae. **F**: *Austrozele* sp. **G**: *Phanerotoma novahebridensis*. (Photos A. Touret-Alby).

many of which are widespread agricultural pests. The Ichneumon female searches the soil surface for host pupation sites and burrows into the pupal gallery to oviposit into the host pupa.

Three other Ichneumonid families collected in Santo comprise species often overlooked because of their small size and discrete brown colours. As Cryptinae and Metopiinae, they were reported for the first time from Vanuatu. Orthocentrinae (three genera, five species) are parasitoid of Diptera larvae leaving in decaying vegetation while Tersilochinae (*Diaparsis* sp.) parasitize small phytophagous Coleoptera larvae, and Mesochorinae (*Mesochorus* sp.) are generally hyperparasitoid that develop on Hymenopteran larvae parasitizing caterpillars. Among the Vanuatu's identified Ichneumonids, 16 species or subspecies can to date be considered as endemic to this archipelago while 15 have a wider Pacific distribution that include mainly Queensland, New Caledonia, New Guinea or Solomon Islands, and five others were recovered until Indonesia or India.

• Braconid wasps

Braconidae are generally smaller insects than Ichneumonidae. Before Santo 2006, *Phanerotoma novaehebridensis* (Cheloninae) (Fig. 166G) and *Utetes cheesmanae* (Opiinae) were the only two species recorded with certainty from Santo, whereas 16 other species have been quoted in Vanuatu mainly from the material brought to the British Museum par E. Cheesman. The specimens collected in 2006 had not yet been studied by a specialist, but at least 74 species (about 40 genera, 15 subfamilies) thus considerably increasing the diversity of the Vanuatu's Braconid fauna. In Santo, about fifteen species (three subfamilies) are parasitoids of Coleoptera larvae, mainly belonging to Doryctinae (12 spp.) that attack stem-boring beetles. The parasitoids of Diptera larvae and pupae (Alysiinae, 12 spp.; Opiinae, five spp.) show the same species diversity but are much more abundant, as are their hosts which mostly develop in leaflitter. Other species (three Braconinae, four Euphorinae) poorly represented in Penaoru may parasitize various hosts, while Aphidiinae (only one sp. recorded) parasitize Aphids. Finally, as in Ichneumonids, the most diverse and numerous Braconidae are parasitoids of Lepidoptera larvae (27 spp., seven subfamilies), among which Microgastrinae (13 species) and even more Cheloninae (five spp.) are dominant by their population density. The most common Braconid in the Penaoru region is the entirely orange *P. novaehebridensis*, an endemic species that has been recorded from the major part of the archipelago while tens *Austrozle* sp. (Macrocentrinae) (Fig. 166F) have bee recorded in a unique sample. Actually, very little is known about the geographic distribution of Vanuatu's Braconids. In the present state of our knowledge, half of the 18 identified species are endemic while the other are much largely distributed around Pacific countries, some of them reaching Indonesia and Indochina. The most widespread are *Apanteles briareus* (Microgastrinae), an enemy of the Macadamia Nutborer *Cryptophlebia ombrodelta* (Tortricidae) that damages macadamia and litchi orchards, and *Ontsira palliata* (Doryctinae) that develops on coffee trees' boring-beetle larvae. However, the presence of these two species in Santo needs to be confirmed by an identification of the Santo's material to a specific level.

- Chalcidoid wasps

The Chalcidoidea superfamily, probably includes several hundreds of thousands species around the world, although less than one tenth is described. Largely ignored because of their small to minute size, Chalcidoid parasitoids exhibit amazingly diverse life histories and profoundly influence all types of ecosystems, from rainforests to agriculture. Many species are natural enemies of insect pests that are used instead of pesticides to control them. Before Santo 2006, only 13 species (six Eulophidae, two Eucharitidae, one Agaonidae, one Eupelmidae, one Leucospidae, one Ormyridae, one Trichogrammatidae) where recorded by various authors from Vanuatu, whereas none was quoted in the literature from the Santo Island. Most were native or introduced parasitoids that probably occurred in cultivated area of Santo with their hosts, i.e. the palmivorous beetle *Promecotheca opacicollis* (Chrysomelidae) or various leaf-miner flies (Agromyzidae) and fruit-flies (Tephritidae). The chalcidoids collected with Malaise traps in Penaoru were identified by Gérard Delvare who recognized at least 90 species in 14 families, among which seven (Aphelinidae, Encyrtidae, Eurytomidae, Mymaridae, Perilampidae, Pteromalidae and Torymidae) are new to Vanuatu. At least two species, *Cirrospiloidelleus* sp. (an Australian Eulophinae genus hitherto known by a single species) and *Neocalosoter* sp. (a tropical Pteromalidae genus), could probably be considered as new, although a thorough study of the material collected in Penaoru would undoubtedly increase the list of Santo's species new to Science. This author recognized fifty genera, of which 15% have a Pacific distribution while most of the others are cosmopolitan. The most diverse subfamilies are Eulophidae (16 gen., 24 spp.) and Pteromalidae (12 gen., 12 spp.) that parasitize a wide range of hosts, and Encyrtidae (14 gen., 20 sp.) that are mainly parasitoid of scale insects.

Fig wasps (Agaonidae) are another well represented chalcidoid family whose species are always associated with figs; 11 species (four subfam.) have been already collected in Penaoru. Many fig wasps (Agaoninae) act as pollinators of various *Ficus* spp., whereas the others are probably parasitoids of the fig-pollinators or gall-formers in other parts of the fig. Female of pollinating wasp that carries pollen from another fig enters the fig via its narrow ostiole. Within the fig, the reproductive flowers are pollinated while the female lays eggs in the sterile flowers where its larvae will develop and pupate. Wingless male offspring emerge first and mate with the females inside the fig before boring in the fig wall a hole that allows females to escape. Figs and their associated species of pollinators are totally dependent on one another, since fig flowers can be pollinated only by the appropriate species of wasp, and no wasp can produce progeny outside the appropriate fig. In a given region, the diversity of the pollinating wasps and their associated species is thus depending on the fig-tree diversity.

Another peculiar lifestyle is that of the Eucharitidae which parasitize mature ants larvae or pupae. Eggs are laid in large numbers on plant parts and the young larva either waits on the leaf or falls to the ground (depending on species) awaiting a passing ant. Once attached to an ant, the larva is taken back to the nest where it attaches to any available ant larva. Further development does not usually occur until the host has reached the pupal stage. Wasp pupation takes place inside the host cocoon or naked inside the ants' nest. Emerging adults leave the nest to reproduce outside. Metallic purple Eucharitid adults were commonly observed in the vicinity of the Penaoru camp while standing on leaves of undergrowth plants (Fig. 167). They belong to the genus *Chalcura* that had not yet been identified in Vanuatu. Another Eucharitid species caught in Malaise trap belong to the genus *Orasema* from which one species, *O. koghisiana* was described from New Caledonia and Vanuatu.

Finally, adults of *Podagrion* sp. (Torymidae) were obtained from a mantis ootheca.

Figure 167: A male of *Chalcura* sp. (Eucharitidae) on a herbaceous plant near the Penaoru camp.

• Cynipoid wasps

This superfamily is represented in Vanuatu only by Eucoilinae (Figitidae), minute, black brown shining wasps that are mostly parasites of small fly pupae (Diptera). No Eucoilin wasps had been recorded from Vanuatu before Santo 2006. The specimens collected in Malaise traps were identified by Mattias Forshage who recognized at least 10 species in seven genera. Two genera (*Diglyphosema* and *Rhoptromeris*) are new to Pacific Islands and three others (*Leptopilina*, *Hexacola*, *Ganaspis*) are cosmopolitan. Another genus, *Maacynips*, occurs only in Australia and the West Pacific Islands; its hosts are yet unknown. *Maacynips* sp. from Penaoru, is close to a Papuan species and new to Science. As in Australian fauna, two widespread genera dominate: *Hexacola* (two spp.) that attacks various fly larvae (i.e. Chloropidae, Drosophilidae…) and *Leptopilina* (four spp.) that are parasitoid of Drosophilidae. The most common species in Penaoru are *Leptopilina boulardi* and *L. heterotoma*, the natural enemies of *Drosophila* flies that are, as their hosts, widespread throughout the world.

• Platygastroid wasps

Platygastroids are minute brown or black shining wasps, with elbowed antennae, that are fairly common all around the world. They include thousands of species representing two distinct morphological groups which were recently gathered into a single family, Platygastridae. Most of the world records on Platygastrids are available on the Hymenoptera Online Database developed by Norman Johnson and Luciana Musetti, who identified the genera collected in 2006. Before, 14 genera and four species were known from Vanuatu, including three *Trissolcus* species and *Scelio flavicornis* that are spread in the Pacific region (but unknown in Santo), *Paratelenomus matinalis*, the only endemic species hitherto recorded from Santo, and the endemic *Fusicornia dissita*, recently described from Tanna. Santo 2006 led to the recognition of 19 platygastrid genera (in four subfamilies), among which 13 are new to Vanuatu. Platygastrinae and among them *Leptacis*, the most common Platygastrid genus in Santo, parasitize gall midges eggs, while Scelioninae and Telenominae attack eggs of very diverse groups of insects. For example, *Scelio* and at least five other genera develop in grasshopper and cricket's oothecae, while *Trissolcus* attack bugs, *Telenomus* moths and *Idris* spiders. However, the hosts of many other species (notably in Teleasinae) are still unknown. Most of the Platygastrid genera recorded from Santo and Vanuatu have a tropical distribution and are largely represented in Australia, New Caledonia and New Guinea. Few are quoted in other Pacific Islands but their fauna is still too poorly known to draw any conclusion.

• Ceraphronoid wasps

They include Ceraphronidae and Megaspilidae, two small families of minute wasps that are commonly found although few in numbers during insect collecting, notably in ground and leaflitter samples. The distribution data of Ceraphronoids already recorded on the Hymenoptera Online Database are largely incomplete. Not much is known of their biology, although some species are parasitoids of the larvae of beetles, flies, bugs, or other hosts, while others are hyperparasitoids developing on insects parasitizing various arthropods. Nothing was known on Vanuatu's Ceraphronoid fauna before Santo 2006 and very few is known from other Pacific Islands. Only five Ceraphronid species and one Megaspilid were collected in Santo in Malaise traps; however their number will certainly increase after sorting the material caught with other sampling devices.

• Diaprioid wasps

They include a single family of wasps (Diapriidae) found nearly everywhere in the world, although rarely observed due to their minute size. Most of them parasitize larvae and pupae of fly larvae, while few others attack beetles, ants or parasitoid wasps. Diapriids show considerable diversity of form with aptery (lack of wings) being fairly common, sometimes in both sexes. Many species exhibit a noticeable sexual dimorphism with males and females often mistaken for separate species. As far as we know, Diapriidae had never been recorded from Vanuatu in the litterature. In Santo, 22 morphospecies were recognized from Malaise trap samples collected in Penaoru in 2006. The largely dominant genus *Aclysta* (two spp.) parasitize fungus gnat flies (Mycetophilidae) larvae that feed on decayed leaves and fungi. Further studies are needed to identify the other diapriid species at least at the genus level. Other species will be certainly obtained by sorting the material caught in Santo with other traps.

• Evanoid wasps

Evanoids include in Vanuatu two families of medium sized wasps: Gasteruptiidae that parasitize the larvae of solitary bees'nests and Evaniidae that are natural enemies of cockroaches (egg parasitoids).

The only Gasteruptid species known from Vanuatu, *Pseudofoenus ritae*, was described by E. Cheesman from Malekula and Tanna. Also known from New Caledonia, it is still unrecorded in Santo. The genus *Pseudofoenus* is mainly distributed in the South Pacific region with a majority of species endemic to a single country.

The Evaniids or Ensign Wasps have a very distinctive body shape since no other Hymenoptera have a so small and laterally compressed gaster attached so high on the thorax, giving the appearance of a small hand flag, hence their common name. Female wasps oviposit into the concealed egg cases (ootheca) of the cockroach; their larvae feed on the eggs and pupate inside the ootheca. The two widespread spe-

Figure 168: **A**: *Szepligetella sericea* female (Evaniidae). **B**: *Foenatopus* sp. female (Stephanidae). (Photos A. Touret-Alby).

cies of Evaniidae known throughout Vanuatu were recorded for the first time by E. Cheesman in the 1930s. *Evania appendigaster* is a black stout species that has achieved an essentially worldwide distribution in warm and temperate climates, having been introduced along with various cockroach species (*Blatta* and *Periplaneta* spp.) associated with man. *Szepligetella sericea* (Fig. 168A), the Lesser Ensign Wasp, was widespread from Australia, throughout Pacific Island and Southern Asia. It parasitizes cockroaches of the genus *Periplaneta*.

- Stephanoid wasps

They include a single family, Stephanidae, primarily tropical and subtropical in distribution. They have a slender and oddly elongated body with a subspherical head, set out on a long neck, and highly modified hind legs. Females bear a long ovipositor they insert in the wood to lay an egg on wood-boring beetle larvae. Stephanids are not easily spotted in their natural habitat, and are rarely collected by most of the usual sample devices. The single specimen collected by sweep netting in Santo was seen flying around the Penaoru camp. It belongs to the genus *Foenatopus* (Fig. 168B) as the two species already quoted in Vanuatu but seems to be different. None of these species was previously recorded from Santo. The endemic *F. unistriatus* was described by Cheesman from Malekula while *F. salomonis*, from Salomon Islands, seems to have been erroneously reported from Vanuatu by its descriptor.

••• The affinities of Santo's Hymenoptera fauna

One of the objectives of the Santo 2006 expedition was to investigate the origin of the Santo's fauna by comparison with that of the surrounding regions. As 80% of the Hymenoptera species collected mainly in Penaoru were hitherto identified only to genus or morphospecies level, it is difficult to have a faithful view of the geographic distribution of this fauna. However, in the best investigated taxa (bees and predatory wasps, ichneumonids) representing almost 25% of the species already recorded in Vanuatu (ants excepted), about 47% are represented by species or subspecies endemic to this archipelago, while a majority of the remaining species is reported from other parts of the Pacific region. The most significant similarities

occur inside Vanuatu with Malekula, and outside with New Caledonia and Salomon-New Guinea. Relationships with the other Pacific Islands are difficult to assess until more detailed surveys lead to a better knowledge of their Hymenoptera fauna. Finally, about 3% of the identified species are exotic species generally introduced involuntarily through maritime trade or willingly for biological control. With increasing tourism and trade, accidental introductions of additional species can be expected in the future.

••• Further studies

The hymenoptera diversity of Santo assessed here remained largely underestimated knowing that many other samples have still not been sorted yet, notably those including soil and litter fauna. Beside ants, these samples will provide a large amount of other species mainly belonging to microhymenopteran parasitoids. A great part of these samples has been collected in Penaoru by the "Forest, mountain and river" team although others have been gathered by the scientists involved in the "Karst" team or the "Wasteland and aliens" team. Nevertheless, new surveys in other parts of the Santo Island and during other seasons are needed to get a better overview of its Hymenoptera fauna.

Only few of Santo's species have been already described as new. However, taking into account the huge diversity of Hymenoptera, it is difficult to assess in many case which species is new to Science without comparing it with material of the Pacific, Australian (mainly Queensland) and Indo-Papuan fauna. The possible introduction of species through maritime transport compels to search also for at larger geographic scale. Such investigations would thus imply for a long period a number of taxonomists that should examine a wide range of material housed in numerous museums and perform various taxonomic revisions.

Nevertheless, the increased knowledge of Vanuatu Hymenoptera biodiversity attained by the Santo 2006 expedition is actually significant, making of Santo one of the best known islands of the Pacific for its entire Hymenoptera fauna. Moreover, this survey has not only highlighted the diversity of this incredibly diverse order but also given a better idea of the diversity of other arthropod orders which serve as prey and hosts to predatory and parasitoid wasps.

Hydnophytum cf. *longistylum*
In situ artwork by Roger Swainston/anima.net.au

MYRMECOPHILY IN SANTO:
A CANOPY ANT-PLANT AND ITS EXPECTED AND LESS EXPECTED INHABITANTS

Bruno Corbara

A few ant species are known to live in close association with particular plants that have therefore been named ant-plants or "myrmecophytes" (from the Greek words *myrmecos* for ant and *phytos* for plant). Ant-plants are widespread in the tropics where they belong to very diverse families. They develop hollow structures (such as leaf pouches, hollow trunks, spines or cavernous tubers) that shelter ants, and are often called "domatia" (from *domus*, the Latin word for house). In Southeast Asia, the Pacific Islands and northeastern Australia, the Rubiaceae family comprises some unusual epiphytic myrmecophytes belonging to five genera: *Myrmecodia* Jack, *Hydnophytum* Jack, *Myrmedoma* Becc., *Myrmephytum* Becc. and *Squamellaria* Becc. The ants live in large chambers in these epiphytes that develop in tubers derived from the hypocotyls[7].

7 - The hypocotyl is the part of a plant embryo or seedling that is found between the radicle and the cotyledons.

Nesting sites are quite rare in arboreal environments and may be considered a limiting factor for many species. By living inside of the hollow tubers of epiphytic Rubiaceae, ants have found an original solution to this widespread ecological problem.

Plant-ants are often known to protect their host from herbivores. In the case of the myrmecophytic Rubiaceae, previous studies conducted in Papua New Guinea indicate that ants provide no protection from phytophageous insects; however, according to the same studies, ants are considered to benefit the epiphytic plants for they provide them with nutrients, especially nitrogen, by means of fecal material and debris which are brought into the chambers.

Ant societies are well known to tolerate many "squatters" or myrmecophiles (i.e. "species who like ants") living in close association with them. Moreover, regardless of the presence of ants, many species may be attracted to the hollow structures of the Rubiaceae in order to use them as shelter or a nest site. Indeed, previous studies have shown that invertebrates (insects or other arthropods) as well as vertebrates (amphibians or reptiles) live inside of the chambers of *Myrmecodia* and *Hydnophytum* tubers.

••• A myrmecophytic Rubiaceae in Santo

The first written mention of a myrmecophytic Rubiaceae on Santo appears, as far as I know, in the report of the 1997 Tsukuba Botanical Garden (TBG) Expedition. Tatsuo Konishi from the TBG and his colleagues report having collected *Hydnophytum* sp. "*between 2nd Camp and Mt. Vutimena, on the steep slope of a montane forest, 700 m. alt.*"; according to their itinerary described in the Annals of the TBG, the 2nd camp was situated at 500 m a.s.l. on the south ridge of Mt. Vutimena (1 446 m a.s.l.). To the best of my knowledge, this is the only species of myrmecophyte reported on the island.

During the IBISCA-Santo project, part of the Santo 2006 Global Biodiversity Survey, a group of participants conducted a detailed inventory of the vegetation and of the arthropod fauna along an altitudinal gradient in the Penaoru area. The different study sites from the base camp (100 m a.s.l.) to the 600 m, 900 m and 1 200 m a.s.l. were attainable by means of a trail that had been opened by our local guides, and that was used daily. The guides, who were interviewed about the local flora and fauna, confirmed —based on our description— the presence of a myrmecophytic Rubiaceae on high branches in the area. We therefore asked them to look for these plants on each side of this path, and to mark the host trees. Thereafter, the myrmecophytes were collected *in situ* with the help of professional climbers and opened up in place or at the base camp in order to identify the ants inhabiting their hollow tubers.

••• Ants and other animals dwelling inside of Hydnophytum tubers

In total, we collected 65 myrmecophytes, between 550 m and 950 m a.s.l., on 11 different trees that grew inside of a well-preserved kauri-tamanu forest (*Agathis-Calophyllum* forest). Many trees bearing *Hydnophytum* epiphytes were not climbed, and for each tree inspected the climber only collected a small proportion of them. These myrmecophytes may therefore be considered as very common in the area.

Those we collected were located at a mean height of 16.5 m (+/- 4.5 m) from the ground. For a non-specialist, they seemed to belong to the same species; the specimens sampled for botanical collection were identified as *Hydnophytum* cf. *longistylum*. The 65 myrmecophytes sheltered 12 different species of ants (Figs 169 & 170). Unsurprisingly, the most common species, whose colonies inhabited the hollow tubers, belonged to the genus *Iridomyrmex* (subfamily: Dolichoderinae), and is probably *I. cordatus*. Indeed, this ant is the regular occupant of *Hydnophytum* and *Myrmecodia* chambers in Borneo, Java, the Moluccas, the Bismarck Archipelago, New Guinea, the Solomon Islands and northern Queensland. We found *Iridomyrmex* colonies in 27 *Hydnophytum* out of 65; i.e. in 42 %

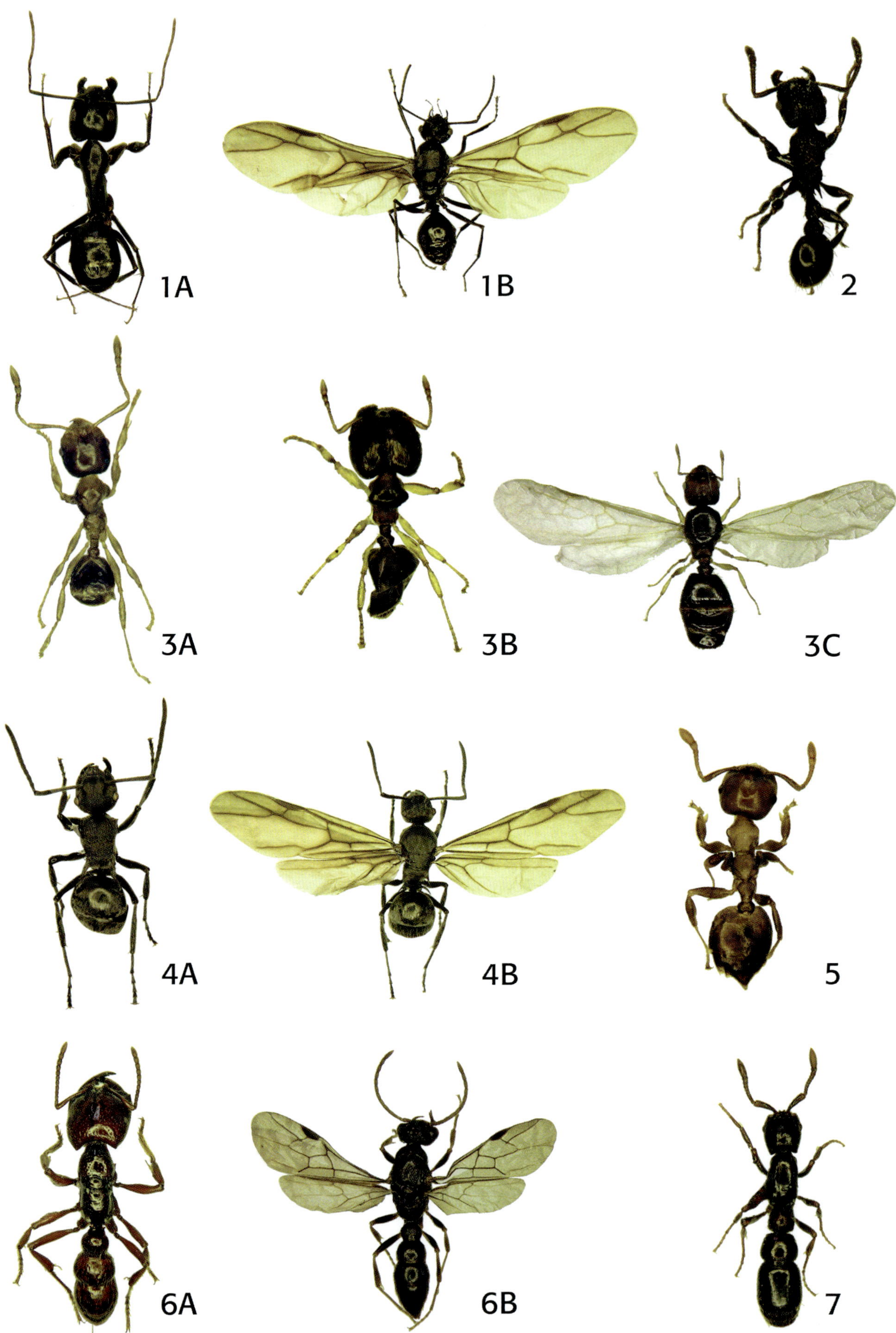

Figure 169: Ants inhabiting *Hydnophytum* cf. *longistylum* tubers from the Penaoru area. **1A**: *Iridomyrmex* sp., worker. **1B**: *Iridomyrmex* sp., alate gyne. **2**: *Tetramorium* sp., worker. **3A**: *Pheidole* sp., minor worker. **3B**: *Pheidole* sp., major worker. **3C**: *Pheidole* sp., alate gyne. **4A**: *Polyrhachis* sp., worker. **4B**: *Polyrhachis* sp., alate gyne. **5**: *Crematogaster* sp., worker. **6A**: *Amblyopone* sp., worker. **6B**: *Amblyopone* sp., alate gyne. **7**: *Cerapachys* sp., worker. (Photos B. Calmont).

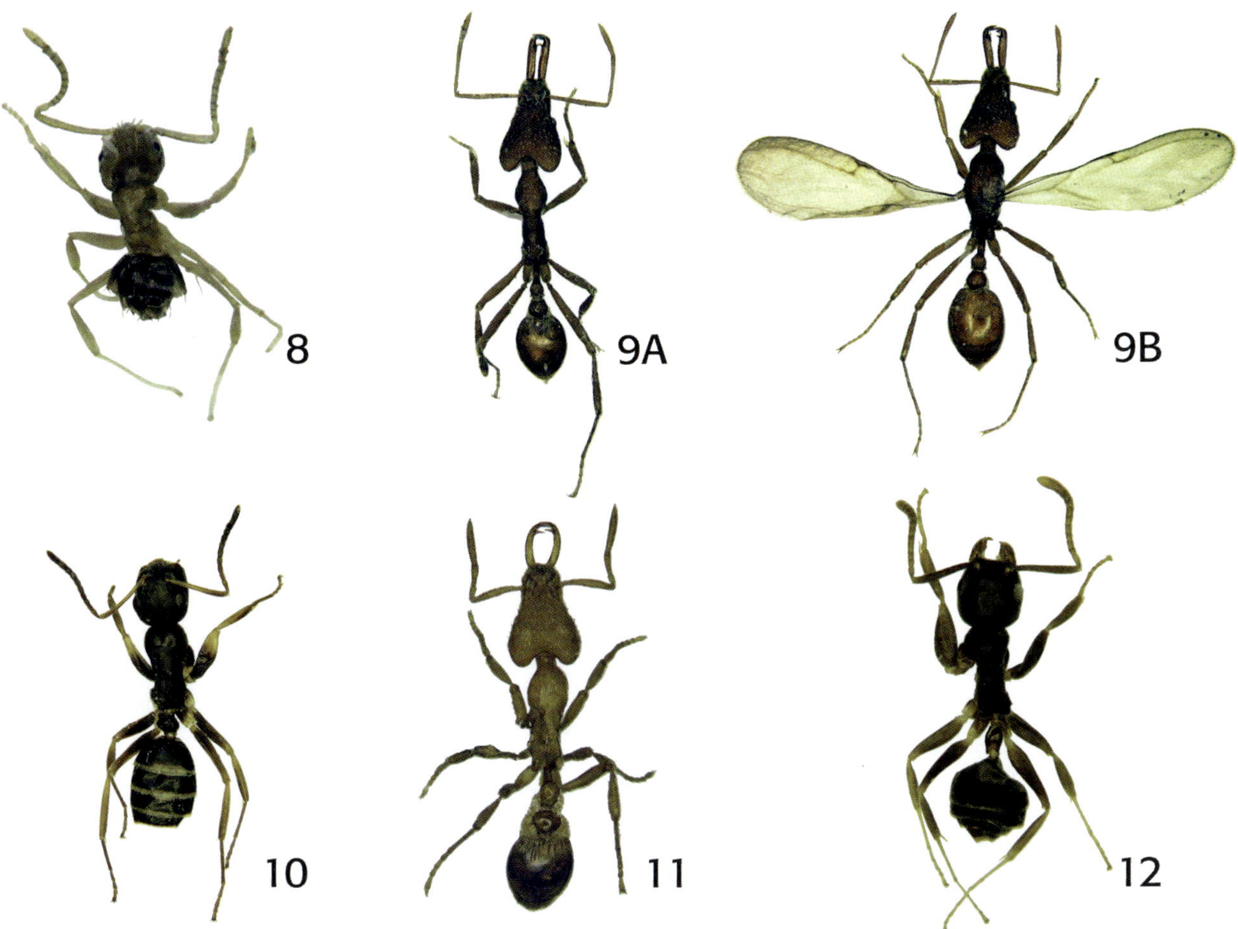

Figure 170: Ants inhabiting *Hydnophytum* cf. *longistylum* tubers from the Penaoru area. **8**: *Paratrechina* sp., worker. **9A**: *Strumigenys* sp.1., worker. **9B**: *Strumigenys* sp.1., alate gyne. **10**: *Camponotus* sp., worker. **11**: *Strumigenys* sp.2., worker. **12**: *Turneria* sp., worker. (Photos B. Calmont).

of them. However, *Iridomyrmex* was not at all the only plant-ant living in the *Hydnophytum*, which contrasts with data gathered from other locations and for other myrmecophytic Rubiaceae. The tubers commonly sheltered two other species of Myrmecinae, *Tetramorium* sp. (22 cases; 34%) and *Pheidole* sp. (17 cases; 26%), and, less commonly, six others: *Polyrhachis* sp. (Formicinae), *Crematogaster* sp. (Myrmecinae), *Amblyopone* sp. (Amblyoponinae) (in seven, four and three cases, respectively), *Cerapachys* sp. (Cerapachinae), *Paratrechina* sp. (Formicinae) and *Strumigenys* sp.1 (Myrmecinae) (in two cases for each). Moreover, three species were found only once: *Camponotus* sp. (Formicinae), *Strumigenys* sp.2 (Myrmecinae) and *Turneria* sp. (Dolichoderinae). Different colonies of different species (a maximum of three) of ants could inhabit the same *Hydnophytum*. Only very young *Hydnophytum* tubers were devoid of any ants (six cases).

The outer appearance of the *Hydnophytum* leaves in Penaoru did not support the hypothesis that the ants protect the plant. Those inhabiting the tuber were not aggressive; very few were patrolling on the surface of the tuber or a fortiori the leaves (at least during the day); in most cases, it was necessary to open the tuber in order to confirm the presence of ants. Moreover, the level of herbivory on the leaves was generally very high.

As has been noted in other locations for other myrmecophytic Rubiaceae, the hollow tubers of *Hydnophytum* cf. *longistylum* shelter other animals, invertebrates as well as vertebrates. These invertebrates include other insects such as Dermaptera, Blattodea, Orthoptera (most notably, the large katidid cricket (Tettigoniidae) *Salomona redtenbacheri* (Fig. 171), as well as spider eggs and Mollusca (*Aneita* sp. slugs, Athoracophoridae).

Interestingly, the hollow tubers also appeared to be the laying site of females of two geckos species: *Gehyra vorax* Girard (which we found only in the tuber of one myrmecophyte) and a *Lepidodactylus*, which appears to be new to science and has recently been described and named *L. buleli* by Ivan Ineich (Fig. 172). While opening one *Hydnophytum* tuber during this survey, I accidentally killed a gravid *L. buleli* female. Moreover a total of 50 living eggs (25 x 2, as these sticky eggs are laid two by two) was inventoried in 10 other *Hydnophytum* tubers. Therefore, in total, 17% (11 out of 65) of the sampled myrmecophytes showed presence of

L. buleli. Some of the big tubers especially appear to be regular laying sites for this species. We found, for example, 16 and 12 eggs in (large or neighbouring chambers of) two *Hydnophytum* tubers, respectively; moreover, numerous pieces of egg shell, evaluated to correspond in each case to a minimum of 20 hatched eggs, were present in the same chambers. When possible, after opening the tubers, eggs were put back into the field inside of the pieces of tuber sheltering them.

This limited study of the ants and other inhabitants of the only known myrmecophyte on Santo shows that *Hydnophytum* tubers represent a very important dwelling and egg-laying site for a large spectrum of taxa, vertebrates included. Epiphytes are often considered as keystone species for arboreal animals in tropical forests. In the well-preserved kauri-tamanu forest of Penaoru, where it is very frequent due to its role in biodiversity enhancement, *Hydnophytum* may be considered a plant of very high value in terms of conservation.

Figure 171: The large Tettigoniidae *Salomona redtenbacheri* may be found inside *Hydnophytum* chambers.

Figure 172: *Hydnophytum* tubers are laying sites for two species of gecko (Reptilia, Gekkonidae). **A**: Two *Lepidodactylus buleli* eggs, new species. **B**: One *Gehyra vorax* egg.

BEETLES IN SARATSI RANGE, SANTO

Alexey K. Tishechkin & Jürgen Schmidl

The task of writing about such diverse group of animals like beetles on a remote, tropical, rather large oceanic island like Santo is analogous to an attempt to describe a black cat in a dark room. You know it is supposed to be there, but have a quite vague idea of its size, age, sex and character, even if you are holding it by the tail. More, you still are not sure if there are any other cats inside, how many, and even if the creature you are holding is a cat at all.

Santo, as a relatively large composite island with its warm and wet climate, lush rainforests, high mountains and limestone plateaus, looks as a good candidate for being a "tropical paradise" for beetles, these champions of diversity and adaptability in animal kingdom. However, there are several important factors of uncertainty in respect to the state of local beetle fauna and our knowledge about it.

First, Santo is a remote oceanic island which has never been connected to a large landmass. This fact brings the issue of colonization, dispersal limitations, survival and diversification of immigrants. Although its major potential colonization sources are represented by such landmasses as Australia and New Guinea which harbor rich and diverse faunas, the water gaps dispersers have to cross are still long several hundreds of kilometers and the nearest lands (Fiji, New Caledonia and Solomon Islands) are remote oceanic islands themselves. These islands are only stepping stones for dispersers from mainland situated more like thousands kilometers away. There is a very limited knowledge available to predict which beetles and how many of them have reached, survived and diversified on an island like Santo.

Second, the remote nature of the island makes it hard to get there not only for animals, but for people who are interested in discovering and studying them. Vanuatu and Santo in particular have never been a popular destination for beetle enthusiasts. Beetle fauna of less remote, larger, more actively used by Europeans islands in the region, such as New Guinea and New Caledonia, had received much more attention from naturalists. So, the available written knowledge and even unsorted museum collections are minimal, especially outside abundant, economically important and appealing to collectors groups, i.e. for "little brown beetles" representing the vast majority of species in this mega-diverse insect order.

And third, there is no way to tell how reliable even this limited knowledge is. The task of obtaining representative picture of local fauna and its natural history for a diverse group of tropical insects at a large and relatively complex site (case of beetles of Santo) is notoriously difficult. In the tropics, most of the beetle species are rare. It could be a true rarity, meaning small populations and small localized ranges, or rarity in collections, when you need a piece sheer luck or very special time window and/or collecting method to detect a specialized species. At a given site, "to get them all" you need to employ a huge effort of mass-collecting methods or apply a multitude of collecting tricks in variety of microhabitats, or better combine two or more approaches. Most of the biotic inventories, even nowadays, fail to do that consistently. That means there is almost no way to tell how complete a given inventory is, and which group failed to reach the island and which was just not netted or trapped. Given all the above, we would not even mention the task of learning the natural history of the beetle fauna as a whole. So, here we are after five weeks of traversing a single range in island's mountains, holding the tail of our black cat in the dark and writing about beetles of Santo.

• • • The background.
What do we actually know about them

Anyway, a big picture gets started with a few first touches by a brush. A natural way to begin is to review the available knowledge. In an attempt to get around at least some of above uncertainties, the most promising way is to consult the up-to-date, complete regional or global systematic treatments dealing with particular groups of beetles. This kind of sources would provide the best possible picture. The table 12 lists known beetle species counts for Vanuatu, its three major neighboring island groups, and New Guinea as the major source for Melanesian island faunas. Although these data are sketchy, some patterns are evident. Known New Guinean beetle faunas are approximately three times richer the New Caledonian ones, the latter emerging as the most diverse among Melanesian archipelagoes analyzed. There is no clear picture in relative richness regarding Vanuatu, Fiji and Solomons. The relatively lower diversity for the

Table 12: Catalogued diversity of selected Melanesian beetle taxa by islands/archipelagoes. Dash means no representatives of corresponding taxon are listed, n. a. stands for "not available". Espiritu Santo expedition 2006 data not included.

Taxon	Number of Species					Source
	New Guinea	New Caledonia	Solomon Islands	Vanuatu	Fiji	
Anthribidae	177	58	20	9	29	Rheinheimer 2004
Apionidae: Myrmacicelinae and Rhadinocybinae	245	71	-	1	1	Wanat 2001
Chrysomelidae: Alticini	n. a.	22	4	16	41	Samuelson 1973
Chrysomelidae: Cassidini	29	2	7	-	-	Borowiec 1999
Histeridae	160	50	16	9	14	Caterino 2006, 2007; Caterino & Dégallier 2007; Gomy 2007; Gomy & Aberlenc 2006; Kanaar 2003; Mazur 1997, 1999
Hybosoridae (including Ceratocanthidae)	23	2	-	1	-	Ocampo & Ballerio 2006
Hydrophiloidea: Hydrochidae and Hydrophilidae	75	22	8	13	8	Hansen 1999

latter archipelago, evident for every taxon but Anthribidae and Histeridae, apparently is due to less collecting and less material available, given the fact that their proximity to New Guinea and large size and complex nature makes the Solomons a logical candidate to be the richest among these three island groups. The case of Histeridae is illustrative in that respect, since the species count is based on a very recent work of Gomy and Aberlenc which reviews several collections done in the 1990-2000s.

Not all the sources consulted to get these data discriminate between island records within Vanuatu, but Santo does not emerge as a popular collecting destination within the archipelago. The highest proportion of Vanuatuan fauna recorded from Santo is just above the half (Anthribidae, five species out of nine). It is worth of noting that two leaf beetle datasets provide an odd picture in comparison with the rest of data, with New Caledonian fauna being relatively poor for both tribes and the fauna of Solomons being uncharacteristically rich for cassidines and Fijian fauna —for alticines (and also anthribiids).

••• The first impressions from the field

Armed with the results of homework reading, you reach the camp in Santo forests and begin your part in the task of local beetle exploration. With the past experience in tropical rainforests on major landmasses, you are somewhat cautious in your expectations about an oceanic island beetle fauna. But local forest looks lush and tall and diverse, and… beetles are out there. You can observe them easily, catches by standard trapping techniques do not look poor in terms of specimen numbers, and all major groups are present. Phytophages, weevils (Curculionidae in a broad sense) and leaf beetles (Chrysomelidae) are abundant and diverse on vegetation. Cut and damaged trees at campsite attract armies of wood-boring weevils and powderpost beetles (Bostrichidae), and many longhorns (Cerambycidae, Fig. 173) and few buprestids (Buprestidae) are prominent (although not so abundant) on this freshly dead wood. Closer look at that substrate reveals plentiful selection of fungus weevils (Anthribidae, Fig. 174), darkling beetles (Tenebrionidae), fancy-looking brentids (Brenthidae, Fig. 174) and representatives of several smaller families of Cucujoidea and Tenebrionoidea on and under tree bark and on dead wood-infesting fungi. At nights, camplights attract numerous ground beetles (Carabidae), scarabs (Scarabaeoidea) and longhorns, including some huge, very "exotic"-looking representatives of Dynastinae and Prioninae (Figs 175 & 176), as well as literally swarms of bark and ambrosia beetles (Scolytinae and Platypodinae). Small and not so small rove beetles (Staphylinidae) are everywhere, making up the bulk of flight intercept trap catches, and several rains bring out flushes of flying water (Dytiscidae and Hydrophilidae) and

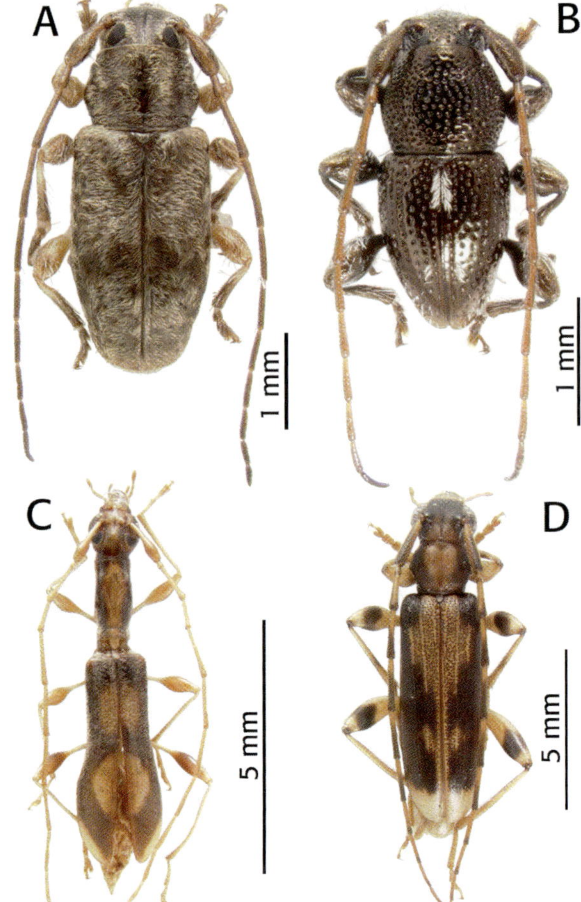

Figure 173: Representatives of longhorn beetles (Cerambycidae). **A**: Unidentified species of *Acanthocinini* (Lamiinae). **B**: *Diastosphya* sp. **C**: unidentified species of a genus near *Tsusuia*. **D**: *Glauscytes* sp. (Photos A. Tishechkin).

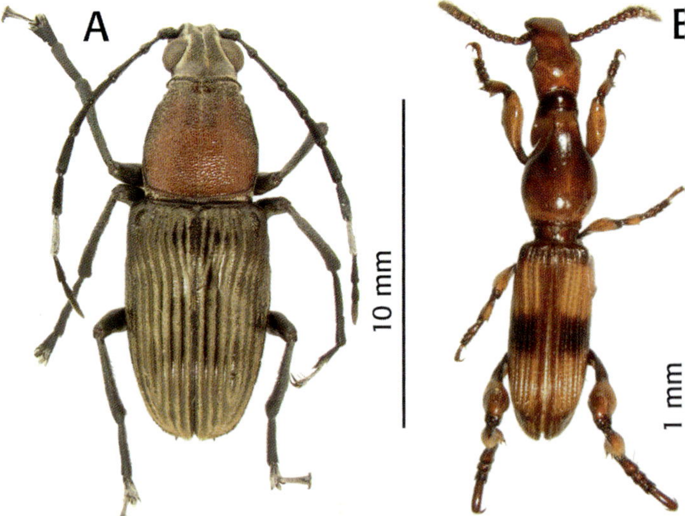

Figure 174: Representatives of Anthribidae (**A**) and Brenthidae (**B**), unidentified species. (Photos A. Tishechkin).

sap (Nitidulidae) beetles. Forest floor leaf litter at different elevations is full of life, with pselaphids (Pselaphinae), scaphidiids (Scaphidiinae), scydmaenids (Scydmaeninae) and other subfamilies of Staphylinidae abundant and diverse, rivaling in numbers ubiquitous ants of various sizes and shapes. And as usually in the tropics, every new day brings multiple discoveries of species not seen before.

Figure 175: *Xylotrupes ulisses* (Scarabaeidae: Dynastinae).

Figure 176: *Olethrius tyrannus* (Cerambycidae: Prioninae).

Initial impression about local beetle fauna does not bring a feeling of an inbalanced or poor fauna, most of its components seem to be in place. However, after closer inspection the picture begins to look different. Leaf beetles are common and rather abundant, but represented by no more than nine species, most of them in Eumolpinae. Dead rotten trunks of fallen trees are beaten by galleries of Passalidae, prominent subtropical and tropical scarabaeoid group in this habitat, but there are only one common large species (*Gonatas hebridalis*) and a rare small one; the latter after a look through

the scope turns out to be a stag beetle, *Figulus foveicollis* (Lucanidae, Fig. 177), the only species we found. Another common group, click beetles (Elateridae), is out, but the first days it looks like only three species (small, larger and medium) are represented in catches until one night several large elaterids with spectacular luminescent spots on the pronotum (*Photophorus bakewelli*) show up at the

Figure 177: *Figulus foveicollis* (Lucanidae). **A**: imago (Photo A. Tishechkin). **B**: larva.

Table 13: Beetle species counts per family at Penaoru camp site and Saratsi Range, November 2006. Columns represent results of dead wood hand collecting (**HC**) at study plots from 100-1 200 m (HC deadwood Saratsi range), miscellaneous hand collecting at camp site 100 m (HC Camp 100 m Penaoru) and total species count (sum HC, not necessary additive, since some species are shared between two categories).

	HC deadwood Saratsi range	HC Camp 100 m a.s.l. Penaoru	sum HC
Anobiidae	1	0	1
Anthicidae	0	1	1
Anthribidae	17	5	21
Bostrichidae	0	3	3
Brenthidae	1	5	6
Buprestidae	0	2	2
Byrrhidae	0	1	1
Callirhipidae	1	1	1
Carabidae	2	9	11
Cerambycidae	16	12	27
Cerylonidae	1	0	1
Chrysomelidae	9	1	9
Ciidae	12	0	12
Cleridae	1	0	1
Coccinellidae	0	1	1
Colydiidae	7	5	9
Corylophidae	1	2	3
Cryptophagidae	1	0	1
Curculionidae	35	24	57
Dermestidae	1	1	2
Elateridae	4	5	7
Erotylidae	2	2	3
Eucnemidae	3	2	5
Histeridae	2	4	4
Hybosoridae	0	1	1
Laemophloeidae	2	3	5
Lucanidae	1	0	1
Lyctidae	0	1	1
Melandryidae	1	0	1
Monommatidae	1	1	1
Monotomidae	0	2	2
Mordellidae	3	4	7
Mycetophagidae	3	4	4
Nitidulidae	4	6	9
Oedemeridae	0	2	2
Passandridae	1	0	1
Phloeostichidae	0	1	1
Propalticidae	1	1	2
Ptinidae	1	0	1
Rhysodidae	2	0	2
Scarabaeidae: Aphodiinae	2	2	3
Scarabaeidae: other subfamilies	2	5	7
Scarabaeidae: Scarabaeinae	0	3	3
Scolytidae (without Platypodinae)	3	5	6
Scolytidae: Platypodinae	4	3	5
Silvanidae	0	3	3
Staphylinidae: other subfamilies	9	4	13
Staphylinidae: Scaphidiinae	6	0	6
Tenebrionidae	15	11	25
Thoscidae	1	0	1
Trogossitidae	1	2	2
Sum	**180**	**150**	**304**

expedition dinner table, making themselves (and explaining beetle experts) a center of the expedition crew's attention. Closer inspection reveals later multiple species of click beetles in above-mentioned categories, but total species count stays nevertheless low. We collected only seven species of click beetles from the camp site and from the dead wood collecting along the Saratsi mountain gradient, namely *Abelater* (*Melanoxanthus*) *hebridanus*, *Abelater* sp., *Agrypnus glirinus*, *Calais carinulatus*, *Photophorus bakewelli*, *Simodactylus buxtoni* and *Simodactylus risbeci*.

··· Good and bad ocean travelers: who's there and who's not

These examples represent a clear pattern: most of the beetle families are present in the Santo's forests, but some only by limited diversity of species (see Tables 13 & 14). Staphylinids, anthribids, bark and ambrosia beetles, weevils and, with some reservations, carabids seem to approach their corresponding dominant diversities in terrestrial mainland communities. The last two families are classical examples of groups prone to successful colonizations and diversifications in remote oceanic islands. But not all habitats may have been colonized successfully by these groups. For example, we could not record one specialized carabid from tree bark, despite thorough sampling efforts, whereas in Australia these are very speciose on bark.

Speaking of beetle groups which were unsuccessful in their dispersion and/ or apparently failed to make it to Santo, very few of the major beetle lineages come to mind. Despite bark and ambrosia beetles being abundant and diverse, one of their prominent specialized predators in Australasia, the genus *Trypeticus* (Histeridae: Trypanaeinae), apparently failed to reach Santo. Our collecting efforts and methods employed would most probably detect these beetles if present at the site and behaving in a typical way. Another example includes Endomychidae, a widespread family of fungivorous beetles, with brighly colored and relatively large representatives, prominent in tropical forest both in Old and New Worlds. The

Table 14: Beetle species counts per family at different altitudes along the transect at Saratsi Range, November 2006. Data were collected by tree trunk bark spray method at 100 m, 300 m, 600 m, 900 m and 1200 m. BS100 = Bark spray plots 100 m, other abbreviations similarly refer to different altitudes.

Family	BS100	BS300	BS600	BS900	BS1200	BS spe sum
Aderidae	0	0	0	0	1	1
Anthribidae	3	1	1	0	0	5
Cantharidae	0	1	0	0	0	1
Carabidae	0	0	1	0	1	2
Chrysomelidae	1	0	3	4	4	7
Ciidae	1	0	0	0	0	1
Coccinellidae	1	1	1	1	1	1
Colydiidae	0	0	0	2	1	3
Corylophidae	1	1	0	2	1	5
Curculionidae	0	1	3	16	1	21
Elateridae	1	0	0	0	1	1
Endomychidae	0	1	0	0	0	1
Lathridiidae	0	0	0	2	0	2
Mordellidae	0	0	0	1	0	1
Nitidulidae	0	1	1	0	3	5
Propalticidae	1	0	0	0	0	1
Ptiliidae	0	0	0	1	1	2
Rhizophagidae	0	0	0	1	0	1
Salpingidae	0	0	0	1	0	1
Staphylinidae: Scaphidiinae	0	0	1	0	1	2
Staphylinidae: Scydmaeninae	0	0	0	1	2	2
Staphylinidae: Pselaphinae	0	0	0	2	6	7
Staphylinidae: other subfamilies	1	1	1	4	4	10
Tenebrionidae	0	0	0	1	1	2
Throscidae	0	0	0	0	1	1
sum	**10**	**8**	**12**	**39**	**30**	**86**

only endomychid record at Saratsi Range was a few specimens of *Holoparamecus*, tiny (2 mm long) brown leaf litter-dwelling beetles, apparently feeding there on some microfungi, found on nearly all elevations. No large, flashy, macro-fungi-feeding endomychids were found despite of extensive special searches. Checkered beetles (Cleridae) appeared to be very rare, only one species, *Omadius santo* (already described as new, Fig. 178), was recorded in numbers on dead wood, a very common habitat for the family. One extra species, *Necrobia* sp., most probably introduced, was observed foraging on drying cow bones at our field kitchen dump. Many others families seem to be with low species numbers (Table 13), but final conclusions should await analysis of other habitat types (e.g. canopy, coastal zone, etc.) and sampling methods.

Some help from humans

Another case of naturally missing beetle lineage has been a subject for a manipulation by the humanity. Oceania does not have any native mammals other than bats, so the lack of dung beetles (Scarabaeinae in strict sense) is a regional faunistic feature related to this lack of food source regardless of dung beetles' dispersal ability per se. The only known representatives in Vanuatu are four established exotic species (*Copris incertus*, *Liatongus militaris*, *Euoniticellus intermedius* and *Digitonthophagus gazella*), the latter three (plus *Sisyphus spinipes* which obviously did not establish) intentionally introduced to cope with the dung of brought and abundant now cows, horses etc. At Penaoru camp site, near the sea level and human settlement, we collected (Table 13) three species of coprophagous Scarabaeinae: *L. militaris*, *D. gazella* and, as a

Figure 178: *Omadius santo* (Cleridae), a new checkered beetle species from Santo. (Photo J. Reibnitz).

surprise, *Onthophagus sagittarius*, not recorded from Vanuatu so far, and unclear in its origin.

• • • Mountains matter

One of the important components of Espiritu Santo's beetle diversity might be the island's mountainous topography. Since climatic conditions and habitats are different along the gradient from coastal plain to high montane cloud forests, one expects to find at least some specialization in respect to elevation in animals, beetles in our case. The extent of such specialization in relatively small and not too high oceanic islands (below 2 000 m) is poorly known, and there is some evidence, at least for birds, that on Pacific islands altitudinal specialization could be quite low. However, several examples from the results of our preliminary specimen sorting and identification at Saratsi Range represent the presence of clear altitudinal limits in distribution of some beetle species. Extensive flight intercept trapping has allowed to uncover rather precise upper limit in distribution of *Phaeochrous* sp. (Hybosoridae). This species is one of the dominants at 100-300 m of

Figure 179: Representatives of Pselaphinae (Staphylinidae), including a species of undescribed myrmecophilous genus (right). (Photos A. Tishechkin).

elevation, a few specimens were collected at 500 m and none at 600 m (where collecting effort was the highest along the transect) and above. Out of nine species of Pselaphinae (Fig. 179), small, low vagile staphylinids associated with leaf litter and dead wood, represented by longer series of specimens to allow at least some quantitative assessment, four are restricted to 1 000-1 200 m, two to 100-300 m of elevation and the rest have broader altitudinal niche between 300 and 1 000 m.

An illustrative reflection of altitudinal changes in Saratsi montane forest ecosystem is the correlated altitudinal shift in beetle composition on tree trunks as portrayed by barkspray sample data (Table 14). These data were collected by spraying about a dozen of tree trunks with arthropod-specific insecticide at several plots along the transect between 100-1 100 m. A total of 86 beetle species was identified from this collecting, and there is a clear diversity increase from dryer forests at 100-600 m to moist forest formations within the cloud zone at 900-1 200 m. Increased epiphyte cover (mosses, lichens and algae) on moist tree trunks presumably provides much better trophic basis for a more speciose foodweb (where beetles and mites play the most important role) than mostly bare bark at the lower altitudes. Mite feeding beetle taxa like Scydmaeninae, Pselaphinae and specialized other groups of Staphylinidae clearly prefer the cloud forest tree trunks, but also "classic" phytophagous groups like leaf beetles and weevils (Fig. 180) do, reflecting the higher food supply there for these plant feeders.

Different collecting methods yielded different species and family compositions. Hand collecting (HC) of dead wood along the Saratsi mountain altitudinal gradient (Table 13) yielded 180 beetle species, and four-week occasional collecting at Camp site Penaoru (100 m a.s.l.) another 150 species, summing up to a total of 304 species. Bark spray (Table 14) added another 86 species, almost not recorded by the hand collecting. Abundant records at campsite lights proofed that many beetles species are attracted by lights, so we suspect light and Malaise trap catches yet to be analyzed will reveal a lot of extra diversity, making estimates of total species richness an interesting future task.

... Much, much more species then we knew

The mentioned preliminary results of partial specimens sorting and identification from different methods shed some light on the quality of the existing knowledge about the beetles of Vanuatu. The general trend is that the available diversity data for this island fauna in literature are quite incomplete. In fact, the only case when observed diversity was lower than reported represents flea beetles, where only one species was found so far, while five are known from Santo. For the rest of groups with some sorting results available the situation is the opposite. Only two species of Scaphidiinae, small, fast-running, compact-bodied fungivorous staphylinids, are reported from Vanuatu, none of them from Santo. We already sorted out six species. At least 16 pselaphine species were collected at Saratsi Range, more than known so far for the entire archipelago of New Caledonia. This statistics however may be more illustrative in respect to the poor knowledge of New Caledonian Pselaphinae.

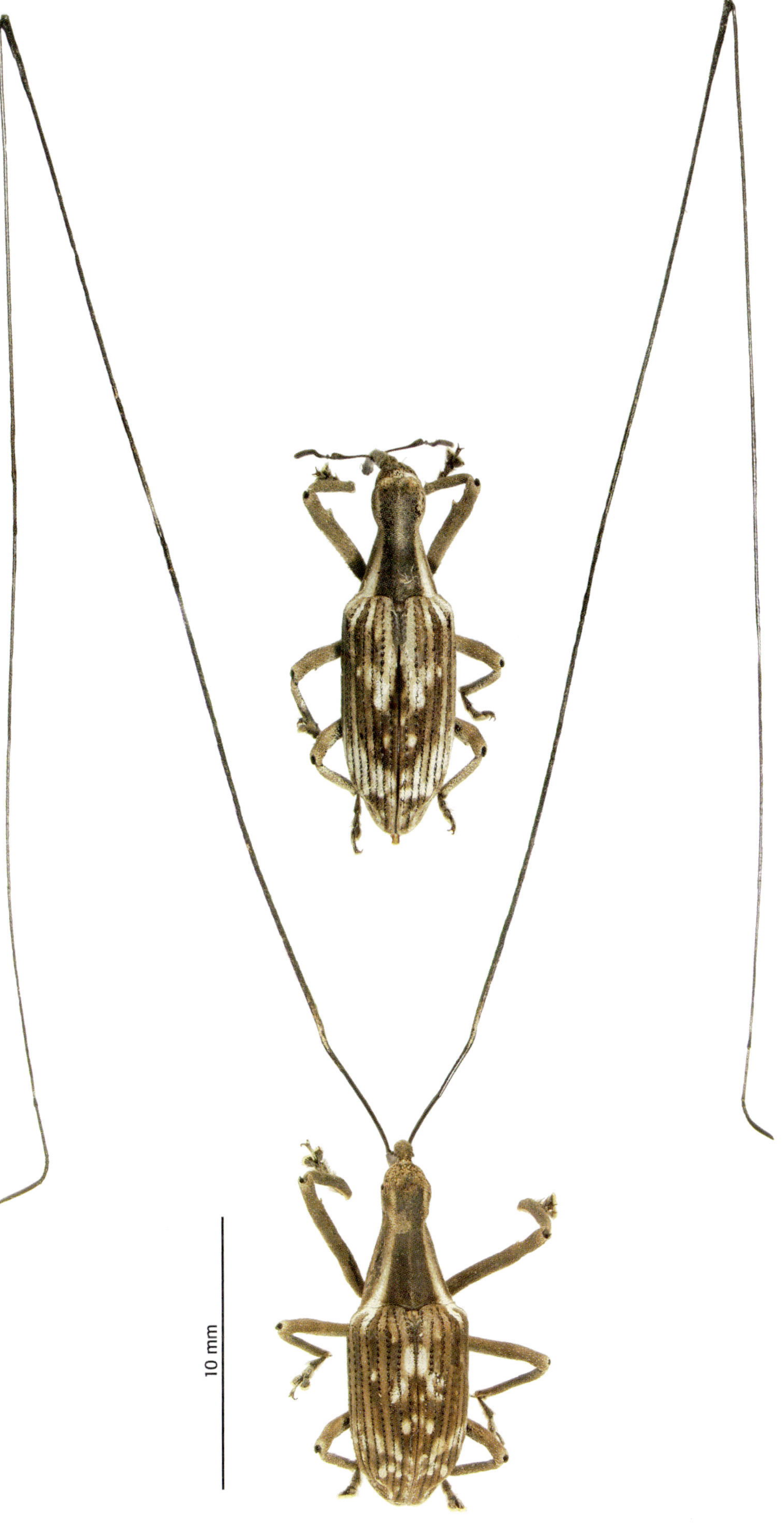

Figure 180: *Mecomastyx* sp. (Curculionidae), attracted to freshly dead wood at 600-900 m, male (with long antennae) and female. (Photos A. Tishechkin).

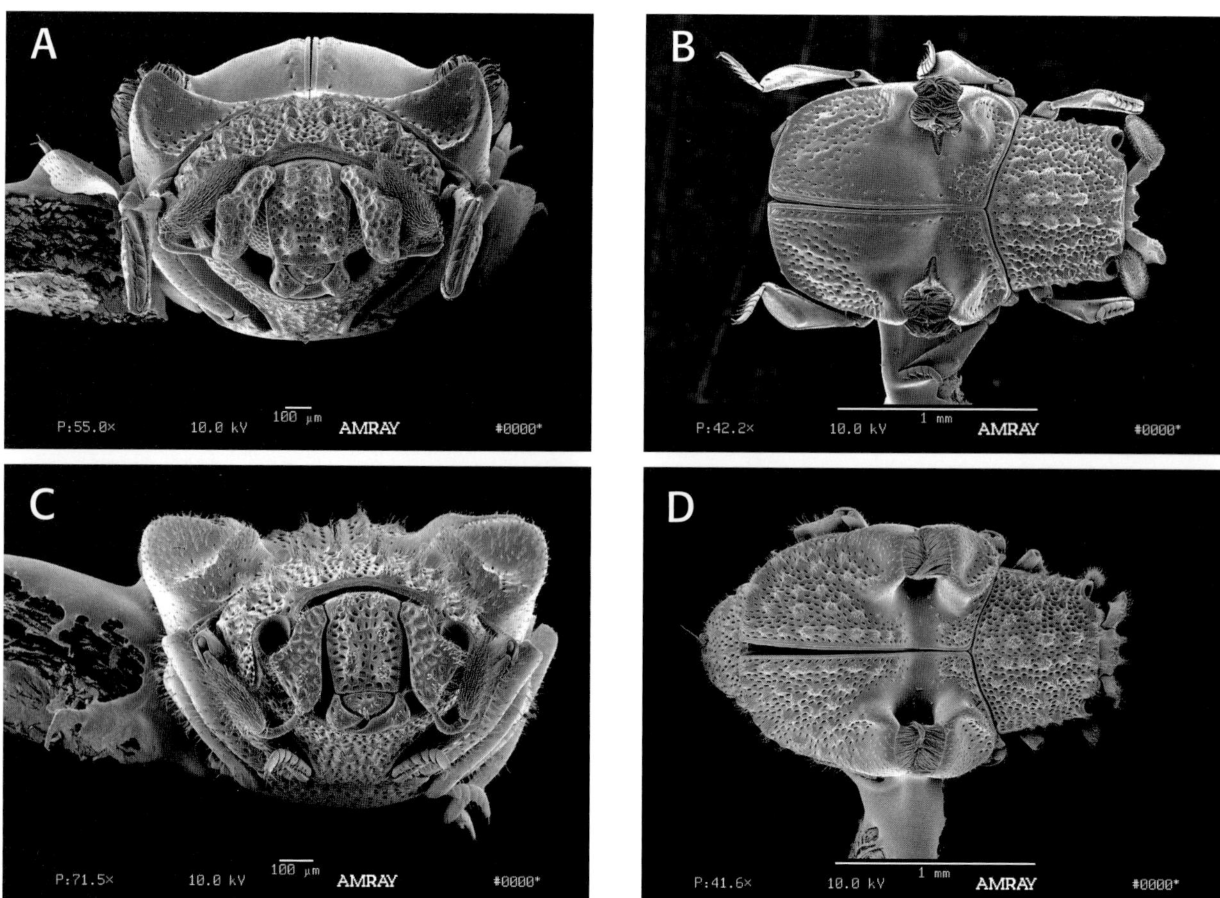

Figure 181: Specialized myrmecophiles. **A-B**: *Eucurtiopsis pascali*. **C-D**: *E. kitchingi*. (Histeridae: Chalmydopsinae). (Photos A. Tishechkin).

Vanuatan ground beetle tribe Platynini has been revised recently by Liebherr as represented by 11 species, five of them recorded on Santo. We identified so far six species, three of which are new island records of decribed species and one —of apparently undescribed one. Twenty one species of anthribids from Saratsi Range and camp site have been sorted so far. The entire Vanuatan fauna is represented by nine species, and only five of them are known from Santo.

The final example includes the Histeridae, a family for which we can provide now the best taxonomic expertise. Histerids are known to be represented in Vanuatu by nine species (Table 12), and at least three are known from Santo (two more is a possibility as island localities for them are not available). We apparently have collected five of these species (although some identification verification is still necessary). One of the undiscovered species belongs to the genus *Bacanius*, tiny dwellers of rotten wood and leaf litter, which are widespread in Oceania and represent a substantial component of otherwise impoverished local faunas. Along with the tribe Acritini, tribe Bacaniini comprise so-called "microhisterids", a group with apparent substantial dispersal and adaptation potential and so an important part of histerid faunas on oceanic islands. The former tribe is also widespread in Oceania, but was not reported in Vanuatu until we found six species

at Saratsi Range. One of those species, *Acritus komai*, has almost cosmopolitan range, with known Oceanian records in Mariannas and Hawaii. The total number of histerid species collected in 2006 on Santo is 23 (25 if two extra species collected by different parties of the expedition are included), nearly three times the number known for the entire Vanuatu before the expedition.

Special example: the case of ant "pets"

Eight new species from the histerid subfamily Chlamydopsinae represent quite exciting discovery. This Australasian subfamily consists of obligate social insect colonies' associates. These beetles live exclusively in colonies of ants (10 genera in four subfamilies) and termites (one genus) and possess unusual defensive and glandular morphological modifications (Fig. 181) associated with their symbiotic ways of life. Until the early 2000's they were considered to be an almost exclusively Australian group, with no more than seven species (out of about 50 in total) were known outside this continent, from Japan, India, Taiwan, New Guinea and Fiji. The recent surge in systematic research on the group during the last decade has increased the known diversity more than three times and revealed the existence (or potential for existence) of rich local faunas on New Guinea, Borneo, Sulawesi and New Caledonia. The discovery and description of Chlamydopsinae in New Caledonia has doubled the size of local histerid fauna, considered prior to this study well known. Our collecting by flight intercept traps at Saratsi has produced eight species in three genera (*Ceratohister*, *Chlamydopsis* and *Eucurtiopsis*), the number currently known for the entire New Guinea. Biogeographic affinities of the genera suggest both New Caledonian and New Guinean (apparently via Solomon Islands) sources of emigration to Vanuatu. Santovian species of *Chlamydopsis* is closely related to the only two, pretty characteristic New Caledonian species of the genus, while two other genera are known from Greater Sundas, suggesting island hopping through New Guinea and Solomons (where representatives of both these genera however remain to be discovered).

This chlamydopsine example is characteristic in several points. The sophisticated symbiotic system of myrmecophilous beetles and their hosts appears to have dispersed to and diversified at a remote Melanesian Island. This discovery also illustrates our poor knowledge not only of the biotas of small remote islands like Santo, but of a major island like New Guinea. Who knows how many big discoveries of poorly known creatures, obscure or not so, await their students in the field or museum storage rooms. Such possibility drives people behind inventories like Santo 2006, united by the goal to add more pieces to the great puzzle of the World's natural history and to preserve the land and its inhabitants for generations to come.

LEPIDOPTERA IN VANUATU: FAUNA, GEOGRAPHY AND THE IBISCA-SANTO PROJECT

Roger L. Kitching

The butterflies and moths, the Lepidoptera, are one of the four great insectan megadiverse Orders. Exact numbers are impossible to predict but there could be as many as a million species world-wide. Further, because of the fascination and efforts of generations of mainly amateur scientists the Lepidoptera are undoubtedly the best known and taxonomically most tractable of all the insect Orders.

This relative wealth of knowledge, however, should not lead to complacency. In many families of Lepidoptera, especially within the so-called micro-Lepidoptera, levels of description and knowledge are very poor with an estimated fewer than 50% currently described in many instances. Even within the three super-families constituting the so-called butterflies (Papilionoidea, Hesperioidea, Hedyloidea) new species are constantly being discovered and described.

The butterflies, as already mentioned, are but a small fraction of this mighty Order, totalling perhaps 100 000 species organised, usually, into six families. These largely day-flying species with knobbed antennae and a generally upright resting posture are usually contrasted with the moths. Moths, mainly nocturnal species with wholly filamentous antennae and a flat, more or less triangular, resting posture, make up the remaining 126 families and dominate the Order both in terms of abundance and species richness.

The Lepidoptera have become widely used as a tool for detecting environmentally-driven change at the community level and have been incorporated into all mixed protocols for detecting beta-diversity. Indeed their only rival in this regard is the Formicidae which, although well known, represent but a single family of the Order Hymenoptera. There are several interconnected reasons supporting the use of Lepidoptera within baseline and monitoring surveys of insect assemblages:

- As already mentioned, compared with all other Orders, the Lepidoptera are well known taxonomically and any survey, particularly of the macro-Lepidoptera may reasonably expect to be able to put species names on many of the morphospecies encountered;
- Identification of most species can be made based on wing and body patterns with dissection required only for confirmation of identity within "difficult" groups;
- More than 99% of Lepidoptera are herbivores (mostly, but by no means exclusively, of living foliage) and a representative sample will reflect, to a greater or lesser extent, the phylogeny and diversity of the plant assemblage within which it is collected;
- A large majority of species of Lepidoptera are attracted to light and large samples can be collected and quantified in this fashion.

For all of these reasons quantitative samples of Lepidoptera based on light trap catches have been a key part of the four IBISCA Projects carried out to date: in Panama, Queensland, Vanuatu and France.

••• The Western Pacific: a biogeography laboratory for Lepidoptera

Biogeography is the science which attempts to understand how the current distributions of animals and plants arose. Its ambit stretches from study of the evolutionary processes of diversification through to the ecological processes of movement and community assembly. As a science it has an old and distinguished history stemming from nineteenth century giants such as Alfred Russel Wallace and his co-worker, Charles Darwin.

Ever since the foundational works of Wallace and Darwin, islands have been a key tool in biogeography. One of the reasons why both Wallace and Darwin came up with the theory of evolution by natural selection independently and contemporaneously was because both had examined and pondered upon the faunas of islands — the Galapagos in Darwin's case and the Sundas in Wallace's. The islands of the Western Pacific — Melanesia, Micronesia and Polynesia — have provided further input to the development of biogeography more recently. Of particular importance in this regard has been the work on bird faunas studied by Jared Diamond, Ernst Mayr and others. Because of the popularity and accessibility of data on birds this proved both profitably and do-able. These authors drew the attention of biogeographers to the importance of island size, geological history and topography and how these interacted with different bird lineages, levels of vagility and dietary specialization to produce the island faunas that we encounter currently.

It was against this background that there arose a new group of biogeographers interested in using the Lepidoptera rather than the birds as their tools. Pre-eminent among these has been J.D. Holloway with many other authors making additional important contributions. This has translated into key monographic works on the biogeography of the Lepidoptera of Fiji, New Caledonia and Norfolk Island. Other authors have restricted themselves both geographically and/or taxonomically. De Jong has written extensively on the Hesperioidea. Tennent has produced key handbooks and checklists on the butterflies (the Solomons 2002; the island arcs 2006). Other handbooks with much

more global remits also provide key resources from which data on the Lepidoptera of the islands can be "mined". Key examples are D'Abrera (*Butterflies of the Australasian region* in 1971), D'Abrera & Hayes (*Hawkmoths of the World* in 1986), Scoble (*Geometridae* in 1999) and Brown (*Tortricidae* in 2005).

In all of this fervour of activity Vanuatu has been relatively neglected. A series of early papers by Butler drew initial attention to the lepidopteran fauna by describing new species and species records from the archipelago. Viette built on these early records with an important series of papers on particular groups. In 1975 Robinson also generated a useful checklist of the lepidopteran fauna as a spin-off from his extensive doctoral studies on Fiji. A Royal Society/Percy Sladen Expedition to the New Hebrides generated much useful biological information and the relationships of the invertebrates are discussed at length by Gross in 1975. That paper also presented species-level checklists for the butterflies and the Heteroptera (Gross's speciality). Throughout this period there has been a trickle of other studies generally drawing attention to species of economic importance. These and other more scattered resources allowed Holloway to discuss key comparisons between the fauna of Vanuatu with that of New Caledonia (and other neighbouring land-masses).

Gross discusses the earlier conjecture that, in invertebrate terms at least, the fauna of Vanuatu is depauperate. In fact Gross shows quite convincingly that when analysed using the standard species/area curves of the theory of island biogeography that this is not the case. Faunistically, he suggests, the islands of Vanuatu contain almost exactly the species richness that more general area and topographic considerations would predict. This conclusion is supported further in a short but insightful paper by Robinson.

Table 15: The Lepidoptera of Vanuatu and adjacent territories. Note "endemic" refers to both species and subspecies. Geographical information extracted from Wikipedia™ and Google Earth™.

	Vanuatu		New Caledonia		Norfolk Island		Fiji	
	Total	Endemics	Total	Endemics	Total	Endemics	Total	Endemics
Total Macro Lepidoptera	364	106 (29%)	444	138 (31%)	98	19 (19%)	395	107 (28%)
Papilionoidea	86	28 (32%)	61	12 (20%)	12	2 (17%)	42	4 (9.5%)
Sphingidae	9	0	18	2 (11%)	6	0	13	1 (8%)
Tineidae	16	8 (50%)	9	5 (55%?)	11	6 (54%)	59	13 (22%)
Total Land Area (km²)	12 189		18 575		34.6		18 274	
Distance (km) to New Guinea	2 685		2 949		3 610		3 510	
Distance (km) to Australia	1 984		1 338		1 428		2 797	
Distance (km) to New Zealand	2 341		1 773		772		2 112	
Highest Point (m)	1 879		1 628		319		1 300	

The Lepidoptera of Vanuatu: what do we know?

Table 15 summarises some of the information available about the Lepidoptera of Vanuatu and compares it with that of nearby insular land-masses — New Caledonia, Fiji and Norfolk Island. Key geographical statistics for each of these islands (or island groups) are also included in the table. Reasonable estimates of total faunas are available only for the macrolepidoptera (that is: the Zygaenoidea, Hesperioidea, Papilionoidea, Geometroidea, Sphingoidea and Noctuoidea plus a number of smaller and less widespread superfamilies). Infraordinal level information is available only for the Papilionoidea, Sphingidae and Tineidae.

Gross notes 53 species of butterflies from Vanuatu, at least one of which he records as a distinct subspecies on different subsets of the archipelago. A very recent check-list by Tennent describes 86 species from Vanuatu including two endemic species and 26 endemic subspecies.

In 1979 Holloway records a total of 364 species of macro-Lepidoptera from Vanuatu based on Robinson. Holloway cautions that these figures are based on only a "preliminary analysis" of Robinson's 1971 collections. Robinson noted that the largest proportion of the non-endemic 259 species had generalised south-western Pacific distributions occurring, variously, in New Caledonia, the Solomons, Fiji, Samoa and/or Tonga. Much smaller numbers of these non-endemics had restricted distributions in which their Vanuatan distributions were shared with only New Caledonia (14 species), the Solomons (eight species) or Australia (three species). Robinson notes that Santo Island, undoubtedly because of its size, had the largest number of endemics. Endemicity of macrolepidoptera in Vanuatu, though, is generally at the level of the single species or with (according to Robinson) few if any of the radiations observed elsewhere in the region. The possible exception to this rule among the macrolepidoptera is the hypenine noctuid genus *Schrankia* which exhibits different species on different islands of the archipelago. Robinson concludes that the fauna of Vanuatu is relatively recent reflecting the youthfulness of the landmass itself. He suggests that Vanuatu has undergone considerable increase in land area through the Quartenary era.

Robinson's analyses, insightful as they are, do not cover any "micro-" groups and these remain poorly known. In his final work, Robinson discusses the global distribution of one family of "micros-" — the Tineidae. He summarizes known species from each of the island groups of the Pacific (and indeed for all other regions of the World). His comments on the fauna of Vanuatu reiterate the very poor level of collecting that has been carried out. However, of the 16 known species eight (50%) are endemic suggesting, as would be expected on first principles, that higher levels of endemism may be expected in the micros than is apparent among the macrolepidoptera.

The IBISCA-Santo project

Background

The IBISCA model of biodiversity research in forests was developed by Yves Basset, Bruno Corbara, Hector Barrios and many other colleagues. In essence the approach relies on a centrally developed experimental design against which a large international team carry out sub-projects targetting a wide range of arthropod taxa and associated ecological processes. To date the three tropical and subtropical IBISCA Projects have been, respectively, in Panama, Queensland (Australia) and Vanuatu (Santo). IBISCA Panama examined the contrasts between the ground zone fauna and the canopy fauna within the San Lorenzo forest reserve on the Caribbean slope of the Republic of Panama. Both the Queensland and Vanuatu projects examined sites at different altitudes along a continuous forest gradient. The last two projects were designed against a goal of identifying which taxa or sets of taxa would be most sensitive to projected climate changes. In particular we hope to identify appropriate future monitoring targets.

In all four IBISCA Projects considerable investment has also been made in collecting spatially and temporally contemporaneous vegetation and other environmental data both as an end in itself and to aid in the interpretation of the arthropod data.

Hypotheses

The Lepidoptera sub-project within the IBISCA-Panama project has focussed on the families of macro-Lepidoptera plus the Pyraloidea and a few other minor families of the larger "micros". In this connection the "macrolepidoptera" was also taken to include other large moths conventionally included with but not being of the macrolepidopteran clade: the Hepialoidea, the Zygaenoidea and Cossoidea.

The following hypotheses have driven the project:
- There will be characteristic assemblages of moths at different altitudes along the sampled transects;
- These characteristic assemblages will mirror, to a greater or lesser extent, the vegetation and land use changes along the transect;
- The levels of local endemism will increase with altitude (reflecting the greater specialisation seen in the vegetaton);
- Different taxa of moths will respond to altitudinal changes in different ways;
- A set of taxa can be identified which can be used as surrogates for the whole assemblage (of Lepidoptera or, perhaps, all herbivorous insects) in future monitoring activities.

Figure 182: The author setting light traps at 1 200 m.

••• Methods

Moths were sampled in November 2006. The trap design was a modified Pennsylvania light-trap in which a vertical, battery driven actinic tube is held above a funnel by three vertical, tranparent vanes (Fig. 182). Moths attracted to the light strike the plastic vanes and drop into the funnel. They are collected in a bucket or bag held beneath the funnel. Resin strips impregnated with the insecticide dichlorvos™ are placed among cardboard fragments within the collecting vessel. These have the advantage of killing moths quickly yet without the active principle evaporating rapidly.

This particular design of trap has been used extensively when the goal is to obtain a local sample of moths representative of the species assemblage flying in the immediate vicinity of the trap. It has been shown by others that even in ideal conditions light traps probably divert insects over a distance of no more than about five to ten metres.

A total of six trap nights was accumulated using three traps over two nights at each site being sampled. Traps were suspended at about 2 m above the

Figure 183: Pinned moths, in the field.

ground and when more than one trap was in operation at the same elevation, traps were located at least 50 m apart and with no line-of-sight between them.

In this fashion the faunas at each of six elevations were sampled: 100 m a.s.l., 300 m a.s.l., 600 m a.s.l.,

Table 16: Preliminary results of moth studies during IBISCA Vanuatu (macromoths + Pyraloidea).

Altitude	100 m	300 m	600 m	900 m	1 200 m	Overall
Total moths caught	662	604	642	1 219	1 747	4 874
Macrolepidoptera Total Abundance	275	316	235	467	783	2 076
Estimated Species Numbers Noctuidae	44	28	31	20	8	129
Estimated Species Numbers Geometridae	9	10	19	15	8	38
Actual Counts:						
Arctiidae - Total Abundance	11	9	29	68	209	326
Arctiidae - Species Richness	3	4	5	10	8	20
Sphingidae - Total Abundance	4	3	11	3	0	21
Sphingidae - Species Richness	1	1	2	3	0	4
Pyraloidea - Total Abundance	387	288	407	752	964	2 798
Pyraloidea - Species Richness (based on sorting 60 % of the entire sample)	77	45	48	47	14	131

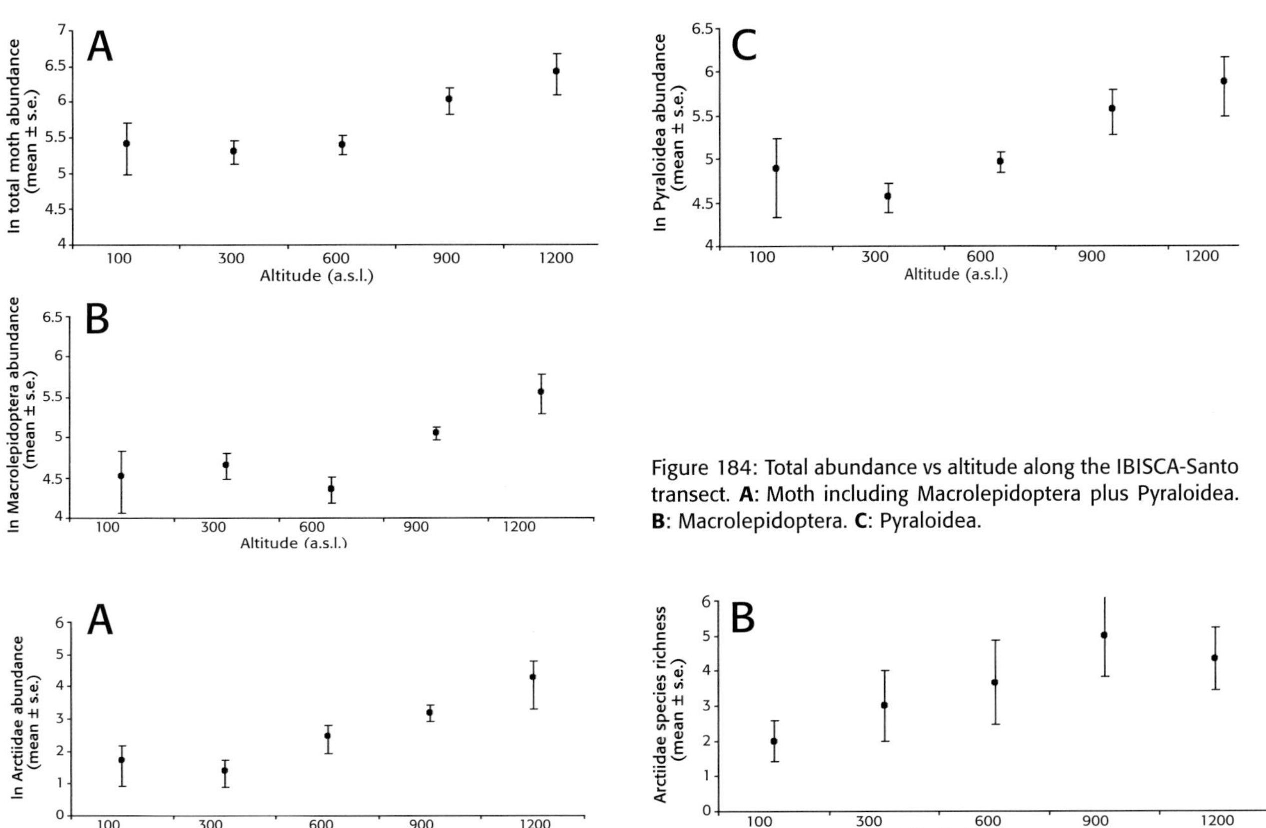

Figure 184: Total abundance vs altitude along the IBISCA-Santo transect. **A**: Moth including Macrolepidoptera plus Pyraloidea. **B**: Macrolepidoptera. **C**: Pyraloidea.

Figure 185: Abundance and richness of Arctiidae vs altitude along the IBISCA-Santo transect. **A**: ln Total abundance. **B**: Species richness.

900 m a.s.l. and 1200 m a.s.l. Traps were emptied daily and collections roughly sorted and pinned in the field (Fig. 183).

These samples were subsequently sorted further and representative examples of the morphospecies used have been set and fully labelled, ready for identification. In line with the protocols used in the IBISCA-Queensland project all moths with forewing length greater than one centimeter were counted plus any smaller Pyraloidea. In general this restricted our target groups to the conventional macrolepidoptera, the Pyraloidea and a few easily recognised "micro-"families such as the Yponomeutidae and Ethmiidae.

••• Preliminary Results and Interpretation
• Composition of the Samples
Identification and analysis is on-going. Identification of the collections made is hindered by the absence of reference works and the scattered nature of useful collections (not to mention the author's location in Australia).

To date total counts of the moths have been made and estimates of the total numbers of species encountered made, based on extrapolations from the substantial fraction of specimens (> 50 %) that have been processed to date. In addition actual totals and species counts have been made for the Sphingidae and Arctiidae. A special focus has been placed on the largely unknown Pyraloidea (Pyralidae and Crambidae). These results are summarized in table 16.

A total of almost 5000 moths was sampled. Of these more than half were Pyraloidea, confirming the relative dominance of this family in general and in the Australasian region in particular. In the relatively well-known Australian fauna the number of species of Pyralidae (*s.l.*) is exceeded only by the much better worked Oecophoridae, Geometridae and Noctuidae. Although the total number of species encountered is yet to be determined, I estimate that there were about 129 species of Noctuidae and 38 species of Geometridae in the Vanuatu samples. Absolute counts of the noctuoid family, Arctiidae, have been made and a total of 20 species was encountered.

Within the Pyraloidea, I estimate the existence of about 131 species within the samples. This is based on an extrapolation from a thorough sorting of about 60 % of the totals sample. Species belonging to some subfamilies, such as the Phycitinae, are virtually impossible to sort to species without dissection and the quality of the samples, after transportation, is such that further certainty as to their identity is unlikely.

• Cross-altitudinal Patterns
Figure 184A presents the average number of moths encountered at each site with the associated standard error plotted on a semi-logarithmic scale against altitude. There is effectively no differences in catch sizes across altitudes 100 m, 300 m and 600 m. Above this level I encountered significantly more moths and this increase continued to the 1200 m

levels (although the differences between means at 900 m vs 1200 m are probably not significant). This pattern is clearly reflected in figures 184B & 184C which break the results down into Pyraloidea (Fig. 184C) and the remainder (Fig. 184B). The sites at 100 m exhibited very high variability reflecting the very disturbed nature of the environment at that altitude as well as the inadequate replication in the sampling regime (as it turned out).

In figures 185A & 185B I have presented comparable results, but exclusively for the family Arctiidae. There is a very clear increasing trend in numbers of Arctiidae from the 100/300 m elevation to the higher elevations reflecting a manyfold increase in numbers within the samples. This is partly reflected in the actual number of species encountered at each elevation (Fig. 185B —note this is not a logarithmic plot) although the trend is broken at the highest elevations. Note that means are low and variability high, yet the trend is clear.

Although our analyses are far from complete there is clearly a cohort of species particularly associated with the high elevation forests. Several species of pyraloid and of arctiid achieve very high numbers within these high elevations and are virtually absent lower down the transect.

• General Discussion
So how may these trends be interpreted?

The simplest interpretation is that there is a naturally occurring gradient from lowland to highland reflecting increased ecological complexity, greater species packing and finer subdivision of host-plants resources (perhaps even a greater variety of food-plants available). If the environment surrounding the transect was undisturbed forest then this might be an appropriate, if puzzling, explanation. Yet it is naïve in the extreme. Fundamentally the lower elevations, from 100 m to as high as 600 m show extensive signs of on-going recent human disturbance. Much of the forest is post-slash-and-burn regrowth and shows all the signs of long-term, on-going disturbance.

There is evidence that the higher elevations also experienced some disturbances relatively recently. Folk memory has it that the villages, including Penaoru (or its predecessors) moved (or were moved) to their present coastal locations only with the arrival and dominance of Christian mores. Members of the IBISCA team noted several old habitation sites high in the mountains. So even the 900 m sites cannot be presented as somehow pristine. The 1200 m sites are so wet and inhospitable to humans that the case for only occasional visitation may be sustainable. The higher elevation sites, whatever their history, now present extensive stands of more or less mature forest.

Accordingly, I interpret the patterns of moth diversity that I encountered as representing the consequences of the extent and history of human disturbance over as much as several centuries. I add the caveat that the fauna at 1200 m probably represents the long-term assemblage characteristic of that altitude and the forest it contains. Again many features of the forest at 900 m can also be seen as representing the pre-human background biodiversity of the island. At lower elevations without much more, currently unavailable, information on the identities and life-histories of the species of moth encountered, I take the fauna encountered to represent substantially modified, anthropogenic assemblages.

The one exception to this rather negative interpretation may be found in the patterns of diversity exhibited by the Arctiidae. The Arctiidae in general is a small, rather specialized family. It is generally divide into three subfamilies, the Arctiinae, Lithosiinae and the Ctenuchiinae (also known as the Syntominae). Of these three subfamilies, the latter two feed almost exclusively on lichens, algae or mosses. Members of the nominate subfamily are generally polyphagous feeding on herbaceous plants. In the Vanuatu samples the numbers of individuals of the arctiids in general increased geometrically from 12 at 100 m progressively to 64 at 900 m and a colossal 218 at 1200 m. Not only that but the percentage of these represented by Lithosiinae and Ctenuchiinae within my samples moved from 18% at 100 m, through 66% and 64% at 300 m and 600 m respectively (all based on small samples) to 81% at 900 m and 93% at 1200 m. The higher elevations were dominated by two species of lithosiines, one a species of *Eilema*. The highest elevations along the IBISCA-Santo transect are converging on true cloud forest with trees literally festooned with mosses and foliose lichens, with bark encrusted with algae, liverworts and crustose lichens. They undoubtedly represent a veritable cornucopia of food resources for Lithosiinae. Such patterns are found in high elevation rainforests elsewhere.

••• **Further Work**
Much further work is needed on the collections already to hand.

That having been said it is also clear that Vanuatu and Santo in particular represent a much under-worked region still, in spite of the best efforts of the IBISCA team. Within the Lepidoptera the large micro families of Pyralidae, Crambidae, Tortricidae and, probably, Oecophoridae (that characteristic family of Australasia) will be the most fertile fields for further study. They would well repay further expeditionary work both from a taxonomic and ecological viewpoint.

What is most needed, though, is the training of a cohort of Vanuatan specialists who can continue the work begun by the IBISCA team and its predecessors.

Invertebrates

DIVERSITY OF SPIDERS

Christine Rollard

Spiders constitute an important group of invertebrates nearing 41 000 known species in the world, and distributed over 109 families according to the latest version of *The World Spider Catalog* (Platnick 2010). Spiders are present on all continents except Antarctica, and are found in all kinds of habitats, at different levels of vegetation, and ranging from the ground, under rocks, among leaves, to the tops of the trees...

The hunting techniques used and the structure of their environment determines the distribution of each individual species of spider. All spiders are carnivorous and either active or passive hunters; their predatory behaviour is highly variable along with the means used, sometimes with a silk trap and sometimes without. Their main prey are insects, on which spiders have a significant impact.

Body sizes ranges from 0.4 mm to about 130 mm and there is much variation in their biology. Spiders are really good models for the study of biodiversity in general and the evaluation of natural habitats in particular.

••• Historical knowledge
Data on arachnids and especially spiders from Vanuatu were old and scattered. Simon in 1897 was the first person to study some material collected on Malicolo and Vanikoro. From 1930, knowledge improved owing to the trips of several naturalists such as Aubert de la Rue and his wife, and M. Risbec.

Slightly less than 100 species have been described from the archipelago and Berland in 1938 drew up a list (Table 17) with 12 cosmopolitan species and 40 present on more than one island. Only 23 species, distributed over 13 families, were identified from Santo of which five species are only found on the island. Berland notes that endemism amounts to 35%, a lower number than that of other archipelagos which are closer to 50%; which should mean that there are more common species in Vanuatu than on other islands elsewhere in the Pacific. This conclusion will be checked by this new collection because it will be interesting to compare this fauna more thoroughly with that of the nearest archipelago — namely New Caledonia — and with the other Pacific Islands. According to Berland, the different parts of the archi-pelago of Vanuatu have more affinity to each other than they have with the surrounding landmasses, and that there really exists a distinct Vanuatu fauna, with the archipelago composing a zoogeograhical unit. This zoogeographical hypothesis needs to be verified.

••• Discovering the spider fauna
Spider specimens were collected by several teams who sampled in diverse sites on the island at different altitudes, in particular the entomologists confined to the base camp at Penaoru. The majority have been collected by the author helped by malacologists and botanists, during one month (November 2006) of itinerant prospecting at four main sites: Tasmate and Penaoru on the west, Matantas and the surroundings of Luganville (south) to Butmas and Port Olry (north-east) as well as the traverse of the Cape Cumberland from west to east. Complementary collections have also been done among other orders of arachnids such as scorpions, opiliones (harvestmen), pseudoscorpions, etc. —groups on which there has hitherto not been a lot of information. This material has already been described or is being studied.

Various sampling methods are necessary for collecting spiders, such as opportunistic collection, sweep netting in low vegetation, beating of trees, extraction or sorting of leaf litter...

The number of species collected during this expedition is still unclear today; however the diversity can be estimated now at 150 morphospecies in just under thirty families in which both principal suborders are represented: the Mygalomorphae and the Araneomorphae.

We cannot list them all in this paper, the majority of the specimens being still under study in collaboration with other colleagues around the world but a few can be presented as part of this work.

••• Mygalomorphae
There are currently 15 families of mygalomorph spiders in the world, comprising around 2 600 species.

Three species were collected on Santo with only a few individuals. No Ctenizidae (trapdoor spiders) were found. Two of the species belong to one of the already known families, the Barychelidae (represented by 300 species with a pantropical distribution), they include

Figure 186: Barychelidae found under a dead trunk on the ground.

an *Encyocrypta* and a new species from the same genus that awaits description, and maybe the male of another genus, *Nihoa*, known only from a female described from Vanuatu. Barychelids (Brushed trapdoor spiders) live in burrows of various forms such as a small silk retreat attached to the underside of fallen trees (Fig. 186) or complex retreats closed with a trapdoor.

The third species, of the genus *Masteria*, belongs to another family not listed above, the Dipluridae (165 species in the world with a wide distribution). Diplurids are ground-dwelling spiders that construct sheet webs with a funnel-shaped retreat in existing crevices, or within the litter in the case of the Masteriinae.

Figure 187: Araneidae of the genus *Arachnura* (scorpion-spider).

The work on these small tarantulas (between 5 to 12 mm) is currently under preparation with P. Maréchal (MNHN).

••• Araneomorphae

The majority of the spiders are araneomorphs, that comprise almost 39 000 species. Among them we find in particular the web spiders that often include species of medium to large size and are thus more easily spotted on the ground. Most of the other species of interest do not exceed 2-3 mm, even though some are very visible at different levels in the vegetation.

On Santo, 22 families had been listed and representatives of most of these were collected during the expedition, and possibly also specimens from three or four others families but the identities of these need to be confirmed.

Five families are currently under investigation with others arachnologists: the Araneidae, Nephilidae, Salticidae, Sparassidae, Tetragnathidae, and some genera of Theridiidae.

••• Araneidae

The best known category are the orb web spiders, for which the web is a very effective passive means of capturing prey that come towards them. Several genera in this family have reduced webs or have entirely abandoned web building. This large family comprises almost 3 000 species occurring worldwide and occupying a wide range of habitats.

A strange spider called the "scorpion spider" —probably from the genus *Arachnura*— has been found during the traverse of the Cape Cumberland from the west to the east; unfortunately the only specimen collected is immature (body length around 10 mm) and we are not able to identify the species (Fig. 187). The prominent tail is somewhat similar to a scorpion, though harmless. The web is constructed near the ground, at an angle or sometimes horizontal. Small flying insects are the usual prey. No species of this genus (found especially in Australia or even in New Caledonia) had previously been reported in Vanuatu.

Another genus found at several locations on the island is *Cyrtophora* or the tent-web spider. Their tent-like, highly complex non-sticky web is sometimes considered a precursor of the simplified orb-web. These webs are aligned horizontally, with a network of supporting threads above them. These spiders often live in colonies. Females usually have a body length of about 10 mm.

The hunting strategy of the group of Gasteracanthinae is different —they wait upside down at the center of the web constructed between trees, and we found several species such as *Gasteracantha* (Fig. 188); they are also called kite spiders because of their

Figure 188: *Gasteracantha* (Araneidae), ventral view of spider at the center of its web.

special form: the abdomen is much wider than it is long and usually has spurs (often black or red) on the sides, rendering this spider unmistakable.

Species in the genus *Argiope* (Fig. 189) often have a prominent abdomen with stripes in alternating colors (yellow, black, white, red…) and spin large webs, usually in tall grass or low bushes. Their size is between 5 to 15 mm. In those species, the web shows a characteristic silk zigzag called the stabilimentum. It is assumed that the role of this conspicuous white silken structure is a protective device that warns birds of the presence of webs in their flight path. Webs endowed with artificial equivalents of stabilimenta tended to survive intact the early morning period when birds are on the wing; unmarked webs showed a high incidence of destruction.

••• Nephilidae

This family of 58 pantropical species contains sticky orb web weavers that occupy many habitats, their webs being characterized by the permanent auxiliary spiral and the cupped hub against the substrate. *Nephila* are famous for their huge aerial webs, and are groups of spiders that can reach 5 cm in body size. They weave a yellowish silk geometrical web from which they derive their name of the golden orb web spiders; they are visible in broad daylight in the centre of their very resistant silk traps, waiting peacefully upside down for the prey that collides with it. *Nephila plumipes* was often seen while travelling around on the island (Fig. 190).

Nephila are common and colonize different habitats ranging from very wet areas such as the edges of forest groves to dry bushes. They are often abundant on buildings, boxes or between poles. They often have company on the outskirts of the web with many small spiders qualified as kleptoparasites such *Argyrodes* (silver dew-drop spiders of the Theridiidae family containing 2 297 species). The study of the genus *Argyrodes* is currently ongoing —these are spiders live in the webs of other spiders, either preying on the host spider or feeding on prey remains in the web.

Figure 189: *Argiope* (Araneidae) on its web bearing a stabilimentum.

Figure 191: Tetragnathidae of the genus *Leucauge* sp. (silver marsh spider).

Figure 190: *Nephila plumipes* (Nephilidae) at the center of its web.

••• Tetragnathidae (947 species)

A number of individuals belonging to this sedentary family have been captured, which generally prefer quite moist habitats. *Tetragnatha*, the long-jawed water spiders, make typical wagon-wheel webs constructed near or above a stream, usually set horizontally. The spider hangs upside-down on the web with its first and second pair of legs stretched forward. Unlike many of the other orb weavers, tetragnathids do not stay permanently on the web and may be found at rest on grass stems or twigs in their characteristic lying position. Other genera such as *Leucauge* (Fig. 191) are often found a short distance away from the web where they hide by stretching out along a grass stem. There are usually located near fresh water.

So far we have only dealt with families of web-bound spiders but many araneomorphs are ground-dwelling or plant-dwelling. The wandering spiders do not build webs and they are very numerous in the tropics. Some families have few species such as the Oonopidae, Pisauridae, Miturgidae but other families that we observed on Santo have many species.

••• Sparassidae (Fig. 192) (1 090 species)

These large (15-35 mm) wandering crab spiders are free-running, living either in built-up areas (many specimens are found on trees close to village homes) or under bark, stones, in crevices of rocks, or at the base of low vegetation. They have long robust legs, turned outwards in crab-like fashion. They are nocturnal in the wild, coming out at night to hunt. In houses, they may also be active during the day, and are most often seen indoors on walls, attracted by the insects that are brought in by the light. Resting with their long legs outstretched, they will move away at great speed if disturbed.

••• Salticidae

This family is the world's largest with more than 5 000 described species. Jumping spiders are diurnal, and apparently have the best vision of all the hunting spiders. Most genera are smaller than 5 mm and often very colourful. They are active, moving around in quick darts or long leaps. They also occupy a wide variety of habitats. They are free-roaming hunters that "jump" onto their prey. For this they require the stereoscopic vision given them by the large median eyes. Some species mimic ants with which they may or may not live. Such spiders are called myrmecomorphs; holding their first pair of legs aloft to mimic feelers, they take on the coloration and "look" of the ants and may even mimic the ants' typical searching gait. They do not spin a web but do make a sac-like nest in the vegetation or under leaves or stones.

The study of this family of jumping spiders is already well advanced under the responsibility of B. Duhem (MNHN). Around 90 % of the species reported by Berland in 1938 for the archipelago have been recollected. Twenty species have been identified at the moment, with maybe three or four new species for Santo and two new genera for Vanuatu.

These new data therefore considerably increase our knowledge of this zoological group with some species having a wide distribution and others being much less common or even endemic. Overall diversity seems to be high, but a large number of species have been found in very small quantities and not always at the adult stage, so determining them will not be easy.

Figure 192: Sparassidae female on the ground carrying a cocoon.

Table 17: Updated list of species from Vanuatu by Berland (1938).
∗ on Santo. ∘ on Santo and other islands of the archipelago.

Name currently used according to Platnick	Name indicated in Berland
Mygalomorphae	
Barychelidae	
Encyocrypta meleagris Simon, 1889	Idiommata meleagris Simon, 1889
Ctenizidae	
∘ Conothele hebridisiana Berland, 1938	Conothele hebridisianus Berland, 1938
Araneomorphae	
Araneidae	**Argiopidae**
Araneus aubertarum Berland, 1938	
Araneus finneganae Berland, 1938	
∘ Araneus neocaledonicus Berland, 1924	
Argiope appensa (Walckenaer, 1841)	Argiope crenulata Doleschall, 1857
∘ Argiope picta L. Koch, 1871	
Argiope trifasciata (Forsskal, 1775)	
Cyclosa litoralis (L. Koch, 1867)	
∘ Cyrtophora moluccensis (Doleschall, 1857)	
Eriophora transmarina (Keyserling, 1865)	Araneus capitalis L. Koch, 1871 & Araneus productus L. Koch, 1867
Gasteracantha hebridisia Butler, 1871	
Gasteracantha regalis Butler, 1873	
Gasteracantha silvestris Simon, 1877	
Gasteracantha westringi Keyserling, 1864	
Larinia phthisica (L. Koch, 1871)	
Neoscona theisi (Walckenaer, 1841)	Araneus theisi (Walckenaer, 1841)
Neoscona nautica (L. Koch, 1875)	Araneus nauticus (L. Koch, 1875)
∗ Poecilopachys australasia (Griffith & Pidgeon, 1833)	Cyrtarachne bispinosa Keyserling, 1865
Clubionidae	
Clubiona alveolata L. Koch, 1873	
Clubiona hystrix Berland, 1938	
Corinnidae	**Clubionidae**
Oedignatha canaca Berland, 1938	
Desidae	**Amaurobiidae**
Epimecinus sp.	
Filistatidae	
∗ Pritha bakeri (Berland, 1938)	Filistata bakeri (Berland, 1938)
Gnaphosidae	**Drassidae**
Anzacia insulana (Rainbow, 1902)	Leptodrassus insulanus (Rainbow, 1902)
Linyphiidae	**Argiopidae**
Laetesia oceaniae (Berland, 1938)	Bathyphantes oceaniae (Berland, 1938)
Lycosidae	
Artoria berenice (L. Koch, 1877)	Lycosa ambrymiana Berland, 1938

Name currently used according to Platnick	Name indicated in Berland
Lycosa caenosa Rainbow, 1889	
Lycosa konei Berland, 1938	
∘ *Lycosa tanna* (Strand, 1913)	
Lycosella hebridisiana Berland, 1938	
Lycosella minima Berland, 1938	
Miturgidae	**Clubionidae**
Cheiracanthium longimanum L. Koch, 1873	
Cheiracanthium mordax L. Koch, 1866	*Cheiracanthium gilvum* L. Koch, 1866
Systaria insulana (Rainbow, 1902)	*Hebrithele longicauda* Berland, 1938
Nephilidae	**Argiopidae**
Nephila maculata (Fabricius, 1793)	
∘ *Nephila plumipes* (Latreille, 1804)	
Oecobiidae	
Oecobius annulipes Lucas, 1846	
Oonopidae	
Xestaspis loricata (L. Koch, 1873)	*Gamasomorpha loricata*
Pholcidae	
Artema atlanta Walckenaer, 1837	*Artema mauriciana* Walckenaer, 1837
∘ *Pholcus ancoralis* L. Koch, 1865	
Smeringopus pallidus (Blackwall, 1858)	*Smeringopus elongatus* Vinson, 1863
Pisauridae	
* *Dolomedes naja* Berland, 1938	
Dolomedes titan Berland, 1924	
Salticidae	
Araneotanna ornatipes Berland, 1938	*Tanna ornatipes* Berland, 1938
Ascyltus pterygodes (L. Koch, 1865)	
Athamas whitmeei O.P.Cambridge, 1877	*Athamas univittatus* Berland, 1938
Bavia aericeps Simon, 1877	
Bavia sexpunctata (Doleschall, 1859)	*Bavia dulcinervis* L. Koch, 1879
∘ *Cosmophasis chlorophthalma* (Simon, 1898)	
* *Cosmophasis risbeci* Berland, 1938	
Cytaea fibula Berland, 1938	
Cytaea flavolineata Berland, 1938	
Efate albobicinctus Berland, 1938	
Hasarius adansoni (Audouin, 1826)	
Jotus arci-pluvii Peckham (?)	
∘ *Lycidas nigriceps* (Keyserling, 1882)	*Saitis nigriceps* Simon, 1901
Marengo bilineata Peckham in Rainbow, 1902	
Menemerus bivittatus (Dufour, 1831)	

Name currently used according to Platnick	Name indicated in Berland
∘ *Mollika microphthalma* L. Koch, 1881	
Muziris wiehlei Berland, 1938	
Plexippus paykulli (Audouin, 1826)	
Saitis auberti Berland, 1938	
Sobasina amoenula Simon, 1898	
* *Tatari multispinosus* Berland, 1938	
Scytodidae	**Sicariidae**
Dictis striatipes L.Koch, 1872	*Scytodes striatipes* L.Koch, 1872
∘ *Scytodes lugubris* (Thorell, 1887)	
Sparassidae	
Heteropoda nobilis (L. Koch, 1875)	
Heteropoda venatoria (Linnaeus, 1767)	
Prychia gracilis L. Koch, 1875	
Tetragnathidae	**Argiopidae**
Leucauge celebesiana (Walckenaer, 1842)	
∘ *Leucauge granulata* (Walckenaer, 1841)	
Leucauge hebridisiana Berland, 1938	
∘ *Opadometa grata* (Guérin, 1838)	*Leucauge grata* (Guérin, 1838)
Tetragnatha keyserlingi Simon, 1890	
Tetragnatha panopea L. Koch, 1872	
Tetragnatha protensa Walckenaer, 1841	
Tylorida striata (Thorell, 1877)	
Theridiidae	
∘ *Argyrodes miniatus* Dol.	
∘ *Argyrodes neocaledonicus* Berland, 1924	
Argyrodes samoensis O.P.Cambridge, 1880	
Argyrodes miniaceus (Doleschall, 1857)	*Argyrodes miniatus* & *Argyrodes walkeri* Rainbow, 1902
Dipoena pacificana Berland, 1938	
Nesticodes rufipes (Lucas, 1846)	*Theridion rufipes* (Lucas, 1846)
Steatoda atrocyanea (Simon, 1880)	*Lithyphantes atrocyaneus* (Simon, 1880)
Theridion epiense Berland, 1938	
Theridion hebridisianum Berland, 1938	
Theridion ondulatum Berland, 1938	
∘ *Theridion piriforme* Berland, 1938	
Theridion setosum L. Koch, 1872	
Thomisidae	
∘ *Diaea bipunctata* Rainbow, 1902	
Ozyptila heterophthalma Berland, 1938	
Stephanopis cheesmanae Berland, 1938	

Name currently used according to Platnick	Name indicated in Berland
Uloboridae	
Daramulunia gibbosa (L. Koch, 1872)	*Uloborus gibbosus* (L. Koch, 1872)
Daramulunia tenella (L. Koch, 1872)	*Uloborus bistriatus* (L. Koch, 1872)
Tangaroa dissimilis (Berland, 1924)	*Uloborus dissimilis* (Berland, 1924)

Name currently used according to Platnick	Name indicated in Berland
Zozis geniculatus (Olivier, 1789)	*Uloborus geniculatus* (Olivier, 1789)
Zodariidae	
Storena lesserti Berland, 1938	
○ *Storena parvula* Berland, 1938	

SOME ARTHROPODS
AS EXPRESSED IN THE WORDS OF PENAORU VILLAGERS

Bruno Corbara

During the 45 days I worked in close contact with the inhabitants of Penaoru, some of them shared with me what is really just a small part of their vast knowledge of the biodiversity of their forest. They provided this information either in response to a question I had asked, or spontaneously like, for instance, when I was invited to eat the small, yellow fruit of the "*nakatambol*" tree (*Dracontomelon vitiense*, Anacardiaceae) found on the path leading to our study sites. The minutes of these informal conversations — that were held in the only language we had in common, basic English— have no real scientific value (indeed, I am not an ethnologist). They only are a testimony. Below is a small collection of some of these trivia directly or indirectly related to forest arthropods.

This information reminds the tropical biologist and naturalist that I am that our guests — unsurprisingly— know their natural environment very well (and not only from a utilitarian perspective), and that they have a very different perception of it than the Western, scientifically-biased one. And as the French anthropologist Philippe Descola —author of the recent book *Par-delà nature et culture*— stresses, I tend to consider that, in Vanuatu or elsewhere: "*la préservation de la biodiversité ne pourra être pleinement efficace que si elle tient compte de cette pluralité des intelligences de la nature*" ("the preservation of biodiversity will be totally efficacious only when it takes into account this plurality of ways to understand nature"; translation mine).

• • • Aranea

In Penaoru, the word for spider is "*parr'*", and "*tepuarr'*" means "*spider web*". The word "*parr'*" is also used to designate a hairy, urticant caterpillar which is usually seen hanging to a silk thread and which is known as "*parr'tutun'*", as "*tutun'*" means 'hot' or 'burning'. This caterpillar was really not considered a lepidoptera for when I asked if it would become a butterfly, the answer was "No, not this one".

According to my informants, some very solid and sticky spider webs, especially those that are spun in the fork of a tree branch, are sometimes used for a peculiar kind of fishing. The fork has to be cut at its three ends without destroying the spider web. This unusual device, which might look like a primitive tennis racket, has to be held parallel and very close to the surface of the water; some grated pieces of coconut are spread onto the upper surface of the spider web. The fish, which is attracted by the coconut, tries to catch the bait and consequently sticks its mouth on the spider web; then, the fisherman only has to lift up the device to catch the fish. Unfortunately, I was unable to see this demonstrated as we did not find the appropriate spider species.

• • • Isoptera

In Penaoru, termites are called "*vi'ô*" and the carton nest of some species "*pâtut'*". The nest of one common species, *Nasutitermes novarumhebridarum* (Termitidae), was present near our base camp. While opening it, my informants told me that there are different inhabitants inside of the termite nest. They could distinguish the "*lot'lot'*", which are numerous, white and tiny, from the "*viô u'*" which are fat and have a black head, and the "*panis'*" which are brown and winged. These different categories clearly fit with the notion of caste polymorphism which distinguishes workers, soldiers and sexuals.

• • • Diptera

One unidentified species of Bibionidae was common at the base camp and near the Penaoru River, flying around people at dusk. The villagers called it "*talman'oro*" which means the spirit of the flying fox ("*oro*" is the word for "flying fox" and "*talman'*" is the spirit). Indeed the shape of the head and the colour of this fly are reminiscent of the fruit bats on Santo.

• • • Hymenoptera

There is only one species of social wasp in Santo, *Polistes olivaceus*, which is locally known for its vicious sting, and which is called "*o'net*" (derived from the English word, "hornet"; should the use of an anglicized word for this species be seen in light of the fact that this invasive wasp probably only recently colonized the island?). Their nests (Fig. 193) are called "*nukun' o'net*". "*Nukun'*" is a generic word for "nest" ("*Nukun'man'*" is also the correct way to say "bird nest", as "*man*" means "bird") and "*nukun' u'u'*" means "ant nest".

Mud wasp nests were very common on the rocks along the Penaoru River, especially near our base camp (Fig. 194). They belong to the Sphecidae *Sceliphron laetum* (the common Pacific mud dauber) known locally as "*ururr'ui*". In the case of these mud daubers, the word that designates the nest is different; people do not use "*nukun*", but "*puêan*" which means "bed". The explanation given was that "*puêan*" is the place where the "*ururr'ui*" sleep; the "*ururr'ui*" correspond in fact to the very inhabitants of the animal construction: the larvae and the ("sleeping") nymphs of the sphecid wasp. According to my informants, there are two differents "*ururr'ui*"; some are brown and others reddish. This is consistent with the fact that the mud nests are sometimes occupied by other solitary wasps. Indeed, Claire Villemant reports to have seen the potter wasp (Eumenidae) *Anterhynchium alecto* reusing the vacant cells of *S. laetum*. *A. alecto* is reddish in color.

••• Ants (Hymenoptera, Formicidae) and ant-plants

Ants are known locally as "*u-u*"; their nests, consistent with the above-mentioned term, are generally called "*nukun'u-u*". The inhabitants of Penaoru —who live near the sea shore— know the myrmecophytic Rubiacea *Hydnophytum* well, although in

Figure 193: "*Nukun' o'net*", the nest of the polistine wasp, *Polistes olivaceus*. Note that the pedicel is covered with a glossy, dark substance which acts as an ant repellent.

this area the epiphyte does not seem to occur at altitudes lower than 500 m a.s.l., nor does it seem to grow on low branches. Two local terms were given to me to refer to *Hydnophytum*: one, "*nukun' u-u*" for ant nest, and more specifically "*u-u missal*", with "*missal*" corresponding to "something to eat"; the meaning here being that the ants eat the flesh of the plant tuber which consequently is hollowed out (this information is based on interviews with seven villagers; five males, and two females). One particular use of these epiphytes —other than cutting their flowers to decorate houses— was also exposed: in the past, in case of a major conflict ("*jealousy*") between people, *Hydnophytum* tubers were used as a kind of "*poison*". They were hidden in the irrigation canals on the taro plantation (*Colocasia esculanta*, Araceae) cultivated by the targeted victim; consequently, the taro tubers progressively acquired a very bad taste which seemingly rendered them inedible.

Figure 194: Locally called "*ururr'ui*", the common Pacific mud daubers *Sceliphron laetum* (Sphecidae) build their nests on the rocks along the Penaoru River.

INDIGENOUS LAND SNAILS

Benoît Fontaine, Olivier Gargominy & Vincent Prié

Vanuatu is one of the Pacific countries where the terrestrial malacofauna has received most attention, with works by Alan Solem, published in 1959-1962, that remain the standard reference on the land snails of Vanuatu. A. Solem (who actually never visited Vanuatu himself), compiled and revisited all the previous literature and added results of his own studies of museum collections. As a result, 59 indigenous or cryptogenic land snails species are known from Santo (a cryptogenic species is defined as a species whose status —indigenous or introduced— is undecided), including seven which are restricted to the neighbouring islands of Malo and Aore. Most species reviewed by A. Solem had been collected from the southeastern part of the island, which is the most impacted by agriculture. A handful of species originated from the Tabwemasana range, but almost nothing was known of the mountainous and forested areas of the island (particularly the northern half of Cape Cumberland).

Solem's work updated the somewhat poor knowledge on some groups. Indeed, species are often described twice or even more by researchers who are not aware of previous descriptions: as a result,

several names will turn out to represent a synonym only. The genus *Partula* for instance was given as having 12 species in Santo. According to Solem, five only are valid, and there might be even fewer true biological species. On the other hand, our own sampling revealed species new to science, and several new records for the island.

... Fauna characteristics

Although land snails are famous for their slow motion, some of them have succeeded in reaching the most remote islands by passive dispersal (rafts torn off by hurricanes and migratory birds are hypothetized as possible dispersal agents). The isolation and further evolution of these species, once anchored in a new island, results in a real taxonomic signature. Usually, the more isolated the island is, the highest the level of endemism is. Santo is part of a large archipelago, with several neighbouring islands; the level of endemism of Santo itself (26 species, i.e. 44%) is thus not as high as in more remote islands, although 41 species (69%) are endemic to Vanuatu. The island malacofauna has three Vanuatu endemic supraspecific taxa: the genera *Diplomorpha*, *Pseudosesara* and *Reticharopa*, representing six species.

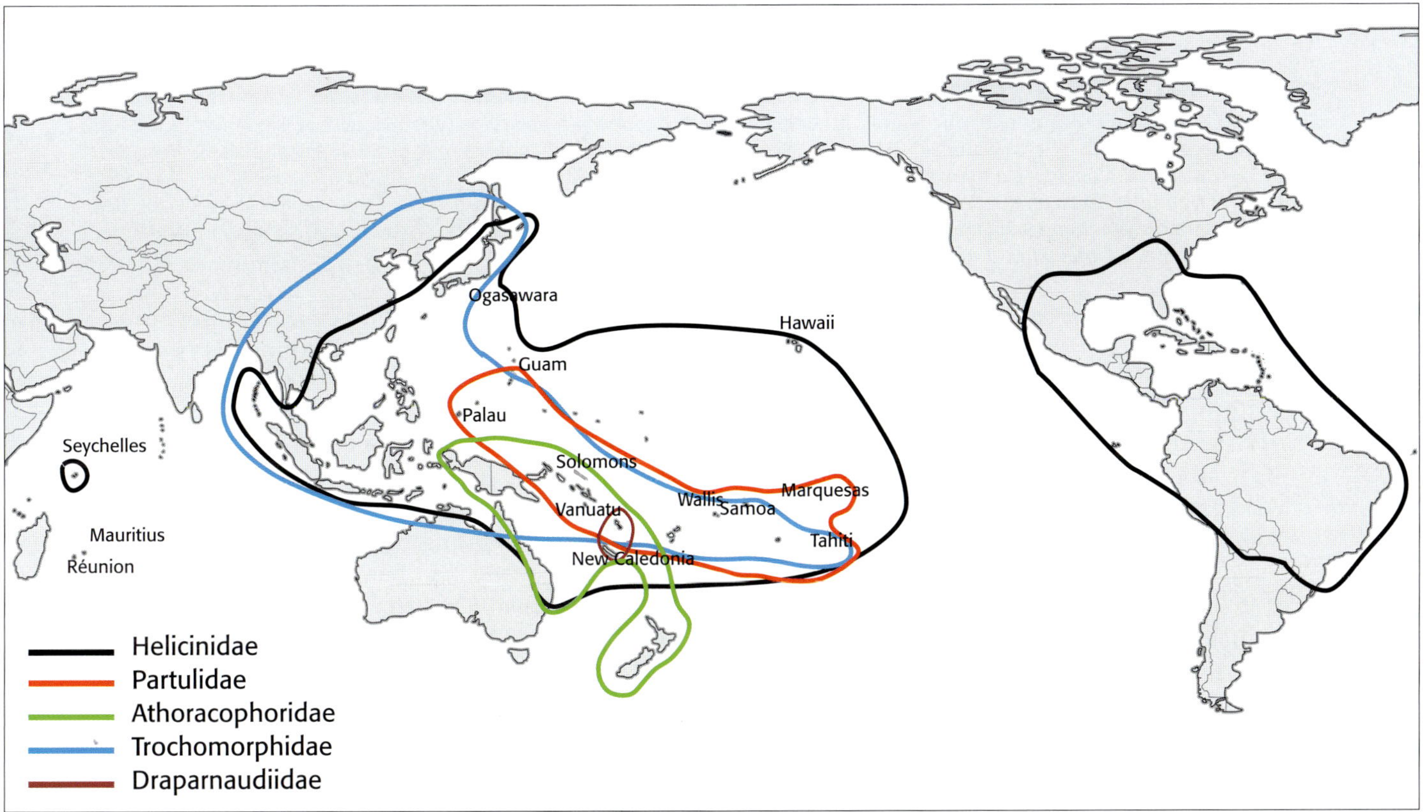

Figure 195: Range of selected indigenous molluscan families found in Santo. Beside large range families (such as Helicinidae), the fauna is at the crossroad of several influences: Polynesian (such as Partulidae), southwestern Pacific (Athoracophoridae), Asiatic (such as Trochomorphidae), New Caledonian (Draparnaudiidae).

Figure 196: Cryptogenic species of Pacific Islands, including Santo. **A**: *Gastrocopta pediculus* (Gastrocoptidae), Linua, Torres Islands. **B**: *Gastrocopta pediculus* (Gastrocoptidae), Matantas, Santo, h=2.5 mm. **C**: *Coneuplecta microconus* (Euconulidae), Matantas, Santo, h=2.8 mm. **D**: *Coneuplecta microconus* (Euconulidae), Sakao Island, Santo. **E**: *Lamellidea pusilla* (Achatinellidae), Matantas, Santo, h=3.0 mm. **F**: *Elasmias apertum* (Achatinellidae), Tegua, Torres Islands, with one brooded juvenile visible inside the last whorl. **G**: *Elasmias apertum* (Achatinellidae), Matantas, Santo, h=2.15 mm. **H**: *Liardetia* cf. *discordiae* (Euconulidae), Matantas, Santo, h=2.15mm. **I**: *Liardetia* cf. *discordiae* (Euconulidae), Butmas, Santo. (Scale bar: 1 mm, x12).

As a result of several colonizing events, the land snail fauna of Vanuatu in general, and Santo in particular, exhibits an interesting mixture of influences from various regions (Fig. 195):

• Western Pacific Islands (particularly Papua New Guinea and Solomons) and more broadly, Asia: families Trochomorphidae, Rhytididae and some genera among Euconulidae;

• Australia, New Zealand and New Caledonia: Placostylidae, Athoracophoridae, Draparnaudiidae;

• Eastern Pacific (Polynesia): Partulidae, Charopidae, Endodontidae and some genera among Euconulidae and Assimineidae.

These three components form relatively equal portions of the fauna. To this clearly indigenous fauna must be added seven broad-range cryptogenic species (Fig. 196).

Endemism within islands exists in Santo. Although some species can be found everywhere, others show a limited range. These range restrictions can result of geology (*Gonatoraphe fornicata*, very abundant on limestone, is completely absent from the western —volcanic— part of the island) or altitude (one species of *Pythia* and another belonging to Euconulidae were only found in the Butmas uplands, *Pseudosesara tabwemasanana* is restricted to the Tabwemasana summit). A striking characteristic of the mollusc fauna in Santo is its poverty in high altitude cloud forests, both in terms of species diversity and of individual abundances. In such habitats above 1 000 m a.s.l., a search by two persons during one hour would typically yield one to three species each represented by one or two individuals, which is very poor compared to other Pacific volcanic islands.

Considering the number of already known species and the species found during Santo 2006 which had not been recorded before (a dozen species), it is probable that more than 80 species should be expected on the island. Alpha diversity (the number of species in a given area) seems to be similar to New Caledonia but lower than in the Solomons: in Santo, a rich station on limestone would yield 20-25 species, or 15-20 species in forest on volcanic soil (vs 30-40 in the Solomons).

••• Conservation

During the 2006 expedition, *c.* 30 indigenous species were found alive, but knowledge on the native fauna is too scarce to assess the conservation status of each single species. In the Cumberland range, the degradation of the forest by introduced plants might be a source of trouble for species living at low altitude. In the eastern and southern parts of the island, the habitat has been deeply modified for agriculture (cattle and coconut plantations), and the fauna has probably been impacted. However, in these areas, there are still many limestone outcrops covered with vegetation that harbour a rich indigenous mollusc fauna, and act as refugium for limestone-dwelling snails. One family seems to have disappeared from Santo, the Partulidae: we did not find a single *Partula* shell, and a survey in villages in Santo and neighbouring islands, showing partulid shells to inhabitants, revealed that they do not know these species, whereas they are well aware of Placostylidae. These extinctions might be linked to the introduction of *Euglandina rosea*, which is well-known for its devastating impact on partulids and achatinellids in Polynesia (see "Focus on alien land snails" in section "Man and nature").

••• Collecting

A visitor familiar with faunas of temperate regions coming to Santo would immediately be struck by the numerous arboreal snails he would find in the forest. Indeed, the Santo fauna, as in most tropical areas, is characterized by families being partly or completely arboreal (Assimineidae, Helicinidae, Placostylidae, Partulidae). They favour trees with large and smooth leaves, and are often found on the underside of the leaves, probably as a protection against predators and heavy rains. Other species may live on the floor, in the soil, under rotten leaves, on rocks or on grass in wet places. Population density and species diversity are usually higher on limestone. However, the local abundance of species depends greatly on microhabitats: damp, hidden places at the base of rocks or trees are often much richer than flat ground a few meters away.

For sampling, once a good habitat and microhabitat have been found, the main problem that should be overcome is the size of these animals: most snail species are minute (less than 5 mm when adult), brownish, and finding them by eye is not easy. Sieving the leaf litter is the solution. The first sieving is done in the field, with a 1 cm mesh Winkler sieve where litter that has accumulated in depressions, below rocks or between buttresses is processed. The fine fraction is stored to be processed later: after having been dried, the leaf litter is sieved once more through several mesh sizes, down to 0.66 mm. The dust which passes through the smaller sieve is discarded, and the remaining fractions are thoroughly searched for snails, with the help of a stereomicroscope for the smaller ones. This allows getting small species in abundance, together with rare species.

However, collecting by hand-picking in the field is still necessary, for several reasons:
- large species are usually not very abundant, and will not be found in the sieving
- for arboreal species, only empty shells are found in the leaf-litter
- as the leaf-litter has to be dried before sieving, the snails are dead by the time they are extracted.

For these reasons, it is necessary to spend at least one hour on a station to look for live specimens, to take pictures, get an idea of their habits and collect live specimens for anatomical and molecular studies.

Litter collecting can be standardized by sampling a standard amount of litter. This allows to assess the relative abundance of snail species and thus characterize communities in different habitats.

During Santo 2006, sampling was done by all these methods (hand picking, standardized and non-standardized litter sieving). Altogether, 96 stations were sampled, including 25 in the most anthropized area (eastern and southern parts of the island), 23 above 500 m a.s.l. (Cumberland, Butmas) including four above 1 000 m a.s.l., and 60 on limestone.

••• Family accounts

Land snail faunas comprise two basic types of snails: those with an operculum (a horny or calcareous plate situated on the tail of the body, which closes the aperture when the animal retracts into the shell), which have the eyes at the base of the tentacles; and those which lack an operculum and have their eyes at the tip of the tentacles. The families occurring in Santo are briefly presented below:

••• **Families with operculum:**
- Assimineidae (Figs 197G & 197H)

The family Assimineidae comprises species smaller than 10 mm, with a shell predominantly conical. Three species are known from Santo, two arboreal, the other one living on the forest floor. They can be very abundant, especially in disturbed ecosystems

Figure 197: Operculate land snail species of Santo. **A**: *Sturanya* cf. *albescens* (Helicinidae), Piarao, Santo. **B**: *Sturanya* cf. *albescens* (Helicinidae), Sakao Island, Santo, d=6.2 mm. **C**: *Pupina brazieri* (Pupinidae), Butmas, Santo. **D**: *Pupina brazieri* (Pupinidae) (Holotype), Erromanga, h=5.7 mm. **E**: *Sturanya sublaevigata* (Helicinidae), Sakao Island, Santo, d=7.2 mm. **F**: *Sturanya* cf. *sublaevigata* (Helicinidae), W Luganville, Santo. **G**: *Omphalotropis conella* (Assimineidae), Sakao Island, Santo, h=5.5 mm. **H**: *Omphalotropis poecila* (Assimineidae), W Luganville, Santo. **I**: *Gonatoraphe fornicata* (Poteriidae), Butmas, Santo, d=8.4 mm. **J**: *Gonatoraphe fornicata* (Poteriidae), Sakao Island, Santo. Note the operculum on the tail. (Scale bar: 5 mm, x8).

Figure 198: Small to minute land snails of Santo. **A**: *Nesopupa* sp. (Vertiginidae), Sakao Island, Santo. **B**: *Nesopupa* sp. (Vertiginidae), Sakao Island, Santo, h=1.8 mm. **C**: *Palaina sykesi* (Diplommatinidae), left: probably a female, h=4.0 mm; right: probably a male, h=3.3 mm, Matantas, Santo. **D**: *Palaina sykesi* (Diplommatinidae), Matantas, Santo. **E**: *Reticharopa* sp.1 (Endodontidae), Matantas, Santo. **F**: *Reticharopa* sp.2 (Endodontidae), Matantas, Santo, d=2.8 mm. **G**: *Reticharopa latecosta* (Endodontidae), Matantas, Santo, d=2.0 mm. **H**: *Reticharopa* sp.3 (Endodontidae), Port Olry, Santo, d=2.0 mm. **I**: *Phrixgnathus glissoni* (Endodontidae), Port Olry, Santo, d=4.3 mm. **J**: *Phrixgnathus* n.sp. (Endodontidae), Port Olry, Santo, d=4.4 mm. **K**: *Phrixgnathus* n.sp. (Endodontidae), Butmas, Santo. (Scale bar: 1 mm, x12).

where they represent up to 75% of the shells in standardized samples.

• Diplommatinidae (Figs 198C & 198D)
Diplommatinids are small to minute snails, with an ornamented shell. In Santo, two species are known. They have a ribbed shell, and have the unusual characteristic of being left-handed: when viewed from the front, the aperture is on the left (in most snail species, it is on the right). They are ground-dwelling species. In both species, shells are distributed in two size class: it is

believed that this represents sexual dimorphism, the larger shells probably belonging to females. They are more common in undisturbed forests, where they account for 20-60% of shells in standardized samples.

• Helicinidae (Figs 197A, 197B, 197E & 197F)
Species belonging to this family share shell characteristics that render them immediately recognizable: a little less than 1 cm in diameter, they are usually dome-shaped, the suture between whorls is poorly marked, and the umbilicus (axis hole at the base of

the shell) is hidden by a heavy callus. They are often brightly colored, pink, yellow or brown, sometimes with darker spiral bands. However, if identifying a snail as a helicinid is easy, the lack of sculpture on the shell and the variability of color patterns make species discrimination very difficult without dissection. Two species are arboreal. They are among the most obvious indigenous land snails in Santo, though they are less abundant than smaller, less visible species.

- Hydrocenidae

If it was not for their vivid orange color, these tiny (less than 2 mm long) land snails would be almost impossible to find by eye on the rocks and dead leaves where they live. However, they can be amazingly abundant in sieved leaf-litter, especially in humid environment (up to 50% of the shells collected in standardized samples). They where unknown from Santo prior to the 2006 Expedition.

- Poteriidae (Figs 197I & 197J)

In Santo, the only member of this family, *Gonatoraphe fornicata*, is unmistakable with its large size (*c.* 1 cm in diameter), flattened shell and round aperture. It is only found in limestone areas, where it can be very abundant. It is a rock and ground dweller.

- Pupinidae (Figs 197C & 197D)

The ovoid, smooth and glossy shell of *Pupina brazieri*, the only member of this family in Santo, is remarkable and unmistakable. A soil-dweller, this small (5-7 mm long) species is often found in numbers in the fallen fronds of palm-trees or tree-ferns.

- Truncatellidae

The sole member of this family in Santo, *Truncatella guerinii*, has an elongated and heavily ribbed shell, which is truncated at the apex in adults (hence the family and genus names). It should be looked for in lowland areas, close to the sea.

••• **Families without operculum**
- Achatinellidae (Figs 196E, 196F & 196G)

This family has radiated all over the Pacific Islands and reaches its maximum diversity in eastern Polynesia. In Western Pacific, the shells are minute to small, thin, conical, and more or less elongated, depending on the species, generally with teeth inside the aperture. Most are arboreal and can be found on the leaves and trunks of shrubs and trees. Both Santo species have a large range in the Pacific ocean and are typically cryptogenic species: they might have been spread all over Pacific Islands by early Melanesians and Polynesians,

Figure 199: Indigenous slugs (Athoracophoridae, *Aneitea* spp.) of Santo. **A**: Ground-dwelling species, Beesel Valley, Cumberland Range, Santo. **B**: Arboreal species, East Coast, Nanda Blue Hole, Santo. **C**: Athoracophoridae can be found in high density in rotten wood, Tasmate, Lalautei ridge, Santo. **D**: Mating, Port Olry, Dionn Island, Santo, and egg laying, Sakao Island, Santo. Note the diversity of color patterns.

Figure 200: Succineidae, Draparnaudiidae and Rhytididae of Santo. **A**: *Succinea kuntziana* (Succineidae), Pwatmel, Santo. **B**: *Succinea kuntziana* (Succineidae), Pwatmel, Santo, h=8.7 mm. **C**: *Draparnaudia walkeri* (Draparnaudiidae), Butmas, Santo. **D**: *Draparnaudia walkeri* (Draparnaudiidae), Beesel Valley, Cumberland Range, Santo, h=7.9 mm. **E**: *Ouagapia santoensis* (Rhytididae), Sakao Island, Santo. **F**: *Ouagapia santoensis* (Rhytididae), Butmas, Santo, d=5.8mm. Note the wide umbilicus. (Scale bar: 5 mm, x8).

or could have reached all these places through passive dispersal without human help.

• Athoracophoridae (Fig. 199)

These large (6-7 cm long) slugs belong to a southwestern Pacific family, ranging from New Zealand and the Australian East coast to Papua New Guinea and New Caledonia. Some species mimic veined leaves, and their color varies from white to dark grey, ground-dwelling species being brownish. In Santo, they are abundant in many areas, even degraded, and often gather in large numbers under the bark of decaying logs. Some species are arboreal, and they seem to be nocturnal, being hidden during the day and wandering on tree trunks at dark. Four species are known from Santo.

• Charopidae

These small ground-dwelling snails have a flat shell, usually with a sculpture of dense radial ribs, somewhat ressembling Endodontidae (Fig. 198). Due to their small size and brownish color, they are mostly found by sieving. One species only is known from Santo, but new species are expected as more material is studied.

• Draparnaudiidae (Figs 200C & 200D)
With only one member in Santo, this family endemic to New Caledonia and Vanuatu is easily recognized, with its medium (6-8 mm long) conical left-handed shell. *Draparnaudia walkeri* lives on the forest floor.

• Ellobiidae (Figs 201H & 201I)
Ellobiidae are predominantly coastal and estuarine snails, living on the high tide mark, or just above it. In Santo, some species live on the shoreline, but a larger one lives in the lowlands and can be found several hundred meters from the sea. Another species was

Figure 201: Euconulidae, Trochomorphidae and Ellobiidae of Santo. **A**: *Dendrotrochus* sp.1 (Euconulidae), Mt Voutmaru, Cumberland Range, Santo. **B**: *Trochomorpha rubens* (Trochomorphidae), Tasmate, Santo. **C**: *Trochomorpha rubens* (Trochomorphidae), Tasmate, Santo, d=12.8 mm. **D**: *Dendrotrochus* sp.2 (Euconulidae), Beesel Valley, Cumberland Range, Santo. **E**: *Dendrotrochus* sp.2 (Euconulidae), Beesel Valley, Cumberland Range, Santo, d=13.1 mm. **F**: *Dendrotrochus layardi* (Euconulidae), Millenium Cave, Santo, h=12.2 mm. **G**: *Dendrotrochus layardi* (Euconulidae), Matantas, Santo. **H**: *Pythia* sp. (Ellobiidae), Sakao Island, Santo, h=27.1 mm. **I**: *Pythia* sp. (Ellobiidae), Tegoa, Torres Islands. (Scale bar 10 mm, x3).

unexpectedly found in the Butmas upland, 600 m a.s.l and probably represents a species new for science.

• Endodontidae (Figs 198E-K)
Endodontidae represent a family which has undergone an important radiation in the Pacific Islands. These small (usually no more than 5 mm in diameter) snails have a flat and rounded shell, often characterized by radial ribs. Due to their small size and dull color, they typically represent the kind of snails that can only be found in large quantity by sieving. They are strictly ground-dwellers, and easily escape snail collectors: as a result, this family is probably a reservoir of species new to science in Vanuatu. Indeed, our samplings revealed a species previously unknown to science which seems to be

Figure 202: Large snails of Santo. **A**: *Placostylus bicolor* (Placostylidae), Pwatmel, Santo, h=25.2 mm. **B**: *Placostylus bicolor* (Placostylidae), old individual, Butmas, Santo, h=43.6 mm. **C**: *Placostylus bicolor* (Placostylidae), Beesel Valley, Cumberland Range, Santo, h=34.4 mm. **D**: *Diplomorpha* sp.1 (Placostylidae) (terrestrial species), Tasmate, Santo, h=34.5 mm. **E**: *Diplomorpha* sp.1 (Placostylidae) (terrestrial species), Tasmate, Santo. **F**: *Diplomorpha delautouri* (Placostylidae), Butmas, Santo, h=23.4 mm. **G**: *Diplomorpha delautouri* (Placostylidae), Dionn Island, Port Olry, Santo. **H**: *Diplomorpha bernieri* (Placostylidae), Butmas, Santo, h=29.1 mm. **I**: *Partula* sp. (Partulidae), Port Olry, Santo, h=25.1 mm. Specimen from museum collections. (Scale bar 10 mm, x2).

widespread on the island, as it was found in Butmas and Port Olry. Around ten species live in Santo, they are mostly found in undisturbed habitats.

• Euconulidae (Figs 196C, 196D, 196H & 196I; Figs 201A & 201D-G)

This large family exhibits a wide array of shapes, from smooth, rounded flat shells to conical and ribbed ones, with or without an angular periphery (keel). Their size ranges from small (4-5 mm diameter) to medium (1 cm diameter), some are ground-dwelling, others are arboreal. Most species have a corneous translucent shell, but some are light colored, whitish to brownish, sometimes with darker spiral bands, and an appendice is generally visible at the end of the tail. A dozen species should be expected on the island. Euconulidae are abundant in most samples, representing 10 to 60 % of collected shells in standardized samples. Some Euconulidae have a restricted range within Santo: a small species, new to science, with characteristic large radial ribs on the shell was only found near Butmas; a large (sub)species was only found on the summital ridge of the Cumberland range.

• Gastrocoptidae (Figs 196A & 196B)

Gastrocopta pediculus is another example of cryptogenic species. It is found all over the Pacific, usually in lowland areas, and can be very abundant on the soil in sandy coastal areas. It has a minute shell, whitish, with a toothed aperture, and is usually only found after sieving the soil and leaf-litter.

• Partulidae (Fig. 202I)

This Pacific Islands family has large (*c.* 2 cm long), elongate oval to conical shells, with the aperture often thickened. They are exclusively arboreal. They are famous for having been decimated in Eastern Polynesia, due to habitat destructions and predation by the introduced *Euglandina rosea*. Indeed, five species were known from Santo and neighbouring islands, but they seem to be extinct now. In Vanuatu, this family still survives at least in the Torres Islands.

• Placostylidae (Figs 202A-H)

In Santo, the five species of Placostylidae are the largest indigenous land snails. This Western Pacific family ranges from northern New Zealand to the Solomons and Fiji. The snails have a thick and elongated shell, and seem to be nocturnal, being hidden under stones and logs during the day. A few species are arboreal. As the shells are quite solid, they remain a long time on the ground, and can still be found in large numbers in degraded areas where no live animals are found nowadays.

• Rhytididae (Figs 200E & 200F)

The Rhytididae are carnivorous, feeding on other small snails or soil invertebrates. Two species belonging to the genus *Ouagapia* are known from Santo. They have a small flat shell and live on the forest floor.

• Succineidae (Figs 200A & 200B)

Succineidae have a peculiar pear-shaped fragile and reduced shell. In Santo, *Succinea kuntziana* is a ground-dwelling species, sometime found on the leaves and stems of low vegetation, and usually not too far from freshwater.

• Trochomorphidae (Figs 201B & 201C)

One species is known from Santo. It has a medium (ca. 1 cm diameter) sized dome-shaped shell, the last whorl keeled, and is often found under logs and dead wood.

• Vertiginidae (Figs 198A & 198B)

This family was previously unknown from Santo. *Nesopupa* spp. are tiny snails, with a toothed aperture, living on the bark of trees trunks and on rocks. Although rare, it was found all over the island, for instance near the village of Tasmate, on the West coast, on the island of Sakao, on the East coast, and near Matantas in Big Bay, suggesting a large distribution on the island. This shows that a widespread species can escape collecting for decades if it is minute enough.

# of Santo

TERRESTRIAL BIRD COMMUNITIES

Nicolas Barré, Thibaut Delsinne & Benoît Fontaine

The avifauna of Vanuatu plays an important role in the understanding of south Pacific biogeography, and has been the subject of attention from scientists and ornithologists since the end of the nineteenth century. Several expeditions during the beginning of the twentieth century improved our knowledge. In 1945 Mayr published the first comprehensive list which was subsequently confirmed by Medway and Marshall in 1975, and more recently by Bregulla in 1992. Santo, the largest (3 959 km²) and highest (1 879 m) island of the archipelago has diverse, mostly forested habitats and has received special attention by all these authors. Thus the objectives of the Santo 2006 expedition were focused more on determining bird distribution, habitat preference, conservation status, and where possible community composition.

... Material and methods

The terrestrial component of the Santo 2006 expedition took place in October-December 2006, the austral spring and the breeding season for most species when the birds are most active, thus making their detection easier. Despite occasional observations of sea, coastal and wetland birds, we focused our attention on terrestrial birds and have restricted this note to these birds. The three authors each spent a total of about one month on Santo, they belonged to three different scientific teams of the Expedition, and worked independently in both the same and different sites on the island. Altogether, 74 days were dedicated to bird record and observation. The survey sites (Table 18 & Fig. 203) were clustered in five classes reflecting a decreasing degree of human disturbance from urban areas to native forests, and increasing elevation, from sea level to the highest sites surveyed at the top of the Cumberland range. The names of the sites are reported, as well as the surveying effort (i.e. number of days spent at each habitat class, number of survey sites and number of surveys for each habitat classes).

Effort was not related to habitat class: nine days in Luganville, a small town of *c.* 3 km length waterfront,

compared with 12 days in the higher mountains. The habitats may be highly diverse especially at low/medium elevation, in sites impacted by humans. Large patches of natural dry/semi-humid forests remain, with secondary forests where logging or clearing has occurred, with Melanesian gardens, old coconut plantations and pastures. This is particularly true at low elevation in the southern part of the island and especially in Luganville and Saraoutou area. Luganville suburbs themselves are immediately in contact, without transition, with these diverse habitats. Small private gardens, open fallow lands, parks and riverside vegetation in the town are replaced rapidly at the periphery by a heterogeneous agricultural landscape dominated by pastures, coconut plantations and patches of forest. Medium elevation (100-300 m) sites are less complex, large pastures and coconut plantations were usually absent or occupied only a limited area. Such sites have considerable habitat diversity, although dominated by different forest types and forest birds can potentially be found

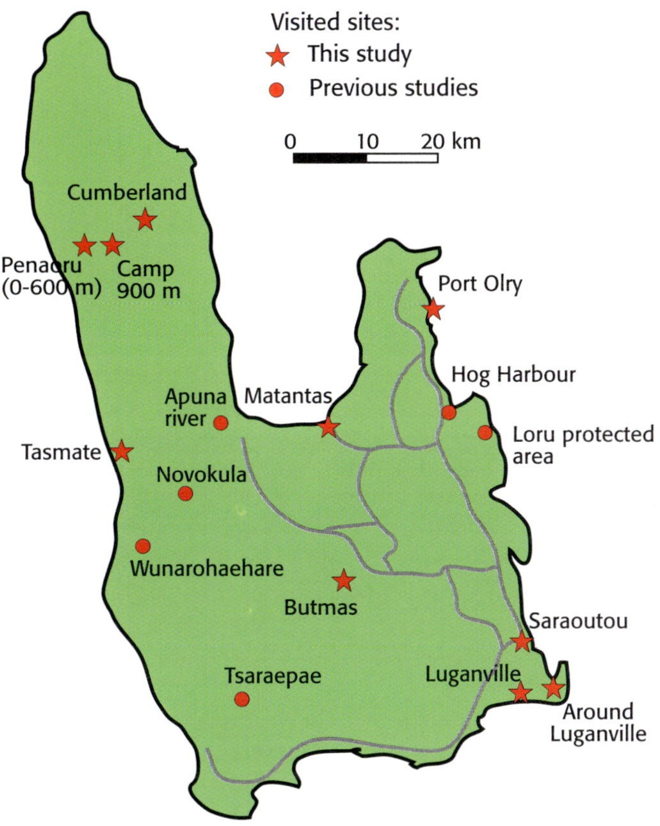

Figure 203: Survey sites of bird communities.

Table 18: Survey sites, description and effort
(Note: Numbers in brackets include surveys by others, in sites in italic characters in the Locality column – refer text)

Habitat Type	Description	Altitude	Days	Sites	Surveys	Localities
Urban site	Urban	Low elevation (0-30 m)	9(9)	1(1)	3(3)	Luganville
Secondary lowland sites	Villages, cultures, pastures, coconut plantations, clearings, secondary forests, coastal dry forests	Low elevation (0-60 m)	31(36?)	5(6)	6(7)	Saraoutou, Tasmate, Port Olry, Matantas, Penaoru, *Loru*
Secondary medium elevation sites	Villages, traditional gardens, clearings, primary (most) and secondary forests	Medium low elevation (100-300 m)	14(22)	2(4)	4(6)	Butmas, Penaoru, *Apuna, Hog harbour*
Primary medium elevation forests	Wet primary forests on slopes and hills	Medium-high elevation (600 m)	8(15)	2(3)	3(4)	Butmas (hills), Penaoru (slopes), *Tsaraepae*
Primary mountain forests	Wet primary mountain forests and ridges	High elevation (900-1 200 m)	12(20)	1(3)	4(6)	Penaoru, Cumberland, *Wunarohaehare Nokovula*
TOTAL			**74(102?)**	**11(17)**	**20(26)**	

in all sites. High elevation habitats (600-1 200 m) consist mainly in well preserved or lightly disturbed forests. Vegetation is described in "Vegetation and Flora" in this book.

In order to increase the data set, we have included data collected, mainly in the southwestern part of Santo, by:
- Medway and Marshall in 1975 who spent four days at each of the following sites: Apuna River (altitude 100 m), Hog Harbor (160 m), Nokovula (1 120 m);
- Kratter and coauthors in 2006 who mist-netted seven days at Tsaraepae (5-700 m) and four days at Wunarohaehare (600-1 200 m);
- Bowen in 1997 who visited the Loru protected area (0-120 m) for several (five?) days (Fig. 203).

Most of these studies were made in September-November. In table 18, we indicate in brackets the total time and the number of sites and prospections for each of the five major habitat types prospected.

Since inventory methods and survey effort between observers were not uniform (Table 18), we record only whether the bird species was observed or not (i.e. presence/absence data) at each survey site. Nevertheless, with the objective of presenting the abundance of the bird species for each habitat class, we calculated the percentage of surveys where the species was recorded.

Except for the Vatthe Conservation Area in the humid forest adjacent to Matantas village (and the Loru Protected Area), none of the study sites we surveyed were under legal or customary protection. Traditional hunting using bows or catapults is in favour among children and men in all tribes visited. Wild birds, especially pigeons and silvereyes, are considered to be an important source of protein and are specifically collected.

Results

Species richness

Forty-five species were recorded during our study (Table 19). This is less than Bregulla who listed 50 species in 1992, but more than Medway and Marshall, and Kratter and coauthors who found 41 species in 1975 and 35 species in 2006 respectively. In 1997 Bowen restricted his survey on low elevation habitats and therefore listed only 25 species (Table 20). The list of Bregulla includes birds which have not been observed by any of the four expeditions (Table 20). These are two native species, the Fan-tailed Cuckoo and the White-rumped Swiftlet and two introduced species, the House Sparrow and the Chestnut-breasted Munia. The first two are well known by one of the authors (N.B.) in New Caledonia and the probability is low that we missed them. Quite intensive and careful surveys were made in Luganville and its vicinity where the two introduced granivorous birds were not seen. They are not established if they were ever introduced and are considered absent.

We add one species to the list of Bregulla: the feral Rock Pigeon which occurs in Luganville (a group of 10 birds established around the market) and in the suburbs. It was not recorded in previous surveys.

Regarding native species, the status of one of them, the Santo Mountain Starling is problematic. This island endemic has not been observed by any of the survey teams despite a total of 20 days being spent by ornithologists in different sites of its preferred habitat, the primary mountain forests (Table 18).

••• Degree of endemism and species on the IUCN (2006) red list

Among native birds recorded by Bregulla in 1992 at Santo, 13 have a wide distribution in the Pacific region, 21 are represented by a specific subspecies in Vanuatu, ten are Vanuatu endemics at the species level (including the Mountain Starling and the Santa Cruz Ground-Dove exclusive of Santo) and one, the Buff-bellied Monarch, at the genus level (Table 19).

Our observations show that four exotic species are naturalised in Santo. They were introduced either by Melanesian settlers during their colonisation of the south Pacific during the last 3 000 years (Red Junglefowl) or by Europeans during the last century (the remaining three species, Table 19). The Common Myna was not recorded in Santo in 1944 by Scott, as quoted by Medway and Marshall, who found the bird widely dispersed in urban, suburban and agricultural areas except in the northern and northeast settled strips in 1971. The Black-headed Munia is believed to have been released in Luganville in 1960. Medway and Marshall saw the bird only in Luganville and Aore, the island in front of Luganville on the other side of the Segond channel, during their 1971 survey. The Rock Pigeon is restricted to Luganville area and has not been reported before. None of these three species were recorded by Kratter and coauthors in 2006. Among the native birds, ten are on the IUCN red list (Table 19): Santa Cruz Ground-Dove (Endangered); Vanuatu Megapode, Baker's Imperial Pigeon, Palm Lorikeet, Chestnut-bellied Kingfisher, Royal Parrotfinch and Mountain Starling (Vulnerable); Thicket Warbler, Rusty-winged Starling and Tanna Fruit-Dove (Near Threatened).

••• Habitats used by bird communities

The 45 species recorded are distributed depending on elevation and habitats. They can be classified according to their site preferences from urban and cultivated habitats at low elevation to undisturbed forests in the Cumberland range. We recognize five groups : lowland birds, ubiquitous birds, ubiquitous birds excluding urban areas, low/medium elevation birds and birds restricted to high elevations (Tables 20 & 21).

• Lowland bird communities

All the sites studied in lowlands (0-300 m) are more or less disturbed and modified by human occupancy and activities. They are distributed from see level to about 300 m (Luganville and villages of Saraoutou, Tasmate, Penaoru, Matantas, Butmas). Five birds, among which three are introduced, are restricted to this level and were not observed in higher sites: natives are the Pacific Swallow and the Swamphen. The Pacific Swallow was recorded breeding in Luganville (two nests) and in the cliffs of Matantas area (where it hunted flying insects with Woodswallows and Glossy and Uniform swiftlets). The three exotics are recently introduced birds: the Rock Pigeon seen in Luganville and the vicinity (two different groups), the Common Myna and the Black-headed Munia. It is in urban and agro-pastoral habitats that these two species are the most abundant. They are particularly common in the harbour site, in gardens and fallow lands of the town and in grassy habitats along roads. The Myna is strongly attracted by cattle and has therefore high densities in all coconut plantations/pastures grazed by cattle, especially at Saraoutou. The Myna was also recorded at Matantas, Butmas (two birds), Penaoru village (one bird) and Tasmate (one bird).
Out of Luganville, the Munia was seen along roads and tracks at Saraoutou, Matantas (large flocks over 50), Butmas (two birds), Penaoru (five), and Port Olry (5-6 birds).
Birds of this lowland group cohabit with the most important group of ubiquitous birds at the lower level of their range.

• Ubiquitous bird communities

This is the most important group, composed of 16 highly adaptable species showing no altidudinal zonation and able to live from urban habitats to mountain forests. Fifteen out of the 16 are natives. However, the aboriginally introduced Junglefowl has the behaviour and the ecological requirements of a native species, restricted like most of them to forest habitats (the bird contacted in Luganville was in a secondary forest and should be the domestic form). This group contains species which favour open cultivated habitats (Woodswallow, Rainbow Lorikeet, swiftlets), secondary scrubs, lowland forests and clearings (Silvereye and White-eye, Gray Fantail, Emerald Dove, Long-tailed Triller, Cardinal Myzomela, Collared Kingfisher), as well as secondary or undisturbed lowlands and premontane forests (Tanna and Red-bellied Fruit Doves, Streaked Fantail, Golden Whister). Most of them are very common at each altitudinal level and each habitat type, including the Tanna Fruit-Dove, the only endemic species of the group. This dove was seen twice in Luganville, and 80% of the surveys from low elevations to premountain forests recorded it (Table 21). Its frequency decreases at higher altitudes where only 50% of the surveys recorded this bird. Despite being common, it is classified as Near Threatened by the IUCN.

• Ubiquitous (excluding urban areas) bird communities

This is also a large group of 14 birds which have the same characteristics than the previous group but avoid urban (Luganville) habitats. They generally show a preference for forest habitats and are

Table 19: Birds observed in Santo by the authors during the Santo 2006 survey or by Medway and Marshall 1975 (**MM**); Bregulla 1992 (**BR**), Kratter *et al.*, 2006 (**KR**). Avian nomenclature follows Howard and Moore, 3d edition, corrigenda 5 (Avibase 2008, Dickinson 2003). Subspecies partly from Bregulla 1992. Bislama names from Anthony Harry, pers. comm. 2006. **Endemism-W**: widespread; **EndGen**, **EndSp**, **EndSSp**: endemic genus, species, subspecies; **Int**: introduced. **IUCN (2006) criteria-NT**: near threatened; **VU**: vulnerable; **EN**: endangered. **H**: high elevations.

Family	Latin name	English name	Bislama name	Endemism	IUCN 2006	MM	BR	KR	Santo 2006
Accipitridae	*Circus approximans*	Swamp Harrier	Big fala hawk, Mala, Namala, Pijin blong fowl	W	-	X	X	X	X
	Falco peregrinus nesiotes	Peregrine Falcon		W	-	X(H)	X	-	X
Megapodiidae	*Megapodius layardi*	Vanuatu Megapode	Skraptak, Scrub duck, Namalu	EndSp	VU	X	X	X	X
Phasianidae	*Gallus gallus*	Red Junglefowl			Int	-	X	X	X	X
Rallidae	*Gallirallus philippensis sethsmithi*	Buff-banded Rail	Nambilak	W	-	X	X	-	X
	Porphyrio porphyrio samoensis	Purple Swamphen	Nambiru	W	-	X	X	-	X
Columbidae	*Ducula bakeri*	Baker's Imperial-Pigeon	Nawemba blong hill, Natutus soraifas	EndSp	VU	X(H)	X(H)	X	X
	Ducula p. pacifica	Pacific Imperial-Pigeon	Nawemba, Nawimba	W	-	X	X	X	X
	Macropygia m. mackinlayi	Mackinlay's Cuckoo-Dove	Long fala tel, Brown pijin	EndSSp	-	X	X	X	X
	Chalcophaps indica sandwichensis	Emerald Dove	Sot leg (Short legs)	W	-	X	X	X	X
	Columba livia	Rock Pigeon		Int	-	-	-	-	X
	Columba vitiensis leopoldii	Metallic Pigeon	Natarua	EndSSp	-	X	X	X	X
	Ptilinopus greyii	Red-bellied Fruit-Dove	Smole fala green pijin, Small green pijin	W	-	X	X	X	X
	Ptilinopus tannensis	Tanna Fruit-Dove	Big fala green pijin	EndSp	NT	X	X	X	X
	Gallicolumba sanctaecrusis	Santa Cruz Ground-Dove	Pimo	EndSp	EN	-	X(H)	-	X
Psittacidae	*Charmosyna palmarum*	Palm Lorikeet	Denga, Dedenga, Maramarei	EndSp	VU	X(H)	X(H)	X	X
	Trichoglossus haematodus massena	Rainbow Lorikeet	Nasiviru, Nasivrou	EndSSp	-	X	X	X	X
Cuculidae	*Chrysococcyx lucidus layardi*	Shining Bronze-Cuckoo		W	-	-	X	X	X
	Cacomantis flabelliformis schistaceigularis	Fan-tailed Cuckoo		EndSSp	-	-	X	X	-
Tytonidae	*Tyto alba delicatula*	Barn Owl	Hoknaet, Hognight	W	-	X	X	-	X
Apodidae	*Aerodramus spodiopygius leucopygius*	White-rumped Swiftlet		EndSSp	-	-	X	-	X
	Collocalia esculenta uropygialis	Glossy Swiftlet		EndSSp	-	X	X	X	X
	Aerodramus v. vanikorensis	Uniform Swiftlet		W	-	X	X	X	X
Alcedinidae	*Todiramphus chloris santoensis*	Collared Kingfisher	Nasiko	W	-	X	X	X	X
	Todiramphus farquhari	Chestnut-bellied Kingfisher	Red Nasiko	EndSp	VU	X	X	X	X
Meliphagidae	*Glycifohia notabilis notabilis*	New Hebrides Honeyeater	Long fala mouth blong hill	EndSp	-	X (H)	X (H)	X	X
	Myzomela cardinalis tenuis	Cardinal Myzomela	Polis	EndSSp	-	X	X	X	X
Acanthizidae	*Gerygone flavolateralis correiae*	Fan-tailed Gerygone	Small nalaklak	EndSSp	-	X	X	X	X
Artamidae	*Artamus leucorhynchus tenuis*	White-breasted Woodswallow		EndSSp	-	X (H)	X	X	X
Campephagidae	*Coracina caledonica thilenii*	Melanesian Cuckoo-shrike		EndSSp	-	X	X	X	X
	Lalage maculosa modesta	Polynesian Triller		EndSSp	-	-	X	X	X
	Lalage leucopyga albiloris	Long-tailed Triller		EndSSp	-	X (H)	X	X	X
Pachycephalidae	*Pachycephala pectoralis intacta*	Golden Whistler		EndSSp	-	X	X	X	X
Rhipiduridae	*Rhipidura albiscapa brenchleyi*	Gray Fantail	Nasiksik	EndSSp	-	- (?)	X	X	X
	Rhipidura verreauxi spilodera	Streaked Fantail		EndSSp	-	X	X	X	X
Monarchidae	*Clytorhynchus pachycephaloides grisescens*	Southern Shrikebill		EndSSp	-	X	BR	XX	X

Family	Scientific name	Common name	Local name	Status	IUCN				
	Myiagra caledonica marina	Melanesian Flycatcher		EndSSp	-	X	X	X	X
	Neolalage banksiana	Buff-bellied Monarch	Zaizari	EndGen	-	X	X	X	X
Petroicidae	*Petroica multicolor ambrynensis*	Scarlet Robin		EndSSp	-	X(H)	X	X	X
Hirundinidae	*Hirundo tahitica*	Pacific Swallow		W	-	X	X	-	X
Sylviidae	*Megalurulus whitneyi whitneyi*	Thicket Warbler	Zizileri	EndSSp	NT	X	X(H)	-	X
Zosteropidae	*Zosterops lateralis vatensis*	Silver-eye	Nalaklak	EndSSp	-	X(H)	X	X	X
	Zosterops flavifrons brevicauda	Yellow-fronted White-eye	Nalaklak	EndSp	-	X	X	X	X
Sturnidae	*Acridotheres tristis*	Common Myna	Pijin blong buluk, Sako	Int	-	X	X	-	X
	Aplonis zelandica rufipennis	Rusty-winged Starling	Woohia	EndSp	NT	X(H)	X(H)	-	-
	Aplonis santovestris	Mountain Starling	Mataweli	EndSp	VU	-	X(H)	-	-
Turdidae	*Turdus poliocephalus vanikorensis*	Island Thrush		W	-	X	X	X	X
Passeridae	*Passer domesticus*	House Sparrow		Int	-	-	X	-	-
Estrildidae	*Erythrura cyaneovirens*	Royal Parrotfinch	Batukira	EndSSp	VU	X(H)	X	-	X
	Lonchura castaneothorax	Chestnut-breasted Munia		Int	-	-	X	-	-
	Lonchura malacca (= *atricapilla*)	Black-headed Munia	Bengali	Int	-	X	X	-	X
TOTAL						**41**	**50**	**35**	**45**

widespread and common from low elevation to upper levels. They include the Melanesian Cuckoo-Shrike, Buff-bellied Monarch, Fan-tailed Gerygone, Mackinlay's Cuckoo-Dove and the quite common Pacific Imperial Pigeon (a group of 20 was seen eating fruits of the "*nakatambol*" (*Dracontomelon vitiense*) at Matantas. Some are less frequently observed at all sites. Amongst these are the Thicket Warbler (IUCN Near Threatened) seen once in dense undergrowth of a clear mature forest at 300 m at Butmas and in mature regrowth at 1120 m in Nokovula but also at low elevation in mesophyllous vine forests at Hog Harbour. If not rare, it is a secretive, inconspicuous bird. At higher elevations the Melanesian Flycatcher (one pair was observed nesting at Penaoru), the Metallic Pigeon, the Swamp Harrier, the Chestnut-bellied Kingfisher (IUCN Vulnerable) and the Shining Bronze-Cuckoo are in this rarely observed category. Two species, the Island Trush (common at Matantas) and the New Hebrides Honeyeater (one seen in degraded forests at sea level at Tasmate) prefer the forests at higher elevations (67% and 83% of the surveys over 900 m recorded them, respectively). In this group, we include the Peregrine Falcon, seen at sea level by us (Saraoutou, Matantas, Tasmate) and Bowen in 1997 (Loru), by us at high altitude at Penaoru (900 m) and by Medway and Marshall in 1975 at 1120 m (Nokovula). It is remarkable that the Harrier appreciates open habitats but also hunts over forests even at high elevations. In this group of 14 species three are endemic at either the genus level (Buff-bellied Monarch) or species level (Chestnut-bellied Kingfisher and New Hebrides Honeyeater). All three are common, at least in one of the habitat types and altitudinal levels, but the Chestnut-bellied Kingfisher is rare at higher elevations.

- ### Low/medium elevation species
Five species are restricted to open (Buff-banded Rail, Barn Owl) or forested (Southern Shrikebill, Santa-Cruz Ground Dove and Vanuatu Megapode) habitats from sea level to the foothill. The endemic Megapode has been seen in the dry forests along the shore at Saraoutou, as well as in the primary undisturbed forests on the coral slopes in the same area. It has been seen and heard also in Matantas forests and is hunted at Butmas (but its dark red meat is not much appreciated). The eggs of this species are also commonly collected, eaten and sold. The other endemic species, the Santa-Cruz Ground Dove is apparently very rare, and is the only Santo bird classified as Endangered by the IUCN. It was not recorded by Medway and Marshall in 1975 nor by Kratter and coauthors in 2006, but was recorded by Bregulla in 1972 (in the highlands at 1000 m and at lower elevations at 300 m) who considered it is scattered in suitable habitats of the mid-mountain forests. This bird was recorded twice (one, then two birds) on the Penarou trail, at 300-400 m near bamboo clumps. This is one of the three rarest terrestrial birds of Santo, and the scarcity of observations did not allow us to identify its ecological requirements and threats.

- ### Species restricted to high elevations
This is a group of seven species (including the Santo Mountain Starling which was not observed by any of the recent ornithological teams), and four endemics at the species level. The endemic Baker's Imperial Pigeon (IUCN Vulnerable) lives in mature forest and has been recorded at 300 m at Butmas, but more regularly at upper elevations up to the mountain summits. It is a shy and elusive bird, possibly because it is a favoured prey species for hunters. The Scarlet Robin seen from 600 m is common at this altitude and may be even more abundant at higher elevations. It lives alone or in small family groups, feeding from 2 m to the upper canopy. The Polynesian Thriller apparently

Table 20: Bird species recorded at sites prospected by the authors during the **Santo 2006** mission or, by Bowen 1997 (**Bow**), Medway and Marshall 1975 (**MM**) and Kratter *et al.*, 2006 (**KR**). Sites are Luganville (**Lug**), Saroutou and vicinity of Luganville (**Sara**), Port Olry (**PO**), Tasmate (**Tas**), Penaoru (**Pen**), Matantas (**Mat**), Butmas (**But**), Loru (**Lor**), Apuna River (**Apu**), Hog Harbor (**Hog**), Tsaraepae (**Tsa**), Wunarohaehare (**Wun**) and Nokovula (**Noko**). Site elevation is precised.

	Urban areas	Anthropogenic low elevations								Anthrop. medium elevations	Primary forest medium/high elevations					Primary forest high elevations	
	Santo 2006	Santo 2006					Bow	MM		Santo 2006			KR		Santo 2006	
	Lug 0-30	Sara 0-50	PO 0-20	Tas 0-200	Pen 0-60	Mat 0-60	Lor 0-120	Apu 100	Hog 160	But 300	Pen 100-300	Pen 600	But 600	Tsa 500-700	Wun 600-1200	Pen 900-1200	Pen 1200-1500
Lowland birds																	
Pacific Swallow	X	-	-	-	-	X	-	-	-	-	-	-	-	-	-	-	-
Rock Pigeon	X	X	-	-	-	-	-	-	-	-	-	-	-	-	-	-	-
Common Myna	X	X	-	X	X	X	-	-	-	X	-	-	-	-	-	-	-
Black-headed Munia	X	X	X	-	X	X	-	-	-	X	-	-	-	-	-	-	-
Purple Swamphen	X	X	-	-	-	-	-	-	-	X	-	-	-	-	-	-	-
Ubiquitous birds																	
White-breasted Woodswallow	X	X	-	X	X	X	X	-	-	-	-	-	-	X	X	-	-
Long-tailed Triller	X	X	-	X	X	X	-	-	-	X	-	-	-	X	-	X	-
Emerald Dove	X	X	X	X	X	X	X	X	X	X	X	X	-	X	X	X	X
Rainbow Lorikeet	X	X	X	X	X	X	X	X	X	X	X	X	-	X	X	X	-
Glossy Swiftlet	X	X	X	X	X	X	X	X	X	X	X	-	X	X	X	-	-
Silver-eye	X	X	-	X	X	X	X	-	-	X	X	X	-	X	X	X	-
Uniform Swiftlet	X	X	X	-	X	X	-	X	-	X	-	-	-	X	X	-	X
Cardinal Myzomela	X	X	-	-	X	X	-	X	X	X	X	X	X	X	X	X	X
Tanna Fruit-Dove	X	X	-	-	X	X	X	X	-	X	X	X	-	X	X	X	X
Gray Fantail	X	X	-	X	X	X	X	-	X	X	X	X	X	X	X	X	-
Collared Kingfisher	X	X	X	X	X	X	X	-	-	X	-	X	X	X	X	X	-
Golden Whistler	X	X	-	X	X	X	X	X	X	X	X	X	X	X	X	X	X
Red-bellied Fruit-Dove	X	X	X	-	X	X	X	X	X	X	X	-	X	X	X	-	-
Streaked Fantail	X	X	-	X	X	X	X	X	X	X	X	X	X	X	X	-	-
Yellow-fronted White-eye	X	X	-	X	X	X	X	X	X	X	X	X	X	X	X	X	-
Red Junglefowl	X	X	-	X	X	X	-	X	X	X	X	X	-	-	-	X	X
Ubiquitous (excluding urban areas) birds																	
Pacific Imperial-Pigeon	-	X	-	X	X	X	X	X	X	X	X	X	X	X	X	X	-
Melanesian Cuckoo-shrike	-	X	-	X	X	X	X	X	X	X	X	-	X	X	X	X	-
Melanesian Flycatcher	-	X	-	X	X	X	X	X	X	X	X	X	X	X	X	-	-
Buff-bellied Monarch	-	X	-	X	X	X	X	X	X	X	X	X	X	X	X	X	X
Fan-tailed Gerygone	-	X	-	X	X	X	-	X	X	X	-	-	X	X	X	X	-
Metallic Pigeon	-	X	-	X	X	X	X	X	-	X	-	-	X	X	X	X	-
Swamp Harrier	-	X	X	-	X	X	-	X	-	X	-	-	-	X	X	-	-
Mackinlay's Cuckoo-Dove	-	X	-	-	X	X	X	X	X	X	X	X	-	X	X	X	X
Chestnut-bellied Kingfisher	-	X	X	-	X	-	X	X	X	X	-	X	X	X	X	-	-
Shining Bronze-Cuckoo	-	X	-	-	-	X	-	-	X	-	-	X	X	X	X	-	-
Peregrine Falcon	-	X	-	X	-	X	X	X	-	-	-	X	-	X	-	X	-
Island Thrush	-	-	-	-	X	-	X	X	X	X	-	-	X	X	-	X	X

Hebrides Honeyeater	-	-	-	X	-	-	-	-	-	-	-	X	-	X	X	X	X	X
et Warbler	-	-	-	-	-	-	-	-	X	X	-	-	-	-	-	-	-	X
Low/medium elevation birds																		
Owl	-	X	-	-	-	X	X	-	-	-	-	-	-	-	-	-	-	-
Cruz Ground-Dove	-	-	-	-	-	-	-	-	-	-	X	-	-	-	-	-	-	-
anded Rail	-	X	-	-	-	X	X	-	-	-	X	-	-	-	-	-	-	-
ern Shrikebill	-	X	-	-	-	X	X	X	X	X	X	X	X	X	X	-	-	-
atu Megapode	-	X	-	-	-	X	X	X	-	X	-	-	X	-	-	-	-	-
Birds restricted to high elevations																		
's Imperial-Pigeon	-	-	-	-	-	-	-	-	X	-	-	-	X	X	X	X	-	X
et Robin	-	-	-	-	-	-	-	-	-	-	-	X	-	-	X	X	X	X
Lorikeet	-	-	-	-	-	-	-	-	-	-	-	-	-	X	-	X	X	X
esian Triller	-	-	-	-	-	-	-	-	-	-	-	-	-	X	X	X	-	-
Parrotfinch	-	-	-	-	-	-	-	-	-	-	-	-	-	-	-	-	X	X
-winged Starling	-	-	-	-	-	-	-	-	-	-	-	-	-	-	-	-	-	X
Birds listed by Bregulla (1992) but not recorded by any of the other surveys																		
ailed Cuckoo	-	-	-	-	-	-	-	-	-	-	-	-	-	-	-	-	-	-
-rumped Swiftlet	-	-	-	-	-	-	-	-	-	-	-	-	-	-	-	-	-	-
nut-breasted Munia	-	-	-	-	-	-	-	-	-	-	-	-	-	-	-	-	-	-
e Sparrow	-	-	-	-	-	-	-	-	-	-	-	-	-	-	-	-	-	-
tain Starling	-	-	-	-	-	-	-	-	-	-	-	-	-	-	-	-	-	-
AL	**21**	**35**	**9**	**21**	**26**	**35**	**25**	**23**	**20**	**34**	**20**	**18**	**16**	**31**	**29**	**23**	**12**	**22**

inhabits the same range as the Robin but appears less common (seen by us only once at about 800 m in a primary forest on Cumberland slopes). The Palm Lorikeet, an endemic small parrot (IUCN Vulnerable) was observed at 800 m (two birds), later at 900 m (one individual flying silently, low in the vegetation) and at 1 100 m (a group of birds in a flowering tree) along the Penarou trail. This nomadic bird was also recorded at 1 100-1 500 m on the Tabwemasana slopes by Medway and Marshall in 1975 who indicate previous observations at Hog Harbour at sea level. A small flock of the Royal Parrotfinch (IUCN Vulnerable) was observed at 1 100 m at Penarou (and also at 1 120 m at Nokovula by Medway and Marshall in 1975). It is certainly very rare, restricted to upper elevation zones only. We did not see either of Santo's two starlings. None of the recent expeditions, which all together spent 20 days in suitable habitats, recorded the Santo endemic Mountain Starling (IUCN Vulnerable), which is obviously very rare and may be restricted to the most remote summits of Santo. The same may also be true for the endemic Rusty-winged Starling (IUCN Near Threatened) recorded by Medway and Marshall in 1975 only, at Nokovula (1 120 m).

••• Conclusion

The terrestrial bird fauna of Santo comprises about 47 species, including two uncertain Starlings that none of the authors saw during this expedition. The avifauna is otherwise well known, with 100 days of field work during four expeditions in the last 30 years, particularly so for low/medium elevation bird communities. The usual rainy weather conditions at higher altitudes make bird survey more difficult and less productive, even though 12 days were spent there. Moreover, some seabirds, especially petrels, nesting in the mountainous interior have never been studied and none was heard during the nights we spent in the mountains. The fauna is diverse with an important guild (eight species) of frugivorous/granivorous pigeons and families well established elsewhere in the Pacific region (Campephagidae, Monarchidae, Meliphagidae). Most of the birds, particularly passerines, are common and do not show altitudinal zonation. Sixty-four percent of the 47 species exhibit a large range of tolerance to altitude, being established from sea level to mountain summits. This is probably an indication of the good state of forested habitats, preferred by most of these birds all around Santo, which have been only partially and locally altered by clearings for pastures and coconut plantations. The situation may change for lowland forests which are potentially threatened by invasive aggressive vines (*Mikania micrantha, Meremia peltata*). For most of the birds the conservation status appears satisfactory despite significant hunting pressure and impact by invasive

Table 21: Proportions (%) of prospections having recorded the different bird species. Proportions are calculated for the five habitat classes. We called prospection the inventory of the bird species of any site belonging to the habitat class under study, by every single observer whatever the time he spent at this site. Several prospections can have been carried out at a same site (i.e. when several observers inventoried the bird species of this site). Name of sites are given in Tables 18 & 20. **Endemism-W**: widespread; **EndGen, EndSp, EndSSp**: endemic genus, species, subspecies; **Int**: introduced. **IUCN (2006) criteria-NT**: near threatened; **VU**: vulnerable; **EN**: endangered.

	Endemism	IUCN 2006	Urban areas	Low elevation habitats (0-60 m)	Medium elevation habitats (100-300 m)	Medium/high elevation habitats (500-700 m)	High elevation habitats (900-1500 m)
Number of survey sites			1	7	4	3	4
Number of surveys			3	7	6	4	6
Lowland birds							
Pacific Swallow	W	-	67	14	0	0	0
Rock Pigeon	Int	-	33	29	0	0	0
Common Myna	Int	-	100	71	33	0	0
Black-headed Munia	Int	-	100	71	17	0	0
Purple Swamphen	W	-	33	29	17	0	0
Ubiquitous birds							
White-breasted Woodswallow	EndSSp	-	33	86	0	25	33
Long-tailed Triller	EndSSp	-	33	57	17	0	50
Emerald Dove	W	-	100	100	100	50	67
Rainbow Lorikeet	EndSSp	-	100	100	100	50	33
Glossy Swiftlet	EndSSp	-	100	100	83	50	17
Silver-eye	EndSSp	-	100	86	50	75	67
Uniform Swiftlet	W	-	100	71	50	25	33
Cardinal Myzomela	EndSSp	-	100	43	83	75	83
Tanna Fruit-Dove	EndSp	NT	67	71	83	75	50
Gray Fantail	EndSSp	-	67	71	50	100	33
Collared Kingfisher	W	-	33	100	33	75	33
Golden Whistler	EndSSp	-	33	86	100	100	100
Red-bellied Fruit-Dove	W	-	33	71	83	50	33
Streaked Fantail	EndSSp	-	33	71	83	75	33
Yellow-fronted White-eye	EndSp	-	33	71	67	75	67
Red Junglefowl	Int	-	33	57	83	25	33
Ubiquitous (excluding urban areas) birds							
Pacific Imperial-Pigeon	W	-	0	86	83	100	33
Melanesian Cuckoo-shrike	EndSSp	-	0	86	83	50	50
Melanesian Flycatcher	EndSSp	-	0	86	67	75	17
Buff-bellied Monarch	EndGend	-	0	71	83	75	83
Fan-tailed Gerygone	EndSSp	-	0	71	50	50	50
Metallic Pigeon	EndSSp	-	0	71	33	50	17
Swamp Harrier	W	-	0	71	33	25	33
Mackinlay's Cuckoo-Dove	EndSSp	-	0	57	67	50	67
Chestnut-bellied Kingfisher	EndSp	VU	0	57	67	25	17
Shining Bronze-Cuckoo	W	-	0	29	17	25	17
Peregrine Falcon	W	-	0	57	0	0	33
Island Thrush	W	-	0	14	50	0	67
New Hebrides Honeyeater	EndSp	-	0	14	0	50	83
Thicket Warbler	EndSSp	NT	0	14	17	0	17

Low/medium elevation birds						
Barn Owl	W		0	43	0	0	0
Santa Cruz Ground-Dove	EndSp	EN	0	0	17	0	0
Buff-banded Rail	W	-	0	57	33	0	0
Southern Shrikebill	EndSSp	-	0	43	67	75	0
Vanuatu Megapode	EndSp	VU	0	43	33	25	0
Birds restricted to high elevations						
Baker's Imperial-Pigeon	EndSp	VU	0	0	17	50	50
Scarlet Robin	EndSSp	-	0	0	0	25	67
Palm Lorikeet	EndSp	VU	0	0	0	25	67
Polynesian Triller	EndSSp	-	0	0	0	25	33
Royal Parrotfinch	EndSSp	VU	0	0	0	0	33
Rusty-winged Starling	EndSp	-	0	0	0	0	17

plants, however, the situation seems critical for some endemic species which have not, or have only rarely, been observed by visiting ornithologists. This is the case for a group of birds restricted to mid-mountain forests (Santa Cruz Ground Dove) or to highlands (Baker's Imperial Pigeon, Palm Lorikeet, Royal Parrotfinch, Santo Mountain and Rusty-winged Starlings). Specific attention has to be paid to the two Santo island-endemics at least, the Santo Mountain Starling and the Santa Cruz Ground-Dove, the IUCN status of the former needs probably to be revised to a higher category of threat. Conversely, species like the Chestnut-bellied Kingfisher should be downgraded and the Tanna Fruit-Dove, very common everywhere, could even be removed from the Red List Threat Categories. As in New Caledonia, endemic and threatened species are mainly restricted to mountain forests (Table 21).

The introduced bird community is very limited with three birds coming in contact with the native fauna (the feral Rock Pigeon is in Luganville town and suburbs). The Common Myna and the Black-headed Munia are strictly restricted to lowland disturbed/modified/open habitats in association with humans, and any potential impact on native avifauna is questionnable. The Junglefowl is established in forests. It may compete with the Megapode which shares the same habitat, but this endemic remains apparently in good health after centuries of cohabitation.

AMPHIBIANS AND REPTILES

Ivan Ineich

••• A brief history of herpetological collections from Vanuatu

The oldest known vertebrate collection from Vanuatu took place during the explorations of Captain James Cook. He discovered Tanna in 1774. That volcanic island was covered with thick forest and the naturalist Forster discovered a pigeon there, *Gallicolumba ferruginea*. The only specimen known in the world is the one he killed on 17 August 1774 and that his son illustrated. Subsequent searches have not found the species again.

Herpetological collections from Vanuatu are located in several European museums among which those in London (BMNH, Natural History Museum, UK), Paris (MNHN, Muséum national d'Histoire naturelle, France) and Basel (NMBA, Naturhistorisches Museum Basel, Switzerland) are the most important. Other European museums holding such collections included Geneva (MHNG, Muséum d'Histoire Naturelle de Genève, Switzerland), Bonn (ZFMK, Zoologisches Forschungsinstitut und Museum Alexander Koenig, Germany), Hamburg (NHMH, Naturhistorisches Museum von Hamburg, Germany) and Frankfurt-am-main (SMF, Senckenberg Museum von Frankfurt-am-Main, Germany).

••• The Natural History Museum, London (BMNH)

The Natural History Museum (London) possesses several specimens collected around 1865 by J.L. Brenchley (six lizards) who visited several Vanuatu islands (Anatom, Tanna, Erromango, Efate and Vanua Lava in the Banks), but not Santo. This historically important collection contains several geckos forming the type series of *Gymnodactylus multicarinatus* (recently revalidated as *Nactus multicarinatus*) described by Günther in 1872. During the 1940's

Edgar Aubert de la Rüe was born in Geneva on 7 October 1901. In 1924 he obtained his geological-engineering diploma from the Institute of Applied Geology of Nancy in France. He later made 15 field trips as a geological-engineer for the French Ministry of Colonies and later French Overseas Ministry, visiting Africa, North and South America, the Indian and Pacific Oceans as well as Antarctica. He was named correspondant of the Muséum national d'Histoire naturelle (Paris) in 1934 and associate researcher on 20 December 1951. By 1963, he had authored about 263 scientific publications.

Philippe François was born in Saumur (France) on 22 November 1852. He studied in different cities according to the military postings of his father. He obtained a Licence in Natural Sciences in 1882 at Poitiers. He achieved a Ph.D. in science in 1885, with his thesis dealing with the nervous system of hirudine flat worms. The same year he was named "Maître de conférences de zoologie" at the University of Rennes. In 1888, he made a trip to the Pacific Islands to study the development of coral reefs on the Great Barrier Reefs of Australia. He was also put in charge of representing France at the Universal Exposition of Melbourne in Australia. He left on 1st August 1888 and returned to Paris only on 1st May 1891. After arriving in Noumea in 1888 he spent a total of eleven months (five different trips) in Vanuatu. François left New Caledonia on 2 August 1894 and visited Vanuatu, including Banks and the Santa Cruz group in the Solomon Islands before returnning through Port Vila on 3 January 1895. He died from a liver infection on 13 March 1908 at Paris.

J.R. Baker (1947) made important collections under the Percy Sladen Expedition, including 238 lizards and 21 snakes. He studied the reproduction of some common skinks (*E. cyanura sensu lato* and *E. caeruleocauda*) at Hog Harbour, Santo. That work was based on data obtained during the Oxford University Exploration Club (1933-34), a trip financed with funds from the Royal Society, the Percy Sladen Memorial Found, the University of Oxford and the New College of Oxford. Miss Evelyn L. Cheesman made collections around 1929 and from 1954 to 1955, totalizing 46 lizards and one snake. Later, in 1971, the Royal Society/Percy Sladen Expedition allowed the Earl of Cranbrook (formerly Lord Medway) and A.G. Marshall to make additional herpetological collections comprising 15 frogs, 345 lizards and four snakes. That expedition also collected an introduced typhlopid snake (not previously reported from the archipelago), and a new skink (*Emoia*). Islands visited were Anatom, Tanna, Erromango, Efate, Malakula and two satellite islets (? Ouri and Suaro), Santo and two satellite islets (Malo and Aore). All herpetological specimens were deposited at BMNH. Several smaller herpetological collections from Vanuatu have been deposited at BMNH among which that of J. MacGillivray (*HMS Herald*) from Anatom, Mr Cuming (around 1860, two lizards), W. Wykeham Perry Esq. (around 1875-1876, five lizards), Dr Corrie (around 1878, one lizard and one snake), "Challenger" collections (around 1882, 14 lizards), E.L. Layard, consul of Noumea (around 1886, three lizards and six snakes), D. McNabb (or MacNab), surgeon of the Royal Navy (around 1892, one lizard and three snakes), R.A. Lever from the Department of Agriculture of Fiji (two lizards). Finally Noel L.H. Krauss made a gift of reptiles from Vanuatu to several world museums (Hawaii, Paris, six lizards given in 1987 to the Univerity of Hamburg, Germany) including the BMNH.

••• Muséum national d'Histoire naturelle, Paris (MNHN)

Even though the New Hebrides was a British-French condominium, herpetological collections from that area are few in the Paris MNHN collections. Angel (1935) first published about a collection made in Vanuatu by Aubert de la Rüe (62 lizards and four snakes) (see above). Later, during his visit to Paris, Lord Medway (1974) examined that collection but also the François collection from Vanuatu. The later collection comprised one crocodile, 35 lizards and 13 snakes (some lizards and snakes were destroyed) but also included specimens from the Loyalty Islands (see above). The François collection was registered in 1894 and later in 1973-1974 (14 lizards and four snakes). Several smaller herpetological collections from Vanuatu were also deposited at MNHN among which are those of Dr Cailliot (six snakes), Deshouillères (one snake) and Dr Joly (four lizards).

••• Other European collections

Dr Félix Speiser spent several years in Vanuatu from 1910 to 1912 and collected many specimens that he send to the Switzerland Natural History Museum of Basel (NMBA). That collection was later studied by Roux in 1913 and comprised two new species of skink, *Emoia nigromarginata* and *Emoia speiseri*, the second being later considered a synonym of the first during a revision of the genus by Brown in 1991. The Speiser collection was re-examined by Lord Medway in 1975. The Natural History Museum of Geneva in Switzerland (MHNG) possesses two major collections from Vanuatu, one made by H. Larsen during the summer of 1958 on Ambrym and Vanua Lava in the Banks (28 lizards and eight snakes) and another made in June 1976 by A. de Chambrier. The Bonn Natural History Museum in Germany (ZFMK) possesses one collection made in Vanuatu in February 1975 by H. Meier. The Senckenberg Natural History Museum of Frankfurt-am-Main in Germany (SMF) has the famous Bregulla unregistered collection, which is unfortunately in a bad state and is of only limited interest.

••• Non European Vanuatu collections

Other important herpetological collections made in Vanuatu are located in the USA. The most important of these are deposited at the American Museum of Natural History (AMNH, New York), United States National Museum, Smithsonian Institution (USNM, Washington), California Academy of Sciences (CAS, San Francisco) and Bernice P. Bishop Museum at Hawaii (BPBM). Minor collections were also deposited at the Field Museum of Natural History (FMNH, Chicago), Museum of Comparative Zoology (MCZ, Harvard University), and University of Michigan (UMMZ, Ann Arbor).

The Whitney South Sea Expedition yielded a huge set of herpetological specimens giving rise to several important scientific publications. That collection was deposited at the American Museum of Natural History and later examined by Lord Medway. Another important collection was recently made on numerous Vanuatuan Islands by George R. Zug and colleagues and deposited at USNM (Washington). Collections deposited at the California Academy of Sciences were made in Vanuatu by Robert N. Fisher (November-December 1988) on Efate and some of its satellite islets (Emao). The Bernice P. Bishop Museum at Hawaii possesses some specimens given by Noel L.H. Krauss and collected from December 1986 to January 1987. Collections deposited at the Field Museum of Natural History of Chicago were made during the years 1952 to 1958. The most important collections were obtained through A.G. Marshall and Elizabeth Laird (August 1952) and B. Malkin (July 1958). The most important collections from Vanuatu held by the University of Michigan at Ann Arbor were made by R. Kuntz in August 1944 on Santo, and by Craig Moritz in 1986 on Efate and Tanna.

Other important herpetological collections from Vanuatu are located at Sydney in Australia (AMS) and at the Auckland Museum and Institute in New Zealand (AMI). The Australian Museum of Natural History (AMS) has several specimens from Vanuatu from collections made by Prof W.A. Haswell (1903), R.J. Etheridge (1910), W.W. Frogatt (1921), A.J. Marshall (1934), R.H. Pickering (1982), L. Gibson and his collaborators (1990), P. German (1992), and McAfee, E. Jacquier, Capt. Braithwaite among others. One other collection was made by Ross Sadlier in August 1983. There is also one more recent collection made in May 1990. The most important Vanuatu collections of that museum were made by Harold G. Cogger from 1983 to 1994. The Auckland Museum and Institute in New Zealand possesses some specimens of reptiles collected on Santo (e.g. AMI 1675).

Including our most recent collections from the SANTO 2006 and the Torres 2007 expeditions, there are certainly about 3000 specimens of reptiles and amphibians from Vanuatu preserved in world museums. Among those MNHN should now possess one of the most important collections.

 Biodiversity of the herpetofauna of Vanuatu

Our actual knowledge of the Vanuatu herpetofauna is based on scattered published papers dealing with some of the collections cited above or parts of them, either totally devoted to Vanuatu or partial reports in more general revisions. *Naika*, the Journal of the Vanuatu Natural Science Society (Efate, Port Vila), was published from 1981 to 1993 [numbers 1-42]. David Dickerson was President of that society in 1982 and Richard Pickering chief-editor, both being authors of herpetological notes published in that journal. New Caledonia, Solomon Islands, Fiji and Hawaii have one or more field guides or monographs devoted to their herpetofauna, but there is unfortunately no such book for Vanuatu.

Several Pacific island reptiles are still undescribed and numerous are known from a few specimens or even considered as extinct or severely endangered. Thus any attempt to make an analysis of that biodiversity is still fragmentary and remains only partial. The Pacific Ocean occupies about one third of the Earth's surface (larger that all continents together). Together there are more than 12 000 islands in the area totaling less than 2% of the earth's landmass. In 1996, Allison counted 672 amphibians and terrestrial or dulcaquicolous reptile species on Pacific islands. The six species of marine turtles are present in the whole area and thus do not carry much biogeographical information. There are also about 30 seasnake species in the area but knowledge on their distribution and even taxonomy is still fragmentary. New Guinea alone occupies more than 80% of the Pacific landmass and possesses at least 472 species totalizing 2/3 of the tropical Pacific herpetofauna. The New Guinean herpetofauna derives from the Oriental region but also to a lesser extent from Australia.

Gekkota (pygopods, diplodactylids, and gekkonids) and scincid lizards are the most visible elements of the herpetofauna on Pacific islands, and to a lesser extent hydrophiine elapids (sea kraits, seasnakes and Melanesian terrestrial elapids) and boids, marginally typhlopid worm snakes, among snakes. The Melanesian herpetofauna serves as the source for the colonisation of islands located north and east by natural trans-oceanic dispersal that began at least during the mid-Miocene. Species diversity clearly decreases eastwards from New Guinea to Easter Island and that decrease is even stronger east of Fiji. Curiously, the Pacific islands have only a low influence on the composition of the New Caledonia herpetofauna, which is mostly related to an Australian fauna of Gondwana origin (e.g. diplodactylid geckos). Note however that the Loyalty Islands, located just beneath New Caledonia, possess a clearly distinct herpetofauna more closely related to that of the eastern tropical Pacific Islands (e.g. Vanuatu and Fiji) rather than to New Caledonia. Gondwana fragmentation began

150 million years ago. Fijian endemic iguanids of the genus *Brachylophus*, Pacific boas of the genus *Candoia*, and numerous Pacific plants show South American affinities that also agree with the dispersion hypothesis that they originated in the neotropics. Distribution patterns of the boid genus *Candoia* are however similar to that of most Melanesian skinks and geckos; this could however be an artifact related to food requirements of the snake needing large lizards for its diet prior to the arrival of the rat. Actual herpetofaunal distribution patterns in the tropical Pacific Ocean are thus a mixture of vicariance and dispersal events, natural dispersion but also accidental and voluntary human introductions.

The description of the 672 Pacific species listed by Allison in 1996 began with the European colonisation of the area. Before 1800, only nine species from the area were described, mostly species also living in Southeast Asia. During the 19th century, 276 additional species were described, and again 371 more during the 20th century. Numerous species with wide distributions in the area are classically considered as recent arrivals, often through man mediated travel. Most herpetologists agree with such an explanation, generally based on weak biological, ecological, biogeographical and even molecular data. In 1985 Gibbons showed that between islands, dispersal could have been largely facilitated during maximal glaciation periods giving rise to important additional exposed land areas. His map shows that 18 000 years ago exposed areas and island numbers were about double the present situation. Allison also reported 49 amphibian and reptile species from the islands located east of the Solomon Islands, including two endemic frog species of the genus *Platymantis* from Fiji. The most diversified tropical Pacific island genera are lizards of the genera *Emoia* (Scincidae) and *Lepidodactylus* (Gekkonidae).

Concerning Vanuatu, Medway and Marshall noted in 1975 that the relatively high number of species that they reported from Efate could be related to a higher collection intensity and more frequent external contacts related to the geopolitical position of that island (hosting the capital Port Vila); the only known specimen of the gecko *Gehyra mutilata* from Vanuatu could have arrived there by boat. The herpetofauna of Efate is as diverse as that of Santo and that island does not show the north-south rarefaction observed for birds and mammals in the archipelago. They also noted that two geckos, *Gehyra oceanica* and *Lepidodactylus lugubris*, were never collected in non-anthropic habitats and that on Santo, the number and diversity of lizards decreases considerably with elevation; on Nokovula at 1 100 m a.s.l, the skink *Emoia caeruleocauda* is the only registred species. Local people however mention a green lizard (probably the green forest skink *Emoia sanfordi*?) and that no geckos (locally called "big eye") occur at

such elevations. This is not accurate since I often observed *Nactus multicarinatus* on Santo at around 900 m elevation.

Some reptile species were reported from Vanuatu as cryptozoological animals. They have not yet been collected and their existence is not confirmed. However these animals are locally reported in folklore and tales about them often exist; local observations and reports could be based on real animals and observations, but sometimes also on the wrong interpretation of an observation. In New Caledonia, a beach stranded banded lizard with short legs was interpreted as a cross between a sea krait (that lizard was found in sea water on the beach and was banded like the snake) and a lizard... It was in fact a nearly dead juvenile of Bocourt's terrific skink (*Phoboscincus bocourti*) that had probably fallen from the claws of a raptor. In 1994, Whitaker and Whitaker mentioned a flying gecko and an arboreal skink with a blue tail reported by some local people from the Matantas area on Santo. Such reports have of course to be considered as serious and attention has to be given to them. It has been noted that on Malakula, in the Uripiv language, there are four different words for lizards: "*mokoblab*" for *Gekko vittatus*, "*mob*" for other geckos, particularly *Gehyra oceanica*, "*negel*" for the green skink (*Emoia sanfordi*) and "*wejur*" for all other skinks. The biodiversity of the herpetofauna of Vanuatu was long considered as impoverished compared to islands north of it (Solomon Islands), south of it (New Caledonia) and east of it (Fiji). The most commonly accepted hypothesis to explain this situation are geologic events that I will discuss at the end of this paper. There is no known extinct species of reptile from Vanuatu. However giant skinks like those of the genera *Tachygyia* (Tonga, extinct?) or *Phoboscincus* (New Caledonia) could possibly have been present in Vanuatu or could still exist. Local people in Vanuatu also mentioned particular snakes from Malakula but they could correspond to uncommon color morphs of *Candoia bibroni*.

••• The Herpetofauna of Vanuatu, with special focus on Santo

••• Seaturtles from Vanuatu

Several species of sea turtles were reported from Vanuatu and all of them can be encountered in Santo waters: *Caretta caretta*, *Chelonia mydas*, *Eretmochelys imbricata* and *Dermochelys coriacea*. *C. caretta*, a carnivorous species, is very uncommon in Vanuatu.

Nesting sites of *C. mydas* have been reported from Vanuatu, and the species is considered as one of the most common sea turtles in the area. *C. mydas* is herbivorous and feeds on seagrasses and seeweed, thus implying active migration from feeding to reproduction areas (beaches). *C. mydas* sometimes concentrates on collective egg laying sites (rookeries).

E. imbricata is considered to be the second most common sea turtle of Vanuatu but its reproduction has never been attested to there, despite it reproducing in New Caledonia, Solomon and Fiji. *E. imbricata* is easy to observe by divers in the reef areas of Vanuatu where it feeds on invertebrates, sponges and soft coral. Egg laying sites of *E. imbricata* are more dispersed than those of *C. mydas*. The occurrence of *D. coriacea* in Vanuatu was confirmed by a specimen from South West Bay, Malakula, caught in a fisherman's net and eaten; the occurrence of that species remains however very unusual in Vanuatu.

D. coriacea is known from Fiji and Solomon. Its shell can reach 2 m and weigh nearly 650 kg. That marine turtle is present in deep water but occasionally occurs along coasts. It is the typical marine turtle of subtropical and temperate waters and feeds almost entirely on jellyfish. It lays its eggs from November to January in the southern hemisphere, generally on isolated beaches with low human disturbance, which are subjected to strong waves and slope steeply into deep water.

Lepidochelys olivacea could be present in Vanuatuan waters but that has never been confirmed. Note that in some parts of Vanuatu local chiefs prohibit the taking of marine turtles and the consumption of their eggs, as I have noted for fishes in some rivers. Shells of the different marine turtle species of Vanuatu are illustrated on the cover of number eight of the Vanuatuan journal *Naika* issued in 1982.

••• Amphibians from Vanuatu – FROGS

Platymantis Günther is a ceratobatrachid genus present in the Philippines, Palau, New Ireland, New Britain, Solomon Islands, with two endemic species from Fiji. The Solomon Islands also possess several endemic ceratobatrachid genera: *Batrachylodes* Boulenger, *Ceratobatrachus* Boulenger, *Palmatorappia* Ahl, and *Discodeles* Boulenger which is a regional endemic also present in the Bismarck Archipelago and the Admiralty Islands. That endemic stock has clearly an ancient Oriental origin, pre-dating the Australo-Oriental collision. Curiously, none of those species or genera, or even a related one, are present in Vanuatu where native frogs are completely absent. Eggs of *Platymantis* have direct development and such a characteristic was classically used to explain its successful colonisation of the area. In Vanuatu, the absence of native amphibians was often explained by the presence of porous volcanic soils that do not retain surface water and thereby prevent most amphibians from laying their eggs. Such an assumption is wrong, as the recent successful introduction (see below) of an Australian frog to Vanuatu has shown. The introduced species adapt to local conditions and some populations now reach considerable densities.

Platymantis is suspected by some experts to have been transported by man but if that is true, the reason for its absence from Vanuatu is unclear. A recent geological hypothesis for the former position of Vanuatu is now more in accordance with such an absence. Vanuatu is geologically younger than Fiji and the Solomon Islands and the archipelago was previously located north/northeast of Fiji (see discussion). *Platymantis*, as well at least one *Eleutherodactylus* species, another insular frog genus from the West Indies, is suspected to have internal fertilisation and a single gravid female can give birth to a new insular population; note however that internal fertilisation has never been confirmed for *Platymantis*. Despite such presumed adaptations, native frogs never reached Vanuatu.

AMPHIBIA, HYLIDAE
Litoria aurea

The Green and Golden Bell Frog, *L. aurea*, is a hylid frog from coastal areas of New South Wales in Australia where it is paradoxically considered a threatened species. It has been introduced by humans to New Zealand, New Caledonia and Vanuatu. Its introduction to New Zealand dates from 1867-1868 from where it was introduced to New Caledonia. The first New Caledonian report dates from 1926. It was also introduced to Hawaii around 1920 but without success. The introduction to Vanuatu was most probably made from New Caledonia by planters around 1960, ostensibly to control mosquitoes. In 1975, Medway and Marshall reported that: "*There are no amphibians native to the New Hebrides, and none was taken by Dr Felix Speiser who collected herpetological material in these islands in 1910-1912*". The first published report of that frog for Vanuatu was made in 1967 by Fischthal and Kuntz in their study of amphibian and reptile parasites. In Vanuatu, the frog is presently known from Efate, Malakula and Santo, which are among the largest islands of the archipelago, and therefore the ones that are best able to retain fresh water for long periods. Note however that the species also exists on the small island of Aore off Santo. On Santo it is encountered as high as 1 132 m a.s.l. It was reported from Santo in 1975 by Medway and Marshall and by Challacombe in 1986. I have observed that frog in the most favourable habitats of Santo, particularly in taro plantations.

The colonization and spread of introduced populations was rapid on Efate, thus showing that habitats on Vanuatu are favourable for amphibian life, contrary to what was formerly believed. The absence of native amphibians on Vanuatu therefore remains a biogeographical mystery that is probably best explained by former geological and tectonic events.

Figure 204: Some specimens of *L. aurea* show a light green dominant colouration that is typical of the species. A typical dark band runs from anterior eye border to the nostril. Matantas, Santo.

Figure 205: Typical colouring is variable in *L. aurea* and not all specimens are light green, some being dark green or even nearly completely brownish. Santo.

Figure 206: *L. aurea* is common in flooded taro plantations. Tasmate, Santo.

The Green and Golden Bell Frog shows a high body size and colour variability. Colour is generally greenish on the back with brown or golden variable sized marks, particularly on the limbs. A distinct light band is seen on the sides of the head extending along the body, often with a fine black band below, mostly visible in front of the eyes. Eyes are goldenish with a horizontal dark colored band typical of the genus. The belly is granular and white. Snout end is sometimes turquoise blue. Some frog sightings, erroneously recorded as unknown endemic frogs, almost certainly refer to unusual color morphs of *L. aurea*.

L. aurea lives mostly in or around water, and is an excellent swimmer often seen where aquatic vegetation is well-developed. When disturbed, the frog dives and swims under water to escape. *L. aurea* is often active and singing after rain.

••• **Amphibians from Vanuatu – TOADS**

AMPHIBIA, BUFONIDAE
Bufo marinus

In 1985 Challacombe reported in the local newspaper *Vanuatu Weekly* the collection of a specimen of the toad *Bufo marinus* [sometimes also called *Rhinella marina*]. It has been widely introduced on tropical Pacific islands and Australia, where its spread has been spectacular. No other observation of that toad has since been reported from Vanuatu. That observation could also correspond to an undescribed (endemic?) species but that is unlikely; more probably it refers to unusual color morphs of *Litoria aurea* that could have been mistakenly confused with a "large brown toad". The same paper also reported a small green arboreal frog. The absence of endemic amphibians in Vanuatu seems to be confirmed by recent collections, however it cannot be excluded that additional field collections will discover such an undetected endemic animal.

••• Reptiles from Vanuatu – CROCODILES

CROCODYLIA, CROCODYLIDAE
Crocodylus porosus

A dwarf fossil crocodile, *Mekosuchus kalpokasi*, has been described from Efate. This recently extinct species was most likely exterminated through human predation since fossil remains are dated about 3 000 BC. The absence of rigorous pre-quaternary and quaternary dating of fossil vertebrates from New Caledonia, Vanuatu and Fiji does not allow us to establish colonisation routes for mekosuchine crocodiles in the south Pacific but they most likely derive from an Australian dwarf crocodile stock.

Crocodylus porosus reaches its eastern distribution limit in Vanuatu. The species is present from southwest India to northern Australia and New Guinea, Micronesia, the Solomon Islands and Vanuatu. The occurrence of occasional specimens on New Caledonia and the Loyalty Islands is confirmed. The species reaches more than 7 m total length and is typical by its nuchal scalation and its bone crest above each eye extending at mid-length on the snout. Commercial exploitation has decimated many populations, except those of New Guinea and Australia that are still exploited under the control of strict management programmes. *C. porosus* is present in marshes and coastal areas but is not exclusively associated with saltwater.

The species was still common at some places on Vanua Lava (Banks, northern Vanuatu) before 1972 and its population was estimated at about 200 specimens at that time. That crocodile was responsible for a significant predation of pigs and cattle. The largest known specimen from Vanuatu reached 5.5 m. The 1972 hurricane was particularly violent and subsequent floodings can certainly explain the elimination of those crocodile populations, either by washing them away or by the destruction of their habitats and nests on the island. Sightings still take place but their frequency is now low. No human attack cases are known from Vanuatu.

Dickenson reported an estimation of the Vanua Lava population at 50 individuals [Ministry of Lands]. Hunting is uncommon but hunters coming from the Solomon Islands around 1973 killed seven individuals and in January 1980 an additional individual of 4.8 m length was killed by an Australian. Dickenson demanded the strict protection of the Vanua Lava populations of Silver River through the creation of a National Park or a Nature Reserve. In 1993, the species was still considered as facing extinction on Vanua Lava. (See: http://www.flmnh. ufl.edu/natsci/herpetology/act-plan/cporo.htm).

No specimen of that species from Vanuatu is present in the BMNH collections. That crocodile has been reported from Vanuatu by several authors. The species was reported from the Santa Cruz Islands, Solomon, not far from northern Vanuatu by Roux. A jaw of that species from Vanikolo (Solomon; sometimes called Vanikoro) is present in MNHN collections (François collection). Several reports also exist from Tikopia (Solomon) not far from northern Vanuatu. When Baker visited Santo in 1927, he observed a large crocodile track near the mouth of Yoro River but did not see the animal that was regarded as rare on the island. Dickenson reported some crocodile observations from Santo and Malo around 1980-1981. It seems likely that the species is now extinct on Santo and other Vanuatu islands and now only survives on the Banks archipelago. However occasional observations of vanished specimens can still be possible on Santo but occurrence as established populations is excluded.

••• Reptiles from Vanuatu – LIZARDS (Iguanidae)

SQUAMATA, IGUANIDAE, IGUANINAE
Brachylophus bulabula

Fiji is considered to have originated from a volcanic geological event dating at least 45 million years ago. Iguanas arrived through floating rafts from South America. However it is also thought that Madagascar iguanids originating from an older lineage arrived before the Gondwana landmass breakup (50-70 million years ago). Iguanine ancestors of that endemic Fijian/Tongan lizard group have formerly travelled more than 8 000 km, which is one of the longest known dispersal events by large terrestrial vertebrates via natural rafting. Among central Pacific island iguanas, two genera and five species are recognised among which two species are extinct, *Lapitiguana impensa* from the Fijian late quaternary, and *Brachylophus gibbonsi* from the Tongan late Holocene. Both were probably eaten to extinction after the arrival of humans about 2 800 years ago.

A recent study has shown that existing species comprise *Brachylophus vitiensis* in northeast Fiji (dry forest), a new species in the centre of Fiji, *B. bulabula* (moist forest), and *B. fasciatus* in the southeast of Fiji (Lau group, dry). Vanuatuan populations have been shown to belong to *B. bulabula*. Tongan populations were also introduced from the Fijian Lau group and they belong to *B. fasciatus* (type locality).

Brachylophus bulabula, which was collected for the first time on Vanuatu in 1976, seems to be a recent introduction originating from different Fijian stocks. It is now established in Vanuatu in the southwest of Efate Island. This population is recent and the first wild individuals were caught in the vicinity of Mele Maat. It is believed that they came from a small zoo on top of Klehm Hill and were relased into the nearby forest sometime prior to 1980 when the zoo closed. Gibbons noted that a German reptile collector named Heinrich Bregulla (who deposited an important reptile collection at the Senckenberg Museum of Frankfurt-am-Main; see above) lived on Efate. H.L. Bregulla authored a field guide on the Birds of Vanuatu (ed. Anthony Nelson) in 1991. He left Fiji, where he was based, for Efate in Vanuatu at the end of the 1960s. It has been suggested that H. Bregulla took the Vanuatu founding stock from Fiji to Efate in the late 1960s. Banded iguana specimens were exhibited at a tourist zoo on Efate. After about 10 years of captivity and public display, the animals were released. The first wild *B. bulabula* was collected on Efate only three months after the departure of Bregulla. The species is now well established in the area of Klemms (or Klem's) Hill on Efate (near MeleMart) and it is the only area on the archipelago where it has been reported. The species is abundant around the waterfall and is sometimes carried by children on their shoulders to attract tourists. Their actual distribution extends at least to Pango. Horrocks considered Vanuatu populations to be expanding. This seems to still be the case.

B. bulabula incubation can last as long as 8-9 months, a long time compared to most other lizard species. Such an incubation period explains the colonisation capacity and raft transport potential for its ancestors' arrival from the neotropics. Hatchlings can survive a long time without eating, living on their fat reserves. Males are larger than females and their white bands are more obvious. In the sun, green dominant coloration rapidly darkens and the white marks become more visible. Only males have dorsal light bands whereas females are nearly uniform green. This species can vary its coloration in a spectacular way according to its physiological state (stress, sexual displays…). This faculty has been attributed to the presence of cutaneous receptors rather than nerve connections to the vision system. Adults as well as juveniles possess functional nasal salt glands. Such glands could have permitted that lizard to undergo its transoceanic travel allowing its colonization of Fiji.

In captivity, banded iguanas readily eat *Hibiscus* sp. flowers. They feed mostly on plants but their food sources can vary from completely vegetarian to almost totally insectivorous. Females lay 3-4 eggs (40 x 30 mm) that are deposited in a hole dug in the soil. In the laboratory, at 30°C, incubation time was 17-23 weeks (119-161 days). Hatchling body size is 65 mm whereas adult size is about 160 mm, males being larger than females. Tail length is about three times body size.

Fijian iguanas (three endemic species) are particularly threatened by the recent introduction of the mangoose *Herpestes auropunctatus* on numerous islands, and by other introduced mammals like cats, pigs, and goats. Note that all iguana species in the genus *Brachylophus* are on the CITES appendix I list and live specimens can reach a very high price in both the legal and illegal pet trade. However it is possible to obtain live specimens on Efate for only a few euros! Local authorities need to take care on the attractiveness of that species for smugglers; the Vanuatu price makes them particularly competitive… However, the occurrence of *B. bulabula* in Vanuatu is problematic because of its CITES status. In Vanuatu it is clearly an alien species and as such has no conservation significance or value. In fact, it could be validly argued that it should be eradicated to protect the local biota. However, because of its present situation in Fiji, where it is seriously threatened by a wide variety of anthropogenic factors, the eradication of the population in Vanuatu would be controversial and certainly make it hard to argue against export for science or the pet trade. Live animals from the Mele population are frequently on sale in the market place in Vila, as pets and not for food.

No specimen of that species from Vanuatu is present in BMNH or MNHN collections. Two specimens from Efate are present in CAS collections.

••• Reptiles from Vanuatu – LIZARDS (Gekkonidae)

SQUAMATA, GEKKONIDAE
Gehyra mutilata

The gecko *Gehyra mutilata* is widespread and broadly distributed in the tropical Pacific and Indian oceans, as well as on the Asian mainland. The Indian Ocean populations, which are generally attributed to that species, certainly belong to a

Figure 207: *Gehyra mutilata* from Toga Island, Torres. Specimens from this population are highly contrasted and patterned. They typically have narrow white rings on their tail and a dark band bordered with white behind their eyes. They can easily lose their skin when handled.

different taxa and the name *Gehyra peroni* should be applied to them. In the same way *Gehyra insulensis* should be applied to the Pacific populations, as the morphology and ecology of both populations are different. The status of geographically intermediate Asian populations has to also be carefully clarified since the type locality (Philippines) of *G. mutilata* is located there. The species is widespread on Pacific islands but generally not abundant.

Only one immature specimen was collected in Vanuatu from Efate by J.R. Baker in 1924 or 1925. The species is considered to have been recently introduced to Vanuatu. Fieldwork by Medway and Marshall on Efate did not locate the species. The only occurrence on Efate seems to be linked to a recent introduction of the species on the archipelago. Note that in 1980 McCoy reported the species from Santa Cruz, southern Solomon, not far from northern Vanuatu.

Our recent surveys on Santo did not succeed in locating this gecko. However, during a follow-up expedition to the Torres Islands I collected numerous specimens of *G. mutilata* in the main village of Toga island (November 2007). The species was collected on house walls in sympatry with another gecko, *Lepidodactylus lugubris*, where both were common. *G. mutilata* was certainly recently introduced to that village and was not found anywhere else on the Torres Islands. Moreover another introduced snake, *Ramphotyphlops braminus* was also collected for the first time in this island group and in the same village. Thus I suspect a recent introduction of both species to this village, probably not from Vanuatu because *G. mutilata* had not been collected there previously. The colour pattern (narrow white bands on tail and back), rapid and significant loss of skin when handled and small size of *G. mutilata* link the Toga population to

Asiatic ones rather than other Pacific populations or even those in the Indian Ocean. The absence of natural populations of that widespread species on Vanuatu remains a mystery. It has been proposed that *G. mutilata* has an ecology on Pacific islands that is consistent with human-mediated dispersal, in support of the conclusions of its genetic data. I do not agree with such conclusions since the ecology of *G. mutilata* is not more reliant on a human-mediated dispersal than the ecology of *G. oceanica*, which is considered as a potential natural colonist of those islands. *G. mutilata* is not a typical forest species on Pacific islands and most frequently occurs along coastal plains and on atolls such as those in French Polynesia.

SQUAMATA, GEKKONIDAE
Gehyra oceanica

This gecko *Gehyra oceanica* has a nearly continuous distribution from northern Australia (occurrence not confirmed) and New Guinea, Mariannas and Palau to Easter Island. *G. oceanica* is curiously absent from Hawaii. According to Bauer and Henle, that species occurs in nearly the whole of Vanuatu. Roux and McCoy reported the species from Santa Cruz and Reef Islands, southern Solomon. The species was sometimes mixed with *G. vorax*. According to scale characters (mostly the number of toe lamellae and femoral pores), Beckon recognized three morphotypes, respectively located in Micronesia, Solomon (Bougainville) and other parts of Oceania; he suggested that these forms could correspond to three distinct species but no recent study has confirmed that point of view.

G. oceanica has 13-20 lamellae on the fourth toe, 12-50 preanal and femoral pores, a snout-vent length varying from 27-102 mm (52-102 mm for

Figure 208: *Gehyra oceanica* is a common gecko of Vanuatu. It is most frequently found in and around houses but also in cultivated areas. However the species is present but rare in forest where favourable habitats are uncommon. The species is more common in small rural villages than in towns like Luganville.

Table 22: Egg size and mass for *G. oceanica* from Tonga, Viti Levu (Fiji) and Rotuma (Fiji), compared to our data for Santo and French Polynesia. **MES**: mean egg size (mm). **MEW**: mean egg weight. * n = 88.

Locality	n	MES	Range	MEW
TONGA	7	12.4 x 14.0	12-13 x 12-15	1.25
FIJI (Viti Levu)	4	12.3 x 14	12-13 x 13-15	1.18
FIJI (Rotuma)	10	12.5 x 13.5	11.8-13.1 x 12.6-14	1.26
VANUATU (Santo)	26	12.4 x 13.6	11.4-13.3 x 12.2-15.2	/
FRENCH POLYNESIA (Moorea)	55	11.6 x 13.1	10.8-12.4 x 12.2-14.9	1.06*

Table 23: Hatchling size and mass for *G. oceanica* from Tonga and Rotuma (Fiji), compared with Moorea Island (French Polynesia). **MHS**: mean hatchling size (mm). **MTL**: mean tail length (mm). **MHW**: mean hatchling weight (g). * n = 62. ** n = 55.

Locality	n	MHS	Range	MTL	Range	MHW
TONGA	7	33.7	33-35	29.7	28-32	0.83
FIJI (Rotuma)		33.6	33.1-34.2			
FRENCH POLYNESIA (Moorea)		30.7*	28-33			0.76**

Table 24: Egg size for *G. oceanica* from two localities on Santo, Peavot village on the east coast of Cape Cumberland and the mangroves of Palikoulo. **MES**: mean egg size (mm).

Locality	n	MES	Range
Peavot	11	12.77 +/- 0.37 x 13.82 +/- 0.38	12.21-13.33 x 13.37-14.59
Palikoulo	12	12.01 +/- 0.33 x 13.30 +/- 0.87	11.38-12.52 x 12.24-15.18

Figure 209: *G. oceanica* is a massive gecko with a large head and heavily built limbs allowing it to easily climb large trees. Adult snout-vent length varies from 60-90 mm and the tail is nearly equal to body size.

Figure 210: *G. oceanica* has enlarged finger and toe pads, the terminal portion of each digit arises from the middle of the enlarged pad and each digit terminates with a well-developed claw. The pinkish band on the posterior part of thigh and leg is typical of the species.

Figure 211: Hatchlings of *G. oceanica* are strongly patterned. Their tails are dorsally annulated with dark and light rings.

Figure 212: *G. oceanica* hatchlings typically have a pinkish wash around the eye. Total length is about 60-70 mm with a nearly equal body and tail length.

Figure 213: Ventral colouration of *G. oceanica* is either lemon yellow (that image) or greyish white; there are no intermediates. Colour polymorphism is not related to sex, is present in both hatchlings as well as adults, and has as yet no explanation.

Figure 214: Adult males of *G. oceanica* are easily distinguished from females by the presence of a large swelling on each side at the base of the tail (red arrow), cloacal spurs and a typical V shaped row of femoral and preanal pores (white arrow).

mature males with femoral and preanal pores). The species occupies a wide spectrum of habitats, including mangroves, inhabited areas, cultivation and to a lesser extent primary forest, where suitable habitats are rare (dead and/or old trees) and generally occupied by a larger species of the genus, *Gehyra vorax*. *G. oceanica* lays its eggs in communal egg laying sites which can sometimes hold more than 40-50 eggs together (each female lays two eggs). Such egg concentrations can be found under the loose bark of live coconut trees or even dead trees and all development stages are present at the same time, including hatched empty eggs.

Data in table 22 clearly shows that the Vanuatu populations of *G. oceanica* have egg sizes similar to those of Tonga and Fiji. There are however significant differences between all of these and those of French Polynesia as shown in tables 22 & 23. Such differences correspond to the significant live colour differences between French Polynesian populations and others and are not in agreement with Beckon's biogeographical separation of potentially distinct species. French Polynesian specimens are less coloured (often uniform greyish) and less patterned, they also have less orange-pinkish colours on their body, even as hatchlings, compared to Melanesian and Tongan specimens. The tails of hatchlings are often ventrally pinkish orange like in other populations. The differences observed between French Polynesia and other islands (Tables 22 & 23) cannot be related to habitat choice since the differences are very low between French Polynesian populations when comparing eggs on atoll and more humid high altitude islands. They really should correspond to a specific or subspecific differentiation between populations.

In the same way if I consider populations from two different habitats on Santo:
- An east Cape Cumberland population from the village of Peavot on Santo,
- The mangrove population of Santo (Palikoulo, 15.498°S, 167.251°E);

the differences are also relevant (Table 24) but slight, and not to the same extent as between French Polynesia and other populations (Tables 22 & 23).

Gehyra oceanica has a wide distribution in Vanuatu. It is reported from Torres and Banks in the north, and to the south on most if not all other islands until Anatom in the extreme south of the archipelago. The species was reported from Santo by Medway and Marshall in 1975. Specimens from Santo are present in several museum collections: AMS, BMNH, MNHN and UMMZ. Some recent data are consistent with the hypothesis that *G. oceanica* was naturally dispersed across the Pacific, prior to the arrival of humans.

SQUAMATA, GEKKONIDAE
Gehyra vorax

For about 75 years after its original description, *G. vorax* was regarded as a valid species, distinct from *G. oceanica*. In 1932, Burt and Burt placed it in the synonymy of *G. oceanica*. That concept was accepted until 1984 due to new data from Gibbons and Clunie. The revision of Beckon in 1992 showed that this species has clear differences from *G. oceanica* in terms of habitat, behavior, and ecology, but also morphology.

G. oceanica has lamellae on about half the toe pad whereas in *G. vorax* lamellae occupy two thirds of the pad. The pad is also distinctly larger. Furthermore *G. vorax* has a mottled pattern and a greater number of subdigital lamellae (more than 18 (19-34) vs 12-15 in *G. oceanica*). For *G. vorax*, Beckon indicated 26-90 pores, a snout-vent length

of 35-156 mm and 90-156 mm for mature males with pores. *G. vorax* is less common in collections than *G. oceanica*. I have also noted that even as a hatchling and juvenile, *G. vorax* can easily be distinguished from *G. oceanica*. The former has only about 5-6 dark dorsal tail bands whereas the later has at least 9-10 such bands.

According to external morphological characters (toe lamellae and preanal/femoral pores), three morphotypes are distinguished in *G. vorax*: Fiji, Vanuatu and New Guinea. Fijian and Vanuatuan specimens do not share the same coloration and can easily be distinguished. The morphotype from New Guinea looks very closely like *G. oceanica* (the later species being absent from there) and could correspond to *Gehyra membranacruralis*. *G. vorax* does not occur in New Caledonia where it is replaced by *Rhacodactylus* sp. It is however present on the Loyalty Islands. Reports from Tonga seemed to be valid but the species is now considered to be extinct on that island group, but of course might still occur there. Beckon supposed the extinction of *G. vorax* on Tonga arose through habitat perturbation or destruction but such a hypothesis is not

Figure 215: In contrast to hatchlings of *G. oceanica* (left), those of *G. vorax* (right) have only 5-6 dark dorsal tail bands. Santo.

Figure 216: Adult male of *G. vorax* on a giant banana trunk at Butmas, Santo.

Figure 217: Red morph (female) of *G. vorax*. Butmas, Santo.

Figure 218: Eggs of *G. vorax* (numbers 3 and 6) can easily be distinguished from smaller ones of *G. oceanica* (numbers 1 and 10) and also from all other species of Vanuatu by their size and circular shape. Santo.

acceptable since primary forests are still present on several islands like 'Eua. A fourth morphotype from the Moluccas could also represent another distinct species. Thus the available biogeographic and morphological data seem to agree with multiple species in the *G. vorax* complex.

G. vorax is poorly represented in world museums and its biology is only poorly known. It is the second largest gecko in the southern Pacific; on Fiji, its total length reaches nearly 25 cm. The body is heavily built with strong muscles; fingers and toes are widely enlarged. Locals fear the species due to the way it mimics vegetation, its capacity to produce sounds, its adherence to human skin associated with its developed subdigital lamellae, strong claws able to cause injuries, and also its capacity to lose skin when handled. I have observed a green muscle colour under the lost skin in one specimen from Butmas on Santo. In Fiji, some specimens show a dorsal colour pattern comprising an alternate brown and grey-blue or greenish bands sometimes causing confusion with endemic iguanas of the genus *Brachylophus*; such colouration was not observed in Vanuatu. In that area the gecko is named the "lizard that barks like a dog" and its cry can be confused with that of the pigeon *Ptilinopus luteovirens*. Its diet is based on fruits and in captivity it readily eats papaya. It is mostly a forest species but it can also be seen on breadfuit trees or even large banana plants, *Pandanus* sp., coconut trees, large palm trees, mangroves, sometimes not far from human habitation. Contrary to *Gehyra oceanica*, it never enters human habitation. *G. vorax* is an arboreal species showing a clear tendency to avoid man, contrary to *G. oceanica*. I suspect that species to be naturally less anthrophilous than *G. oceanica* and not to avoid humans as a secondary adaptation to its fear of being predated for meat. *G. vorax* is hunted for food in Fiji and also on Vanuatu where *G. vorax* is often active during day time. Its skin perfectly matches tree trunks covered with lichen, and it can change its coloration from light green to grey or dark chocolate brown. On Vanuatu, sexual colour dimorphism occurs: adult females are rather brownish and adult males greyish to greenish. The species is difficult to collect or observe in the field since it often lives high on tree trunks and even in the canopy, mostly in primary forest. When sympatric with *G. oceanica*, *G. vorax* is found higher up towards the canopy. *G. vorax* is less common in coastal forest and highly disturbed habitats than in deep humid forests.

Eggs of *G. vorax* are clearly larger than those of *G. oceanica* with which it was long confused. Gibbons and Zug measured two eggs from Fiji (Viti Levu Island) with a mean size of 18 x 20 mm and a mass of 3.9 g. The mean size for three eggs from Santo is 17.57 x 19.24 with ranges of 16.59-18.11 x 18.59-19.75 mm.

G. vorax is particularly sensitive to two threats, deforestation and the international pet trade. The species is regularly and easily available on the internet from where several colour morphs can be obtained from Vanuatu ("red morph", "ornate morph").

G. vorax was reported from several islands of Vanuatu: Erromango, Efate, Epi, Malakula, Pentecost, Aoba and Santo. Specimens from Santo are present in AMS and MNHN. The species was never reported north of Santo and south of Erromango and seems limited to the largest islands of the archipelago.

SQUAMATA, GEKKONIDAE
Gekko vittatus

G. vittatus is present in the Indo-Australian archipelago from Java (Indonesia) to New Guinea, Palau [probably a distinct species], Admiralty, Bismarck Archipelago, and as far as the Solomon Islands (Santa Cruz Group) and northern Vanuatu. The species was reported from Santa Cruz (southern Solomon) by Roux based on one specimen from the Speiser collection (NMBA). He stated: "*C'est, à l'est, le point de dispersion le plus extrême connu pour cette espèce. Notre exemplaire n'appartient pas à la var. bivittatus D.B.*", thus indicating observed

Figure 219: *Gekko vittatus* is an arboreal species living on small branches of small trees and shrubs. Torres Islands.

Figure 220: Like all lizards, *G. vittatus* sheds its skin in pieces. Torres Islands.

10 infralabials, and 23 lamellae on the fourth toe. It shows larger granules on the back separated between them by smaller scales. The throat is covered by enlarged separated granules and followed by true scales on chest and belly, without granules. The legs show enlarged scales anteroventrally, followed by distinctly smaller granules in the posteroventral area. Postmental scales are asymmetric and median. Four successive larger scales lie behind the mental plate. The mouth is elongated by a dermal fold going up to just below the ear. The tail cross-section is typically circular. According to my observations, the Vanuatu and southern Solomon specimens clearly differ from Indonesian populations and might represent a separate taxon.

differences with *Gekko bivittatus* from New Guinea. McCoy also reported the species from the Santa Cruz Islands, Southern Solomon. Cranbrook and Pickering reported the species from the Banks: Vanua Lava, Mota Lava and Gaua (nowadays Santa Maria). Bauer and Henle stated: "*A single, questionable record from the main islands of Vanuatu*". Burt and Burt considered the species unable to migrate south of a limit fixed by Banks and Santa Cruz, a statement that seems correct.

This relative large gecko can be distinguished from other geckos from Vanuatu by the lack of claws on its fifth toe and finger and its flattened and elongated body. The MNHN specimen from southern Solomon (Vanikolo) shows a snout-vent length of 98 mm, tail length 96 mm, 12 supralabials,

In Vanuatu, the species was reported from the northernmost islands: Torres Islands (Hiu, Tegua, Loh, Toga) and Banks Islands (Mota Lava, Vanua Lava, Santa Maria) where it is common. It was also recorded southerly, from Malakula by McKerras but its occurrence there has not been reconfirmed. Our research on Santo and Torres did not confirm its presence south of Banks. Locals consider the species to live in deep forest, which is wrong. It is sometimes eaten by the local people. On Torres Islands, the locals told us that children often use that lizard to make scarifications (tattoos). They get the lizard to bite them on body parts like the arm, then toss it away to produce injuries that later give rise to a kind of "V" shaped scarification. That gecko is the main prey of Barn Owls on the Torres Islands.

Figure 221: *G. vittatus* females deposit two adhesive eggs under bark or more generally under vegetation, like here in the axil of a *Pandanus* sp. leaf. Torres Islands. Detail: Eggs of *G. vittatus* have nearly the same size as those of *G. oceanica* but can be distinguished because they are adhesive.

Figure 222: In this gravid female of *H. frenatus*, eggs are visible through the body wall. Matantas, Santo.

SQUAMATA, GEKKONIDAE
Hemidactylus frenatus

The gecko *Hemidactylus frenatus* is a recent colonist in Vanuatu where it arrived most likely around 1980 through man-mediated introduction. *H. frenatus* presently occurs on almost all tropical and subtropical areas of the Indo-Pacific. It naturally occurs in South and Southeast Asia but also on part of the Indo-Australian archipelago. It is an invasive species particularly adapted to human dispersal. It is generally believed to compete with indigeneous species and displace them. Such species interactions were noted as early as 1985 by Gibbons on Fiji, where populations of *H. garnotii* seem to suffer from such competition. In Vanuatu, the species is common in many villages and towns where it is often syntopic with *Lepidodactylus lugubris* and as yet has not replaced the latter.

Figure 223: Head of *H. frenatus*. In Vanuatu, geckos (Gekkonidae) can be distinguished from skinks (Scincidae) by the presence of small granules on top of the head in the former and large symmetrical head plates in the later. Matantas, Santo.

Figure 224: *H. frenatus* is only active at night and does not hesitate to come to electric lights to capture the attracted insects on which it feeds.

Figure 225: *H. frenatus* can be distinguished from the other house gecko *Lepidodactylus lugubris*, by its larger adult size, the lack of deep black symmetrical spots on the back, and regular rings of erected granules around the dorsal tail vs only lateral small spines in *L. lugubris*. Luganville, Santo.

This introduced gecko was only reported from Efate and Espiritu Santo in 1994 and more recently (2008) from Malakula. Specimens from Santo are present in the Australian Museum (Sydney) and MNHN. AMS collections have specimens from Luganville (Santo) and Port Vila (Efate) collected in 1984. BMNH does not possess any specimens of this species from Vanuatu.

SQUAMATA, GEKKONIDAE
Hemidactylus garnotii

The gecko *Hemidactylus garnotii* has a broad distribution including the entire tropical Pacific, from New Guinea to French Polynesia. It occurs in India and mainland Southeast Asia through Indonesia to the Philippines. It is apparently absent from most of Melanesia and Micronesia. The species is, however, poorly represented by voucher specimens because it is relatively rare.

H. garnotii is a parthenogenetic species. It has a flattened appearance and an entire tail with only lateral spines but no dorsal spine rings. The ventral part of the tail is often coloured reddish pink in juveniles and subadults.

The species has not been previously reported from Vanuatu, but its absence is certainly an artifact, as is the case for *Gehyra mutilata*. *H. garnotii* is confirmed by a voucher at BMNH from Anatom, collected by Miss Cheesman but this record most likely constitutes a wrong identification. During the Santo 2006 expedition, I collected two specimens from the village of Tasmate, on the western coast of Cape Cumberland; these represent the first records of the species from Santo and the first vouchers for Vanuatu. It is certainly just a rare species that is difficult to locate.

Gibbons and Zug (1987) provide some data on eggs and hatchling from 'Eua Island (Tonga): one egg size was 9 x 10 mm, mass 0.6 g; three hatchlings had a mean snout-vent length of 27.3 mm (range of 27-28), a tail length of 20 mm (range 19-21) and a mass of 0.31 g (range 0.28-0.36).

SQUAMATA, GEKKONIDAE
Hemiphyllodactylus typus

The gecko *Hemiphyllodactylus typus* has a wide distribution extending from South Asia to the Mascarenes, through Indonesia, Philippines and the whole tropical Pacific to French Polynesia. In Oceania, the species is represented by only a few records. This poor representation is the result of its rarity. It has not been reported from Vanuatu, but its absence is unlikely, and it will certainly be discovered in the coming years. It is a small gecko with a typical elongated body and short enlarged toe pads. It can be present in forest as well as around human habitation but is not attracted to lights on houses.

SQUAMATA, GEKKONIDAE
Genus *Lepidodactylus*

In his PhD thesis, Russell considered the genera *Gekko*, *Pseudogekko*, *Luperosaurus*, *Ptychozoon*, *Lepidodactylus*, *Hemiphyllodactylus*, *Gehyra* and *Perochirus* as belonging to the same relational group, the *Gekko* group of the subfamily Gekkoninae. Within the genus *Lepidodactylus*, he examined toe morphology and formulated an evolutionary hypothesis related to distal transformations, mostly a reduction of the number of subdigital lamellae, accompanied by a median division of some distal lamellae to form two rows. Later, Brown and Parker indicated that a more compressed tail and members habitus is parallel to that evolution. They distinguished three evolutionary lineages in the genus *Lepidodactylus*. Group I species (**pumilus-oorti group**) still possess numerous undivided "Gekko like" lamellae at all digits. *L. manni*, a Fijian endemic, belongs to that group. Group II members (**guppyi-pulcher group**) also have undivided lamellae but several terminal lamellae are divided. *L. gardineri*, a Rotuma endemic (north of Fiji) and *L. vanuatuensis* (see below), belong to this group. Tails of groups I and II are subcylindrical, without lateral fringes or spines. Group III species (**lugubris-woodfordi group**) are characterised by a reduced number of subdigital lamellae but the terminal and some subterminal lamellae divide, associated with a depressed body and a more flattened and widened tail. Group I is considered as the most primitive of the genus. Species of that group are present from Indonesia to Papua New Guinea, Torres Straits islands, Solomon and Fiji, and on Christmas Island in the Indian Ocean. *L. listeri* from Christmas Island cannot be distinguished morphologically from *L. manni* of Fiji. Group II species are present over a smaller area from north New Guinea, Solomon, Vanuatu, Admiralty, Bismarck and Rotuma. Group III species are the most evolved species regarding their digital structures according to Russell's hypothesis. Several distal lamellae,

Figure 226: *L. buleli* seems to be endemic from the west coast dry forests of Santo. It was only observed in myrmecophilous plants but it is unlikely that this represents its only habitat.

Figure 227: *L. lugubris* is easily distinguished from other Vanuatu members of the genus by its dorsoventrally flattened tail with lateral indentations. This specimen has typical black marks of triploid clone B.

Figure 228: Diploid clone A is the most common clone of Vanuatu. That clone was recently (during World War II) introduced over most tropical Pacific islands. It is typical by its seven pairs of double symmetric fine black points from the neck to the base of the tail.

Figure 229: Ground colouration is variable in most geckos but black marks are always present in that species and do not disappear whether the gecko is active under light or under shelter in the dark during day time. Their size, number and position allow clonal attribution to each individual (diploid or triploid clones).

including the terminal one, are divided and lamellae are in more reduced numbers, with a clear tendency to disappear at the base of the finger; their tail is wider and more compressed dorsoventraly. These species have a wide distribution comprising several Indian Ocean islands and most tropical Pacific islands, reaching both sides of Central America where populations are generally considered to have been recently introduced by man.

SQUAMATA, GEKKONIDAE
Lepidodactylus buleli

The gecko *Lepidodactylus buleli* is known only from the Cape Cumberland west coast, in the forest around the village of Penaoru. It was discovered during the Santo 2006 expedition and is known from two specimens, one in bad condition and the second, the holotype, which arrived in Paris as an egg that was hatched and the juvenile grown to an adult, allowing it to be described. It is a typical group II *Lepidodactylus* (terminal lamellae entire) due to its cloacal spurs at base of the tail in males, its subcylindrical tail, yellowish lips and dorsal and head pattern.

This gecko has a unique habitat and lives in myrmecophilous plants hanging on tree trunks around 20 m high in primary forest of the dry west coast of Santo. It is however not unlikely that it occurs in a wider range of habitats that only such kind of plants.

SQUAMATA, GEKKONIDAE
Lepidodactylus lugubris

L. lugubris is widely distributed from the Indian Ocean to South Asia (India and Sri Lanka) through Southeast Asia, Indonesia, Philippines and almost all tropical Pacific islands. Most authors consider that the species owes its broad occurrence to accidental transport by man; even if the genetical structure of the populations referring to that species reflect a long evolutionary history in the Pacific. The species occurs throughout Vanuatu and was reported from there already since a long time.

The species is common around human habitation and does not hesitate to visit the interior of houses. It occurs in most habitats from mangroves, cultivated areas, villages and cities, and also in forest at low elevation. The populations of Vanuatu are unisexual and females reproduce clonally through parthenogenesis. Males have never been reported in Vanuatuan populations. At least three clones are present in Vanuatu, diploid recently-introduced

Figure 230: *L. vanuatuensis* is typical by its yellowish anterior eye border, short and rounded snout and purple brownish colouration.

Figure 232: This male *L. vanuatuensis* has a swollen tail base that allows easy sex recognition for adults. Tasmate, Santo.

Figures 231: Other diagnostic features of *L. vanuatuensis* are enlarged dark dorsal symmetric blotches and a roundish tail section. Females (this image) do not have a swollen tail base.

Figure 233: Hatchlings of *L. vanuatuensis* are smaller than those of *L. lugubris*. They typically have large symmetric dark dorsal blotches, a rounded snout, and a round tail section without lateral fringes.

clone A, and triploid clones B and C which are rare. In Vanuatu, the species was reported from Torres and Banks groups, Espiritu Santo, Aore, Malo, Aoba, Tutuba, Malakula, Ouri, Ambrym, Tongoa, Efate, Emao, Erromango, Aniwa, Tanna, Futuna, Anatom. Specimens from Santo are present in numerous world collections: AMNH, AMS, BMNH, FMNH, MCZ, MNHN, UMMZ and USNM. The species was reported from Santo by Medway and Marshall in 1975.

Each female lays two adhesive eggs (adhesive to the substrate but also to one another). Eggs are often deposited in communal egg laying sites under the bark of upright living or dead trees. They can also be found in rock crevices or even in pompilid wasps nests. Hatchling snout-vent lengths from Santo vary from 19-21 mm (n = 13, mean 20.1 +/- 0.9) and tail length varies from 16-23 mm (n = 13, mean 19.6 +/- 1.8).

SQUAMATA, GEKKONIDAE
Lepidodactylus vanuatuensis

The recently described gecko *Lepidodactylus vanuatuensis* is a Vanuatu endemic and its type locality is located on Santo (holotype in USNM). It was reported from several Vanuatu islands: Santo, Efate, probably Tanna, and Anatom. Specimens from Santo are present in the following museum collections: AMNH, USNM, MNHN and AMS.

The species often lives in high numbers on the same tree, sometimes in complete syntopy with *L. lugubris*. It is present around villages not far from sea level or on the seashore, e.g. under tree bark like mango trees or *Casuarina* sp. It can be distinguished from *L. lugubris* by its round tail cross section, short and rounded snout and red brownish dorsal colouration with a yellow eye border.

Hatchling snout-vent length varies from 16-17 mm (n = 5, mean 16.6 +/- 0.5) and tail length from 12-15 mm (n = 5, mean 14.2 +/- 1.3). Eggs of this species are smaller than those of *L. lugubris*; they are laid in pairs, stuck together and onto the support were they are laid in communal egg laying sites.

SQUAMATA, GEKKONIDAE
Lepidodactylus guppyi

L. guppyi is a group II *Lepidodactylus* species. It was first reported by Burt and Burt in 1932 from Tinakula (Tinehula) in the Santa Cruz group. In 1980, McCoy recorded the species from Santa Cruz and Vanikolo Islands. I recently collected a single specimen of this species in northern Vanuatu (Hiu, Torres Islands) in November 2007. It is a forest arboreal species observed during the day under loose bark on a dead tree in a garden surrounded by deep forest.

SQUAMATA, GEKKONIDAE
Genus *Nactus*

Despite that some populations of *Nactus* were considered to be distinct species in the past, all populations were subsequently considered as a single, wide-ranging species, *Nactus pelagicus*,

Figure 234: *Lepidodactylus guppyi* (right) can be distinguished from *L. lugubris* (left) by its habitus and colouration. Torres islands.

until the revision by Zug and Moon in 1995. Schwaner found only females in Samoan populations, whereas most Australian and Melanesian populations are bisexual. In 1987 Moritz was the first to review this species complex. He examined 322 specimens from Vanuatu: Banks (Vanua Lava), Espiritu Santo, Aoba, Malakula ("Malekula") and Efate for the northern islands from which populations comprised males and females, Erromango, Tanna and Anatom for the southern islands from where populations only comprised females, except for Anatom. Chromosomal and enzymatic polymorphism analysis shows that bisexual populations are highly polytypic and probably comprised several species that could be separated into four geographic groups based on electrophoresis and cytogenetics. He made no nomenclatural decision to distinguish those morphologically similar populations. Unisexual populations are diploid, highly heterozygous but do not show a genetic segregation between them, thus suggesting parthenogenetic reproduction. The high heterozygosity of unisexuals indicate a probable origin through hybridization with one of both parental species genetically similar to the bisexual populations of northern Vanuatu and the second not identified but

whose genetic characteristics could be predicted. Oceania (comprising the islands located east of a Philippines/New Guinea/Solomon/Vanuatu limit) and New Caledonia have only unisexual populations whereas Vanuatu possesses mixed populations —both unisexual and bisexual at the same time (mixed on Anatom in Vanuatu), whereas elsewhere in southwest Pacific populations are unisexual. Moritz also shows a scalation character concerning post-mentals allowing the separation of unisexual females (small post-mentals separated from infralabials) from bisexual females (enlarged post-mentals in wide contact with infralabials).

During a detailed analysis of populations until then attributed to the *N. pelagicus* complex, Zug and Moon in 1995 showed that north Vanuatu and Solomon populations have males and females, whereas elsewhere in Oceania and New Caledonia populations only comprise females, as in southern Vanuatu (except Anatom). Precloacal pores are only present in males; unisexual populations have a significantly larger size than bisexuals but that character, as with body size, and other statistical differences highlighted by Zug and Moon, do not allow a serious separation of unisexual and bisexual

Figure 235: *N. multicarinatus* is typical by the presence of dorsal lines of keeled enlarged granules and black and white striped lips.

Figure 236: *N. multicarinatus* is often heavy dark when disturbed during day time. Symmetrical black dorsal marks can however be seen on its back. Tail base is typical by the presence of enlarged granules forming a kind of lateral spine line. Torres Islands.

Figure 237: When disturbed and stressed, *N. multicarinatus* often curls up its tail. Note the typical black and white striped lips.

populations. Those authors noted however a clear tendency to a correlation between those variations with latitude, and explained that phenomenon by the influence of climatic factors affecting development. They suggest testing that hypothesis by analysing altitudinal variations on a sample from the same island. Moritz's hypothesis, which supposed that one of both parental species of the unisexual form would be the sexual form of Vanuatu and the Solomon Islands seems to agree with those results. According to Zug and Moon, the distinction of unisexual from bisexual specimens by the size of post-mentals as shown by Moritz is not always reliable; their work also shows that it is impossible to

distinguish unisexual and bisexual females through morphology and scalation. The same authors proposed to distinguish Vanuatu bisexual populations at a specific level (*Gymnodactylus multicarinatus*) from Oceanian unisexual populations (*Heteronota pelagica*). Two binomen are available to name the Oceanian unisexual populations (*Arnouxii* and *pelagica*). The International Commision of Zoological Nomenclature validated *pelagica* with its type locality "Feejee and Navigator Islands". Zug and Moon (1995: 88) designated a lectotype and restrict the type locality to Ovalau, Fiji. The syntypic series of *Nactus multicarinatus* comprised six specimens coming from three different localities. Again, Zug and Moon designated a lectotype and restricted the type locality to Efate, Vanuatu. The electrophoretic and cytogenetical studies of Moritz, as well as the morpho-meristic study of Zug and Moon seem to show that unisexual populations are uniform and certainly arose from a unique hybridisation event between two bisexual species.

SQUAMATA, GEKKONIDAE
Nactus multicarinatus

The bisexual species *Nactus multicarinatus* occurs from the southern Solomon Islands to Vanuatu, except Tanna and Erromango. On Vanuatu, it has been reported from the Torres group, the Banks group (Vanua Lava, Santa Maria), Elephant Island (east coast of Santo), Santo, Malo, Aore, Aoba, Maewo Pentecost, Suwarro, Ouri, Ambrym, Malakula, Epi, Emao (off Efate), Niogriki Island (north coast of Efate), Efate, Anatom. Specimens from Santo are present in several world museums: AMNH, AMS, BMNH, MNHN and UMMZ.

N. multicarinatus is very common and widespread on Santo. It is found in many kinds of habitats and can easily be observed during daytime under stones and logs at soil level. Its colouration is heavy dark brown to black when disturbed during the day. It lives mostly on the ground but climbs on to the bases of trees, generally less than one meter high at night. This gecko is easy to distinguish by its toes, which are not enlarged (no toe pads). *N. multicarinatus* is present above 900 m elevation on Santo were it was the only species found during the day under stones and dead fallen trees.

SQUAMATA, GEKKONIDAE
Nactus pelagicus

The unisexual gecko species *Nactus pelagicus* is present from Micronesia, New Caledonia to the east of Melanesia and Polynesia. In Vanuatu, *N. pelagicus* seems to occur only on Erromango and Tanna, perhaps on Anatom, but it is absent from Santo. The status of Anatom populations has to be checked.

Figure 238: Reproduction of the original plate of *Perochirus guentheri* by Boulenger (1885: pl. 12, fig. 4).

Gibbons and Zug indicated a mean egg size and mean mass for 5 eggs from Viti Levu (Fiji): 7 x 13 mm, range 6.5-8 x 12-14 mm, mass 0.44 g, range 0.39-0.49 g.

SQUAMATA, GEKKONIDAE
Perochirus guentheri

In 1976, W. C. Brown revised the genus *Perochirus* first described by Boulenger in 1885. He recognized three species all with restricted distributions, limited to one island or a small island group. *Perochirus scutellatus* is present in Micronesia on Tetau Island (Carolines) and Greenwich Islands (Kapingamarangi Atoll, State of Pohnpei). *Perochirus ateles* occupies the Truk Islands, Kapingamanrangi Atoll, Guam, Tinian, Ponape, Marcus and Cocos in Micronesia. *P. guentheri* is endemic to Vanuatu and the only species of the genus found outside Micronesia. *Perochirus* was erected by Boulenger with *Perochirus ateles* as the type species. It shows different characteristics including a first rudimentary finger, without a claw and a granular back. In 1954, Underwood placed that genus in Gekkoninae, principally based on its pupil shape. Later, in an unpublished manuscript, Russell considered the genus to belong to a group of genera including *Gehyra* and *Hemiphyllodactylus*. In fossil remains discovered on 'Eua island (Kingdom of Tonga), Pregill noted in 1993 the occurrence of nine species among which two are extinct, one such gecko was placed in the genus *Perochirus* and a large extinct skink in the genera *Emoia* (not *E. trossula*), *Eugongylus* or *Tachygyia*. This endemic species seems to agree with the hypothesis that Vanuatu was formerly positioned north of its current location, near the putative center of origin of the genus in the Central Pacific, in the Micronesian area. However, fossil remains of *Perochirus* reported from 'Eua (Tonga) by Pregill could also indicate that the group previously had a wider distribution. That author considered *Perochirus* to have been eliminated from Tonga by the most recent arrival of *Gehyra oceanica*. An illustration of *P. guentheri* is given in the early work of Boulenger published in 1885. Adult sizes vary from 61 to 71 mm snout-vent length and males possess 10 to 12 preanal and femoral pores extending to the lower thigh.

P. guentheri is a southern Vanuatu endemic. The species is known only from four specimens: two specimens from Erromango collected in 1859 or 1860 (type locality) by Cuming (BMNH) and a third one collected on the same island in August 1971 in Marshall's tent, on the site of Nuangkau camp. Cranbrook considered however that origin to be uncertain and that it could have been transported during the expedition going from island to island. A fourth specimen from AMNH was collected on Anatom but was not reported by Medway and Marshall in 1975. Note however that a recent list of BMNH specimens from Vanuatu does not mention any BMNH *Perochirus* from Anatom. Thus the Anatom locality requires confirmation.

No data are available on the ecology of *P. guentheri*. *P. ateles*, a related species, occupies coconut palms on Kapingamarangi, and one specimen was collected on a bread fruit that had fallen onto a tree fern branch. *P. scutellatus* also was observed on coconut palms on Tetau Island. No data are available on reproduction.

••• Reptiles from Vanuatu – LIZARDS (Scincidae)

Most Santo lizards observed during the day are skinks (family Scincidae). They can be distinguished from geckos by the presence of symmetrical head plates and smooth overlapping scales. Some species are terrestrial and others arboreal.

SQUAMATA, SCINCIDAE
Caledoniscincus atropunctatus

The genus *Caledoniscincus* is a regional endemic comprising about 11 recognized species of which all except one are New Caledonia/Loyalty area endemics. The exception, *C. atropunctatus*, occurs throughout New Caledonia and the Loyalty islands, but also in the southern islands of Vanuatu. Its broad distribution in these latter islands suggest that it arrived in Vanuatu prior to humans; however, a human mediated introduction cannot be excluded since the species has probably also been recently introduced to Surprise Island off New Caledonia. The species is widely distributed in New Caledonia and its arrival in Vanuatu is sometimes considered to be recent. It occurs on Ouvéa, Lifou and Maré islands in the Loyalty Islands, and in Vanuatu it is restricted to forest on Erromango, Tanna, Aniwa, Futuna and Anatom. Bauer and Sadlier recently add Efate but that record arises from a misidentification referring to *Emoia* sp. There is no record of that species for Santo and it is clearly a southern Vanuatu skink species.

The species was present in Vanuatu before the Second World War and also occupies forest, thus arguing against a recent human mediated introduction. Medway and Marshall considered it to be restricted to forest; it however only occurs in dry open-canopy forest but is more abundant in larger patches of coastal forest or interior mixed hardwood forest. Its distribution in Vanuatu is however wide and encompasses native dry open-canopy forest, thus suggesting an older arrival.

I have examined MNHN specimens from Tanna which belong to that species. They are small skinks with three dorsal keels and 28 midbody scale rows. Their flanks are dark brown from nose to tail, the

Figure 239: *C. novohebridicus* is a small skink less than 38 mm snout-vent length. Its eyes are covered by a fixed transparent eyelid thus explaining standard name of Vanuatuan snake-eyed skink. Santo.

Figure 240: Specimens of *C. novohebridicus* are usually dark and always possess a pair of lighter lateral bands. They are the only skink of Vanuatu lacking moveable eyelids. Santo.

Figure 241: *C. novohebridicus* is a typical littoral species that regularly forages on sandy beaches when stones and vegetation are nearby, sometimes in syntopy with *Emoia cyanura*. Santo.

back is light brown with spots on a line corresponding to each scale line (like interrupted lines). The belly of both specimens show a pigmentation on each scale. There is a sexual dimorphism in coloration and both specimens are females. Supranasals are separated, prefrontals largely separated, the interparietal is large, and there is a single frontoparietal.

SQUAMATA, SCINCIDAE
Cryptoblepharus novohebridicus

Burt and Burt regarded *Cryptoblepharus boutonii novohebridicus* (as it was originally named) as a synonym of *Cryptoblepharus boutonii poecilopleurus*. The taxonomy and systematics of the genus *Cryptoblepharus* have been recently reviewed for Indian Ocean as well as for Pacific Island populations by Horner. Fuhn in 1969 and Greer in 1974 considered *Cryptoblepharus boutonii* as a super-species with about 36 morphological and/or geographical types. Mertens in 1931 previously attributed a subspecific rank to all these 36 "forms". *C. novocaledonicus*, as *C. novohebridicus*, is now considered as a valid species. In 1931, Mertens recognized affinities between the faunas of New Caledonia and Vanuatu. He also noted affinities between *C. novohebridicus* and *C. virgatus* (northern Australia) from which the former could derive.

C. novohebridicus is a small species with a maximum snout-vent length of 37 mm. The species is sometimes present in mangroves but generally frequents upper sandy beaches covered with vegetation. Coastal populations usually occur on coral litter, palm trunks and dead timber and can display high densities in such habitats. It is also sometimes abundant in dry open forest areas a short distance (approximately 3 km) from the coast as I observed in the Penaoru area, Santo. However, that population was certainly accidentally introduced by the Santo 2006 expedition team from the sea shore to that area.

This species occurs throughout the entire Vanuatu archipelago. It was reported from Banks (Merig), Espiritu Santo, Malo, Aore, Aoba, Pentecost, Ambrym, Malakula, Efate, Emao (off Efate), Erromango, Tanna, Futuna, and Anatom. I confirmed its occurrence in the Torres Islands during a November 2007 trip. It was reported from Santo by Medway and Marshall but not found again. The species can be considered to be present over the whole of Vanuatu from the extreme north to the extreme south.

SQUAMATA, SCINCIDAE
Genus *Emoia*

The genus *Emoia* has numerous insular endemic species, as well as some widespread species. Many species remain to be discovered or described. For example, prior to 1980 Fiji had a single known endemic species. After that date, intensive field-work allowed the recognition of five additional taxa new to science in the genus from Fiji, most of them from the *E. samoensis* species group. The systematics of that species group is being clarified, including the Vanuatuan populations from which additional species remain to be described.

SQUAMATA, SCINCIDAE
Emoia aneityumensis

The skink *Emoia aneityumensis* is an Anatom endemic, in southern Vanuatu. It was described based on three immature specimens collected by Lord Medway during the 1971 Royal Society Expedition and three additional specimens collected by Miss Evelyn Cheesman on the same island and deposited in the BMNH collection. They were first catalogued as *Emoia nigra*. The material collected by Cheesman during 1954-1955 has no precise collection locality. Medway did not find additional specimens in the different collections he examined elsewhere (AMNH, BMNH, MNHN, Senckenberg Museum). The species belongs to the *Emoia samoensis* group comprising 14 described species among which 11 are geographically limited to the Vanuatu, Fiji, Tonga and Samoa subregion, one species is endemic to the Solomon Islands, one to Cook Islands, and another occupies nearly the whole of Melanesia.

The *E. samoensis* group comprises two species sub-groups that seem to be closely related and three species with unclear affinities to other members of the group. It is the only group in the genus where females lay more than two eggs. *Emoia aneityumensis* belongs to the *Emoia concolor* subgroup, which comprises eight species (several more to be described): *Emoia campbelli* (Fiji), *Emoia concolor* (Fiji), *Emoia erronan* (Futuna, Vanuatu), *Emoia aneityumensis* (Anatom, Vanuatu), *Emoia loyaltiensis* (Maré, Loyalty, New Caledonia), *Emoia mokosariniveikau* (Vanua Levu, Fiji), *Emoia tongana* (Samoa and Tonga) and *Emoia nigromarginata* (Vanuatu). *E. aneityumensis* can be distinguished from *E. samoensis* by its higher number of medio-dorsal scales (40-42) and its lower number of subdigital lamellae (36-42); its coloration is identical to that of *E. nigromarginata* (but can be distinguished from the former by its larger size at sexual maturity (up to 92 mm SVL) and its increased number of eggs. Total length is about 2.74 times snout-vent length. Prefrontals are in weak contact and separate nasals from frontals. The interparietal is distinct. The 5th or 6th supralabial is the largest and located under the eye. Twelve specimens of *E. nigromarginata* from different localities in Vanuatu studied by Medway show 28-34 midbody scale rows,

Figure 242: *Emoia atrocostata freycineti* is a littoral species that prefers rocky shorelines. Torres Islands.

38-48 subdigital lamellae under the fourth toe, and a snout-vent length of 60-72 mm. *E. aneityumensis* is only known from Anatom and specimens are conserved at BMNH and FMNH.

The coloration of the holotype (after 18 months preservation in alcohol), is dark brown on the back, with a discontinuous and irregular row of black spots and punctuations extending in the shape of a discontinuous dorso-lateral line from ear to tail base extending above legs and feet. Flanks are brownish, crossed and spotted by intrusions of the whitish ventral coloration. In three live juveniles the back was grey-brown or olive-brown, marked by a broken row of punctuations or black spots, mixed with Nile water flecks extending from the ear to the base of the tail. Flanks are olive-brown, crossed and spotted by the dirty white ventral colouration. The tail is the same brown as the dorsal colouration but regenerated tails are lighter. Medway supposed, according to Cheesman's description, that adults have a general grey blue coloration.

All available specimens originated from a disturbed and partially regenerated forest located about 1 mile northeast of Anelgauhat on Anatom. Medway erroneously considered two reported specimens of *E. samoensis* as valid for Erromango and thus allows for a competitive exclusion between *E. aneityumensis* and *E. samoensis* on Erromango and Anatom, but he excluded the possibility of considering them as subspecies of a same species.

SQUAMATA, SCINCIDAE
Emoia atrocostata freycineti

In 1913, Roux reported the occurrence of the lizard *Emoia atrocostata freycineti* from the Santa Cruz group, southern Solomon, and from Pentecost in Vanuatu (both NMBA collections). He noted: "*Elle n'avait pas encore été signalée aux Nouvelles-Hébrides*". In 1991 Brown distinguished three subspecies and refers Solomon and Vanuatu populations to *E. atrocostata freycineti*. The type locality of that subspecies is Vanikolo Island, Santa Cruz group, southern Solomon Islands. The nominal subspecies occurs in Micronesia (Mariannas, Carolines), Palau and in the Bismarck Archipelago along the Pacific and Australian plates west to New Guinea, Torres Strait islands, Indonesia, Christmas Island, the Malay Peninsula, Indochina, Borneo, Philippines, Taiwan and only one island south of the Ryukyus in Japan. The subspecies *E. atrocostata australis* only occurs in the Cape York area of Queensland, Australia. Cranbrook and Pickering noted that the species is known from Vanuatu by two specimens respectively from Malo and Pentecost.

It is a common specialised species but restricted to some habitats. It is mostly present on rocky

Figure 243: *E. caeruleocauda* has two color morphs in Vanuatu and elsewhere. They are not related to sex or age as classically believed. One morph is characterized by its emerald green tail and the other by its dark brown colour with more or less visible golden stripes on the back. Some specimens can sometimes be nearly completely melanistic.

Figures 244: The most useful character for easy recognition of *E. caeruleocauda* is the presence of a typical axilla black or darker spot. This trait has not been noted previously but it is really useful for identification in the field as in museum.

beach margins where densities can be high. It is an active and agile lizard, which does not hesitate to enter saltwater puddles at low tide to escape from predators. It can easily stay submerged for several minutes. It is active even in the wave-washed areas where it can maintain itself firmly on rocks with its strongly clawed fingers. It feeds in inter-tidal areas. In 1980, McCoy noted that two eggs are deposited on beach litter, under floating driftwood logs or in limestone rock cavities.

E. atrocostata freycineti was reported from several Torres islands, Aore, Malo, Espiritu Santo, Pentecost, Malakula, and Efate. Specimens from Santo were reported by Medway and Marshall in their general distribution table but later in their paper they noted that this report refers to Malo —a small peripheral island of Santo, and not to Santo itself; these authors however mention an observed specimen not collected by Medway on the reefs of

Figure 245: Like *E. impar*, *E. caeruleocauda* lack the black parietal spot of *E. cyanura* located at the posterior part of the frontoparietal head plate (see arrow).

Hog Harbour, Santo. One specimen from Santo is also deposited in the Australian Museum.

SQUAMATA, SCINCIDAE
Emoia caeruleocauda

The skink *Emoia caeruleocauda* belongs to the *E. cyanura* group, and *E. caeruleocauda* subgroup. It was long confused with *Emoia cyanura sensu lato* (see below). Parker in 1925 was the first to provide clear characters to distinguish between them, mostly based on lamellae appearance and number. He called both the Vanuatu species respectively *Lygosoma (Emoa) cyanurum* and *Lygosoma (Emoa) lessoni*. The first taxon comprises the actual *Emoia cyanura* and *E. impar* and the second corresponds to *E. caeruleocauda*. Several wrong names were attributed to the Vanuatu populations of that species: *Lygosoma kordoanum* by Baker, *Emoia werneri* by Burt and Burt in 1932, Baker in 1947, Tanner in 1952, and Medway and Marshall in 1975, *Lygosoma (Emoa) Werneri* by Angel in 1935, *Emoia caeruleocauda* (= *E. werneri*) by Cranbrook and Pickering in 1981, *E. [moia] werneri* by Cranbrook in 1985, and finally *Emoia caerulocauda* [sic] by Goldberg and coauthors in 2005.

E. caeruleocauda, as presently defined, is widely distributed, although it certainly consists of a species complex. Its type locality is located on Sudest Island (formerly Tagula Island) in the Louisiade Archipelago, Papua New Guinea, but unfortunately its holotype is lost. The species is reported from the Mariannas, Carolines, Marshall, Palau, Fiji, Vanuatu, Solomon, Bismarck Archipelago, New Guinea, Moluccas, Sulawesi, northern Borneo and southern Philippines. Brown in 1991 and Zug in 1991 are certainly wrong when they state that the Fijian populations were recently introduced by man; I found them often in deep primary forest on Santo as well as on Fiji.

E. caeruleocauda is a terrestrial species even though some individuals can be seen on tree trunks or bushes, generally no more than one or two meters above ground. In Vanuatu, like on the Solomon

Islands and New Guinea, the species occurs in gardens, open areas and in deep forest; it is however most frequent in forested areas. In Papua New Guinea that skink is present from sea level to almost 1 500 m elevation but I did not observe it above 700 m on Santo.

E. caeruleocauda occurs as two colour morphs over most of its distribution: one emerald green tailed form and one completely dark brown form sometimes with one to three more or less visible golden dorsal back stripes. That later morph can easily be confused with *Emoia cyanura*. They differ however by toe lamellae and other scale counts. One practical way to distinguish between them in the field is the occurrence of a dark blackish axilla spot in *E. caeruleocauda* but absent in *E. cyanura* and other Vanuatu *Emoia* species. That useful character was never used before our study but it is really helpful and reliable. Anyway there is no relationship between colour pattern and sex in that species, contrary to what has often been stated in the past.

Gravid females usually bear two eggs. Eggs are deposited in soil debris and vegetation. In Vanuatu, reproduction is continuous all year round with a peak from November to February and a slower period from May to June. I suspect that rainfall rather than day length is the factor that causes the observed variations. Curiously Baker in his 1947

Figure 248: *E. cyanogaster* has a typical uniform yellow light greenish belly.

work does not report the occurrence of totally dark brown specimens and it is not clear under which binomen he considered them (*E. caeruleocauda* or *E. cyanura sensu lato*?).

In Vanuatu, *E. caeruleocauda* occurs throughout the entire archipelago: Banks and Torres groups, Santo, Malo, Aore, Pentecost (formerly sometimes called Raga), Malakula, Ambrym, Tongoa, Efate, Erromango, Aniwa, Tanna, Futuna, and Anatom. The occurrence of that species on Santo was reported by Baker in 1947, Medway and Marshall in 1975 and Brown in 1991. Specimens of that species from Santo are conserved in AMNH, AMS, BMNH, and MNHN.

Snout-vent length of adults range from 39-52 mm. Tail length varies from 147-178 % of SVL but is generally about 160 %. Prefrontals are well separated in all specimens. About 40 % (n = 47) of our Santo sample had skin parasites (mites). The frequency of the brown morph compared to the emerald green tailed morph on Santo is about 50 % (n = 47); both morphs include males and females as well as adults and juveniles. Midbody scale rows vary from 31 to 38 and fourth toe lamellae from 31 to 41.

Figure 246: *E. cyanogaster* is a slender species with an elongate body and tail. Such characteristics allow this lizard to move easily on small branches of shrubs and short trees, even on large leaves.

Figure 247: *E. cyanogaster* is typical by the strong dark/light colour contrast between the back and the light yellow belly. Four medio-dorsal scale rows are clearly enlarged.

SQUAMATA, SCINCIDAE
Emoia cyanogaster

The skink *Emoia cyanogaster* belongs to the *Emoia cyanogaster* group containing five species distributed across the Moluccas, New Guinea, Bismarck Archipelago, Solomon and Vanuatu. *E. cyanogaster* occurs in the Bismarck Archipelago, Solomon (including Santa Cruz and Reefs Islands) and Vanuatu. The species is absent from Fiji, like *E. atrocostata* and both absences certainly have an important zoogeographical significance.

In 1975, Medway and Marshall reported the species from seral and climax forest to elevations of about 500 m; it is absent from forest at 1 100 m elevation in Vanuatu. The species also occurs in gardens and other degraded areas with introduced vegetation. It is an arboreal species mostly restricted to small trees and bushes. It searches for its food at ground level as well as in trees but does not occur in deep forest. It seems to need more open and sunny areas.

I carefully examined two specimens from Malakula deposited in the MNHN and report below their respective morphological and scalation characters: snout-vent length 65 and 64 mm; tail length 131 and 156 mm; midbody scale rows 24 and 24; scale lines on mid back 55 and 56; fourth toe lamellae 79 and 78. Both show separated supranasals, separated prefrontals, parietal and nucal plates with a contact oriented left, single interparietal (epiphysal spot visible) and single frontoparietal. They have eight supralabials with the 6th enlarged and under eye, infralabials 5-6.

In the same way I observed MNHN 1934.0088: snout-vent length 70 mm, tail length 166 mm, midbody scale rows 24, scale lines on middle back 56, fourth toe lamellae 78, supralabials right/left 8(6)/8(6), infralabials 6/6, small interparietal with a small epiphyseal spot, parietal and nucal suture oriented left, supranasal well separated, prefrontals in left contact, supralabial under eye very large and clearly different in size compared to that of *E. sanfordi* and *E. nigra*,

toe lamellae numerous and bladelike, mediodorsal scales enlarged (mostly the four median rows), irregular rows of dorso-lateral white spots, dark brown band from nostrils to midbody which disaperared progressively posteriorly.

Females lay two eggs that are deposited in loose soil, litter or rotting vegetation.

E. cyanogaster was reported from Vanuatu by McCoy in 1980 and Brown in 1991. The species is present on the following islands or island groups: Torres and Banks, Espiritu Santo, Malo, Aore, Aoba, Pentecost, Ambrym, Malakula, Epi, Efate, Erromango, Tanna. The occurrence of the species on Santo was reported by Medway and Marshall in 1975, and by Brown in 1991. Specimens from Santo are conserved in the following museums: AMS, BMNH, FMNH, and MNHN.

SQUAMATA, SCINCIDAE
Emoia cyanura or *Emoia impar* without precision

Most species of the *E. cyanura* species complex are difficult to distinguish and all together were until recently considered under the unique binomen *E. cyanura*. This is also true for all species records from Vanuatu prior to 1990. Some species of the complex still remain to be described and several of the available older names need to be revalidated. Three morphotypes can be clearly recognized in the complex, blue tailed and brown-greenish tailed striped specimens, but also melanistic specimens that are almost completely dark brown. All of those morphotypes comprise several distinct species. The names that should be applied to those different species

Figure 249: *E. cyanura* (all four pictures) has a variable tail colour but never shiny blue like that of *E. impar*. Both species share a similar size and habitat, but numerous other characters allow them to be distinguished.

are not clearly defined and numerous doubts still exist. The current attribution of names to most populations is not acceptable. Eastern brown tailed populations were generally referred to *Emoia cyanura* or *E. pheonura* and blue tailed specimens from the same area to *Emoia impar* or *E. cyanura*. The name *E. impar* should certainly be applied and restricted to western populations of the Bismarck Archipelago area, but this has still to be demonstrated through molecular studies. In the same way the name *E. cyanura* should be restricted to eastern Polynesian blue tailed populations (nowadays erroneously called *E. impar*), in accordance with the original description of that taxon by R.P. Lesson in 1826. The main problem is how to name the Polynesian brown tailed populations. Their valid name is either *E. arundeli*, *E. arundeli pheonura* or *E. pheonura*, but in no way *E. cyanura* as currently used. Valid names will be defined according to the fact that:

- Clipperton Atoll populations will be considered as belonging to a monotypic species identical to eastern Polynesian populations (thus *E. arundeli* will be valid for all eastern Pacific populations);
- Clipperton Atoll populations will be considered different at specific level from eastern Polynesian species (*E. arundeli* and *E. pheonura* will both be valid and distinct at specific level);
- Clipperton Atoll populations will be considered as belonging to a polytypic species (*E. arundeli arundeli* for Clipperton populations and *E. arundeli pheonura* for eastern Polynesian populations).

Beside these nomenclature problems, several other populations still have to be named, probably as new taxa: e.g. the southern Bougainville blue tailed populations, and some Vanuatu blue tailed populations. Anyway I here still consider only two distinct morphotypes from Vanuatu from what was previously called *E. cyanura*, the brown tailed morph that I will (provisionally) call *E. cyanura* and the blue tailed morph that I will (provisionally) call *E. impar*.

E. cyanura and *E. impar* belong to the *Emoia cyanura* group and the *E. cyanura* subgroup as defined by Brown in 1991. This author considered *E. cyanura* as a superspecies but was wrong in not accepting the recognition of both morphs (blue tailed and brown-greenish tailed morphs) as distinct species. In 1986, Brygoo designated MNHN 7069A as the lectotype of *E. cyanura* but that decision cannot be accepted since the type series, having originated from Tahiti (French Polynesia), does not include blue tailed morph specimens

Figure 250: *E. cyanura* can easily be distinguished from *E. impar* and *E. caeruleocauda* by its dark simple or double parietal spot (black or bluish grey) which is always present.

such as the one illustrated on Lesson's original iconotype plate published in 1826, but only brown tailed morph specimens and some specimens of *E. caeruleocauda*, a species not occurring in French Polynesia (see above). Thus in 1987 Ineich recommended considering that type series as invalid and the syntypes of *E. cyanura* as having been lost (or mixed with other specimens in MNHN collections). That position was not followed by Ineich and Zug in 1991 but should be recommended now. In 1947, Baker did not distinguish *E. cyanura sensu stricto* from *E. impar* but however noted about *E. cyanura sensu lato*: "*The tail is dull olive green during life. In preserved specimens its colour sometimes changes to a blue resembling that of* E. werneri *[actually* E. caeruleocauda*]. It is unfortunate that the name* cyanura *(sic) should refer to a species in which the tail is in life so far from*

Table 25: Frequency variation of gravid adult females for both *Emoia* species of French Polynesia.

	October-December	January-March	June-August
Emoia cyanura s.s.	60.87 % (n = 92)	38.74 % (n = 111)	21.54 % (n = 130)
Emoia impar	56.72 % (n = 134)	64.44 % (n = 135)	31.25 % (n = 80)

Figure 251: *E. impar* is characterized by a shiny sky blue tail. Contrary to most eastern Polynesian populations, melanistic specimens are not present in Vanuatu.

Figure 252: Both *E. impar* (right) and *E. caeruleocauda* (left) have shiny blue tails. *E. impar* does not display the middorsal fusion of dorsolateral dark stripes into a Y shape on the middle of tail base seen in *E. caeruleocauda* (left).

Figure 253: Both *E. caeruleocauda* and *E. impar* (this picture) lack a dark parietal spot typical of *E. cyanura*.

Figure 254: *E. impar* is also distinguished by middorsal scale fusions. Scale fusions can involve one scale but more often several scales and can be repeated on the back separated by double scale rows. Such fusions are often present in *E. impar* but never in *E. cyanura*.

glossy blue". His remark clearly applied to the species that I here erroneously call *E. cyanura* (brown tailed morph) since actual *E. impar* has a distinct shiny blue tail in accordance with the specific epithet *cyanura*.

E. cyanura sensu lato occupies the whole of the tropical Pacific from the Admiralty Islands, Bismarck Archipelago, Solomon and Vanuatu, Micronesia to the whole of Polynesia, including Hawaii and Clipperton Atoll in the eastern Pacific. *E. cyanura sensu lato* was reported from the Santa Cruz Islands (southern Solomon): Vanikolo, Lomlom, Tikopia and Matema by Roux in 1913, by Burt and Burt in 1932, and by Brown in 1991. This species group was also reported from Vanuatu by Brown in 1991. *E. cyanura sensu lato* was reported from the Torres and Banks groups, Santo, Malo, Aore, Aoba, Pentecost, Malakula, Ambrym, Tongoa, Efate, Emao (off Efate), Erromango, Aniwa, Tanna, Futuna, and Anatom. The occurrence of *E. cyanura s.l.* on Santo was reported by Roux in 1913 (as "Spiritu Santo"), Medway and Marshall in 1975, Burt and Burt in 1932, and Brown in 1991. Specimens of *E. cyanura s.l.* (registered as "*E. cyanura*") from Santo are located in the following museums: AMNH, AMS, BMNH, FMNH, MCZ, and NMBA.

Species from that group are terrestrial and always found in relatively high densities. Syntopy between species is common and on Santo all three species, *E. cyanura sensu stricto*, *E. impar* and *E. caeruleocauda*, can be found together. However, their ecology is clearly distinct concerning microhabitat requirements. Baker in 1947 studied the reproduction of *E. cyanura sensu lato* on Santo (collections made during 1933-34) but does not distinguish *E. cyanura sensu stricto* from *E. impar*. He noted the occurrence of one unique egg in each oviduct. Despite the relative constancy of climate in the study area, he noted a reproduction peak in November-December (60% of adult females were gravid) and a clear decrease in May-June (only 15%). Unfortunately his data are not reliable since they mix both of the actually recognized species, *E. cyanura sensu stricto* and *E. impar*. Seasonal variations in the proportion of gravid females seems however constant throughout the distribution range of both species. In French Polynesia, I also observed a clear reproduction peak in November-December and a clear decrease during the austral winter with clear differences between both species. *E. impar* demonstrates a maximum during the full rainy season whereas *E. cyanura s.s.* shows a maximum just before (Table 25).

SQUAMATA, SCINCIDAE
Emoia cyanura [*sensu stricto*]

Specimens referred with confidence to *E. cyanura* from Vanuatu without more precise location are conserved in several museums (AMS, FMNH, MNHN). *E. cyanura s.s.* occurs at the following localities in Vanuatu: Torres and Banks groups, Santo, Malo, Aoba, Pentecost, Malakula, Ambrym, Epi, Tongoa, Efate, Emao (off Efate), Erromango, Tanna, Futuna, and Anatom. Specimens from Santo were deposited at AMS, AIM, BMNH, FMNH, MNHN, and UMMZ.

Adult specimens from Santo have a snout-vent (SVL) length of 45 to 54 mm with a tail of about 150 % to 170 % of SVL. The number of scale rows at midbody varies from 27 to 30 but is most often 28.

SQUAMATA, SCINCIDAE
Emoia erronan

The skink *Emoia erronan* belongs to the *Emoia concolor* subgroup of the *Emoia samoensis* group. It is often mixed with the latter binomen in collections. It is only known from Futuna Island, southern Vanuatu, an island formerly called Erronan. It can be readily distinguished from the other species of the *E. concolor* subgroup (except with *E. aneityumensis*) by its number of middorsal scale lines from the frontoparietal to the base of the tail (77 to 84). It can then be distinguished from *E. aneityumensis* by its number of fourth toe subdigital lamellae (36 to 42 in *E. aneityumensis* vs 47 to 53 in *E. erronan*). The species is endemic to the southern Vanuatu island of Futuna and has never been reported from other islands. No information is available on the habitat and reproduction of this species. Specimens are present in the following museums: AMNH and FMNH.

SQUAMATA, SCINCIDAE
Emoia impar

Bruna and co-authors in 1996 considered the blue tailed *E. impar* as originating in Vanuatu or a neighbouring area. I rather suggest the blue tailed *E. impar* clade to have originated in the Bismarck Archipelago area (Papua New Guinea). A recent molecular study by the aforementioned authors has shown that Vanuatu *E. impar* populations present a 13 % genetic divergence in their mitochondrial DNA haplotype compared to those from Cook, Tahiti, Fiji, Arno and Kosrae in Micronesia. *E. cyanura* populations from Vanuatu only present less than 0.25 % genetic divergence with regard to Hawaiian, Takapoto (Tuamotu, French Polynesia), Cook, Fiji and Clipperton Atoll (melanistic form) populations. Thus blue tailed populations from Vanuatu (at least some, i.e. the ones studied by them) should belong to an undescribed blue tailed taxon distinct from the eastern Polynesian one.

Figure 255: *E. nigra* is a semi-arboreal species that most often forages at ground level. Torres Islands.

Figure 256: Most adult Vanuatuan *E. nigra* are dark black with reddish eyes. Torres Islands.

E. impar was reported from the following islands of Vanuatu: Aore, Tatuba, Malo, Santo, Aoba, Pentecost, Wala (off Malakula), Malakula, Ambrym, Epi, Tongoa, Efate, and Tanna but I collected it recently in northern Vanuatu (Torres Islands) where it is sometimes common. Specimens of *E. impar* from Santo are present in the following museums: AMS, AIM (Auckland), FMNH, and MNHN.

Adult specimens of *E. impar* from Santo have a snout-vent length (SVL) of 41 to 50 mm with a tail of about 145 % to 169 % of SVL; they are on average smaller than *E. cyanura*. The number of scale rows at midbody varies from 27 to 31 and is usually above 28. Specimens without at least one fused medio dorsal scale row are very scarce (about 5 %; n = 137) on Santo whereas the absence of such fusion concerns about 12 % (n = 1266) of specimens in French Polynesia.

SQUAMATA, SCINCIDAE
Emoia nigra

E. nigra belongs to the *E. samoensis* group whose distribution extends from the Bismarck Archipelago to Solomon, Vanuatu, Loyalty, Fiji, Samoa, Tonga and Cook. That skink group comprises numerous species, almost all endemic to single islands or small island groups.

Emoia nigra has a small interparietal plate located just below the posterior edge of the frontoparietal. That interparietal shows a clear tendency to disappear and its anterior border is often linear and not V shaped as in most *Emoia* species. The parietal spot is distinct. The tail is transversally flattened. Adult males and females often present a dark to black throat whose function is unknown. Anterior and posterior loreals are generally of same size and shape. The tympanum is enlarged. Supranasals are sometimes nearly in contact but generally largely separated. Prefrontals are narrowly to moderately separate. Subdigital lamellae are broad and rounded. Snout-vent length of adult males and females (no observed dimorphism) varies from 103-116 mm and the tail is about 164-196% of SVL. Fourth toe lamellae vary from 33-36 and midbody scale rows from 35-44.

E. nigra is distributed on southern Pacific islands from Samoa to Tonga and west to Fiji, Vanuatu, Solomon and the Bismarck Archipelago (uncommon). *E. nigra* was reported from the Santa Cruz group, southern Solomon by Roux in 1913 and by Brown in 1991 from several localities. *E. nigra* was reported from Vanuatu by Burt and Burt in 1932, McCoy in 1980, and Cranbrook and Pickering in 1981. It was reported from the following islands of Vanuatu: Torres and Banks groups, Santo, Malo, Aore, Aoba, Maewo, Pentecost, Efate. Specimens from Santo were reported by Burt and Burt in 1932, Medway and Marshall in 1975, and Brown in 1991. Specimens from Santo are conserved in the following museums: AMNH, AMS, BMNH, FMNH, and MNHN.

On Santo this species clearly occupies open habitats, even if forested, but does not occur in deep dark forest like e.g. *Emoia nigromarginata*. It sleeps under stones and trunks at night, not on tree branches like some other similar large sized *Emoia* (*E. concolor*, *E. trossula*, *E. sanfordi*…). Such sleeping behaviour can explain the disappearance of *E. nigra* when mongoose or pigs are introduced. Competition between *E. nigra* and other large species of the *E. samoensis* group during daytime could have given rise to habitat separation: with *E. nigra* on the ground vs other *E. samoensis* species inhabiting tree trunks and branches. A similar kind of competition could occur during the night for shelter: branch forks for other *E. samoensis* group species vs soil level under logs and stones for *E. nigra*. Some *Emoia sanfordi* specimens are nearly totally black or at least greyish black and thus can be mistaken for E. nigra. They can however be easily distinguished since *E. sanfordi* prefrontals are in narrow to broad contact vs being separated in *E. nigra*, the last supralabial is dorso-ventrally entire and unique in *E. sanfordi* vs separated in *E. nigra*, finger and toe lamellae are broad and rounded in *E. nigra* vs thin and bladelike in *E. sanfordi*, and finally the interparietal plate is small in E. nigra vs larger in *E. sanfordi*. Juvenile *E. nigra* are also characteristic, as their ventral tail is whitish with a typical fine

Figure 257: *E. nigromarginata* is an endemic species from Central and Southern Vanuatu. Tasmate, Santo.

Figure 258: *E. nigromarginata* is a semi-arboreal forest species living on small tree trunks and lower vegetation. It searches for shifting sunny patches where it can thermoregulate. Butmas, Santo.

median black line. Juvenile *E. nigra* have generally strong keels, a dark brown back and black flanks. Medio dorsal scales are larger than those on the flanks (about twice the size) on 8-10 rows.

Eggs of *E. nigra* were found on Santo under stones and under logs, including six eggs under the same stone, which certainly corresponds to two clutches, thus attesting to communal egg laying in that species. Mean egg size for 10 eggs was 15.1 +/- 0.6 mm x 10.3 +/- 0.2 mm (range 14.6-16.3 x 9.9-10.6). However egg size varies considerably for skinks according to the development state of the embryo: one additional egg measured 18.5 x 13.0 and it certainly corresponds to an egg ready to hatch. One of our collected females was gravid with three eggs

Figure 259: *E. sanfordi* is a beautiful Vanuatu endemic arboreal skink.

Figure 260: Complete green morph of *E. sanfordi*.

Figure 261: Black headed green morph of *E. sanfordi*. Torres Islands.

Figure 262: Black headed green morph of *E. sanfordi*. Santo.

Figure 263: Grey headed morph of *E. sanfordi*. Santo.

Figure 264: Marbled morph of *E. sanfordi*. Santo.

Figure 265: Marbled morph of *E. sanfordi*. Santo.

Figure 266: Uniform grey morph of *E. sanfordi*. Santo.

Figure 267: Uniform grey morph of *E. sanfordi* with greenish dots.

in the genital tract (20.5 x 11.1, 18.9 x 12.1, and 21.2 x 11.5 mm). One egg hatched in Paris yelding a juvenile with a snout-vent length of 29 mm and a tail of 53 mm. I have observed an adult *Candoia bibroni* eating a large adult of that lizard in the field after both were placed in the same bag.

SQUAMATA, SCINCIDAE
Emoia nigromarginata

After comparison of the holotypes of *E. nigromarginata* and *E. speiseri* and an examination of a complementary series of 30 specimens, Brown in 1991 showed that the midbody scale count and coloration

differences on which Roux had based the separation of both species in 1913 correspond to the extremes of a continuous variation within the same species. He placed *E. speiseri* in the synonymy of *E. nigromarginata* since the description of the later species is situated on page 154 whereas that of *E. speiseri* is on page 155 in the same publication. It has been proposed that *E. parkeri* (endemic to Fiji) could be a sister species of *E. nigromarginata*, thus implying faunal exchanges between Vanuatu and Fiji. This hypothesis has of course to be demonstrated with modern molecular technics before it can be given any credit. I have observed important colour differences between west coast (Tasmate) and central Santo (Butmas) populations (see pictures).

E. nigromarginata is a medium-sized species (adult snout-vent length varies from about 50-80 mm). It is a forest species that never frequents open habitats. It is endemic to Vanuatu. Cranbrook in 1985 noted that the species was not found again during the important collections made in Vanuatu in 1971. It was reported from Vanuatu from the following islands: Santo, Pentecost, Malakula, Ambrym, Epi, Efate, and Anatom. In 1975, Medway and Marshall considered that the species could be present on Erromango and Tanna; this has not been assessed since that time but seems realistic to us. Only few specimens from Santo (reported by Brown in 1991) are deposited in museums: FMNH, and additional specimens from the Santo 2006 expedition in MNHN.

SQUAMATA, SCINCIDAE
Emoia sanfordi

Before its description, specimens of *Emoia sanfordi* from Vanuatu were refered to *Emoia samoense* by Roux in 1913, Baker in 1928, and Angel in 1935. *E. sanfordi* was described based on specimens from Vanuatu (including Banks Islands) and two specimens from the Solomon Islands (Fauro Island) of doubtful origin.

In 1974, Medway noted: "E. samoensis, *on the other hand, has been found in the New Hebrides only on Erromanga [Erromango], some 180 km north-northwest of Aneityum [Anatom] and the next large island but one. The specimens (BM 1860.3.18.8 & 1860.3.18.11) were collected more than a century ago*

Figure 268: *E. sanfordi* often shows a typical lighter eye border.

Figure 269: Dark museum specimens of *E. sanfordi* can be distinguished from *E. nigra* by their typical blade like toe and finger lamellae (vs rounded in *E. nigra*).

by Mr Cuming (Boulenger, 1887). In 1971 we did not find the species, but our stay on Erromanga was brief and collecting not intensive; there are no grounds to doubt the record. The third of Cuming's specimens allocated to this species by Boulenger (BM 1860.3.18.12) is in fact E. sanfordi, *an identification which confirms at least that the collection derived from the New Hebrides. ... All New Hebrides specimens identified by Roux (1913: 155) as* Lygosoma (Emoa) samoense, *a synonym of* E. samoensis, *prove to be* E. sanfordi *and thus provide a fallacious basis for comparison. ...* E. samoensis *and* E. aneityumensis *may replace each other ecologically on their respective islands in the New Hebrides, but there are no grounds for treating the two taxa as geographical races of one species".* In 1975 Medway and Marshall stated that *E. sanfordi* was not distinguished from *E. samoensis* in the previous publications of Boulenger (1887), Roux (1913) and Angel (1935). Specimens attributed to *E. samoensis* by those authors were re-examined by Medway and Marshall in 1975 and with only two exceptions all correspond to *E. sanfordi*. The two specimens they refer to as *E. samoensis*, among which one is from the same locality as *E. sanfordi*, are of uncertain location (Erromango). The species was not found again by Medway on Erromango in 1971, but his sojourn was short and collections done in low elevation areas only. In 1981, Cranbrook and Pickering noted about *E. sanfordi*: "*Doubfully recorded from the island of Fauro in the western Solomons*". They also mention both specimens of *Emoia samoensis* collected on Erromango in 1860.

The specimen BMNH 1860.3.18.12 was later determined as *E. concolor* by W.C. Brown in 1983 in the British Museum catalogues, and its collecting locality questioned. Both of the other problematic Cuming specimens from Erromango were regarded as *E. samoensis* by Lord Medway in 1972, and later as *Emoia trossula* by Brown and Gibbons during their description of that species. The collection locality of both lizards is highly uncertain, thus excluding them from the type series of *E. trossula* during its description. The University of Hamburg (Germany) specimen R01974 from Ovalau (Fiji) was identified as *E. sanfordi* by W.C. Brown in 1983 regarding *E. sanfordi* from Vanuatu. That specimen was different from other Fijian species and Brown supposed its location to be a mistake.

In his 1991 revision of the genus *Emoia*, Brown restricted the distribution of *E. samoensis* to Samoa and that of *E. sanfordi* to Vanuatu (including Banks in the north), with a doubtful record for Solomon, but he did not examine the important specimens mentioned above. The Genova Museum in Italy (Museo Civico di Storia Naturale) possesses one or more paratypes of *E. sanfordi* according to their catalogues.

E. sanfordi is large with a maximal snout-vent length of about 115 mm and a tail about two times snout-vent

Figure 270: *L. noctua* is a small terrestrial skink that often lives in leaf litter and dead fallen tree trunks. It can climb on tree trunks but seldom above one or two meters above ground. Santo.

Figure 271: *L. noctua* is the only viviparous lizard of Vanuatu. Its black and white barred lips are unique among Vanuatuan skinks. Santo.

Figure 273: Another unique trait of *L. noctua* is the yellow to white head spot bordered by black located at the anterior end of the light middorsal stripe. Santo.

Figure 272: Like some other skinks, *L. noctua* can change its ground colour from light to dark. It is also found in the detritus in the axila of *Pandanus* leaves. Torres Islands.

length; midbody scale rows 30-32, fourth toe lamellae 62-72; always five supralabials in front of the subocular; interparietal always present; parietal suture and nuchal suture oriented left. All examined specimens show a high uniformity for scalation characteristics but not for coloration going from uniform green to almost completely and nearly uniform greyish. When few black marks are present they are restricted to the head area and to some parts of the back and flanks, with black punctuations only concerning some scales. Prefrontals are always in wide contact. Anterior and posterior loreals are identical, flat and elongate but the anterior is higher; supranasals are largely separated and prefrontals fused. Juveniles are sometimes uniform green or uniform dark brown to almost black. The throat is sometimes more or less marbled. Differences between *E. nigra* and *E. sanfordi* can be seen above (see *E. nigra*). I can add that the tail is rather circular or subcircular in section in *E. sanfordi* whereas it is rather laterally flattened in *E. nigra*. During my trip to Santo I observed that uniform grey specimens were usually associated with deeper and dense forest.

In 1975, Medway and Marshall noted that: "*Of large* Emoia, E. sanfordi *is recorded from every island visited except Tanna; the species* E. samoensis, E. aneityumensis *and* E. nigra *occur with* sanfordi, *but never with each other on any island. It is notable that among each pair, one species has a high count of subdigital lamellae and the other low. This feature is possibly a scansorial adaptation. Certainly* E. sanfordi *is highly arboreal in habit."* On Santo *E. sanfordi* occurs as several clearly distinct colour morphs that are certainly related to habitat characteristics: a complete green morph, a green body with black top of head morph, a marbled body morph, a complete grey headed green morph and a totally uniform grey body morph.

This species, a Vanuatuan endemic, was reported from Banks (Vanua Lava, Santa Maria), Santo, Aore, Malo, Aoba, Pentecost, Malakula, Ambrym, Epi, Tongoa, Efate, Erromango, and Anatom. In 1975, Medway and Marshall recognized that *E. sanfordi* could occur on Tanna where they only briefly sampled. I collected it recently on the Torres Islands. Collections from Santo were reported by Medway and Marshall in 1975, and by Brown in 1991. Specimens from Santo are deposited at AMS, BMNH, FMNH, and MNHN. Note that recent molecular investigations showed that the true *E. sanfordi* does not exist south of Efate, an area occupied by other cryptic species.

E. sanfordi is the most widespread endemic species in Vanuatu. Such a distribution probably relates to

Figure 274: *Ramphotyphlops braminus* is often called the flower pot snake. It was easily accidentally introduced, probably with soil, on most tropical and subtropical Pacific islands. It is also called the two headed snake since tail and head are similar in shape. The tail, however, ends in a spine that does not inject venom. This snake is completely harmless. Santo.

Figure 275: *R. braminus* feeds on ants and termite eggs and larvae. Populations are composed only of females since the species reproduces by producing clones through parthenogenesis. Santo.

the age of this skink species. The diversified colour polymorphism observed on Santo was not seen on the Torres Islands.

SQUAMATA, SCINCIDAE
Lipinia noctua

The genus *Lipinia* belongs to the *Sphenomorphus* group of lygosomine skinks and comprises 21 species. It has two main radiation centers, the Philippines (eight species) and New Guinea (seven species). In 1975, Medway and Marshall noted that the species is rare and secretive in Vanuatu. I can add that this is the case throughout its distribution range and in Vanuatu it is as abundant as in other places. *L. noctua* occupies most of the area where the genus is distributed; it can be found from the Papuasian area to Oceania, including Hawaii and Pitcairn, but the species is not present on Easter Island. It is also absent from Clipperton Atoll, thus showing clearly reduced colonisation abilities compared to other similar sized skinks (e.g. *Emoia cyanura* and *E. impar*); such characteristics certainly have to be associated with its partly fossorial habits. Its type locality is located on Kosrae Island, State of Kosrae, Federated States of Micronesia. Classically the genus was considered as having a Papuasian origin but molecular studies conducted by Austin and published in 1998 have shown that its origin is instead located on the Asian mainland or on the Philippines.

L. noctua was reported in Vanuatu from Santo, Malo, Aore, Pentecost, Malakula, Ambrym, Epi, Efate, Tanna, and Anatom. *L. noctua* also occurs north of the archipelago since I collected it recently on most of the Torres islands, so it is widespread on the whole archipelago. The species was reported from Santo by Medway and Marshall in 1975. Specimens from Santo are conserved in the following museums: AMS, BMNH, MNHN, and UMMZ.

••• Reptiles from Vanuatu – SNAKES (Typhlopidae)

SQUAMATA, TYPHLOPIDAE
Ramphotyphlops braminus

Fossil remains recovered from presumably pre-human levels in the Marianas (Micronesia) suggest that this species occurs naturally there. *R. braminus* was first reported from Vanuatu on Efate by Medway and Marshall in 1975 and later by Cranbrook and Pickering in 1981. The first specimens were collected in the gardens of White House, British Paddock, Independance Park, Port Vila and in plantations in the Agricultural College of Tagabe on Efate. According to Medway and Marshall, on Efate the species seems only to occupy cultivated areas highly modified by man. The Australian Museum collections also possess a specimen from Efate collected by I.W.B. Thornton in 1976.

The species was until now reported from Vanuatu only from both main islands, Santo and Efate, and more recently from Ambrym. I collected it recently (November 2007) in the very far north of the archipelago, on Toga Island (Torres Group), thus clearly indicating that the snake has largely expanded its distribution range in Vanuatu. On that island I also collected *Gehyra mutilata*, another recently introduced gecko (see above). The species arrived in Vanuatu probably shortly after 1970 but it seemed to be absent in 1971 during the Percy Sladen Expedition. Specimens from Santo are conserved in several natural history museums: AMS, MHNG and MNHN. The oldest record known for Santo was collected at Luganville by A. de Chambrier between 9 and 26 June 1979.

Figure 276: *C. bibroni* is the most arboreal species of its genus. Its prehensile tail allows it to climb high in trees and to hang by only its tail, when necessary, e.g. to explore bird nests. Torres Islands.

Figure 277: The typical snout shape (flattened rostrum) of *C. bibroni* explains its common name of the shovel nosed snake. Santo.

Figure 278: Dorsal colour polymorphism is considerable in *C. bibroni*, varying from red to grey or even nearly black. Santo and Torres Islands.

Some specimens can have a light electric grey blue colouration and they can thus be confused and considered as belonging to another species. In fact such coloured specimens correspond to the same species when shedding its skin. All the specimens I observed in Vanuatu were from human disturbed habitats, but they can be found far from human habitation (e.g. roadsides and cattle parks). They can locally reach high densities and several snakes can be seen under the same small stone. Local people often consider them as worms rather than snakes.

••• Reptiles from Vanuatu – SNAKES (Boidae)

SQUAMATA, BOIDAE
Genus *Candoia*

The position of the genus *Candoia* inside boine snakes has always been controversial. Three distinct hypotheses can be considered:
- A basal position;
- A sister group of the Neotropical forms (e.g. *Boa constrictor*);
- A sister group of the Madagascan forms (*Sanzinia* and *Acrantophis*).

These hypotheses are either related to dispersion or to the biogeographical mechanisms of vicariance. An additional difficulty to resolve is that the genus only occurs on former Gondwana fragments. Three species are classically recognized. *C. bibroni* occupies the largest distribution from eastern Solomon and the Loyalty Islands east to American Samoa. *C. carinata* occupies the whole of Solomon, including Santa Cruz, western New Guinea, Sulawesi and north of Palau. *C. aspera* is a New Guinea species, living below 1 500 m elevation, also present in the Bismarck Archipelago and Manus Island. Despite important morphological differences, those three species show a similar diet mostly based on rodents. A recent molecular analysis, albeit based on limited material, demonstrates the basal position of *C. bibroni* in the genus and close links between *C. aspera* and *C. carinata*. It also shows that a clade (*Sanzinia/*

Figure 279: Juvenile *C. bibroni* usually are yellowish orange and mistaken for a different species by non familiar observers. Torres Islands.

Candoia) diverged from Neotropical boas at least 40 million years ago. The basal position of *C. bibroni* is controversial but in 2000, Austin formulated the hypothesis that it is in competition with species of the genus *Python*, while also sympatric with both other species of *Candoia* but absent from the distribution of *C. bibroni*. Other interpretation problems also arise from absence of calibration through fossils; it is in fact not possible today to distinguish vertebrae of Boinae from those of Pythoninae inside boid snakes. Complementary works are necessary to obtain a clear view of the evolutionary history of the genus *Candoia* in the Pacific.

SQUAMATA, BOIDAE
Candoia bibroni

According to osteological characters, McDowell in 1979 distinguished two forms in *Candoia bibroni* that he considered as valid subspecies, one eastern (*C. bibroni bibroni*: Loyalty, Fiji, Samoa and extending to the east as far as Tokelau in Polynesia) and one western form (*C. bibroni australis*: Vanuatu, southeast Solomon). The specimens used in Austin's study (see above) unfortunately only comprised the eastern form. In 1985, Gibbons noted: "*Based on distribution the genus [Candoia] is Papuan in origin, and resembles that of* Emoia *skinks and* Gehyra *geckos, both of which are present in the Loyalty Islands, but absent from New Caledonia.*"

C. bibroni is a highly polymorphic species concerning its colouration in Vanuatu, with blackish, reddish, brownish, orange and grey blue morphs. It seems that orange-yellowish specimens are largely predominant in northern Vanuatu (Torres) where its size seems smaller. Juveniles are often without patterning and uniformly orange, thus giving them the look of a different species for people who are not aware of this. Specimens of about 1.5-2 m in length were observed around Port Vila. I also have seen a specimen well over 2 m (unfortunately not measured) at Penaoru (Santo).

That snake is arboreal in habit but can sometimes be observed resting at ground level. It seems to be more terrestrial in the north of Vanuatu (Torres). It shows a clear sexual dimorphism for size (females larger than males) and cloacal spurs (absent in about 50 % of females). Maximum published snout-vent is 1 460 mm for females and 1 190 mm for males but larger specimens clearly exist and are not uncommon, except perhaps in museum collections. Each gravid female gives birth to 15 to 18 live youngs but reproduction is not annual since about only 1/3 of the females in collections are gravid. Juveniles eat lizards (skinks) and adults mammals (rats and mice). In 1935, Angel noted that one skink (erroneously called *Emoia samoensis* but in fact *E. sanfordi*) was found in the digestive tract of one *Candoia* from Vanuatu. Burt and Burt also noted the occurrence of the skink *Emoia cyanogaster* regurgitated by such a snake from Tapua Island (Santa Cruz Group, Solomon). I also have observed predation by that snake on adult *Emoia nigra* on Santo.

C. bibroni was reported from the Santa Cruz group (southern Solomon) by Roux in 1913, and Burt and Burt in 1932. In Vanuatu, the species is reported from Torres and Banks groups, Santo, Malo, Aore, Aoba, Maewo, Pentecost, Malakula, Ambrym, Efate, Emao (off Efate), Erromango, Tanna, and Futuna. The species was reported from Santo by Roux in 1913 (on "*Spiritu Santo*"), by Burt and Burt in 1932, Medway and Marshall in 1975 and Horrocks in 1989. Specimens from Santo are present in numerous world museums (AMNH, AMS, BMNH, MNHN, NMBA, and UMMZ).

SQUAMATA, BOIDAE
Candoia carinata

In 1932, Burt and Burt proposed the eastern distributional limit for the boa *Candoia carinata* as the Banks and Santa Cruz groups. They report a specimen from Santa Cruz. In 1989, Horrocks reported a curious observation: "*As I look at the colour picture of* C. carinata *in the magazine* [Solomon Airlines inflight magazine *1988 or 1989] I instantly see the Boa which we kept for several weeks in an observation tank at Malapoa. Our Malapoa specimen came from the Vila* [Port Vila, Efate] *rubbish tip.* C. carinata *is described as a very adaptable and variable species. It seems highly likely that we have both* C. bibroni *and* C. carinata *Boas in Vanuatu.*" Recently A. H. Whitaker kept a captive boa collected on Efate and which he identified as *C. carinata*. That specimen could have been a recent accidental introduction, but its genetic analysis by C. Austin revealed a unique pattern among other studied Pacific populations. On the web, it is easy to find *Candoia carinata paulsoni* from Vanuatu in the international pet trade (September 2006; e.g. Society "*World Wild Fauna*"). It seemed highly probable that this snake was also deliberately introduced to Vanuatu (at least on Efate) through the German reptile dealer Bregulla, as he certainly did for *Brachylophus* iguanas.

In 1979, McDowell demonstrated the existence of two morphotypes inside that species, one "short tailed" and one "long tailed". Austin's molecular studies showed clear differences between populations of southern and northern New Guinea, the central mountain chain being an important biogeographical barrier. This suggests that the colonisation of the Solomon Islands occurred through a southern route via the d'Entrecasteaux Archipelago rather than via a northernly route through the Bismarck Archipelago.

C. carinata has supralabials in contact with the eye whereas they are separated from the eye by suboculars in *C. bibroni*. The snake is nocturnal and terrestrial. It can be observed during daytime in tree trunk holes and, like *C. bibroni*, is common in and around caves. It feeds on lizards and rodents, but sometimes also on frogs and bats. Austin (2000) considered the species to be semi-arboreal and an ecological intermediary species between *C. bibroni* and *C. aspera*.

••• Reptiles from Vanuatu – SEA KRAITS or AMPHIBIOUS SEASNAKES
(Elapidae, Hydrophiinae)

SQUAMATA, ELAPIDAE, HYDROPHIINAE
Laticauda colubrina

In 1913, Roux reported a juvenile specimen from Malo (Vanuatu) (deposited in NMBA) and noted that: "*Les taches noires ne forment pas des anneaux complets comme chez les exemplaires trouvés en Calédonie*". He thus made an interesting observation recently taken into account by Heatwole and co-authors when revalidating *L. frontalis* from Vanuatu (see below). The specimen examined by Roux certainly refers to *L. frontalis*. Old records of *L. colubrina* from Vanuatu could refer either to the true *L. colubrina* (*sensu stricto*) or to *L. frontalis*, since both Vanuatu species were previously mixed under the binomen *L. colubrina*. Thus old records are not really relevant without a re-examination of the concerned specimen.

In 2005, Heatwole and coauthors examined a large sample and proposed the following distribution of *L. colubrina sensu stricto* in Vanuatu: Santo, Elephant Island (= Ais Island) near Santo, Ngoriki Islet near Kakula Island, Malakula, Efate, Tanna, and Anatom. Numerous specimens of *L. colubrina* were reported from Santo by Heatwole and coauthors in 2005; they are conserved in AMS, FMNH, MNHN, UMMZ, USNM.

Table 26: Morphometric and meristic comparison of *L. frontalis* and *L. colubrina*. **MBS M/F**: Midbody scale rows for males (M)/females (F). **V M/F**: Number of ventrals for males (M)/females (F). **SVL max M/F**: maximum snout-vent length for males (M)/females (F).

	MBS M/F	V M/F	SVL max M/F
Laticauda frontalis	19-21/21-23	192-208/199-211	654/783
Laticauda colubrina	23-24/23-27	216-234/202-239	893/1450

Pickering in 1984 supposed that this seasnake could feed on littoral skinks of the species *Cryptoblepharus novohebridicus*, a hypothesis that is completely wrong since these seasnakes only feed on marine fish. Even though skinks are common on many small islands where sea kraits are abundant, they don't feed on lizards; no explanation is available for this.

SQUAMATA, ELAPIDAE, HYDROPHIINAE
Laticauda frontalis

L. frontalis can be distinguished from *L. colubrina* by the following combination of characters: midbody scale rows, number of ventral plates, smaller size (Table 26), and in sympatry by the absence of a lower lateral connexion between the black cephalic band and the first nucal black band, and the absence of black bands contacting at least the anterior ones at mid-belly. *L. frontalis* occurs in Loyalty and Vanuatu (from Santo south to Anatom) but is reported from New Caledonia by a single specimen without a precise collection locality (also Loyalty Islands?). Food is based on moray eels, mostly Muraenidae and Congridae as in other species of the genus. That species is less common than *L. colubrina* in Vanuatu.

In 2005, Heatwole and coauthors reported *L. frontalis* from several Vanuatu localities. It was cited from Santo, Elephant Island (east coast of Santo), Efate, Ngoriki Islet near Kakula Island (northern coast of Efate), and Anatom. Specimens of *L. frontalis* from Santo are conserved in AMS, FMNH, LACM, MCZ, and UMMZ.

SQUAMATA, ELAPIDAE, HYDROPHIINAE
Laticauda laticaudata

In 1913, Roux reported the occurrence of *Laticauda laticaudata* from southern Solomon (Santa Cruz group). The MNHN François collection comprises one specimen from Vanuatu without a precise locality. The species is known from the following islands of Vanuatu: Ambrym, Efate, and Anatom. The species was not previously reported from Santo but I collected a specimen from Tasmate (west Cape Cumberland coast) during the Santo 2006 expedition.

••• Reptiles from Vanuatu – TRUE SEASNAKES (Elapidae, Hydrophiinae)

SQUAMATA, ELAPIDAE, HYDROPHIINAE
Hydrophis coggeri

The seasnake *Hydrophis coggeri* is present from northern Australia to Fiji, through the Philippines, the north coasts of Sulawesi (Indonesia), the Coral and Timor seas, New Guinea, New Caledonia and Vanuatu. In 2005, Kharin reported the occurrence

of the species from Vanuatu and placed it in the genus *Leioselasma*, a position not always followed. In 1986, David reported the shed skin of a seasnake found near a mangrove on Malakula and identified as a *Hydrophis*. It could correspond to *H. coggeri*. No specimen of *H. coggeri* from Vanuatu is present in BMNH or MNHN collections.

SQUAMATA, ELAPIDAE, HYDROPHIINAE
Pelamis platura

The pelagic seasnake *Pelamis platura* is represented from Vanuatu by only a few museum specimens. It was only reported from Malakula and Efate, but the species certainly occurs in waters around Santo and even throughout the whole archipelago. No specimen of that species from Vanuatu is present in either the BMNH or in the MNHN collections.

••• Potential and doubtful species

Several lizard and snake species are reported from the southern Solomons (Santa Cruz group) and thus are potentially present in northern Vanuatu (Torres and Banks). The Santa Cruz Islands are the closest of the Solomon Islands to Vanuatu. However, fresh water is absent on the Torres Islands, which are composed of coral limestone and rarely exceed 300 m elevation.

••• Skinks

Potential Santa Cruz lizards that could occur in Vanuatu include several skink species. In 1913, Roux reported the occurrence of *Lamprolepis smaragdina* as *Dasia smaragdina* on the Santa Cruz group and noted: "*Cette espèce n'avait pas encore été signalée aussi loin à l'est en Mélanaisie (sic)*". In 1932, Burt and Burt also reported that species from Duff and Santa Cruz islands of the Santa Cruz group. In 1980, McCoy also reported *Lamprolepis smaragdina* from Santa Cruz and Reefs Islands (Solomon) where it is common. That arboreal species is present in forest but also in coconut plantations, mostly on tree trunks but rarely on branches. *Emoia trossula* was formerly considered as a regional endemic (Fiji, Tonga, and Cook) and its situation is still incompletely resolved (species complex). Two specimens at the BMNH are registered from Vanuatu, under the binomen *E. samoensis* [note however that one is reported as a syntype of *Perochirus guentheri*?]. That mention is considered as a locality mistake because those specimens are actually *E. trossula*, a species not present in Vanuatu. In 1986, Brown and Gibbons stated: "*(these specimens), which were purchased from Mr. Cuming, are stated to be from Eumonga and Vanuatu. However, since they agree in most characters with examples from populations of* E. trossula *from the Fijis and not with samples from any of the species known from Vanuatu, we assume that the locality data are probably in error and have referred these two specimens to* E. trossula, *but have not included them in the paratypes"*. Another skink species of the genus *Emoia*, *E. rufilabialis*, is an endemic skink from Santa Cruz Island, southeastern Solomon. That *E. cyanura* group and *E. cyanura* subgroup species could be present in northern Vanuatu but our recent searches on the Torres Islands proved that this was not the case.

Another large skink, *Eugongylus albofasciolatum*, was reported by Roux in 1913 from the Santa Cruz Islands, by Burt and Burt in 1932 on Tikopia Island, Santa Cruz Group, and by McCoy in 1980 from the same island group. *E. albofasciolatum* is particularly shy and difficult to observe. It could be potentially present in northern Vanuatu (Torres and Banks). Our recent searches on Torres Island showed however that one is unlikely to encounter this skink in Vanuatu. One smaller skink, *Prasinohaema virens*, is reported from Santa Cruz and Reefs Islands by McCoy in 1980. One specimen from Vanikolo (Santa Cruz Group), labelled as *"Lygosoma anolis, Leiolopisma* or *Lipinia anolis* Boulenger" corresponds to the subspecies *Prasinohaema virens anolis*. That subspecies was described from Treasury and Santa Ana Islands. Its status is uncertain and the taxon was not recently compared with the nominotypical subspecies. This arboreal species can be observed as high as 2-20 m in forested areas. It is easily recognisable since the head and body stay close to the tree trunk when the lizard moves. The species can be observed at night when sleeping on accessible branches. Its blood pigment is typically green, as are its mouth and tongue, and also the egg shell. This skink has toe lamellae like geckos and can easily climb glass walls.

••• Snakes

Finally one snake species, *Dendrelaphis salomonis* [sometimes referred to as *Dendrelaphis calligaster*, or *D. calligastra*] was reported by Roux in 1913, based on several specimens from the Santa Cruz Islands. He noted: *"Cette espèce n'avait pas encore été signalée à l'est des îles Salomon"*. In 1932, Burt and Burt collected that species on Nendö (Santa Cruz) and Utupua (Tapua) Islands, Santa Cruz Group; in 1980, McCoy reported the species (as *D. salomonis*) from this island group. This snake could be present in northern Vanuatu but it is unlikely. It is a diurnal and active snake present in forests and cultivated areas. It feeds on amphibians and lizards.

••• **Discussion**

••• Pacific herpetofauna origin

Pacific island herpetofaunas have Oriental, Gondwanian and Neotropical origins. Natural recent arrivals originated from the Papuan area (animal flotation for short distances or drifting rafts) and by more recent man mediated introductions during pre-historical and historical events originating from the Papuan area or through exotic

alien species introductions from different origins (Asia, Australia). Short distance flotation implies air pockets and impermeable skin permitting the animal to remain on the surface without sinking. Sea currents can be totally reversed during climatic anomalies like hurricanes, thus explaining the scarcity of Neotropical elements even in eastern Polynesia, despite trade winds going from east to west. Neotropical elements are however clearly present in southern Pacific herpetofaunas with two endemic genera containing several species each, the iguanid lizard genus *Brachylophus* and the boid snake genus *Candoia*.

Biogeographical analyses are still tentative since numerous species have yet to be described or even discovered, some of them being cryptic species. Also the phylogenetic relationships between species are still unresolved for the most important speciose genera like the geckos *Lepidodactylus* and *Gehyra* or the skinks *Emoia*. Furthermore, several important areas have only been weakly explored. Nonetheless, three factors can explain the biogeographical patterns observed for South Pacific herpetofaunas:
   • Vicariance [separation of populations previously in contact through plate tectonics and/or orogenesis];
   • Dispersion [natural through rafts for reptiles];
   • Accidental or voluntary human transportation.

In human transportation we have to distinguish between ancient human transportation through early Pacific settlers (mostly Polynesians and the first European explorers) that I call **historic introductions** and more **recent introductions**, some of them clearly being intentional like the frog *Litoria aurea* or the iguanid *Brachylophus bulabula* on Efate. *Gehyra oceanica* and *G. vorax* are generally considered as having been mostly transported by man but I disagree with that hypothesis proposed by Beckon. In his 1992 work, this author concluded that his morphological data on *Gehyra oceanica* and *G. vorax* are in accordance with human transportation but with different time scales between *G. oceanica* and *G. vorax*, the later being a forest species with a limited distribution. Brown in 1956 and Gibbons in 1985 also considered geckos and skinks as recent arrivals constituting the main element of the current herpetofauna; no endemic genera are present east of Solomon. Most of these species have a distribution to the west, adjoining their center of origin, the Papuan area. Specific endemism stops at a line joining American Samoa/Niue and only few endemics are present to the east. I think that this scarcity of endemism, even at the specific level, is rather explained by a strong selection for island efficient colonists east of Fiji, thus not allowing populations to become isolated on any island. Pacific lizards east of Fiji are all ecological generalists that are biogeographically specialized to efficient island colonization.

A recent genetic study of Pacific Islanders has shown the same for *Homo sapiens*: "*At some point, prehistoric Oceanic mariners apparently became so accomplished that the inter-island water crossings in the central Pacific were often no more of an impediment to travel than the (already occupied) rugged terrain of the larger island interiors in the western Pacific. In many areas the ocean was transformed from a formidable barrier into a highway*". They also have shown that shore-dwelling Oceanic-speaking human groups are more intermixed since dispersal along the shorelines was easier. The sailing capabilities of the ancestors of the Polynesians transformed the nature of their diaspora and kept them relatively homogenous through constant gene mixing and flow. Thus the situation seen for widespread lizard species on Pacific Islands is exactly the same as the one observed for earlier island human settlers: they all are homogenous since they had specialized to favourise frequent inter-island travels long before the human colonization of those islands, not because they arrived recently.

Pacific island reptile distribution was and still is controlled by: geology (including plate tectonics and volcanism), sea level variations, hurricanes, natural dispersion (rafts) and the role of man. It is classically accepted by most herpetologists that Pacific island species with large distributions have benefited from recent man mediated transportation, but if one endemic species has strong affinities to another geographically separated species then their separation is due to natural colonisation. Such an interpretation seems to not be really logical! In the same way the biology of some alien reptile species like the gecko *Hemidactylus frenatus* or the parthenogenetic snake *Ramphotyphlops braminus*, which recently colonised almost all tropical or subtropical areas of the world, cannot explain why they were not present earlier on Pacific tropical islands through human transportation. Also, how could some shy species like the skink *Lipinia noctua* have been transported by man on such a large scale as the whole tropical Pacific while both the former recently introduced species were not?

Since 1957 and the studies of Darlington, the following dichotomous paradigm has been accepted:
- Natural dispersion with breaks giving birth to micro-differentiations and geographical variations eventually producing endemism;
- Recent human introduction, which has been rapid and thus gives rise to uniform populations.

The reality is of course not so dichotomous. One important faunal break between the Oriental region and the Pacific islands is the lack of several genera such as the skinks *Carlia*, *Sphenomorphus* and *Tribolonotus*, as well as the monitor *Varanus* on most islands. Another second break can be observed more to the east with the lack of *Candoia*

boid snakes and all larger *Emoia* species east of a line through Niue/Samoa (but reaching the Cook Islands). Interestingly, terrestrial elapids have arrived not far from Vanuatu at Santa Cruz and Reefs Islands. Fijian endemism of old faunistical elements (*Platymantis* frogs, *Brachylophus* endemic iguanid lizards and the endemic genera *Ogmodon*, an elapid snake) could be related to a mixture of island size and former position through plate movements, island age, colonization hazards but perhaps also former volcanic events.

Other species biological characteristics like parthenogenesis (*Hemidactylus garnotii*, *Hemiphyllodactylus typus*, *Lepidodactylus lugubris* complex, and *Nactus pelagicus*, the introduced snake *Ramphotyphlops braminus*), saltwater tolerance, delayed incubation time, continuous reproduction, sperm storage, communal egg laying, adhesive eggs… have clearly favourised older natural or man mediated more recent island colonisations. Those characteristics can be in accordance with both hypotheses (old natural colonisations or recent man mediated colonisation) and thus cannot help choose one theory against the other.

••• The role of plate tectonic movements around Vanuatu

Plate tectonics provide an explanation for part of the current distribution, at least for the most ancient elements. Eustatic movements resulting from plate tectonics and sea level variations are also important biogeographical factors. The border between the Pacific and Australo-Asiatic plates is unstable and irregular with several important subduction areas according to age and in different directions. Micronesia, Samoa, and the whole of Eastern Polynesia (east of Tonga) are located on the Pacific plate whereas New Caledonia and New Zealand are located on the Indo-Australian plate. Fiji, Vanuatu, Solomon and Tonga are located at a dynamically unstable area called the Outer Melanesian Arc (OMA). That Arc stretches from New Guinea, through the Solomon Islands, Vanuatu Archipelago, Fijian Islands, the islands of Samoa, and southeast to Tonga. It is a major geological feature of the Pacific basin, west of the Tongan Trench. Components of the OMA are the result of tectonic activity over the last 11.2 to 2 MY, and the biota of this region results from speciation within the OMA as well as dispersal from Southeast Asia, New Guinea, Australia, and New Caledonia.

Strong interactions between both plates (Indo-Australian and Pacific) occur at OMA. Vanuatu was located on that arc between the Solomon Islands (north) and Fiji (south) in mid Oligocene (30 MY) times. The Bismarck Archipelago is located on a different arc. Fiji and Tonga are ancient and estimated to about 40 MY. The Pacific plate is moving under the Indo-Australian plate and the Tongan Trench is located on one end of that subduction area, with the Vanuatu Trench on the other end. On the Melanesian border of that subduction area several movements have taken place, including partial rotations explaining the variable relative positions of some of those archipelagos during time. The Australo-Oriental collision is dated from the mid-Miocene, about 17 MY. Note also that between New Guinea and Australia there existed a now disappeared insular chain that was absorbed to the north of New Guinea during the Australo-Oriental collision at mid-Miocene. That chain permitted the connexion of the southwestern Pacific islands with Southeast Asia.

••• **Geologic movements of Vanuatu through time**
Differences in the herpetofaunal composition between the Loyalty Islands and the New Caledonia mainland are considerable and constitute a clear biogeographical indication of a more important geographical separation in the past. Loyalty possesses mostly Papuan elements (skinks of the genera *Emoia* and *Lipinia*; gecko genus *Gehyra*; snake genera *Ramphotyphlops* and *Candoia*). The presence of *Cryptoblepharus* in New Caledonia can certainly be explained by its strictly littoral life. In the same way *Emoia impar* is absent from Loyalty whereas *E. cyanura* (*sensu stricto*) is present, a situation uncommon on Pacific islands were both species most often occur together. One explanation of such considerable faunal differences should be related to a different past position of Loyalty and the only recent arrival at its present position. Thus Loyalty could have been closer to Vanuatu, explaining its long separation from New Caledonia but also the strong observed faunal affinities between Loyalty and Vanuatu (*Caledoniscincus*, *Emoia loyaltiensis* which is related to Vanuatu endemics of that genus, *Gehyra vorax* [absent from Solomon but present on Loyalty and Vanuatu], *Nactus pelagicus*, *Candoia bibroni* [probably arrived in Vanuatu from the south through Loyalty], and *Laticauda frontalis* [a regional endemic from Vanuatu and Loyalty, sister species of the New Caledonian endemic *Laticauda saintgironsi*]); it would thus not be surprising that an endemic forest gecko species of group II of the genus *Lepidodactylus* will one day be discovered on the Loyalty Islands.

The current position of Vanuatu was only reached recently. In 1996, Allison noted that the emergence of Vanuatu is associated with that of the Vitiaz Arc (= OMA), an intense subduction area formed by the convergence of Indo-Australian and Pacific plates south of the Ontong Java Plateau and the dilation of the north Fiji basin. At the beginning of the Oligocene, the Vitiaz Arc represented an eastern extension of the North Solomon Arc to west of actual Fiji. Relictual islands formed by that arc comprise Santa Cruz and Duff archipelagos (politically the most eastern Solomon Islands) and Banks Islands located north of Vanautu. During the late Miocene (10-8 MY), the New Hebrides arc began to form itself as an eastern extension of the southern Solomon Arc in conjunction with the reverse subduction reported above in relation to the Solomons. Later during the Miocene (about 6 MY), the area between both arc systems began to inflate, thus forming north of Fiji basin and giving rise to a rotation of the New Hebrides Arc of about 30° clockwise. Volcanism began during that time and later moved during the Pliocene to the west to form the central chain currently active in Vanuatu and the active part of Santa Cruz. Following previous tectonic movements, 6-8 MY ago, Vanuatu was still close to north or northwest of Fiji. Colonisation of Fiji and Vanuatu by *Gehyra vorax* from New Guinea is old and predates their separation. Later Vanuatu drifted counter clockwise to progressively reach its current position. Such movements can explain the current distribution of Micronesian elements such as *Perochirus*, if such an origin is confirmed. The Vanuatu herpetofauna is of Papuan origin with a low level of local endemism. Bats are the only native mammals and about half of Vanuatu bird species are shared with New Guinea. The Vanuatu herpetofauna has no old endemic elements compared with Solomon, Fiji or New Caledonia. In 1957, Cheesman admitted that the flora of Vanuatu's southernmost islands (Anatom, Erromango and Tanna) probably originated from a land connexion between Vanuatu and New Caledonia during the Cretaceous before the breaking away of both elements from northern Queensland. However, current data shows rather that Vanuatu has never been in contact with New Caledonia. An important ceratobatrachid frog speciation exists on Solomon (comprising several endemic genera) and Fiji (two endemic species), as well as an ancient snake group, elapid genus *Ogmodon*, an endemic Fijian genus.

Vanuatuan faunas are in all likelihood less than 2 MY old. The oldest Vanuatu rocks are dated to 14 MY but the whole archipelago only emerged out of the sea about 2 MY. The present position of Vanuatu is a result of counterclockwise rotational movement since the Miocene; prior to this rotation the islands of Vanuatu were located to the north of Fiji and Tonga, and were more isolated than at present. Its recent emergence and position could explain the lack of *Platymantis* frogs and *Ogmodon* elapid snakes and the presence of *Perochirus* geckos. The biogeographic patterns highlighted by *C. atro-punctatus* show in particular a biotic break betwen the islands of southern Vanuatu (Erromango, Tanna, Futuna, Aniwa and Anatom) and the islands of central and northern Vanuatu. That break is congruent with patterns recovered from a literature review made by Hamilton and co-authors in 2009 for a broad range of diverse taxonomic groups including vertebrates, invertebrates, and plants. That biogeographical line is defined as Cheesman's line by these authors.

••• Can fossil help us to understand the biogeographical history of Vanuatu's reptiles?

With the exception of one mekosuchine dwarf crocodile, no native vertebrate species seem to have been eliminated by human activity in Vanuatu. According to fossil records, the past distributions of most Pacific island birds were clearly larger than at present. Numerous bird species considered as island or archipelago endemics today correspond in fact to relictual distributions of previously more widely distributed species. Lizard fossil remains are not very common but it seems unlikely that situation was similar. Bird extinctions are classically attributed to man's arrival but sea level movements during glaciation times also produced important habitat variations on islands (increasing salinity, plant mortality, expansion and contractions of littoral plains). During the sea level changes related to glaciations, evolution for survival and colonisation of new habitats mainly took place on smaller islands. On those smaller islands, larger endemic lizard species are rare or absent but this can also be explained because they are more dependent on favourable physical conditions and food than smaller skinks or geckos such as the eastern Polynesian species. In a 1986 paper, Gibbons and Clunie suggested that the major human migrations in the west Pacific could be correlated to migrations for food during such sea level variations resulting from glaciations and dated -700 000 to -2 MY. The situation could be the same for reptiles. Those sea level variations also had a considerable impact on coral reef faunas producing mass extinctions and recolonisation events from refuges as shown for bivalve molluscs by Paulay in 1990. Gibbons and Clunie propose that the current sea level was reached 4 000 years ago and that the lowest sea level was reached 18 000 years ago and increased progressively thereafter. For example, the exposed area of the Seychelles during the glacial maximum was about 600 times greater than the current area. Concerning Vanuatu, almost all islands were separated from each other, even during the glacial maxima, except Santo and Malakula, which were attached together. Endemism on Anatom and Futuna, which is difficult to explain regarding the small size and relative isolation of both islands, could perhaps be explained by their increased size during glacial maxima and perhaps their more diversified habitats.

There is a possibility that fossil bird remains aged more than 3 000-3 500 years could have been present on the islands, but that the areas where these might have been found have been inundated by the sea level rises that started 18 000 years ago. That active period of sea movements could have favourised speciation events through the considerable modification of island ecosystems. Speciation could have taken place through different ways: adaptation to new habitats arising from the increasing salinity of littoral habitats after 18 000 years ago when the sea slowly rose again, increasing numbers of dead trees allowing egg deposition and producing favourable microhabitats, possible recontact between separated species, subspecies, or populations giving rise to parthenogenesis or speciation through hybridisation. In the same way, sea level rises might also have separated populations giving rise to an important subsequent vicariant speciation on Pacific islands.

During glacial periods, sea levels were low, exposing large dry calcareous areas around the edges of islands that were impossible to colonise by elapid snakes and amphibians originating from the wet forests of Fiji and Solomon, whereas smaller lizards could better support such xeric conditions, even for *Lipinia* which needs humidity but can survive in open areas if small wet patches can be found under stones or in trunks, in dead trees or in *Pandanus* sp. crowns. The study of fossil remains from Vanuatu, with the exception of the recent discovery of an extinct mekosuchine dwarf crocodile, have unfortunately not provided significant biogeographical information on the herpetofaunal history of the archipelago.

••• Biodiversity of the herpetofauna of Vanuatu and Santo

The number of reptile species listed for Vanuatu is regularly being increased. Baker (1947) reported 14 lizard species from Vanuatu and stated that the two most common were *E. cyanura* (*sensu lato*) and *E. caeruleocauda* (named *E. werneri*). In 1975, after having surveyed six islands of the archipelago, Medway and Marshall listed 20 species of terrestrial reptiles (five gekkonids: *Gehyra oceanica*, *G. mutilata*, *Perochirus guentheri*, *Lepidodactylus lugubris*, *Nactus pelagicus* [as *Cyrtodactylus pelagicus*]; 13 scincids: *Emoia cyanura*, *E. caeruleocauda* [as *Emoia werneri*], *E. nigronarginata*, *E. speiseri* [later considered a synonym of *E. nigromarginata*], *E. atrocostata*, *E. cyanogaster*, *E. sanfordi*, *E. samoensis* [mostly *E. sanfordi*], *E. aneityumensis*, *E. nigra*, *Lipinia noctua*, *Caledoniscincus atropunctatus* [as *Lampropholis austrocaledonica*], *Cryptoblepharus novohebridicus* [as *Cryptoblepharus boutonii*], and two snakes: *Ramphotyphlops braminus* [as *Typhlops braminus*] and *Candoia bibroni*). They considered four species to be endemic and four as introduced, and one introduced amphibian from Vanuatu. Thus a total of 18 native species were reported from Vanuatu in 1975. They noted that the origin of the terrestrial vertebrate fauna of Vanuatu has clear Indo-Australian affinities with some species having an exclusively insular distribution and others that are endemic. They also noted that 95 of the 98 terrestrial vertebrate species known at that time from Vanuatu were present on the six islands they

Table 27: Current knowledge on amphibians and reptiles of Vanuatu reported from Islands excluding Santo (Vanuatu excl. Santo), previously reported from Santo and collected during the Santo 2006 expedition. *Crocodylus porosus* has not been recently observed on Santo (?); *Hemidactylus garnotii* was previously only reported from Anatom from a BMNH specimen that I have not checked; during the Santo 2006 expedition I have not explored sea margin and mostly collected inland in forested areas, thus common sea margin species like *Emoia atrocostata* and sea kraits (except *Laticauda laticaudata*) were not observed. I do not consider *Gehyra mutilata* as an introduced species except the recently arrived population of the Torres Islands. Species that I consider as introduced by man (historic or recent) are indicated with an *.

	Vanuatu excl. Santo	Santo	Santo 2006 exped.
AMPHIBIA			
Hylidae			
Litoria aurea	+	+	+
REPTILIA, CROCODYLIA			
Crocodylidae			
Crocodylus porosus	+	(?)	-
REPTILIA, SQUAMATA			
Iguanidae			
Brachylophus bulabula	+	-	-
Gekkonidae			
Gehyra mutilata	+	-	-
Gehyra oceanica	+	+	+
Gehyra vorax	+	+	+
Gekko vittatus	+		-
Hemidactylus frenatus	+	+	+
Hemidactylus garnotii	+	-	+
Hemiphyllodactylus typus	-	-	-
Lepidodactylus buleli	-	-	+
Lepidodactylus guppyi	+	-	-
Lepidodactylus lugubris	+	+	+
Lepidodactylus vanuatuensis	+	+	+
Nactus multicarinatus	+	+	+
Nactus pelagicus	+	-	-
Perochirus guentheri	+	-	-
Scincidae, Lygosominae			
Caledoniscincus atropunctatus	+	-	-
Cryptoblepharus novohebridicus	+	+	+
Emoia aneityumensis	+	-	-
Emoia atrocostata freycineti	+	+	-
Emoia caeruleocauda	+	+	+
Emoia cyanogaster	+	+	+
Emoia cyanura	+	+	+
Emoia erronan	+	-	-
Emoia impar	+	+	+
Emoia nigra	+	+	+
Emoia nigromarginata	+	+	+
Emoia sanfordi	+	+	+
Lipinia noctua	+	+	+
REPTILIA, SQUAMATA, SERPENTES			
Typhlopidae			
Ramphotyphlops braminus	+	+	+

Boidae			
Candoia bibroni	+	+	+
Elapidae, Hydrophiinae			
Laticauda colubrina	+	+	-
Laticauda frontalis	+	+	-
Laticauda laticaudata	+	-	+
Hydrophis coggeri	+	-	-
Pelamis platura	+	-	-

visited. The most diverse vertebrate fauna was on Espiritu Santo but the most diverse herpetofauna was on Efate. They could not attribute the faunal decrease observed on the southern islands of Vanuatu to isolation or to supposed distance from potential sources, not even to diminution of island area or elevation. They also noted that all endemic species were found in mature seral or primary forest and that few species, all cosmopolitan or with large repartition, are restricted to open habitats. Forest faunas show a clear elevation zonation and a vertical stratification under the canopy. The distributions of the three larger skink species [*E. sanfordi*, *E. samoensis* and *E. nigra*] were considered to be mutually exclusive and complementary by Medway and Marshall in 1975, a situation not seen for birds or bats; recent distribution data however disagree with that old observation. Distributions of some bird species indicate that complete archipelago colonisation has not been achieved. Later, in 1994, Whitaker and Whitaker listed 26 reptile species from Vanuatu, including introduced species. Nonetheless the Vanuatu herpetofauna is generally considered to be poor, Allison in 1996 numbered a total of 27 native terrestrial and freshwater species, a number nearly identical to that of Fiji (29 species): *Crocodylus porosus*, 11 gekkonids in six genera, 13 skinks in four genera, one boid and one typhlopid probably introduced but for which he indicated that natural dispersion cannot be ruled out.

There is currently no synthesis of all the available information on the Vanuatu herpetofauna. Our field and literature survey allowed the recognition of 27 native reptile species from Vanuatu (Table 27), four recently introduced species, three sea kraits and two sea snakes (elapids), excluding sea turtles. Native species include one crocodile, 12 geckos, 13 skinks, and one boid snake; note however that the gecko *Hemiphyllodactylus typus* is certainly present in Vanuatu. Thus a total of 37 amphibian and reptile species (32 native species, one potential species [*H. typus*] and four introduced species) are confirmed for Vanuatu today. This diversity of course does not reflect a depauperate herpetofauna when compared with neighbouring archipelagos.

Concerning native species (excluding sea turtles and elapid snakes), Santo has 18 of 27 species, which is about 2/3 of the Vanuatu species. Differences (lacks) are mostly related to southern endemics of the skink genus *Emoia* and northern endemics with clear Solomon affinities (*Gekko vittatus*, *Lepidodactylus guppyi*), all lacking from Santo probably due to its intermediate geographic position.

The herpetofauna of Vanuatu is unique for its lack or scarcity of several widespread Pacific island gecko species (*Gehyra mutilata*, *Hemidactylus garnotii* and *Hemiphyllodactylus typus*). Other Vanuatu species include widespread Pacific species (*Gehyra oceanica*, *Emoia caeruleocauda*, *E. cyanura*, *E. impar*, *Lipinia noctua*, *Laticauda colubrina*, *L. laticaudata*, and *Pelamis platura*), southern species which are absent north of the archipelago (*L. buleli*), species with Solomon affinities (*Gekko vittatus*, *Lepidodactylus guppyi*, *Nactus multicarinatus*, *Emoia atrocostata freycineti*, *E. nigra*, and *E. cyanogaster*), species with Micronesian affinities (*Perochirus guentheri*), species with Loyalty affinities (*Gehyra vorax*, *Nactus pelagicus*, *Caledoniscincus atropunctatus*, *Candoia bibroni*, *Laticauda frontalis*) (note however that *G. vorax* was reported from Tonga and that fossil remains attributed to *Perochirus* sp. were also found on Tonga), and southern Vanuatu with possible Loyalty affinities (*Emoia aneityumensis*, and *E. erronan*) or strict or nearly strict Vanuatu endemics (*E. nigromarginata*, and *E. sanfordi*), all probably also with Loyalty affinities.

••• **Originality of southern Vanuatu islands: Cheesman Line**
Anatom Island is particularly important from a biogeographical perspective owing to its intermediate position between New Caledonia in the west and other oceanic islands in the east. Anatom experienced volcanism but possesses a rich biota thus suggesting less perturbation than Tanna and Erromango. Erromango, Tanna and Anatom, the three southernmost islands of Vanuatu, probably constituted a unique landmass in the past and their connexion to New Caledonia has been suggested regarding their Australasian flora and fauna. Anatom is located at the crossroads of two important migratory routes. The first coming from north Vanuatu, Solomon and New Guinea where Papuan affinities can be found, including the occurrence of some Australasian elements, the second

coming from southern Vanuatu, New Caledonia and Australia where Australasian affinities can be found with again the occurrence of some Papuan elements. Botanists have established strong affinities between the flora of the southern group of islands and that of New Caledonia. Species considered as New Caledonian endemics are present in southern Vanuatu (*Caledoniscincus* for lizards but also e.g. *Araucaria columnaris*, nine other flowering plants and seven fern species). Entomological samples collected on Anatom by Cheesman allow for the recognition of 11 species with Australian affinities related to New Caledonian forms. In 1957, Cheesman proposed that the continental mass comprising Australia (Queensland) and New Caledonia should also comprise southern Vanuatu, a hypothesis no longer considered valid today. The Australia/New Caledonia separation is dated from the end of the Cretaceous. Cheesman however does not reject the second hypothesis of natural colonisation: "*But if the colonization of islands could be considered due mainly to adventitious means of introduction, there would not exist such distinct faunal areas*". I agree with that point of view since all Vanuatu endemic *Emoia* are clearly biogeographically related to southern Vanuatu. Even though Anatom is only 150 km from New Caledonia, it shows considerable faunal dissimilarities and a land connexion is refuted by modern data. Vanuatu is also separated from New Caledonia by the deep Vanuatu Trench (6 400 m). Cheesman suggested a zoogeographical line in southern Vanuatu, analogue to the Wallace line but not as clearly defined. This biogeographic line was recently defined and named the Cheesman Line; all my data agree well with its validity.

• • • Endemism of Vanuatu herpetofauna

In 1996, Allison noted that only seven reptile species are endemic to Vanuatu but that number now has grown to eight species regarding the new *Lepidodactylus* species that was recently described. All genera, except two (*Perochirus* and *Caledoniscincus*), are present in New Guinea and the Solomon Islands. The absence of iguanids, elapids and native ceratobatrachid frogs from Vanuatu is generally explained through the young age of the archipelago and the plate rotation hypothesis indicating a different position of the archipelago over time. Solomon and Fiji had secondary radiations after the dispersal of ancient elements like elapids, iguanids or ceratobatrachids and ranid amphibians from continental landmasses. In 1985, Gibbons considered *Brachylophus* iguanid lizards to be an old Fijian element since they show important morphological and ecological differences compared to actual Neotropical forms considered as ancestral. Curiously, Vanuatu that is located between them (south of Solomon and north of Fiji on the Outer Melanesian Arc, but near Fiji) has no such elements. Ceratobatrachid frogs of the genus *Platymantis*

certainly migrated from Solomon to Fiji when Vanuatu was located elsewhere, north of Fiji. In the same way the occurrence of the Micronesian genus *Perochirus* in Vanuatu, located near Micronesia during its migration to its current position, can be explained, as can the occurrence of some flowering plants or birds. Endemism is low in Vanuatu, particularly endemic genera. Gibbons considered that most species derived from Fijian forms and reverse is not true; however it would be interesting to know if southern Vanuatu endemic *Emoia* species are more closely related to the Loyalty endemic *Emoia* or to the Fijian endemic *Emoia*, implying an east-west migration in the second case. Gibbons noted that most Vanuatu fauna is recent since it is rich in geckos and skinks with at least two endemics; such an assumption has of course no biological reality. Dispersal to the east continued to operate after Vanuatu reached its current position between Solomon and Fiji. Absence of the ancient skink genera *Leiolopisma* [Fiji] and *Tachygyia* [Tonga] on Vanuatu and Samoa, which are young islands, suggest an older occurrence of those genera in Fiji and Tonga, both of which are old archipelagos.

Historically it was believed that the low Vanuatu biodiversity [compared to Solomon and Fiji] is related to the young age of the archipelago, the presence of volcanoes modifying habitats and the influence of hurricanes giving birth to major habitat perturbations. However, several Solomon Islands or island groups are in the same situation but possess a much more diverse fauna. Thus the varying geographical position of Vanuatu over time seems to have had a much more important impact on levels of endemism. Vanuatu was also a site for local speciation and diversification, not at generic but at species level. Among the 13 native gecko species four (including *N. multicarinatus*) are endemic (33%) and among the 13 native skink species five are endemic (38%), a situation similar to most other Pacific archipelagoes of that area.

Factors classically believed to influence insular fauna diversification are isolation (distance from the source area), island area, elevation range, number of plant species and number of insect species. There is a progressive increase of isolation from Santo to Anatom and a decrease in elevation range and island area. The vegetation shows a clear southern diversity decline; a break in that sense seems to appear at a latitude of around 18° south. All vertebrate classes show an impoverishment of their faunas from Santo to the south. Differences in the geological ages of the islands are also obviously correlated with variations in latitude. Southern islands are younger. During the period before the Pleistocene, exposed areas above sea level were limited and restricted to northern Vanuatu. One cannot exclude that the endemic southern species require an ecology that does not allow them to

Figure 280: Small village cultivated areas surrounded by deep forest are not unfavourable to lizards and herpetofaunal biodiversity is often high. It is easier in such places than in deep forest to observe forest species at the interface between forest/opened cultivation zones or even on trees that have been killed and lie on the soil. Dead trees have to be preserved as long as possible (not burned) in such cultivations to ensure lizard dynamics.

colonise the northernmost islands with different climatic or edaphic conditions.

The Efate fauna is exceptionally rich for all vertebrate classes. Such an anomaly seems to be simply related to survey intensity on that island which is the political center of Vanuatu hosting the capital Port Vila and the main trading centre thus explaining why recent introductions have begun on Efate. Some herpetologists are reluctant to prospect northern islands like Santo where malaria is present and thus focus on Efate. Vegetation is also more diverse on Efate. Inversely the fauna of Tanna, for all vertebrate classes, is abnormaly low (birds and plants) and that could be related to particular edaphic conditions. Most tropical islands, in both the Indian and Pacific Oceans, present a physiography based on the same model. They have primary rainforest on the east coast that is subjected to trade winds, a central mountain chain constituting a barrier to that rain, and a drier western side often covered with typical dry forest. Such habitat differences constitute the first speciation dynamic on most large islands like New Caledonia or New Guinea (north/south separation rather than east/west), or even larger islands like Madagascar. Such differences could certainly exist on Santo, particularly for species like *E. sanfordi*. However available data have not yet shown their occurrence.

Vertebrate fauna is richest in low elevation areas, a rule on all Indo-Australian islands. On Santo it is clear that a significant number of birds and reptiles of low elevation forest are excluded from higher elevations. A specialized montane fauna is only present for birds among which few species occurs only at higher elevations on Santo. The vertebrate fauna restricted to nonforested areas only comprise Indo-Pacific or cosmopolitan or widespread species. Two lizards show a distribution centered on coastal habitats (*E. atrocostata* and *C. novohebridicus*), thus explaining their wide distribution. Among introduced vertebrates only two species occur in primary forest (*Gallus gallus* and *Rattus exulans*), other introduced species do not occur there. If such a man modified habitat restriction is generated by a competitive exclusion phenomenon by native forest species, a limited ecological range or simply recent arrival through man-mediated transport is unknown. *E. caeruleocauda*, *E. impar* and *E. cyanura* are sympatric but not syntopic, with different ecologies and competition between species has not been demonstrated. On the other hand, Medway and Marshall noted competitive exclusion of *E. samoensis* [*E. sanfordi*], *E. nigra* and *E. aneityumensis*. They never found these species in sympatry on one island; the occurrence of one species seems to exclude the others. However, I observed what probably could be considered to be an ecological separation between *E. nigra* and *E. sanfordi* on Santo (see above). Medway and

Marshall also argued that skinks are in dietary competition with insectivorous birds at ground and lower vegetation levels. Thus skink diversity on Anatom Island could be related to bird scarcity in these ecological strata of the vegetation. We however find such an explanation unlikely.

Gibbons considered snakes to be rare on Pacific islands since most micromammals on which they normally feed are lacking. *Candoia* is a large boid snake reaching over 2 m length. It is generally considered to feed on birds and bats, less frequently on lizards. Rodents are not native on tropical Pacific islands, thus those snakes could not always have preyed on them, even if they represent the most frequent prey found today in the stomachs of adults. Recent studies have shown that juveniles feed mostly on lizards whereas adults eat mainly rodents. There is certainly something to conclude about the absence of *Candoia* east of Samoa, in relation to the time of arrival of lizards on those islands or more relevantly in relation to the occurrence/absence of larger prey lizards like *Emoia* of the *E. samoensis* group prior to the arrival of rats. It is certain that adults of that species undertook a dietary shift from lizards to rats once those rodents were introduced to the Pacific islands; rats could be more abundant and easier to catch relative to large lizards. Other snakes, *Ogmodon* and *Ramphotyphlops* feed on invertebrates. Solomon endemic snakes also feed on invertebrates, except one species that preys mainly on amphibians and small lizards rather than invertebrates.

In 1995, Adler and coauthors made a biogeographical analysis of skink occurrence on southern Pacific islands. They distinguished three kinds of species:
- **Continental** [species present on a continent (Australia, New Guinea, Asia) or a set of large islands not far from a continent (Sunda Islands or Philippines)];
- **Pacific** [species endemic to the southern Pacific but occupying more than one archipelago];
- **Endemic** [species present on only one archipelago].

They performed a multivariate analysis with nine variables reflecting archipelago size, isolation and maximal elevation. Their results show that island size and isolation are the main factors explaining the nature of faunal composition whereas elevation is less important. No endemic skink is present beyond Samoa/Tonga in Polynesia (except one recently described species from Cook Islands) and the Carolines/Mariannas in Micronesia. Generally, skink endemism is similar to that observed for birds. That is curious since the reduced mobility of lizards compared to that of birds should give rise to an increased endemism by the former. Comparison with birds shows that the most important differences are seen in Hawaii and the

Marquesas Islands (French Polynesia) where birds are nearly all endemic whereas skinks are only represented by few widespread Pacific island lizards. That clearly shows a different colonization potential between them. However, they noted that Pacific island skinks shows a morphological conservatism formerly demonstrated for other groups of the same family where the existence of cryptic species was evident and demonstrated by genetic analysis. Several cryptic species could exist within the wide distribution of many species. The low diversification of Pacific skinks in the easternmost area of their distribution could also be related to undetermined ecological interactions. Adler and co-authors show a high endemism of south Pacific skinks (66%) where 79% of the species do not occupy defined continental areas and 87% have a distribution limited to three archipelagos or less. Vanuatu possess 13 skink species among which four are continental [they considered the Vanuatu form of *Cryptoblepharus* as belonging to the continental species *C. boutonii* (sic) but that form is presently considered to be an endemic species, *C. novohebridicus*], five Pacific and four endemic, whereas the Santa Cruz Islands (nine species) comprised respectively four, three and two and Fiji (11 species) three, five and three. Island size is correlated with diversity in all those three categories, whereas geographical isolation is mostly correlated with total diversity and to the diversity of Pacific elements. I think that if man is responsible of such transportations then these species should occur randomly and there should be only weak correlations with source distance and also abnormal occurrences should be seen since the Polynesians were island settlers. Their travels were not always from one island to the nearest island but also from one archipelago to other more distant ones. Abnormalities should thus exist but they do not; the Pacific herpetofauna east of Samoa is uniform and widespread, without abnormalities due to stochastic colonisation. Adler and coauthors' analysis is interesting but omits some important factors to explain faunal composition with strongly biased conclusions, notably on species size and ecological requirements of each species. If man had intentionally transported lizards he certainly would have selected large species that he could eat (Monitor, *Gehyra vorax*, *Gekko vittatus* used for tattooing, *Brachylophus* from Tonga to Marquesas). Moreover species size decreases clearly in skink faunal compositions with increasing distance from source areas from the east and thus seems to me to be an important factor that cannot be overlooked. On another hand, species inhabiting sea margins are more easily transported accidentally than primary forest inland endemics, a habitat not present on all islands and probably scarce during sea level movements during glaciation events, a critical period for inter-island lizard movements. They consider the biogeographical origins for the

13 Vanuatu skink species to be continental (*Emoia atrocostata*, *Emoia caeruleocauda*, *Lipinia noctua*), Pacific (*Caledoniscincus atropunctatus*, *Emoia cyanogaster*, *Emoia cyanura*, *Emoia impar*, *Emoia nigra*) or endemic (*Cryptoblepharus novohebridicus* [they considered in fact *C. novocaledonicus* as endemic and *C. boutonii* (sic) as continental, thus implicitely *C. novohebridicus*, whose specific status is now accepted, as endemic], *Emoia aneityumensis*, *Emoia erronan*, *Emoia nigromarginata*, and *Emoia sanfordi*).

Some Santa Cruz Group islands, like Tikopia and Anuta, were clearly colonised by Polynesians as stated by Steadman and co-authors in 1990: "*The contemporary inhabitants of both islands are Polynesians, whose oral traditions indicate that their ancestors came from islands to the east, including 'Uvea, Futuna, Samoa, and Tonga. Culturally and linguistically, the people of Anuta and Tikopia are closely related and maintain regular inter-island contact through canoe-voyaging. Their languages are mutually intelligible, though distinct*". Those islands are geologically recent, 80000 years old, but still possess endemic bird subspecies. They however have no reptile species that are not found on surrounding islands, suggesting that Polynesians are not at the origin of the introduction of lizards to these islands.

••• Dynamic of faunal changes

All tropical Pacific islands possess a common set of small lizards. Their common size (around 50 mm SVL) is not accidental and can certainly be related to their capacity to survive and to travel on floating rafts during trans-oceanic travel. Most herpetologists believe that these lizards were recent colonists on those islands and that their colonisation is mostly related to man mediated introduction. In fact, I believe that most of these reptiles (except of course the introduced *Hemidactylus frenatus* and *Ramphotyphlops braminus*) were present on the Pacific islands long before human colonisation. Their morphological and genetic uniformity has to be linked to their significant colonisation abilities rather than to recent arrival. Such colonisation abilities do exclude populations becoming isolated on any island (except partial isolation on the most remote eastern islands like Easter Island or Clipperton Atoll), thus explaining the scarcity of endemism east of Tonga. That "*reptile eastern rush*" could certainly have been favoured during important sea level movements related to glaciations, long before human arrival in the area. Small sized skinks and geckos could have benefited from such ecological modifications. Major salinity variations and modifications of the fresh water lens related to sea level variations certainly had considerable repercussions on littoral trees, perhaps even on rainfall in the area (through accelerated evaporation on increased littoral seawater covered areas during sea level rise), thus allowing easier colonisations (increased island size) or favouring lizard dispersal (unsatisfactory conditions found on an island). On the other hand, such vegetation dynamics, particularly in increasing the number of littoral dead trees, had a positive impact on lizard demography in producing a high number of favourable egg laying and shelter sites (dead trees). Volcanism and hurricanes certainly also played an important role in the faunal dynamics of Vanuatu.

In 1975, Medway and Marshall noted that 53% of Vanuatuan reptile species are forest species. All endemic species are present in closed habitats. On Santo at least two endemic bird species are exclusively montane species. Most species, however, also occupy modified habitats where not all trees have been destroyed. Natural regeneration following man made vegetation perturbations (mostly for agriculture) are probably comparable to those that happen after natural cataclysms like earthquakes, volcanic eruptions and hurricanes, all frequent events in the area and for which the fauna as well as flora seemed to be adapted. The ecological niche of most vertebrate species on Vanuatu includes the vegetation that covers those islands in the observed conditions without human interventions other than local agriculture at small village level, with small and well separated cultivated areas around each village, distributed like spots on the forest surface. Agriculture as practised in small villages and without modern techniques certainly has no negative impact on the Vanuatu herpetofauna. In contrary it should have a positive impact in creating egg laying sites, shelters (dead trees) and small forest openings useful for thermoregulation of forest species on the forest edge. The threat is from the changing scale of agriculture, e.g. large uniform pineapple fields, which completely eliminate the local herpetofauna.

The Vanuatu herpetofauna is fragile and sensitive to habitat change and plant and animal introductions. It is unique, with particular regard to numerous endemic species or others with a restricted distribution. That herpetofauna is unique among the whole Pacific and has to be preserved for the coming generations but also as a national heritage for the local Vanuatu Melanesian culture.

Rivers and Other Freshwater Habitats

coordinated by Philippe Keith

Habitat Types

Philippe Keith & Clara Lord

Most islands of Vanuatu, like Santo, show a dense hydrographic network. Santo's freshwater system can be divided into two categories reflecting the island's geological features. Thus, on the western and north-western Cumberland coast, the streams are rapid mountain rivers; whereas the eastern coast's aquatic system is characterised by slow rivers, deep pits called "blue holes", and swamps on the calcareous plain.

MOUNTAIN STREAMS AND LONGITUDINAL ZONATION

Mountain streams in Vanuatu tend to be short and steep and generally lack low gradient alluvial floodplain reaches. Where these are present (e.g. on Efate and Santo) the floodplain is usually heavily developed. The river flow depends on various factors, such as, climate, soil, vegetation and catchment basin morphology. Some streams are ephemeral or have sub-surface flows during the drier months.

On Santo's north-western coast, the orientation of streams and catchment areas is usually perpendicular to the coast. Because the Cumberland peninsula is extensive and owing to its mountainous relief, the catchment areas are small and do not allow large streams to develop. The run-off varies considerably, depending on the site's orographic characteristics, the seasons, and tree cover. It is also liable to change when specific weather episodes take place, such as, cyclonic floods, or, on the contrary, droughts.

Western and north-western coast streams can be divided into five zones, depending on altitude and current speed:
- **Spring zone**, located at altitudes of over 800 m.
- **Higher course**, located at altitudes of between 450 and 800 m.
- **Middle course**, located at altitudes of between 150 and 450 m.
- **Low part of the stream**, located at altitudes of between 50 and 150 m.
- **Lower course**, located at altitudes of under 50 m.

The latter four all have an additional sub-zonation: calm zone (0-30 cm/s), medium current (30-75 cm/s), and rapid current (75 to > 100 cm/s).

The five zones described are usually grouped into three larger zones defined according to the river slope, the average current speed and the grain-size of the substrate: **higher course**, **middle course**, and **lower course**.

Specific criteria define these three zones:
- The **higher course** (Fig. 281) is characterised by a steep slope (generally more than 10%); thus current speed is high. The substrate is usually composed of large blocs directly coming from

Figure 281: Pelouva River (higher course).

the parent-rock. The delimitation with the middle course often corresponds to a topographical accident, like a cascade. The distance between this reach and the river mouth is highly variable; it largely depends on the catchment area's geological characteristics.

• The **middle course** (Fig. 282) has an average slope generally under 10%. The riverbed is covered in pebbles and rocks. Sometimes, sandy bottoms can be found in slow current reaches. The length of this zone depends on the geological origins of the catchment area.

• The **lower course** (Figs 283-285) is the part of the stream located in the coastal plain; its length is thus generally reduced. Two areas can be distinguished in this zone: the estuary, immediately under marine influence, and the upstream part, where the water's conductivity is very low. Some estuaries can be very broad (i.e. the Jordan River), and the intrusion of salt water can go relatively far upstream. The slope and the current speed are low to nil; it is a high accumulation zone. In estuaries, the sediments are composed of sand and silt, but higher upstream the grain size is coarser (gravel, pebbles and blocks). This zone is not present in all streams. For some, it is related to the middle course because of the average current speed and the grain size; in this case, the marine influence is somewhat lesser.

There is a relation between the flowing facies and the species found within each zone. The majority of species is found in facies where the current speed is not too rapid. On the contrary, populations found in facies where the current is very strong (rapids or steps) are characterised by the presence of species having specific adaptations to this type of environment; this is, for example, the case for gobies of the *Sicyopterus* genus that are capable of resisting very strong currents by sticking to the substrate with their ventral suction cup, or for the larvae of several fixed aquatic insects.

It is important to point out that this classification is merely an instrument to characterise flowing units for a portion of river at a given time. In these mountainous streams, one facies can quickly be transformed into another because of the flow variability

Figure 282: Peavoho River (middle course).

Figure 283: Penaoru River (lower course, alt. 10 m).

Figure 284: Peavohori River (lower course, alt. 3 m).

Figure 285: Peavohori River mouth.

and the torrential regime, and this is particularly so in these mountains areas where rainfall is high. Nevertheless, the distribution of all populations of aquatic species (fish, crustaceans, molluscs, insects) reflects the altitudinal gradient and their ecological preferences. Indeed, some species favour living exclusively in the lower course, whereas others are found only in the higher course.

PLAIN STREAMS, BLUE HOLES AND SWAMPS

There are few natural lakes of any size on any of the islands, except those associated with volcanic cones. Santo's south-eastern and eastern coasts are characterised by few streams as the relief is low in this region. When present, the streams are only a few kilometres long, comprised of lentic zones, vegetation-rich, and enable the tide to run far up stream. The species found in these zones are typical of lower courses and can withstand high salinity variations. This region is essentially calcareous, and has many "blue holes". These lentic zones are generally found near the shore and harbor euryhaline taxa. A connexion may also exist between the lentic zone of the lower course and flood-plain swamps. The latter can be either of natural origin or anthropogenic when coming from developed channels used to irrigate dasheen (taro) fields. Dasheen fields are also found in some mountainous zones thus favouring the settlement of species usually found in calm zones, particularly Odonata.

Figure 286: Plain stream. Peilapa River.

Biota

FOCUS ON FISH, SHRIMPS AND CRABS

Philippe Keith, Clara Lord, Philippe Gerbeaux & Donna Kalfatak

••• Biodiversity and inventories

Until 1998, Vanuatu's freshwater ichthy-ofauna and carcinofauna was relatively unknown, and the literature was poor. Baker in 1929 had revealed the presence of a few freshwater fish in the Yoro River on Santo Island, but with no further information. Later, Ryan described a new Gobiidae species from the genus *Stiphodon*, caught in a river on Santo.

In October 1998, the Environment Unit of Vanuatu sampled a few freshwater fish and crustaceans on several islands. From July 2002 to 2006, the MNHN has carried out several campaigns on several of the archipelago's islands for the census of freshwater species. Several of these campaigns took place on Santo Island; however the tip of the Cumberland had never been surveyed. It was not until the Santo 2006 expedition that the eastern and western slopes of this cape were prospected (Fig. 287). Further to MNHN's successive inventories, knowledge of freshwater fauna (fish and crustaceans) has greatly increased, as there are now about 100 species indexed for Vanuatu; less than forty were known at the beginning of the 1990s. In addition to the new occurrences of certain known species, a new fish genus and several new species have been discovered

Figure 288: *Sicyopterus aiensis*.

Figure 289: *Schismatogobius vanuatuensis*.

and dedicated to Vanuatu, to its rivers and to its people. Such is the case for *Sicyopterus aiensis* Keith, Watson & Marquet, 2004 (dedicated to the Ai creek in Efate) (Fig. 288); *Schismatogobius vanuatuensis* Keith, Marquet & Watson, 2004 (dedicated to Vanuatu) (Fig. 289); *Stiphodon kalfatak*, Keith, Marquet & Watson, 2007 (dedicated to D. Kalfatak from the Environment Unit of Vanuatu); *Akihito vanuatu* Watson, Keith & Marquet, 2007 (new genus and new species dedicated to Vanuatu).

••• Tolerance to salinity and biogeography

••• Tolerance to salinity

Santo's species need to be classified according to their tolerance to salinity, a major adaptation to insular systems, in order to understand their biogeographical distribution and their ecology. According to Myers' studies on fish in 1949 and to their adaptation by McDowall in 1997 and Keith in 2003, the following classification is used:

• Primary fish are strictly intolerant to saltwater and therefore cannot cross any salty zone.

Figure 287: Freshwater fish sampling in Santo.

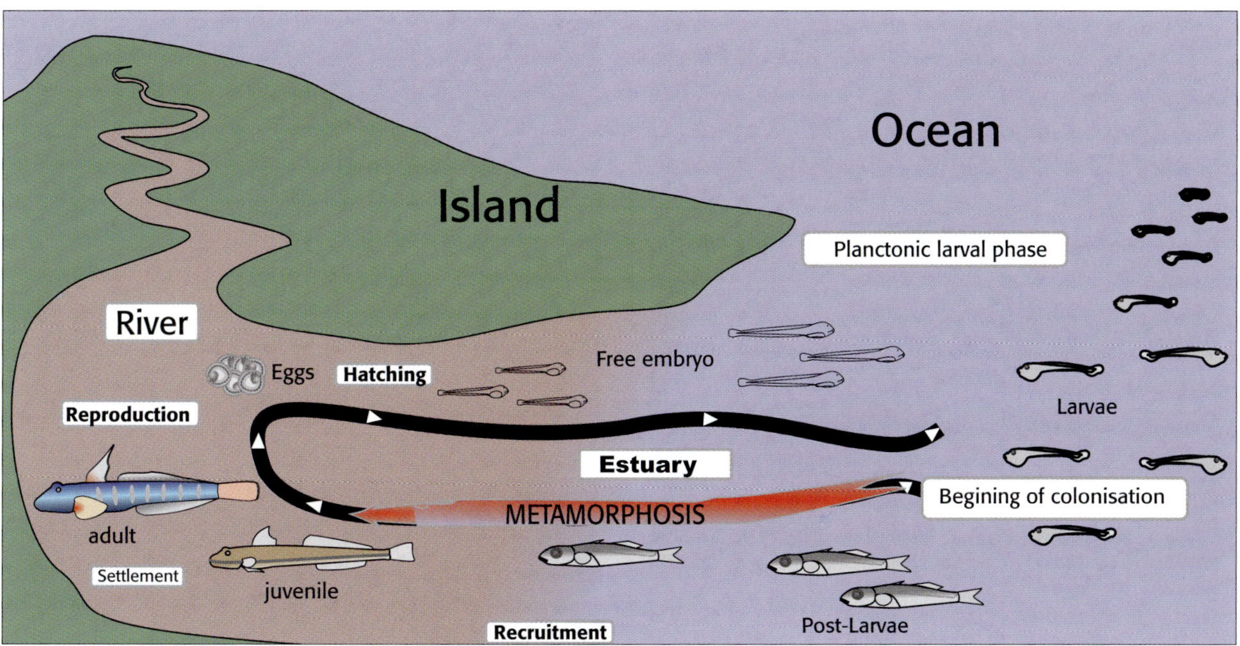

Figure 290: Amphidromous cycle for freshwater gobies (P. Torres & P. Keith).

- Secondary fish are mainly found in freshwater but can bear to cross narrow salty barriers.
- Diadromous fish are migratory and alternate between freshwater and saltwater according to their life cycle. Diadromous fish are classified in three sub categories:
  - Anadromous fish spend the majority of their life in salt water and migrate to freshwater to reproduce.
  - Catadromous fish spend the majority of their life in freshwater and migrate to saltwater to reproduce.
  - Amphidromous fish (Fig. 290): females spawn many ova in freshwater, which are then fertilised by the males. After hatching the larvae are carried by the current out to sea where they spend a variable amount of time. The young fry then go back to freshwater to resume their growth. The migration has no reproductive goal, unlike the two former categories. **Amphidromy is a major adaptation to insular environments**; it will be described in detail in the next paragraph ("Life traits, amphidromy and insular adaptation").
- Vicarious fish are thought to be non-diadromous and are freshwater representatives of originally marine families.
- Complementary fish are freshwater fish, usually diadromous and belonging to marine groups. They become dominant in rivers if categories "Primary", "Secondary" and "Vicarious" are absent.
- Sporadic fish indifferently live in fresh or saltwater or entering sporadically freshwater although there is no true migration.

"True" fresh water species are truly distributed in two groups: primary fish, belonging to families strictly found in freshwater, and secondary fish, belonging to families capable of crossing narrow salted barriers to reach another basin. The capacity to disperse and to colonise, varying according to the different species, is the basis to the understanding the ichthyological populations' variations according to the environment and the epoch.

Like most Pacific islands, Vanuatu is characterised by the absence of indigenous primary and secondary fishes that are intolerant to saltwater. Indeed, the only secondary fish present were introduced in

Figure 291: *Kuhlia marginata*.

Figure 292: *Kuhlia rupestris*.

Figure 293: *Rhyacichtys* cf. *guilberti.*

Figure 294: *Stiphodon sapphirinus.*

Figure 295: *Stiphodon rutilaureus.*

Figure 296: *Sicyopterus aiensis.*

the 20th century: tilapia *Sarotherodon occidentalis*, gambusia *Gambusia affinis* and guppy *Poecilia reticulata*. Santo is home to these three species; they are found more particularly on the eastern and south-eastern coasts.

Vicarious fish are mainly represented by carp (*Kuhlia marginata* and *Kuhlia rupestris*) (Figs 291 & 292), the latter being found only in running water from the estuary to the medium course.

As a consequence of the absence of primary and secondary native fish, the rivers are then mainly colonised by diadromous fish and Crustaceans (migrant amphihaline species performing a part of their biological cycle in freshwater). The fish are represented by two categories: catadromous (Anguillidae, i.e. eels) and amphidromous fish (Gobiidae, Eleotridae and Rhyacichthyidae Fig. 293). The crustaceans are represented by amphidromous species (Palaemonidae, Atyidae).

One of the factors explaining the biodiversity of amphidromous species in such islands is the particularity of the distribution of the different species along the river.
The distribution of amphidromous species in all islands depends essentially on two closely bound factors: the altitude and the strength of the flow. Within the Gobiidae, *Awaous* species, *Glossogobius* cf *celebius*, *Schismatogobius* species and *Stenogobius yateiensis* are restricted to the lower parts of the rivers and to littoral stagnant waters. *Stiphodon* genus (Figs 294 & 295) colonises the lower and middle courses, essentially where the current is average to weak. *Sicyopterus aiensis* (Fig. 296) and *S. lagocephalus*, as well as *Lentipes kaaea*, *Sicyopus zosterophorum* and *Sicyopus chloe*, are able to climb up waterfalls. The Eleotridae can be found in running water (from the estuary to mid-stream) where the current is weak, as well as in stagnant water. The higher the altitude

of the island and the greater the variation of the river flow, the more amphidromous species there are.
The fact that the ecological conditions become increasingly constraining with altitude (strong current, unavailability of food) explains why 75 % of the species are confined to only one zone (the lower course), while only some (*Sicyopterus lagocephalus*, *Macrobrachium lar*, *Anguilla marmorata* Figs 297 & 298) can be found from the lower course to the higher course of the river, and that a few species live only in the higher course (*Lentipes kaaea*, *Macrobrachium latimanus*, *Sicyopus sp*). The

Figure 297: *Macrobrachium lar.*

Figure 298: *Anguilla marmorata.*

number of freshwater species falls gradually working from the estuary upstream. This reduction is well marked after a waterfall.

Inventoried crustaceans are essentially diadromous amphidromous or vicarious and their zonation is also very well marked. Thus, *Macrobrachium equidens* and *M. grandimanus* are strictly found in the brackish water of estuaries; however, these latter species can also be found in caves with brackish water or in the blue holes. *M. australe* (Fig. 300) is found in freshwater in the lower courses. *M. gracilirostre*, *M. placidulum*, and *Atyoida pilipes* favour the rheophile zones of the lower and mid-courses. Finally, *M. latimanus* is only found in the medium and higher courses.
Utica and *Ptychognathus* crabs are found in estuaries and in the lower course, whereas *Varuna* (Fig. 301) and *Labuanium* can be found in the medium and higher courses.

••• Biogeography

Vanuatu's fauna, and particularly that of Santo, corresponds to a blend between Asian-Indonesian elements and Indo-Pacific recent marine intruders. Indeed, Nivan aquatic environments were originally mainly colonised by marine species ascending estuaries or by amphihaline species part of whose biological cycle takes place in freshwater (Eels, Gobiidae, Eleotridae). (See paragraph "Life traits, amphidromy and insular adaptation".) Due to the isolation and adaptations of some populations, endemism has developed for fish and, at a smaller scale, for decapod crustaceans. Some species are thus geographically limited to islands and even only to certain catchment areas.

Vanuatu is close to a suspected biodiversity centre situated in the Indonesia-Philippines area. As it stretches from north to south, it has over time favoured the dispersion of amphidromous species, as they go through their marine larval phase, from this centre towards Vanuatu. This is confirmed by the presence of a high number of West Pacific Gobiidae species (24 %) in Vanuatu that originated from the Indonesian area (West Pacific).

As some endemic species are common to New Caledonia and Vanuatu, we assumed that there are dispersion and exchange of larvae between these two areas, depending on the species and the duration of the larval phase. This duration is variable from one species to another. In a recent study Keith et al. have shown that an amphidromous species (*Sicyopterus lagocephalus*) (Fig. 299), present in New Caledonia and Vanuatu, has a broad distribution and occurs over a range of 18 000 km in the Indian and Pacific Oceans. The overall pattern of distribution and phylogenetic relationship suggests that the lineages leading to endemic species in this genus originated earlier than *S. lagocephalus*. This latter seems to be a secondary migrant species, having gradually colonized both Indian and Pacific Oceans.

Figure 299: *Sicyopterus lagocephalus.*

Other factors occur to control the dispersion of the species and the specificity of each island in Vanuatu and in particular in Santo. The factors affecting genetic structure and larval dispersion of coral reef population in the Pacific Ocean have been divided into three majors groups:
  • functional factors which are related to the ecology and the biology of each species (this

includes parameters such as reproductive behaviour or duration of the pelagic larval phase);
• physical factors including the effect of oceanic currents and/or the geomorphological structure of the reef;
• historical factors such as colonisation or extinction linked to sea level variation, which will directly modify the population structure.

In the case of amphidromous species, it is necessary to consider all of these factors to explain the existence of many endemic species and the broad or small distribution of the species in the Indo-Pacific area. It is indeed probable that a complex system including marine currents, duration of the larval phase, paleohistory of the area (period of emergence of the islands, vicariousness and variation of the sea level, etc.), in addition to the particularities of the islands (altitude, substratum, age, river velocity, etc.), explain the structure observed.

Numerous overlapping factors could explain the biodiversity of amphidromous species in the rivers of Santo, and this needs to be studied, but it seems that the most important factors are more generally, for the Vanuatu archipelago, its long geographical shape and the proximity of a biodiversity centre.

• • • Life traits, amphidromy and insular adaptation

In Santo, river systems are colonised by fish (Gobiidae, Eleotridae…) and crustaceans (Palaemonidae, Atyidae…) with a life cycle adapted to the conditions in these distinctive habitats, i.e. young oligotrophic rivers, subject to extreme climatic and hydrological seasonal variation. These species spawn in freshwater, the free embryos drift downstream to the sea where they undergo a planktonic phase, before returning to the rivers to grow and reproduce, hence they are called amphidromous. These species contribute most to the diversity of fish communities and have the highest levels of endemism.

At certain times of the year, the biomass of fish larvae (in particular Gobiidae) migrating upstream is so great that they become a major source of food for local human populations in Vanuatu and particularly in western Cumberland. However, harvesting this food resource is highly unsustainable, on account of the complexity of the species' life cycle and the hydrological specificities of the Vanuatu islands. Hence, there is an urgent need to understand the biology and ecology of amphidromous species (reproduction, larval life, dispersion, colonisation, etc.) in order to form the basis on which management and conservation policies are formulated. A summary of the most recent knowledge about the main fish family (Gobiidae) present in the rivers of Santo is present below.

• • Spawning and larval life stages

After spawing in the river, the newly hatched gobioid larvae are better adapted physiologically to life in seawater, while prolonged exposure to freshwater postpones development and greatly increases mortality.

Gobioid larvae are known to drift to sea during the night. Free embryos of *Sicyopterus stimpsoni* (Gill, 1860) from Hawaii and of *S. lagocephalus* from Vanuatu "swim" repeatedly upwards until they reach the water surface, then for a while, cease "swimming" and sink and then move towards the surface again. This reaction facilitates their transport to the sea.

In normal or low river flow, embryos from the upstream reaches starve to death before they reach the sea. As a result, early starvation of larvae could limit reproductive success of fish located far from the sea, which would have a strong selection for fish reproduction in certain regions of rivers within a given distance from the sea.

The free embryos are carried passively via river transport to the sea where they will lead a planktonic life for 90 to 240 days.

• • • Recruitment into rivers

On some oceanic islands, hundreds of miles away from any other freshwater habitat, amphidromy necessitates special migratory abilities in larvae and juveniles. The post-hatch embryos appear in the marine environment when they have reached 1-4 mm in length where they spend only a few months before they must locate a source of freshwater.

Recruitment of post-larvae at the river mouth is possible all year round, but seems to occur mostly from December to February.

Soon after entering freshwaters, the juveniles undergo several changes in colour and fin shape. The body of these post-larvae is mostly translucent and light coloured when they enter the estuary, except for the eyes and internal organs. The tail is mainly forked but it gradually becomes truncated and the body acquires pigments as it metamorphoses and changes from a larval planktonic life to an adult benthic life.

Pectoral and tail fins grow bigger, together with the pelvic disc needed for upstream movements. The mouth and the jaw structure are also modified, as well as the digestive system. The juveniles must switch from a planktonic feeding mode to a benthic feeding mode, when they "graze" on the substratum.

When the metamorphosis is complete, the goby is able to climb over waterfalls by using alternately its pelvic suction cup and its lips, and these are also used to graze on the diatoms growing on rock surfaces. This metamorphosis is correlated to the changes in diet and spatial distribution between larvae and juveniles.

The recruitment of gobies is often associated with heavy rain. The freshwater flow seems to trigger their migration inland. Even intermittent rivers that flow only after major storms attract post-larvae.

The movement of these young fish from the ocean to the rivers is the key link in completing their amphidromous cycle. The maintenance of this system is crucial to the long-term supply of freshwater resources in insular systems.

Once in the river, the juveniles then migrate upstream, towards the adult habitat. The timing and direction mechanisms required for the success of this migration remain mostly unknown. Rheotaxis might not be the only phenomenon involved.

To access the rivers, the larvae most often choose to swim close to the shoreline, where the current is weaker and can even be reversed (i.e. a counter-current) with the tide or the swell. Some larvae travel with their spindle shaped body facing the current and climb from one rock to the next using their pelvic suction cups. Others can reach freshwaters following passages through cracks in the rocks, thus avoiding turbulence. Postlarvae first congregate near the shore. They will begin migrating into the estuary at first light and at various moments during the day, depending on the species.

For *Lentipes*, found at high altitude in Santo, the post-larvae enter the rivers mainly during the daylight hours, with the waves of the incoming tide. Migration peaks in the hours following sunrise, when the post-larvae are carried by the rising tide thereby reducing their energy expenditure. When they enter freshwater, *Lentipes* post-larvae swim immediately upstream. No amphidromous goby species (whether adults or young) appears to develop strict nocturnal activities, thus a morning initiated migration enables larvae to reach higher up into the river system.

Sicyopterus lagocephalus post-larvae are strong swimmers. Their well-developed fins and highly effective pelvic suction cup allow them rapidly to access the upper reaches of the catchement area.

For *Awaous guamensis*, the post-larvae were transported to the river mouth by waves and they entered freshwater either by day or by night, with a peak early in the night. *A. guamensis* is a slower climber, and it depends on tidal inundation to move upstream. For *Stenogobius*, the post-larvae appear to move with the incoming tide during the day, or, at night, using cracks in the rocks. In comparison, for *Eleotris fusca* (Forster, 1801) (the most common Eleotridae in Santo), the post-larvae have difficulty manoeuvring and swimming in turbulent water, and this determines recruitment periods. They are transported mainly at night by the waves of the incoming tide.

••• Upstream migration

The juveniles must migrate from the lower reaches of river habitats to the habitats they will occupy as adults. Out of the three genera living highest up the river systems of Vanuatu (*Lentipes*, *Sicyopterus*, *Sicyopus*), *Lentipes* is the only one to swim directly towards the upper reaches, while the other two take more time to complete their migration.

Beyond the first major waterfall which acts like as barrier against predators, schooling and escape behaviours are replaced by agonistic interactions between juveniles competing for space and food.

In the genus *Lentipes*, the adults are territorial (especially the males). These territories have strong currents, good water clarity and vary in size from 0.1-4 m^2.

••• Preferential habitats

As a rule, whatever the island studied, the first major waterfall is a crucial factor for species' distribution along a river.

Predatory species of the genus *Kuhlia* (Kuhliidae) are abundant in the lower reaches, below the first waterfall, but are absent above it. This limitation in the movements of predators of the genus *Kuhlia* affects the distribution of other organisms. Gobies, Palaemonid crustaceans of the genus *Macrobrachium*, or Atyids are often more abundant in the areas situated above cascades where predators are less numerous.

The type of habitat (riffle, pool, etc.) and the type of substratum determine the species' distribution and abundance. The positive correlation between gobies' density and the proportion of hard substratum is the result of the diet of these herbivorous species. These gobies feed by scraping diatoms or algae from hard surfaces in the river. Many of these species have teeth adapted for scraping. Although the site with the highest density of gobies is that with the highest proportion of hard substratum, the wide differences reported even for areas with large amounts of hard substratum show that other factors have to be considered.

The amount of plant cover on the banks of the river is another important factor where gobies' habitat is concerned. Fish density is highest where plant cover on the banks is poorest, thus enabling sunlight to penetrate. This is probably because primary productivity on rocks is higher in sunny locations, providing more algae for the gobies to feed on. However, this might not be the only reason: it is possible that light plays a part in propagating the iridescent colours of displaying males. Whatever the cause, vegetation cover on river banks is a contributing factor in the distribution of gobies within streams.

As a general rule, all the studies carried out show that many species, and particularly those for which the status is critical, prefer sectors where the current speed is higher than 30 cm/s and where the substratum is composed of pebbles and rocks. The higher the river's altitude, the faster the current and the coarser the substrate, but the fewer the species found; however these species tend to be more specific and are generally endemic.

Understanding the role of these factors (density/plant cover, and density/hard substratum) is the first step towards identifying the main habitat factors that are important for amphidromous species, and towards predicting the ecological impact of any environmental disturbance or modification in Vanuatu.

••• Freshwater fish and crustaceans of Santo

Table 28 shows all of the 75 known species from Santo (26 decapod crustaceans, 49 fish). Thus far, it is the richest island of Vanuatu in terms of specific diversity. However, all the freshwater communities found in Vanuatu are not encountered in Santo, some known patrimonial species do not occur on this island.

Table 28: species of fish and decapod crustaceans from Santo (*: endemic to Vanuatu or to the Vanuatu/New Caledonia area. I: Introduced).

Fish	
Anguillidae	**Eleotridae**
Anguilla marmorata	Butis amboinensis
Anguilla megastoma	Belobranchus belobranchus
Anguilla obscura	Bunaka gyrinoïdes
Ophichthidae	Eleotris fusca
Lamnostoma kampeni	Eleotris acanthopoma
Lamnostoma orientalis	Hyseleotris guentheri
Syngnathidae	Ophieleotris sp
Microphis brachyurus	Ophieleotris aporos
Microphis leiaspis	Ophiocara cf porocephala
Microphis manadensis	**Gobiidae**
Microphis retzii	Awaous ocellaris
Microphis cf spinachoides	Glossogobius cf celebius
Poeciliidae	Lentipes kaaea *
Poecilia reticulata I	Psammogobius biocellatus
Gambusia affinis I	Redigobius bikolanus
Mugilidae	Schismatogobius vanuatuensis *
Cestraeus plicatilis	Sicyopterus lagocephalus
Liza subviridis	Sicyopterus aiensis *
Cichlidae	Sicyopus chloe *
Sarotherodon occidentalis I	Sicyopus zosterophorum
Scatophagidae	Stenogobius yateiensis *
Scatophagus argus	Stiphodon astilbos *
Tetrarogidae	Stiphodon sapphirinus *
Tetraroge niger	Stiphodon mele *
Ambassidae	Stiphodon kalfatak *
Ambassis miops	Stiphodon atratus
Kuhliidae	Stiphodon rutilaureus
Kuhlia marginata	**Rhyacichthyidae**
Kuhlia munda	Rhyacichthys cf guilberti *
Kuhlia rupestris	
Krameriidae	
Gobitrichinotus radiocularis	

However, it seems from what is currently known that the tip of the Cumberland, and particularly its western coast, has a fundamental role to play in the maintenance of several "patrimonial" species from Vanuatu. this is particularly the case for the species *Rhyacichthys cf guilberti* for which one of the main existing population is found on this coast. It also the case for the species *Sicyopterus aiensis* (endemic to Vanuatu), as very healthy spawner populations are found in the western catchment areas, particularly that of the Penaoru River.

The western Cumberland also houses flourishing populations of emblematic amphidromous species, such as *Lentipes kaaea*. In total, Santo houses 10 species endemic to Vanuatu or to the Vanuatu/New Caledonia area, and three introduced species (see list below). It has also been noted that the Penaoru catchment area houses a high diversity of species and that all the specimens caught had sizes close to previous records. This can be explained by the fact that this catchment area is taboo; fishing there is forbidden.

It has been commonly found that, in the sectors prospected, the number of endemic species is greater in rivers flowing under natural vegetation cover and where the flow is unmodified. This result can be easily explained from current knowledge about amphidromous species and on the way the river-forest system functions.

Indeed, the vegetation cover maintains a certain river flow, cool temperatures, and thus well-oxygenated water; it produces exogenous food inflows for the aquatic species, an important factor as insular river systems are generally poor in nutritional elements. The vegetation cover thus raises the river's trophic potentialities, whilst favouring habitat diversity and water filtration.

Crustaceans	
Atyidae	Macrobrachium gracilirostre
Atyoida pilipes	Macrobrachium grandimanus
Atyopsis spinipes	Macrobrachium lar
Caridina brevicarpalis	Macrobrachium latidactylus
Caridina gracilirostre	Macrobrachium latimanus
Caridina longirostris	Macrobrachium placidulum
Caridina serratirostris	Macrobrachium spinosum
Caridina typus	Palaemon concinnus
Caridina weberi	**Grapsidae**
Caridina gueryi	Labuanium trapezoideum
Palaemonidae	Ptychognathus riedelii
Macrobrachium australe	Utica gracilipes
Macrobrachium bariense	Utica borneensis
Macrobrachium cf microps	Varuna litterata
Macrobrachium equidens	

Amphidromous species colonising the rivers are distributed along the river from the estuary to the higher reaches according to their ecology. Some are therefore only found at a certain altitude according to the water temperature, its physical and chemical parameters and its hydrological properties. The majority of the species encountered are rheophile (they live in strong currents); in order to maintain a high level of biodiversity, it is therefore necessary to maintain high flow rates. The seasonal variability favours massive freshwater flow in estuaries, thus allowing post-larvae from the sea to colonise the rivers.

Moreover, the shorter the river and the steeper its slope, the higher the success of the downstream migration of larvae to the sea as, according to several authors, larvae have less than three days after hatching to reach the estuary. In these kinds of rivers, the colonisation of the rivers by post-larvae will also be more successful: as they return, they must climb upstream as fast as possible in order, on the one hand, to flee predators that are in greater number in the lower course, and, on the other hand, to find a suitable territory.

Figure 300: *Macrobrachium australe*.

Figure 301: *Varuna litterata*.

Macrobrachium gracilirostre male
In situ artwork by Roger Swainston/anima.net.au

... Identification Keys

The main families or genus found in Santo could be identified using the keys below.

••• Crabs

1-a: Anterior-lateral border shaped in 3 teeth. Front with a straight border ⟶ **2**
1-b: Anterior-lateral border shaped in maximum 2 teeth. Front with a wavy or lobed border ⟶ **3**

2-a: No H shaped drawing on the cephalothorax. Absence of seta fimbriae on the extremity of the legs or, if present, only on the extremity of legs 4 and 5 ⟶ *Utica* **sp.**
2-b: H shaped drawing on the cephalothorax. Presence of seta fimbriae on the extremity of legs 2, 3, 4 and 5 ⟶ *Varuna litterata*

3-a: Front with wavy border; dactylus of the locomotive legs P2 to P5 is about as long as or longer than the propodus ⟶ *Ptychognathus* **sp.**
3-b: Front with 4 prominent equally sized lobes; the dactylus of the locomotive legs P2 to P5 is shorter than the propodus ⟶ *Labuanium trapezoideum*

••• Shrimps and prawns

1-a: Presence of dense seta tufts at the extremity of the legs' P1 and P2 fingers. Branchiostegal and hepatic spines absent on the cephalothorax ⟶ **Atyidae: 2**
1-b: Absence of dense seta tufts at the extremity of the legs' P1 and P2 fingers. Branchiostegal and hepatic spines present on the cephalothorax ⟶ **Palaemonidae: 4**

2-a: Legs P1 and P2 carpus identically shaped ⟶ **3**
2-b: Legs P1 and P2 carpus differently shaped. P2 carpus is more elongated than that of P1 ⟶ *Caridina*

3-a: Cephalothorax anterior border bearing a strong spine; straight rostrum rounded at its extremity with 1 to 8 small ventral teeth ⟶ *Atyopsis spinipes*
3-b: Cephalothorax border acute bearing no spine; curved rostrum with 0 to 2 small ventral teeth ⟶ *Atyoida pilipes*

4-a: One branchiostegal spine on each side of the cephalothorax. Hepatic spine absent. P2 moderately developed ⟶ *Palaemon*
4-b: One hepatic spine on each side of the cephalothorax. Branchiostegal spine absent. P2 very developed ⟶ *Macrobrachium*

••• Fish

1-a: Fish very elongated and filiform. The total length is more than 12 times the height of the body ⟶ **2**
1-b: Fish not very elongated and non-filiform. The total length is at most 10 times the height of the body ⟶ **4**

2-a: Filiform fish with dorsal, caudal and anal fins welded and continuous ⟶ **Anguillidae** (*Anguilla* sp.)
2-b: Filiform fish with non-welded dorsal, caudal and anal fins, or lacking a caudal fin ⟶ **3**

3-a: Smooth, cylindrical fish without any bony rings. Pointed snout. Caudal fin absent ⟶ **Ophichthidae** (*Lamnostoma* sp.)
3-b: Fish composed of bony rings. Tube-like snout. Lanceolate caudal fin ⟶ **Syngnathidae** (*Microphis* sp)

4-a: A unique dorsal fin, sometimes divided into two distinct parts, but joined ⟶ **5**
4-b: Two separate dorsal fins ⟶ **8**

5-a: No spines on dorsal and anal fins ⟶ **Poeciliidae**
5-b: Dorsal and anal fins with spines ⟶ **6**

6-a: More than 14 spines on the dorsal fin; one pair of nostrils only ⟶ **Cichlidae**
6-b: Less than 14 spines on the dorsal fin; 2 pairs of nostrils ⟶ **7**

7-a: 7 spines on the dorsal fin followed by 9 to 11 soft rays; translucent body ⟶ **Ambassidae** (*Ambassis* sp.)
7-b: 10 spines on the dorsal fin followed by 9 to 13 soft rays; non-translucent body ⟶ **Kuhliidae** (*Kuhlia* sp.)

8-a: The distance between the two dorsal fins is greater than the length of the first dorsal fin's basis ⟶ **Mugilidae**
8-b: The distance between the two dorsal fins is smaller than the length of the first dorsal fin's basis ⟶ **9**

9-a: Free pelvic fins; 6 branchiostegal rays ⟶ **10**
9-b: Pelvic fins welded together; 5 branchiostegal rays ⟶ **Gobiidae**

10-a: Pelvic fins close together; lateral line absent ⟶ **Eleotridae**
10-b: Pelvic fins very distant from each other; lateral line present ⟶ **Rhyacichthyidae** (*Rhyacichthys* sp.)

... Conservation and management

The current state of knowledge on the life cycle of the amphidromous species (biology, ecology) of Santo, the length of the larval phase and the part it plays in the dispersal of larvae, is of direct relevance to management and conservation. The management and the conservation of species must take into account both the dependency of adult populations on the larval pool for replacement, and the contribution of each reproductive population to the larval pool. The length of the marine phase might increase the probability of finding a river for colonisation, as will the strength and the direction of marine currents. The survival of the species in Vanuatu depends also on the ability of existing populations to provide enough larvae to maintain appropriate adult numbers.

Seasonal variables (e.g. rainfall, drought, floods, typhoons, etc.) have a major impact on the survival of populations: biological events such as reproduction, spawning, and the dispersal of larvae, are dependant on these events and are synchronised with them. On the islands, the impact of humans on aquatic habitats is highly significant, particularly on estuarine habitats, which are crucial to amphidromous species. These have to undertake two migrations between freshwaters and the sea. The success of such a life cycle, i.e. production of larvae and restocking rivers, depends on maintaining the mountain-ocean corridor in Santo open to allow movement between both habitats.

Several important facts need to be pointed out for the management and/or the conservation of Santo's freshwater fish and crustaceans.

- It is essential to allow species to move freely between the upstream and downstream reaches for trophic or gamic migrations or for the larvae's downstream migration; and between the downstream and upstream reaches for river colonisation by the post-larvae and the juveniles. To ensure the free circulation of these species requires that there be no barriers in the river that cannot be crossed both up and downstream (the ecological and biological characteristics of all the species involved need to be studied). It is also important to note that fishways are effective only if adapted to the context, to the fish species and if there are few on the migration pathway.

- The different ecological studies carried out show that a minimum flow has to be maintained in order to maintain rheophile zones (strong current and high water oxygenation) in the river and thus enable the species adapted to such an environment to complete their biological cycle. The flow rates must be high and must follow seasonal variations: the freshwater wave discharging at sea "calls" the post-larvae which then colonise the rivers. The disappearance of these rheophile areas would rapidly lead to the extinction of the endemic species.

- The vegetation cover must be maintained over rivers which still have a catchment area with a forest. This forest cover ensures the water remains cool and is well oxygenated; it also ensures regular rainfall thus supplying the catchment area with water. Forest cover provides a high diversity of habitats and therefore of species. It also supplies exogenous elements for the nourishment of certain species.

- The installation of structures modifying the flow rate , degrading habitats or causing pollution should be avoided. River eutrophication would lead to the disappearance of rare and/or endangered species because of the modification of the physical and chemical parameters of the water; moreover, the proliferation of filamentous algae would restrain the development of amphidromous species, as they are usually grazing species scraping short algae off pebbles and rocks. Several sectors of the Cumberland high plateau have suffered recent deforestation, leading to significant eutrophication in these areas and to a decrease in local biodiversity.

- Estuaries must be preserved as they represent areas where certain species transit, where larvae of amphidromous species exit to sea, and where post-larvae and juveniles enter to colonise the rivers. It is essential that they will be kept in there natural state.

FOCUS ON AQUATIC INSECTS

Arnold H. Staniczek

Aquatic insects are a heterogeneous assemblage comprising several insect orders which have independently evolved in aquatic habitats. The larvae of mayflies (Ephemeroptera), damselflies and dragonflies (Odonata), stoneflies (Plecoptera), alderflies and dobsonflies (Megaloptera), and caddisflies (Trichoptera) are obligatory bound to freshwater habitats. Furthermore, the true bugs (Heteroptera), beetles (Coleoptera) and flies (Diptera) each include many large families present only in aquatic habitats. In fact, almost every insect order has at least a few aquatic representatives. In all kinds of continental freshwater environments, aquatic insects are usually the largest single group of organisms, by species as well as specimen numbers, and are of paramount importance to the community structure and function.

However, in isolated tropical archipelagos the faunal composition of freshwater environments may entirely differ from continental conditions. The dispersal of aquatic freshwater insects from continents to isolated archipelagos is severely hampered by long distances over sea. Adult mayflies, stoneflies, and caddisflies are poor flyers and delicate insects, so their success in establishing island population is rather limited. A colonisation through larval dispersal can entirely be ruled out, as freshwater insect larvae generally do not survive in marine conditions. While crustaceans, e.g. shrimps, may secondarily conquer freshwater habitats from the seas, all species of insects must bridge the sea either by aerial dispersal or by clinging to driftwood.

This is mainly the reason for an entirely different freshwater species composition on Vanuatu compared to the Australian or Asian mainlands. In the streams of Santo only a very limited amount of freshwater insects species is present, and regarding arthopods the streams are rather dominated by freshwater shrimps.

• • • Mayflies (Ephemeroptera)

Larvae of mayflies are easily recognized by the presence of abdominal gills and three tail filaments. The larvae of most species mainly feed on detritus. Adult mayflies do not feed anymore and have a very short life span ranging from several hours to a few days. Adult mayflies are characterized by the presence of two or three long tail filaments. The hind wings of mayflies are reduced; in some species they are even completely lost.

Kimmins described in 1936 a single species of mayflies from Erromango, but mayflies have never been reported from Santo. During our field work we were able to record three different mayfly species. The larvae and female adult of *Caenis vanuatensis* (Caenidae) have been described recently. The description of a new species of *Labiobaetis* as well as the description of the larva of *Cloeon* sp. is currently in press.

• • • Key for the identification of larvae of the mayfly families of Vanuatu

1-a: All abdominal gills leaf-like, situated on sides of abdomen at least on segments II-VII ➞ **Baetidae**
1-b: Abdominal gill I finger-like, abdominal gills II situated dorsally on abdomen and modified to an operculum covering gills III-VI ➞ **Caenidae**

• • • Baetidae (small minnow mayflies)

Cloeon sp.
Labiobaetis sp.
Larvae of Baetidae are only of few centimetres length. They have a spindle-shaped body to which six or seven leaf-like pairs of abdominal gills are inserted laterally. The outer tail appendages, the cerci, are only medially fringed with long swimming hairs. Baetidae mainly feed on detritus. Usually small minnow mayflies have two to three generations within one year.

A single species of Baetidae, *Cloeon erromangense*, is known from Erromango and was described by Kimmins (1936) from adults. We encountered larvae of two species of small minnow mayflies in several streams of western Santo. One species can be attributed to *Cloeon* sp., but as the adults remain unknown it is not clear if these larvae are conspecific with *C. erromangense* or if they represent a different species. The other collected species from Santo belongs to *Labiobaetis*, a genus that is widespread in the Australasian realm. Both genera are very similar to each other and thus not easy to distinguish for laymen. The larvae of *Labiobaetis* (Fig. 302A) usually have a long and broad medial projection on the second segment of the labial palp and lack gill I. Both species were only collected in calm river parts clinging to the riverine vegetation.

Figure 302: Ephemeroptera. **A**: *Labiobaetis* sp. **B-D**: *Caenis vanuatensis*. (Photos A.H. Staniczek).

Caenidae (small square-gilled mayflies, angler's curses)

Caenis vanuatensis Malzacher, 2007

Caenidae belong to the smallest mayflies, with only a few millimetres of body length. Their adults are characterized by the total loss of their hind wings, and the forewings lack marginal intercalary veins. The larval abdominal gills are modified: Gill I is reduced and fingerlike, gill II is modified to an operculum that covers gills III-VI, gill VII is lost. The larvae of *C. vanuatensis* (Figs 302B & 302C) can be distinguished from other species of *Caenis* by the significantly reduced and shortened dorsal ridge of the operculate gill II.

There are more than 100 species of Caenidae described worldwide. While some species are also reported from Indonesia, Papua New Guinea, Australia, and Fidji, they are neither present in New Zealand nor in New Caledonia.

C. vanuatensis is abundant in many of the investigated streams in both western and eastern Santo. Numerous larvae were collected under stones and on the ground where they feed on detritus. All larvae found were of similar size which may point to a synchronised life cycle.

All collected larvae and reared adults (Fig. 302D) were of female sex. This makes it likely that *C. vanuatensis* is a parthenogenetic species. Parthenogenesis is often assumed to occur more likely on islands, and this phenomenon is also known from other species of Caenidae.

Dragonflies and Damselflies (Odonata)

Adult Odonata are generally more robust and mobile than other aquatic insects. Especially dragonflies are known as agile perchers and thus are very successful in the colonisation of islands. This may be the reason that a comparatively higher number of species could be found on Santo. The vast majority of Odonata found on Vanuatu is widely distributed throughout Southeast Asia and the Australasian region, but the species of *Vanuatubasis* are considered endemic to islands of Vanuatu. Until now 25 species of Odonata were known from Vanuatu. In our field studies we recorded 13 species, adding seven new records for Espiritu Santo. Five of these are also new for Vanuatu. Our study thus increases the total number of dragonflies for Vanuatu to 29, and now 17 species are known from Santo (Table 29).

Key for the identification of adults of the odonate families of Vanuatu

1-a: Fore- and hind wings similarly shaped and petiolate. Eyes well separated on top of the head. Discoidal cell four-sided (quadrilateral), never traversed by crossveins ⟶ Suborder **Zygoptera 2**

1-b: Forewings narrower than hind wings and never petiolate. Eyes touching on top of the head. Discoidal cell three-sided (triangle), sometimes traversed by crossveins ⟶ Suborder **Anisoptera 3**

2-a: Pterostigma distinctly longer than broad. Several wing cells five-sided. Veins IR3 and R4+5 originate far basal to the node. At least one supplementary, intercalated vein several cells long between branches of Rs in distal portion of wings ⟶ Family **Lestidae** with one genus: *Austrolestes*

2-b: Pterostigma not or scarcely longer than broad. Most wing cells four-sided. Veins IR3 and R4+5 originate beneath the node. No supplementary longitudinal veins between branches of Rs in distal portion of wings ⟶ Family **Coenagrionidae** with six genera: *Agriocnemis*, *Ischnura*, *Melanesobasis*, *Vanuatubasis*, *Pseudagrion*, *Trineuragrion*

Table 29: Seventeen species of dragonflies are known from Santo.

Agriocnemis exsudans Selys, 1877	Kimmins (1936), also as *A. vitiensis*
Ischnura aurora (Brauer, 1865)	Kimmins (1936), as *I. torresiana*
Vanuatubasis santoensis Ober & Staniczek, 2009	New for Espiritu Santo and Vanuatu
Vanuatubasis sp.	New for Espiritu Santo and Vanuatu
Pseudagrion microcephalum (Rambur, 1842)	Kimmins (1936)
Pseudagrion sp.	New for Espiritu Santo and Vanuatu
Anax guttatus (Burmeister, 1839)	Kimmins (1936)
Hemicordulia fidelis McLachlan, 1886	New for Espiritu Santo
Diplacodes bipunctata (Brauer, 1865)	Kimmins (1936)
Diplacodes haematodes (Burmeister, 1839)	New for Espiritu Santo
Diplacodes trivialis (Rambur, 1842)	Kimmins (1936)
Neurothemis stigmatizans (Fabricius, 1775)	Kimmins (1936)
Rhyothemis phyllis (Sulzer, 1776)	Kimmins (1936)
Orthetrum serapia Watson, 1984	Kimmins(1936), as *O. sabina*
Orthetrum villosovittatum (Brauer, 1868)	New for Espiritu Santo and Vanuatu
Tramea propinqua Lieftinck, 1942	New for Espiritu Santo and Vanuatu
Pantala flavescens (Fabricius, 1798)	Kimmins (1936)

3-a: Discodial cells in fore- and hind wings similarly shaped and more or less equidistant from arculus. Middle lobe of labium large, lateral lobes widely seperated ⟶ Family **Aeshnidae** with four genera: ***Adversaeschna, Anaciaeschna, Anax, Gynacantha***

3-b: Discodial cells in fore- and hind wings differently shaped and much nearer to arculus in hind wing than in forewing. Middle lobe of labium small and concealed by the contiguous lateral lobes ⟶**4**

4-a: Posterior eye margin with prominence ⟶ Family **Corduliidae** with one genus: ***Hemicordulia***

4-b: Posterior margin of eye without a conspicious prominence ⟶ Family **Libellulidae** with seven genera: ***Agrionoptera, Diplacodes, Neurothemis, Orthetrum, Pantala, Rhyothemis, Tramea***

••• **Annotated species list of Odonata from Santo**

Damselflies (Zygoptera) - Coenagrionidae

Agriocnemis exsudans Selys, 1877 (Figs 303A, 303B & 303C) (no common name)

A. exsudans (syn. *A. vitiensis*) is a small species of damselflies with hind wings shorter than 14mm. It is found throughout the southwestern Pacific. Adult males are very characteristic, as they are covered with whitish pruinosity on thorax and legs. Young males as well as females do not exhibit any pruinescence, but have an orange-red colouration on abdominal segments VIII-X. Otherwise they are coloured green and black. *A. exsudans* has been previously recorded from Shark Bay, Espiritu Santo, by Kimmins. It is also recorded from Vila by Tillyard in 1913. We encountered *A. exsudans* in the area of Tasmate at Taro fields and near Tasmate River. It was also recorded from Penaoru near Penaoru River.

Ischnura aurora (Brauer, 1865) (Fig. 303D) (Aurora bluetail)

This species was commonly encountered at temporary puddles and ponds in the Tasmate area. Its males are characterized by a green thorax, reddish abdomen with a black tail, additionally marked with blue on segments VIII-IX. It is also a small to tiny species. It occurs from India to the Central Pacific.

Pseudagrion is a large genus with more than 150 described species worldwide. *P. microcephalum* occurs from the Indian subcontinent through Southeast Asia and northeastern Australia into the western Pacific. It was previously recorded from Shark Bay, Santo. We did however not encounter this species during our field work. *P. microcephalum* is a medium-sized black and blue damselfly.

Pseudagrion sp. (Fig. 303F)

So far this species was unknown from Vanuatu and is recorded herein for the first time. The specimens are presently investigated and cannot yet be clearly assigned to a certain species. Males of the species found at Penaoru River are predominantly black with dark metallic green mesothorax and blue metathorax. Abdominal segments VIII and IX are almost completely blue.

Female specimens are blackish with green thoracic and and abdominal sterna.

Vanuatubasis santoensis Ober & Staniczek, 2009 (Fig. 303E) (no common name)

Kimmins (1936, 1958) described *Nesobasis malekulana* from Malekula and *Nesobasis bidens* from Aneityum. Ober & Staniczek 2009 restricted *Nesobasis* to 21 species described from Fiji, established the new genus *Vanuatubasis* for the two known species from Vanuatu, and added another new species *Vanuatubasis santoensis* from Santo to the new genus. *V. santoensis* was recorded from the lower courses of Penaoru River, Mamasa River and Paé River.

The males of this endemic species have an extraordinarily long abdomen, its general colouration is mainly olive green with brown longitudinal thoracic stripes. Also characteristic is a blue labrum.

Vanuatubasis sp.

There is probably at least another species of *Vanuatubasis* present on Santo. Female specimens were collected that are probably not conspecific with any known species of *Vanuatubasis* and thus remain undescribed for the moment.

Dragonflies (Anisoptera) - Aeshnidae

Anax guttatus (Burmeister, 1839) (Lesser green emperor)

This genus represents large to very large dragonflies. *A. guttatus* is distributed from India throughout the Pacific. Its light green thorax and abdominal segment I is very distinctive.

Abdominal segments II-III are of light blue colour, the remaining abdominal segments are black with pale brown spots. The species has previously been described from Hog Harbour, Santo by Kimmins in 1936. We were not able to catch the species, but have numerous sightings from Tasmate.

Dragonflies (Anisoptera) - Hemicorduliidae

Hemicordulia fidelis McLachlan, 1886

The distribution of this species is restricted to New Caledonia, The Loyalties, and Vanuatu. Kimmins reported in 1936 this species from Tanna and (under *H. assimilis oceanica*) from Aneityum, Malekula, and Efate. We record this species for the first time from the west coast of Santo. Three female specimens were caught from Tasmate River and Penaoru River. Most females of the species can be distinguished from all other *Hemicordulia* females by the presence of a diffuse amber cloud covering the distal part of the forewing near the pterostigma. Its general appearance is dark green metallic.

Figure 303: Odonata. **A-C**: *Agriocnemis exsudans*. **D**: *Ischnura aurora*. **E**: *Vanuatubasis santoensis*. **F**: *Pseudagrion* sp. **G**: *Diplacodes bipunctata*. **H**: *Diplacodes haematodes*. **I**: *Diplacodes trivialis*. **J-K**: *Neurothemis stigmatizans*. **L**: *Orthetrum villosovittatum*. (Photos M. Pallmann).

Dragonflies (Anisoptera) - Libellulidae

Diplacodes bipunctata (Brauer, 1865) (Fig. 303G) (Wandering Percher)
Diplacodes ranges from Africa and Asia through to the Western Pacific. *D. bipunctata* is found in a variety of still waters and was frequently observed in Taro fields near Tasmate. It was also collected at Tasmate River and Penaoru River. The species was previously also recorded from Banks Island. It is a small to medium-sized, red species. It has a small amber spot at the hind wing base that does not reach the rear margin of the hind wing.

Diplacodes haematodes (Burmeister, 1839) (Fig. 303H) (Scarlet Percher)
D. haematodes is very close to the previous species, but differs from *D. bipunctata* in having an elongate, light yellow tinged area on the hind wing base, which extends to its rear margin and to the triangle. It is widely distributed and occurs in Australia, Sri Lanka, New Guinea, New Caledonia, and Vanuatu. It has previously been only recorded from Aneityum and Erromanga. *D. haematodes* was frequently observed in West Santo at Tasmate River, Pai River, and Penaoru River.

Diplacodes trivialis (Rambur, 1842) (Fig. 303I) (Chalky Percher)
The distribution of *D. trivialis* reaches from the Seychelles to Fiji. This species is recorded in Santo for the first time, though it was previously known from Malekula and Erromanga. It inhabits ponds, taro fields, and swampy areas. This species is of predominantly black colour marked with yellow, becoming bluish grey in the mature male.

Neurothemis stigmatizans (Fabricius, 1775) (Figs 303J & 303K) (Painted Grasshawk)
This species is frequently observed throughout Santo. It is distributed from India to the western Pacific and inhabits a wide range of still waters. We found this species frequently at Taro fields near Tasmate, and also at backwaters of Mamasa River and Penaoru River. The mature males of *N. stigmatizans* are easily recognized by their characteristic, large, brownish red wing markings that cover the basal two thirds of the wings. In young males these markings are of brownish yellow colour. The colour of the pterostigma also changes from whitish yellow to red during maturation. In females, the pterostigma is also red, but there the otherwise clear wings have only lightly tinged wing markings at the tips and basally of the pterostigma.

Rhyothemis phyllis (Sulzer, 1776) (Yellow-striped Flutterer)
This species is distributed from Myanmar to Australia and through the western and southern Pacific. It has been previously recorded from Malekula and with one male and female specimen from Shark Bay, Santo, by Kimmins. We were not able to encounter the species during our field work. *R. phyllis* has colourless or tinged yellow wings marked with brownish to black patches. A major brownish black area at the base of the hind wings is traversed by a yellow cross-band. The wing tips are also darkened.

Orthetrum serapia Watson, 1984 (Green Skimmer)
O. serapia was found near Tasmate at Mamasa River. Kimmins recorded the closely related *O. sabina* from East Santo at Shark Bay, but the latter species is otherwise only known from the Palearctic to Australia and New Guinea. Kimmins' specimens most probably represent *O. serapia*, which is possibly distributed from the Philippines to the southwestern Pacific. The species is characterized by the presence of three light green stripes on the synthorax edged with brown. The abdomen is basally swollen and constricted from segment III on.

Orthetrum villosovittatum (Brauer, 1868) (Fig. 303L) (Fiery Skimmer)
This species is recorded for the first time for Vanuatu. It was found at Tasmate, Mamasa River, and Penaoru. It is a very distinctive, medium to large sized species with a red abdomen in the male. Females are entirely coloured yellow to orange. The species is distributed in North and Eastern Australia, New Guinea and adjacent islands.

Tramea propinqua Lieftinck, 1942 (Northern Glider)
Two mature males of his large dragonfly were found at Tasmate River. These findings represent the first record for Vanuatu. Like all other species of the genus, *T. propinqua* has conspicuous dark brown patches at the base of the hind wings. In this species these markings are narrow, not reaching the base of the triangle. Additionally, the top of the male frons is metallic purple. *T. propinqua* is also recorded from Northeast Australia, New Guinea, and the Bismark Archipelago.

Pantala flavescens (Fabricius, 1798) (Wandering Glider, also Globe Skimmer)
This species has a large distribution throughout the tropics and warmer temperate regions of the world. It is known to be a strong migrant, and its dispersal is associated with the Inner Tropical Convergence Zone. Kimmins (1936) recorded this species from Hog Harbour and Shark Bay, Espiritu Santo. The medium sized, dull orange to brownish species inhabits a wide range of still waters.

• • • Other aquatic insect orders
Plecoptera (stoneflies) have never been recorded from Vanuatu. We have neither been able to confirm the presence of any species of stoneflies during

our field work. It is most likely that Plecoptera have not been able to colonise Santo. Stoneflies are also not reported from New Caledonia and many other Southwest Pacific Islands. The same applies to dobsonflies (Megaloptera) who most probably do not occur on Santo.

We were also not successful in collecting any caddisflies (Trichoptera) at Santo, although Kimmins, Mosely, and Neboiss described several species of Trichoptera from Vanuatu.

In some temporary puddles near Penaoru there were also some specimens of water scavenger beetles (Hydrophilidae) collected, but their taxonomy has not yet been worked out.

FOCUS ON FRESHWATER SNAILS

Yasunori Kano, Ellen E. Strong, Benoît Fontaine, Olivier Gargominy, Matthias Glaubrecht & Philippe Bouchet

• • • Freshwater nerites (Yasunori Kano)

The Neritidae, also called neritids or "nerites", represent one of the commonest members of freshwater snails on Santo and other tropical islands. Neritids seem to have their origin in the sea, comprise roughly 100 living species in a few marine genera (e.g. *Nerita*) and abundantly inhabit intertidal rocky shores. However, they are more diverse in freshwater and brackish-water environments in terms of the numbers of species and genus. There are some 200 limnic species worldwide in eight or more genera (e.g. *Neritodryas, Clithon, Vittina, Neritina, Neripteron* and *Septaria*). The marine species have rather uniformly round shells, while the freshwater lineages have evolved various shell morphologies: they are spined, winged or limpet-shaped (Figs 304 & 305). "Nerites" may refer collectively to the members of Neritiliidae along with the Neritidae. Regardless of their similar names, habitats and shell shapes, the two families are totally different in soft part anatomy and in evolutionary history. Neritiliidae consists of about 30 living species of minute round snails that inhabit oceanic (e.g. *Pisulina*) or limnic (*Neritilia* and *Platynerita*) environments. The text below deals with the species of the two neritimorph families in freshwater and brackish water habitats, including streams, rivers, springs, estuaries and mangrove swamps (see chapter "Habitat types").

Fully-grown shells of limnic neritids may attain a diameter of 4 cm, but many species do not reach 2 cm at maximum. All neritiliids as well as some *Clithon* species are minute with their shell diameters less than 1 cm. Neritid shells often exhibit within-species variation in their shape, colour and markings (Figs 304 & 305). These considerable intraspecific variations accompanied with sometimes quite subtle differences between species make identification difficult without knowing the exact ranges of the variations. Worn shells may further prevent species identification. The apexes of nerite shells are often eroded with the acidic waters of streams (rain is naturally slightly acidic) and mangrove swamps. The extreme examples of erosion occur in long-lived freshwater species of *Clithon* and *Neritina*, where the apical one-third of the shell is eroded away and secondarily plugged by an inner shell layer. Eroded shells are rare in calcium-rich waters flowing through karst terrain.

Nerites are all herbivorous and graze on micro-algae, but habitat partitioning sustains dozens of species in a single river system. Distance from the river mouth, water velocity, salinity and riverbed condition are the primary factors that determine species composition at a certain area of rivers and streams (see "Mountain streams and longitudinal zonation"). At a smaller scale, a different set of species occupies a different type of microhabitats in a short stretch of rivers (Fig. 306). Yet, several species commonly occupy exactly the same type of microhabitat in the same stream, and the surprisingly high species diversity of limnic nerites cannot be solely explained by the resource partitioning. Natural disturbance by flooding in the rainy season may play a significant role in maintaining a heterogeneous environment and hence the species richness. Nerites require hard substrata for their grazing and they cannot be found on mud bottom in estuaries.

Mating and spawning occur year around. Female snails store sperm capsules (spermatophores) received from males for continuous spawning. Neritids lay flat, elliptical egg capsules on hard substrata including rocks and shells of other snails (Fig. 305B). The capsules are 1-3 mm long depending on species and are covered with sand grains sorted from mother's faeces. Newly laid capsules are white and then become cream to brown in a few days. Each capsule contains up to 300 eggs that hatch several weeks after being laid. Most freshwater nerites in the Indo-Pacific region have an amphidromous life cycle (Fig. 307). Amphidromy is a strategy involving migration of juveniles from the sea into freshwater, where growth from juvenile to adult, attainment of sexual maturity, and spawning all occur. With the presence of the obligatory marine phase, nerites do not occur in pools and dams without connection to the sea. However,

Figure 304: Neritid and neritillid gastropods from rivers, streams and mangrove swamps on Santo, Vanuatu. Note that some 10 species found on Santo are not included here. Morphological and genetic criteria were employed for rigorous species classification, but scientific names used are provisional and these assignments may be infirmed by thorough investigation on type specimens. **A**: *Neritodryas* sp. cf. *notabilis*, 41.8 mm. **B**: *Neritodryas cornea*, 26.5 mm. **C**: *Neritodryas* "*subsulcata*", 35.3 mm. **D**: *Clithon siderea*, 5.7 mm. **E**: *Clithon paulucciana*, 5.2 mm. **F-G**: *Clithon oualaniensis*, 8.8 mm (F), 7.5 mm (G). **H**: *Clithon chlorostoma*, 7.0 mm. **I**: *Clithon diadema*, 15.5 mm. **J**: *Clithon aspersus*, 27.0 mm. **K**: *Clithon olivaceus*, 24.0 mm. **L-M**: *Clithon corona*, 24.7 mm (L), 19.8 mm (M). **N**: *Neritilia rubida*, 4.8 mm. **O**: *Neritilia vulgaris*, 2.9 mm. **P**: *Platynerita rufa*, 4.6 mm. Sizes denote maximum shell diameters of figured specimens, excluding spines if present. The spine development of *Clithon* shells sometimes varies greatly within a single species. Very long spines are most frequently found in young shells from calm waters, while adults with and without spines often co-inhabit a fast flowing stream (L & M).

Figure 305: Neritid and neritillid gastropods from rivers, streams and mangrove swamps on Santo, Vanuatu (continued). **A**: *Neritina pulligera*, 39.2 mm. **B**: *Neritina petitii*, 36.2 mm. **C**: *Neritina asperulata*, 25.6 mm. **D**: *Neritina canalis*, 22.5 mm. **E**: *Neritina powisiana*, 25.3 mm. **F**: *Neripteron florida*, 10.5 mm. **G**: *Neripteron siquijorensis*, 10.9 mm. **H**: *Neripteron spiralis*, 8.2 mm. **I**: *Neripteron auriculata*, 22.4 mm. **J-K**: *Neripteron subauriculata*, 20.5 mm (J), 10.3 mm (K). **L**: *Septaria porcellana*, 30.7 mm. **M**: *Septaria tesselata*, 32.0 mm. **N-O**: *Vittina adumbrata*, 31.3 mm (N), 23.9 mm (O). **P**: *Vittina coromandeliana*, 14.0 mm. **Q-S**: *Vittina variegata*, 17.7 mm (Q), 19.4 mm (R), 19.9 mm (S). **T**: *Vittina communis*, 11.1 mm. Shells of *Neripteron* species from protected mangrove swamps are generally thinner, flatter and wider with more prominent wing-like projections (J) than those from relatively exposed estuaries (K). For species with variable shell colours, opercula are sometimes more useful in identification: *V. adumbrata* always bears a white operculum (O), while *V. variegata* has a black one (R & S).

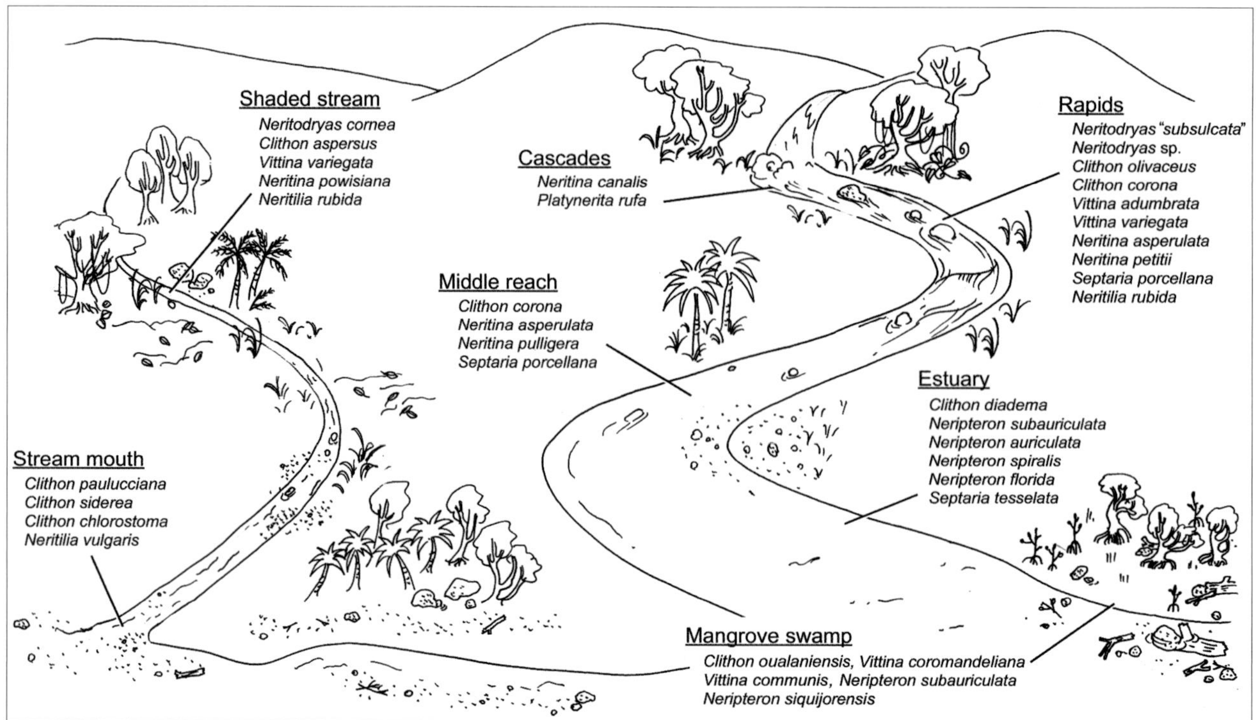

Figure 306: An idealized landscape with various freshwater and brackish-water habitats of the Neritidae and Neritiliidae. Representative species are shown for each habitat. Cascades are often inhabited by *Neritina canalis*; mountainous rapids with rocks harbor quite a few species of *Neritodryas*, *Clithon*, *Vittina*, *Neritina* and *Septaria*. The middle reach of rivers with stones is typically inhabited by *Clithon corona* and a few *Neritina*; brackish estuaries by *Clithon diadema* and *Septaria tesselata*; the mouth of streams at pebble beaches by several small species of *Clithon*; mangrove swamps by *Vittina coromandeliana* and some *Neripteron*. Exceptions to this macrohabitat segregation include the presence of such omnipresent species as *Septaria porcellana* and *Neritilia rubida* from both upper and lower reaches of rivers and streams. Habitat segregation also exists at a smaller scale. For example, a middle reach of a stream may harbour *Neritina petitii* on the lateral sides of rocks in fast-flowing water, *Neritilia rubida* under deep buried boulders, *Vittina adumbrata* and *Neritodryas* "subsulcata" on walls above the level, and *Neritodryas cornea* even on the branches of nearby bush. *Neripteron spiralis* and *Septaria tesselata* prefer drift pieces of wood on the riverbed to rocks and stones as their habitat in an estuary, while preference is reversed in the sympatric *Clithon diadema*. (Illustration by Miyuki Fuji).

some species crawl surprisingly long distances to mountain rapids more than 10 km away from estuaries where they metamorphose and settle.

The longevity of their larvae in the ocean may exceed two months, and such a long duration of the larval phase makes widespread geographic distribution possible. Although adults inhabit strictly freshwater environments, many amphidromous species are distributed widely in the Indo-Pacific. For example, specimens of *Septaria porcellana* on Santo are surprisingly similar to those from Japan and eastern Africa, both morphologically and genetically. Of the approximately 40 species of non-marine nerites on Santo, some 30 have fairly wide distributions stretching to either South-East Asia or French Polynesia, and less than 10 species are restricted to the Melanesian islands. In contrast, few direct developers (those hatch from egg capsules as crawl-away juveniles without a marine phase) show understandably higher levels of endemism. An unidentified *Neritodryas* (Fig. 304A) may represent a new species endemic to Vanuatu, supposedly with a direct development.

The downstream and upstream migration of amphidromous animals can easily be hindered with such human activities as water withdrawals, channel modifications, deforestation and sewage inputs at any reach of a river. Water withdrawals should not dry up rivers, or nerites there disappear for many years; extensive logging may also reduce dry season flow to zero. Dams and certain types of stream crossings (e.g. culverts with pipes) can impede migration and eliminate upstream population. With their large biomass, elimination of the snails may result in food-web alternation in a river system. Santo has been a good representative of high islands in the region in sustaining the natural conditions of various habitats and the snail diversity, while large species are rare in some places, suspectedly due to over-harvest by locals for food.

••• Thiaridae and other freshwater snails (Ellen E. Strong, Benoit Fontaine, Olivier Gargominy, Matthias Glaubrecht & Philippe Bouchet)

The Thiaridae is a family of snails distributed throughout the tropics and sub-tropics around the world, inhabiting inland freshwaters as well as the lower, tidally-influenced reaches of coastal rivers and streams. Members of the group are playing a variety of roles in aquatic ecosystems as primary herbivores, bioturbators, vectors of disease,

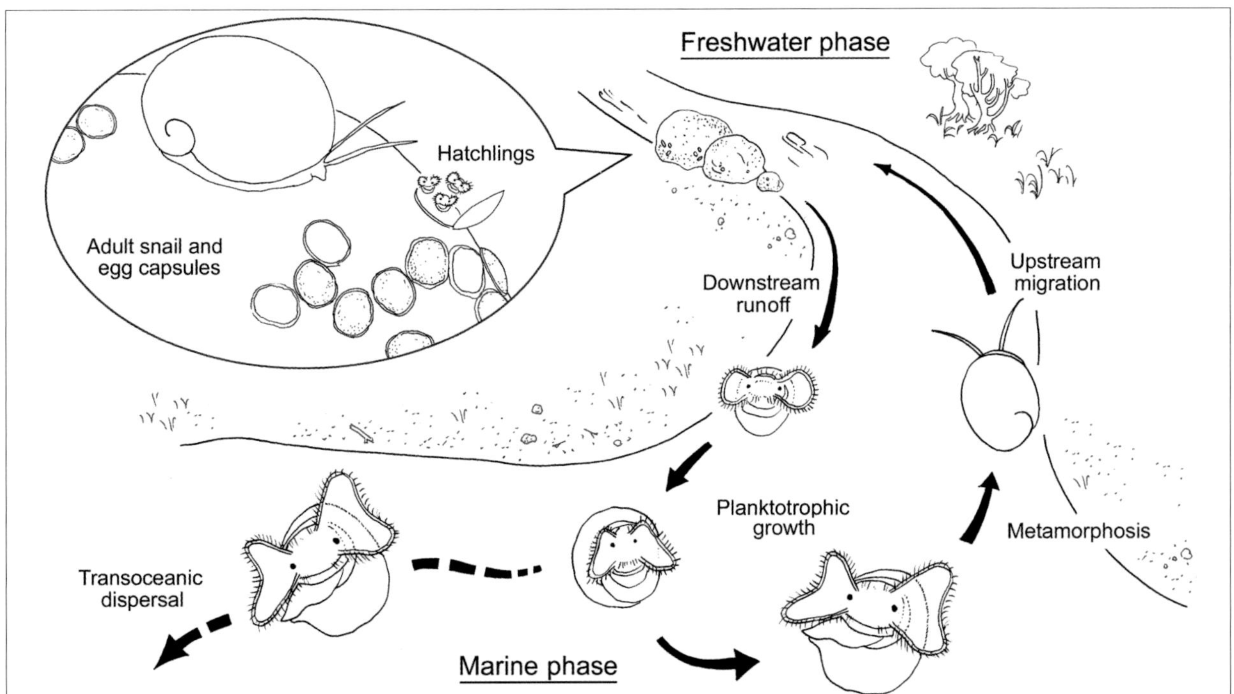

Figure 307: Schematic illustration for the amphidromous life cycle of the freshwater Neritidae and Neritiliidae. (**1**) Mother snails lay egg capsules on rocks in the freshwater reaches of rivers. (**2**) Hatched larvae, 0.1-0.2 mm in size, are swept downstream to the ocean. (**3**) The swimming marine larvae feed on diatoms and other phytoplankton and grow up to 0.3-0.5 mm in several weeks. Most larvae apparently remain in the vicinity of the river mouth, but some may travel great distances with ocean currents and the luckiest of them may happen to drift to an estuary on a remote island. (**4**) Metamorphosis to crawling juveniles occurs at estuaries or brackish reaches of rivers. (**5**) Young snails crawl upstream to the freshwater where they spend the rest of their life. Certain species migrate surprisingly long distances with over 10 (or possibly 20) years of longevity. (Illustration by Miyuki Fuji).

invasive pests, agents of biocontrol and sometimes substantial contributors to biomass. Thiarids are recognized by their often large (up to ~10 cm), high-spired shells that can be smooth but may also bear grooves, nodules, ribs or spines. Their shells are simply coloured and usually appear dark brown or black due to the presence of an organic periostracum on the outer surface of the shell; once this layer is removed, the shells are revealed to be light tan to brown or green in colour, and may be ornamented with reddish-brown "flames", spots or spiral bands. Colour, as well as shell shape and ornamentation patterns often vary considerably within species.

Like other groups of freshwater snails, this extensive intraspecific variation in features of the shell has produced widespread confusion surrounding species recognition. Dozens of species have historically been described from South Pacific islands as "*Melania*", based on minor variation in shell morphology, reflecting differences between juveniles and adults, ecophenotypic variation, or individual pathological variants. In this respect, it is significant to quote Edgar Smith when he described three new species from the "New Hebrides" island of Epi back in 1884: "It is with some reluctance that I name forms so variable and puzzling as the Melaniae of South-Sea Islands". As a consequence, it is unknown how many species of freshwater thiarids currently really exist, and there is little consensus in the scientific literature regarding the correct names of those that have been documented. This situation confounds efforts to communicate effectively about these species and hinders efforts to understand global patterns of distribution and diversity.

Thiarids are ovoviviparous and brood their young in a specialized brood pouch extending into the head-foot from a brood pore on the side of the neck. Reproductive strategies range from species that release hundreds to thousands of swimming (veliger) larvae, to those that brood a few large juveniles. In some species (e.g. *Melanoides tuberculata*), the epithelium of the brood pouch has been thought to provide nutrition for developing embryos; but this claim is however controversial with adelphophagy (brood cannibalism) being the most likely source of nourishment. Males are rare, and in some species completely unknown, indicating that reproduction is primarily (and sometimes exclusively) parthenogenetic (development of unfertilized diploid eggs) and only rarely sexual. When sperm transfer does occur, it is accomplished with elongate sperm packets (spermatophores); however, successful mating has never been observed and is thought to be facilitated by water currents as males are aphallic (lack a penis).

The distribution of thiarids on scattered Pacific islands, including Santo, suggests that development

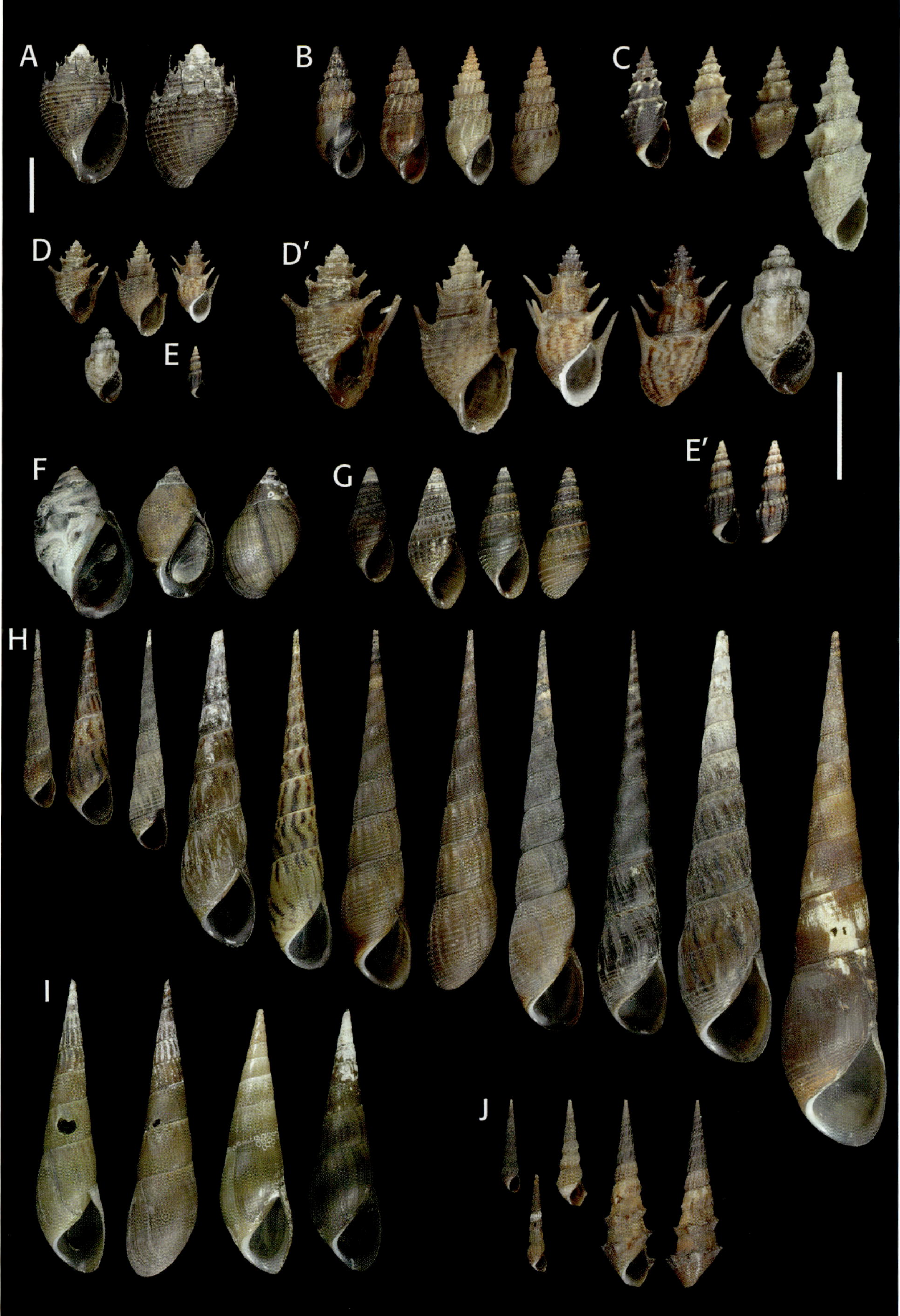

Figure 308: Thiaridae from the rivers of Santo. **A**: *Thiara cancellata*. **B**: *Thiara mirifica*. **C**: *Thiara winteri*. **D-D'**: *Thiara scabra*. **E-E'**: *Thiara terpsichore*. **F**: *Balanocochlis glans*. **G**: *Tarebia granifera*. **H**: *Stenomelania funiculus*. **I**: *Stenomelania plicaria*. **J**: *Stenomelania fastigiella*. Scale bars: 10 mm. All shells are at the same scale, except D' and E' (scale bar on the right) which are enlarged for clarity.

Figure 309: Different forms of Thiaridae that correspond to *Melanoides tuberculata* morphotypes. However, unpublished molecular data (COI, 12S and 16S) indicate strong divergence between the four forms depicted. Scale bar: 10 mm.

is amphidromous (see text on Neritidae, above), allowing the larvae to disperse in the sea. Although the release of veligers has been observed, trans-oceanic dispersal has never been directly confirmed. The observation that juveniles and adults of some species have different ecological tolerances, with juveniles dominating brackish influenced areas and adults in freshwater ones, is further circumstantial evidence of their amphidromous life cycle. Regardless, their inferred oceanic larval development, parthenogenetic reproduction and brooding habit render thiarids effective dispersers and colonizers. Two species in particular — *Melanoides tuberculata* and *Tarebia granifera*—are highly invasive and have become established throughout the tropics primarily through the trade in aquatic plants (e.g. aquarium industry, agriculture). Their effectiveness as colonizers is heightened by their tolerance of low oxygen conditions and disturbed habitats, their rapid growth, high fecundity and ability to maintain high population densities. For example, while native and naturalized thiarids typically do not maintain extremely large populations, some invasive species have been recorded at densities up to 10000/m². Although invasive snail species have had sometimes devastating effects on the local mollusc fauna of Pacific islands (particularly land snails), invasive thiarids usually only have local impacts resulting in displacement of native populations through competition and overcrowding and are not known to have contributed to the wholesale extinction of endemic snail species. Nevertheless, the introduction of thiarids has been cause for some concern as some species can serve as the first intermediate host for the Oriental lung fluke *Paragonimus westermani* and the Chinese liver fluke *Clonorchis sinensis* as

well as for trematode infections of mammals, birds and other wildlife. However, transmission requires raw consumption of the second intermediate host (usually crustaceans) and/or poor water sanitation and in most areas, substantial risks to human health have not materialized.

Genetic studies have demonstrated that invasive populations are typically structured by repeated introductions from different source populations. This alleviates the dramatic and potentially detrimental decrease in genetic diversity that can accompany founder events — a phenomenon heightened among parthenogenetic clonally reproducing species, as only a single female is necessary to establish a viable population. Furthermore, within these populations different shell morphs representing phenotypically and genetically distinct clonal lineages can persist for quite some time owing to the rarity of sexual reproduction that acts to homogenize the gene pool. When different lines do occasionally interbreed, they produce hybrids that become established as independent lines with their own population dynamics. The long term persistence of these distinct shell morphs has been yet another confounding influence in the attempt to understand species diversity in this group.

On Santo, roughly 17 species of thiarids (some not shown) have been found inhabiting coastal and inland rivers and streams (Figs 308 & 309). Together with Fiji and Samoa, Santo represents the eastern extent of native populations of thiarids in the Pacific; in Eastern Polynesia, only the widespread alien species *Melanoides tuberculata*, and, to a lesser

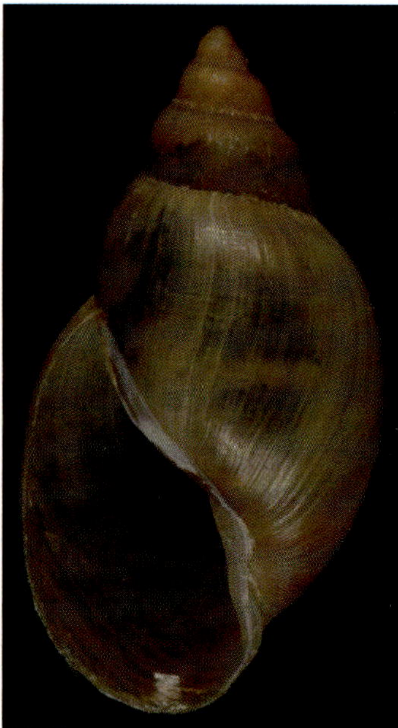

Figure 310: Contrary to most gastropod genera, *Physastra* is left-handed. Specimen from Fapon area, H = 11 mm.

extent, *Tarebia granifera*, are present. Some Santo species have been documented at elevations as high as 700 m above sea level (a.s.l.) in the Cumberland range, or at 570 m a.s.l. in the Butmas area, some 25 km away from the closest coast. Individuals of most species are found in the main channel of slow flowing streams, or in side channels, sandy banks and pools of fast flowing streams, while others prefer more stagnant waters, like small pools, marshes, mangrove swamps and irrigated fields. They occupy many different substrates including mud, sand, pebbles or stones where they graze on diatoms and filamentous algae with some species in tidally influenced areas tolerating prolonged exposure to air. Some species are active primarily at night, burrowing into the sediment away from potential predators by day. Santo thiarids can be quite abundant, reaching densities of several dozen individuals per square meter, and in some areas up to five species of thiarids may co-occur syntopically, and also with various neritids (see text on Neritidae, above).

Although it is not possible to provide an unambiguous identification of each species that occurs on Santo given the confusion that exists in the literature, the distinctness of these species has been confirmed by molecular sequence data, and some may prove to be new species. The variation in shell form and colour pattern that has produced this confusion is evident particularly among individuals of *Thiara scabra* and *Stenomelania plicaria*. This variation is more pronounced when comparing individuals from different islands, further confounding efforts to circumscribe populations as belonging to the same species. Of those species that can be positively identified, published records confirm that these are indeed widespread species, occurring throughout the western Pacific and Indian Oceans in Africa, Madagascar, India, Sri Lanka, Thailand, Java, Borneo, continental southeast Asia, Sulawesi, the Philippines, the Ryukyus of Japan, New Guinea, northern Australia, New Caledonia, Vanuatu and many smaller tropical islands of the western Pacific.

Other than thiarids and neritids, the single most abundant freshwater gastropod species in Santo is a species of *Physastra* (Planorbidae), which is found in taro fields, irrigation ditches and slow streams. The systematics of South Pacific *Physastra* is in considerable need of revision and it is not even known whether the Santo species — traditionally identified as *Physastra layardi* — is endemic, native or introduced. Springsnails in the family Tateidae [formerly classified in Hydrobiidae] represent another important constituent of the aquatic gastropod fauna. The genus *Fluviopupa* has radiated throughout the South Pacific, from New Caledonia to the Austral Islands, and seven species were collected during the Santo 2006 expedition, all of them new to science and endemic to the island. These tiny (less than 3 mm high) gastropods can be very abundant locally in springs, forest streams and seepages, mostly on limestone substrate, and never in large rivers. One species is probably exclusively stygobiont, i.e. it lives in underground aquifers. Most have very restricted ranges, and their habitat can easily be damaged, e.g. by cattle trampling.

Figure 311: Tiny *Fluviopupa* species are sometimes found in large numbers on fallen leaves in seeps and springs. Here, *Fluviopupa espiritusantoana* from a stream near Penaoru. Scale bar: 3 mm.

Caves and Soils

coordinated by Louis Deharveng

Team

Louis Deharveng & Anne-Marie Sémah

With more than 1 600 km² of limestone terrain, i.e. nearly half of the total area of the island, the karst of Santo was an important target of the Santo 2006 expedition. In a larger context, Santo karst was an exciting objective in the Pacific for three reasons. First, it has the largest extent of karstified limestone in the region after New Guinea. Second, it had the least known subterranean fauna among the large islands of the region (Fig. 312). Third, basic sedimentological and archaeological data were already available.

The goals of the karst team were to inventory as completely as possible the biodiversity of karst habitats, and to collate data regarding present and past environments in order to better understand the biodiversity patterns of the island.

Before sending a large team of scientists to a remote part of the world like Santo, the biodiversity potential of the area had to be evaluated, more particularly in its most characteristic and hidden component, i.e. cave fauna. Since literature about regional biodiversity was very scarce, the project leaders funded two prospective expeditions in 2005, one focusing on caves and terrestrial cave fauna, the other one on underground waters, diving and aquatic fauna. Results were very promising, so we rapidly set up the karst team for the Big Trip of 2006.

The team in charge of the scientific study of the Santo karst (Fig. 313) included 20 people from Australia (1), France (12), New Caledonia (3), Indonesia (1), Spain (2) and UK (1). Two Ni-Vanuatu (Charley and Faustin) and Rufino Pineda provided invaluable assistance in the field. Altogether, we had a broad panel of scientific and technical expertise covering human, physical and historical aspects of the karst, as well as soil, cave and aquatic karst invertebrates. A large network of specialists were at least associated with our work, providing additional taxonomic expertise.

We operated in the field as three core teams, with specific targets and investigation tools:
- A team of sedimentologists, palaeontologists and archaeologists;
- A team of hydrobiologists, with the support of experienced cave divers;
- A team of terrestrial biologists, with the support of experienced cavers.

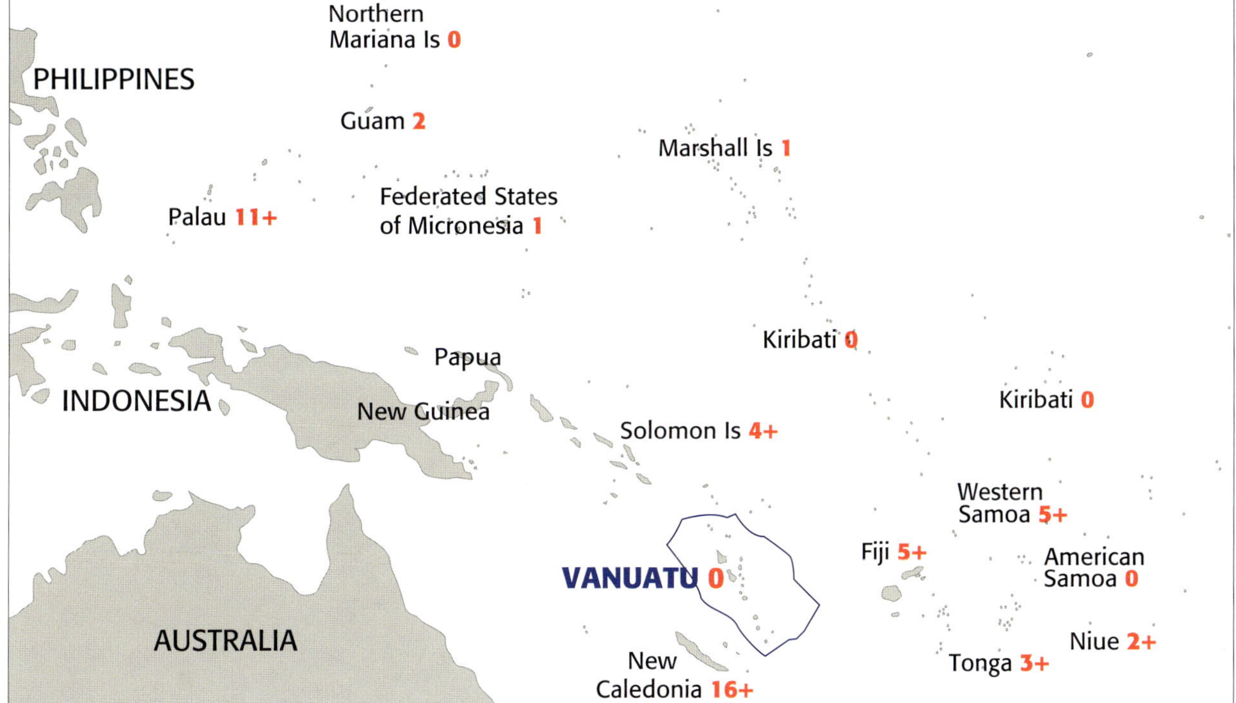

Figure 312: Subterranean fauna of the western Pacific, state of the art before Santo 2006. The approximate number of cave-obligate species is given for the main island-groups; + indicate records of additional not identified cave species.

Figure 313: The Karst team.

Figure 314: Boat is often the most efficient transportation to reach many coastal areas.

Figure 315: "*En route*" for the Malo island caves, by truck.

Figure 316: In the lab of the base camp.

Field work focused on the large limestone block of eastern Santo, with a flash-trip to Cape Cumberland. We mostly operated in one-day trips, made possible by a relatively easy access to most parts of the karst. Technically difficult caving and diving explorations as well as remote areas of the island required overnight and sometimes an expedition of several days (Figs 314 & 315).

The collected material, i.e. bones, sediments, animals, water and soil samples, was labelled, processed and sorted at a gross level at the end of each trip, in the CETRAV building north east of Luganville, our base camp (Fig. 316).

The karst team benefited from an unexpectedly favourable weather in September 2006, which allowed uninterrupted activities. After a month of field work, more than 650 samples from cave, soil, terrestrial and aquatic habitats as well as hundreds of sediment, palaeontological and archaeological remains had been obtained. A large part of this material has been dispatched for study to specialists worldwide, and several papers have been already published, including new taxa among crickets, springtails and Crustacea.

The karst component of Santo 2006 was a great success. Soil and cave karst biodiversity of the island changed from the least known to the best known of the large Pacific islands. The reader will find in the following pages a presentation of the main results obtained during this exceptional expedition.

Karst and Caves

Bernard Lips, Franck Bréhier, Denis Wirrmann, Nadir Lasson,
Stefan Eberhard, Josiane Lips & Louis Deharveng

THE KARST OF SANTO: GEOLOGICAL SETTING

Limestone facies account for almost half of the surface of Santo Island. The Papatai limestone is the most extensive areally, covering *c.* 90% of the karst area east of a line drawn south from the mouth of the Jourdan River (see Fig. 325B in "Caves as Archives" by Wirrmann and coauthors). It overlies the Tawoli calcarenite-calcilutite formations (ranging from middle Miocene to Pliocene), and forms widespread plateaus in eastern Santo as well as isolated plateau remnants in the western part of the island (for example the coastal plateau in Cape Cumberland). The Papatai formation consists of massive Pleistocene coralline limestone, very porous, and made up entirely of high-magnesium calcite: it is an uplifted complex of coral terraces and flat, coralline islands. The coral terraces related to glacio-eustatic sea-level changes during late Pleistocene and Holocene time are constantly uplifted above modern sea-level (see "The late Quaternary reefs" by Cabioch & Taylor). The pure Quaternary reef limestones comprise the East Santo Plateau, its highest elevation being at 348 m in the Butmas Plateau.

The oldest Quaternary limestones, older than the last interglacial, present a pronounced residual karst morphology with well developed conical hills. They formed at former reef crests and along interfluvial ridges on tilted limestone plateaus and terraces in the eastern central part of the island at Butmas Plateau, Mt Tanakar and Mt Tiouri. The conical hills have developed along the reef crests as the result of fluvial erosion perpendicular to the crests rather than through pure dissolution effect. Only in a few locations, more particularly at the northwestern end of Butmas Plateau, are conical hills associated with typical over-deepening of solution dolines. The hills are most pronounced near the local base level of the Sarakata River, suggesting a fluviokarst morphology mainly related to erosional processes. Karst is totally absent in Tertiary calcarenites, where fluvial erosional forms dominate the relief, due to their high content of volcaniclastic impurities which induces better mechanical strength.

The last interglacial terraces (130-115 ka BP, Eemian Stage [1 ka = 1000 years; years BP: before 1950]) form a broad compound terrace surface, termed the Luganville Surface. This formation exhibits a variety of solutional topography: small solutional and shallow depressions (dolines) and residual hills, without large scale dissection into residual conical hills. This topography has been interpreted as the initial stage of a cone and cockpit karst landscape related to over-deepening of preexisting depressions on the former lagoon floor. Numerous collapse features, many of them with maximum depths extending to more than 150 m below the Luganville surface, are associated with the dolines. They are always associated with absence of roofing limestones over underground streams. On topographical arguments, the related subterranean river courses have been considered as very young features, younger than 60 ka BP. Collapse features are most pronounced in the areas where streams from the volcanic basement enter the limestones. Along the tectonic lineaments the allogenic rivers were provided with paths of easier and faster penetration and solution so that extensive underground channels could develop. On the isolated plateaus of the Luganville Surface west of Hog Harbour, north of Port Olry and around the Walroul Plateau, collapse features are generally absent, while surficial solution features are well developed, however.

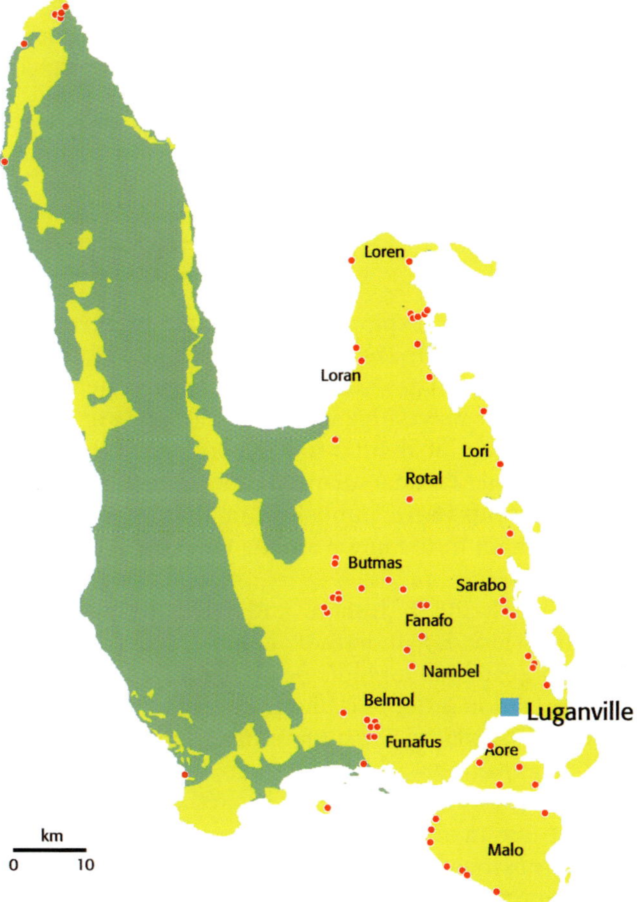

Figure 317: Karst (yellow area) and caves (red dots) of Santo.

The karst forms on the uplifted Holocene coral reefs correspond to circular or elongated collapse dolines of 4 to 6 m deep, sometimes extensive like at Hog Harbor and further north along the east coast and south of Mt Tomebou (See Fig. 325B in "Caves as Archives" by Wirrmann and coauthors). Most of them occur within a kilometre-wide strip parallel to the coastline. Their development has been related to areas where subterranean water reaches the piezometric surface near the present coast. Nevertheless, more especially between Mt Tomebou and Wambu River, other depressions, which are not related to collapse over underground streams, are also observed on this Holocene surface. Their origin probably reflects "premature" emergence of incomplete reef development, rather than a unique solution processes like that of the Luganville surface. Rapid tectonic uplift and a rising Holocene sea-level must have resulted in mainly vertical growth, which has overtopped a former and older submerged reef, explaining how the extensive Holocene reef surface could have formed in such a short time.

This brief overview based on published works shows that, in combination with a perhumid tropical climate, the karst relief on the Quaternary limestones of Santo is the result of facies-controlled lithology and purity of parent material (lack of karst feature on the Tertiary calcarenite formation), tectonic conditions (the linear topographical features defining preferential paths for vertical and lateral solution), influence of soil and plant cover (the deeper the soil, the stronger the dissolutional process) and Quaternary sea-level changes.

THE CAVER PERSPECTIVE

For cavers, the Santo karst is first a huge and compact block of limestone, 60 km long by 25 km wide (Fig. 317), i.e. one of the largest karst blocks in the Pacific. The few outlying islands and small outcrops in the North and the West of the main island have much less speleological potential. Elevation is moderate (784 m at the highest point, Mt Tanakar) and undulating terrain is the dominant landscape, with few large dolines and rare cliffs except along the coast. The archetypical features of tropical karsts, pinnacles or steep karst towers, are absent in Santo. Nevertheless, the karst is amazingly rich in other karst features, including caves, deep dolines and large coastal springs, enough to be very attractive for cavers.

Finding caves in Vanuatu can only be done in tight connection with local people. Not only the lush vegetation (Fig. 318) and absence of prominent landscape features would make direct prospection long and uncertain, but also permission is mandatory for underground access, from the owner of the cave entrance and the tribal chief of the area where it is located. Local guides also are necessary to reach the entrances, and young (and less young) villagers often enthusiastically accompany the cavers into the darkness. At least, in many cases, sleeping in the villages is the best way to optimize the time spent underground —and is an unforgettable experience.

HISTORY OF EXPLORATION

The first cave explorations on Santo were done by Australian divers in search of sumps to dive (1996-2000). They surveyed the large system Mt Hope-Sarakata resurgence (with a total horizontal development of about 3 400 m), and made the first trip to the Patunar giant doline. These pioneer explorations were published in diving journals, and remain little known among cavers.

In August 2005, Josiane and Bernard Lips and Rufino Pineda, undertook a preliminary trip to Santo, searching for biologically significant caves in the island in order to evaluate how much emphasis should be placed on subterranean biology for the 2006 expedition. The results were beyond expectation, with 54 karstic features located (mostly caves) and 5 886 m of underground passages mapped. The most obvious biological features observed were the omnipresence of bats, often in large colonies, and the frequent occurrence of swiftlets, in association with large guano piles supporting abundant animal communities in numerous caves.

A second preliminary reconnaissance targeting subterranean water habitats was made from 17 October to 13 November 2005 by Franck Bréhier, with three objectives: diving some inland caves to detect their richness in subterranean species; exploring karstic areas not seen by the preliminary reconnaissance, specially the northern cape of the island; searching for anchialine habitats and fauna. In total, Franck explored 30 karstic features, of which 10 potential anchialine habitats, 2 000 m cave passages were recognized, and five caves were dived. Aquatic fauna was present in most caves, and in several sites pottery as well as human bones were observed.

Based on these promising results, we set up a strong team of skilled cavers and divers to continue the exploration and mapping of the cave systems of the island, and to assist the biologists in the field.

Figure 318: Cave and shaft entrances.
A: Lavav Aven near Port Olry. **B**: Avorani shaft on Malo. **C**: Bottom of Patunar doline. **D**: Patunar resurgence. **E**: Fapon Cave.

Figure 319: Passing a "voûte mouillante" in Fioha Aven.

MAIN RESULTS

The 2006 expedition allowed us to push exploration and mapping of the caves recognized in 2005, but also to discover 28 additional caves or karstic features. In one month, 13 961 m of subterranean passages were explored, of which 7 961 m were mapped. Eighty-six subterranean karstic features are recognized today on Santo: 10 blue holes, 41 caves less than 50 m length, 15 caves from 50 to 100 m, 21 more than 100 m, 12 more than 500 m and five more than 1 km long. With more than 19 km of explored underground passages in total, Santo is the richest of the Pacific islands in number of caves and length of explored passages — outside New Guinea.

In lowlands, most caves were dry entrances and did not give access to underground streams, with the noticeable exception of the anchialine Loren Cave (see "Focus on Loren cave"). Even the spectacular blue holes did not open onto significant underground passages.

The most interesting cave systems of the island were discovered in two upland areas: the Funafus area, which gave access to the largest caves of Santo, and the Butmas region, the most promising area, which was the most intensively studied biologically.

••• The Funafus system (Fig. 320)

The largest subterranean system of Santo explored so far is located near the village of Funafus, between 100 and 250 m of altitude, in the southern part of the Santo karst. Six caves, awaiting interconnection, belong to this complex:

- Amont Cave (length: 259 m northwestern part of the system, not represented on the map);
- Kafae Aven (length: 3 702 m);
- Streamsink and Tarius Caves (length: 2 139 m mapped + 380 m

not mapped);
- Riorua Cave (length: 400 m);
- Tchawak Cave (length: 58 m);
- Patunar resurgence and doline (length: 742 m).

The total known passages of the Funafus system reach 7 662 m, i.e. more than 1/3 of the total length of subterranean passages currently known from the island. The Patunar doline is probably the most impressive karst surface feature of Santo: it is a large depression with subvertical cliffs on three sides, 100 m deep and 100-150 m in diameter. In

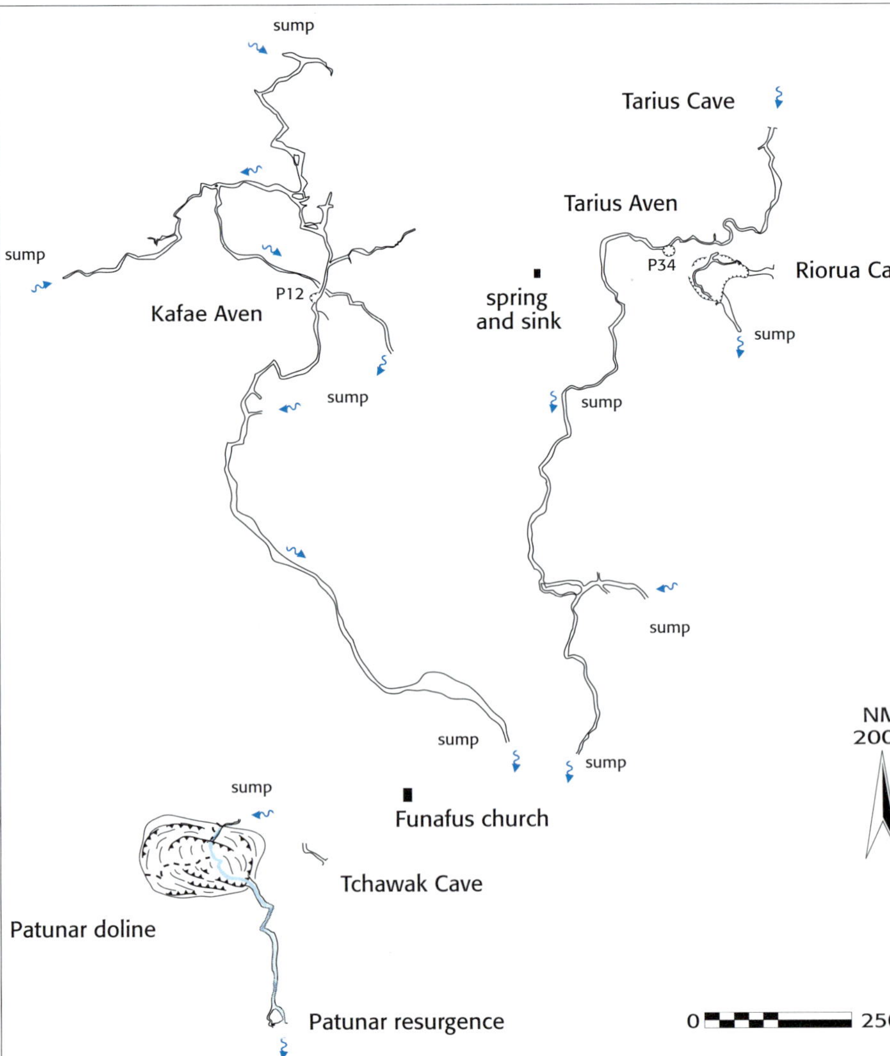

Figure 320: The Funafus system (Funafus-Santo-Vanuatu, Santo 2006 expedition).

Figure 321: Underground exploration.
A, B: Underground river in large passages of the Kafae Aven. **C**: Descent in Kafae Aven. **D**: Amarur Cave. **E**: Fapon Cave.

Figure 322: Underground exploration.
A: Kafae Aven, clay formation. **B**: Calcite formation deposited by a little inlet in Kafae Aven. **C** & **D**: Stalactites in Fapon Cave.
E: Stalactites and stalagmites in Fapon Cave. **F**: Amarur Cave, large fossil clam (*Tridacna* spp.).

Table 30: Longest (more than 300 m) and deepest (more than 25 m) caves of Santo and Malo Islands.

Name of the caves	length in m	depth in m	explored by
Kafae Aven (Funafus) (Figs 321A-C & 322A-B)	3 702	-77	Santo 2006
Tarius Aven (Funafus)	2 139	-90	Santo 2006
Fapon Cave (Butmas) (Figs 318E, 321E & 322C-E)	1 400	-44	Santo 2006
Sarakata resurgence (Fanafo)	1 200	-30	Australians (Harris 2006)
Loren Cave (Lotoror)	1 013	-40	Santo 2006
Mt Hope System: Fifty Four, Champagne Hole, Pump Sink, Three Way Sink, Three sisters, Tourist Blue Hole (Fanafo)	900	-54	Australians (Harris 2006)
Amarur Cave (Nambel) (Figs 321D & 322F)	874	-21	Santo 2006
Mt Hope System: Drinking Hole, Clam Shell (Fanafo)	800	-67	Australians (Harris 2006)
Patunar resurgence and Doline (Funafus) (Figs 318C-D)	791	-106	Santo 2006
Fioha Cave (Belmol) (Fig. 319)	550	-29	Santo 2006
Wanror (Butmas)	525	-40	Santo 2006
Mt. Hope System: Bush Rope Hole (Fanafo)	500	-36	Australians (Harris 2006)
Millenium Cave (Nambel)	432	-41	Santo 2006
Riorua Cave (Funafus)	400	-25	Santo 2006
Mba Aven (Butmas)	387	-59	Santo 2006
Vobananadi shaft (Malo: Avorani) (Fig. 318B)	209	-87	Santo 2006
Lavav Aven (Loran) (Fig. 318A)	60	-31	Santo 2006
Tchawak Cave (Funafus)	58	-31	Santo 2006

the Kafae and Tarius shafts, the longest caves of Santo recognized at this time, passages may reach up to 15 m wide by 15 m high with huge underground streams. In several leads, exploration stopped simply by lack of time, or on big sumps (there were about ten, all of them easily divable). The different parts of the system are likely interconnected by underwater passages, so future exploration is now in the hands of divers, and probably would encompass more than 10 km of passages given the number of galleries running in parallel.

••• Butmas: Fapon Cave and Mba Aven (Figs 323 & 324)

The village of Butmas is set in the middle of Santo in the jungle near the western border of the karst and at the foot of Mt Tanakar. This area is overgrown with dense secondary forest and tangled vegetation, and is the wettest of Santo. Several biological surveys were made in the largest cave of the area (Fapon Cave), which turned out to host a rich and original fauna in its dark passages as well as in its deep dolines which open on the course of the subterranean stream.

The upstream entrance of Fapon Cave is a narrow passage, where a small stream sinks among blocks. The subsequent galleries are moderate in size, some well decorated, and connect three successive deep dolines. Several passages are

aquatic: the last one, downstream of the third doline gives onto a shallow sump 42 m-long. It was dived during the last field trip of the diving team and gave access to a much larger and very different passage, with a big river running northward, perpendicular to the first part of the cave, i.e. towards the Jourdain valley. Interestingly there were bats flying in this post-sump gallery indicating that another direct access exists to this part of the cave. Alone for this diving, Nadir Lasson had to turn back by lack of time after 400 m of easy walk, after observing huge continuations upstream and downstream.

The same day, in the Mba Aven, another team of cavers got access to another large river where exploration stopped upstream on a waterfall, and downstream on a winding, low ceiling passage.

These last-day discoveries of major underground collectors highlighted the Butmas area as the most promising site for future explorations in the Santo karst. Indeed, finding such big underground streams suggests that a huge network of hydrological passages are developed in the three or four hundred meters thick limestone terrain above Butmas. The potential is even higher downstream, with suspected water resurgences at the coastal blue holes, several kilometers away and more than 300 m below).

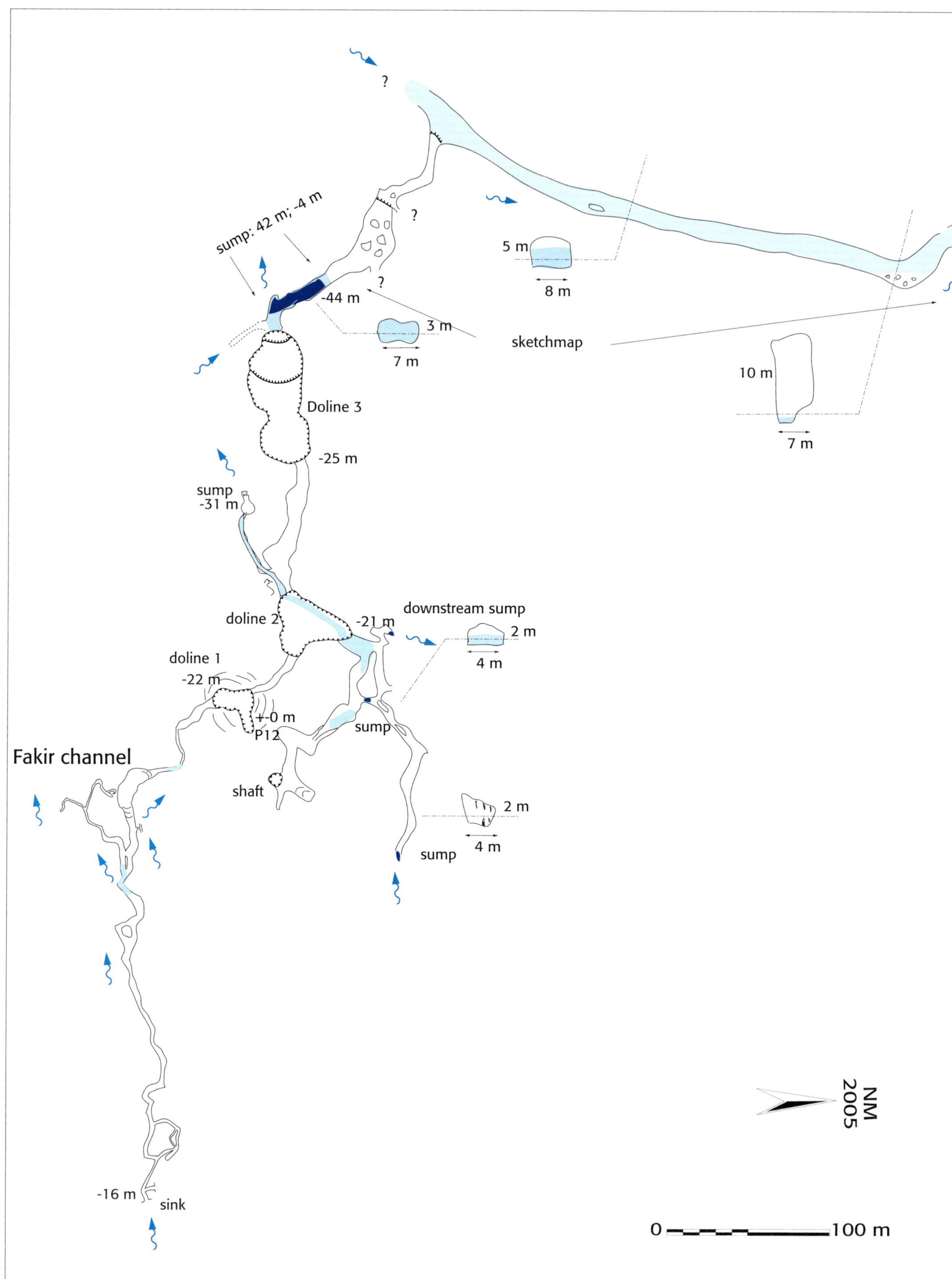

Figure 323: Fapon Cave (Butmas-Santo-Vanuatu, Santo 2006 expedition).

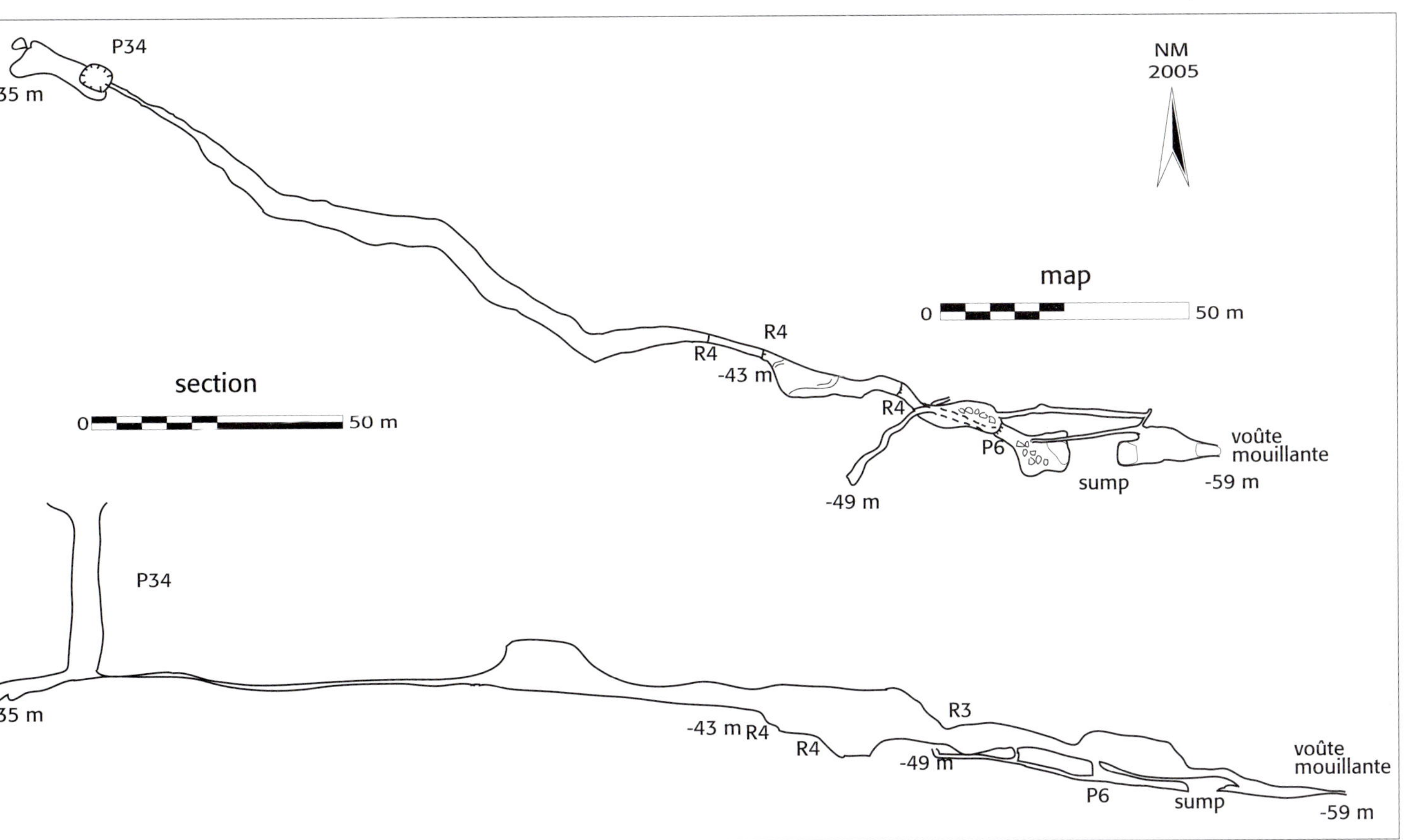

Figure 324: Mba Aven (Butmas-Santo-Vanuatu, Santo 2006 expedition).

THE FUTURE OF CAVE EXPLORATION IN SANTO

One of the important and unexpected finding of Santo 2006 has been the existence of a well organized underground drainage, with very large underground passages in spite of the relative recentness of the karstification (younger than the Upper Miocene). Kilometers of beautiful galleries have already been recognized, but they clearly represent only a small part of what exists in the heart of the limestone, both in terms of length and depth of the karstic features. Furthermore, we have no idea of the hydrological course of the largest subterranean rivers discovered in Butmas: these questions are the exciting challenge that Santo 2006 has raised for future expeditions.

as Archives

Caves

Denis Wirrmann, Jean-Christophe Galipaud, Anne-Marie Sémah & Tonyo Alcover

Hereafter we will present a few examples of the archives which we have retrieved in different sites (caves, rock shelters and dolines) from the limestones realm of Santo (Fig. 325).

The caves from a karst system play a very important role as sedimentary archives, because they are the place of very large sedimentation, and accumulate various kinds of deposits, like clastic fragments in association with calcite secondary precipitates

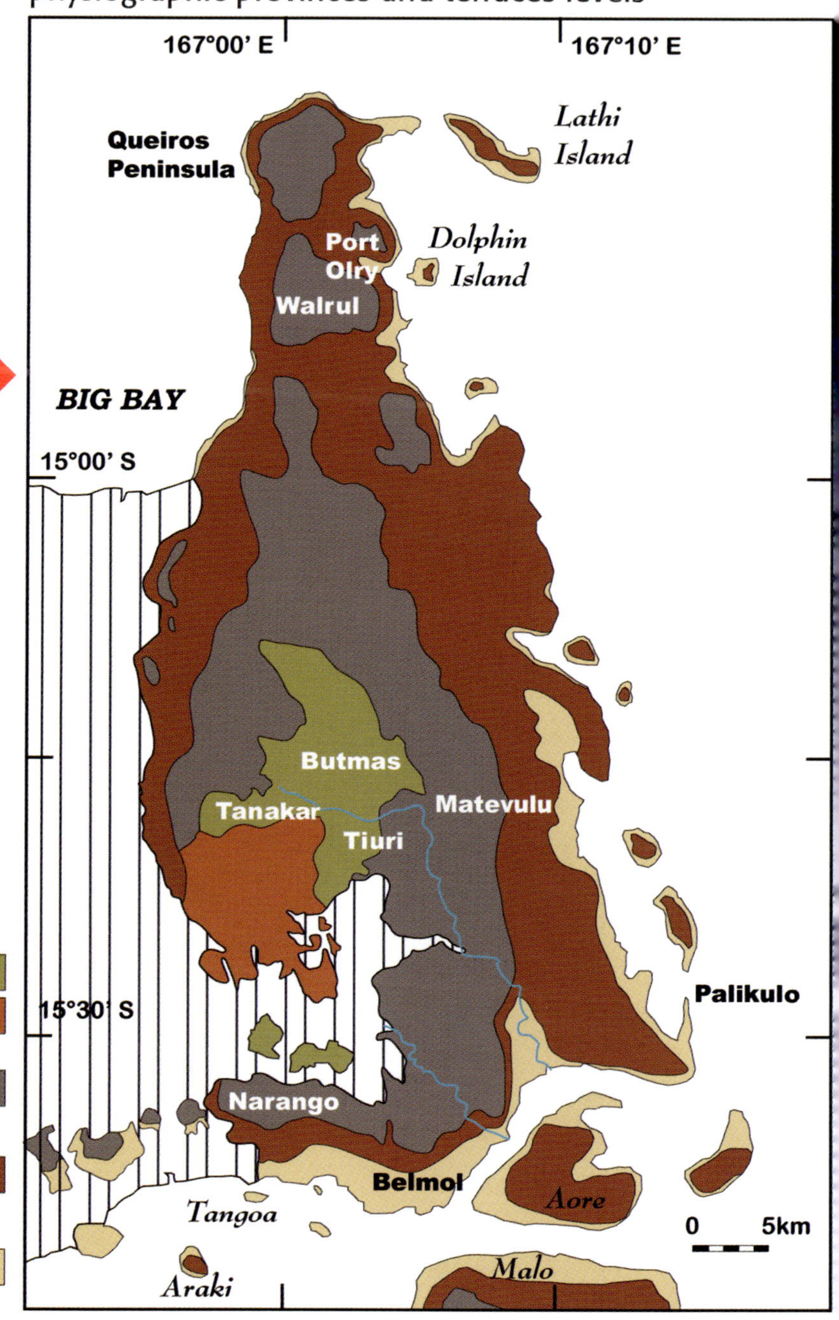

Figure 325: The main prospected sites (**A**) and (**B**) detailed geological pattern of the karst of Santo, redrawn by D. Wirmann after Strecker *et al.* (1987).

and other chemical and/or organic debris. Due to their natural environment (absence of light, difficult access) the caves tend to be preserved from degradation and aerial erosion. Morever, their deep interiors remain mostly stable in temperature and atmospheric composition, representing often particularly rich undisturbed sites. They represent protected repositories amongst the other continental environments and preserve a large spans of time medium for palaeoenvironmental studies. They are also very valuable for archaeological research, the caves or assimilated structures having frequently played and playing still a role

of shelters, accomodations or habitats. That is why, for example, human artefacts are often very common into cave entrances and rock shelter deposits (burials, rock drawings, cave paintings, potsherds, etc.) and in a lesser extent in cave interior deposits.

One can ascertain that all the clastic, organic or precipitated sediments formed within a cave or the deposits transported into it, will allow a multiproxy study of the karst environmental fluctuations, which will encompass hydrogeology, geology (e.g. palaeoseismology), climatology, biodiversity and human palaeontology.

STALAGMITES

A few stalagmites and stalactites from the Fapon Cave (Butmas Plateau) have been sampled during a three-month field trip (Sémah and Wirrmann

in 2005) before the Santo 2006 expedition. Their study is in course and their dating by Thermal Ionisation Mass Spectrometry (TIMS, U-series

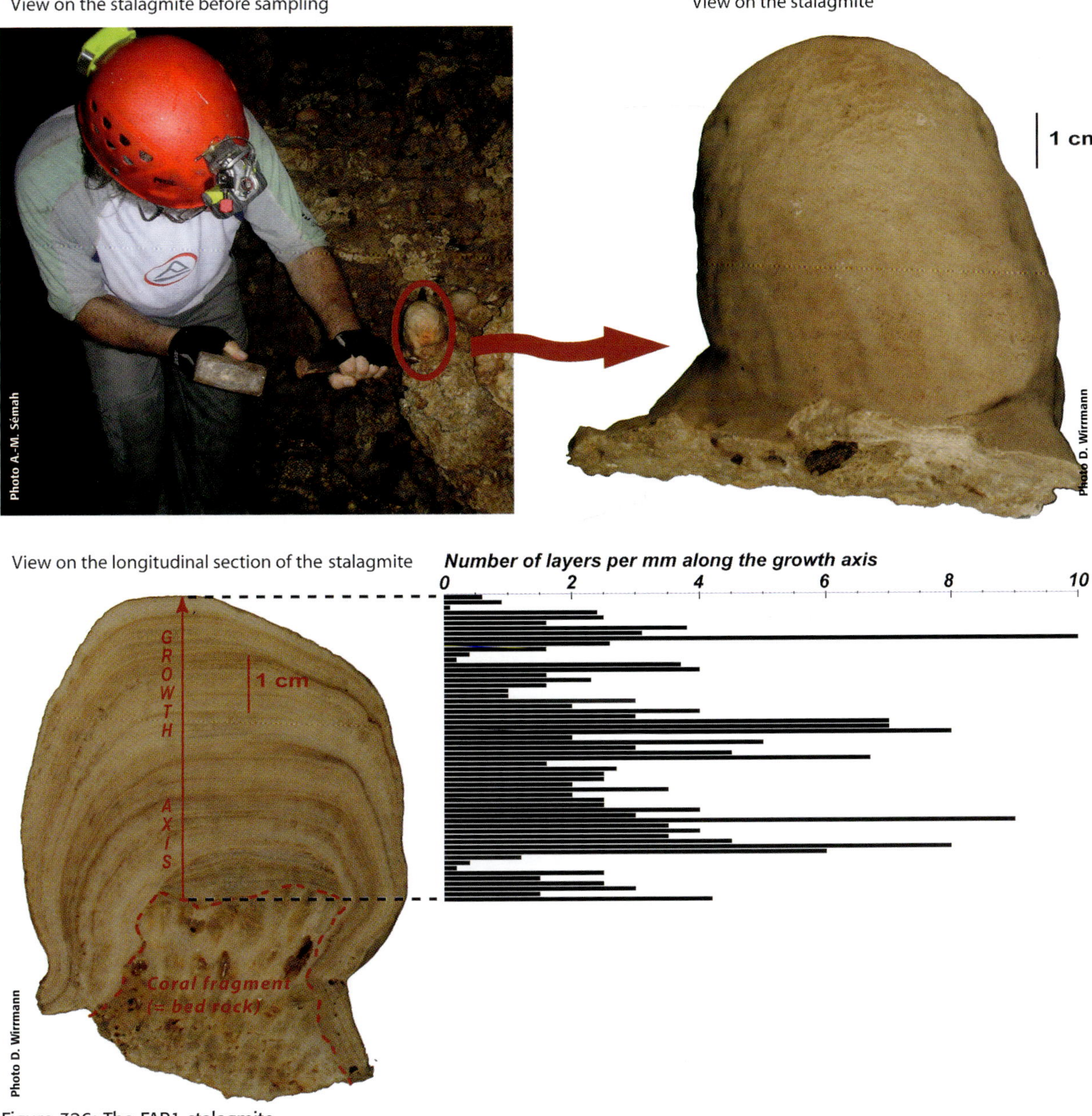

Figure 326: The FAP1 stalagmite.

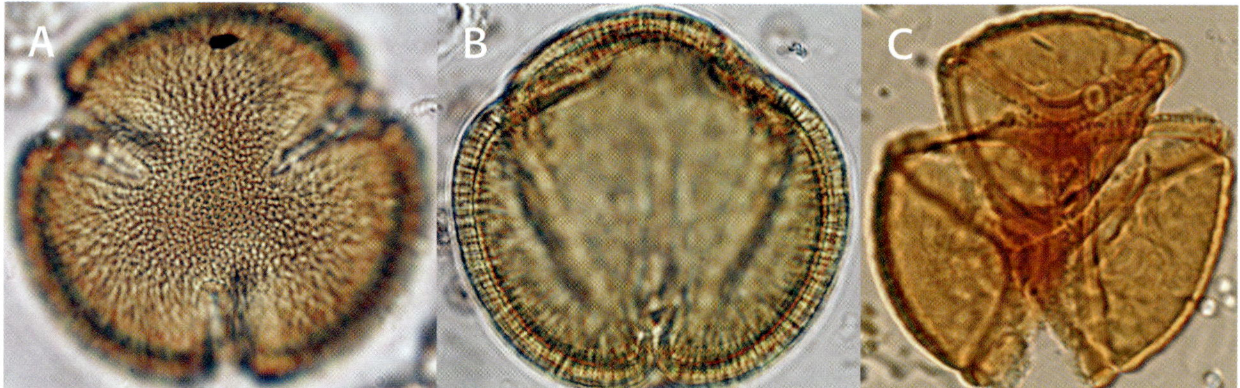

Figure 327: Example of grains of pollen from Santo sediments. **A-B**: *Merremia peltata* (Convolvulaceae), the invasive creeper (diameter size 60 µm). **C**: *Barringtonia novae hiberniae* (Lecythidaceae), endemic to the Vanuatu Archipelago (size 54 x 45 µm). (Photos A.-M. Sémah).

methods) will provide the basis for the respective relative age models.

Petrographic thin sections were prepared all along the stalagmites growth axis, in order to analyze the carbonate fabrics and the laminae under an optical polarizing microscope. Enlarged photographs of the thin sections were also used as analytical aid for mineralogical fabric observation and laminae counting. Along its growth axis the FAP1 sample (Fig. 326), a stalagmite of 9 cm length, shows many layers with distinctive widths and colors: the fluctuation of their number per millimeter illustrates the variation of the feeding conditions during the time of growth, in relation with the available dissolved calcite and water at the origin of the precipitation of secondary calcite into the cave.

The study of pollen, the reproductive male cell (mean size about 30 µm) from spermatophytes (flower plants), is a valuable bio-marker used in "classical" sedimentology since the 1950s for environmental reconstitution. As each plant species is characterized by a specific grain of pollen, the reconstitution of a palynological spectra and its variations during the time will in turn allow the reconstitution of the climate. Nevertheless, it is only from the 1970s that it has been used for such a purpose in stalagmites, flowstone and spring travertine. Usually, the deduced representation of the contemporary vegetation of the deposit does not correspond solely to the true live picture. It will privilege for example the local vegetation at the expense of the regional one, knowing that the sedimentation of the pollen may be more or less hampered by the wind circulation, or by change in aquifer plumbing, or by the percolating system over the cave and/or transformed by allogenic contributions (anthropic activities).

It will be particularly interesting to analyse the samples retrieved in Santo (speleothems, dolines and caves fillings) for precising the apparition of the alien *Merremia peltata* (Convolvulaceae), this invasive creeper which alters the structure and functioning of the insular ecosystem and about which we have no data for it first introduction in Vanuatu. It will be important too, to understand the distribution in stratigraphy of some endemic (*novae hiberniae*) vs pandemic (*asiatica*) species in the *Barringtonia* genus, for instance (Figs 327 & 328).

CAVE FILLINGS

In almost all the twenty caves and rock shelters whe have prospected (Lori, Luri, Lonepré and several sites from Cape Cumberland, Fig. 325A), the artefact rich levels ly often under a more or less thick superficial bats or swiftlets guano layer.

We have put at the day anthropic levels. These levels, still not dated but probably recent according to their stratigraphical position, contained airfall tephras, charcoals, ceramic potsherds and tiny animal bones remains. Beside these sites, a few other places corresponded to burial caves, rich in human remains and pig canines (the well known "tusks" of Vanuatu) as burial accompaniments. In all these places we have sampled superficial and deeper sediments in order to analyse their pollen content.

••• **The North Cumberland area**
The karst environment of the Cape Cumberland area has been for many centuries an important

Table 31: Datings of Hokua shelters (after Galipaud, 1997).

Sample n° (Lab. Miami, USA)	$\delta^{13}C$ ‰ vs standard Pee Dee Belemnite	Conventional Age BP
Beta 96570	-27.4	1 110 ± 80
Beta 97558	-25.3	340 ± 60

population center for the west coast controlling the exchanges with the Banks islands to the North-East. This is a very active environment with one of the bigger uplift rate in Santo, between 2.1 and 5.5 mm/year. Several caves with important anthropic levels have been previously surveyed near the actual village of Hokua. Datings of stone ovens (Table 31)

View from the rim of the doline.

View on the hillside.

Bottom of the doline.

in two shelters near the irrigated taro gardens of Hokua have allowed to attest an occupation of this area during the last 1000 years.

Several new shelters revealing further human activities in the area have been localized during the Santo 2006 expedition.

The most surprising and unexpected discovery was of a burial site well hidden in a very small cavity, partially blocked by accumulated rocks near the ancient village of Salea (14°40.421'S; 166°37.202'E). The site was arbitrary named "Abri aux dents de cochons".

In this cavity we discovered human bones from more than three individuals, in association with a child skull (younger than ten years), suidae's teeth (22 tusks) and maxilla, and pottery pieces (Figs 329 & 330). The site is probably the tomb of a man of high rank as attested by the number of well developped pig tusks. A decayed piece of woven cotton as well as the relatively well preserved pig tusk suggested that the burial was not of great antiquity. Half of an open bowl with red slip further confirmed a inhumation during the last millenium. We took a pig molar for datings and submitted the sample to the Waikato laboratory (New Zealand). The result, (Sample n° Wk 20296, C14 Age BP = 209 ± 10, modern activity = 97.6%) indicates that this inhumation dates to the last phase of the Santo prehistory, between the 17th and the 19th century.

In the same area, the "Abri Palatin" (14°40.178'S; 166°37.407'E) includes three adjoining chambers

A 65 cm deep borehole sampled in the doline bottom.

Figure 328: Example of a doline, the doline of Arifos.

Figure 329: View on human bones and tusks in "*Abri aux dents de cochons*".

Figure 330: Fragment of a red slipped bowl, "*Abri aux dents de cochons*".

containing human bones (Fig. 331). In the right chamber, more than five individuals were discovered, while on the left side, there was only one post cranial skeleton remains associated with two potsherds.

The third cavity contained another skeleton and faunal remains plus charcoals. In several other shelters from the same area, blocked chambers or scattered human bones suggest that such burial practices were frequent in the region during the second millenium. No such practices had been evidenced in Santo prior to this survey.

Further west along the same area, a dry cave named Malaostitir in the local language (14° 39.673' S; 166° 37.291' E; Figs 332 & 333), revealed traces of red and black drawings partly covered by calcite. Fragments of pottery with red slip on the surface suggest an occupation at the same period as the previously described burial sites. the drawings which are difficult to see correspond to geometrical forms of squares and crosses.

••• **The Cape Quiros area**

This region encompass a well developed underground karst system. We will focuse here on one of its largest cavity, the Lonepré Cave (14° 59.227' S; 166° 59.160' E). Fallen rocks block partly its entrance, wich is oriented north-west south-east and makes a porch which opens in the first coastal crest over the forested littoral belt. A survey in this shelter made by J.-C. Galipaud and R. Pineda in late 2005, following its discovery by F. Brehier on 4 November 2005 uncovered large pieces of red-slip pottery under a the large and small coral blocks scattered on the surface. A one square meter

pit near the back wall of the shelter, revealed a homogenous dark clayish sediment with scatters of charcoals.

In 2006, a one square metre pit, reaching 60 cm depth, has been done in the middle of the entry. No well marked stratigraphy is observed. Nevertheless, different remains have been collected according to the depth of sampling: between 20 and 40 cm many mollusks fragments are present, the following level 40-45 cm is very rich in microcharcoals and below the sediment becomes dustier, ash-like and sterile. After 50 cm there are again numerous microcharcoals

Figure 331: Human remains in "*Abri Palatin*".

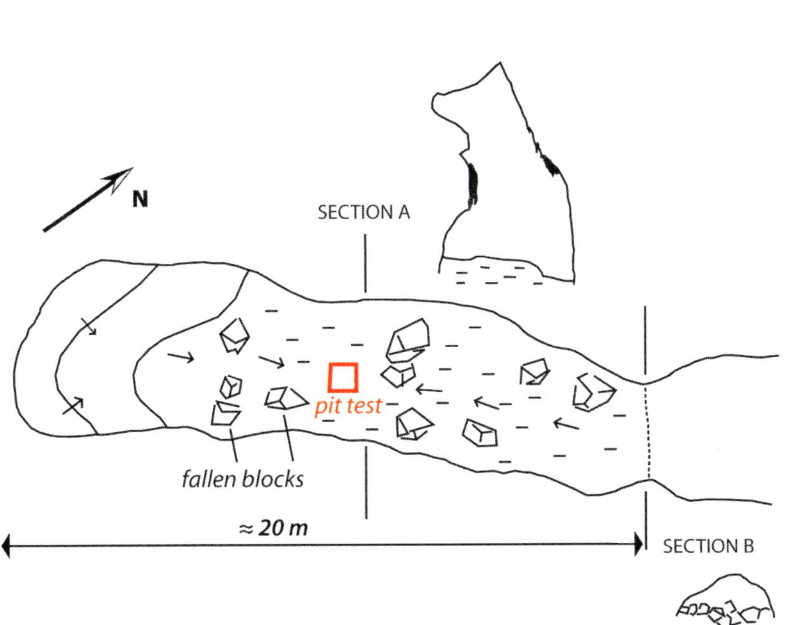

Figure 332: Sketch of The Malaostitir Cave. (15/09/2006, redrawn by D. Wirrmann after Lips & Lips, 2007)

Figure 333: The Malaostitir Cave. Cave entry.

and fallen calcareous blocks appeared from 60 cm depth. As the time allowed to this field trip was ended, we stopped the excavation and filled it with calcareous blocks and branches to protect it and to be able to excavate again easily in a next future. A total of 51 potsherds, 20 charcoals, twomicrofaunal bone fragments and seven mollusks has been collected in this excavation.

CONCLUSION

Many small caves were discovered during two preliminary reconnaissance trips undertaken in 2005. After further deep explorations and sump diving during the expedition itself, several of these caves subsequently turned out to be very large indeed. The longest of Santo, the Funafu system, develops several kilometers of underground passages, and exploration is far from being completed. Another surprise came at the end of our stay, when we discovered huge subterranean galleries with big underground river near Butmas. Today, Santo has become the most promising karst in the Pacific islands outside New Guinea for the exploration of large underground systems.

The twenty rock shelters or caves sampled in the karstic zone of Cape Cumberland, Cape Quiros, east coast and Aore Island provided plenty of archaeological and palaeontological rests. Megapod birds, thought to have possibly existed in Santo, as in New Caledonia, were searched for, but no remains were found. On the other hand, we found, at almost every site, layers with ceramics (more than 1 000 years old), fire marks, sometimes tools and fossil remains (more than two species of rattus, birds, bats and lizards). Human bones were also discovered, sometimes associated with jaws and canines (tusks) of Suidae (pig family), indicating the importance of burial in these shelters.

Information extracted from the Santo karst is exciting in several respects. After a month of intensive sampling, Santo has become the best known island of the Pacific for cave and soil fauna. By the size of its karst, the easy access to its subterranean habitats, and the relatively low disturbance and high richness of its soils, Santo is a model island for testing key questions about the origin and evolution of biodiversity, well beyond its regional interest.

The use of multi-proxy studies of speleothems, involving combinations of geochemistry (isotopes and traces elements), mineralogy and palynology, integrated with parallel proxy materials from the same region (lake or swamps sediments, dolines fillings, coral cores, archaeology) will provide very useful data for identifying process controls and for working out the mechanisms of climate change and the exact timing of connections between climate and environmental changes.

Perception and Attitudes Vis-à-Vis the Karstic Environment

Florence Brunois

One of the many objectives of the cross-module "Perception Plurielle of Biodiversity" was to study the perception that the Ni-Vanuatu population has of the Santo karstic landscape, especially dry or drowned caves, avens, dolines and resurgences. Our intention was to document the local representations and practices related to these typical karstic features, as well as local knowledge of their fauna and flora. Conducted in parallel with the "Karst module" of the expedition, our investigation lasted three weeks and ranged across the territories of six communities. Thanks to the precious contribution of "field workers" —designated by the communities or their leaders themselves— the results highlight a kind of cultural constancy. Uninhabited by humans or inhospitable for man, almost all caves and cavities, are instead inhabited by spirits, guardians of places and holders of its inhabitants (swifts, bats, eels, etc.). This spiritual occupation confers on the cavities a sacred character that naturally has some impact on the way human populations apprehend them and act inside these particular environments.

● ● ● The perception of the karstic environment

To clarify the logic governing the local perception of karstic environments and the impact it would exert on practices and knowledge associated to them, we first had to study the local conception of the world in order to appreciate the exact place that karstic environment occupies within it.

●●● The Ni-Vanuatu conception of the World

As described in detail in the chapter "Ni-Vanuatu perceptions and attitudes vis à vis biodiversity", our study revealed the tremendous complexity of the daily world of Ni-Vanuatu people. Although converted to Christianity, all communities encountered maintain their traditional cosmologies. Beside the Christian God, the Ni-Vanuatu allow the co-presence of several categories of spirits whose identities may vary from one community to another: spirits of ancestors, spirits of the clan and spirits investing certain areas or species (animal and floral). However, as diverse as they are, all these entities have in common the power to materialize in living beings and express themselves through natural phenomena. Finally, all hold the power to harm humans (by illness or death) if they fail to fulfil their obligations to them. The Ni-Vanuatu are therefore forced to maintain a certain relational harmony by practising rituals, sacrifices and offerings or by observing many taboos toward the spirit, living beings or spaces in which they are custodians.

The permeability of the borders between the human and spiritual worlds, the danger it represents for humanity and the role allotted to the spirit beings explain the extraordinary overlap between the phenomenological world and the spiritual one.

This perception does not mean, of course, that the thoughts of the Ni-Vanuatu are unclear. As a demonstration, if necessary, their taxonomy of the fauna and flora that we studied corresponds, with few exceptions, to Linnean species (see "Ni-Vanuatu perceptions and attitudes vis à vis biodiversity"). It simply reflects the fact that local cosmologies do not erect a border between human and non-human beings.

Understanding this unique conception of the world was crucial to our study because, as we shall see, it governs and gives meaning to the karstic landscape representations and the activities of the local populations.

●●● The place of the karstic environment in the Ni-Vanuatu world: myths and realities

With knowledge of the local cosmologies and their application, we analyse how the karstic landscape and its inhabitants form part of this very complex world. This research has led to an examination of the mythological corpus of local populations in order to detect whether the karstic environment enjoyed a special treatment in the Ni-Vanuatu imagination and, if so, the reasons for such a privilege.

Our study has helped to show how karstic features like mountains and marine landscapes do not leave local people indifferent. Quite the contrary is the case: whether they are dried or drowned, followed or not by a resurgence, located on the coast or in the green mountains, all karstic caves visited or identified are the subject of a myth recounting their origin or their properties and their remarkable owners.

● From their human origin…

The mechanisms underlying the formation of karstic landscapes are similar in all myths listed. It is people

Figure 334: An eel captured close to the cave. A fabulous animal, half snake, half fish returning groundwater, feeding in the creeks only to disappear again in the deep caves. A fabulous representative of spirits Masters of the caves.

who are at the root of their formation or, more precisely, it is their subsistence activities —hunting, gardening and fishing— that cause a sudden collapse of walls, the creation of a shaft or a deadly rush of water. The landscape is abruptly and fundamentally altered —the birth of a cave, an abyss, or a bay (that of Matantas)— by the actions of humans, who metamorphose themselves into an eel, snake or rock and are invested these new places (Fig. 334).

The Santo mythology thus focuses on the formidable geological instability governing the karstic landscape and wards off mortal dangers that humans face when carrying out even the most ordinary daily activities.

Subject to unexpected metamorphosis, these places can only be inhabited by exceptional beings —the spirits— that are able to metamorphose themselves to survive erosion.

- …To their spiritual occupation

This close association between the karstic and spiritual world is found in the whole mythological corpus studied. But, as mythical as it is, this association is still seen as real by the Ni-Vanuatu interviewed. All cavities, without exception, are known to be occupied by an exceptional being, often irascible and in all cases with the power to kill men, even all humanity. On the island of Aore, it is the spirit of an ancestor named "toy" that reigns in the cave called "toy house" in the form of a snake that is sometimes red, sometimes green. At Okwa, the cave paplow ship is inhabited by the spirit "mata", mother of snakes, while the cave tasipo is guarded by the spirit "to usus". Taking the form of a stone, "to usus" is responsible for guiding spirits of the recent dead along the path they must take to enter the antrum land. At Butmas, it is the spirit "sosol", capable of turning into

birds or eel, that occupies the drowned cave totororo, while in Fioanumu and Batunarr cavities, it is the spirits of humans (who led to their own creation), which occupy these places in the form of a snake and eel. Masters of karstic places, these spirits are also commonly regarded as the guardians of white-bellied swifts and bats (*Miniopterus* sp.), who live "in good intelligence" in their home. Indeed, these two species not only share the cave depending on its depth and darkness, but also they adjust the timing of their activities to co-protect their respective offspring. This "over-protected" karstic wildlife is not therefore characterized by its formidable species diversity. However, it is distinctive for its high population densities which is a sign, for the local people, of the formidable fertility associated with these places. Water dripping stalactites betrays this: assimilated with breast-milk, the bats from Okwa come in the cave just to "suck" the concretions, while at Aore, where the water is regarded as semen, it fattens pigs.

Apprehension, practices and knowledge related to the karstic features

The study of the Ni-Vanuatu's modes of comprehension, activities and knowledge related to the karstic environment formed the second part of our research. Our aim was not to raise limited uses that draw local populations to this environment, but rather to clarify their perception of the karstic landscape and understand how this original perception might have an impact on the activities of local people and on their knowledge.

Attitudes observed and to be observed on entering a cave

Observed or orally described, attitudes underlying the entering of a cave are under a code of conduct that is very rigorous and often similar in all communities. Broadly speaking, entering a cave arouses strong apprehension, with some individuals even refusing to enter. This emotion makes sense in light of their representations, but also in terms of the

Figure 335: Scientists in discussion with village leaders to obtain permission to descend into the caves without creating damage or disrupting the masters of the place.

traditional code governing access to caves. First, if the entrance to a cave is not simply forbidden, not all humans necessarily have the right to access it (Fig. 335). This case concerns all menstruating women from Butmas, Aore, Patuna and Fioanumu, but also all foreigners[8] at all sites. Violation of that rule would lead to the collapse of the cave in the first case and a deadly flood in the second. In fact, only men who are indigenous to the territory in which the caves are situated have the right to access them, but again only under certain conditions. For all the communities, it is thus necessary to call up the spirits present in order to inform of them of one's arrival. At Butmas, men must remain silent or use another language in its approach and present offerings of meat at the entrance of the cave. If they do not observed this rules they face the risk of provoking the anger of the spirits, which could have fatal consequences.

8 - This law explains why, at Butmas, leaders bring chicken and meat as offerings to the spirit of the caves, with the aim to protecting the life of the "Kars team".

••• **Activities within the caves**
Unlike the rock cavities in which populations of Okwa and Vate freely shelter or stay with their family during hunting parties, fishing or gardening, activities within the caves are more codified, sometimes even ritualized. During our investigation, we found four activities traditionally performed within these karstic environments:

• The Hunt
All communities hunt swifts and bats in caves once or twice a year. At Aore and Fioanumu, hunting swifts is performed exclusively before the celebration of the closing mourning ceremony: the meat of these birds[9] is the main dish, offered to guests in a basket manufactured solely for this purpose. However, whatever their motivations are, these predatory activities are subject to certain conditions. For example, at Fioanumu the Chief must precede the other hunters or else risk the death of his companions. At Butmas, women are prohibited from visiting gardens during the hunting party: otherwise plants would rot immediately. The hunting techniques are also varied. Illuminated by a torch of bamboo, men can use a spear, a stone or just their hands. In Okwa, hunters take care to cover their arms and their hands with an anointment of a magical plant to ensure the capture of prey; some sing ballads to attract birds. But all these hunters respect the same mandatory law: never overkill the population of animals, else the whole community will die.

9 - This ceremony, which serves to definitively separate the dead spirit from the living human community, use this bird because of its black and white feathers, colours that symbolize death and life.

• The oracles
A world between two worlds, caves are the preferred place to practice oracles. This is the case at Okwa and Butmas, where, following a dream, a man will go in the cave with wood and the heart of a pig to identify the sorcerer trying to cannibalise his soul or that of his loved one. On the island of Aore,

before going to war, no men would ever fail to seek advice from the Toy spirit. If the snake interviewed buries his head in the ground or under a stone, the men immediately renounced their attack.

• Witchcraft
Only at Okwa, caves can be bewitched by human beings for criminal purposes: it suffices to smell the odour of the hidden poison (unidentified) to die[10].

10 - It is for this reason that Okwa people did not inform J.-C. Gallipaux of the existence of certain caves: they were afraid for his life.

• A refuge
All dry caves can be used as refuge in extreme situations like war, tidal waves or cyclones. Today, it seems that young unmarried lovers also shelter in caves away from prying eyes! (Fig. 336).

Figure 336: Some children preferred to stay outside the cave rather than accompany us, especially when it was narrow and deep, while others were not impressed.

••• **Knowledge of the karstic environments**
The study of the Ni-Vanuatu's knowledge of the karstic environment revealed how their perception of that environment, and the practices resulting from it, had exercised a strong influence on the formation of their knowledge. Indeed, as in many non-scientific societies, their understanding presents itself as an integrated knowledge in the sense that it does not dissociate the object from the socio-cosmic context, in which man participates fully. Yet this context, as we have seen, is very complex because it adds a spiritual dimension to the phenomenological one, entailing a certain ambivalence and thus posing a constant threat to humans. For caves, it seems that this cosmological principle has exercised a certain constraint, not only on the acquisition of knowledge but also on their development. Firstly, the opportunities to get there are, as we have seen, limited: the duration of visits is curtailed, and in general, the fear of an encounter with karstic beings and its consequences discourage Santo people from satisfy any curiosity they might have. To this epistemological explanation can be added that of the erosion of knowledge resulting from

Figure 337: I was very concerned to understand why the local knowledge on karst environment was no longer transmitted to children. Christianization, education, modernization were, of course, the most obvious reasons. The caves belonged to another time, another space inhabited by the traditions and spirits...

Christianization and schooling, processes that induce local populations to no longer transmit traditional knowledge, precisely because their knowledge is not decontextualized, i.e. dissociable from the traditional spirit world! (Fig. 337).

This study has also highlighted another, quite exciting phenomenon concerning the denominations of bats and swifts and the confusion that this is likely to produce in the minds of both the Ni-Vanuatu and the ethnologist. Although in all communities

Figure 338: Karstic environment, a world between two worlds, like these men who seem able to oscillate between tradition and modernity.

the two animals are distinguished and named differently, we have found that the term "*kalakala*" is used to designate the swift in some languages, but refers to *Miniopterus* sp. in others. This generalization of the term "*kalakala*" poses a question: is it due to migration of human populations within the island or migration of the animals themselves?

••• The karstic environment: a world between and beyond two worlds

In the light of cosmology and mythology, practices and knowledge, the karstic environment seems to represent, for the people of Santo, a world between two worlds —the phenomenological world of humans and the invisible world of spirits— but also —and there lies no doubt his remarkable singularity— a world able to accommodate this ontological duality or even to transcend it (Fig. 338). Unstable by essence, the karstic environment is the domain of metamorphosis, associated with the passage from life to death, from the land to the underground, from day to night, from human to spirit beings. This exceptional property, which can link these dualities and sometimes even combine them, explains the fabulous fertility that the Ni-Vanuatu ascribe to the karstic world. A fabulous fertility that echoes that which the researchers of "module Karst" actually discovered.

Habitats of Santo

FOCUS ON SOILS

Anne Bedos, Vincent Prié & Louis Deharveng

As the place where organic matter breaks down and is recycled, soil is an essential component of terrestrial ecosystems. Earthworms and nematodes are all functionally important in this habitat, but in terms of diversity, arthropods are far ahead. During the Santo expedition, we focused on this group and on snails, which are also an important component of biodiversity in limestone terrains.

Figure 339: Map of Santo. Limestone terrain in yellow. Sample sites for soil and litter shown by red circles.

••• Collecting soil fauna in the Santo karst

The sampling effort we expended on the Santo karst was significant, with 287 samples of litter and soil processed in one month. We followed a simple sampling design, combining standardized and *ad hoc* sampling. Standardized sampling consisted of five sample points in each of 16 forested areas scattered across the large karst block of Eastern Santo

(Fig. 339). At each point, we took a sample of 500 cm³ of litter and 500 cm³ of deeper soil layer down to 10 cm. To capture the largest possible diversity of fauna, we used the following five efficient and well-tried collecting methods.

As a good way to soak up the local habitat characteristics, we performed whenever possible hand collecting on arrival at a new site, catching medium-size arthropods (above 2 mm) and snails in the litter with a brush or a pooter.

Non-baited pitfall traps were operated in order to catch the fauna foraging at the soil surface, especially during night, mainly springtails, ants, spiders and beetles (Fig. 340). This fauna is often poorly extracted from the litter itself. Small vials with a 2 cm diameter aperture were buried with the opening at the soil level, with 95 % ethanol inside, and left for one or two nights. Ethanol has a negligible repulsive effect, and may even attract large number of some species like bark beetles (Scolytidae); in Santo, two springtail species of the genera *Microgastrura* and *Brachystomella* were also obviously attracted.

When the content in organic matter was low, the granulometry of the substrate very fine and the fauna sparsely distributed, i.e. in crude sand and deep layers of the soil, washing was the only simple way to catch the arthropod fauna (Fig. 341). This technique (immersing a soil core in a basin of water) allows the collection of most arthropods that float

Figure 340: A pitfall trap. Arthropods foraging at the surface of the litter fall into the bottle half-filled with ethanol. At Santo, in one night, several tens, sometimes hundreds, of specimens, mainly ants and Collembola, may be collected.

Figure 341: Washing soil in a plastic basin is an efficient and inexpensive way to concentrate and catch the minute and rare arthropods of deep soil layers. Washing and collection of floating arthropods with a brush and a lens may be done in the field or in the lab.

and concentrate at the surface of the water, except a few groups like woodlice.

The most productive method to extract arthropods from soil and litter is the Berlese-Tullgren device (Fig. 342). Fresh soil cores were placed, directly or after sieving, in a thin layer on the surface of a grid, itself placed on a funnel of 30 cm diameter. A vial filled with 95 % ethanol is placed at the base of the funnel. Soil core drying causes the animals to migrate downwards through the grid mesh and to fall into the vial. The Santo Berlese model used a 2 mm grid mesh, with a few bigger holes laterally to allow largest arthropods to fall through. Fifty such funnels were operating simultaneously in a small abandoned greenhouse, well ventilated and warm, at our base camp near Luganville. Drying lasted four to six days, without using a heating bulb above the device, which would have been superfluous in these favourable extraction conditions.

Catching snails requires special techniques, which combine sieving, washing and hand collecting.

••• What did we know?
Information about the diversity of soil invertebrates on Santo was extremely scarce before the Santo 2006 expedition. Basic taxonomic descriptions of a few arthropods formed the bulk of our knowledge of the Santo soil fauna, and nothing was known about the composition of the communities or on the spatial patterns of biodiversity. As an example, only two species of terrestrial isopods were recorded from Vanuatu Islands before the Santo expedition: *Lobodillo hebridarum* Verhoeff, 1926 from the Western Pacific, and *Xestodillo lifouensis* Verhoeff, 1926, originally described from the Loyalty Islands in New Caledonia. In comparison, 57 species of woodlice are reported from New Caledonia, and 20 species from the Solomon Archipelago, the largest islands near Vanuatu. Our collections, which are currently under study, have produced at least 13 additional soil species from the Santo karst alone.

Figure 342: The Berlese-Tullgren funnels in the extraction house at the CETRAV field station. (Photos L. Deharveng).

••• Composition
A wide range of classical soil invertebrate groups were collected in Santo: snails, spiders, false-scorpions, scorpions, whip-spiders, whip-scorpions, micro-whip-scorpions (Palpigradi), short-tailed whip-scorpions (Schizomida), various mites, centipedes, ants and several beetle families (with Staphylinidae dominant) among predators; various mites, millipedes, woodlice, springtails, Diplura, Protura, termites and many insect larvae among the decomposers. As in most regions of the world, mites (Acari) and springtails (Collembola) were numerically dominant.

For instance, Collembolan densities estimated from standardized litter samples reached 300 to 1000 specimens per litre of substrate in litter, and about 10 times less in soil. These values are similar to those recorded in temperate regions, and depart significantly from those often encountered in the tropics, where mite and springtail relative abundances are usually lower, while the reverse is true for mesoarthropods (termites, ants, beetles). Unexpectedly, aquatic Crustacea, which often colonize soils in hyperhumid tropical islands, were either absent (Copepoda) or poorly represented (Talitridae shrimps, Ostracoda) in Santo, in spite of favourable climatic conditions.

••• Diversity

The vegetation is rather homogeneous on the karst, because the altitudinal range is limited (0-784 m), bioclimate is hyperhumid all along the year, and the relief is mostly gentle — nothing like the rough scenery of many tropical karsts. The mangrove and littoral vegetation, present as minute patches along the coastline, host a very peculiar fauna linked to seashore rather than to limestone, and was only marginally sampled. Outside this narrow fringe, diversity in habitats is mostly linked to freshwater occurrence, and to the disturbance gradient from secondary forest to gardens and coconut tree plantations. Though all these habitats were sampled, we mainly focussed on less disturbed forest and bush, which cover most of the karst. Soils differed from place to place by their depth and humidity, determined by bedrock and drainage. Drier and shallower soils were found on limestone terraces along the coast, wetter and deeper soils occurred in the center of the karst near Butmas. The composition of the soil communities shows differences, even if moderate, apparently related to this gradient. Drier soils host, for instance, the large springtail *Dicranocentrus* sp. as a characteristic species, while Talitridae shrimps, Polychaeta and *Orthonychiurus* sp. (Collembola) are only encountered in the wettest soils. There is also faunal heterogeneity in a given type of forest soil that cannot be explained by the depth/humidity gradient or by disturbance levels, but — not unexpectedly for an habitat as complex as soil— probably relies on undetected differences in food resources.

At the present stage, taxonomists are working hard at describing the species collected on Santo. Names for the majority of them are not yet published in the literature and we therefore have to talk in terms of morphospecies, i.e. forms that have been recognized by the specialists, but are not yet described.

••• Tramps and widespread species

As is typical in the tropics, the fauna encountered in Santo includes a group of pantropical species and a larger group of more narrowly distributed species. Most pantropical species (introduced, invasive, tramps, or simply widespread) have long been known to the taxonomists, often being the first species recorded historically from islands in the Pacific. They represent a significant part of the fauna, but seem to be less important in Santo than in more remote Pacific Islands — at least in secondary forest. All large invertebrate groups have such representatives: e.g. snails, millipedes, woodlice, springtails and ants. In many cases, and especially in human-disturbed habitats, these pantropical species dominate the communities. Thus, introduced Subulinidae form the bulk of the soil malacofauna in disturbed areas, comprising up to 90% of the collected shells (coastal forests near Luganville or grazed forest in Aore), as compared to 10% in better-preserved inland landscapes (Fapon, Amarur, Malo...). Our knowledge of the Arthropod fauna is more limited because heavily disturbed habitats were hardly sampled, but available evidence reveals patterns strongly influenced by widespread or tramp species. For instance, the communities of soil Collembola, even in well-preserved secondary forests, were frequently dominated by two subcosmopolitan species, *Folsomides parvulus* and *Hemisotoma thermophila*, which may, however, represent clusters of cryptic species. The two species of millipedes (*Orthomorpha coarctata* and *Leptogoniulus sorornus*) recorded from Santo in the literature were both pantropical, while endemics (probably five or six species) were undetected, even though they largely dominate the forest soil community. Broad distribution ranges of species with very poor dispersal abilities raise questions about their presence across the whole Pacific. It is generally assumed that many were introduced unintentionally by man.

••• Deep soil fauna (Fig. 343)

We discovered a relatively well-diversified arthropod fauna in the deep soil layers in Santo. Its representatives exhibited the typical morphological traits of endogeous species compared to their epigeous relatives: smaller size, shorter appendages, lack of pigment, eyes reduced or absent. Characteristic euedaphic species, i.e. those living almost exclusively in deep soil layers, were detected among several groups: mites (unstudied); minute Fuhrmannodesmidae-like millipedes; Palpigradi (*Koeneniodes* sp.); woodlice (Eubelidae: *Elumoides monocellatus*); springtails (Odontellidae: *Axenyllodes* sp.; Neanuridae: blind *Cephalachorutes* sp., *Friesea santo* and a probably new genus of Pseudachorutinae; Tullbergiidae: *Mesaphorura* sp. and a probably new genus of Tullbergiidae; Onychiuridae; Isotomidae: *Isotomodes* sp.; Entomobryidae: *Pseudosinella* sp.); Diplura (Parajapygidae); and beetles (Anillini Carabidae, a blind Ptiliidae).

Noteworthy is the fact that several species present in both soil and litter, i.e. several minute and blind Armadillidae woodlice, have the morphology of euedaphic species.

Figure 343: Deep soil arthropods. **A**: *Koeneniodes* sp. (Palpigradi). **B**: Siphonophoridae with sucking mouthparts (Diplopoda). **C**: cf. *Myrmecodillo* sp. (Armadillidae, Isopoda). **D**: Armadillidae (Isopoda). **E**: Parajapygidae sp. (Diplura). **F**: Anillini (Carabidae, Coleoptera) on the left and Oribatida on the right. (Photos L. Deharveng).

Despite being significant, deep soils of the Santo karst nevertheless have a slightly impoverished fauna compared to those of the few regions of Southeast Asia studied in this respect, with fewer blind micro-beetles and the absence of Projapygidae (*Symphylurinus* spp.), which are characteristic of this habitat in the surrounding tropical regions.

••• Litter fauna (Figs 344-346)

Dramatic changes in abundance, species richness and taxonomic composition occur across the few centimeters of the interface between soil and litter, which acts as a very efficient barrier. On Santo, faunal differences between a soil sample and a litter sample taken just above it were much greater than the differences between litter samples collected several kilometers apart.

As expected, litter hosted a larger range of arthropods than soil, most of them more colourful and of larger size. Aside from Collembola and Acari, the numerically dominant groups among organic matter decomposers were millipedes of the families Haplodesmidae, Pyrgodesmidae, Glomeridesmidae, slender Opisotretidae and Fuhrmannodesmidae, woodlice (at least three species of Armadillidae, a Philosciidae and a Styloniscidae), beetles (Ptiliidae), Diptera larvae, and even very locally in a Fapon doline, Talitridae shrimps.

Figure 344: Diversity of litter invertebrates: a partial view of the arthropods in a single sample from the Santo karst collected using Berlese-Tullgren extractors.

Many predators exploit these dense populations of decomposers on Santo. Geophilomorpha are remarkably diversified; predaceous mites, false-scorpions and spiders are present in every sample; but beetles are poorly represented, mostly by a few species of Staphylinidae and Scydmaenidae.

The overall abundance and diversity of the litter communities of Santo —decomposers and predators combined— was surprisingly high for the tropics, recalling the temperate fauna. For instance, the number of Collembolan species contained in a standard 500 cm³ litter sample was frequently 40% higher than for similar habitats in Southeast Asia. Not all groups, however, followed this pattern. In particular, beetles were obviously less diversified than in other tropical litters, where huge number of Pselaphidae and Staphylinidae often coexist.

Few termites - many ants (Fig. 347)

Two groups of arthropods have a special importance in the tropics: termites and ants. For unknown reasons, termites were rare in the karst of Santo, except in the most disturbed coastal areas, mainly planted with coconut trees. Fewer termites means more time for wood to decay, and hence a larger array of microhabitats available for other arthropods. Indeed, decaying wood represents a considerable part of the forest litter (probably as much as dead leaves) on Santo. The high abundance and diversity of decomposers, especially springtails and mites, in the soils of Santo might be due, at least in part, to this rarity of termites.

Ant abundance is the second characteristic of lowland tropical soils. Ants were not found in very large number in the forests of the Santo karst, but they were present everywhere, with many species contained in a single sample (usually from five to six for a half-litre core). This ant diversity is probably related to the diversity of their potential prey, and, again, to the absence of termites. Virtually nothing is known about the biology of the indigenous ants of Santo, except that almost all are predators. Species of the tribe Dacetonini, many of which are known to prey on Collembola, were unusually frequent, with often two or three species present in a half-litre sample: this is consistent with the high abundance and diversity of their potential prey.

Snails

Snails were sorted by sight from a part of the dry soil sample after arthropod extraction by Berlese funnel. Most were minute species, only a few millimeters long. Actually, the diversity of the soil malacofauna is measured almost exclusively from shells, although live individuals can sometimes be trapped as well. As a consequence, the species found are not necessarily living in the litter itself, but might be arboreal or cliff dwellers. Furthermore, because shells last longer that invertebrate corpses in the soil, they constitute an overestimated part of a soil sample compared to that of live invertebrates. This introduces biases when comparing snail to arthropod diversity, but may also provide complementary ecological data such as information about the dynamics of local environment. For instance, in the Fapon dolines, aquatic snails form part of litter samples, suggesting occasional flooding of the sampled soil.

Over 40 species belonging to 20 families were identified from eight sites sampled in a standardized way. They represent almost 70% of the Santo land snail fauna (see "Indigenous land snails" in section "Terrestrial fauna" and "Focus on snails" in section "Rivers and other freshwater habitats"). On this basis, using classical statistics, we estimate the total biodiversity to be between 50 and 80 species (Fig. 348).

As one would expect, big species appear to be less common in standardized samples than the smaller ones. Draparnaudidae, Placostylidae and Succineidae for example, believed to be common species as they are easily found in the field, are each represented by a single shell out of about 3000 shells identified.

Amongst the native species, the families Assimineidae, Euconulidae, Diplomatinidae and Hydrocenidae represent altogether more than 70% of the collected shells. These families, best represented in preserved forests like those of Fapon and Malo Island, were not recovered in disturbed ecosystems such as the surroundings of Luganville. Malo Island hosted the richest and most diverse malacofauna: 21 species, with only two of them introduced. The family Endodontidae was mostly found at this site. In disturbed areas native snails can drop down to about 10%.

The future

Thanks mostly to the productive Berlese-Tullgren funnel method and exceptionally favourable sampling conditions —a one month stay between two typhoons— we were able to gather what is probably

Figure 345: Detritivorous invertebrates of the litter in the Santo karst. **A**: *Namanereis* cf *malaitae* (Polychaeta). **B**: Phthiracaroidea (Acari Oribatida) and (upper right) Ostracoda. **C**: Lohmanniidae (Acari Oribatida). **D**: Glomeridesmidae (Diplopoda). **E**: *Eutrichodesmus communicans* (Haplodesmidae Diplopoda). **F**: *Lobiferodesmus vanuatu* (Pyrgodesmidae Diplopoda). **G**: Talitridae (Amphipoda). **H**: Armadillidae (Isopoda). **I**: Styloniscidae (Isopoda). **J**: Protura. **K**: Diptera larva. **L**: Ptiliidae (Coleoptera). (Photos L. Deharveng).

Figure 346: Predaceous arthropods of the litter in the Santo karst. **A**: *Ideobisium* (Pseudoscorpionida). **B**: Chthoniidae (Pseudoscorpionida). **C**: Lycosidae (Araneae). **D**: *Cryptothele* sp. (Araneae). **E**: Schizomida. **F**: Geophilomorpha (Chilopoda). **G**: Staphylinidae (Coleoptera). **H**: Lampyridae (Coleoptera). (Photos L. Deharveng).

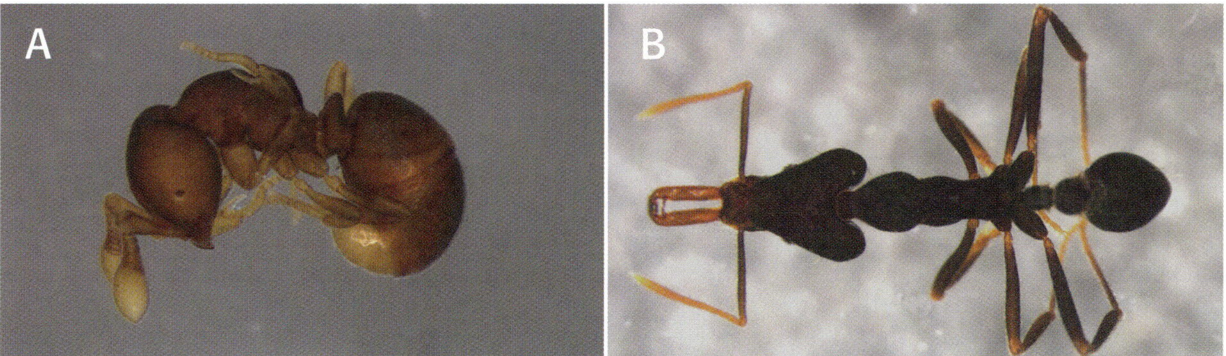

Figure 347: **A**: *Discothyrea* sp., a rare ant in the forest litter of Santo. **B**: A springtail-eating ant (cf. *Strumigenys* sp.), frequent in Santo. (Photos L. Deharveng).

today the most important soil invertebrate collection made in the western Pacific islands outside of New Caledonia, with about 100 000 specimens. The real number of species present in the soils of the Santo karst cannot be ascertain with precision as we have no way to take into full account the mites, which are often dominant but desperately in need of specialists for identification. Under the conservative assumption that mites and insects (as adults or larvae) each have two times as many species as springtails in the limestone soils of Santo, 500 species would be a conservative estimate. A huge work of identification and taxonomic descriptions has begun and will obviously last for many years.

We expected to find the characteristic features of tropical soil biodiversity in Santo, i.e. a lower diversity and abundance of microarthropods versus a higher diversity and abundance of mesoarthropods, compared to the temperate soil fauna. The oceanic nature, relative isolation and youth of the island should also have resulted in a depleted biodiversity in all groups, with some disharmony in faunal composition. In short, we expected a poor fauna. The results have been far from meeting our pessimistic expectations. To our surprise, we discovered that the karsts of Santo support a high diversity of mites and springtails living in unusually dense populations, and that there was no significant gap in the taxonomic composition of the communities.

Furthermore, it is clear that the biodiversity of Santo karst soils has not been completely exhausted by the Santo 2006 expedition. Many taxa were collected only once, implying statistically that others remain to be found. Beyond the karst itself, which was reasonably well sampled, the non-karst part of Santo, representing more than half the surface of the island, remains virtually unsampled. In particular, the fauna has not been sampled at high elevation, and only very superficially in the mangroves, two habitats that always host remarkable species. Regarding the soil fauna at least, there is room for a further expedition…

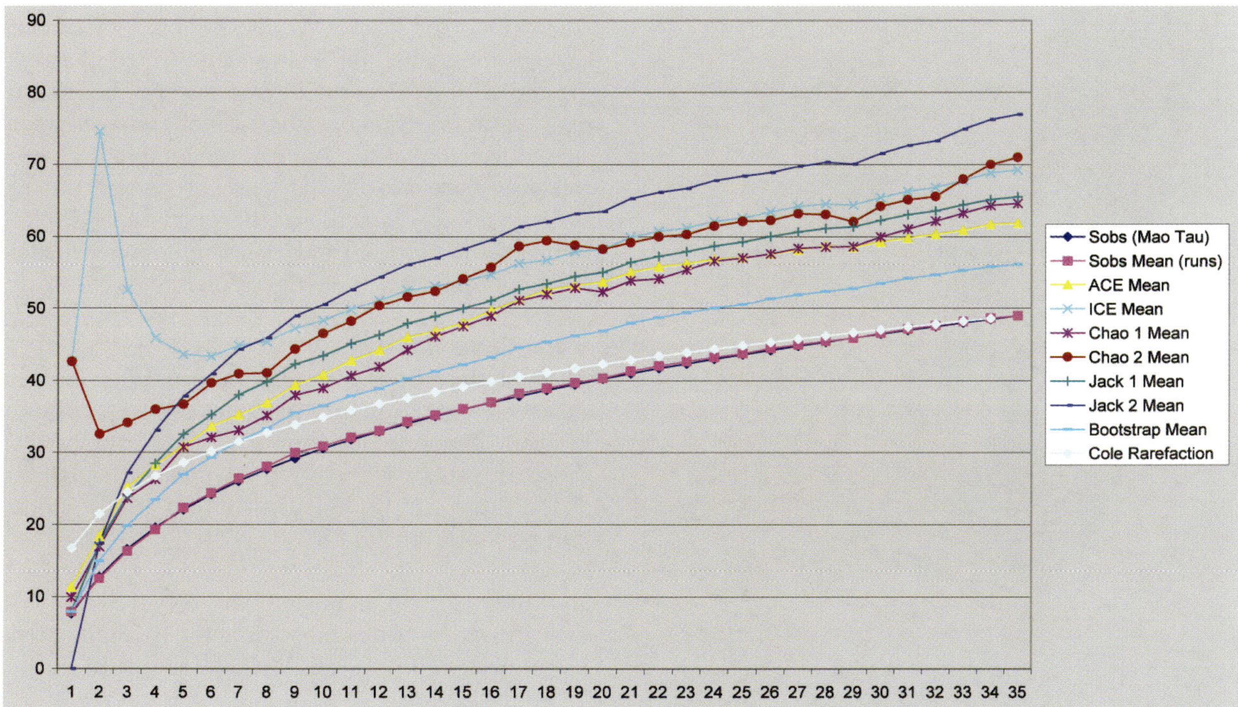

Figure 348: Various indicators converge on an estimation of 50-80 species of land snails on Santo and neighboring islands of Aore and Malo.

FOCUS ON CAVE TERRESTRIAL HABITATS

Louis Deharveng, Anne Bedos, Vincent Prié & Éric Queinnec

Santo's caves host an interesting fauna: this was the first discovery of the karst team during the Santo 2006 expedition. In a way, it does not have much merit. Any karst in the tropics hosts subterranean animals, and nobody had searched for them in Vanuatu before, contrary to other large islands of the Pacific. However, access to subterranean fauna is never certain. It requires the presence of entries into the underground domain, which caves may provide if they are large enough for man, and dark and humid enough to host subterranean life. Only a handful of such caves were known on the island before the expedition, far from enough to provide the conditions needed for a successful biological trip. In order to evaluate the potential of Santo in this respect, two preliminary reconnaissances were launched in 2005 — one for terrestrial cave fauna, the other one for aquatic subterranean habitats, including blue holes. These trips were extremely productive, revealing a large number of subterranean passages with both aquatic and terrestrial habitats favourable for troglobionts, as well as plenty of guano. The way was thus opened for the 2006 expedition itself, during which extensive collections of the fauna were made: 255 samples in 42 caves scattered across the whole karst area of Santo

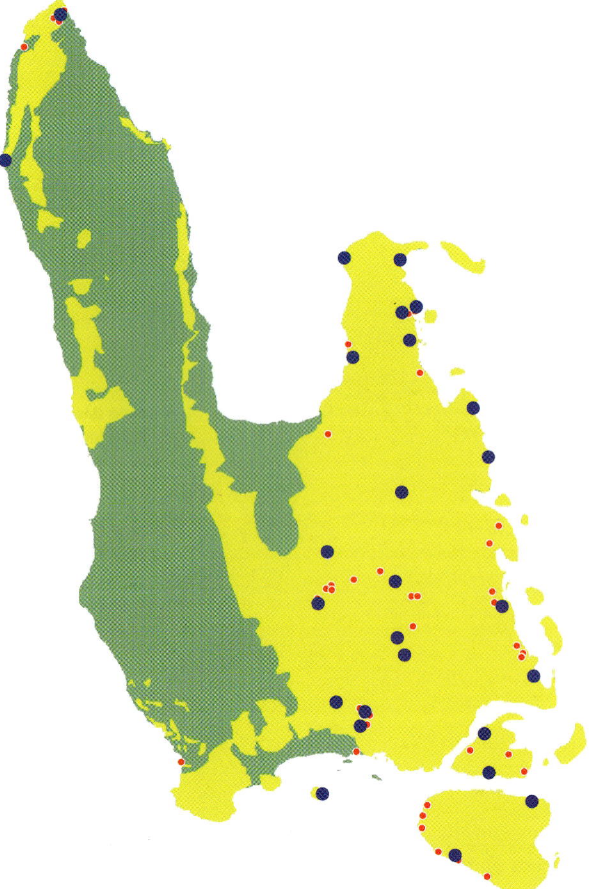

Figure 349: Caves sampled biologically (blue circles) and not sampled (red circles).

(Fig. 349), as well as 40 samples in interstitial habitats, and 41 samples in springs and blue holes.

••• Collecting in Santo caves

Basic techniques for collecting the fauna are the same underground and above ground. They just have to be adapted to the particular conditions of subterranean environments: a limited diversity of habitats, a patchy distribution of trophic resources and the co-occurrence of extremely food-rich habitats (guano) and extremely food-poor habitats (the so-called oligotrophic habitats). This pattern makes standardized sampling and faunal comparisons between-caves difficult: the distribution of caves in the karst is very uneven and the different cave habitats do not occur in all caves. Oligotrophic habitats in particular are absent or difficult to access in most caves of Santo because of the widespread occurrence of guano. Ideally, they are present behind sumps (which bats cannot cross), but reaching these habitats requires diving skills and equipment, and sumps of interest were limited to a few large caves near Butmas and Funafus. As a result, oligotrophic habitats were strongly undersampled in comparison with guano habitats.

We used various sampling methods in caves (Figs 350A-F), three in particular for the terrestrial fauna: direct hand collecting with forceps, brush or aspirator (Figs 350A & 2B), sometimes near baits; pitfall trapping using small vials filled with alcohol or beer, sunk into the soil flush with the aperture and emptied after two to four days; and substrate extraction of the terrestrial fauna from cave soil (Fig. 350C) and guano. We also sampled soil outside caves at the same time. This greatly helped in assessing the ecological status of cave species, because in many cases, morphology alone cannot tell us if a species collected in a cave is a real troglobiont; its status can only be ascertained if it is not present outside caves, especially in soil. For this purpose, 50 Berlese funnels from both soil and cave samples were operating simultaneously (see "Focus on Soil").

••• Cave habitats

The stability of temperature and humidity, and the absence of green plants (roots excepted), strongly limit the diversity of habitats in caves compared to those outside. It is mostly the type and availability of food that control the distribution and organisation of subterranean life.

Six kinds of habitats hosted cave life in Santo:
• Entrance zone, mostly in the twilight zone, inhabited by an obscuricole fauna of rather large species;
• Oligotrophic habitats in the dark zone, where food is provided by scattered faeces of bats or

Figure 350: Collecting cave fauna. **A**: Collecting terrestrial invertebrates with a pooter, after a sump in Loren Cave. **B**: Collecting microarthropods on a wall of Amarur Cave. **C**: Taking a soil sample in the "Grotte du Golf", a shallow cave near Luganville. **D**: Placing fish nets in an underground stream. **E**: Collecting in large gours of Kafae. **F**: Small gours that trap terrestrial micro-arthropods at their surface.

swiftlets, or dissolved organic matter from soil percolating water; they typically host cave-obligate species;

• Roots, which sometimes penetrate into caves, especially when they are shallow, and host a mixed fauna of soil root-suckers and troglobiont animals;

• Vegetal debris brought into caves by sinking streams, especially during floods, with a large number of outside species. The number of these trogloxenes decreases with time as many are unable to cope with subterranean life; flood debris deposited by exceptionally high flooding may slowly decompose for years, providing habitats for troglobionts;

• Guano habitats (including swiftlet nests), where food is plenty and very concentrated, are exploited by a peculiar fauna; they are described in "Focus on guano";

• Aquatic habitats, which are also the subject of separate contributions.

The first four habitat types are considered in this part, "Focus on cave terrestrial habitats".

● ● ● The composition and diversity of the non-guanobiont terrestrial cave fauna

In terms of species numbers, subterranean voids are never rich compared to vegetation or soil habitats. Among other reasons, the absence of light prevents the growth of green plants and most of their associated fauna, except for a few root-sucking bugs. However, caves are unsurpassed among terrestrial environments in terms of endemism levels, which are particularly high among obligate cave species. In this respect, cave habitats have long fascinated biologists. Much effort was expended during Santo 2006 to discover the endemic cave species that we expected for a relatively ancient island like Santo.

In fact, we found that the main characteristic of terrestrial cave fauna in Santo was that such troglobionts are very few in number, being restricted to oligotrophic habitats that are difficult to sample. Subterranean communities were largely dominated by guanobionts (see "Focus on guano") and by non-troglobiotic species, i.e. trogloxenes and troglophiles. Among the troglophiles, two categories are present in Santo: those using caves as shelter during the day, such as bats and, probably, some Amblypygi, or during the night, such as swiftlets; and those with reproducing populations both inside and outside caves, including most of the troglophilic invertebrates.

> The following terms are used to describe the link of a species with the subterranean environment:
> •Troglobiont/stygobiont: a species living exclusively in terrestrial/aquatic subterranean habitats.
> •Troglophile/stygophile: a species living part of its life cycle in caves, or with both epigean and subterranean populations.
> •Trogloxene/stygoxene: a species whose main habitat is outside caves, encountered accidentally inside caves.
> •Guanobiont: a species living exclusively in guano.
> •Troglomorphic: a species or a morphological trait modified in relation to cave life.

● ● ● A large number of outside and obscuricole species

Even if many species and groups do not usually penetrate into caves, the overall number and abundance of outside species in the subterranean habitats of Santo is high. These invertebrates do not exhibit troglomorphic characters and are also present in soil and litter outside caves. Many are carried passively into caves by sinking streams, and are found among flood debris deposited along subterranean banks. Water percolating through the edaphic layer and limestone cracks also brings minute arthropods, which are often trapped on the surface of micro-gours at the top of stalagmites (Fig. 351A). Large obscuricole species are also abundant; attracted by darkness, but not adapted to cave climatic conditions, they are typically seen on the walls of the entrance zone, in partial or total darkness (Fig. 351B). The most conspicuous species are spiders of different families, particularly Sparassidae and Pholcidae.

Interestingly, the abundance of trogloxenes and troglophiles in caves was not proportional to their relative abundance in surface habitats. There is a variety of situations that require a variety of adaptive abilities, from entering passively into caves to establishing short to long-term viable populations underground. This diversity of troglophily levels can be highlighted by three examples:

• In Santo, like anywhere in the world, deep-soil species do not colonize caves, though their habitat is in direct connection with the bedrock, and they are already adapted to darkness and limited food supplies. Minute endogeous arthropods —micro-woodlice, Symphyla, micro-Collembola or endogeous Acari (see "Focus on soil")— are present in caves only as isolated specimens, whatever the group considered, except in places where soil "flows" into the cave through large connections with the surface. It is not known which biological traits prevent these arthropods from colonizing caves.

• Two species of relatively large Neanurinae springtails are found both in litter and in caves. The first, *Telobella* sp., is more abundant in litter than the second, *Yuukianura* sp. Conversely, *Yuukianura* sp. is far more abundant than *Telobella* sp. in caves, despite their similar ecology and size, and frequent co-occurrence in litter. The *Yuukianura* show abundant guano-adapted populations that seems to be conspecific with litter populations according to the current state of our analyses, whereas *Telobella* are never found on guano. Both are otherwise prone to be carried into caves by percolating waters, in spite of their large size (Fig. 351A).

• All terrestrial snails collected in caves were also present outside, in litter or rocky environments, but a few were found alive underground. Among them, Hydrocenidae species may be observed on cave walls, surviving on leaves and organic matter carried in by flood waters.

● ● ● Few terrestrial troglobionts

Four species were found exclusively in caves and can be qualified as troglobionts, living mostly in resource-poor habitats and absent in large guano accumulations:

• A Paradoxosomatidae millipede;
• A woodlouse;
• A Sminthuridae springtail;
• A Nocticolidae cockroach.

All are probably new to science. Except for the cockroach, these species may also be found occasionally or marginally on small guano accumulations.

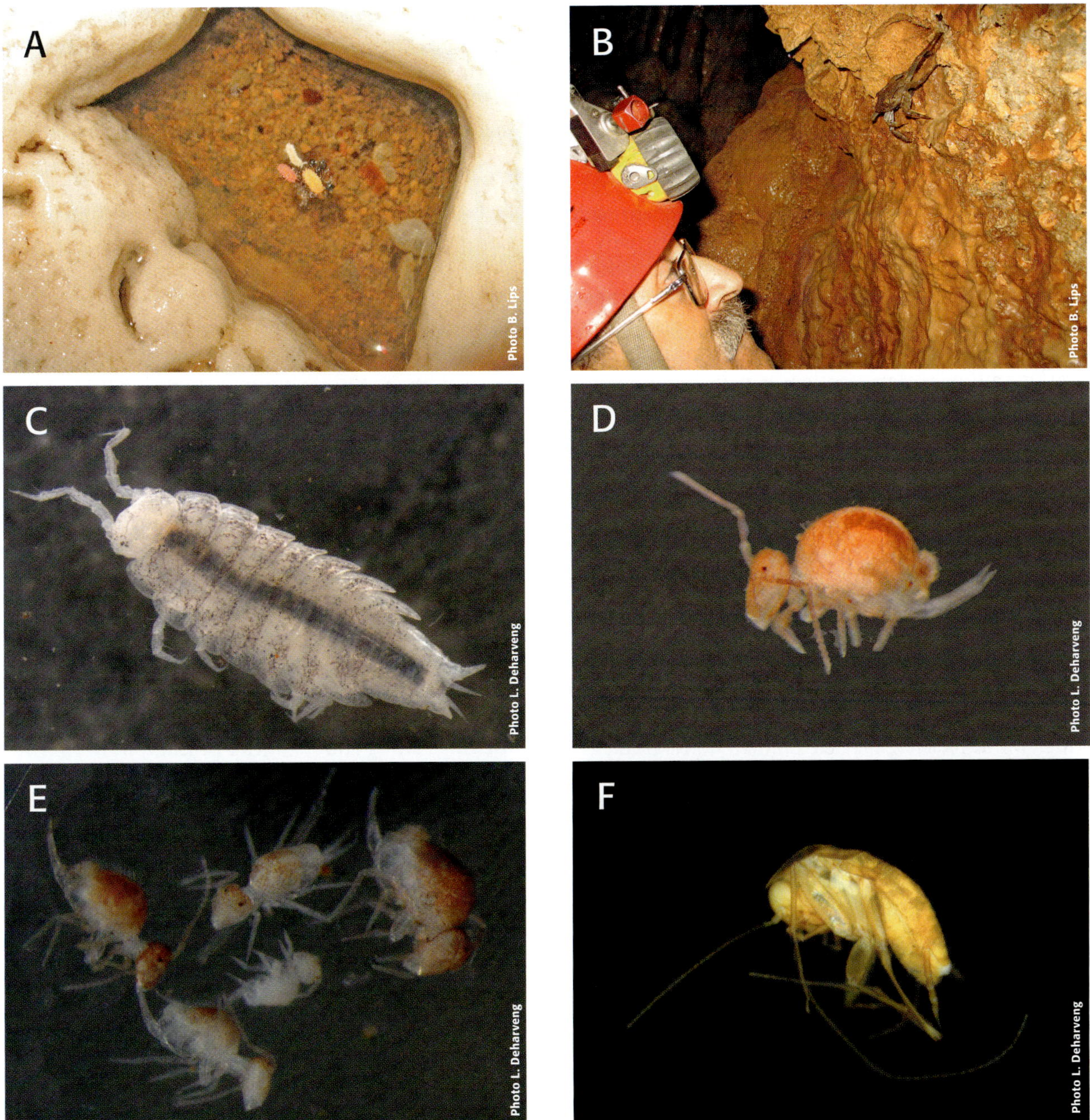

Figure 351: Some cave species of Santo. **A**: Three Neanurinae species (one or two species of *Telobella* and one *Yuukianura*) trapped on the surface of a small gour in Fapon Cave. **B**: Observing a probably trogloxenic land-crab on the wall of Amarur Cave. **C**: A troglobiotic Styloniscidae, from oligotrophic, post-sump habitats of Patunar. **D, E**: *Pararrhopalites* sp. from Fapon Cave (D) and Autabelchiki Cave in Aore (E). **F**: A blind Nocticolidae from Fapon Cave.

At least three Styloniscidae species in the caves and soils of Santo have reduced eyes. One is a blind guanobiont, abundant in the Mba Aven, rare in Riorua Cave and absent elsewhere. A second one has a single small eye on each side and was collected in two caves and two soil samples. The third species is the only woodlice that could be considered as a troglobiont, as it was found in several caves of the Funafus system, but not in guano or outside (Fig. 351C). This species has three small, poorly pigmented eyes per side and retains a vestigial pigmentation. Interestingly, the only significant population was collected from a sump in Patunar in the absence of swiftlet or bat colonies; it was associated there with a large population of troglomorphic Nocticolidae.

A large, pale caramel millipede of the genus *Eustrongylosoma* is frequent in Santo caves (Amarur, Fapon, Kafae, Riorua and Tarius), roaming on small guano piles or scattered organic matter. Its absence outside caves, where other Paradoxosomatidae millipedes are frequent, suggests a troglobiotic status.

Collembola are not rare in oligotrophic habitats, feeding on microorganisms growing on bat droplets or on scattered pieces of wood, but all these "cave" springtails are also present outside, or are isolated specimens of guanobiotic species. The only exception is a species of the genus *Pararrhopalites* (Figs 351D-E), known from five caves (Aore, Fapon, Kafae, Tarius and Sarabo), but which does not exhibit obvious troglomorphic characters compared to soil

Pararrhopalites from other Australasian regions. The *Pararrhopalites* were collected from guano in three caves, but not in the two other caves, so we assume it could be a troglobiont. The genus has not been found in the soils of Santo analyzed so far.

The Nocticolidae species of Santo caves (Fig. 351F) is a small, fragile, blind and long-legged cockroach, which runs rapidly on cave walls, preferentially in confined galleries. Such troglomorphic Nocticolidae are characteristic cave inhabitants in Southeast Asia and Australasia. But, while eyed relatives of blind cave Nocticolidae are present outside caves in these other regions, none was collected in Santo, raising the question of how Nocticolidae may have colonized caves of the island.

It cannot be ruled out that some of the collected spiders —frequent cave colonizers in the tropics and numerous in Santo caves— are troglobionts, but at the current stage of sorting, none has been recognized as troglomorphic. The only species found to have strongly-reduced eyes is an Oonopidae from the "Grotte du Golf", but it looks like a soil rather than a cave species.

In terms of its diversity, the obligate terrestrial cave fauna of Santo therefore has two characteristics:
- Except for Nocticolidae, the terrestrial troglobionts of Santo are at most only moderately troglomorphic;
- With four troglobionts, the obligate non-guanobiotic cave fauna has a low diversity compared to that of many tropical karsts of Southeast Asia or Australasia, where up to 15 troglobiotic species may be found in a single cave. Nevertheless, it is "rich" when compared to that of other small Pacific Islands. In fact, guanobionts, like troglobionts of oligotrophic cave habitats, are cave-restricted and should equally be considered as well adapted troglo-

bionts. In this respect, the cave fauna of Santo would appear to be reasonably rich, even in comparison to that of Australasia.

The study of the insular cave fauna of Hawaii and the Canary Islands has demonstrated that speciation on islands may occur in an amazingly short space of time, i.e. a few thousand years. The oldest Santo limestones are of Pleistocene age, but it is likely that subterranean fauna colonized the deep karst habitats after having evolved in volcanic voids, since the non-limestone terrains of Santo were emerged several millions years before coral reef uplift. This would be enough time to have allowed cave adaptations in arthropods. Surprisingly, however, the troglobiotic fauna of Santo is poor: the recent age of the Santo Island karst cannot fully account for the rarity of non-guanobiotic troglobionts and the low levels of troglomorphy. Indirect evidence suggests that local bioclimate and the connection of karst voids with the surface may partly explain this moderate level of biodiversity. In the subterranean systems surveyed in Santo, interconnections between surface and caves are numerous, as caves are only developed for a few tens of meters at most under karst surface, with frequent openings along subterranean passages. This permeability, combined with a similar bioclimate inside and outside caves (characteristic of a hyperhumid tropical climate), allows easier faunal exchanges between caves and the surface, as illustrated by the massive presence of trogloxene species, hence limiting opportunities for speciation. A further reason may be that most, if not all, caves of the island host bat colonies with large guano accumulations, which repel "true" troglobionts. It is quite possible that habitats favourable to troglobiotic species (in particular galleries beyond sumps, or cracks in limestone that bats cannot enter) have not been sufficiently sampled, as they are rare and often of difficult access in Santo.

FOCUS ON GUANO

Louis Deharveng, Josiane Lips & Cahyo Rahmadi

Bats and swiftlets are the most familiar inhabitants of Santo caves. Their guano is found in almost all caves we visited (Fig. 352), as isolated droplets or accumulated in piles of varying sizes (Fig. 353). Bats were observed in almost all the caves surveyed in Santo, with nine species being recorded during the expedition (see "Focus on bats"). Large colonies of more than 50 000 bats were not exceptional, mostly comprising *Miniopterus australis australis* and *Asellicus tricuspidatus*. Swiftlets were also very common. Three species are recorded from Vanuatu, all nesting in caves (Fig. 354): *Collocalia esculenta*, *C. vanikorensis* and *C. spodiopygius*.

A rich fauna was associated with the bats and swiftlets, including parasites and free-living species found in the guano. Though parasites were not our target, it would have been difficult to miss the very conspicuous yellow Diptera of the families Nycteribiidae and Streblidae crawling through the fur of the bats (Fig. 355). In some colonies of *Miniopterus australis*, nearly every specimen carried these large flies, visible even on animals in flight. On very rare occasions, parasitic flies were found on guano, having fallen from bats.

Our efforts focused on free-living invertebrates associated with the guano. Guano is an unique habitat,

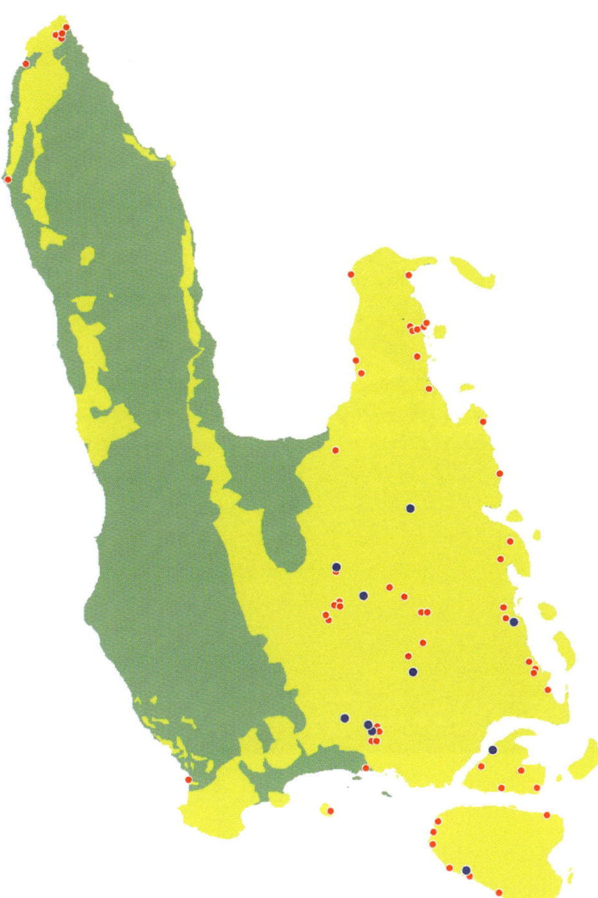

Figure 352: Localisation of the guano caves. Blue circles, most extensively sampled guano caves on Santo; red circles, other caves of Santo.

Figure 353: Searching for guanobionts in small guano piles below swiftlet nests in Kafae Cave.

Figure 354: Swiftlet in its nest in Fapon Cave.

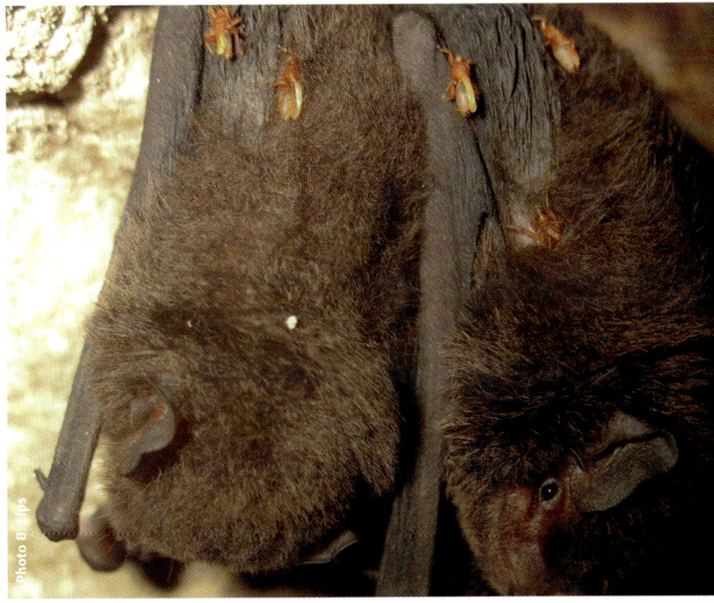

Figure 355: *Miniopterus australis* bearing large parasitic flies in Amarur Cave.

made of faeces, urine, feathers and corpses of bats and swiftlets, accumulating daily and decomposing in situ. It often constitutes islands of very high energetic resources in a mineral matrix of very poor resources. Its only links with the exterior are bats and swiftlets that import nutrients into the caves. In Santo, both these groups are insectivorous, as no frugivorous bats roost in the caves of the island. When scattered, their droplets provided the main food resources supporting troglobionts in oligotrophic habitats, as mentioned in "Focus on Cave terrestrial habitats". The true guano fauna was mostly limited to small piles under swiftlet nests (Fig. 354) and to larger accumulations under bat colonies (Fig. 356). These guano accumulations were inhabited by an abundant fauna that we sampled intensively, following a standardized protocol, consisting of five guano replicates per pile in each of 10 caves distributed across the largest karst block of Santo (Fig. 352). Intensive non-standardized hand collecting was also carried out in order to create an inventory of invertebrates living in guano, and to characterize their community structure in relation with life in this very peculiar habitat — two fields in which our knowledge remains extremely weak.

••• The fauna

Species of several invertebrate groups are regular inhabitants of guano in the caves of Santo. These are:

Figure 356: Large bat guano accumulation in Sarabo Cave.

••• Gastropoda

The subulinid *Allopeas gracile* is the only terrestrial snail living in the cave of Santo and seems to be restricted to guano in this habitat (Fig. 357). It is also widespread outside caves (see "Focus on Alien Land Snails" in section "Man and Nature").

••• Arachnida

• Mites —both decomposers (Uropodina and Oribatida) and predators (Gamasina)— were present in and on the guano, often in very dense populations. They accounted, on average, for three quarters of the guano arthropods in terms of abundance, and were dominant in most cases, often forming huge swarms covering in of the guano. They have not been identified to species level, but 56 morphospecies were recognized, probably a world record for a karst the size of that of Santo.

• Whip-spiders (Amblypygi) of the genus *Charinus* were common in caves and surface litter of Santo. In caves they were not strictly linked to guano, but mostly found on walls near accumulations of organic debris. A single species (*C. australianus*) is present, possibly represented by an undescribed subspecies with geographical varieties, which is currently under study (Fig. 358).

• False scorpions were represented by isolated specimens of epigean species from a surprising variety of families (Atemnidae, Chthoniidae, Lechytiidae and Syarinidae), but also by true guanobionts of the Cheiridiidae and Chernetidae.

Figure 357: *Allopeas gracile* in Fapon Cave a common snail on the guano of Santo caves.

Figure 358: A whip-spider, *Charinus* sp. from Fioha Cave.

Centipedes

• Scolopendromorpha centipedes were present, though not common, in the caves of Santo; they are found exclusively on guano, where they are active predators (Fig. 359).

Millipedes

• Some guano piles supported abundant populations of millipedes of the genus *Hypocambala* (Fig. 360). Interestingly, the species was identified as *H. anguinea*, which is considered to be a pantropical tramp species. Its absence outside cave guano habitats in Santo raises interesting questions about the way it might have colonized the island. *Hypocambala anguinea* is the only true guanobiotic millipede found on Santo, but other species of Paradoxosomatidae and Polydesmidae are occasionally present on guano. Like other cave Cambalopsidae, this millipede excretes a smelly liquid that probably acts as a deterrent to predators.

Woodlice

• Woodlice were represented by a guanobiotic Styloniscidae limited to two caves: Mba Aven where it was abundant, and Riorua Cave where it was rare. Though typically guanobiont, the species lacks eyes and pigment, and is not closely related either to the troglobiotic species of the Funafus sytem (see "Focus on cave terrestrial habitats") or the other troglophilic or epigean species of the family.

Springtails

Springtails are the second most abundant arthropods of the guano, after mites. Only one species (*Psammisotoma* sp., see "Focus on Springtails") was strictly linked to guano, though it was rare. At least three others (*Xenylla yucatana*, *Yuukianura* sp. and *Coecobrya aokii*) reached their highest densities in this habitat.

• *Xenylla yucatana* is a pantropical species that can be encountered in very large populations on guano, but it is also abundant in soil outside caves.

• *Yuukianura* sp., a species of Neanurinae new to science, lives in dense populations on the surface of guano piles, often aggregating around fresh dung (Fig. 361). It also seems to be widespread in litter habitats on the island. This is the first time that representatives of the otherwise halophilous genera *Psammisotoma* and *Yuukianura* have been reported from inland habitats (see "Focus on Spingtails").

• *Coecobrya aokii*, an entomobryid described from a single site on Santo, is in fact common in soil, litter and caves around the island. In guano it is sometimes found in huge populations, but this is never the case in outside habitats (Fig. 362).

• Other springtail species (*Orthonychiurus* sp., *Rambutsinella* sp., *Isotomiella symetrimucronata*) are occasionally found in guano piles, but their preferred habitat is forest litter.

Figure 359: A centipede eating a young cockroach in Autabelchiki Cave.

Figure 360: *Hypocambala* swarm on a guano pile in Mba Cave.

Figure 361: *Yuukianura* springtails around a fresh dung on a guano pile of Rotal.

Figure 362: *Coecobrya aokii*, a jumping springtail abundant on guano in Fapon Cave.

Figure 365: Dermaptera among *Coecobrya aokii* and other microarthropods in a guano collection from Vobananadi shaft.

Figure 363: Bent-wing bats and *Periplaneta* cockroaches co-habiting on the roof of Autabelchiki Cave.

Figure 364: *Periplaneta* cockroaches eating a dead bat in Autabelchiki Cave.

••• Insects

• Blattodea: The most disturbed or warmest lowland caves, in which guano is abundant, may host dense populations of large cockroaches of the genus *Periplaneta*. They are often found on cave walls and roofs, amongst bats (Fig. 363), and they may be active scavengers (Fig. 364).

• Dermaptera: earwigs were regularly found in guano piles, but never in large numbers. At least two species, both with eyes, were present in Santo, of which one — paler and with slightly smaller eyes— was not found outside the cave guano habitat (Fig. 365).

• Nemobiinae gryllid crickets which turned out to be new to science (*Cophonemobius faustini* Desutter, 2009) were the only Orthoptera regularly associated with guano in Santo. They were not collected outside caves, though they are pigmented and fully oculated. Two closely related forms were recognized and described, and another one is possibly present. *Cophonemobius* was so far only known from Samoa. Rhaphidophoridae, which are widespread in caves in surrounding regions (Australia, Southeast Asia), are absent from Santo caves.

• Coleoptera. Tenebrionid beetles are regular inhabitants of guano worldwide, with both widespread and endemic species. In Santo, large swarms of a rather large, black species (probably the widespread *Alphitobius laevigatus*) were found locally, strictly restricted to guano piles (Fig. 366); the species was not encountered outside caves. Other beetles were also present in the guano, including many unidentified larvae. The adult specimens collected belong to several families, of which Histeridae, Hydrophilidae, Scydmaenidae (Fig. 367) and Staphylinidae were the most common.

• Diptera were abundant as larvae and adults, but remain unstudied.

• Lepidoptera. Tineidae are a major component of guano communities in all cave systems on Earth. Santo is no exception, with tineid larvae sometimes occurring in very dense populations (Fig. 366). The species present in Santo caves have not yet been identified.

Figure 366: A large swarm of Tenebrionidae mixed with encased Tineid larvae on a guano pile in Autabelchiki Cave.

Figure 367: A Scydmaenidae (probably *Scydmaenus* sp.), abundant in the guano of Rotal shaft.

••• The characteristics of guanobionts

In contrast to troglobionts, guanobionts are often morphologically similar to species found outside caves: of similar shape and size, eyed, pigmented, often winged, without any adaptive morphological features that distinguish them from non-guanobionts. This traditional image of guanobionts has, however, to be nuanced for tropical caves. In Santo, guanobiotic millipedes or beetles have eyes and pigment, like their closest counterparts outside caves. What links guanobionts to guano is probably rooted in their physiological ability to cope with extremely rich organic nutrients. Most guanobionts also share a high reproductive potential and often live in extremely dense populations. Because their food source is permanently available, different life stages usually occur together.

In terms of their geographical distributions, guanobionts contrast with troglobionts because they include a large number of widespread species and relatively few endemics. Special dispersal abilities may partly account for this pattern. Thus, some guano species belong to families in which dispersion by phoresy is common (gamasid mites and false-scorpions of the families Atemnidae, Lechytiidae, Cheiridiidae and Chernetidae) while others are flying insects. Among the smallest guano springtails, several (*Isotomiella symetrimucronata*, *Folsomides centralis*) also live in soil and are apparently parthenogenetic. They are not confined to fragmented subterranean and karst habitats and so are less prone to endemism.

••• Guano foodweb

The food-web of guano is based on the organic debris formed by the dung, urine, corpses and nests produced day after day by bats and swiftlets. These debris accumulate on the cave floor where they are rapidly decomposed by bacteria and fungi, providing very rich —and very smelly— nutrients for invertebrate decomposers. In Southeast Asia, the fauna of guano is organized into two almost independent assemblages: the giant arthropod community and the meso-/microarthropod community.

The giant arthropod community includes the most conspicuous invertebrates to be seen in caves: large crickets and cockroaches living in dense populations, and their predators: huge sparassid spiders, whip-spiders and centipedes. In Santo, we have only a pale reflection of this giant fauna. Large cockroaches (Figs 363 & 364) were locally present, but crickets were only represented by small Nemobiinae. Large Sparassidae were rarely seen. Only whip-spiders of the genus *Charinus* were widespread, but their size is moderate compared to that of many tropical members of this group (Fig. 358). The large, long-legged Scutigeridae, which are regular and spectacular inhabitants of Southeast Asia caves, were rarely encountered in Santo caves, possibly because their favorite prey, the large rhaphidophorid crickets, were not present.

Conversely, meso- and microarthropod communities, numerically dominated by mites, were strongly diversified. Decomposers of these communities often live in very dense populations in guano. They include mites, springtails, tineid larvae in their cases, tenebrionid beetles, and a great number of less abundant groups.

Predators feeding upon these large swarms of decomposers are rather diverse. None is supposed to be specialized, but evidence of what eats what is starkly lacking in Santo caves. The predators are mostly arachnids: mites (Gamasina), whip-spiders (*Charinus* sp.) and false scorpions. Ants were often abundant, as in most tropical caves where their importance in guano fauna has been largely overlooked. They belong to epigean species, but some probably have permanent populations in cave guano. Centipedes of the groups Geophilomorpha and Scolopendromorpha are uncommon in guano (Fig. 359), whereas they are very common in soils.

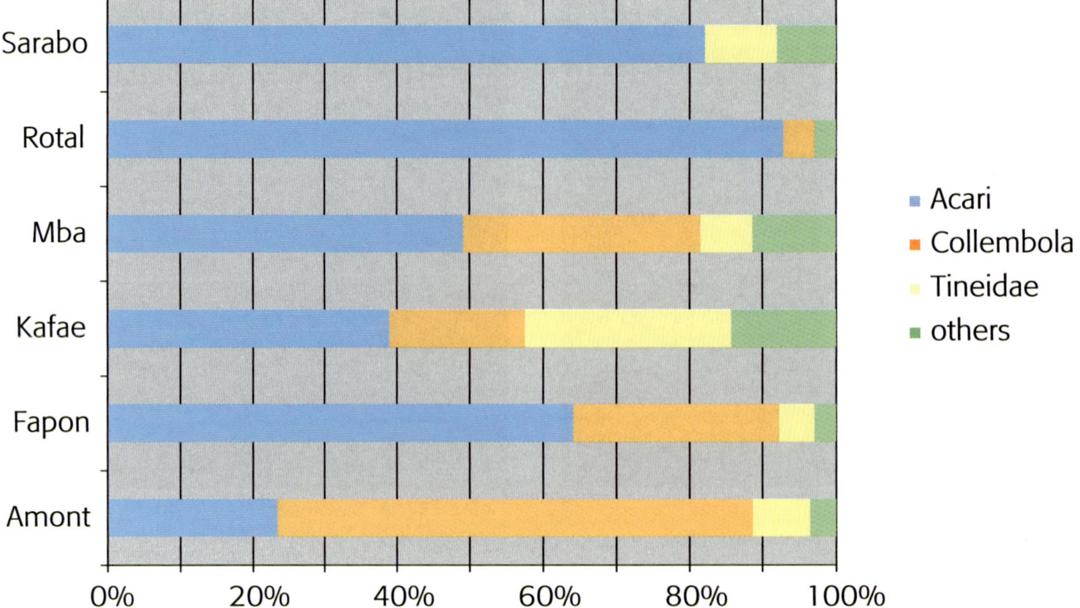

Figure 368: Relative abundance of different Arthropods in the guano of six caves of Santo.

••• General patterns of biodiversity

In terms of alpha-diversity (i.e. for a given guano pile), richness of invertebrates ranged from 7 to 36 species and morphospecies, with mites always the most diverse before springtails. One to three species of decomposers dominated the guano community. They were often homogeneously distributed through the pile. Their density was much higher than that observed for any outside soil species for their respective groups. Similar patterns have been described from many guano communities worldwide, but they have rarely been quantified. Here, the density per square meter may roughly reach 500 000 individuals for some Uropodina, 100 000 for *Coecobrya aokii*, 50 000 for Tenebrionidae and Tineidae, 20 000 for *Yuukianura* sp., 10 000 for *Hypocambala* and 1 000 for large *Periplaneta*.

Unexpectedly beta-diversity was remarkably high. From one cave to another, and even from one guano pile to another inside the same cave, the proportion of dominant taxa were often different (Fig. 368), not only taxonomically, but also in their size and ecology, in spite of strong similarities in habitat characteristics. We did not detect any clear relationship between biodiversity pattern and habitat parameters. No evident guano invertebrate diversity patterns was found linking guano community structure or composition

with the species (bats or swiftlets) producing the guano. This result is in line with the large trophic requirements of most guano-inhabiting species. It is humidity, age of the guano and continuity of guano piles over the years that seem to play the essential roles in structuring guano assemblages.

Some of the most abundant guanobionts, though devoid of adaptations to subterranean life, were never encountered outside caves (e.g. Cambalopsidae millipedes and Tenebrionidae beetles), while others (e.g. the springtails *Coecobrya aokii* and *Yuukianura* sp.) were also present outside caves, but at lower population densities.

On the whole, the fauna of the guano in Santo caves shows obvious similarities to that of other regions in tropical Asia, notably in the frequent occurrence of noodle-millipedes (Cambalopsidae, here species of the genus *Hypocambala*), as well as in the overall alpha-diversity pattern (low species numbers, with strong dominance by a few species). The most unexpected feature, thus far not observed elsewhere, is the high level of beta-diversity between caves and guano piles. This may, however, reflect our currently poor knowledge of the organisation of invertebrate guano communities in caves in other parts of the world, rather than being truly exceptional.

FOCUS ON BLUE HOLES

Stefan Eberhard, Nadir Lasson & Franck Bréhier

Whereas the western part of Santo is characterised by high and steep mountains of volcanic origins, the eastern part has a more subdued topography with marine-deposited limestones lying on top of

the underlying volcanic rocks. As often occurs on limestone terrains, most rainfall disappears underground into caves and few streams flow on the surface for any great distance. The underground

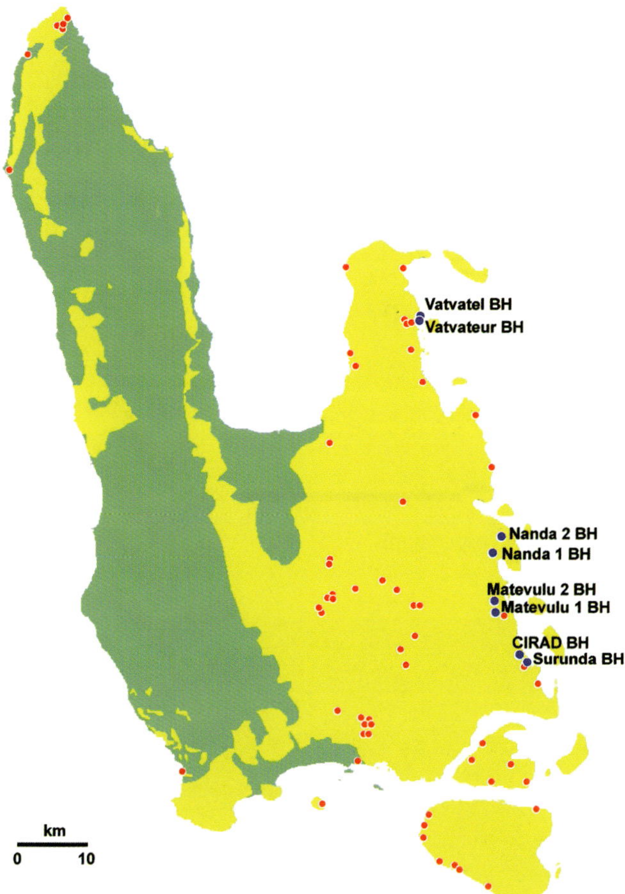

Figure 369: Location of the blue holes of Santo Island eastern coast. Red circles: caves. Blue circles: blue holes. Vatvateur and Vatvatel Blue Holes were not explored.

the coast. As it penetrates further inland, the wedge of seawater is pushed deeper beneath the freshwater lens which is thinnest near the coast and which gradually thickens further inland. Fresh groundwater flows along the hydraulic gradient towards the coast where it is forced to discharge to the surface by the underlying wedge of seawater pushing inland. Discharge of the fresh groundwater to the surface is focused at a series of points forming springs, which are sometimes formed into blue holes, and from where the water continues its course to the ocean as a stream on the surface.

••• Why blue holes are "blue"

As light passes through water the different wavelengths forming the visible light spectrum are absorbed at different rates. Wavelengths at the red end of the spectrum are absorbed preferentially while wavelengths at the blue end of the spectrum are least absorbed. Thus, deep clear water usually appears blue. The molecular structure of water also confers a typically blue colour, often enhanced by reflection of a blue sky. The natural blue colour exhibited by bodies of water may be altered by dissolved substances or suspended particles. On Santo, the deep spring-ponds discharge water that is crystal clear, and when there is a clear blue sky overhead, the "blue holes" are at their most spectacular.

••• The blue holes: a target for the Santo 2006 expedition karst team

Four of the blue holes were mapped and sampled for subterranean fauna by the karst team during the Santo 2006 expedition. These were:

- Matevulu 1 and 2;
- "Trou Bleu du CIRAD";
- Nanda Blue Holes.

The blue holes mapped on Santo ranged in diameter from 10 to 50 m, and reached up to 19 m in depth (Figs 370, 372 & 374). At the base of the blue holes the spring water discharges from the sandy bottom, or small caves and fissures in the soft and friable limestone. The caves cannot be entered far by divers before further passage is blocked by collapsed rocks. Entering the caves is extremely dangerous due to the risk of silt-out and collapse of the soft and friable rock. Cave diving should only be undertaken by properly trained and experienced cave divers. For most divers the caves are of little interest and the open water in the ponds provides the most interesting experience with the plants and abundant fish and animal life.

Another aim of the Santo 2006 expedition was to attempt to collect aquatic subterranean fauna from the blue holes which may be "washed-out" from deeper aquifer habitats. In this aim we were partially successful in collecting a new species of subterranean amphipod from Nanda Blue Hole, however we had more success in sampling the

streams resurge at the surface near the coast, sometimes into large and deep circular ponds known as blue holes. Blue holes occur in similar situations on limestone islands in the tropics elsewhere, however the blue holes of Santo are spectacular. The blue holes of Santo, somewhat hidden in the jungle but still of easy access from Luganville, the main town of Santo, are large and deep pools of clear water with a beautiful turquoise hue (Fig. 369). They are popular with tourists for swimming, canoeing, and diving. The blue holes are also an important source of freshwater, and a habitat for many species of plants and animals, including some which are unique to the blue holes.

••• A brief description

The blue holes are actually a type of spring that discharges groundwater from the limestone aquifer which is fed by rain falling in the hinterland catchment. Most cave systems are formed by the corrosive action of rainwater, which is naturally acidified by carbon dioxide from the atmosphere and soil, and that is capable of dissolving carbonate rocks such as limestone. The portion of rainfall that is not lost by evaporation or transpired by plants seeps underground and recharges the groundwater aquifer. The aquifer forms a lens of less dense freshwater that "floats" on top of a "wedge" of underlying seawater that penetrates inland from

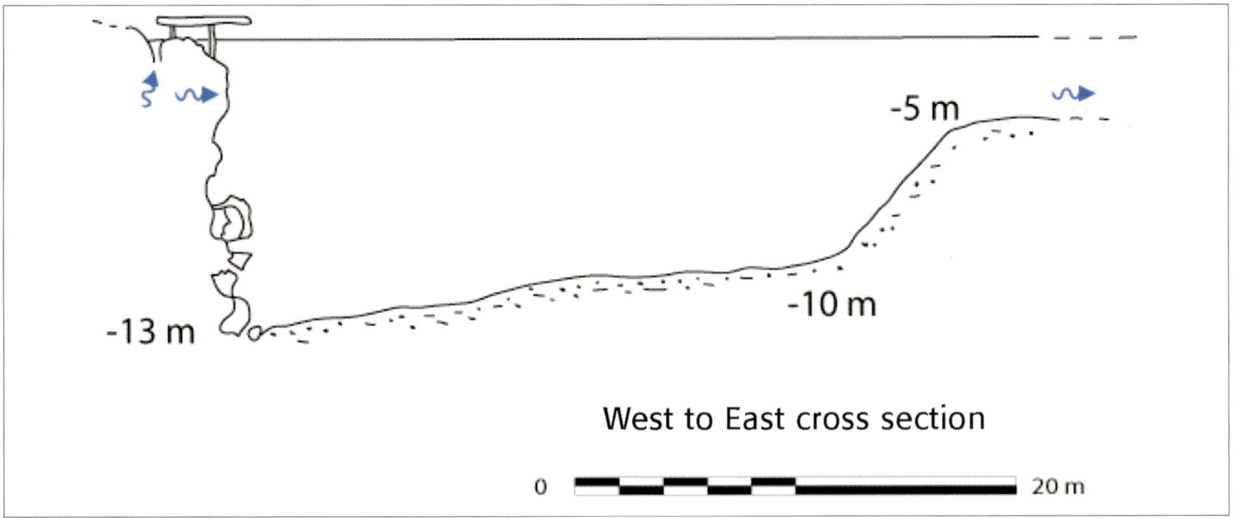

Figure 370: Topography of Nanda Blue Hole (Natawa-Santo-Vanuatu, Santo 2006 expedition).
Topographical plan: N. Lasson (16/09/2006).

deeper aquifer habitats by pumping from the wells which provide the villages with freshwater.

••• Nanda Blue hole

With its crystal-clear water, it is one of the most striking blue holes on Santo (Figs 370 & 371). Most of the strong flow comes from a crack that opens close to the surface. It is a habitat for many species of plants, fish and invertebrates including the freshwater hermit crab, *Clibanarius fonticola*.

••• Matevulu blue holes 1 & 2

Matevulu is a large and deep (-19 m) pool of fresh water from which discharges a strongly flowing river that reaches the sea along a winding course. It is a very popular tourist attraction and local guides offer visitors a swim or a peaceful descent of the river. We dived and mapped this blue hole, and found a major outflow of water in the bottom (Fig. 372). Although a passage could be observed, it was too narrow for further exploration. We put a net to sample fauna in the outflow point of the spring, but did not capture any subterranean aquatic species.

Close by to Matevulu 1, lies Matevulu 2, which is smaller, shallower, and with no obvious spring outflow at the bottom. However, the deep blue of its water, well hidden amongst a luxurious forest, attracts many visitors (Fig. 373). Diving failed to reveal any submerged passages in this hole.

Figure 371: Diving in crystal waters of blue holes: Nanda blue hole.

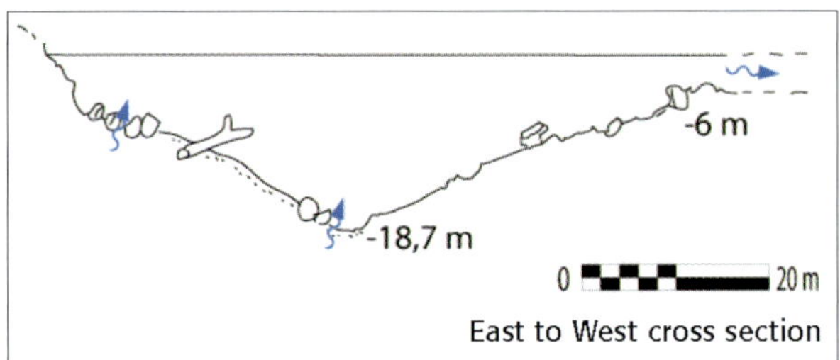

Figure 372: Topography of Matevulu 1 Blue Hole (Matevulu-Santo-Vanuatu, Santo 2006 expedition).
Topographical plan: N. Lasson (23/06/2006).

Figure 373: Matevulu 2 blue hole.

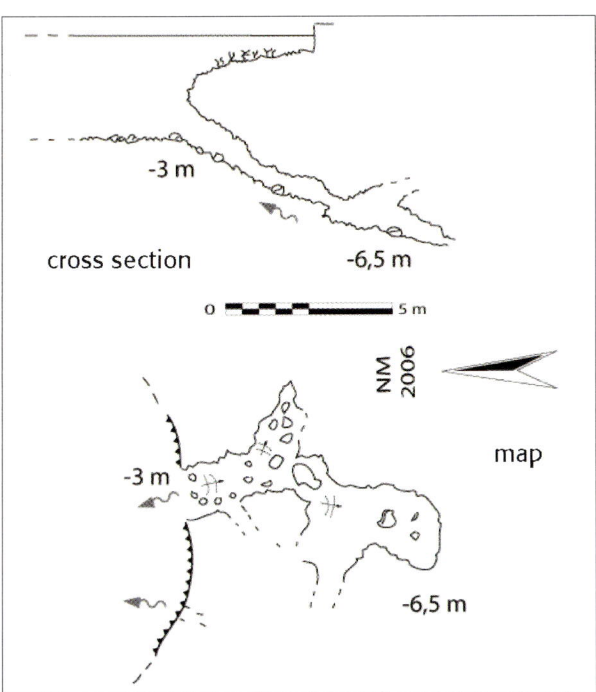

Figure 374: Topography of "Trou Bleu" du CIRAD (Surunda-Santo-Vanuatu, Santo 2006 expedition). Topographical plan: N. Lasson (03/09/2006).

Figure 375: "CIRAD" blue hole.

••• "Trou Bleu du CIRAD" ("CIRAD" blue hole)

It is 40 m in diameter but only reaches a depth of about 3 m (Figs 374 & 375). The spring outflow occurs at several points along one side of the hole, one of which is just large enough to be entered by trained cave divers, however the passage soon becomes too narrow after only a few meters. We sampled for fauna inside this cave and collected several species which had obviously originated from nearby surface waters.

••• Why are the blue holes important?

The blue holes on Santo are a spectacular natural feature popular with locals and tourists for swimming and diving. They are an important source of freshwater, as well as a habitat for plants and animals, including some rare and significant species such as the freshwater hermit crab, *Clibanarius fonticola*, known only from Nanda and Matevulu Blue Holes. Hence it is important to protect and conserve the blue holes from any human activity that may threaten them.

Possible threats to the blue holes include pollution or eutrophication in the spring-lakes or the groundwater which feeds them. Pollutants may enter the lakes directly, or less obviously, by flow of groundwater originating in the hinterland catchment areas. Eutrophication is an increase in chemical nutrients containing nitrogen or phosphorus, which typically originate from human sources such as fertilisers, sewage, etc. Lakes subject to eutrophication typically experience excessive algal blooms, resulting in poor water quality including lack of oxygen and severe reductions or death of fish and other animal populations. There is some evidence that eutrophication has already impacted CIRAD Blue Hole (Fig. 376).

Another possible threat to the blue holes may be tourism, if visitors and visitor impacts are not carefully managed and monitored. Visitor impacts may include for example, trampling and damage to vegetation, disturbance of habitats and animals, pollutants (e.g. sunscreen) and rubbish. The cumulative impacts of visitors over many years may lead to gradual degradation of the natural and aesthetic qualities which visitors come to see, and ultimately lowering the value to tourism of the blue holes. Thus, tourism at the blue holes needs to be carefully managed and monitored, so that future generations can continue to appreciate and enjoy them in pristine condition.

Figure 376: Eutrophication of the CIRAD blue hole.

FOCUS ON THE LOREN CAVE

Franck Bréhier, Stefan Eberhard & Nadir Lasson

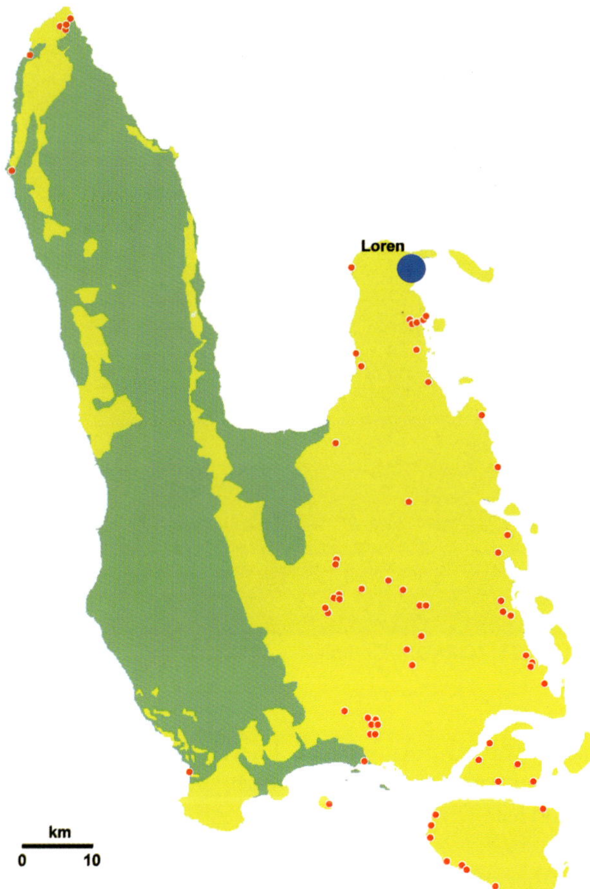

Figure 377: Loren Cave location.

In this gallery, the height above the water is around 4 m whereas the depth reaches -8 m. The section is narrow above the water but wider below. The walls are very sharp due to intense dissolution, and there are almost no accretions. Another "dry" gallery of some 150 m ends up in a restriction on one side, and in a sump that connects with Sump 2 on the other side. Sump 2 is 150 m long and -10 m deep and emerges in a 5 m air-bell. Sump 3 begins just behind this. It has been explored for 190 m and down to a maximum depth of -28 m (Fig. 380). Exploration at the furthest point is completed, but several other side passages were not explored to a definite conclusion. In total, more than 1 km of cave passage were surveyed, of which more than 700 m was submerged and explorable only by diving. There are still opportunities of discovering further passages throughout the cave system, especially in Sump 3.

Figure 378: On the other side of Sump 1.

One of the main goals of the "Karst Team" was exploration and sampling of anchialine caves, since these unusual and special caves, situated close to the coast and under the influences of both marine and fresh waters, are well known in other places for their high diversity of rare and relict species. In 2005, a major reconnaissance effort was put in to locate such caves, but only one site, Loren Cave (Fig. 377), proved promising enough for further exploration and sampling in 2006. Throughout the month of September 2006, a team of cave divers and cave biologists explored and sampled Loren Cave, which proved to be an extensive anchialine cave system containing several significant and interesting species of aquatic fauna.

••• A brief description

Loren Cave (14° 58.850' S; 167° 03.553' E) is situated on the east coast of Cape Queiros (Northern Santo) some 30 m from the coastline. The entrance, 2 m wide by 1 m high, leads to a wide chamber occupied by a pool of water. From there a small gallery leads to Sump 1 (Figs 378 & 379). This gallery is 60 m long and -6 m in depth, and has a restriction (although enlarged during the first explorations in 2005, the passage is still narrow). Behind, a 100 meter-long gallery leads to Sump 2.

Figure 379: Stefan Eberhard and Nadir Lasson emerging from Sump 1.

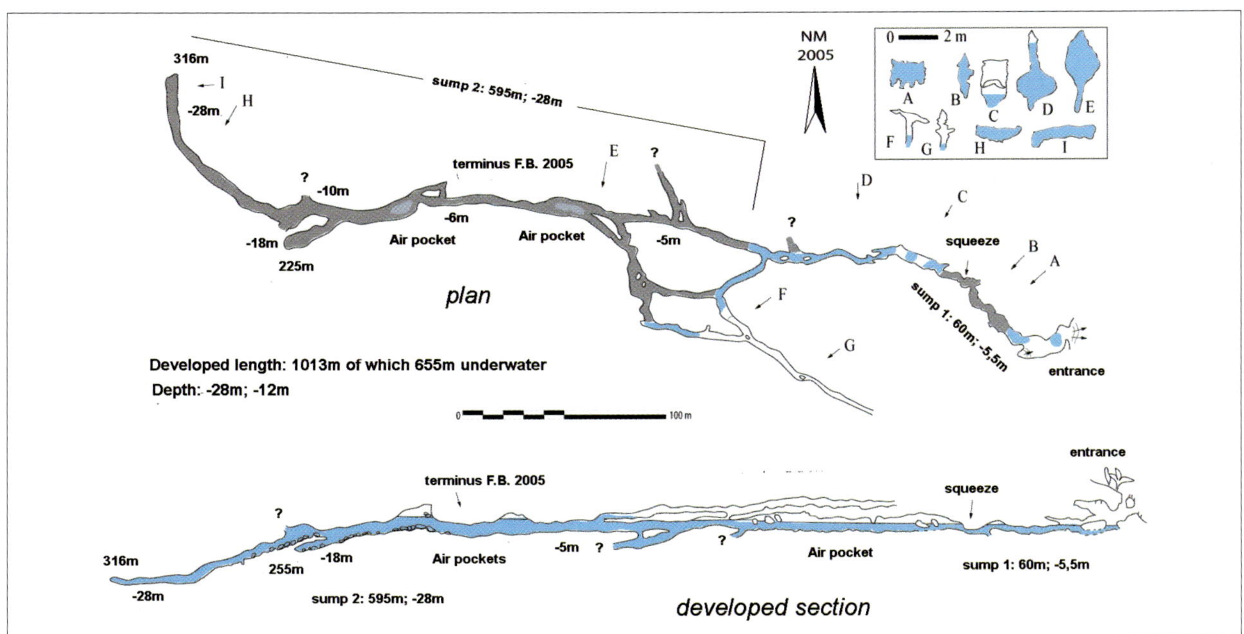

Figure 380: Topography of Loren Cave (Lototor-Santo-Vanuatu, Santo 2006 expedition).
Topographical plan: F. Bréhier (October 2005). S. Eberhard, F. Bréhier, N. Lasson (September 2006). Report: N. Lasson.

Loren Cave: an anchialine cave

Loren cave has all the characteristics of a true anchialine cave: it has a superficial freshwater layer which gradually increases in salinity with increasing depth and eventually approaches the salinity of sea water. The salinity of seawater is about 35 ppt (parts per thousand). Salinity was measured throughout the water column in the third sump of Loren Cave by collecting water samples at discrete depths then measuring the salinity with an Oaklon tester kit. A first halocline can be observed in the first few meters of the water column, depending on the tide level and distance into the cave. In Sump 3 there is a first weak halocline at approximately 8-10 m depth, then a second abrupt halocline at 25-27 m depth where the salinity jumps from 10 to 28 ppt, near full seawater (Fig. 381). Associated with the haloclines are slight temperature gradients (thermoclines) from 24-25°C. There are marked tidal influences in Loren Cave —with an estimated amplitude of about 1 m observed near the entrance— and tidal flow is evident throughout the cave.

Biology of Loren cave

Catching aquatic animals in flooded caves can be difficult. One successful method is to use plastic traps baited with dried meat or fish. Cave fish and crustaceans use their well developed chemosensors to home in on the odour trail diffusing out from the bait into the cave lake.

Deep into the cave we laid baited traps in the first cave lake (Fig. 382), which were left in position at depths from about 5-7 m, for between two and four days. In these traps we caught numerous decapod crustaceans

(Fig. 383): small crabs (provisionally identified as *Orcovita* sp. and *Laubuanium trapezoideum* by Peter Ng), and shrimps belonging to the families Palaemonidae (two species) and Atyidae (three species). The presence of large shrimps in the traps can be problematic since their frenzied feeding movements can damage each other and the smaller animals that have been attracted into the traps.

Beyond the first sump, traps were placed and retrieved by the divers and these produced more shrimps and crabs. Several specimen of *Macrobrachium* cf *microps* were caught in the traps, most of them in the deeper and more saline waters. Hand-net samples were also taken by the divers and sampling was performed to a depth of 28 m, into the deep saline water below the halocline. These deep samples caught one spectacular copepod (see "Focus on the anchialine fauna").

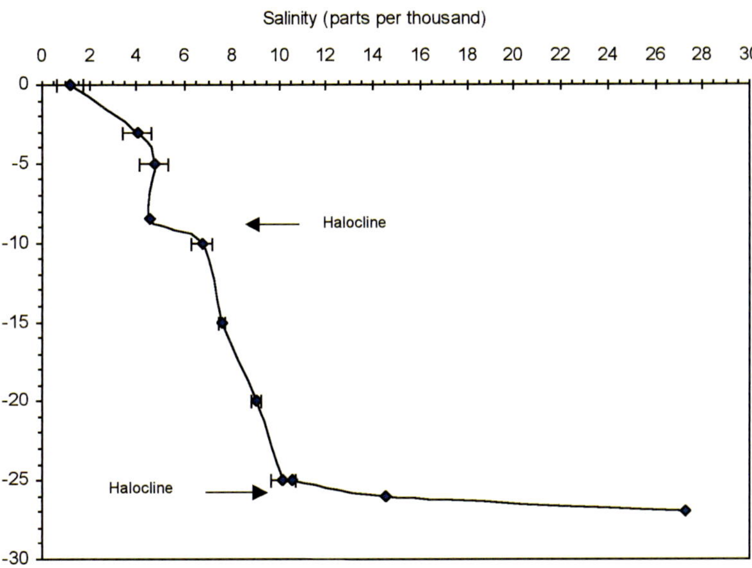

Figure 381: Salinity profile of Loren Cave showing halocline and mixing zone.

Figure 382: The cave diver and cave biologist Stefan Eberhard retrieves traps laid in the deepest part of Loren Cave, at a depth of -28 m.

Figure 383: The cave divers and cave biologists Stefan Eberhard and Franck Bréhier looking at just- caught decapods between Sump 1 and Sump 2.

During a last dive in the deepest parts of the cave, several amphipods were seen swimming in the water column, but unfortunately the divers were unable to catch them. A fish of the same species as those at the entrance, but partly depigmented, was also caught.

FOCUS ON ANCHIALINE FAUNA

Geoff Boxshall, Damià Jaume

Around the coastal zone of tropical and subtropical islands like Santo it is possible to find flooded marine caves, particularly in karstic areas that represent raised fossil coral reefs. Some of these flooded caves show a unique layering of the water within them, and have a surface layer of fresh water sitting above a deeper layer of dense salty water that penetrates through the porous rock separating the cave from the adjacent sea. Habitats which have this double layer of fresh water on top of sea water are known as anchialine habitats and they tend to be inhabited by unusual organisms. On Santo we found only one such habitat, Loren Cave, located just South of the village of Lotoror. Loren Cave has many of the characteristics of a true anchialine cave: it has a superficial freshwater layer which overlies a deeper seawater layer. There is also a marked tidal effect — with tidal rise and fall of about one meter, and there seems to be considerable exchange of water during the tidal cycle. The physical characteristics of the cave and the structure of the water column of this cave system are described in the earlier text "Focus on the Loren Cave".

Cave lakes in anchialine caves are inhabited by unique organisms that have become accessible for study only since the advent of specialized cave diving techniques and equipment (Fig. 384).

Caves are extreme environments for aquatic organisms for several reasons; firstly there is no light, so there is no photosynthesis by green plants. This means the only sources of energy to support the community of organisms in the cave lake are either bacterial production or the inflow of organic material carried into the cave in the inflowing water. These sources of energy usually cannot support dense populations of animals. Secondly the layered water column is often characterised by low oxygen levels, especially in the deeper marine water layer, and oxygen is required by most organisms for their normal respiration. Animals adapted for life at low oxygen levels typically do not survive well outside of these conditions. So aquatic animals that are cave adapted, often have highly localised or restricted distribution patterns.

Anchialine organisms are highly specialized yet many retain extremely primitive features and some, especially those living in and around the Caribbean region, have been classified as living fossils. Specializations that are common in anchialine animals include loss of eyes, enhancement of chemosensory systems, tolerance of low oxygen levels and physiological adaptation to extremely low energy systems.

Figure 384: Member of cave diving team in Loren Cave preparing baited traps before diving.

Figure 385: Copepods are very small and it is best to pick them out from samples while they are still alive, using a field microscope.

Just inside Loren Cave is a shallow pool of fresh water which is dimly lit by light from the cave entrance. It contained a few shrimps belonging to a crustacean genus called *Macrobrachium*. Its Latin name means "long arms" and this shrimp has the most amazing elongate claws, often brightly coloured. There are more than 200 species of *Macrobrachium* known and they can be found in warm tropical and subtropical fresh waters around the world. In addition to some small fish, which were not caught, the pool contained some other small crustaceans known as copepods. These are tiny relatives of the shrimps and crabs, but their adult size is often only one millimeter or even less (Fig. 385). The pool contained copepods belonging to the genus *Halicyclops*. The name *Halicyclops* refers to the fact that these species prefer to live in very slightly salty water (known as brackish water) and that they belong to the group typified by Cyclops, the mythical monster with only one eye.

Halicyclops species, like most copepods, have only one eye, a tiny spot located in the middle of the front of their heads. The eyespot is a simple structure which only allows the copepod to detect light and dark, but enables them to see a predatory fish by its passing shadow and to make the appropriate escape reactions. *Halicyclops* frequently swims away from the bottom of the pool and up into the water. This behaviour helps to distinguish *Halicyclops* from the harpacticoid copepods which were present in the same pool. Harpacticoids have elongate slender bodies with short antennules and move rapidly over the sediment and stones on the floor of the pool, rarely venturing up into the water. Harpacticoid copepods typically consume small particles of organic material that they find in the sediment; some scrape off and feed on the film of bacteria that is present on the surface of sediment particles.

Using baited traps laid in the cave lake we caught numerous larger crustaceans, mainly crabs and shrimps, which were attracted to the smell of the bait. Two kinds of crabs were found in the

submerged passages of the cave, a species of *Orcovita* and *Laubuanium trapezoideum* both provisionally identified by Professor Peter Ng, a crab expert from Singapore. There was also a surprising variety of shrimps species present; we found at least two species of the family Palaemonidae and three more species of the family Atyidae. Cave shrimps typically have extremely long antennae which are supplied with sensors that allow them to detect traces of chemicals as well as mechanical disturbances in the water caused by swimming of other organisms. They use these sensors to find their food and to detect possible predators.

The divers also swam holding hand-nets and collected samples down to a maximum depth of 28 m, into the deep saline water below the halocline (the zone of rapid salinity change marking the boundary between the upper freshwater and lower seawater layers). In the deep samples we found a single adult male copepod nearly 2 mm in length, belonging to the order Calanoida. Calanoid copepods dominate the plankton community of the world's oceans and this male is a member of the family Centropagidae. Centropagid copepods, particularly those belonging to the genus *Centropages*, are common in shallow coastal waters around the world but members of some other genera in the same family have colonised freshwater habitats in Australia and South America. The male we found in Loren Cave represents a new species of the marine genus *Centropages* (Fig. 386), most closely related to species described from Australian waters. This is the very first *Centropages* to be found in an anchialine cave, and it is an interesting discovery as it leads us to conclude that this *Centropages* probably colonised Loren Cave from the coastal waters surrounding Santo.

Centropages copepods feed on small particles like single celled algae and protozoans which they detect in the water and catch by grasping them with their finely branched mouthparts.

They typically have long paired antennules armed with fine sensitive hairs (setae) for detecting their food as well as for detecting the vibrations made by potential predators, such as swimming fish. This male has asymmetrical antennules, with the left hand antennule normal but the right one modified for grasping onto the female during mating.

About 50 m away from the entrance of Loren Cave is a small pool into which an active spring discharges water. Although it is about 30 m from the sea, at high tide the water flow is greatly reduced and the water is slightly brackish (3.2 parts per thousand). As well as being indirectly connected to the sea, this pool is probably also connected to the Loren Cave system. The floor of the pool is formed from coarse coral rubble and it is surrounded by trees. Samples taken by passing a hand net through the water in the pool (Fig. 387) and brushing it firmly across the surface of the stones contained a small amphipod shrimp, in large numbers. Amphipod shrimps are typically flattened from side to side and can be found on and under rocks and stones in most freshwater habitats around the world. The species found here, belongs to the family Sebidae and was described as *Seborgia sanctensis* in 2009.

The adult female of the new *Seborgia* species (Fig. 388) is just less than 2 mm in length. Specimens were observed alive under the stereomicroscope in the field laboratory and we found that the animals moved ventral side down, not upside-down or on one side. This posture is quite unusual for an amphipod, more closely resembling a typical isopod due to their slightly dorso-ventrally depressed body. The females also tended hold the

Figure 387: Fishing for tiny amphipod and tanaid shrimps using fine mesh plankton nets in the pool near Loren Cave.

hind end of the body (the pleon or abdomen) bent forwards underneath the anterior part (the pereon), giving the animal a short and "tail-less" appearance when seen in dorsal view. Female amphipods carry their developing embryos in a marsupium, or brood pouch, located ventrally under the body, and in this species each brooding female carried only two embryos in its brood pouch.

The biogeography of *Seborgia* is remarkable. Even though this genus displays an extremely broad but discontinuous distribution, its members live only in subterranean waters. The ten species known thus far appear scattered in very localised sites across tropical-subtropical latitudes, stretching from an inland aquifer in Texas (USA) to anchialine habitats of the Andaman Islands, Vietnam, the Salomons, Loyalty Islands and Vanuatu. In trying to understand the distribution pattern of these species, we assume that they were all derived from shallow-water marine ancestors and that they are unable to disperse across wide and deep oceanic

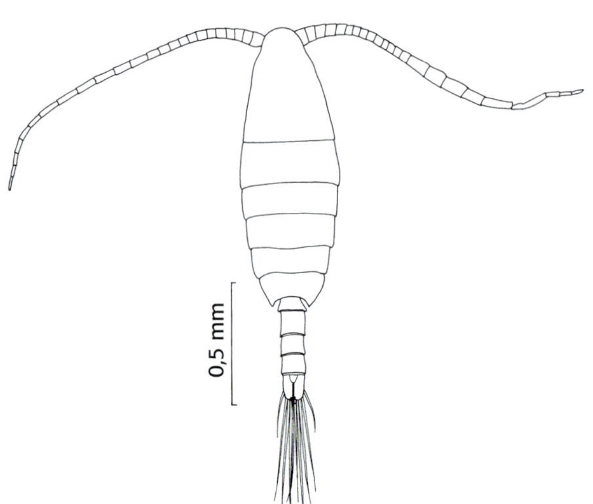
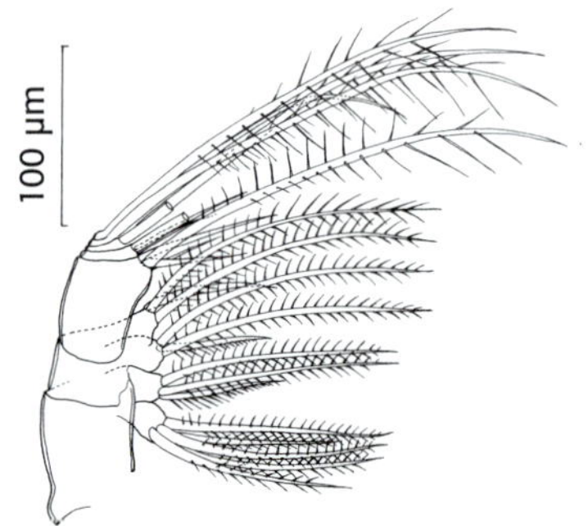

Figure 386: The adult male of a new species of *Centropages* — a calanoid copepod — with inset figure showing one of its specialised mouthparts used for catching the small particles that it feeds on. (Drawings by G. Boxshall).

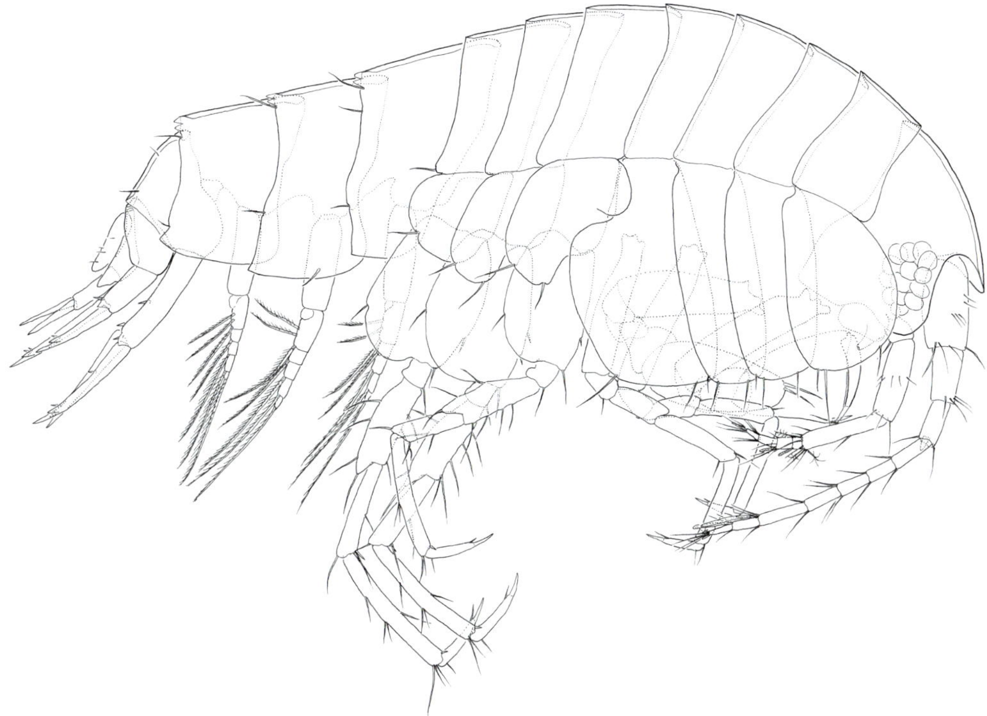

Figure 388: Side view of an adult female of the new species of tiny amphipod shrimp *Seborgia*, collected in the pool near Loren Cave. This female is 1.9 mm long. (Drawing by D. Jaume).

basins or continental subterranean waters. The fact that there are species located around the Gulf of Mexico occupying areas not covered by the sea at least since the Late Cretaceous, should be suggests this genus has an ancient origin. However, the existence of several species in the Indo-West Pacific leads us to infer that the origin of the genus is even older, perhaps Early Cretaceous, prior to the opening of the Atlantic, when the fragmentation of the ancient, shallow-water Tethys Sea commenced. This vanished sea once connected the South West Pacific through to the Caribbean Sea, and separated the former northern and southern supercontinents of Laurasia and Gondwana. Its shores were probably already populated by members of the genus *Seborgia*. It leads us to suggest that modern members of *Seborgia* are probably relicts of this ancient tropical shallow-water fauna.

Living in this same pool were many tanaid shrimps. Tanaids are small shrimps which, like the amphipods are characterised by having a ventral marsupium or brood pouch, within which the females carry their eggs and developing young. Tanaids, despite being so small (2-3 mm), have interesting sex lives with each adult male attempting to gather together a harem of numerous females with which to mate. In the pool the tanaids lived within a thick blackish mat of algae or filamentous bacteria that coated the surface of the submerged stones. Tanaids are rarely found in non-marine waters so this is an unusual and interesting discovery.

Loren Cave and its nearby pool contained a fascinating aquatic fauna including several species new to science. In the dry passages of the cave deep into the cave we also found a terrestrial crab (*Discoplax longipes*) which was eventually caught by hand. Its name refers to its relatively long legs (Fig. 389).

Figure 389: A terrestrial crab *Discoplax longipes* was found in dry passages deep inside Loren Cave.

FOCUS ON BATS

Vincent Prié

Bats are the only native mammals from Vanuatu (marine species excepted). Based on literature and recent observations, twelve species are known from Santo (Table 32). Despite the insularity context, the bat fauna exhibits low endemism levels: bats are successful colonizers and species are generally widespread in vast biogeographic regions. Santo shares most of its bat fauna with neighbouring islands: Papua New Guinea and Solomon Islands, Fiji, New Zealand, New Caledonia and Northern Australia. Nevertheless, Santo hosts two Vanuatu endemic species (Vanuatu's flying fox and Bank's flying fox), and two near-endemic species: the Fijian mastiff bat, restricted to Vanuatu and Fiji Islands and the Long-tailed fruit bat found only in Vanuatu, Fiji and New Caledonia.

These endemic and near-endemic species are more vulnerable than widespread species because their habitat is more reduced and a single event such as a natural or a human induced disaster can affect the whole individuals. The Long-tailed fruit bat and Banks flying fox are categorized as vulnerable in the IUCN Red List. Although more widespread, Polynesian sheath-tailed bat is considered endangered in the IUCN Red List. A striking fact for a naturalist used to temperate countries is the abundance of common species (Fig. 390). Almost any cave visited by the Karst team during the Santo 2006 expedition would host colonies of bats, sometimes huge ones housing several tens of thousand of individuals. To some extent, the abundance of bats in caves would make speleological progress difficult because of collisions with flying bats or dangerously thick guano piles. Bats can be detected at any time and in any place at night in Santo with a ultrasound detector. Although disturbed by human activities and introduced species, Santo's ecosystems still support important bat populations.

Table 32: Bat species' checklist for Vanuatu and IUCN Red List categories (*: Vanuatu endemic species).
LR: low risk. **lc**: least concerned. **VU**: vulnerable. **EN**: endangered.

PTEROPODIDAE	
Vanuatu flying fox *Pteropus anetianus* Gray, 1870*	LR/lc
Banks flying fox *Pteropus fundatus* Felten & Kock, 1972 *	VU
Insular flying fox *Pteropus tonganus* Quoy & Gaimard, 1830	LR/lc
Long-tailed fruit bat *Notopteris macdonaldi* Gray, 1859	VU
HIPPOSIDERIDAE	
Trident horseshoe bat *Aselliscus tricuspidatus* (Temminck, 1835)	LR/lc
Fawn horseshoe bat *Hipposideros cervinus* (Goul, 1854)	LR/lc
VESPERTILIONIDAE	
Large-footed bat *Myotis adversus* (Horsfeld, 1824)	LR/lc
Little bent-wing bat *Miniopterus australis* Tomes, 1858	LR/lc
Small melanesian bent-wing bat *Miniopterus macrocneme* Revilliod , 1913	LR/lc
Nyctophilus sp.	
MOLOSSIDAE	
Fijian mastiff bat *Chaerephon bregullae* (Felten, 1964)	LR/nt
EMBALLONURIDAE	
Polynesian sheath-tailed bat *Emballonura semicaudata* (Peale, 1848)	EN

••• Bats in caves

Bats preferably roost in dry caves. This could be due to flood risks in river-flowing caves. Almost every dry cave in Santo hosts bat colonies (Table 33). Colony sizes range from a few hundred to about 50 000 individuals for the largest ones (Fig. 390).

The importance and extent of bat colonies in Santo give them a key role in subterranean ecosystems' functionality. While aquatic subterranean ecosystems rely mainly on water's infiltration to bring in organic matter, dry caves are very poor in nutrients unless colonies of bats roost there. In such caves, the only organic income would come from guano and dead corpses. Indeed, bats guano can form huge amounts of several meters deep in some caves (Autabelchiki in Aore Island for instance). This huge amount of organic matter supports diversified and highly specialised life forms: springtails, millipedes and insects are often found feeding on guano or bats corpses.

••• Methods

Several methods were used for the inventory of cave dwelling bats in Santo: direct observation

Table 33: Record of bats and birds in caves explored during the Santo 2006 expedition.
(X): less than 100. **XX**: >10 000. **XXX**: >50 000. **?**: may be present but hasn't been formally identified.

Site name	Bat colonies	*H. cervinus*	*A. tricuspidatus*	*M. a. australis*	*M. macrocneme*	*C. bregulae*	Swiftlets
Sarabo 1 (Luganville)	XXX		XXX	XXX			X
Sarabo 2 (Luganville)	X	X		(X)	X		
Autabelchiki (Aore)	XXX		XXX	XXX	X		(X)
Amarur	X	(X)	(X)	X			(X)
Fioha	X		(X)	X	(X)		(X)
Rotal	XXX	XX	XX	?	X		
Fapon (Butmas)	X			X			X
Vobananadi (Malo)	XXX			XXX			
Riorua (Funafus)	X			(X)		X	X
Kafae (Funafus)	X					X	X
Millenium Cave	X					X	X
Cape Cumberland Cave	X	X					
Silova (Akari)	XX		X	X			
"Puits sans Nom" (Butmas)	X			X			X
Amarirua (Malo)	XX		X	X		X	
Lemeloc 3 (Port Orly)	XXX		XX	XX			
Yekavon	X						
Yet Veun	X			X			X
Luvuth Yet (Port Olry)	X	X					
Raïa (Aore)	XX	X		X	X		
Gouffre Lavav (Loran)	X			X			
Sanuarav							X
Tarlisengo (Malo)	X						
Mba		X		X			X
Tari Boi	X						
Lori (Kole)	X			X			
Ukupo (Tasiriki)							X
Santo Cave (Butmas)	XXX						
Patunar (Funafus)	X						X

Figure 390: A huge colony of Little bent-wing bat in the island of Aore.

Figure 391: Ni-Vanuatu father and son hunting bats with catapults.

(mostly in caves), capture and biometry, echolocation recording and genetic studies. All means have been put together in.

Direct observation was used to inventory bat colonies in caves. It is difficult to estimate the number of individuals in a colony: most colonies reached several thousands individuals, many of them flying during the time of observation. As a result, the numbers given here are only rough estimates. Biometry datas include forearm and tibia lengths, taken in the field. For each species, pictures were taken with a digital camera. These include head side and wing, in order to measure the length of each finger bone with accurate morphometric software. Echolocation calls were recorded as 10 time-expanded signals with a Pettersson D240x ultrasound detector and a MiniDisc recorder. Time expansion records allows to play the recorded sound 10 time slower, thus 10 times deeper and audible to human ears. Signals were subsequently re-digitalized with the software BatSound 3.31 at a sampling frequency of 44.1 kHz (16 bits per sample).

••• Species accounts

••• Flying foxes (PTEROPODIDAE)

Flying foxes are large fruit-eating species that roost in trees and lack echolocation calls. They are hunted for meat by natives (Fig. 391) and are even sometimes tamed and live around houses (Fig. 392). Among the four species found in Santo, only the Long-tailed flying fox roosts in caves.

The Insular Flying fox *Pteropus tonganus* is widespread in the Pacific region, ranging from New Guinea to the Cook Islands. Three subspecies are currently recognised : *P. t. basilicus* from Karkar and Koil Islands off Papua New Guinea, *P. t. geddiei* from the Solomon Islands, Vanuatu and New Caledonia

Figure 392: A tame Insular flying fox in Tutuba Island.

and *P. t. tonganus* from Fiji, Tonga, Kiribati, Tuvalu, American Samoa, Western Samoa, Niue and Cook Islands. The subspecies present in Santo is *P. t. geddiei*. It was recorded from sea level up to 1 000 m and was found in a variety of habitat types including lowland, intermediate and highland forest as well as mangroves and agricultural areas.

Little is known about Vanuatu's endemic Banks flying fox *Pteropus fundatus* and Vanuatu's flying fox *Pteropus anetianus*. The latter has been recorded from Vatthe Conservation Area and collected from the West coast, Wusi village (Australian Museum).

It has been observed near Kerepoa (Tabwemasana) during the Santo 2006 expedition. Among the six recognised subspecies, one was described from Aore Island (*P. a. aorensis* Lawrence, 1945).

The Long-tailed flying fox *Notopteris macdonaldii* is a distinctive species that occurs in Fiji and Vanuatu. Whereas it roosts in cave in large numbers, it has not been found in the caves investigated during Santo 2006. It is easily hunted in roosting places and is believed to be threatened in most parts of its range.

••• Cave dwellers and other small species

HIPPOSIDERIDAE

These bats use peculiar echolocation calls: call is emitted at a very high frequency from the nose, which generally has a very weird shape (Fig. 393). The signal consists of an high-pitched regular frequency, with an abrupt drop at the end. The fundamental is filtered out while the harmonics are accentuated, allowing bats to emit high frequency calls with a reasonable effort. This type of calls is very directional and short ranged but gives a more precise picture of the environment. This is consistent with Hipposideridae's hunting behaviour: they generally hunt insects in foliage or close to the ground.

Figure 393: Trident horseshoe bat with its typical nose shaped for the ultrasound emission.

Two species occur in Santo: The Trident horseshoe bat and the Fawn horseshoe bat. Although they can be found in the same caves, they never roost in the same rooms. The Trident horseshoe bat is often seen together with the Small bent-winged bat, while the Fawn horseshoe bat is found alone, or with the Small melanesian bent-winged bat.

Trident horseshoe bat
Asellicus tricuspidatus

As its name suggest, this species exhibit a characteristic nose with three cuspids. "*A-sellicus*" means "without a saddle", a reference to the typical "saddle" that characterises Hipposideridae's nose. The hair is ginger. It is a small species, with forearm about 42 mm (41.3-45) and tibia 16.0 mm (15.0-17.0). It is known from the Pacific Islands only, from Vanuatu to New Guinea.

Trident horseshoe bat is found in most caves in Santo, with colonies of several thousand individuals. Females are pregnant in October, give birth in December and are lactating from March to May. In the Santo 2006 expedition, individuals were paired in Aore in late September (Fig. 394), when reproduction probably takes place.

Figure 394: Trident horseshoe bats looked paired in Aore in late September. This could be premises of mating period.

Fawn horseshoe bat
Hipposideros cervinus

Less common than the previous one, but has been found regularly in caves. It is larger than Trident horseshoe bat: forearm 48.5 mm (47.5-49.3), tibia 21 mm (20-22) and easily distinguished by its simple nose leaf (Fig. 395). The species occurs from northern Australia to the Philippines, New Guinea, Borneo, Sumatra and Sulawesi. Colonies of hundreds of individuals have been observed in Santo, together with Small Melanesian bent-wing bat. Juveniles are found from January to early March.

Figure 395: The simple nose leaf of the Fawn horseshoe bat allows emission of very high-pitched calls (up to about 150 kHz).

VESPERTILLIONIDAE

Large footed mouse-eared bat
Myotis adversus

The only mouse-eared bat known from Santo, it is widespread, ranging from Northern and Eastern Australia to south-east Asia (Malaysia and possibly Vietnam) and from Malaysia to New Guinea and Vanuatu. It has not been encountered in caves during the Santo 2006 expedition but records in literature mention it from Aore Island. Some sound samples recorded during the exprdition could belong to this species.

Large footed mouse-eared has a brown-grey fur above with a paler colour below. It has conspicuous large feet (more than 8 mm long), used to forage over water and catch preys such as small fish. Wingspan is about 28 cm. It is known to roost in various sites ranging from caves to pots and other utensils in houses, generally in groups of 10-15, close to water. It is difficult to find in caves as it retreats into deep cracks and crevices when disturbed. It roosts with Bent-winged bats and Trident horseshoe bats. Lactating females were recorded from June to November in Papua New Guinea.

Little bent-wing bat
Miniopterus australis australis

This species (Fig. 396) is found in Australia, Indonesia, Philippines, and Vanuatu. It is by far the most common species of cave-dwelling bat encountered in Santo with colonies of more than 50 000 individuals. It is distinguishable from the Small Melanesian bent-wing bat by its small size, the tibia being less than 18 mm long. In September, individuals were not sexually active. Pregnant females have been caught in November.

Small Melanesian bent-wing bat
Miniopterus macrocneme

It is quite similar to the previous species but is distinguished by its larger size, with tibia over 18 mm long (Table 34). It occurs in Indonesia, Papua New Guinea, Philippines, Solomon Islands, and Vanuatu. Colonies are smaller than those of *M. australis* and less frequent.

Figure 396: Little bent-wing bat resting during the day.

Table 34: Biometry of Little bent-wing bat and Small Melanesian bent-wing bat.

Species	Forearm (mm)				Tibia (mm)			
	Mean	Min	Max	N =	Mean	Min	Max	N =
M. a. australis	37.5	36.3	40	128	15.6	14.5	17	85
M. macrocneme	42.5	41.3	43.5	22	19.4	18.6	21	22

Nyctophilus
Nyctophilus sp.

A specimen probably belonging to the Greater Nyctophilus *Nyctophilus timoriensis* (Geoffroy, 1806), a widespread species found from Australia to Papua New Guinea, has been caught during the Santo 2006 expedition near the Tabwemasana summit. This is the first record of the genus *Nyctophilus* for Santo.

MOLOSSIDAE

Fijian mastiff bat
Chaerephon bregullae

This molossid was formerly classified as *Chaerephon jobensis* and thought to occur from the Solomon Islands to Australia. *C. jobensis* has recently been split into three separate species, based on bio-chemical analysis: *C. jobensis* (Australia), *C. solomonis* (Solomon Islands) and *C. bregullae* (Vanuatu and Fidji Islands), based on molecular analyses. However, individuals measured in Santo during the 2006 expedition are much smaller than the biometry given in literature for individuals from Fiji Islands (Table 35).

The Fijian mastiff bat is the only molossid bat found in Santo. It is characteristic with its strong muzzle and ears pointing forward. It is a high flying species and the wings are long and narrow (Fig. 397).

Very few roost sites are known. In Santo, the species is known only from Millenium Cave, Riorua Cave, Kafae and Amarirua Caves in Malo Island. Sexually active males have been observed in September in Riorua Cave. Youngs are probably born around December, females are lactating up to April when the young are about to be weaned. *C. bregullae* individuals bear hundreds of acarid parasites that fall like rain from the colony when disturbed. As in Fiji, natives use to collect hundreds of specimen for meat. In Riorua, custom harvesting is done with a fire lighted under the colony. Bats seem to collapse from the fire smoke and are caught by hand. People from Funafu estimate that *c.* 300 individuals are eaten each year.

Figure 397: The Fijian mastiff bat with its rounded ears.

EMBALLONURIDAE

Polynesian sheath-tailed bat
Emballonura semicaudata

It has a wide but disjunctive distribution that extends from Palau and the Northern Mariana Islands in the western Pacific to Vanuatu, Fiji, Tonga and Samoa in the east. It is clearly linked to forest habitat, whether natural or degraded and forages in forest understorey from ground level to the canopy.

Although preferably a cave-dweller, no colony has been observed in Santo during the 2006 expedition. Despite the relatively preserved forest habitat of the island, this species is believed to have been extirpated from Santo in 2008, according to IUCN. Polynesian sheath-tailed bat is thought to have declined drastically in all its range during the last 60 years, probably due to habitat degradation by cattle.

Table 35: Compared biometry of Fijian mastiff bats from Fiji Islands and from Santo.

	Forearm (mm)				Tibia (mm)			
	Mean	Min	Max	N =	Mean	Min	Max	N =
C. bregulae, Males, Fiji (Flannery 1995)	52.9	51.3	53.6	5	21.3	20	22.3	5
C. bregulae, Females, Fiji (Flannery 1995)	52.2	51.4	52.8	5	20.8	20	21.4	5
C. bregulae, Males, Santo 2006	48.6	48.5	48.7	3	17.3	17	17.8	3
C. bregulae, Females, Santo 2006	47.1	45.9	48	3	16.6	16.3	16.9	3

••• Echolocation calls

The following standard measurements have been done: Maximum energy frequency (**ME**), Terminal frequency (**TF**), Call duration (**CD**), Band width (**BW**) and time between calls (**TBC**).

••• Constant frequency calls

Trident horseshoe bat
Aselliscus tricuspidatus

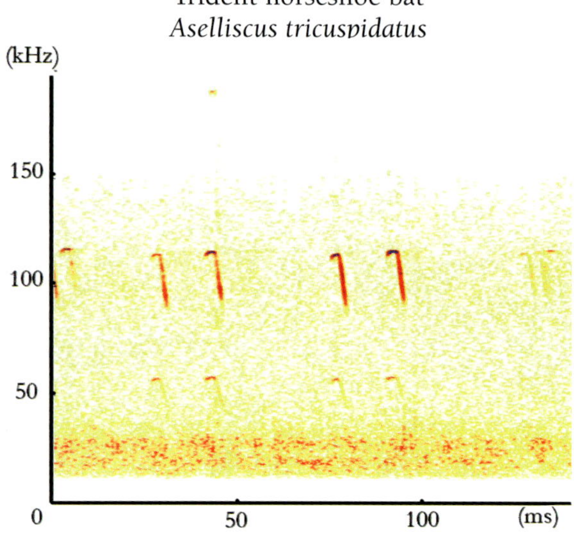

Call pattern typical of bats emitting with the nose: constant frequency with a "comma"; an abrupt drop of frequency at end of call. Fundamental is masked and only the harmonic is perceptible. The fundamental can often be distinguished on sonograms at about 60 kHz, as well as the second harmonic at about 190 kHz.
ME: 117 kHz. When flying in a confined space, Trident horseshoe bat can high the pitch up to 122 kHz.
TF: 96 kHz.
CD: 5.3 (3.8-7) ms.
TBC: 13-60 ms. Calls are often paired with about 15-20 ms between two calls and from 30-60 ms between pairs.

Fawn horseshoe bat
Hipposideros cervinus

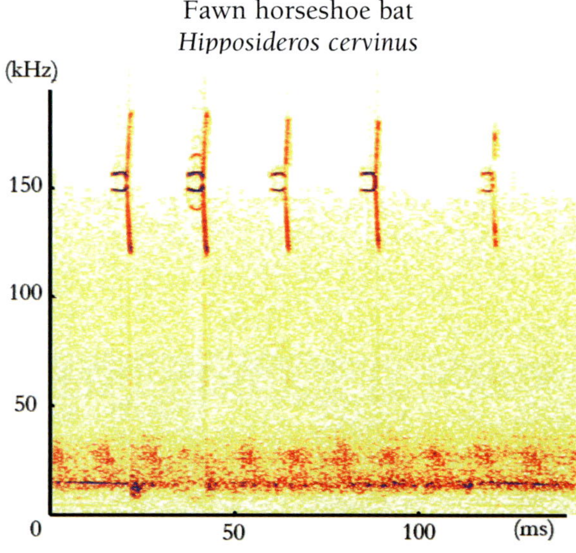

The call is very high pitched, with a fundamental at about 78 kHz and a first harmonic at about 155 kHz.

The main call is the result of two regular frequencies, one at 150 and the other at 157 kHz. Sometimes, only the fundamental is heard as a constant frequency at 78 kHz with terminal frequency at 75 kHz.
ME: 156 kHz.
TF: 121 kHz.
CD: 5.3 (4.4-6) ms.
TBC: 14 ms.

••• Modulated to quasi-constant frequency calls

Little bent winged bat
Miniopterus australis

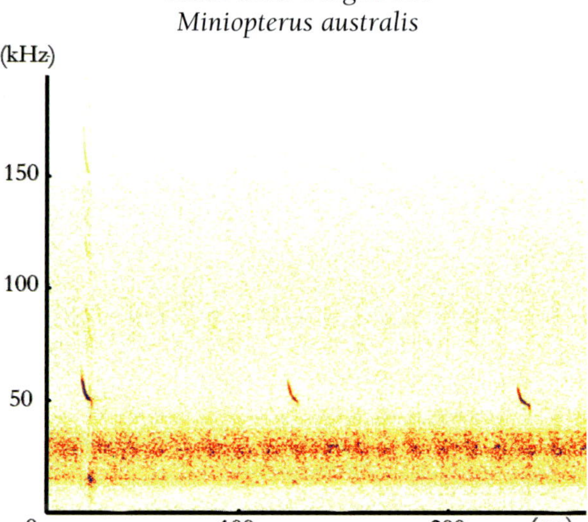

Quasi constant frequency calls typical from the genus with a maximum energy at about 55 kHz.
ME: 52 kHz.
CD: 6 (4-8) ms.
BW: 40 (25-55) kHz.
TBC: 83 (50-130) ms.

Small Melanesian bent winged bat
Miniopterus macrocneme

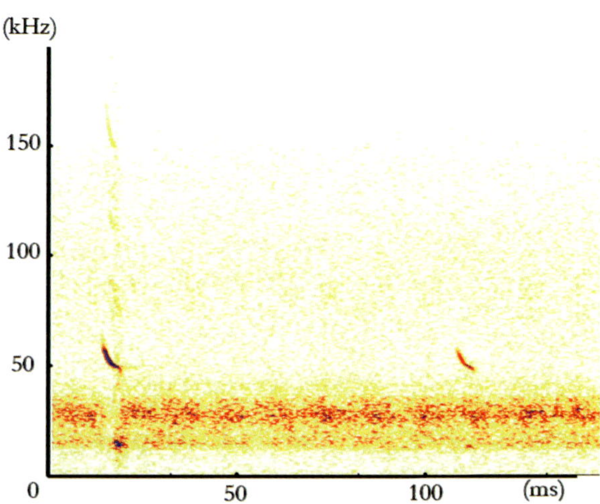

Typical quasi constant frequency, lower than the previous one, as size would suggest.
ME: 36 kHz.
CD: 10 (5-13) ms.
BW: 25 kHz.
TBC: 150 (50-250) ms.

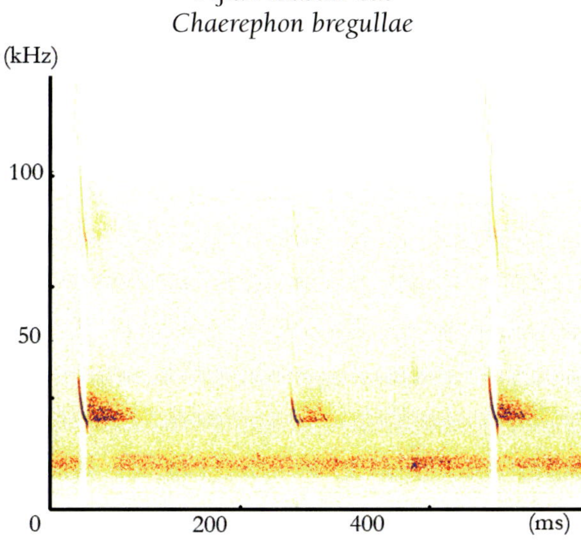

Fijian mastiff bat
Chaerephon bregullae

Only a few sequences have been recorded from caught animals. The call is a quasi-constant frequency, much deeper and more spaced than that of any other bat from Santo as size would suggest.
ME : 14-17 kHz.

Emballonura sp. ?

Some unidentified calls have been recorded in the wild. Their characteristics cope with what is known about the genus *Emballonura*: Emballonurid bats use the second harmonic as main call, the first harmonic being masked. The ME could correspond to the Polynesian sheath-tailed bat, but it is a modulated to quasi-constant frequency, low and spaced.
ME: 28 kHz
CD: 12 (8-15) kHz
BW: 14 (9-21) kHz
TBC: 270 ms

••• Frequency-modulated calls

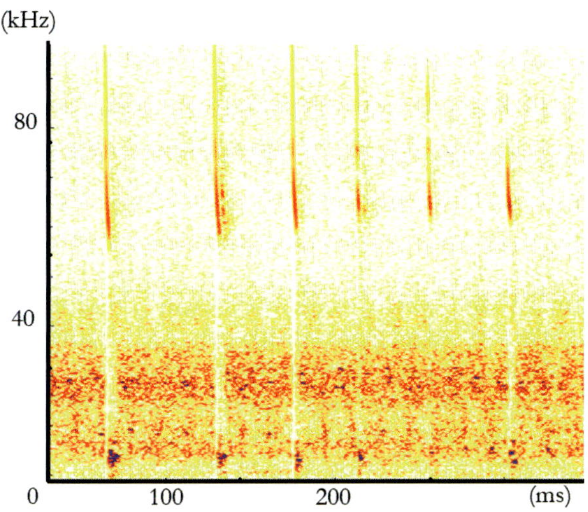

Myotis/Nyctophilus sp.

Abrupt frequency modulated calls, with a somewhat "explosive" beginning and a high terminal frequency, were recorded in forest habitat. This type of call is characteristic of *Myotis* species. They were tentatively attributed to the Large footed mouse-eared bat *Myotis adversus*.
Type 1: **TF**: 38 kHz (37,9-38,9)
       CD: 4,5 ms (3-6)
       BW: 13 kHz (12-14)
       TBC: 136 ms (117-179)
Type 2: Similar to type 1,
       but **TF** higher, up to 42 kHz (N = 15).

FISH AND SHRIMPS OF SANTO KARSTIC SYSTEMS

Marc Pouilly & Philippe Keith

The eastern part of Espiritu Santo Island is constituted by a coral karst, which nowadays presents an important subterranean system with freshwater habitats such as sinks, lakes and rivers. These specific habitats increase the local diversity of ecological conditions and may be colonized by species with ecological and behaviour predispositions to living in lightless environment. Espiritu Santo is around 22 millions of years old, making it one of the oldest islands of the Vanuatu Archipelago, old enough to have led to local speciation (appearance of new species through a specialisation from an ancestor) adapted to subterranean conditions.

Crustacean decapods (crabs and shrimps) are regularly found in subterranean freshwater or brackish habitats. Due to their ecological predispositions, they can survive in such habitat with or without genetic adaptation. More than 90 subterranean fish species are now described from around the world. Directly related to the recent increase in karst system exploration, this number has risen continuously from 1970. The majority of known species comes from the Asian (especially China) and South American continents. However, some species have been described from different regions of the Indo-pacific: Madagascar (three species), Australia (two species), Papouasia, Malaysia, Sulawesi and Galapagos (one species in each area). Two of the principal fish families that colonized freshwater habitats of Vanuatu Islands show species adapted to subterranean conditions elsewhere: Gobiidae (*Glossogobius ankaranensis* from Madagascar) and Eleotridae (*Typhleotris pauliani, T. madagascariensis, Oxyeleotris caeca*). Many species of the family Anguillidae (eels), which is also common in the Vanuatu Islands, are able to survive in subterranean habitats without any adaptation. It is therefore possible that local species of Espiritu Santo Island may have colonized subterranean rivers, sinks or lakes and developed adaptations to maintain such populations. However, two important factors may limit this colonization and, moreover, prevent adaptation of the species to subterranean conditions. Firstly, the local species have a predominantly herbivorous diet. In order to survive in subterranean habitats, such species would have to shift their diet toward resources that are not light-dependant, such as detritus or small prey. Second, most of these species present a priori a diadromous migration cycle (they spend part of their biological cycle in freshwater habitats and the rest in marine habitats). For example, eels are generally catadromous. They carry out their reproduction in the ocean, the larvae migrate to rivers and, once adult, the eels return to the ocean for reproduction. On the contrary, Gobiidae and Eleotridae are generally amphidromous. In this case, the species spawn in freshwater and the free embryos drift downstream to the sea, where they undergo a planktonic phase before returning to the rivers to grow and reproduce. A complete shift to subterranean live would required a total change of the reproduction strategy and biological cycle of these species, with a shift from a migration to a sedentary cycle, because subterranean life requires profound morphological and physiological adaptations (regression of ocular system, pigmentation loss, etc.) that are incompatible with an oceanic existence.

••• Study sites, material and method

There are only a few records of freshwater fish and decapods crustaceans (crab and shrimp) available for Vanuatu Island. Sampling carried out since 2000 by the National Museum of Natural History, Paris (MNHN) and the Santo 2006 expedition are under evaluation and should rapidly improve this situation. Currently, 29 species are known from Vanuatu and 26 from Santo.

For fish, Keith and collaborators have inventoried 67 species on the archipelago, the majority of which are amphidromous. Most of these species are restricted to the lower parts of the river (altitude below 50 m) and in the estuarine area.

Espiritu Santo Island is the largest of the 80 islands of the Vanuatu republic archipelago. The western part of the island is volcanic and the eastern part is covered by calcareous coral that culminate at an altitude of *c.* 600 m. In this landscape rivers are short and steep. Only the bigger ones (Sarakata, Jourdain) present a lower part with a small alluvial floodplain.

During the Santo 2006 expedition we sampled different karst freshwater aquatic habitats (rivers, sinks and blue holes) to identify crab, shrimp and fish species that colonize these habitats, and to look for evidence of subterranean adaptations. We carried out a comparative sampling of the aquatic fauna (fish and decapods) of superficial (EXT) and subterranean (INT) freshwater aquatic habitats belonging to four systems:
- Fapon system (Butmas region): river upstream of Fapon Cave (EXT) and in the Fapon doline (INT);
- Amarur system (Nambel region): Amarur River in the cave (INT), at the resurgence (EXT) and downstream near Nambel village (EXT);
- Patunar system (Funafus region): river in Kafae Cave (INT), upstream (EXT) and downstream (INT) of the Patunar doline, resurgence (EXT) and downstream reaches at altitudes of 90 m and 15 m;
- Fioha system (Funafus region): Fioha River in the cave (INT) and at the resurgence (EXT).

Figure 398: Electrofisfing in the subterranean river of the Amarur Cave in Santo, during the Santo 2006 expedition.

Most of the sampling sites were situated at altitude above 50 m and correspond to small streams, less than 10 m wide and 1 m deep, with gravel or rock substrate. All present an alkaline pH (8.0 to 8.2) in the upper reaches and neutral pH (7.0 to 7.4) in the lower reaches. Conductivity increased with the downstream gradient from (130 µS.cm⁻¹) to salted water. Most of the sampling sites had a conductivity between 200 and 350 µS.cm⁻¹. Other occasional observations were made in sink parts of the same systems by divers of the Santo 2006 karst team and in four blue holes (Sarabo, CIRAD, Nanda and Porpor spring) situated near the ocean.

Due to these characteristics and difficulties of access, electrofishing is likely to be an efficient sampling method (Fig. 398). It was the most used and was sometimes, when possible, complemented by dipnets and traps.

Results

The karstic rivers of Espiritu Santo Island are colonized by 20 fish species, belonging to 10 different families (Table 36). The most diverse family (that showing the highest number of species) is Gobiidae (Fig. 399), represented by nine genera and 14 species. Eleotridae is the second most diverse family, with four genera and six species. The family Poecilidae is only represented by *Gambusia affinis*, which is a non-native species introduced for mosquito control.

Most of the native species colonize only the lower parts of the rivers (< 50 m), or only the estuary and river mouth: Ambassidae, Kraemeriidae, Kuhliidae, Ophichthidae and Syngnathidae. Only three families (Anguillidae, Eleotridae and Gobiidae) colonize the rivers at higher altitude and include species typical of these rivers. Species richness decrease drastically with altitude, falling from 30 species in river reaches below 50 m in altitude to eight species in upper reaches.

Decapods are represented by two shrimp families, Palaemonidae (genus *Macrobrachium*) and Atyidae, and one crab family, Grapsidae. *Macrobrachium* shrimps are widely distributed in the Pacific region and eight species are present on Santo Island. As for fish, species richness decreased with altitude and there are only two or three species that colonize the rivers above an altitude of 50 m.

Globally, at higher altitudes (> 50 m) rivers present a high abundance of shrimps and only a few fishes (< 10 individuals/100 m²). This tendency is reversed in rivers at low altitudes and in habitats near the estuaries (blue holes and river mouths).

No fish or decapods species captured or observed showed clear evidence of strong adaptation to subterranean conditions (depigmentation, ocular system regression). Shrimps, especially of the family Atyidae, sometimes showed pigmentation differences between individuals captured in superficial and in subterranean rivers. However theses differences fade with the conservation of the specimens, so that it is likely that they correspond more to a response to light stimulus than to a genetic difference. Only few individuals of Atyidae captured

Table 36: List of freshwater fish, shrimp and crab genera captured in karst systems of Santo during the Santo 2006 expedition.

Family	Genus	Species	> 50m	Subterranean
Fish				
Ambassidae	*Ambassis*	1	0	0
Anguillidae	*Anguilla*	2	1	1
Eleotridae	*Butis*	1	0	0
	Eleotris	1	0	0
	Hypseleotris	2	0	0
	Ophieleotris	2	0	0
Gobiidae	*Awaous*	1	0	0
	Lentipes	1	1	0
	Psammogobius	1	0	0
	Redigobius	2	0	0
	Schismatogobius	1	0	0
	Sicyopterus	2	1	0
	Sicyopus	1	1	0
	Smilosicyopus	1	1	
	Stenogobius	1	0	0
	Stiphodon	3	2	0
Kraemeriidae	*Gobitrichinotus*	1	0	0
Kuhliidae	*Kuhlia*	3	0	0
Ophichthidae	*Lamnostoma*	2	0	0
Syngnathidae	*Microphis*	3	0	0
Shrimp				
Palaemonidae	*Macrobrachium*	8	7	3
Atyidae	*Caridina*	3	3	2
	Atyoida	1	1	1
Crab				
Grapsidae	*Utica*	1	1	1

Figure 399: *Sicyopterus lagocephalus*. (Gobiidae). A common fish species in Santo freshwater rivers (male is coloured and female drab).

presented a strong depigmentation accompanied by a slight microphthalmy (reduction of the size of external ocular system) (Fig. 400).

All the aquatic subterranean habitats (sinks, rivers and small lakes) are colonized by decapods (crabs and shrimps). The same species colonize both superficial and subterranean rivers (five of the eight *Macrobrachium* species are present in both conditions). Habitats with low velocities (small lakes, sinks) appeared to be more favourable for Paleomonidae shrimps, although lotic habitats are more suitable for Atyidae shrimps. Some family density variations exist between the different systems. For example, the Amarur system has a high density of Paleomonidae, whereas the Fioha system has high density of Atyidae. These variations remain unexplained and require a more detail study to understand the environmental factors that determine the quality of an habitat for each type of organisms.

Figure 400: Specimens of *Atyoida pilipes* (decapods shrimps). The first (**A**) was captured in a superficial river reach and the second (**B**) from a subterranean reach. The pigmentation difference is obvious, but in almost all the case it did not appear to correspond to an adaptation to subterranean life.

Figure 401: Eels (here *A. marmorata*) are the only fish that colonize the subterranean rivers of Santo. Due to their nocturnal behaviour and carnivorous diet, they can colonize indifferently both superficial and subterranean rivers without any kind of adaptation. However, their biological cycle requires a migration to the ocean for reproduction, thus limiting the probability of an adaptation to a strictly subterranean way of live. The eel is one of the few freshwater fish appreciated by local people.

Fish are absent in the subterranean habitats, with exception of eels (*Anguilla marmorata* and *A. obscura*) (Fig. 401) that were observed in almost all the explored sinks and in many sampled rivers at different altitudes.

The presence of eels can be explained by their carnivorous diet and nocturnal activity. This is in contrast to most of the other species of other groups, which are limited in their colonization of subterranean habitats by their herbivorous diets, as well as by local physical conditions, such as the openness and connectivity of the subterranean systems (river or sink vs groundwater), altitude, high inclinations and waterfalls that characterize the upper river reaches. However, even if these fish species have not colonized the subterranean habitats, it should be noted that most of them use it at least to pass through and attain the river upper reaches. Indeed, because most of these species are amphidromous, they sometimes need to pass through long subterranean sections (e.g. the Patunar system, which is more than 1 km long) to get to the upper reaches.

FOCUS ON SPRINGTAILS

Louis Deharveng & Anne Bedos

Formerly an order of Insecta, Collembola (springtails) are now considered to be a class of their own, representing the sister group of either Crustacea or Insecta. These tiny animals (0.3 to 8 mm long) are among the most abundant Arthropods in soils, where they regulate fungal growth and contribute to organic matter breakdown. Together with mites, they dominate the animal communities of soils and related habitats, such as decaying wood or mosses on rock. They also live in all strata of the vegetation up to tree canopy, where they are among the most numerous arthropods in some areas (e.g. New Caledonia).

In spite of their importance in terrestrial habitats, Collembola are only moderately diverse —about 7 000 described species in the world. However, as one of the lesser-known groups of terrestrial arthropods, their relative contribution to global biodiversity is rapidly increasing. From what is observed in the tropics (where at least 60 % of the species are new to science in almost any sample of an unstudied region), and the overall high levels of endemicity in the group, it can be safely assumed that at least 50 000 species of Collembola exist worldwide.

• • • History of research on Vanuatu Collembola

The first work on the springtails of Vanuatu was published by Womersley in 1928. Three species were briefly described in that paper, including one from water lying in an opened coconut, from which the first copepod of Santo was also described. Remarkably, none of these Collembola was retrieved among the biological material sorted so far from the soils and caves collections of the Santo 2006 expedition, and we assume they live on the vegetation, a habitat that we did not sample. Sixteen additional species were recorded, by Womersley (1937, one species), Da Gama (1976, two species) and Yoshii (1995, 13 species including three newly described). A total of 19 species of Collembola, mostly from the karstic part of Santo, were therefore known in the island before the Santo 2006 expedition.

Three species had also been recorded from islands of Vanuatu other than Santo: the Neanurinae *Penelopella pacifica* from Efate (Cassagnau in 1987), retrieved in our samples; a Neanurinae (*Achorutes rosaceus*) of uncertain placement from Malekula Island, cited by Womersley in 1937; and *Xenylla cavernarum salomonensis*, cited by Da Gama in 1976 from both Santo and Malekula.

The first results of the Santo 2006 expedition regarding Collembola were published in 2009, bringing to 36 the total number of named species for the island.

• • • Springtails sampled during the Santo 2006 expedition

Springtails were collected independently in three areas of Santo during the expedition: in the Penaoru

Figure 402: A partial view of the arthropods collected in a single litter sample from the Santo karst, illustrating the dominance of Collembola among soil arthropods.

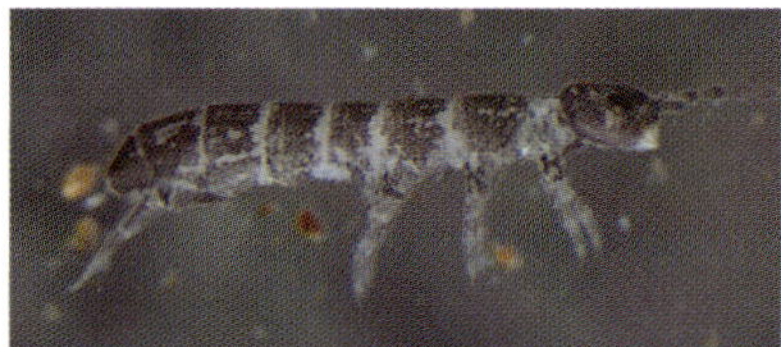

Figure 403: Some pantropical species of Collembola collected in the Santo karst. From left to right: *Hemisotoma thermophila* (Isotomidae); *Isotomiella symetrimucronata* (Isotomidae); *Folsomides parvulus* (Isotomidae); *Megalothorax minimus* (Neelidae). (Photos L. Deharveng).

area, on the coast and in the karst. The Penaoru Collembola, obtained from Malaise traps and litter samples, became available for study too recently to be included. The littoral Collembola are interstitial species, mostly collected from sand beaches of the karstic part of Santo. They were listed by Thibaud in 2009, and include 18 species, of which 16 were identified to species level.

To give an idea of how poorly-known the springtail fauna of Santo was before the expedition, the number of Collembola collected by Thibaud in the sand beach habitat alone is as high as the number of Collembola previously known from the whole island. Our own collections from various habitats in forest and caves included four times more species than previously known on Santo. A single sample of 500 cm³ of litter from Santo karst forest hosted on average 20 to 30 springtail species, i.e. more than the total number of species known from the island in 2006.

Thus, the increase of our knowledge brought by the Santo 2006 expedition is very significant, making Santo the best known island of the Pacific after New Caledonia and Hawaii for its Collembolan fauna.

In the karst alone, 86 species were collected, of which 11 are pantropical and at least 35 are new to science. The remaining 40 species correspond to species insufficiently described in the literature, or belonging to genera with a confused taxonomy. More detailed statistics on the diversity of Collembola in the Santo karst are as follows:
- Total number of genera on Santo: 70
- Total number of species on Santo: 110
- Number of species known before Santo 2006: 19 (18 named)
- Number of littoral species collected during Santo 2006 (Thibaud 2009): 18 (16 named)
- Number of species collected by the karst team during Santo 2006: 93
- Total number of taxa identified at species or species group level: 49
- Number of new species from the karst samples (minimal estimate): 35

••• Santo compared to other Western Pacific Islands

No other region of similar area in the Pacific, except New Caledonia, has benefited from a sampling effort as intensive as the one we applied to Santo Island.

The remote Polynesian archipelagoes are very poor in terrestrial arthropods. Carpenter in 1935 and Yosii in 1967 mentioned 10 species in the "Iles de la Société" and 17 species in the Marquises. In the minute atoll of Fangataufa, in Tuamotou, 16 species were recorded by Thibaud & Najt in 1993, of which at least six are pantropical parthenogenetic forms. Tahitian soils host a very poor fauna (three to five times less species than the Santo average for equivalent volumes of soil).

Conversely, a few samples that we recently examined from the Torres Islands hosted a fauna as rich as, and very similar to, that of Santo. New Caledonia, which is now well known in terms of its collembolan fauna, is also very rich.

A general pattern of collembolan diversity distribution in the Western Pacific is still lacking, but the limited evidence available is in agreement with the pattern of species richness described for coral reefs, now well documented across the Pacific, showing a rapid decrease in diversity from the Indonesian Islands towards the isolated archipelagoes to the East.

••• Widespread and common species

The Collembola of Santo include a number of taxonomically well known species that are widespread in the tropics or around the world. These often dominate collembolan communities in the most disturbed areas, suggesting the possibility of human introduction. They include representatives of Isotomidae (one *Folsomia*, two *Folsomides*, one *Hemisotoma*, three *Isotomiella*, one *Folsomina*), Tullbergiidae (*Mesaphorura yosii*) and Neelidae (*Megalothorax minimus*). With the possible exception of *Hemisotoma thermophila*, all these species are parthenogenetic and small (usually less than 1 mm), allowing much higher colonization success; most are also unpigmented and blind or microphthalmic, traits without an obvious link to their ability to colonize.

Their density in litter may be stupendously high, amounting locally to three hundred million individuals per hectare (*Hemisotoma thermophila*, extrapolated from available samples).

••• Endemics and new taxa

More interesting for biodiversity and historical biogeography are taxa with a narrow distribution area, i.e. ranging from the whole Pacific region to a single site in the Santo karst. All species of this category are bisexual. Like pantropical forms, some species assumed to be endemics reach very high densities (*Coecobrya aokii* on guano and an undescribed species of *Subisotoma* in litter of lowland bush), but most are patchily distributed. Three genera and at least 35 species — most of them new to science— belong to this category. Two of them (*Friesea santo* and *F. septem*) have already been described from the Santo 2006 collections.

The Neanurinae genus *Penelopella*, with its unique species *Penelopella pacifica*, previously known only from Ffate, was present in our Santo collections. This "big" species (up to 2.5 mm) lives in forest litter, always in small populations. Nine other representatives of Neanurinae are present in Santo, two of which, already described, have a wider distribution in the Pacific region (*Australonura rennellensis* from the Solomon Islands, *Paleonura louisi* from the Fangataufa Atoll), and seven are new to science —a very good score at the world level, given the size and isolation of the island. Two genera of Poduromorpha, both probably new to science, are present in our samples. One is a deep-soil Tullbergiidae found at only three sites. The second is a minute Pseudachorutinae, also from deep soil, with reduced eyes, no furca and no pigment, morphologically very similar (by convergent regressive evolution?) to *Paranurida* Skarzynski & Pomorski, recently described from Europe. Neanuridae other than Neanurinae, Isotomidae, Entomobryidae, Paronellidae and Sminthuridae include additional species in already described genera, which are often new to science and endemic as well.

Figure 404: Some endemic Collembola new to science collected in the Santo karst. **A:** *Cephalachorutes* sp. **B:** *Ceratrimeria* sp. **C:** *Pseudachorutes* sp. cf *longisetis*. **D:** *Pseudachorudina* sp. **E:** *Friesea santo*. **F:** *Telobella* sp. **G:** *Paranura* sp. **H:** cf *Willowsia* sp. **I:** cf *Seira* sp. **J:** *Lepidonella* sp. **K:** *Parasphyrotheca* sp. **L:** Dicyrtomidae. (Photos L. Deharveng).

••• The affinities of the Collembolan fauna of the Santo karst

Investigating the origin of the springtail fauna of Santo by comparing it to that of surrounding regions was one of the objectives of the project. Two large archipelagoes are less than 1 000 km away from Santo and were partly emerged before it: Solomon-New Guinea (referred to here as Solomon), and New Caledonia. Both have significant faunistic similarities with Santo.

In the present state of our knowledge, and excluding pantropical and littoral species (several species shared by Santo and New Caledonia actually live in the beach interstitial, a habitat unsampled in Solomon), Santo shares less than 10% of its identified species with New Caledonia or Solomon. Significant information is drawn from Neanuridae, which are excellent biogeographical markers due to the low vagility and restricted distributions of their species. Thus, the *Australonura* species present in Santo is *A. rennellensis*, described from Rennell Island (900 km northwest). In contrast, the representative of the genus in New Caledonia (about 650 km southwest of Santo), is a different, though closely related species (*A. novaecaledoniae*). A new *Telobella* species, very abundant in the forest litter of Santo, is closely related, both morphologically and genetically, to another, undescribed species from Torres Islands (200 km north). This genus has never been recorded in New Caledonia. Among Isotomidae, the species *Folsomina lawrencei* is distributed from New Guinea to Vanuatu, but is absent from New Caledonia. Faunal relationships also exist with other Pacific Islands, as illustrated by the Neanurinae *Paleonura louisi*, described from the Fangataufa Atoll, which was found in Santo. In contrast, the genus is absent from New Caledonia and has not been reported from Solomon.

Thus, the available data point to a closer relationships with Solomon than New Caledonia, agreeing with the few other arthropods investigated in this respect (e.g. the decapod fauna —crabs and shrimps— analyzed by Marquet and his collaborators in 2002), as well as to the existence of "Pacific" lineages. However, this is insufficient for proposing any general scenario about the origin of the Santo Collembola.

••• Colonization of inland soils and caves by littoral terrestrial species

It is widely accepted that terrestrial habitats of oceanic islands are mostly colonized by long-distance and passive dispersal of organisms on rafts, or by humans or birds. In contrast, the colonization of island freshwaters, especially in interstitial and subterranean habitats, has often been the result of ecological shifts in marine species, as documented for Santo in the "Focus on Microcrustaceans". Surprisingly, colonization of inland terrestrial habitats by marine littoral organisms has been very rarely reported. Among Collembola for instance, genera and species-groups whose species are living on the sea-shore generally have no inland representatives, except for a few large ubiquitous genera. It was therefore exciting to discover inland species of two genera (*Yuukianura* and *Psammisotoma*), previously considered to be typically littoral.

Yuukianura was represented by two species in soils and caves, with local adaptation to life in guano. Indeed, the eight known species of *Yuukianura* are basically restricted to marine littoral habitats, where they may be very abundant, except for a Himalayan species of uncertain taxonomic status. In Santo, *Yuukianura* has colonized with great success a large array of inland habitats: litter, deep soil layers and cave guano, from the coast to the heart of the Santo karst, more than 10 km inland. In all these habitats, *Yuukianura* are frequent and abundant. Preliminary molecular data from COI genetic analysis furthermore indicate that two closely allied, but well distinct forms are present on the island, which we were able to recognize morphologically afterwards, both being new to science. One form is spread across the different habitats mentioned above, while the other was only found at a single site near Belmol, where it lives in large populations among plant debris on the surface of sweating rocks. The assumed ecological shift of *Yuukianura* may not have occurred locally on Santo, as we did not obtain specimens of the genus from coastal habitats, but at the moment there is no report of non-littoral *Yuukianura* anywhere else in Eastern Asia.

The isotomid genus *Psammisotoma* was described from littoral habitats in 1986, and four additional species were subsequently recognized, all from the

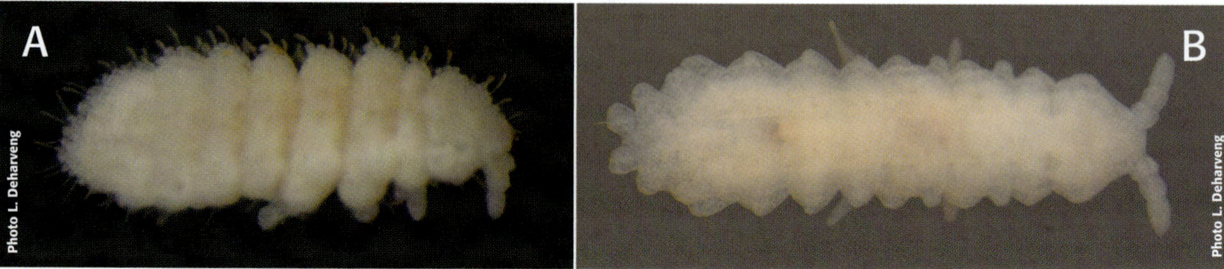

Figure 405: **A**: *Australonura rennellensis*, widespread on Santo Island, especially in moderately disturbed habitats, is also present on Rennell Island (Solomon), from where it was originally described. **B**: *Yuukianura* sp. from forest litter (Matantas).

same kind of habitat. To our surprise, a small species of this genus is present, though rare, in Tarius and Mba Caves, both far inland. Futhermore, it was only collected from guano.

Hence, in both these cases, the ecological shift to inland habitats has been associated with an adaptation to life in cave guano.

••• The future

Biodiversity exploration of the Santo karst provided an unexpectedly large number of springtail species from soils and caves, which are currently under study. After such intense sampling, it can be safely assumed that most collembolan species in the soils and caves of the Santo karst have been collected. But the karst represents only part of Santo Island, and soil is only one of the habitats colonized by Collembola. Collembola of the forest canopy and litter in the Penaoru site —in a non-karst area of Santo— have just been made available to collembologists, and will produce a large but unpredictable number of new taxa. The remarkable results obtained during the Santo 2006 expedition make clear today the work that remains to be done in order to obtain a comprehensive overview of the collembolan diversity of Santo Island: sampling the vegetation in the karst and the soils in Penaoru on the one hand, setting up a sound sampling coverage of the non-karst part of the island, particularly at highest altitudes, on the other hand. This would give an invaluable insight into the magnitude and origins of the biological diversification of Collembola on the island.

Whatever the future holds, the Santo karst is already the best documented site in the western Pacific for springtail biodiversity.

FOCUS ON MICROCRUSTACEANS

Damià Jaume, Geoff Boxshall, Eric Queinnec

The freshwater lakes, streams and rivers of oceanic islands are prone to be colonised by typically marine groups that seldom penetrate far inland on continental landmasses. The process of colonisation of these island habitats is presumed to have been facilitated by the reduced competition, if any, posed to the newcomers since the island hydrographic networks (inland water systems) have never been connected to the continents and are assumed to offer plenty of vacant niches. In addition, and since many oceanic islands are fringed by karstified zones consisting of fossil coral reef terraces that have been raised, their running waters typically exhibit raised calcium concentrations, which can lessen the osmotic barriers preventing colonisation by marine animals. A classic example of colonisation of island freshwaters by a typically marine taxon is the hermit crab *Clibanarius fonticola*. This is the only strictly freshwater anomuran crab known to science thus far, and dwells in karstic coastal springs and blueholes adjacent to the shore on the island of Espiritu Santo. The name of the species, *fonticola*, means living in springs, and refers to this unique aspect of its biology.

The exploration of Santo surface freshwater habitats undertaken in 2006 revealed a second example of the colonisation of island freshwater habitats by a predominantly marine group: this time an isopod belonging to the basically marine family Sphaeromatidae. Isopods are a diverse group of crustaceans containing terrestrial forms, such as woodlice, as well as freshwater and marine forms. Some species are even parasitic on shrimp hosts and exhibit highly modified body shapes. Most have dorso-ventrally flattened bodies but, like amphipods and tanaids, they belong to the Peracarida which are characterised by carrying their developing young in a ventral brood pouch.

The new species of Sphaeromatid was discovered in a karstic stream and associated cave sink located on a raised coral reef terrace about 390 m a.s.l. and covered by lowland rainforest. This site was situated near Butmas village, about 23.5 km inland from the east coast of the island. The Sphaeromatidae is a large family, currently comprising about 655 species, and the vast majority of these are marine.

The new species was named *Exosphaeroides quirosi* (Fig. 406) by two members of the expedition, Damià Jaume and Eric Queinnec in 2007. The name of the new species honours the Portuguese explorer Pedro Fernández de Quirós, who was the first European to reach the island on 1606, exactly 400 years before the date of our expedition.

This is a particularly exciting discovery because *Exosphaeroides quirosi* is the only truely freshwater sphaeromatid known to occur in the Pacific Islands outside of New Zealand. It is only the third species described in the genus *Exosphaeroides*, and the other two are known to live in brackish waters in New Zealand and in Easter Island.

Another important habitat to search for microscopic crustaceans is the subterranean waters. We were allowed to sample water from wells in villages around the islands of Santo and Malo and in most cases there were pumps so we were able to collect

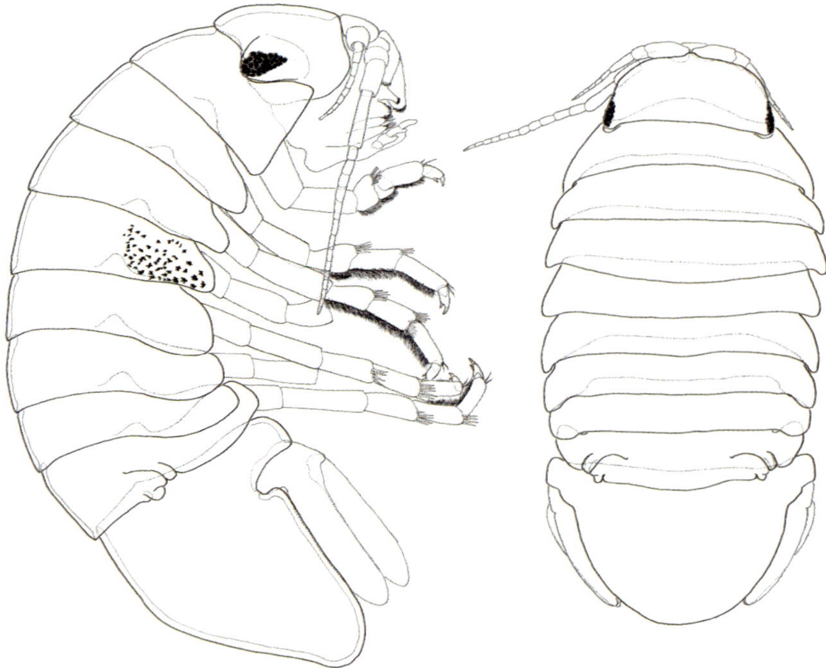

Figure 406: Adult specimens of *Exosphaeroides quirosi* shown in side view and dorsal view. These adults are about 11.5 mm in length. (Line drawings by Damià Jaume).

discovery was a blind and unpigmented amphipod shrimp, possibly a new species of *Josephosella* belonging to the family Melitidae. It was collected in virtually fresh water, salinity was only 0.3 parts per thousand, at pumps in the villages of Natanara and Matevulu. It has not been fully studied yet, but apparently the same amphipod species was caught by Stephan Eberhard at Nanda Blue Hole at a depth 13 m, and also at the spring at Porpor, this time entangled in submerged tree roots.

Most of the pump samples taken around Santo also contained copepods, but they were particularly common at the pumps at Matevulu and Natanara. The copepods inhabiting subterranean waters are typically tiny, with a body length of around half a millimetre (0.5 mm). They have slender bodies and usually have short sensory antennae at the front of the head (Fig. 408). They are usually white and lack the single mid-dorsal eyespot present in forms living above ground.

samples simply by pumping water through a hand held net (Fig. 407) supplied with a fine mesh that retained even minute crustaceans less than 1 mm in body length.

Around the Eastern and Southeastern coast of Santo, these wells provide access to parts of the anchialine aquifer fringing the island that otherwise could not be sampled for aquatic fauna. The most interesting

Early developmental stages of copepods were also found in many pump samples. Copepod eggs hatch

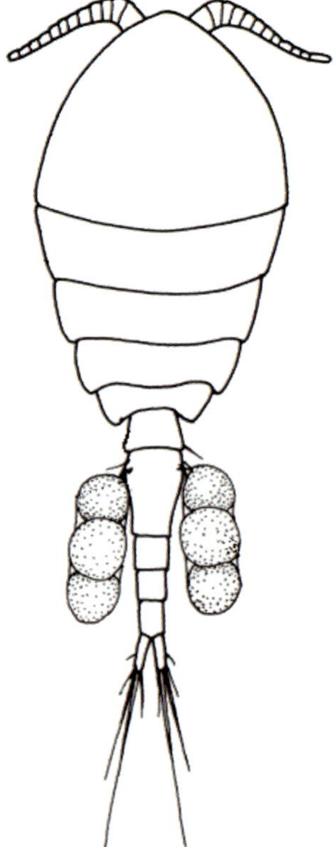

Figure 407: Filtering water from a pump near Matevulu. Using a fine mesh plankton net allowed us to catch interesting microscopic crustaceans.

Figure 408: A typical copepod from the family Cyclopidae, shown in dorsal view. This is an adult female bearing a pair of egg sacs, each with only few eggs, as is typical for subterranean forms. This female is 0.6 mm long. (Line drawing by Geoff Boxshall).

as nauplii, a tiny swimming larval stage with only three pairs of limbs (unlike the 12 pairs of limbs present in the adult). Like all arthropods, copepods moult as they develop and grow up to become adults, and they typically add one pair of limbs at each moult until the adult total of 12 pairs is reached. The presence of larvae representing different stages of the life cycle indicates that there are viable populations of the microcrustaceans living in the subterranean waters.

The subterranean copepod fauna of the southeastern coast is quite diverse, consisting of at least two different species of the family Cyclopidae and three different species of the order Harpacticoida. All of these species are probably new to science. Only one species of the Cyclopidae has previously been reported from the island of Santo, and that was not a subterranean form. It was *Bryocyclops anninae* and was found back in 1927 from water caught in empty coconut shells collected near Hog Harbour. This tiny species has an adult body length of 0.4 mm and was first described from Java in Indonesia. We were unable to find this same species again, perhaps because the original collections were made during February and we were collecting in September. However, we did find other members of the Cyclopidae in several surface streams and swamps on Santo and Malo. We collected species of the genera *Mesocyclops* and *Eucyclops* in various localities on both islands and these are still being studied.

In addition to the amphipod *Josephosella* and the cyclopid copepods, other samples taken at pumps contained representatives of different types of crustacaea. One pump sample taken just east of Luganville contained two specimens of an unidentified microparasellid isopod, although this time water was slightly brackish with a salinity of 1.10 parts per thousand. Finally, another pump in the same area rendered a single specimen of a juvenile tanaid. Because it is a juvenile it is not possible to identify or describe the species until more material becomes available, but the discovery is remarkable because the water was virtually fresh (salinity only 0.4 parts per thousand). The Tanaidacea is a strictly marine group with only a handful of species ever reported from fresh water anywhere. Interestingly, we can mention here that another tanaidacean has just been collected from a cave stream of the coastal Maros karst in Sulawesi, in running fresh water. This new species has been recently described as a new species of *Pseudohalmyrapseudes*, a genus with four species reported from freshwater habitats on Pacific Islands.

Marine Ecosystems

coordinated by Philippe Bouchet

Algal and Seagrass Communities from Santo Island in Relation to Habitat Diversity

Claude E. Payri

The coral reef communities of Vanuatu have been little studied and nothing has been previously published on the benthic algal flora from Espiritu Santo Island. Some marine algae from Vanuatu have been found in the British Museum collections (BM) and are mainly *Sargassum* species and *Turbinaria ornata*. In their report on Vanuatu's marine resources, Done and Navin (1990) mentioned *Halimeda opuntia* as occurring in most of the sites investigated and noted the high encrustation of coralline algae in exposed sites. More work has been done on seagrass communities; earlier authors reported a total of nine species from Vanuatu including five species from Santo, i.e. *Cymodocea rotundata*, *Halodule uninervis*, *H. pinifolia*, *Halophila ovalis* and *Thalassia hemprechii*.

The present algal flora and seagrass investigation of Santo was conducted during August 2006 as part of the "Santo 2006 expedition". This work is a companion study to that of the Solomons and Fiji and is intended to provide data for ongoing biogeographic work within the West and Central Pacific.

Extensive surveys have been carried out in most of the habitats recognized in the southern part of Santo Island and in the Luganville area, including islands, shorelines, reef flats, channels and deep outer reef slopes.

SAMPLING SITES AND METHODS

The 42 sites investigated are shown in figure 409 and are distributed from Palikolo in the northernmost part of the study area, down to Urepala islet located in the southern part including the Segond Channel, the Malo passage and Abokisa Island on the east coast of Aore Island. Sites were selected to include the largest possible range of environments.

Most of the sites were surveyed by SCUBA divers from the surface down to 60 m deep. The shallow areas, including fringing reef flats, reef channels and rocky shorelines, were sampled by snorkelling and walking on the reef. The sampling effort was standardized and inventory duration at each site was fixed to 80 minutes. A species inventory was compiled in order to create a more comprehensive species list for the southern part of Santo. Specimens were sampled to make a taxonomical collection for the area.

All the collected specimens were pressed and dried for herbaria; fragments of specimens were preserved in a solution of buffered formalin in seawater (5%) for further anatomical studies. Tissues from various taxa were also preserved for further phylogeny and molecular analysis and all the herbarium specimens were air dried (without formalin) which allows for extra DNA analysis if necessary. The collection is currently housed in the phycological herbarium at IRD Nouméa (IRD-NOU) and will be transferred to the *Muséum national d'Histoire naturelle* in Paris (PC). Part of the collection will be deposited at USP in Suva, Fiji.

MARINE MACROPHYTES IN SANTO: GENERAL INSIGHTS

Macrophyte communities on coral reefs are generally distributed in assemblages that more or less reflect reef zonation. The distribution of the marine flora on a coral reef is influenced by various factors including sunlight, salinity, water turbulence and currents, the nature of the substratum, depth, exposure to air, geomorphology, topography, herbivore pressure and biological competition with other benthic organisms.

In addition, Vanuatu's reefs have a complex tectonic history, having experienced several emergence and subsidence events. These have resulted in some features that are typical of many reefs with rocky shorelines, and recent tectonic displacements and uplifts may have affected some of their benthic assemblages. However, we did not observed a recent influence of tectonic displacement on benthic communities in the study areas and the most significant disturbances we

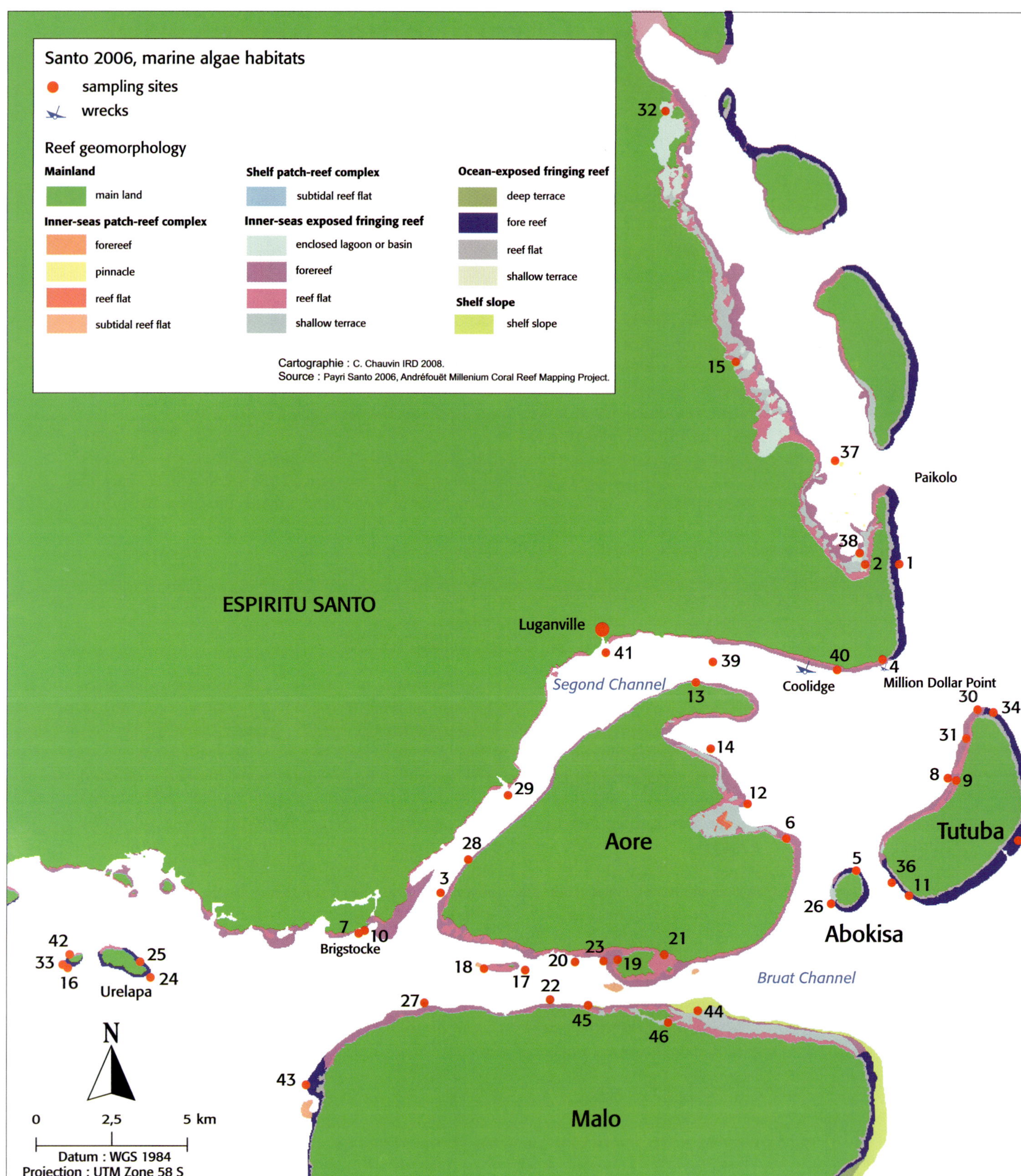

Figure 409: Map of sampling sites in the Luganville area, southern part of Santo Island.

observed were the result of recent cyclone activity and bleaching events over the past few years.

However, environmental factors are not homogeneous across the reefs, and gradually change from the shoreline to outer reef slopes through lagoons and reef flats. The rather patchy zonation patterns are generally distributed parallel to the shorelines and reef margins.

This synthesis comprises an overview of the representative biotopes investigated and describes the different macrophyte (algae and seagrasses) communities associated with the different identified habitats.

REPRESENTATIVE BIOTOPES

The main features of the Santo coral reef complex are the absence of a barrier reef and associated habitats. Most of the structures comprise narrow fringing reefs, outer reefs, patch reefs in shallow water, sheltered and open embayments, deep channels and shallow passes, exposed outer reef slopes and reef walls or drop-offs.

Most of the sites have reef habitats compressed into narrow coastal margins and exposed to ocean influences. On unsheltered coasts the coral reefs are wave-beaten structures that are heavily encrusted by coralline algae as well as by coral species that are well-adapted to strong turbulence.

The great ocean depths, large fetches and the refraction of swells around the small islands adjacent to Santo mean that the open coasts on all sides are subject to strong wave forces, and this limits the types of reefs that can establish. Less robust forms of corals and other benthic communities are however able to develop in more sheltered embayments. The islands around the Segond Channel provide significant shelter from the open ocean, and the channel accumulates siltation originating from the terrestrial erosion of the adjacent island of Santo; the channel supports a range of habitats with conditions ranging from intermediate to abundant shelter, and muddy substrata.

The 42 sites surveyed have been classified into 12 major habitat groups which include geomorphology, topography and major benthic communities. Schematic diagrams (profiles) are given in figures 411-422, the list of the symbols used in the profiles are given figure 410. Descriptions of the profiles are as follows.

••• Segond Channel (Fig. 411)

This long channel runs between Santo and Aore Island. Around the Luganville area and down to the south there are few reef formations and coral communities and these are mainly developed on sandy slopes and rubble. Narrow reef flats are present, mostly at both entrances to the Channel and along the Aore Island coast. Coral communities are mostly Acroporidae and occasional massive *Porites*. There is also evidence of damaged coral in the high proportion of coral rubble. Large beds of the green macrophyte *Halimeda opuntia* intermixed with sponges colonise the hard substratum. In the northern area, the middle part of the channel is deep (70 m depth) and muddy, marked with ghost shrimp (Calianasseae) hummocks and small benthic communities including some Nephtheidea, *Dophleina* and *Asthenosoma* urchins. The benthic communities of the channel environment are dominated by sponges and octocorallians (soft corals and sea fans). On the shallow muddy flats adjacent to the shore of Santo island seagrasses such as *Halodule* and *Cymodocea* form sparse patches.

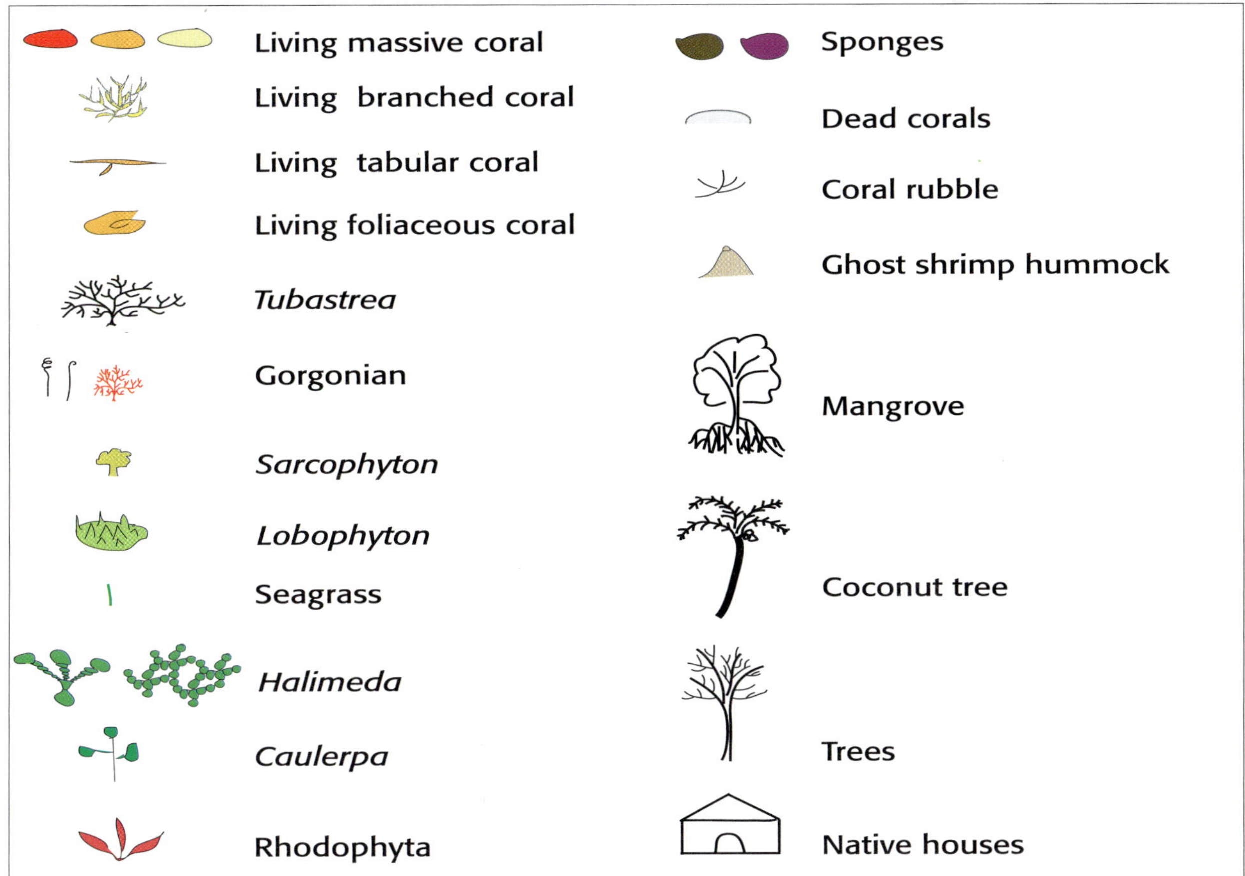

Figure 410: List of symbols used for figures 411-422.

Figure 411: Schematic diagram of the topography and reef communities of the Segond Channel.
A: Mangrove trees on the shoreline. **B**: Typical gentle slope with dense *Halimeda* bed. **C**: Typical mixed community on channel slope dominated by Neiphteidae. **D**: Diffuse seagrass bed of *Halodule pinifolia* and *Halophila ovalis*. **E**: *Halimeda minima* forming thick mats with sponges and other organisms on the channel slope. **F**: Typical community of invertebrates and algae on the deep channel edge. **G**: Mixed seagrass

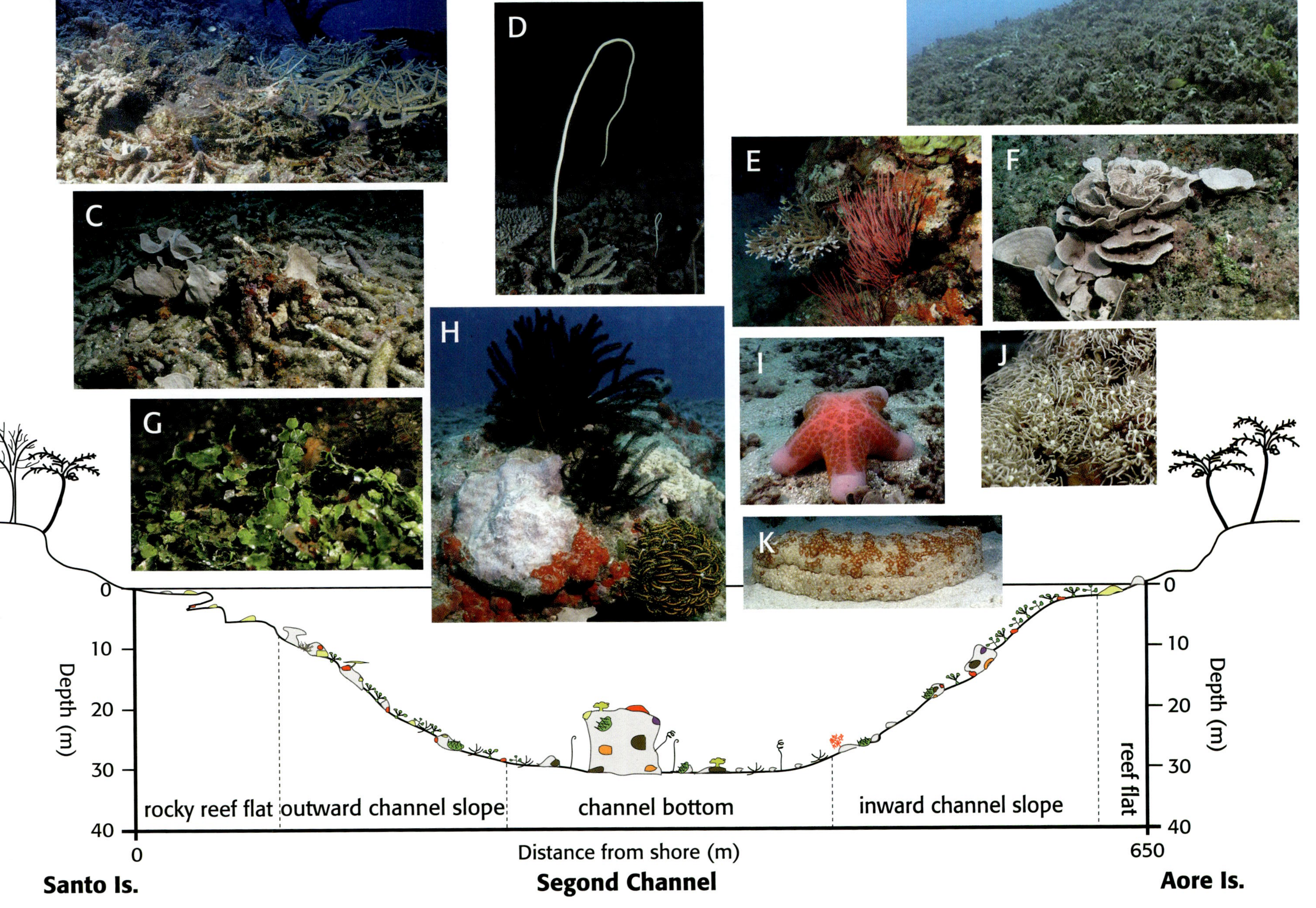

Figure 412: Schematic diagram of the topography and reef communities of Channel Pass (southern entrance of the Segond Channel).
A: Branching *Acropora* community on channel slope. **B**: Typical *Halimeda* beds on the edge of the slope. **C**: Rubble of *Acropora* branches. **D**: Gorgonian. **E**: *Ellisella*. **F**: *Phyllospongia lamelosa* with *Halimeda*. **G**: *Halimeda minima*. **H**: Crinoids *Comantheria briareus* and *Comanthus bennetti*. **I**: *Choriaster granulatus*. **J**: *Tubipora musica*. **K**: *Thelenota anax*. (Photos J.-L. Menou & J.-M. Boré IRD Nouméa).

Channel Pass (southern entrance of Segond Channel) (Fig. 412)

The topography and the environment of the Segond Channel changes from the north to the south with the increasing current. The coral communities are well developed on the slopes with massive corals in the shallow areas while a rich branched coral community is associated with green calcareous algae *Halimeda opuntia* and the coralline algae *Amphiroa* on the gentle sandy slope. The bottom of the channel is a hard substratum supporting large coral patches with flourishing communities of benthic invertebrates including octocorallians and numerous crinoids. This area experiences strong tidal currents. The reef flat on the Aore side is narrow with a steep slope dominated by *Halimeda macroloba* anchored in the muddy sand and various red algae entangled in dead *Acropora* branches.

Sheltered fringing reef (Brigstocke point, SW Luganville) (Fig. 413)

Most of the reefs fringing the southern corner of Santo are characterized by a narrow area of subtidal grooves adjacent to the rocky shoreline, with dense macroalgal vegetation of red algae *Callophycus* spp, *Portieria hornemanii* and *Amphiroa crassa*. The slopes are mostly covered with coral debris with some large blocks of dead *Porites*. The green calcareous alga *Halimeda opuntia* develops spectacular beds from 15-30 m deep, while *Halimeda macroloba* and the red foliaceous algae belonging to the Peyssonneliaceae form aggregations (soft nodules) up to 10 cm in diameter that are locally abundant on the top of the slope.

Malo water passage (between Aore and Malo Islands) (Fig. 414)

The Aore site is fringed by small *Rhizophora* clumps that grow on the beaches along a narrow and shallow depression where *Acroporids* and the seagrass *Enhalus acoroides* are well developed on sand flats with moderate tidal currents. The shallow reef flat on the side of the islet is covered with many massive *Porites*; on the side of Malo Island the reef is deeper and the corals are more massive and have developed into large patches with abundant encrusting coralline algae. The coral cover is high on the reef slope along the water passage. Foliose and branched corals are abundant on the reef slopes while they decrease further down to the bottom of the passage where strong currents limit the development of a benthic community. Large spurs parallel to the sea floor that support octocorallians are the main feature of the base of the slope on the Malo side.

Sheltered sandy slope (Malo passage) (Fig. 415)

The Malo passage has typical sandy slopes with little reef formation along the shores of Aore and Malo islands. Fringing reef flats are narrow, shallow and protected. No seagrass beds were found, only the delicate paddle seagrasses *Halophila* spp. were observed on the sandy slope down to 50 m deep. Sparse coral blocks and rubble are the main features from the top down to the mid-slope while coarse sand and debris are dominant further down beyond 30 m deep. Species diversity is low except for starfish and holothurians with various species such as *Nardoa gomophia*, *Echinaster callosus*, *Choriater granulatus*, *Holothuria* (*Microthele*) *fuscogilva* and *M. fuscopunctata* and the red algae that display several gelatinous species in the deeper part of the slope.

Sheltered embayment (Palikolo Bay) (Fig. 416)

From shore to open ocean, this bay contains several biotopes including:
- Prolific seagrass beds in the fringing muddy flats intermixed with a green macroalgal complex of *Halimeda* and *Caulerpa*;
- Shallow sheltered reefs on sand dominated by acroporids;
- Large areas of coral-construction on deeper (6 m) patch reefs;
- A steep slope from 15 m down to 60 m deep. The patch reefs here support a high diversity of species with a very rich coral community including many fungids and octocorallians. Large mounds of rubble covered by the brown alga *Lobophora variegata* indicate an accumulation of coral skeleton fragments that have broken in recent decades. The diversity decreases down the slope; some holothurians including *Thelenota anax* and *Neoferdina cumingii* (50 m) have been observed along with green algae *Cladophora* and *Halimeda* in the deeper zone.

Open embayment, partially sheltered (east Aore) (Fig. 417)

This habitat occupies the north eastern part of Aore Island. The coral community is developed on a gentle sandy slope down to 25 m deep and looks like the silty bottom of a lagoon. *Porites* with abundant soft corals, sponges and various branching *Acropora* form large beds down to 15 m deep. Various coralline and red fleshy algae were recorded on hard substrata. Corals such as *Polyphyllia*, *Herpolitha limax*, *Sandalolitha robusta* and *Cynarina lacrymalis* were also observed in these sheltered areas. The accumulation of fine carbonate sand in the deeper part is a typical characteristic of lagoons and supports large patches of mixed green algae including *Halimeda*, *Udotea*, *Avrainvillea* and *Caulerpa*. Visibility was reduced in this environment due to the abundance of fine carbonate particles in the water column.

Sheltered fringing reef (West Tutuba Island) (Fig. 418)

On the west side of Tutuba Island adjacent to the beach there is a narrow and patchy fringing reef fronting in some places an enclosed lagoonal gutter (10 m deep) and then a gentle outer slope that nonetheless

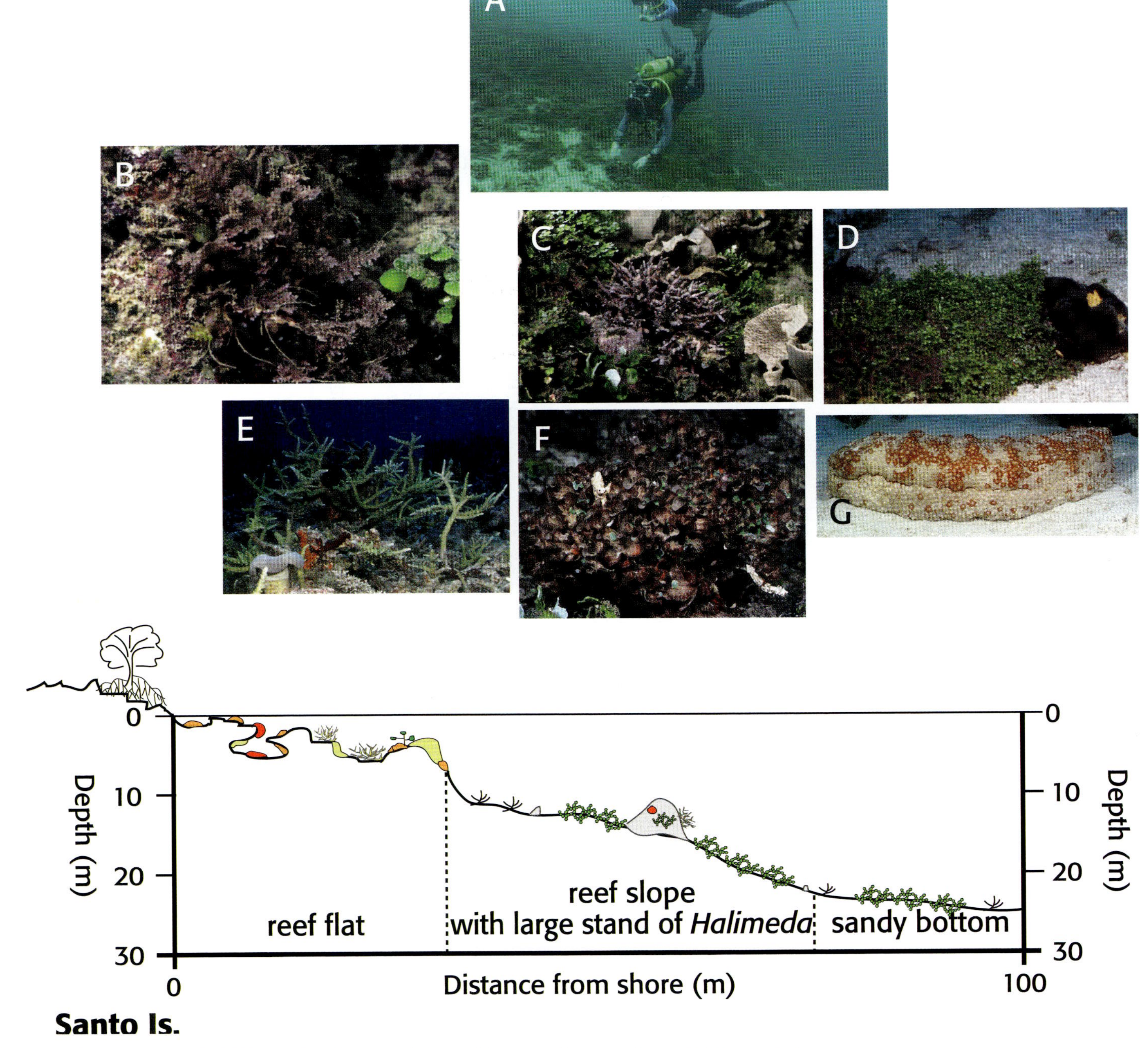

Figure 413: Schematic diagram of the topography and reef communities of a sheltered fringing reef (Brigstocke point, SW Luganville).
A: *Halimeda* beds on the gentle sandy slope. **B**: *Callophycus serratus*. **C**: Branched coralline algae. **D**: *Halimeda distorta* on sandy bottom. **E**: Branching coral community. **F**: Ball-like Peyssonneliaceae. **G**: *Thelenota anax*. (Photos J.-L. Menou & J.-M. Boré IRD Nouméa).

Figure 414: Schematic diagram of the topography and reef communities of the Malo passage (between Aore and Malo Island).
A: Flourishing community on the hard channel bottom. **B**: Spurs at the top of the reef slope. **C**: Branching corals on the reef flat. **D**: *Steronephtya*. **E**: Reef slope community composed of branching and massive corals. **F**: Seagrass *Enhalus acoroides*. **G**: *Aglaophenia*. **H**: Gorgonian community on the bottom of the reef slope. **I**: Candle-like coralline algae on the reef edge. **J**: Massive corals **K**: Branching

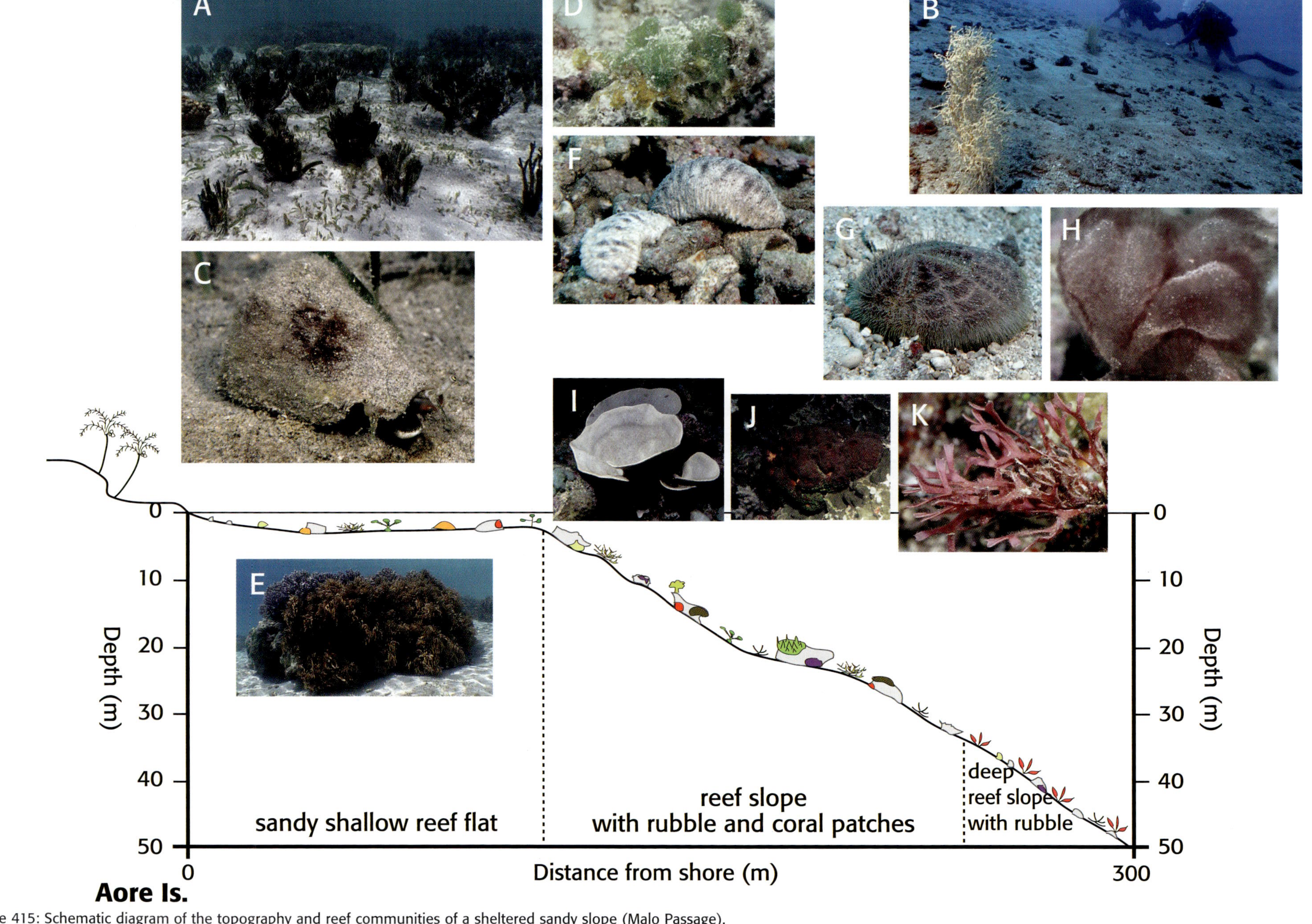

Figure 415: Schematic diagram of the topography and reef communities of a sheltered sandy slope (Malo Passage).
A: Typical mixed beds of *Halimeda cylindracea* and *Halophila ovalis*. **B**: Sandy slope. **C**: *Strombus luhuanus* on shallow sandy bottom. **D**: *Rhipilia crassa* on rubble. **E**: Soft coral on sandy bottom. **F**: Holothurian on rubble on the upper part of the slope. **G**: *Metalia sternalis*. **H**: *Gibsmithia hawaiiensis* on deep rubble. **I**: *Phyllospongia lamelosa*. **J**: Sponge *Melophlus*. **K**: *Dichotomaria marginata*. (Photos J.-L. Menou & J.-M. Boré IRD Nouméa).

Figure 416: Schematic diagram of the topography and reef communities of a sheltered embayment (Palikolo Bay).
A: Invertebrate assemblage on dead corals. **B**: *Distichopora*. **C**: Deep sandy bottom with massive coral patches. **D**: *Cymodocea serrulata* (large) mixed with *Halodule universis*. **E**: *Tydemania expeditionis*. **F**: Soft coral *Rumphella aggregata*. **G**: Rubble covered by *Lobophora variegata*. **H**: *Halimeda macroloba* growing among corals on sandy bottom. **I**: Fungids growing among rubble. **J**: Branching Acropora

Figure 417: Schematic diagram of the topography and reef communities of an open and partially sheltered embayment (east Aore).
A: Sponges growing among rubble on the upper slope. **B**: Holothurian *Bohadchia graeffei*. **C**: Red algae *Titanophora* and *Padina*. **D**: Sandy lagoon floor. **E**: *Polyphylla*. **F**: Branching *Acropora*. **G**: *Halimeda* and *Padina* on gentle sandy slope. **H**: Sponge *Phakellia cavernosa*. **I**: Fungids growing among rubble. **J**: *Acanthaster plancii* feeding on coral. **K**: Green algae *Udotea argentea*. (Photos J.-L. Menou & J.-M. Boré IRD Nouméa).

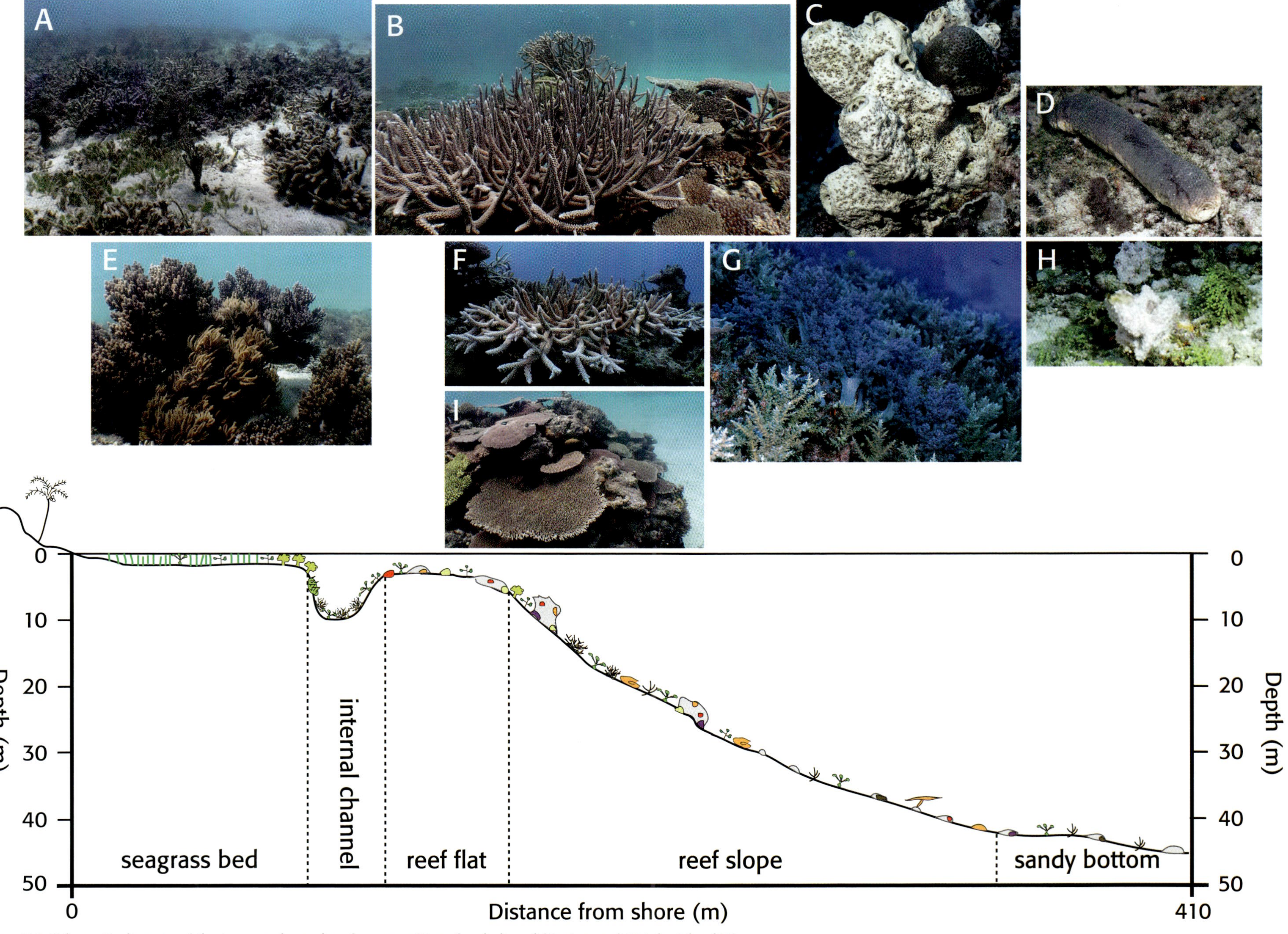

Figure 418: Schematic diagram of the topography and reef communities of a sheltered fringing reef (Tutuba Island W).
A: Mixed *Acropora* and *Halimeda* community. **B**: Attractive coral community. **C**: Sponge *Liosina*, abundant on hard corals. **D**: *Holothuria edulis* on the floor. **E**: Soft corals. **F**: Large branching *Acropora*. **G**: Mixed

Figure 419: Schematic diagram of the topography and reef communities of a windward fringing reef and exposed outer reef slope (Tutuba Is. N & E).
A: Spurs. **B**: Top of reef. **C**: Typical reef edge community composed of massive and tabular corals. **D**: *Culcita novaeguinea*. **E**: *Caulerpa fergusoni*. **F**: Red algae *Predaea laciniosa*. **G**: Coralline algae. **H**. *Halimeda* on rubble. **I**: Octocorallian community on spur edges. **J**: Crinoid assemblage. **K**: Sponge *Melophlus*. **L**: Abundant large sponges. **M**: Octocorallian community on the reef edge. **N**: *Heliopora* on slope. **O**: Luxuriant coral community on mid slope. **P**: Massive corals on slope in deep water. **Q**: Green alga *Cladophora obukhoana* on coarse sand. (Photos J.-L. Menou & J.-M. Boré IRD Nouméa).

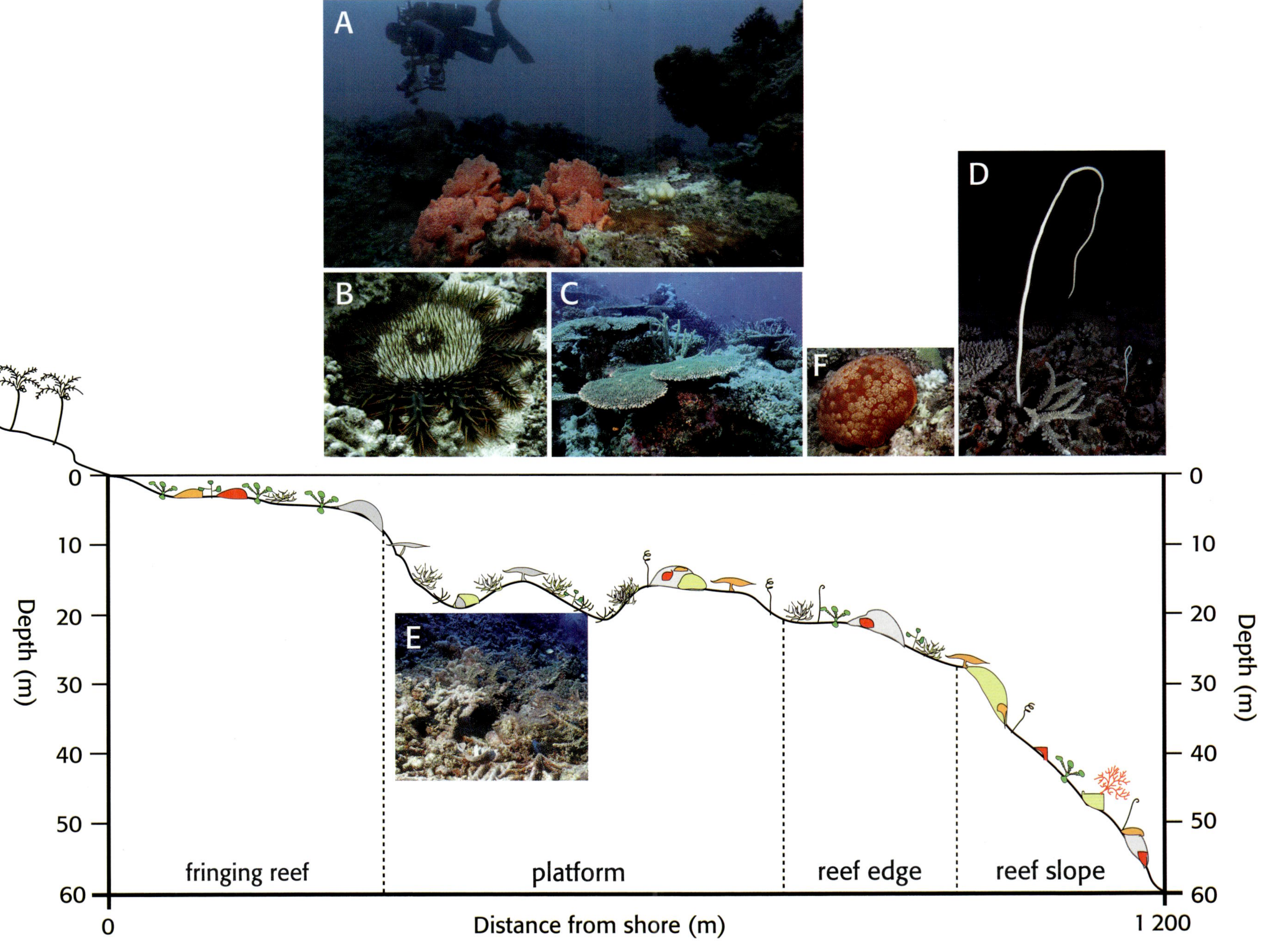

Figure 420: Schematic diagram of the topography and reef communities of an outer reef platform (Malo Is. W coast).

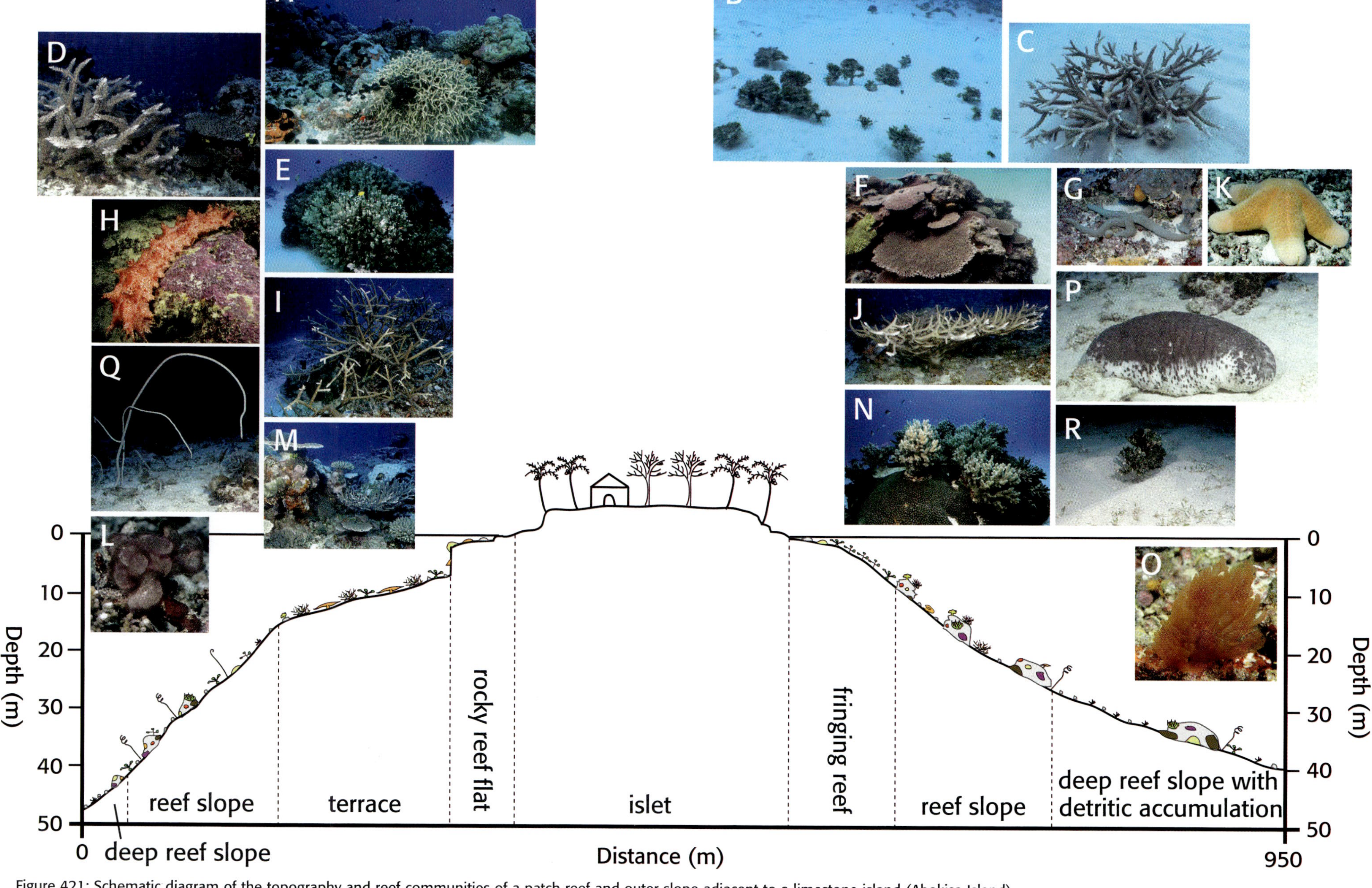

Figure 421: Schematic diagram of the topography and reef communities of a patch reef and outer slope adjacent to a limestone island (Abokisa Island).
A: Attractive coral community on the reef flat. **B**: *Halimeda cylindracea* on shallow sandy bottom. **C**: Branching *Acropora* on shallow sandy bottom. **D**: Massive tabular and branching corals. **E**: Massive coral heads on sandy bottom. **F**: Tabular *Acropora* community on patch reef. **G**: *Linckia* guild. **H**: *Thelenota rubrolineata*. **I**: Branching *Acropora* on sandy slope. **J**: Branching *Acropora* on slope. **K**: *Choriaster granulatus*. **L**: *Gibsmithia hawaiiensis*. **M**: Massive coral community on the terrace. **N**: Soft corals on massive coral head. **O**: *Predaea weldii*. **P**: *Holothuria (Microthele) fuscogilva*. **Q**: Gorgonian on the bottom of the outer slope. **R**: *Halimeda cylindracea* and *Halophila ovalis* on deep sandy bottom. (Photos J.-L. Menou & J.-M. Boré IRD Nouméa).

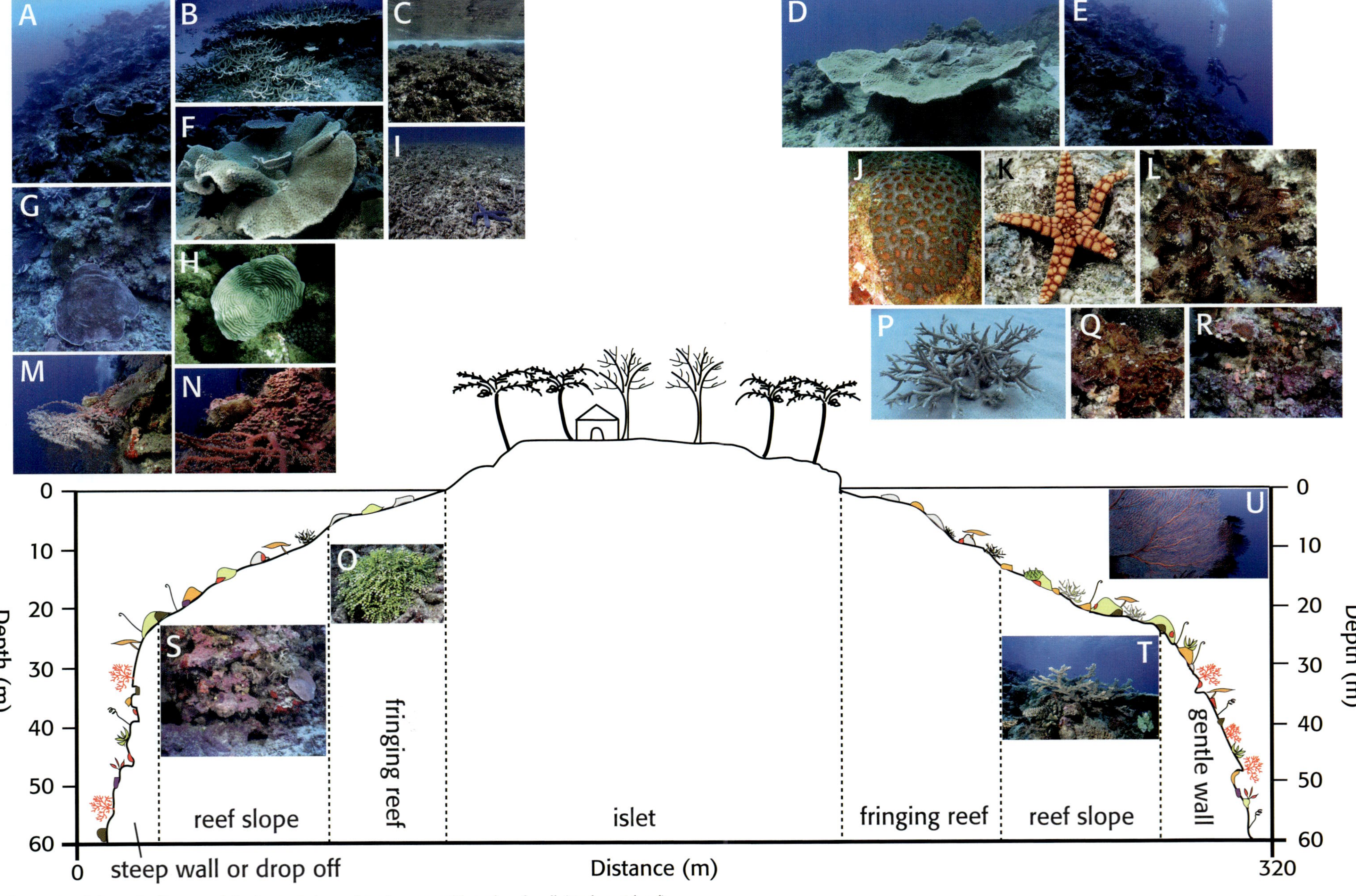

Figure 422: Schematic diagram of the topography and reef communities of reef wall (Urelapa Island).
A: Typical outer slope at 30 m depth. **B**: Branching *Acropora* community on the sandy terrace. **C**: Top of the reef. **D**: *Diploastrea heliopora*. **E**: Reef slope and wall. **F**: *Foliacea coral* cf. *Turbinaria*. **G**: Foliaceous corals on the wall. **H**: *Pachysiris speciosa*. **I**: *Acropora* rubble on the reef crest. **J**: *Montastrea annuligera*. **K**: *Celerina*. **L**: *Asteronemia anastomosans*. **M**: Gorgonian and crinoids on the drop-off. **N**: Coralline algae and octocorallian on the wall 2. **O**: *Halimeda minima* on rubble. **P**: *Acropora* on sandy bottom with ripple marks. **Q**: *Peyssonnelia*. **R**: Cryptic community of coralline algae on the outer slope. **S**: Coralgal

goes down to 50 m in depth. The reef flat area is very shallow and supports a diffuse seagrass bed composed mainly of *Cymodocea serrulata*, intermingling with algae and small coral colonies. The front of the stretches of reef and edges of the lagoon are overgrown by a dense covering of erect soft corals such as *Sinularia* and *Sarcophyton*, while the narrow silty lagoon contains isolated massive *Porites*. The steep internal slope and the outer reef flat are covered by thick swathes of rubble without coral communities. This entire zone is exposed at low tide. The outer reef slope is similar to other sheltered slopes with large areas of broken *Acropora* branches and plates on white coral sand down to 8 m deep; further down there is an accumulation of coral branches and other carbonate debris with few coral colonies and occasional fleshy algae.

In the area outside the lagoonal depression the sandy bottom supports a mosaic of patch reefs dominated by robust massive corals that are highly dissected with spectacular communities of green and red algae. Several *Holothuria edulis* have been observed on the shallow sandy bottom.

● ● ● Windward fringing reef and exposed outer reef slope (north and east Tutuba Island) (Fig. 419)

On the windward side, fringing reefs are deeply dissected with massive spurs and narrow grooves from 3 to 12 m deep that are littered with coarse sand and coral rubble. Heavy crusts and candle-like coralline algae are well developed in this exposed area along with numerous small species in the overhangs and reef interstices. Corals are sparse on the reef top and mostly consist of massive *Pocillopora*, *Acropora* and *Millepora*. The outer slope is steep from 15 to 30 m in depth, with scattered low spurs and large patches of coarse sand with abundant *Halimeda* segments. Beautiful sea fans and other octocorallian fauna are present on the top of the reef with numerous crinoids making this a very attractive area. From 30-60 m deep the slope is less steep and comprises rubble and scattered coral heads. The deeper part of this seaward slope is typical of many deep slopes, especially with respect to the associated red gelatinous algae (*Predaea*, *Dudresnaya* and *Gibsmithia*), green algae *Caulerpa fergusonii* and *Cladophora ohkuboana* and *C. dotyana*.

The reef slopes facing the open sea are less steep from 30 m down to at least 50 m deep. Coral cover is reduced; *Halimeda minima* coverage is high and contributes to sand accumulation from their calcified segments. *Seriatopora* cf *histrix*, and black coral *Cirrhipathes anguineus* have been observed at 45 m deep.

● ● ● Outer reef platform (west coast of Malo Island) (Fig. 420)

Reef formation on the northwest coast of Malo Island provides an example of a platform that was not seen elsewhere during the survey. This reef is totally subtidal with broad, irregular and meandering spurs and grooves. The site has a high proportion of rubble and corals that have been dead for several years. The coral communities were dominated by plate and branching forms. At the time of the survey coral recovery was observed with several living colonies of the same size (20-30 cm in diameter). The inshore reef and outer slopes were not studied.

Numerous *Culcita novaeguineae* were observed, along with one specimen of *Acanthaster planci*.

● ● ● Patch reef and outer slope adjacent to limestone island (Abokisa Island) (Fig. 421)

The small limestone islet located between the larger Tutuba Island and Aore is surrounded by an intermittent fringing reef developed on coral sand to about 6 m in depth with an attractive coral community. Structurally they are dominated by stands of *Acropora* in plate (*A. danai*) and branched forms, both living and dead. The dead skeletons provide the substrata for a complex and beautiful coral community in shallow sandy water and include coralline algae. The adjacent slope is dominated by massive *Porites* down to 15 m deep with numerous *Halimeda* and encrusted rubble as well as rare echinoderms such as *Holothuria* (*Microthele*) *fuscogilva* and *Linckia guildingi*. Further down the slope drops off to 40 m deep in the north and more than 60 m on the southwestern side with a steeper declination. The coral community is replaced by rubble and a few small colonies (< 1 m high). In deep water, species diversity is low with some red gelatinous algae (*Predaea* and *Gibsmithia*), the green algae *Caulerpa* and *Cladophora ohkuboana* and the echinoderms *Choriaster granulatus* and *Thelenota rubrolineata*.

● ● ● Reef wall (Urelapa Island) (Fig. 422)

Fringing reefs on limestone islands adjacent to deep water such as Urelapa and Tuvana islets located off the southern part of Santo have vertical underwater cliffs. These reef walls are distinct features that represent one of the more spectacular biotopes for species diversity. Stretches of fringing reef are found adjacent to limestone and coral sand beaches, which change gradually to a reef slope dominated by a mixture of massive corals such as *Diploastrea*, *Goniastrea* and *Montastrea magnistellata* and branching Acroporidae down to the cliff precipice. The coral walls start beyond 20 to 25 m deep down at least 60 m deep and are present around the islets where the coast is neither sheltered nor exposed. Coralline algae in association with numerous fleshy red algae (large patches of *Halichrysis irregularis* and *Asteromenia anastomosans*) are dominant components with octocorallians. Corals are encrusting or foliaceous such as *Pachyseris speciosa*.

REPRESENTATIVE MACROPHYTES COMMUNITIES

••• Algal vegetation

The species list of the benthic marine algae and sea-grasses collected from Santo is shown in tables 37 & 38. The classification follows *The catalogue of the benthic marine algae of the Indian Ocean* by Silva and coauthors (1996). The 271 listed species of algae consist of 163 Rhodophyta (red algae), 83 Chlorophyta (green algae) and 25 Pheaophyceae (brown algae). A selection of species is illustrated in figures 423-428. About 150 specimens of coralline algae are under study and are not included in this work; only the most common encrusting coralline algae are considered here. The species belong to 12 orders and 45 families (Figs 429 & 430). Most of the specimens have been identified to species level and these represent 90% of the collection; the 10% remaining unidentified species comprise taxa that could be new to science. Among the identified species, three of them are newly described from Solomon Islands, Fiji and New Caledonia; while at least five taxa including four species of *Martensia, Rhizophyllis, Rhodymenia, Hypoglossum* and *Dudresnaya* and one new genus belonging to the Dumontiaceae are being studied to describe new taxa or establish them as belonging to existing species. The study of the coralline algae will probably reveal new taxa as well.

The algal flora is typically tropical and most of the species belong to the Indo-Pacific biogeographic province. Comparison with flora from adjacent archipelagos is limited due to the difference in sampling effort in the various regions. However, 55% and 53% of the species of Santo are present in Solomons and Fiji respectively.

The Rhodymeniaceae *Asteronemia pseudocoalescens* described from Lord Howe Island was observed for the first time outside of its type locality, suggesting that its geographic distribution is broader than originally thought; this discovery enhances the known biogeographic affinities of the Santo marine flora with the tropical west Pacific.

Algal assemblages are characterized within the biotopes as shown in the following sections.

Table 37: List of Rhodophyta, Chlorophyta and Phaeophyceae species from Santo waters.

Class Rhodophyta			
Order Bonnemaisoniales			*Wrangelia argus* Montagne
Family Bonnemaisoniaceae	*Asparagopsis taxiformis* (Delile) Trevisan		*Wrangelia elegantissima* R.E. Norris
Order Ceramiales		Family Dasyaceae	*Dasya anastomosans* Weber-van Bosse
Family Ceramiaceae	*Aglaothamnion boergesenii* (Aponte & D.L. Ballantine) L'Hardy-Halos & Rueness		*Dasya baillouviana* (S.G. Gmelin) Montagne
	Anotrichum tenue (C. Agardh) Nägeli		*Dasyphila plumarioides* Yendo
	Antithamnionella elegans (Berthold) J.H. Price & D.M. John		*Heterosiphonia crispella* (C. Agardh) M.J. Wynne
	Balliella repens Huisman & Kraft		*Thuretia* sp. nov.
	Centroceras clavulatum (C. Agardh) Montagne	Family Delesseriaceae	*Frikkiella searlesii* M.J. Wynne & C.W. Schneider
	Centroceras minutum Yamada		*Haraldia lenormandii* (Derbès & Solier) Feldmann
	Ceramium flaccidum (H.E. Petersen) Furnari & Seiro		*Hypoglossum simulans* M.J. Wynne, Price & Ballantine
	Ceramium marshallense Dawson		*Martensia* cf. *australis* Harvey
	Ceramium upolense South & Skelton		*Martensia elegans* Hering
	Corallophila apiculata (Yamada) R. Norris		*Martensia flabelliforme* Harvey ex J. Agardh
	Griffithsia heteromorpha Kützing		*Martensia fragilis* Harvey
	Haloplegma duperreyi Montagne		*Martensia* sp. nov.
	Monosporus indicus Børgesen		*Myriogramme melanesiensis* N'Yeurt, Wynne & Payri
	Spyridia hypnoides (Bory de Saint-Vincent) Papenfuss		*Nitophyllum adhaerens* M.J. Wynne
	Tiffaniella saccorhiza (Setchell & Gardner) Doty & Menez		*Vanvoorstia spectabilis* Harvey
		Family Rhodomelaceae	*Acanthophora pacifica* (Setchell) Kraft

	Acanthophora spicifera (Vahl) Børgesen
	Amansia rhodantha (Harvey) J. Agardh
	Bostrychia tenella (J.V. Lamouroux) J. Agardh
	Chondria armata (Kützing) Okamura
	Chondria dangeardii Dawson
	Chondria minutula Weber-van Bosse
	Chondria ryukyuensis Yamada
	Chondria simpliciuscula Weber-van Bosse
	Chondria bullata N'Yeurt & Payri
	Chondria sp.
	Chondrophycus parvipapillatus (C.K. Tseng) Garbary & Harper
	Chondrophycus succisus (A.B. Cribb) K.W. Nam
	Exophyllum wentii Weber-van Bosse
	Herposiphonia nuda Hollenberg
	Herposiphonia tenella (C. Agardh) Ambronn
	Laurencia brachyclados Pilger
	Laurencia cf. *distichophylla* J. Agardh
	Laurencia decumbens Kützing
	Laurencia sp. 1
	Laurencia sp. 2
	Neosiphonia apiculata (Hollenberg) Masuda & Kogame
	Polysiphonia scopulorum Harvey
	Polysiphonia sertularioides (Grateloup) J. Agardh
	Polysiphonia sp
	Polysiphonia triton P.C. Silva
	Spirocladia barodensis Børgesen
	Tolypiocladia glomerulata (C. Agardh) F. Schmitz
Order Corallinales	
Family Corallinaceae	*Amphiroa anceps* (Lamarck) Decaisne
	Amphiroa crassa Lamouroux in Quoy & Gaimard
	Amphiroa foliacea Lamouroux in Quoy & Gaimard
	Amphiroa fragilissima (Linnaeus) Lamouroux
	Amphiroa sp. nov.
	Amphiroa tribulus (Ellis & Solander) Lamouroux
	Amphiroa valonioides Yendo
	Cheilosporum acutilobum (Decaisne) Piccone
	Cheilosporum spectabile Harvey ex Grunow
	Hydrolithon onkodes (Heydrich) D. Penrose & Woelkerling
	Hydrolithon orthoblastum
	Hydrolithon reinboldii (Weber-van Bosse & Foslie) Foslie
	Jania adhaerens Lamouroux
	Jania rubens (Linnaeus) Lamouroux
	Lithophyllum pygmaeum (Heydrich) Heydrich
	Lithothamnion proliferum Foslie
	Neogoniolithon fosliei (Heydrich) Setchell & mason
Order Gelidiales	
Family Gelidiaceae	*Gelidiella acerosa* (Forsskål) Feldmann & G. Hamel
	Gelidium cf. *crinale* (Turner) Gaillon
	Gelidium isabelae W.R. Taylor
	Pterocladiella sp.
Order Gigartinales	
Family Caulacanthaceae	*Caulacanthus ustulatus* (Turner) Kützing
Family Corynocystaceaea	*Corynocystis prostrata* G.T. Kraft
Family Dicranemataceae	*Pinnatiphycus menouana* N'Yeurt, Payri & Gabrielson
Family Dumontiaceae	*Dudresnaya capricornica* Robins & Kraft
	Dudresnaya hawaiiensis R.K.S. Lee
	Dudresnaya sp. nov.
	Dumontiaceae gen. nov.
	Gibsmithia dotyi Hoyle
	Gibsmithia hawaiiensis Doty
	Gibsmithia larkumii Kraft
Family Hypneaceae	*Hypnea cervicornis* J. Agardh
	Hypnea nidulans Setchell
	Hypnea pannosa J. Agardh
	Hypnea saidana Holmes

	Hypnea spinella (C. Agardh) Kützing
	Hypnea valentiae (Turner) Montagne
Family Nemastomataceae	*Predaea laciniosa* Kraft
	Predaea weldii Kraft & I.A. Abbott
Family Peyssonneliaceae	*Peyssonnelia* cf. *boergesenii* Weber-van Bosse
	Peyssonnelia inamoena Pilger
	Peyssonnelia sp. 1
	Peyssonnelia sp. 2
Family Rhizophyllidacea	*Portieria hornemannii* (Lyngbye) P.C. Silva
	Rhizophyllis sp. nov.
Family Schizymeniaceae	*Titanophora weberae* Børgesen
Family Solieriaceae	*Callophycus densus* (Sonder) G.T. Kraft
	Callophycus serratus (Harvey ex Kützing) P.C. Silva
	Eucheuma horizontale Weber-van Bosse
	Eucheuma sp.
	Meristotheca procumbens P. Gabrielson & Kraft
	Wurdemannia miniata (Sprengel) Feldmann & G. Hamel
Order Gracilariales	
Family Gracilariaceae	*Gracilaria dotyi* Hoyle
	Gracilaria sp.
Order Halymeniales	
Family Halymeniaceae	*Cryptonemia* cf. *lomation* (Bertoloni) Agardh
	Cryptonemia cf. *umbraticola* Dawson
	Cryptonemia crenulata (J. Agardh) J. Agardh
	Cryptonemia umbraticola Dawson
	Grateloupia ovata Womersley & J.A. Lewis
	Halymenia maculata J. Agardh
	Halymenia porphyraeformis Parkinson
	Halymenia stipitata I.A. Abbott
	Prionitis angusta (Okamura) Okamura
Order Halymeniales	
Family Sebdeniaceae	*Sebdenia cerebriformis* N'Yeurt & Payri
	Sebdenia flabellata Zablackis
Order Nemaliales	
Family Galaxauraceae	*Actinotrichia fragilis* (Forsskål) Børgesen

	Dichotomaria australis (Sonder) Huisman, J.T. Harper & G.W. Saunders
	Dichotomaria marginata (Ellis & Solander) Lamarck
	Dichotomaria obtusata (Ellis & Solander) Lamarck
	Galaxaura divaricata (Linnaeus) Huisman & Townsend
	Galaxaura filamentosa R. Chou
	Galaxaura obtusata (Ellis & Solander) Lamouroux
	Galaxaura rugosa (Ellis & Solander) Lamouroux
	Tricleocarpa fragilis (Linnaeus) Huisman & Townsend
Order Nemaliales	
Family Liagoraceae	*Liagora* sp.
	Yamadaella caenomyce (Decaisne) I.A. Abbott
Order Nemaliales	
Family Scinaiaceae	*Scinaia furcata* Zablackis
Order Plocamiales	
Family Plocamiaceae	*Plocamium sandvicense* J. Agardh
	Plocamium sp.
Order Rhodymeniales	
Family Champiaceae	*Champia compressa* Harvey
	Champia parvula (C. Agardh) Harvey
	Champia vieillardii Kützing
Order Rhodymeniales	
Family Faucheaceae	*Gloiocladia iyoensis* (Okamura) R. Norris
Order Rhodymeniales	
Family Lomentariaceae	*Lomentaria corallicola* Børgesen
Order Rhodymeniales	
Family Rhodymeniaceae	*Asteromenia anastomosans* (Weber-van Bosse) G.W. Saunders, C.E. Lane, C.W. Schneider & Kraft
	Asteromenia pseudocoalescens G.W. Saunders, C.E. Lane, C.W. Schneider & Kraft
	Botryocladia kuckuckii (Weber-van Bosse) Yamada & Tanaka
	Botryocladia skottsbergii (Børgesen) Levring
	Botryocladia spinulifera W.R. Taylor & I.A. Abbott
	Chamaebotrys boergesenii (Weber-van Bosse) Huisman
	Chrysymenia procumbens Weber-van Bosse

	Coelothrix irregularis (Harvey) Børgesen
	Gelidiopsis intricata (C. Agardh) Vickers
	Gelidiopsis repens (Kützing) Weber-van Bosse
	Gelidiopsis scoparia (Montagne & Millardet) De Toni
	Halichrysis irregularis (Kützing) A.J.K. Millar
	Leptofauchea sp.
	Rhodymenia intricata (Okamura) Okamura
	Rhodymenia pacifica Kylin
	Rhodymenia sp. 1
	Rhodymenia sp. 2

Class Chlorophyta

Order Bryopsidales	
Family Bryopsidaceae	*Bryopsis pennata* J.V. Lamouroux var. *secunda* (Harvey) Collins & Hervey
Order Bryopsidales	
Family Caulerpaceae	*Caulerpa biserrulata* Sonder
	Caulerpa brachypus Harvey
	Caulerpa cupressoides (Vahl) C. Agardh
	Caulerpa fastigiata Montagne
	Caulerpa fergusonii Murray
	Caulerpa manorensis Nizamuddin
	Caulerpa microphysa (Weber-van Bosse) Feldmann
	Caulerpa nummularia Harvey ex J. Agardh
	Caulerpa racemosa (Forsskål) J. Agardh var. *clavifera* Turner (Weber-van Bosse)
	Caulerpa racemosa (Forsskål) J. Agardh var. *lamourouxii* (Turner) Weber-van Bosse
	Caulerpa racemosa (Forsskål) J. Agardh var. *peltata* (Lamouroux) Eubank
	Caulerpa sedoides C. Agardh
	Caulerpa serrulata (Forsskål) J. Agardh
	Caulerpa sertularioides (S. Gmelin) M. Howe
	Caulerpa taxifolia (Vahl) C. Agardh
	Caulerpa verticillata J. Agardh
	Caulerpa webbiana Montagne

	Caulerpella ambigua (Okamura) Prud'homme van Reine & Lokhorst
Order Bryopsidales	
Family Codiaceae	*Codium arabicum* Kützing
	Codium geppiorum O.C. Schmidt
	Codium mamillosum Harvey
	Codium ovale Zanardini
Order Bryopsidales	
Family Halimedaceae	*Halimeda borneensis* W.R. Taylor
	Halimeda cuneata K. Hering
	Halimeda cylindracea Decaisne
	Halimeda discoidea Decaisne
	Halimeda distorta (Yamada) Hillis-Colinvaux
	Halimeda gigas W.R. Taylor
	Halimeda heteromorpha N'Yeurt
	Halimeda lacunalis (W.R. Taylor) Hillis
	Halimeda macroloba Decaisne
	Halimeda macrophysa Askenasy
	Halimeda micronesica Yamada
	Halimeda minima (W.R. Taylor) Colinvaux
	Halimeda opuntia (Linnaeus) Lamouroux
	Halimeda taenicola W.R. Taylor
Order Bryopsidales	
Family Pseudocodiaceae	*Pseudocodium floridanum* Dawes & Mathieson
Order Bryopsidales	
Family Udoteaceae	*Avrainvillea erecta* (Berkeley) A. Gepp & E. Gepp
	Avrainvillea lacerata Harvey ex J. Agardh
	Boodleopsis pusilla (Collins) W. Taylor, Joly & Bernatowicz
	Chlorodesmis fastigiata (C. Agardh) Ducker
	Chlorodesmis hildebrandtii A. Gepp & E. Gepp
	Rhipidosiphon javensis Montagne
	Rhipilia crassa A.J.K. Millar & G.T. Kraft
	Rhipilia penicilloides N'Yeurt & Keats
	Rhipilia sinuosa Gilbert
	Rhipilia sp. nov.
	Rhipiliopsis carolyniae Kraft
	Rhipiliopsis echinocaulos (A.B. Cribb) Farghaly

Order	Family	Species
		Rhipiliopsis howensis Kraft
		Siphonogramen sp.
		Tydemania expeditionis Weber-van Bosse
		Udotea argentea Zanardini
Order Cladophorales		
	Family Anadyomenaceaea	*Anadyomene wrightii* Harvey ex J. Gray
		Microdictyon umbilicatum (Velley) Zanardini
Order Siphonocladales		
	Family Boodleaceae	*Phyllodictyon anastomosans* (Harvey) Kraft & M.J. Wynne
Order Cladophorales		
	Family Cladophoraceae	*Chaetomorpha antennina* (Bory de Saint-Vincent) Kützing
		Cladophora dotyana Gilbert
		Cladophora glomerata (L.) Kutzing
		Cladophora liebetruthii Grunow
		Cladophora ohkuboana Holmes
		Cladophora prehendens Kraft & Millar
		Cladophora sp.
Order Siphonocladales		
	Family Siphonocladaceae	*Boergesenia forbesii* (Harvey) J. Feldmann
Order Dasycladales		
	Family Dasycladaceae	*Bornetella nitida* Sonder
		Bornetella sphaerica (Zanardini) Solms-Laubach
		Neomeris vanbosseae Howe
Order Siphonocladales		
	Family Boodleaceae	*Boodlea composita* (Harvey) F. Brand
		Cladophoropsis herpestica (Montagne) M.A. Howe
		Cladophoropsis vaucheriaeformis (J.E Areschoug) Papenfuss
		Struvea elegans Børgesen
Order Siphonocladales		
	Family Siphonocladaceae	*Dictyosphaeria cavernosa* (Forsskål) Børgesen
		Dictyosphaeria intermedia Weber-van Bosse
		Dictyosphaeria versluysii Weber-van Bosse
		Siphonocladus sp.
Order Siphonocladales		
	Family Valoniaceae	*Valonia aegagropila* C. Agardh
		Valonia fastigiata Harvey ex J. Agardh
		Valonia macrophysa Kützing
		Valonia ventricosa J. Agardh
		Valoniopsis pachynema (G. Martens) Børgesen
Order Ulvales		
	Family Ulvaceae	*Ulva intestinalis* (Linnaeus) Nees
		Ulva lactuca Linnaeus

Class Phaeophyceae

Order	Family	Species
Order Dictyotales		
	Family Dictyotaceae	*Dictyopteris repens* (Okamura) Børgesen
		Dictyopteris sp.
		Dictyota bartayresiana Lamouroux
		Dictyota ceylanica Kützing
		Dictyota cf. *canaliculata* O. De Clerck & E. Coppejans
		Dictyota cf. *friabilis* Setchell
		Dictyota cf. *pfaffii* Schnetter
		Dictyota divaricata Lamouroux
		Dictyota friabilis Setchell
		Dictyota grossedentata De Clerck & Coppejans
		Dictyota hamifera Setchell
		Dictyota sp.
		Distromium sp.
		Lobophora papenfussii (W.R. Taylor) Farghaly
		Lobophora variegata (Lamouroux) Womersley ex Oliveira
		Padina boryana Thyvi
		Padina melemele Abbott & Magruder in Abbott
		Padina sp.
		Padina sp. nov.
		Stypopodium flabelliforme Weber-van Bosse
Order Ectocarpales		
	Family Acinetosporaceae	*Hincksia indica* (Sonder) J. Tanaka
Order Fucales		
	Family Sargassaceae	*Sargassum aquifolium* (Turner) C. Agardh
		Spatoglossum asperum J. Agardh
		Turbinaria ornata (Turner) J. Agardh
Order Sphacelariales		
	Family Sphacelariaceae	*Sphacelaria tribuloides* Meneghini

Figure 423: Rhodophyta. **A**: *Amansia rhodantha*. **B**: *Amphiroa crassa*. **C**: *Amphiroa foliacea*. **D**: *Asteronemia anastomosans*. **E**: *Botryocladia spinuligera*. **F**: *Callophycus serratus*. **G**: *Cheilosporum spectabile*. **H**: Corallinales complex. (Photos J.-L. Menou IRD Nouméa).

Figure 424: Rhodophyta. **I**: *Dasyphila plumarioides*. **J**: *Dichotomaria marginata*. **K**: *Dichotomaria obtusata*. **L**: *Galaxaura divaricata*. **M**: *Gibsmithia hawaiiensis*. **N**: *Halymenia porphyraeformis*. **O**: *Halymenia stipitata*. **P**: *Lithothamnion proliferum*. (Photos J.-L. Menou IRD Nouméa).

Figure 425: Rhodophyta. **Q**: *Martensia flabellata*. **R**: *Martensia* sp. nov. **S**: *Neogoniolithon fosliei*. **T**: *Peyssonnelia inamoena*. **U**: *Plocamium sandvicense*. **V**: *Portieria hornemanii*. **W**: *Predaea laciniosa*. **X**: *Titanophora weberae*. (Photos J.-L. Menou IRD Nouméa).

Figure 426: Chlorophyta. **A**: *Caulerpa bisserulata*. **B**: *Caulerpa fergusoni*. **C**: *Codium mamillosum*. **D**: *Halimeda cuneata*. **E**: *Halimeda discoidea*. **F**: *Halimeda lacunalis*. **G**: *Halimeda macroloba*. **H**: *Halimeda minima*. (Photos J.-L. Menou IRD Nouméa).

Figure 427: Chlorophyta. **I**: *Halimeda taenicola*. **J**: *Avrainvillea erecta*. **K**: *Rhipilia* sp. **L**: *Tydemania expeditionis*. **M**: *Cladophora ohkuboana*. **N**: *Bornetella nitida*. **O**: *Cladophorospsis herpestica*. **P**: *Valonia ventricosa*. (Photos J.-L. Menou IRD Nouméa).

Figure 428: Phaeophyceae. **A**: *Dictyota barteyresiana*. **B**: *Dictyota ceylanica*. **C**: *Dictyota friabilis*. **D**: *Distromium* sp. **E**: *Padina boryana*. **F**: *Padina melemele*. **G**: *Padina* sp. **H**: *Stypopodium*. (Photos J.-L. Menou IRD Nouméa).

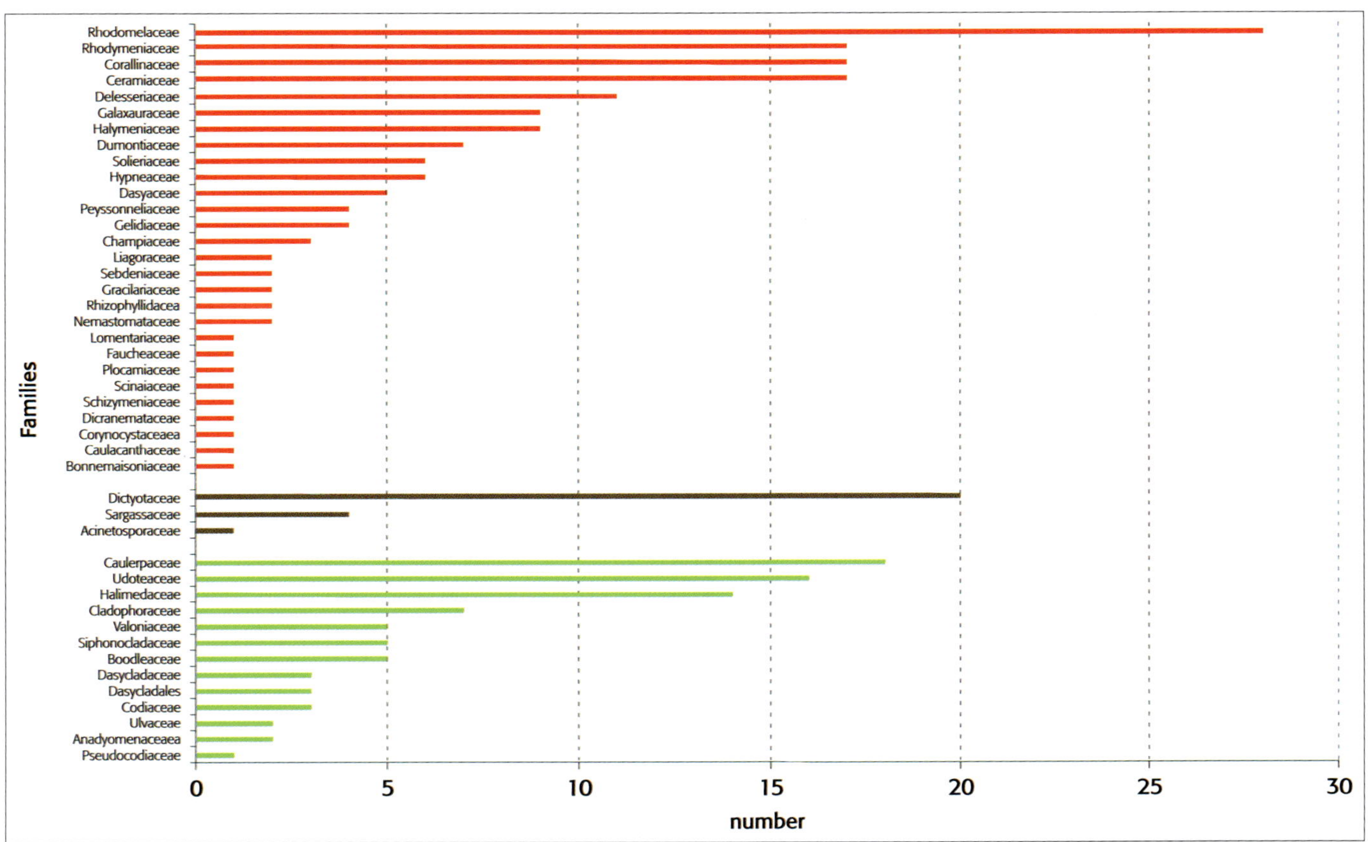

Figure 429: Species richness of the families of Rhodophyta (red), Phaeophyceae (brown) and Chlorophyta (green).

Algal assemblages

Algal vegetation on outer reefs and slopes down to 20 m

The algal community on the outer reef and slope down to 20 m in depth comprises a large number of encrusting coralline algae mixed with several dozen species growing among corals. Near the top of the reef, many species — mainly red algae — grow within the interstices of corals, and include *Chondrophycus parvipapillatus*, *Avrainvillea lacerata*, *Martensia flabelliformis*, *Halymenia porphyraeformis*, *Meristotheca procumbens*, *Champia vieillardii*, *Caulerpa nummularia*, *C. biserrulata* and *Halimeda micronesica*. In the breakwater area coralline algae *Hydrolithon onkodes* and *Neogoniolithon fosliei* develop thick candle-like crusts, with *Hydrolithon orthoblatum* or branched clumps of *Lithophyllum pygmaeum*. The vegetation can vary according to the topography and the presence of gutters and grooves is often associated with large clumps of *Callophycus serratus*, *Cheilosporum spectabile*, *Asparagopsis taxiformis*, *Dasyphila plumarioides*, *Tricleocarpa fragilis*, *Caulerpa* spp. and *Halimeda* spp. and small species such as *Chondria armata*, *Botryocladia* spp., *Chamaebotrys boergesenii* and *Portieria hornemanii*. The pinkish colours of coralline algae contrast with

the very bright green pompom-like morphology of *Chlorodesmis hildenbrandtii* and *Rhipilia penicilloides*. Further down the reef slope, from 8-20 m deep, the motion of the water is reduced and the reefs support a higher coral cover and articulated calcareous algae such as *Amphiroa crassa*, *A. tribulus* and *A. foliacea*, and the green *Halimeda cuneata*, *H. gigas*, *H. minima* and *H. taenicola* dominate some reef slopes. Fleshy algae are less abundant and mostly comprise *Gibsmitha hawaiiensis*, *Amansia rhodanta* and *Valonia*

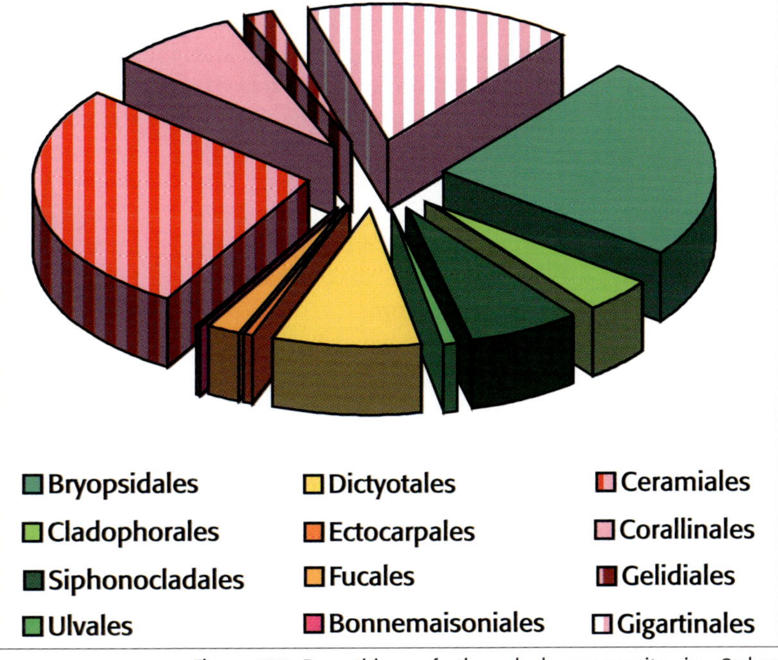

Figure 430: Repartition of the algal community in Orders Rhodophyta (red), Phaephyceae (brown), Chlorophyta (green).

fastigiata. Rubble is often found at the bottom of the slope rupture, at ~ 15 m deep and *Caulerpa serrulata, C. sedoides, Microdictyon umbilicatum, Neomeris van-bossea, Halimeda distorta, Valonia aegagropila, Myriogramme melanesiensis, Stypopodium flabelliforme* and *Padina* spp. grow among the coral debris. Various thin and small fronds of dark green *Rhipilia* spp. and *Rhipiliopsis* spp., *Anadyomene wrightii* form small associations in the shady areas with *Corynocystis protrata* and *Cryptonemia crenulata.*

••• Algal vegetation on deep outer slopes
In the outer slope from 40 m to at least 60 m deep, coral debris and coarse sand dominate the substratum. At the first glance fleshy algae are relatively few in species number and the vegetation is not luxuriant. Most of the gelatinous red algae such as *Dudresnaya capricornica, Predaea weldii, P. laciniosa, Gibsmithia hawaiiensis, G. larkumii,* and the green algae *Caulerpa fergusonii, C. sedoides, C. microphysa, Codium mamillosum, Rhipilia* sp. nov. grow in this environment on and among the coral debris, while the delicate green *Cladophora ohkuboana, C. dotyana* are found on sand. Brown algae are very few and *Dictyota bartayresiana* and *Padina* groupe *melemele* can be observed in this deep habitat.

••• Algal vegetation on coral walls
Coral walls usually start deeper than 30 m on the edge of limestone islands located in deep open water. This environment is often dark due to a heavily variegated surface with numerous interstices, overhangs and small caves. Coral walls are of great beauty with spectacular encrustations by coralline algae and *Peyssonnelia* spp. giving an attractive mosaic of forms and colours. The dominant *Lithothamnion prolifer* is easily recognisable by its pink crust and numerous short knobs. Numerous Rhodymeniales including several species of *Leptofauchea* and *Rhodymenia* live in the caves and interstices with *Cryptonemia crenulata, C. umbraticola, Corynocystis prostrata* and *Callophycus serratus,* while the iridescent *Halichrysis irregularis* and the star-like *Asteromenia anastomosans* grow luxuriantly on the walls with the large foliose *Peyssonnelia inamoena* and *P. capensis.* The golden-yellow *Padina melemele* and the green ball-like *Codium mamillosum* are present in discrete clumps among coral debris with very occasional *Sebdenia flabellata* and *S. cerebriformis.*

All these species can be found in other deep areas but in less abundance. The shady environment and open ocean influences enhance this algal community that is generally sheltered in the reef interstices.

••• Algal vegetation on the sandy bottom of deep lagoons
Various green algae grow together and develop meadows between coral colonies located on the coral sandy bottom at 25-30 m deep in the embayments.

This sheltered and silty environment supports luxuriant vegetation including: *Udotea argentea, Avrainvillea erecta, Halimeda borneensis, H. distorta, Caulerpa verticilata, C. cupressoides, C. racemosa, C. sedoides, C. serrulata, C. taxifolia* as well as some red algae such as *Martensia, Titanophora webera* and the brown alga *Stypopodium flabelliforme* with its fan-like shape and iridescent blue on the thallus surface.

••• Algal vegetation on shallow reef flats
The shallow fringing reef flat along the shoreline to the north of Luganville supports many algae from the beach to the reef front. The flats are exposed at low tide and corals are therefore reduced in abundance, except at the outer part of the reef flat where large stands of staghorn *Acropora* grow in the gutters perpendicular to the reef front. Adjacent to the beach, the reef is covered with a green underwater "turf" mainly composed of *Cladophora glomerata, Boodlea composita* and *Boergesenia forbesii* that is partially buried in the sand. Several *Caulerpa, C. fastigiata, C. racemosa, C. serrulata* along with light green *Chlorodesmis fastigiata, Halimeda opuntia* and the red pompom-like algae *Galaxaura filamentosa* and *G. rugosa* grow on the inner part of the reef. Among the coral branches there are numerous green algae *Dictyosphaeria cavernosa* and *H. micronesica.* The edible red seaweed *Meristotheca procumbens* was abundant within the coral branches and in the interstices on the reef margin. The vegetation on the front part is dominated by nongeniculated coralline algae including crusts of *Hydrolithon onkodes* and the candle-like thallus of *H. orthoblastum.* Various articulated coralline algae such as *Amphiroa* spp. form clumps on the reef top. Surprisingly, no stands of *Sargassum* were observed except occasional young stages of *Sargassum aquifolium.*

••• Algal vegetation in shallow sandy coral communities
The algal vegetation associated with the coral community in shallow sandy environments is mainly represented by patches of the fan-like brown alga *Padina boryana* mixed with another brown alga *Turbinaria ornata* and various species of green algae such as *Caulerpa cupressoides, C. racemosa, C. racemosa* var *lamourouxii, C. fergusonii, Boodlea composita* and *Udotea argentea* and the red algae *Galaxaura rugosa, Hypnea* spp. and *Tolypiocladia glomerata. Microdictyon umbilicatum, Halimeda discoidea* and *Myriogramme melanesiensis* grow among the branches of staghorn *Acropora.* Moreover, the delicate red algae *Martensia fragilis, Neomartensia flabelliforme, Haloplegma duppereyi,* along with *Laurencia* spp., *Exophyllym wentii* and several Rhodymeniales form small associations of a rich algal flora in the interstices of submassive corals *Porites rus* and *Montipora.*

••• Algal vegetation in channel environments
In general, the algal vegetation of the channels is not very rich due to the silty and muddy environment that

limits algal diversity. There is no typical association of algae from this environment except the brown alga *Spatoglossum asperum* which has only been recorded in the Segond Channel and Malo passage. The algal flora has characteristics that are typical of sheltered areas; *Halimeda* spp. and *Caulerpa* spp. can locally cover the substratum and most of the investigated sites showed coral damage. The algal vegetation associated with dead coral communities is described in the next section.

••• Algal vegetation on dead coral communities

Santo coral reefs have experienced heavy damage from successive cyclones, bleaching events and crown-of-thorns starfish (*Acanthaster*) outbreaks in the past decades. Thus on the outer reefs and slopes along the channel, the dead corals are colonized by small prostate algae such as the brown *Dictyota friabilis*, *Lobophora variegata* and large green calcareous *Halimeda distorta* and *H. minima*. In more exposed locations, dead and collapsing branches are overgrown by coralline algae, turfs of filamentous red algae and cyanobacteria assemblages. Depending on the local environmental and reef condition, this pioneer stage of colonization will evolve into a secondary succession of algal-dominated communities or revert to coral recolonization.

••• Remarks on the absence of *Sargassum* beds

Sargassum species are common features of the algal vegetation of tropical islands in the Pacific. However, around Santo this genus is restricted to limited germlings and the reduced thalli of *Sargassum aquifolium* on reef flats, while personal observations in Efate at the same period have shown the presence of large beds of *Sargassum* including several common tropical species such as *S. polycystum*. The lack of suitable habitats such as sheltered shallow lagoons could explain the absence of the species in these biotopes around Santo.

••• Seagrass communities

Seagrasses are flowering plants belonging to the Cymodoceaceae and the Hydrocharitaceae families which are placed in the Alismatales order (nomenclature based on molecular analysis). In tropical regions they are mostly permanently submerged in marine and estuarine biotopes that are generally sheltered from wave action and offer a suitable substratum for rooting in mud, sand or coarse rubble. In many places they can also develop into large meadows or beds in deeper lagoon zones down to 40 m deep, on barrier reefs or surrounding lagoon islands. They are remarkable habitats in tropical shallow waters and they often form a key functioning system on sandy bottoms along shorelines between mangroves and coral reefs.

Most coastal areas around Santo do not have these typical seagrass habitats and only the fringing sandy flats adjacent to estuarine and river catchments, sheltered embayments and inner reef sandy flats provide the necessary conditions for seagrass development. However, deep sandy slopes, sandy channel slopes and bottoms also support the paddle-like *Halophila* seagrasses.

Eight species of seagrass were reported from our survey (Table 38), four of these are new records for Santo: *Cymodocea serrulata*, *Enhalus acoroides*, *Halophila capricorni* and *H. decipiens*, and the two species of *Halophila* had not been previously recorded for the Vanuatu archipelago.

Seagrass diversity and abundance were relatively low in the investigated areas. Plants never form large meadows; they mostly developed in scattered patches except in Palikolo bay where they form dense mats (>75 % coverage) in 70 m wide zones that represent the most extensive bed surveyed. The seagrass communities generally comprised few species; most of the sites had just one to three species growing together. The inner sandy areas such as Palikolo bay, the Aore shoreline in the Malo passage and the estuarine zone adjacent to Luganville showed the highest species diversity with four species growing together. However, most of the time one species was dominant in the bed, i.e. *Halodule uninervis* in Luganville, *Cymodocea rotundata* in Palikolo. In some localities seagrasses form mixed communities with marine algae such as *Halimeda macroloba*, *H. cylindracea*, *H. borneensis*, *Caulerpa serrulata*, *Padina boryana* and *Acanthophora spicifera*.

Table 38: List of seagrass species from Santo waters.

Class	Order	Family	Genus	Species	Authority
Anthophyta	Alismatales	Cymodoceacea	*Cymodocea*	*rotundata*	(Hemprich & Ehrenberg) Aschers & Schweinf
			Cymodocea	*serrulata*	(R. Brown) Aschers & Magnus
			Halodule	*uninervis*	(Forsskål) Ascherson in Boissier
		Hydrocharitacea	*Enhalus*	*acoroides*	(Linnaeus) Royle
			Halophila	*capricorni*	Larkum
			Halophila	*decipiens*	Ostenfed
			Halophila	*ovalis*	(R. Brown) J.D. Hooker
			Thalassia	*hemprichii*	(Ehrenberg) Ascherson

Except for the species of *Halophila*, all the other taxa were confined to very shallow waters although they are known to grow in deeper habitats elsewhere.

It is clear that the coastal physiography of Santo does not provide ideal habitats for seagrass meadows, but it is not clear why seagrasses are so restricted in shallow waters and are not well developed in other areas that appear to be suitable. Part of the explanation could be due to climatic conditions. The high occurrence of cyclones and rough seas can provoke sediment movements and salinity changes, which may have prevented the establishment of seagrasses or removed beds which would both have limited the development or the absence of this key functioning habitat. This situation could turn critical with the predicted increase of threats as a result of human activities and climate change. Seagrass habitats must be considered as associated ecosystems to coral reefs just like mangroves. All these habitats are important and integral components of the natural environment of Santo and they must be considered as priorities in conservation efforts. This study provides information that could aid coastal zone planning and development.

Figure 431: Seagrass species. **A**: *Cymodocea serrulata*. **B**: *Halodule uninervis*. **C**: *Enhalus acoroides*. **D**: *Thalassia hemprechii*. **E**: *Halophila capricorni*. (Photos J.-L. Menou IRD Nouméa).

# of Santo in Relation to the Centre of Maximum Marine Biodiversity (the Coral Triangle) Based on Mushroom Corals and their Associated Mollusc Fauna

Bert W. Hoeksema & Adriaan Gittenberger

The centre of marine maximum marine biodiversity has become increasingly important as a means to draw attention to the conservation of coral reefs. Due to its shape, it has been named the Coral Triangle, which is supposed to encompass all or some of the reefs of the Philippines, Malaysia, Indonesia, Timor-Leste, Papua New Guinea, and the Solomon Islands. The criteria used to define this diversity centre as it is presently recognized, are based on high numbers of species recorded from within this centre. However, data from within and, especially, from outside the centre's hypothetical boundaries are far from complete due to insufficient sampling. Ideally, study areas should be surveyed by the same scientists using the same methods for reaching any conclusions regarding their position in- or outside the centre of maximum marine biodiversity.

In order to get a clear picture of biodiversity patterns for various groups of animals and plants, large research teams are needed to visit many different localities. Another way to investigate species diversity patterns is by selection of representative species groups that have been well studied by means of taxonomic revisions and many field surveys. An example of such a target group is the mushroom coral family (Scleractinia: Fungiidae), which consists of at least 47 species. Most of these species were treated in a taxonomic revision: *Taxonomy, phylogeny and biogeography of mushroom corals (Scleractinia: Fungiidae)* by Hoeksema (1989), whereas additional ones were described later on, or still have to be described. Use of such target groups will give a consistent overview of species compositions over various geographic areas, in which not only the presence of species is known, but also the absence of species should be indicated with a high reliability using species richness estimators.

Reef corals, reef fish, and molluscs are usually recognized as the most important indicator groups of animals that tell us whether a reef area is rich in species or not. However, no reef-dwelling species is living on its own, since each one is part of a network of inter-specific associations such as parasitism, commensalism, predator-prey relations, etc. Some of these interrelations are very explicit, in which, for example, a parasitic snail depends on a single particular host coral. Presence or absence of particular host species is therefore also indicative for the occurrence of the associated fauna, which is demonstrated by various groups of molluscs, such as parasitic wentletrap snails (Epitoniidae), boring coral snails belonging to coral eating Coralliophilidae (*Leptoconchus* spp.), and boring mussels (Mytillidae: *Fungiacava eilatensis*, *Lithophaga* spp.).

METHODS

Mushroom corals and their associated molluscs were studied during the Santo 2006 expedition, at the Southeastern side of Espiritu Santo, in the vicinity of the base camp at Luganville. The mushroom coral species were recorded during 25 dives in a time span of 15 days (September 2006). The molluscs were recorded at the same sites and at additional localities in a period of nearly two months.

CORAL SPECIES RICHNESS

A total of 35 species of mushroom coral species was recorded (Table 39). Of these, 34 were observed during the surveys. The other one, Fungia (*Cycloseris*) cyclolites, could be listed thanks collecting efforts during a separate dive at 40 m depth by Mr. Eric Folcher, professional diver of IRD (Nouméa, New Caledonia). Of the 35 species, 16 are new records for Vanuatu and one of these, a *Sandalolitha* sp., is

Table 39: Mushroom coral species encountered at 25 localities (dive sites) during the Santo 2006 expedition. Per species the number of localities is mentioned at which it was recorded to indicate its relative abundance. One species (*) was observed during a separate occasion. Earlier records for Vanuatu are indicated by their publication.

Species	Number of dive sites	Earlier records
1. *Cantharellus jebbi* Hoeksema, 1993	17	
2. *Ctenactis albitentaculata* Hoeksema, 1989	6	
3. *Ctenactis echinata* (Pallas, 1766)	18	Veron 1990a
4. *Ctenactis crassa* Dana, 1846)	17	Veron 1990a [1]
5. *Fungia (Cycloseris) costulata* Ortmann, 1889	16	
6. *Fungia (Cycloseris) cyclolites* Lamarck, 1815	0*	
7. *Fungia (C.) fragilis* (Alcock, 1893)	4	Veron 1990a [2]
8. *Fungia (C.) hexagonalis* Milne Edwards & Haime, 1848	1	
9. *Fungia (C.) sinensis* (Milne Edwards & Haime, 1851)	7	
10. *Fungia (C.) somervillei* Gardiner, 1909	2	
11. *Fungia (C.) tenuis* Dana, 1846	1	
12. *Fungia (C.) vaughani* Boschma, 1923	3	
13. *Fungia (Cycloseris)* sp.	5	
14. *Fungia (Danafungia) horrida* Dana, 1846	15	Hoeksema 1989; Veron 1990a [3]
15. *Fungia (D.) scruposa* Klunzinger, 1879	13	Veron 1990a [4]
16. *Fungia (Fungia) fungites* (Linnaeus, 1758)	24	Hoeksema 1989; Veron 1990a
17. *Fungia (Lobactis) scutaria* Lamarck, 1801	12	Veron 1990a
18. *Fungia (Pleuractis) gravis* Nemenzo, 1955	9	
19. *Fungia (P.) moluccensis* Van der Horst, 1919	9	
20. *Fungia (P.) paumotensis* Stutchbury, 1833	23	Veron 1990a
21. *Fungia (Verrillofungia) concinna* Verrill, 1864	18	Veron 1990a
22. *Fungia (V.) repanda* Dana, 1846	22	Veron 1990a
23. *Fungia (V.) spinifer* Claereboudt & Hoeksema, 1987	9	
24. *Fungia (Wellsofungia) granulosa* Klunzinger, 1879	19	Veron 1990a
25. *Halomitra pileus* (Linnaeus, 1758)	1	
26. *Heliofungia actiniformis* (Quoy & Gaimard, 1833)	5	Hoeksema 1989
27. *Herpolitha limax* (Esper, 1797)	23	Hoeksema 1989; Veron 1990a
28. *Lithophyllon mokai* Hoeksema, 1989	9	Veron 1990a [5]
29. *Podabacia crustacea* (Pallas, 1766)	3	Veron 1990a
30. *Podabacia motuporensis* Veron, 1990	1	Veron, 1990a [6]
31. *Polyphyllia novaehiberniae* (Lesson, 1831)	16	Hoeksema 1989; Veron 1990a
32. *Polyphyllia talpina* (Lamarck, 1801)	15	
33. *Sandalolitha robusta* (Quelch, 1886)	22	Veron 1990a
34. *Sandalolitha* sp.	13	
35. *Zoopilus echinatus* Dana, 1846	1	Veron 1990a

Notes: (1) As *Fungia (Ctenactis) simplex* (Gardiner, 1905); (2) As *Cycloseris patelliformis* (Boschma, 1923); (3) also as *Fungia (D.) valida* Verrill, 1864, and *F. (D.) klunzingeri* Döderlein, 1901; (4) as *Fungia (Danafungia) danai* Milne Edwards & Haime, 1851; (5) as *Lithophyllon undulatum* Rehberg, 1892; (6) as *Podabacia* sp. (see Veron 1990b).

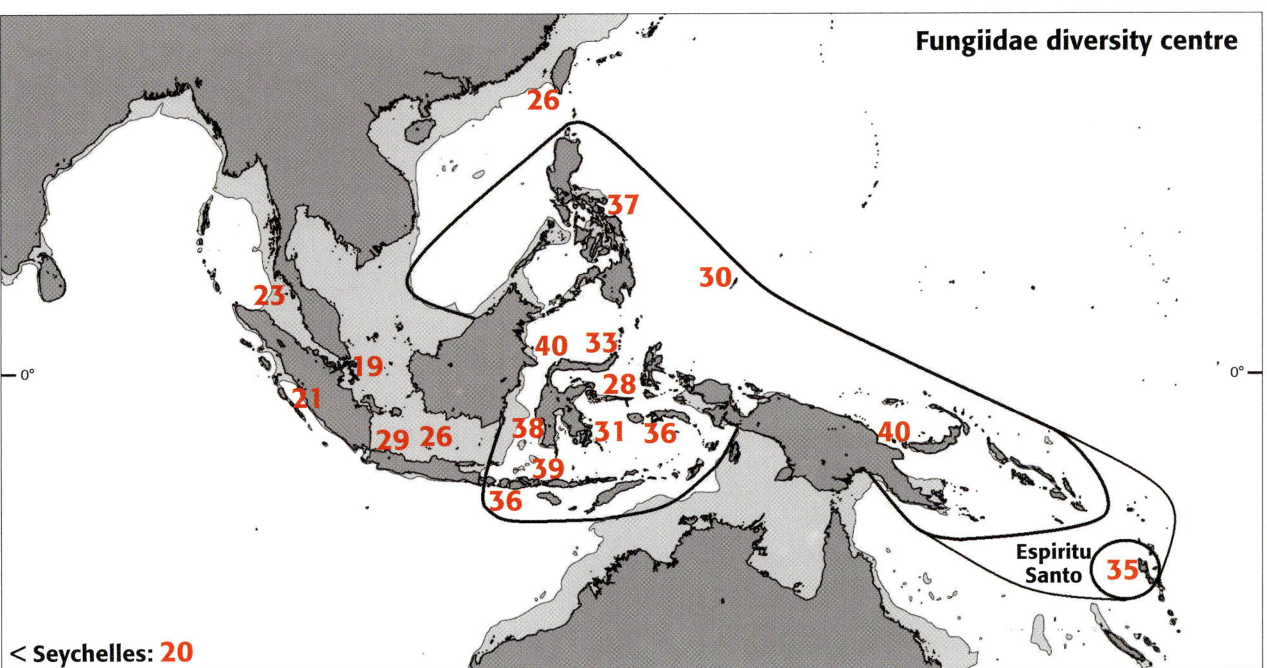

Figure 432: A hypothetical Coral Triangle based on mushroom coral data (Hoeksema 2007), with an additional southeastern extension including Espiritu Santo, based on results obtained during the Santo 2006 expedition.

new to science. It is remarkable that many species were abundant (present at almost each site), whereas other ones were quite rare (Table 39), which may be important for their role as potential host species for associated molluscs. Twelve species were absent:

- *Fungia (Cycloseris) distorta* Michelin, 1842
- *F. (Verrillofungia) scabra* Döderlein, 1901
- *F. (Danafungia) fralinae* Nemenzo, 1955
- *F. (Pleuractis) taiwanensis* Hoeksema & Dai, 1991
- *F. (P.) seychellensis* Hoeksema, 1993
- *Sandalolitha dentata* Quelch, 1884
- *Halomitra clavator* Hoeksema, 1989
- *Lithophyllon undulatum* Rehberg, 1892
- *L. ranjithi* Ditlev, 2003
- *Cantharellus doederleini* (Von Marenzeller, 1907)
- *C. noumeae* Hoeksema & Best, 1984
- *Podabacia* sp.

The 19 species (54%) that have been recorded earlier from Vanuatu were either represented in museum collections or observed during an Australian expedition. The presently reported increase in the number of mushroom coral species of the known coral fauna of Vanuatu, suggests that the number of all reef coral species (296) recorded previously by Veron in 1990, should be much higher also. On the other hand, this number also includes some synonyms (Table 39). Extrapolation of the mushroom coral species numbers projected on all reef coral species, suggests that more than 500 reef coral species are to be expected, which would be enough to consider Espiritu Santo part of the centre of coral species richness. Therefore, the present mushroom coral data suggest that the Coral Triangle should have a southeastern extension, including Espiritu Santo and adjacent parts of Vanuatu (Fig. 432).

ASSOCIATED MOLLUSCS

Just a list of coral species does not give us complete idea of marine biodiversity. Most coral species are known to act as host for symbionts, such as crustaceans and molluscs. We found representatives of various species of boring mussels (Mytilidae: *Lithophaga* spp., *Fungiacava eilatensis*) and snails (Coralliophilidae: *Leptoconchus* spp.) living as endoparasites inside mushroom corals (Table 40). Furthermore, a total of nine species of parasitic wentletrap snails (Epitoniidae) has been found in association with mushroom corals only. This number represents just more than 50% of the 17 epitoniids known to occur with Fungiidae. An additional three species were found on other coral hosts (Table 40). Such observations are only possible when the coral hosts are well known and systematically searched for their associated fauna.

Seven epitoniid species were not found despite the presence of potential coral hosts, i.e.:

- *Surrepifungium oliverioi* (Bonfitto & Sabelli, 2001) known from various host species,
- *Epitonium crassicostatum* Gittenberger & Gittenberger, 2005, known from *Fungia (Cycloseris) costulata*,
- *E. graviarmatum* Gittenberger & Gittenberger, 2005, known from *F. (Cycloseris) vaughani*,
- *Epifungium adgranulosa* Gittenberger & Gittenberger, 2005, known from *F. (Wellsofungia) granulosa*,

Table 40: Stony corals (Scleractinia) and their associated molluscan fauna observed at Espiritu Santo (Santo 2006).

Fungiidae acting as host (with miscellaneous parasites)	Coralliophilidae	n	Epitoniidae	n
Ctenactis crassa	Leptoconchus sp.	1	Surrepifungium costulatum (Kiener, 1838)	4
			Surrepifungium ingridae (Gittenberger & Goud, 2000)	1
			Surrepifungium patamakanthini Gittenberger & Gittenberger, 2005	2
Ctenactis echinata			Surrepifungium costulatum (Kiener, 1838)	1
Fungia (Cycloseris) costulata - 1 Fungiacava eilatensis - 1 Lithophaga sp.	Leptoconchus sp.	5	Epifungium lochi (Gittenberger & Goud, 2000)	1
Fungia (Fungia) fungites	Leptoconchus sp.	2	Epifungium ulu (Pilsbry, 1921)	1
Fungia (Pleuractis) gravis	Leptoconchus sp.	1		
Fungia (Pleuractis) paumotensis	Leptoconchus sp.	2	Epifungium nielsi Gittenberger & Gittenberger, 2005	3
Fungia (Verrillofungia) concinna	Leptoconchus sp.	2		
Fungia (Verrillofungia) repanda	Leptoconchus sp.	4	Epifungium ulu (Pilsbry, 1921)	3
Fungia (Wellsofungia) granulosa - 1 Fungiacava eilatensis - 1 Coralliophila bulbiformis (1st record Coralliophila on Fungiidae, M. Oliverio pers. comm.)	Leptoconchus sp.	2	Epifungium adgranulosa Gittenberger & Gittenberger, 2005	1
Halomitra pileus			Epifungium ulu (Pilsbry, 1921)	1
Heliofungia actiniformis	Leptoconchus sp.	1		
Herpolitha limax - 1 Lithophaga sp.	Leptoconchus sp.	5	Epifungium twilae (Gittenberger & Goud, 2000)	8
			Surrepifungium ingridae (Gittenberger & Goud, 2000)	3
Podabacia novaehibernae	Leptoconchus sp.	4		
Polyphyllia talpina	Leptoconchus sp.	1		
Sandalolitha robusta - 1 Fungiacava eilatensis - Lithophaga sp. in three specimens of S. robusta			Epifungium pseudotwilae Gittenberger & Gittenberger, 2005	7
			Epifungium ulu (Pilsbry, 1921)	1
Zoopilus echinatus			Epifungium pseudotwilae Gittenberger & Gittenberger, 2005	1
Scleractinia: non-Fungiidae		**n**		**n**
Plerogyra diabolotus			Epifungium hartogi (Gittenberger, 2003)	1
Hydnophora rigida	Leptoconchus sp.	1		
Tubularia sp.			Epidendrium aureum Gittenberger & Gittenberger, 2005	1
			Epidendrium sordidum Gittenberger & Gittenberger, 2005	3

- *E. adgravis* Gittenberger & Gittenberger, 2005, from *F. (Pleuractis) gravis,*
- *E. hoeksemai* (Gittenberger & Goud, 2000) only known from *Fungia (F.) fungites* and *Heliofungia actiniformis,*
- *E. lochi* (Gittenberger & Goud, 2000) known from various *Fungia (Cycloseris)* spp.

Only two wentletrap snail species were not observed, wich was due to the absence of their coral host:
- *Epifungium adscabra* Gittenberger & Gittenberger, 2005, known from *Fungia (Verrillofungia) scabra,*
- *E. marki* Gittenberger & Gittenberger, 2005, only known from a *Fungia (Pleuractis)* sp. at Egypt.

Obviously, potential coral hosts were sufficiently available. Some of these hosts were even abundantly present (Table 39). Epitoniid species that were not recorded were either overlooked, restricted by environmental factors, or Espiritu Santo is not included in their distribution range.

on Selected (Micro)Habitats

SULPHIDE RICH ENVIRONMENTS

Yasunori Kano & Takuma Haga

Sulphur is an essential material for all organisms, playing a significant role in the synthesis of protein and enzyme. Regardless of this vital requirement, hydrogen sulphide —a major sulphuric compound in marine ecosystems— is highly toxic and often lethal for animals by preventing cellular respiration. Also called "sewer gas" with its foul odour of rotten eggs, hydrogen sulphide is produced by bacterial decomposition of organic matter in the absence of oxygen. The organic rich, oxygen poor "reducing" sediment in mudflats, mangrove swamps and other fine grained seabed is therefore an inappropriate and unsafe environment for most sediment-dwelling animals due to high concentration of sulphide, in combination with the low concentration of oxygen itself. Burrowing invertebrates including thalassinidean ghost shrimps construct tubular nests that allow water exchange and oxidization very deep in mud bottom. Yet, animal diversity in such sulphide rich sediment is understandably lower than in usual, fully aerobic marine environments.

However, there exist particular species that tolerate, or even prefer, the presence of hydrogen sulphide and shortage of oxygen. Although no animal completes its life cycle in absolute lack of oxygen, many are capable of growth and normal behaviour for extended periods under near-anoxic and sulphide-rich conditions. Nematode worms and other microscopic groups probably represent the hardiest examples, while larger invertebrates including certain molluscs and annelid worms show comparatively less but still amazing tolerance.

This tolerance often attributes to the unique nature and ability of their blood. A good number of bivalve and annelid species in mud sediment and similar harsh environments have crimson blood that looks very much like ours, clearly contrasting to the usual, bluish transparent blood of molluscs and crustaceans in well-oxidized habitats. Their red blood contains haemoglobin and provides greater capacity in transporting oxygen from the gill or other surface of the body to internal organs than that of the haemocyanin-containing, bluish blood. Moreover, some red blood in these invertebrates is less affected by sulphide poisoning and thus maintains cellular respiration in reducing environments. Sulphide tolerance also attributes to protection from symbiotic bacteria, which oxidize and neutralize the toxic compound for their energy. Such symbiotic bacteria are found either externally on the surface of the animal or internally in the gill or other organs.

Sulphide oxidizing bacteria have a more significant role for animals at the interface between anoxic and aerobic environments. The symbiotic bacteria provide the energy from sulphide oxidization to host animals, which partly or completely depend on this supply, with or without a degenerative digestive tract for normal food consumption. On the other hand, free-living members of sulphide oxidizing bacteria in sediment and on hard substrates may constitute the main diet for deposit feeders and grazers in the reducing habitats. Such free-living bacteria are ubiquitous and sometimes form large filamentous colonies that are visible to the naked eyes as white, hairy patches on buried stones and wood in mud sediment. The reducing environments can thus be favourable habitats for the specialized animals, often with a limited number of competitors and predators.

The sulphide-dependent ecosystem is first recognized at deep-sea hot springs called hydrothermal vents. Since the astonishing discoveries of the *Riftia* tubeworms, *Vesicomya* giant clams and *Bathymodiolus* mussels in the late 1970's, the relationship between these deep-sea animals and their symbiotic bacteria have attracted much attention from scientists, and a number of other symbiotic or non-symbiotic invertebrates have been described from the deep-sea vents. The vent ecosystem is truly unique as it is independent of the energy from sunlight and photosynthesis: hydrogen sulphide originates from the ocean plate heated by the hot magma. Soon after the discovery of the vent fauna, symbiotic association was found between sulphide oxidizing bacteria and several groups of clams (e.g. *Solemya, Lucinoma, Anodontia* and *Thyasira*) in mud sediment in intertidal and subtidal waters, where sulphide originates from bacterial decomposition of organic matter. Other reducing environments with the sulphide of organic origin include sunken wood and whalebone communities in the deep sea. Symbiotic animals including mytilid mussels and *Osedax* worms have attracted increasing interest along with a number of non-symbiotic, bacterial feeders in these communities.

Figure 433: Collecting molluscs from sulphide rich, oxygen poor habitats.
A: Turning over a dead coral, deep buried in sand-silt sediment, at a beach on Aore Island. **B**: Under-surface of the dead coral shows the orange colour of oxidized iron, a good sign for the anoxic/aerobic interface and its unique molluscan fauna, along with the smell of rotten eggs. **C**: Gastropods found on the surface (arrows): an unidentified skeneimorph snail (upper left) and a phenacolepadid limpet (right). **D**: *Phenacolepas scobinata*, another species of Phenacolepadidae from the same rock, also shown in E. The shell approximates 10 mm in diameter; arrowheads indicate its egg capsules. **E**: Phenacolepadids show amazing tolerance to oxygen-poor reducing water, thanks to their red blood cells and huge gill (arrows). Their degenerative eyes are useless in the cryptic habitat, but numerous tentacles, two long ones on the head and others on the mantle edge, compensate the lack of vision. **F**: Mangrove swamps provide more strongly reducing habitats, which are preferred by a special assortment of hardy gastropods. A new species of *Phenacolepas* was collected under decaying driftwood (arrow) at Belmoul Lagoon, Santo. **G**: *Cinnalepeta pulchella* is the commonest species of the family in mangrove swamps on western Pacific islands. Photo taken at Matafou, Santo. (Photos Y. Kano).

In contrast, much less attention has previously been paid for apparently non-symbiotic molluscs associated with reducing environments in more easily accessible tidal flats. These molluscs can be found on the under-surface of deeply embedded rocks and decaying wood, which create sealed and stable reducing condition (Figs 433A-B & 433F). The mollusc-bearing surface is often brown to orange in colour due to the activity of iron oxidizing bacteria, while sometimes it is black and covered with white, hairy patches of (possibly sulphide oxidizing) bacteria. The molluscan fauna of deep-buried rocks and wood is diverse but largely unexplored and contains countless of new species. Gastropod species exclusive to this assemblage belong to the family Phenacolepadidae, Rissoidae (some *Rissoina*), Iravadiidae, Elachisinidae, Caecidae, Vitrinellidae (e.g. *Pseudoliotia*, *Circulus* and *Teinostoma*), Costellariidae (some *Thala*), Cornirostridae, Orbitestellidae and Pyramidellidae (some *Odostomia*). Bivalves and polyplacophorans are comparatively rare, but galeommatid clams and leptochitonid chitons are conspicuous members of the fauna. Symbiosis with sulphide oxidizing bacteria has not yet been found in these molluscs. They are mostly grazers of bacteria or deposit feeders, while costellariids and pyramidellids are carnivorous and feed on other molluscs or annelids. Many of the true members of this reducing environment, including all phenacolepadids, some iravadiids and vitrinellids, and galeommatid *Barrimysia* have crimson to pinkish appearance of the body, suggesting the presence of extracellular or intracellular haemoglobin in their blood. Among the entire Gastropoda, erythrocytes or red blood cells containing haemoglobin have been found only in the neritimorph family Phenacolepadidae (Fig. 433E). Phenacolepadid limpets represent the commonest and most characteristic molluscs of the buried environment. Besides the unique blood, they evolved a huge gill to increase the capacity of oxygen intake. They are relatively large (4-20 mm) among the small molluscs of the environment; their flat body and shell, seemingly advantageous under rocks and wood, may possibly have acquired as a result of adaptation.

Phenacolepadids also dwell in various reducing habitats in the deep sea including hydrothermal vents, cold seeps and sunken pieces of wood. Interestingly, many other molluscan groups of the deep-buried objects in the tidal fat, such as Leptochitonidae, Iravadiidae, Elachisinidae, Vitrinellidae and Orbitestellidae, similarly have close relatives in the vent or sunken-wood community, suggesting ecological and evolutionary ties between the shallow- and deep-water biotopes. Future studies on this little known assemblage in the tidal flat may shed new light on the origin and evolution of the now extensively investigated fauna in the deep sea.

MARINE INTERSTITIAL

Timea Neusser

It was not until the beginning of the 20th century that biologists discovered the water-filled interstitial space between the grains of coastal marine sands as a habitat for organisms.

The interstitial environment is characterized by extreme ecological conditions, such as weak light or limited amount of space that determines the size of the organisms and restricts the meiofauna to tiny organisms suited to a lacunar environment. Wind and wave action modify the interstitial biotope by continuous restratification of the surface of the sand. The permanent rearrangement of the particles contributes to a dynamic environment and avoids the colonization by plants. Moreover, the living conditions in the intertidal zone or shallow water are complicated by different physical factors: the temperature is dependent on the daytimes, seasons and the rhythm of tides and thus, fluctuates considerably in the surface sand layers. On the other hand, the salinity decreases by rainfall or by the inflow of coastal freshwater and otherwise increases by evaporation.

Organisms which successfully colonize the marine interstitial, called the mesopsammon, often develop different morphological and biological adaptations. Body sizes are usually very small ranging from 0.5 mm to approx. 5 mm. Vermiform elongated or flat and broad body shapes are commonly favoured. The body wall often is reinforced by cuticle or subepidermal spicules for mechanical protection. Usually, members of the meiofauna have a good contractibility and a high adhesive capacity by epidermal glands to avoid being washed away. Locomotion occurs via ciliary gliding, crawling or writhing. The miniaturization of the species generally involves the reduction of gametes so that only approximately 1-10 oocytes mature per female at a time.

Many phyla of invertebrates are represented in the marine meiofauna, e.g. Cnidaria, Turbellaria, Nematoda, Polychaeta, Echinodermata, Arthropoda and Mollusca. However, the main focus of meiofauna sampling during this expedition was on opisthobranch molluscs.

Interstitial opisthobranch assemblages are subject to seasonal variations and comprise rheophilous species most of them living in clean and oxygenated waters. A long-term study in the Mediterranean Sea demonstrated that they are particularly sensitive to any clogging of their habitat, either by man-made coastal pollution or by decrease of marine hydrodynamism, resulting in an impoverishment or even disappearing of the opisthobranchs. Therefore, they have been proposed as biological indicator organisms in the past.

While the interstitial fauna along the European coast has been extensively sampled, the interstitial

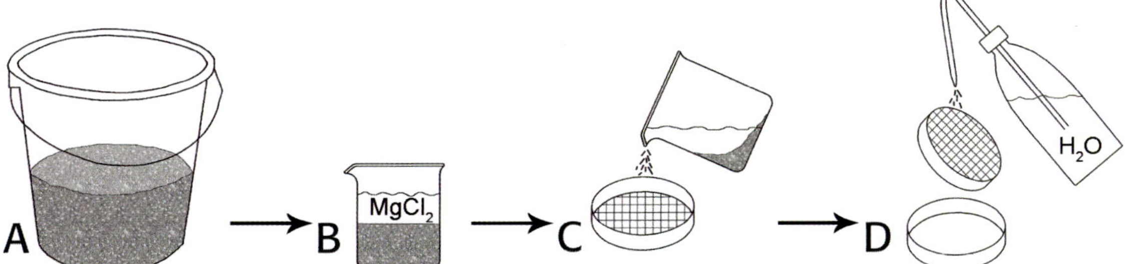

Figure 434: Method for collecting interstitial molluscs. **A**: Collect sand sample with few seawater, let the sand in the shade for 1-2 days. **B**: Take some sand, add $MgCl_2$ solution, mix gently and wait at least 10 minutes. **C**: Shake gently to suspend organic particles, decant quickly through a 80 μm sieve. **D**: Turn around the sieve, wash particles into dish with seawater. Finally look for organisms under a binocular microscope.

molluscs of the South Pacific are almost unexplored. Our knowledge is limited to a few species of Fiji and the Solomon Islands.

Opisthobranch species on Vanuatu were extracted from superficial, subtidal sand samples as indicated in figure 434. The majority of the opisthobranch assemblage discovered are representatives of the Acochlidia (Figs 435A-D), which comprises species with high morphological and biological diversity. At least one of the acochlidian species found is new and was recently described as *Pseudunela espiritusanta* Neusser & Schrödl, 2009 (Fig. 435D) is about a new, yet undescribed species. It inhabits an interstitial brackish-water fauna underside of stones deeply embedded into the sand, where coastal subsoil water mixes with salt water in the intertidal pools. Several opisthobranch species with so far unknown identity (Figs 435E-H) are a sign of the large amount of unknown species and indicate: wherever we have a closer look to the sand, we can discover new opisthobranch species. We are far from estimating the actual species diversity of tropical interstitial opisthobranchs, or knowing about their ecology and evolution —and probably the same applies to most other interstitial groups as well.

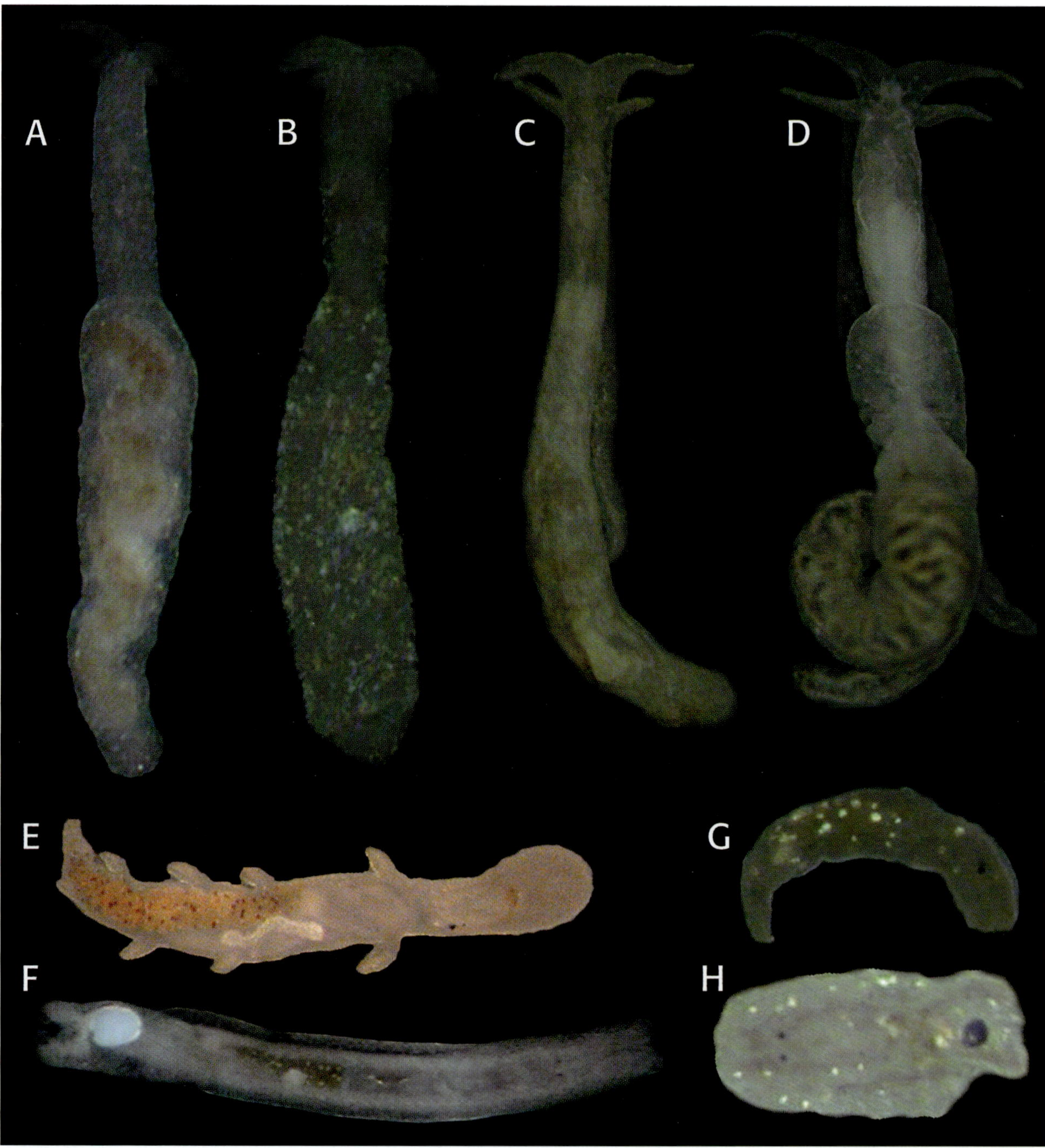

Figure 435: Interstitial opisthobranch molluscs from Santo and surrounding area. **A**: *Paraganitus* sp., 2 mm (collected on Malo Island). **B**: *Microhedyle* sp., 3 mm (collected on Malo Island). **C**: *Pseudunela* sp. 1, 4 mm (collected on Mounparap Island). **D**: *Pseudunela espiritusanta*, 10 mm (collected on Santo). **E**: *Pseudovermis* sp., 4 mm (collected on Mounparap Island). **F**: *Philinoglossidae* sp., 2 mm (collected on Mounparap Island). **G**: Cephalaspidea sp., 1 mm (collected on Mounparap Island). **H**: *Philine* sp., 0.4 mm (collected on Mounparap Island). (Photos T. Neusser).

MANGROVE ENVIRONMENTS OF SOUTH EAST SANTO

Jean-Claude Plaziat & Pierre Lozouet

The volcanic origin and a persistent raising tectonic activity of the Vanuatu archipelago explain the limited extension of mangrove environments, even on the largest island of Espiritu Santo. Along most of its steep shoreline, there is few appropriate areas for forested salt-marsh development: no deltaic area, very limited estuaries and cliffs resulting from a general upheaval of the Quaternary fringing reefs. The adverse ocean energy of exposed coasts nevertheless is reduced by the complex islands topography of SE Santo. Settings of the most interesting mangrove settlements are therefore located behind local accumulations of sediments deposited on protected shores of the mainland or between associated islands (Palikulo, Matevulu, Malo).

Rainfall is not only drained by brooks and small turbulent rivers, but also delivered to the sea by seepage or through innumerable karstic channels of the raised reef terraces. This ubiquitous freshwater delivery lowers the littoral salinity between 20 and 30‰ at low tide, even in the absence of

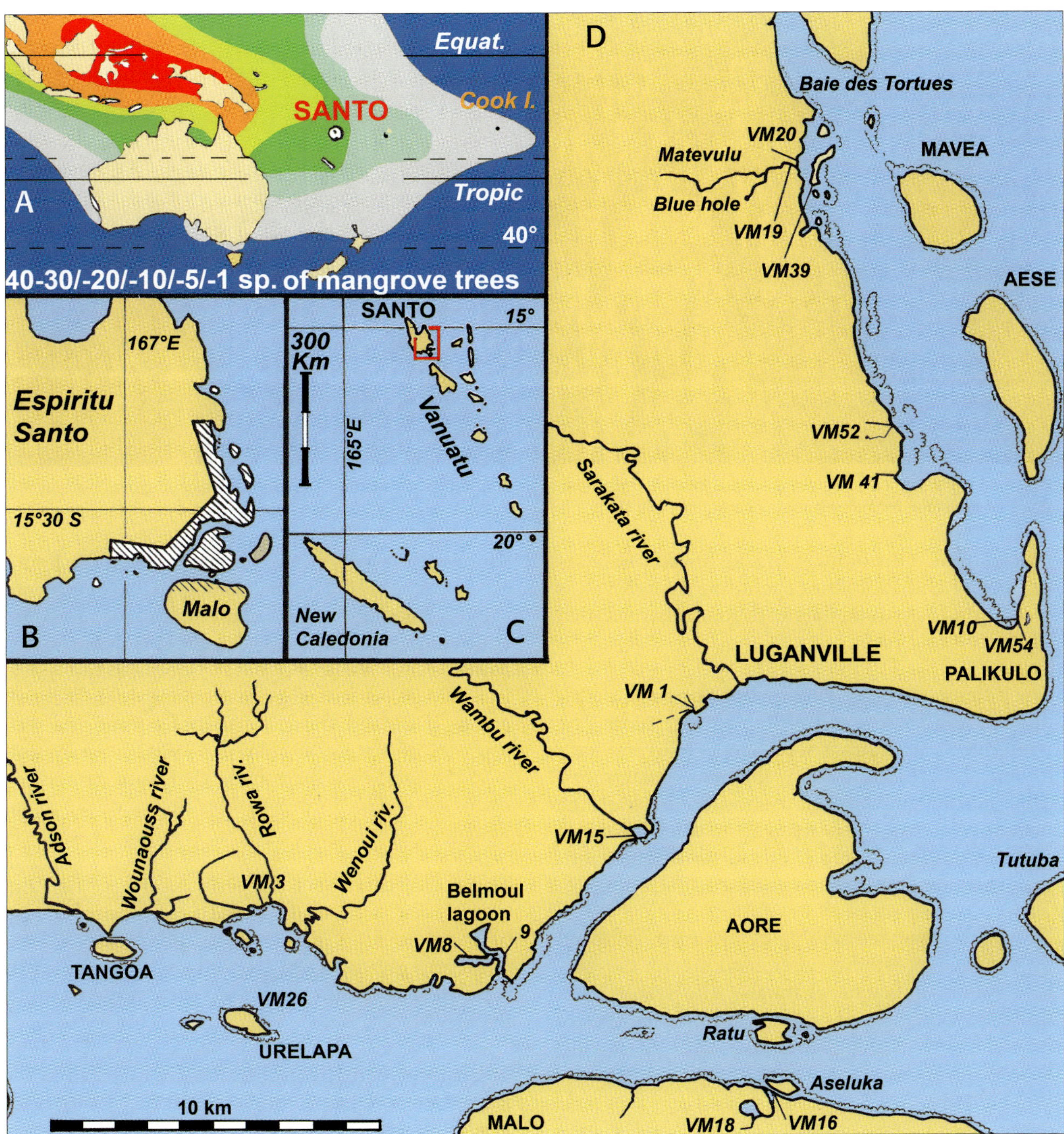

Figure 436: **A**: mangrove trees diversity gradient in the Pacific ocean (according to Duke, 2006). **B, C**: Location of the prospected shoreline on the SW Espiritu Santo area, with respect to latitudes and longitudes. **D**: mangal sites where molluscs were collected.

Figure 437: The conical pneumatophores show the peripheral extension of the buried roots of a *Sonneratia* mangrove tree (on the right) whereas the arched stilt roots of a *Rhizophora* (left) are mainly subaerial. VM3 Mangal established on a fossil fringing reef.

river output. This induces brackish water habitats in front of cliff shorelines excluding mangroves as well as on rocky tidal flats with *Rhizophora* settled on paleo-coral reefs.

The different settings that allowed mangrove development are discrete and of limited extension. The main rivers (Wambu – VM15, Sarakata Rivers) have a very short saline estuary including mangroves. Their bottom gradient and freshwater flow limit the mangrove ecosystem (= mangal) to a few hundred metres from the shoreline. Inland steep banks contribute to the rapid mangroves exclusion.

Other drainage morphologies are now followed by rivulets, possibility after sunking of the surface waters in karsts channels. The results is a drowned valley mainly influenced by tidal waters i.e. a lagoon setting (Belmoul – VM9, Matewulu lagoon – VM19) where mangrove trees fringe the inland banks.

Satellite islands with high relief or flat but extensive reefal islands favour the SE coasts of Espiritu Santo with respect to the protection from the highest ocean energy. The main mangals developed on sheltered shores where coral sands accumulated and built a spit (Palikulo – VM10, 54) or emergent banks (Ratu, Aseluka – VM16, Mounparap-Oyster isl. – VM39). A few other small mangals are located on the mainland shore, in naturally protected settings (N of Matewalu channel – VM20, Saraoutou – VM52) and in a man-made enclosure bathed by open sea waters (Vanuatu Maritime College at la Roseraie – VM1).

The rocky tidal flats are only locally settled by mangroves, where the usual cliff morphology of fossil reefs is dismantled and more or less abraded, giving sand pockets where the stilt roots of *Rhizophora* may creep their way (Nasouli bay – VM3).

Finally, the mangrove substrates are especially varied but not extensive, ranging from rocky shores to muddy sands while excluding the most propicious deltaic/estuarine muds. On the other hand, freshwater delivery and climate equability are favouring

Figure 438: **A**: A narrow marine mangal made by *Rhizophora* encroached on coral rubbles, low tide. The small boat is dredging the bottom of a strait between low islands (R. von Cosel managing). Mounparap island (VM39). **B**: Another marine but unusual setting for a mangal. The tidal channel between Asekula and Malo islands (VM16) is partly filled by a living coral carpet on which the pioneer *Rhizophora* overgrow. The branching coral substrate is emergent at low tide while the arched stilt roots are drown at each high tide. **C**: A rare (unusual?) Holothurian (5/10 cm long) creeping on submerged *Rhizophora* stilt roots (VM16). **D**: A spiny Chiton (3 cm long) below tide level on a *Rhizophora* root (VM39). (Photos J.-C. Plaziat).

factors. The mangrove trees diversity is an intermediate one. *Rhizophora*, *Sonneratia* and *Avicennia* are the most representative, plus local *Bruguiera*, *Xylocarpus* and *Acanthus*. The classic algal muff

Figure 439: **A**: An oyster settlement on the mangal front of a protected shore (VM20). At low tide, P. Lozouet looks at the low diversity molluscs association of *Saccostrea* and *Isognomon* clinging to the *Rhizophora* stilt roots, mostly at the back of their bark cylinder. This reflects the need of protection for larvae settlement of the oysters on the outermost root substrates. **B**: In the more internal root thicket the large *Isognomon ephippium* are byssaly attached by bundles up to tens of shells. **C**: On the muddy sand bottom of the internal part of a larger mangal (Palikulo, VM10) the potamidid *Terebralia palustris* is especially abundant, with a high size diversity that suggests a fairly continuous recruitment (no seasonality). (Photos J.-C. Plaziat).

around *Avicennia* or *Sonneratia* pneumatophores (cf. bostrichietum) is a very local subenvironment (Palikulo – VM10, Saraoutou – VM52). As the mangals never extend on areas more than tens of metres large, except at Palikulo, internal zonation of trees is not evidenced. The only general pattern is the seaward fringe dominated by *Rhizophora*.

The molluscs collected in these mangals reflect the intermediate location of the Vanuatu archipelago: between the impoverished New Caledonia and the empty Pacific ocean core (Fig. 436A) this is not a rich mangal malacofauna.
The most abundant species is *Terebralia palustris*, locally considered as a seafood (resettled from Palikulo to Urelapa isl. – VM26, in an anchialine pond, for example). Its moderate size and the non-seasonality of its recruitments (no excessively dry season) explain an unusual continuity of size distribution ($\leq$ 5-100 mm) on the muddy bottom of shaded sites, the younger specimens living in lower intertidal depressions. The location of *T. palustris* in the inner mangal at Palikulo may be interpreted as a response to the limitation of mud bottom and decaying vegetal detritus to this part of the mangal where permanent puddles favour the first growth stages of this potamid. The outer mangal is conversely enriched in coarse sand, repeatedly washed and spread by storms, adverse to mud-creepers.

The most conspicuous mangal neritid is the big *Nerita planospira*, generally clinging to mangroves trunks and branches above high tide but foraging down to the muddy bottom of the inner mangal. Other species *N. pulicata*, *squamosa*, *reticulata* creep only on the intertidal muddy sand. In the same mangals, the smaller neritinids (*Neritodryas*, *Vitta*, *Clithon*, *Neripteron*) do not move far from the mud and litter bottom.
The littorinid mangal genus *Littoraria* is not unfrequent but not so varied. It has been collected not only on peripheral mangrove trees but also on the bark of terrestrial big trees, rooted on low cliff shores, that low down to the tidal reach. The ellobid family is only represented by a small *Cassidula*, locally crowded on shaded slender trunks, above high tides. Other so-called mangal ellobid genera (*Pythia* and *Melampus*) actually dwell in the coral rubbles associated with drifted plant detritus in the high tide deposits of exposed shores. Another controversial subenvironment is the drifted deposit of the highest tide mark, that may constitute the innermost mangal belt, at the limit of the terrestrial realm. At Palikulo its benefits from a mixture of mangroves and terrestrial leaves and other decaying detritus, under shade of the terrestrial forest. The malacofauna includes *Melampus*, *Pythia* and *Truncatella*.
The *Fissalabia* mangal habitat of Palikulo (with *Terebralia* in an *Avicennia* subenvironment) was

observed only in 2005. This is evidence of the patchy unstable distribution of molluscs in the subtropical mangals.

Bivalves are also sporadically abundant. Oysters (three morphs including typical *Saccostrea cuccullata*) are locally attached to *Rhizophora* roots or less frequently to *Avicennia* or *Sonneratia* pneumatophores of the outermost mangal belt. North and South of Matewulu channel (near Oyster island) oyster settlements are associated with an *Isognomon*. The big *Geloina coaxans*, an estimated seafood, is abundant at Palikulo but also identified in other sites by empty shells abandoned after fire opening in the wild.

We shall insist on the extreme rarity of barnacles on the mangroves barks, in contrast with New Caledonia for example. Mangals crabs are not varied and abundant either, except locally with small red or yellow clawed *Uca*. The infrequent amphibious fish Periophthalmids are less than 10 cm long.
A very special mangal setting deserves comments. In the shallow (intertidal) inlet of Asuleka island – VM16, 18 (N of Malo I.) *Rhizophora*, *Sonneratia* and *Bruguiera* developed fairly well whereas the mangal malacofauna is especially poor. On sand banks emergent at low tide, a dense coral carpet (a Poritid *Montipora* and few other forms), locally associated with patches of sea-grass and *Halimeda*, extends to the *Rhizophora* belt, so that its stilt roots anchor between living corals; It therefore is difficult to exclude intermingling assemblages, especially from wandering animals such as Holothurians and Polyplacophora. Nevertheless spiny species of both groups have been observed only on the *Rhizophora* roots of this marine setting (Fig. 438). Another *Holothuria*, *H. leucospilota* seems to characterize the inner mangal with *Terebralia* at Palikulo.
As far as the animal biodiversity is concerned, we must insist on the distribution between the mangal biotope —the intertidal forest shaded by mangroves trees— and adjacent environments. The varied Terebrid malacofauna associated with asterid echinoderms of Belmoul lagoon (VM9), while the adjacent mangal malacofauna is especially poor, clearly illustrates such a juxtaposition with local contrasting biodiversities.

To sum up, the varied mangals of Santo suggest that the Vanuatu archipelago is a significant benchmark in the general West-Pacific impoverishment gradient from the Indo-malayan core, at the genera as well as the species level. High volcanic islands and raised reef terraces are additional unfavorable local conditions, more valid explanations than anthropic deterioration as most of the observed mangals appear to have not suffer from recent human interferences.

Figure 440: **A**: The landward fringe of a large mangal (VM54, Palikulo). This is the true shoreline, at the limit between the supratidal forest (left) and the tidal swamp (mangal) with *Avicennia* (pneumatophores) and *Rhizophora* mangrove trees. The sandy "beach" slope is underlined by drifted plant detritus (tide marks), a special biotope (**B**) rich in decaying leaves, favouring Ellobids (banded *Melampus*, up to 10 mm) and *Truncatella* (right top).**C**: *Geloina*, the tasty mangal Cyrenid partly buried and excavated (6 cm long) from the muddy sand bottom of the internal mangal at Palikulo (VM10). **D**: A *Littoraria* creeping on a *Rhizophora* leave of the the peripheral mangroves belt, above tide level. **E**: Another tide-mark litter but drier, with *Pythia* (15 mm). (Photos J.-C. Plaziat).

on Selected Biota

CHECKLIST OF THE FISHES
Ronald Fricke, John L. Earle, Richard L. Pyle & Bernard Séret

This checklist includes the fish species recorded from the Espiritu Santo Group, Vanuatu. A detailed, annotated checklist of all Vanuatu fish species including distribution data, literature references and material lists is in preparation by R. Fricke. The checklist comprises all records that could be verified either by museum specimens (including deep water materials of the MUSORSTOM 8 cruise), by confirmation by revising authors, or by the identification of photographs and video sequences taken by Adrian Gittenberger (National Museum of Natural History Naturalis, Leiden), Tan-Heok Hui (University of Singapore), Damien Hinsinger (MNHN), Peter Ng (Raffles Museum Singapore), Richard Pyle and John Earle (BPBM) and Sarah Samadi (IRD, Paris) during the Santo 2006 expedition, are included.

Families are arranged systematically according to the fourth edition of *The fishes of the world* by Nelson (2006), and genera and species alphabetically under the family names. The genus and species classification is mainly based on the Online version of the *Catalog of fishes* by Eschmeyer. In the checklist, reference is given to materials in the collections of the Australian Museum Sydney (AMS), the Academy of Natural Sciences of Philadelphia (ANSP), the Natural History Museum London (BMNH), the Bernice P. Bishop Museum Honolulu (BPBM), the California Academy of Sciences (CAS), the Field Museum of Natural History Chicago (FMNH), the Muséum National d'Histoire Naturelle Paris (MNHN), the Stanford University Collection (SU, now deposited at CAS), the Staatliches Museum für Naturkunde Stuttgart (SMNS), and the National Museum of Natural History Washington D.C. (USNM) in order to document records. Video clips serving as species records are kept at BPBM, along with location data, habitat, depth and abundance for those species recorded by video.

The present checklist of the fishes of the Espiritu Santo Group includes a total of 917 species. Among these, 809 species are recorded from Espiritu Santo for the first time; 514 species are first records for Vanuatu. As the fish fauna of Espiritu Santo is still poorly known, additional species are expected to be added to the list in the future. An ichthyological expedition using rotenone and other collecting techniques and searching in a wider variety of habitats might double the number of species known from the island group.

The 10 largest shore fish families are the Labridae (81 species, 8.8% of the total species) and Pomacentridae (81 species, 8.8%), Gobiidae (72 species, 7.8%), Serranidae (49 species, 5.3%), Apogonidae (37 species, 4.0%), Chaetodontidae (31 species, 3.4%), Acanthuridae (28 species, 3.1%), Blenniidae (25 species, 2.7%), Scaridae and Holocentridae (21 species each, 2.3% each). Small and cryptic species, however, are probably under-represented in these statistics. In 2006, in a checklist of the marine shore fishes of the neighbouring New Caledonia, Fricke & Kulbicki found the Gobiidae to be the most common shore fish family with 169 species (9.4% of the total species), followed by the Labridae (118 species, 6.7%), Pomacentridae (109 species, 6.2%), Serranidae (86 species, 4.9%), and Apogonidae (78 species, 4.4%). Similar percentages, though slightly fewer species, are expected for the Espiritu Santo Group.

••• Fishes recorded from Espiritu Santo group, Vanuatu

Species indicated with an asterisk (*) are new records for the Espiritu Santo Group and for Vanuatu; those marked with a cross (†) are new records for the Espiritu Santo Group, but have been recorded previously from other areas in Vanuatu.

••• Scyliorhinidae
Galeus priapus Séret & Last, 2008 – MNHN material (Recently described by Séret & Last, 2008)
Parmaturus albipenis Séret & Last, 2007 – MNHN material

••• Triakidae
†*Iago garricki* Fourmanoir & Rivaton, 1979 – MNHN material

••• Carcharhinidae
Carcharhinus albimarginatus (Rüppell, 1837) – Video sequence 2006 Vanuatu
Carcharhinus amblyrhynchos (Bleeker, 1856) – Video sequence 2006 Vanuatu
Carcharhinus melanopterus (Quoy & Gaimard, 1824) – Video sequence 2006
†*Galeocerdo cuvier* (Péron & Lesueur *in* Lesueur, 1822) – Underwater photograph 2006
†*Triaenodon obesus* (Rüppell, 1837) – Video sequence 2006

••• Squalidae
Squalus rancureli Fourmanoir & Rivaton, 1979 – First record by Fourmanoir & Rivaton (1979: 437-438, figs 25-26)

••• Arhynchobatidae
Notoraja sp. 1 (to be described by Séret & Last) – MNHN material
Notoraja sp. 2 (to be described by Séret & Last) – MNHN material

••• Dasyatidae
†*Aetobatus narinari* (Euphrasen, 1790) – Video sequence 2006
†*Neotrygon kuhlii* (Müller & Henle, 1841) – MNHN material; photograph 2006
†*Taeniura lymma* (Forsskål *in* Niebuhr, 1775) – Video sequence 2006

••• Halosauridae
Aldrovandia affinis (Günther, 1877) – MNHN material; photograph 2006
Aldrovandia phalacra (Vaillant, 1888) – MNHN material; photographs 2006
Halosaurus ovenii Johnson, 1864 – MNHN material

••• Anguillidae
Anguilla marmorata Quoy & Gaimard, 1824 – Keith *et al.*, this volume
Anguilla megastoma Kaup, 1856 – Keith *et al.*, this volume
Anguilla obscura Günther, 1872 – Keith *et al.*, this volume

••• Muraenidae
Anarchias allardicei Jordan & Starks *in* Jordan & Seale, 1906 – USNM material
†*Echidna nebulosa* (Ahl, 1789) – USNM material
†*Echidna polyzona* (Richardson, 1845) – USNM material
†*Gymnothorax buroensis* (Bleeker, 1857) – USNM material
†*Gymnothorax chilospilus* Bleeker, 1865 – USNM material
†*Gymnothorax enigmaticus* McCosker & Randall, 1982 – USNM material
Gymnothorax eurostus (Abbott, 1860) – Photograph 2006
†*Gymnothorax fimbriatus* (Bennett, 1832) – USNM material
Gymnothorax javanicus (Bleeker, 1859) – Video sequence 2006
†*Gymnothorax margaritophorus* (Bleeker, 1865) – USNM material
†*Gymnothorax melatremus* Schultz *in* Schultz, Herald, Lachner, Welander & Woods, 1953 – USNM material
Gymnothorax meleagris (Shaw *in* Shaw & Nodder, 1795) – USNM material
†*Gymnothorax pictus* (Ahl, 1789) – ANSP, FMNH, SU and USNM material
†*Gymnothorax rueppelliae* (McClelland, 1845) – ANSP material
†*Gymnothorax thyrsoideus* (Richardson, 1845) – Video sequence 2006
†*Gymnothorax undulatus* (Lacepède [ex Commerson], 1803) – FMNH and SU material
†*Gymnothorax zonipectis* Seale, 1906 – CAS, MNHN and USNM material; photograph 2006
†*Uropterygius concolor* Rüppell, 1838 – FMNH material
Uropterygius fuscoguttatus Schultz *in* Schultz, Herald, Lachner, Welander & Woods, 1953 – USNM material
†*Uropterygius marmoratus* (Lacepède, 1803) – SU material

••• Muraenesocidae
Muraenesox cinereus (Forsskål *in* Niebuhr, 1775) – MNHN material; photograph 2006

••• Synaphobranchidae
Atractodenchelys robinsorum Karmovskaya, 2003 – MNHN material
Histiobranchus sp. – MNHN material
Synaphobranchus affinis Günther, 1877 – MNHN material

••• Ophichthidae
Lamnostoma kampeni Weber & Beaufort, 1916 – Keith *et al.*, this volume

Lamnostoma orientalis (McClelland, 1844) – MNHN material; photograph 2006; Keith *et al.*, this volume
Lamnostoma polyophthalmum (Bleeker, 1853) – BMNH material
†*Myrichthys maculosus* (Cuvier, 1816) – MNHN material

••• Nemichthyidae
Nemichthys sp. – MNHN material

••• Congridae
Ariosoma anago (Temminck & Schlegel, 1846) – First record by Castle (1964: 13-14, fig. 2)
Ariosoma scheelei (Strömman, 1896) – First record by Castle (1964: 3-13, fig. 1 A-I, 2)
Bathycongrus aequoreus (Günther, 1887) – MNHN material
Bathycongrus guttulatus (Günther, 1887) – MNHN material
Bathycongrus wallacei (Castle, 1968) – MNHN material
Bathyuroconger sp. – MNHN material
†*Conger cinereus* (Klunzinger [ex Rüppell], 1871); authorship according to Fricke (2008: 15) – USNM material
Conger verreauxi Kaup, 1856 – First record by Castle (1964: 24-28, fig. 8 G-H)
Conger wilsoni (Bloch & Schneider [ex Banning], 1801) – First record by Castle (1964: 19-24, fig. 6 A-H, 9 A-C)
Parabathymyrus sp. – MNHN material
Rhechias sp. – MNHN material

••• Clupeidae
Anodontostoma selangkat (Bleeker, 1852) – USNM material
†*Herklotsichthys quadrimaculatus* (Rüppell, 1837) – Video sequence 2006; photograph 2006
†*Spratelloides delicatulus* (Bennett, 1832) – FMNH and SU material

••• Chanidae
†*Chanos chanos* (Forsskål *in* Niebuhr, 1775) – Video sequence 2006

••• Alepocephalidae
Conocara microlepis (Lloyd, 1909) – MNHN material

••• Synodontidae
Saurida gracilis (Quoy & Gaimard, 1824) – USNM material and video sequence 2006
Saurida nebulosa Valenciennes *in* Cuvier & Valenciennes, 1850 – MNHN material; video sequence 2006; photograph 2006
Synodus jaculum Russell & Cressey, 1979 – Video sequence 2006
Synodus similis McCulloch, 1921 – MNHN material; photograph 2006
†*Synodus variegatus* (Lacepède [ex Commerson], 1803) – Video sequence 2006

••• Chlorophthalmidae
Chlorophthalmus sp. – MNHN material

••• Ipnopidae
Ipnops agassizi Garman, 1899 – MNHN material

••• Neoscopelidae
Neoscopelus microchir Matsubara, 1943 – MNHN material

••• Sternoptychidae
Argyropelecus lychnus Garman, 1899 – MNHN material; photograph 2006
Polyipnus indicus Schultz, 1961 – MNHN material
Polyipnus omphus Baird, 1971 – MNHN material

••• Phosichthyidae
Polymetme sp. – MNHN material

••• Myctophidae
†*Symbolophorus evermanni* (Gilbert, 1905) – MNHN material

••• Polymixiidae
Polymixia japonica Günther, 1877

••• **Bregmacerotidae**
Bregmaceros sp. – MNHN material

••• **Macrouridae**
Caelorinchus acutirostris Smith & Radcliffe, 1912 – MNHN material
Caelorinchus anatirostris Jordan & Gilbert, 1904 – MNHN material
Caelorinchus argentatus Smith & Radcliffe, 1912 – MNHN material
Caelorinchus sereti Iwamoto & Merrett, 1997 – MNHN material
Cetonurichthys subinflatus Sazonov & Shcherbachev, 1982 – MNHN material
Cetonurus crassiceps (Günther, 1878) – MNHN material
Hymenocephalus longibarbis (Günther, 1887) – MNHN material
Hymenocephalus megalops Iwamoto & Merrett, 1997 – MNHN material
Hymenocephalus nascens Gilbert & Hubbs, 1920 – MNHN material
Kuronezumia bubonis Iwamoto, 1974 – MNHN material
Lucigadus acrolophus Iwamoto & Merrett, 1997 – MNHN material
Mataeocephalus cristatus Sazonov, Shcherbachev & Iwamoto, 2003 – MNHN material
Nezumia aspidentata Iwamoto & Merrett, 1997 – MNHN material
Nezumia propinqua (Gilbert & Cramer, 1897) – MNHN material
Ventrifossa nigrodorsalis Gilbert & Hubbs, 1920 – MNHN material

••• **Moridae**
Physiculus longifilis Weber, 1913 – MNHN material
Physiculus roseus Alcock, 1891 – MNHN material
Physiculus therosideros Paulin, 1987 – MNHN material

••• **Bythitidae**
Alionematichthys riukiuensis (Aoyagi, 1954) – USNM material (recorded by Møller & Schwarzhans, 2008: 113)

••• **Aphyonidae**
Barathronus sp. – MNHN material

••• **Ophidiidae**
†*Brotula multibarbata* Temminck & Schlegel, 1846 – USNM material
Dicrolene longimana Smith & Radcliffe *in* Radcliffe, 1913 – MNHN material
Monomitopus garmani (Smith & Radcliffe *in* Radcliffe, 1913) – MNHN material
†*Neobythites malayanus* Weber, 1913 – MNHN material; photograph 2006
Neobythites pallidus Nielsen, 1997 – MNHN material
Neobythites sp. – MNHN material; photograph 2007
Neobythites unimaculatus Smith & Radcliffe *in* Radcliffe, 1913 – MNHN material; photograph 2006
Porogadus miles Goode & Bean, 1885 – MNHN material
Sirembo jerdoni (Day, 1888) – MNHN material; photograph 2006
Tauredophidium hextii Alcock, 1890 – MNHN material

••• **Carapidae**
Carapus mourlani (Petit, 1934) – MNHN material; photograph 2006
Pyramodon ventralis Smith & Radcliffe *in* Radcliffe, 1913 – MNHN material; photograph 2006

••• **Antennariidae**
Antennarius coccineus (Cuvier *in* Lesson, 1831) – First record by Pietsch & Grobecker (1987: 144-153); USNM material
Antennarius commerson Lacepède [ex Commerson] *in* Anonymus, 1798 – MNHN material; photograph 2006
Antenarius dnorehensis Bleeker, 1859 – First record by Pietsch & Grobecker (1987: 166-170), and AMS material
†*Antennarius nummifer* Cuvier, 1816 – BPBM and MNHN material; video sequence 2006
Antennarius rosaceus Smith & Radcliffe *in* Radcliffe, 1912 – MNHN material; photograph 2006

••• **Lophiidae**
Lophiodes naresi (Günther, 1880) – MNHN material; photograph 2006
Lophiomus setigerus (Vahl, 1797) – MNHN material; photograph 2006

••• **Chaunacidae**
Chaunax sp. – MNHN material

••• Ogcocephalidae
Halieutaea nigra Alcock, 1891 – MNHN material
Halieutaea stellata (Vahl, 1797) – MNHN material; photograph 2006
Halieutopsis nudiventer Lloyd, 1909 – MNHN material; photographs 2006
Malthopsis annulifera Tanaka, 1908 – MNHN material
Malthopsis lutea Alcock, 1891 – MNHN material; photograph 2006

••• Mugilidae
†*Cestraeus plicatilis* Valenciennes *in* Cuvier & Valenciennes, 1836 – BMNH material
†*Chelon macrolepis* (Smith, 1846) – BMNH material
Crenimugil crenilabis (Forsskål *in* Niebuhr, 1775) – First record by Thomson (1954: 117-118, fig. 13); AMS and BMNH material
Liza subviridis (Valenciennes *in* Cuvier & Valenciees, 1836) – Keith *et al.*, this volume
Moolgarda seheli (Forsskål *in* Niebuhr, 1775) – USNM material

••• Exocoetidae
Parexocoetus brachypterus (Günther, 1866) – USNM material
Parexocoetus mento (Valenciennes *in* Cuvier & Valenciennes, 1847) – MNHN material

••• Hemiramphidae
†*Zenarchopterus dispar* (Valenciennes *in* Cuvier & Valenciennes, 1847) – USNM material

••• Belonidae
Strongylura urvillii (Valenciennes *in* Cuvier & Valenciennes, 1847) – USNM material
†*Tylosurus crocodilus* (Peron & LeSueur *in* LeSueur, 1821) – USNM material

••• Atherinidae
†*Atherinomorus duodecimalis* (Valenciennes *in* Cuvier & Valenciennes, 1835) – SU material
†*Atherinomorus lacunosus* (Bloch & Schneider [ex Forster], 1801) – FMNH and USNM material

••• Monocentridae
Monocentris japonica (Houttuyn, 1782) – MNHN material; photograph 2006

••• Berycidae
Centroberyx sp. – MNHN material

••• Holocentridae
†*Myripristis adusta* Bleeker, 1853 – USNM material; video sequence 2006
†*Myripristis berndti* Jordan & Evermann, 1903 – USNM material; video sequence 2006
Myripristis chryseres Jordan & Evermann, 1903 – Video sequence 2006
†*Myripristis hexagona* (Lacepède, 1802) – Video sequence 2006
†*Myripristis kuntee* Valenciennes *in* Cuvier & Valenciennes, 1831 – Video sequence 2006
†*Myripristis murdjan* (Forsskål *in* Niebuhr, 1775) – FMNH and USNM material; video sequence 2006
†*Myripristis violacea* Bleeker, 1851 – SU material; video sequence 2006; underwater photograph 2006
Myripristis vittata Valenciennes *in* Cuvier & Valenciennes, 1831 – MNHN material; photograph 2006; video sequence 2006
Neoniphon argenteus (Valenciennes *in* Cuvier & Valenciennes, 1831) – USNM material; video sequence 2006; underwater photograph 2006
†*Neoniphon opercularis* (Valenciennes *in* Cuvier & Valenciennes, 1831) – Video sequence 2006
†*Neoniphon sammara* (Forsskål *in* Niebuhr, 1775) – USNM material; video sequence 2006; underwater photograph 2006
†*Plectrypops lima* (Valenciennes *in* Cuvier & Valenciennes, 1831) – USNM material
Sargocentron caudimaculatum (Rüppell, 1838) – Video sequence 2006
†*Sargocentron diadema* (Lacepède [ex Commerson], 1802) – Video sequence 2006
Sargocentron ittodai (Jordan & Fowler, 1902) – USNM material
†*Sargocentron microstoma* (Günther, 1859) – FMNH and SU material
Sargocentron punctatissimum (Cuvier *in* Cuvier & Valenciennes, 1829) – USNM material
†*Sargocentron spiniferum* (Forsskål *in* Niebuhr, 1775) – USNM material; video sequence 2006
Sargocentron tiere (Cuvier *in* Cuvier & Valenciennes, 1829) – Video sequence 2006
†*Sargocentron tiereoides* (Bleeker, 1853) – SU material; video sequence 2006
Sargocentron violaceum (Bleeker, 1853) – Video sequence 2006

••• Zenionidae
Zenion japonicum Kamohara, 1934 – MNHN material; photograph 2006

••• Poeciliidae
Gambusia affinis (Baird & Girard, 1853) – Introduced; Keith *et al.*, this volume
Pocilia reticulata Peters, 1859 – Introduced; Keith *et al.*, this volume

••• Rondeletiidae
Rondeletia loricata Abe & Hotta, 1963 – MNHN material

••• Aulostomidae
†*Aulostomus chinensis* (Linné, 1766) – Video sequence 2006

••• Fistulariidae
†*Fistularia commersonii* Rüppell, 1838 – USNM material; video sequence 2006

••• Centriscidae
†*Aeoliscus strigatus* (Günther, 1860) – USNM material

••• Solenostomidae
Solenostomus paradoxus (Pallas, 1770) – Photographs 2006

••• Syngnathidae
†*Choeroichthys sculptus* (Günther, 1870) – CAS material
Corythoichthys amplexus Dawson & Randall, 1975 – USNM material
†*Corythoichthys haematopterus* (Bleeker, 1851) – CAS, MNHN and SU material; photograph 2006
†*Corythoichthys intestinalis* (Ramsay, 1881) – USNM material; underwater photograph 2006
Corythoichthys sp. (undescribed, Fricke, MS) – MMHN material; photograph 2006
†*Doryrhamphus melanopleura* Bleeker, 1858 – USNM material
Hippocampus denise Lourie & Randall, 2003 – First record by Lourie & Randall (2003: 286); USNM material
†*Hippocampus novaehebudorum* Fowler, 1944 – USNM material; photograph 2006
†*Micrognathus pygmaeus* Fritzsche, 1981 – CAS material
†*Microphis brachyurus* (Bleeker, 1853) – USNM material; Keith *et al.*, this volume
Microphis leiaspis (Bleeker, 1853) – Keith *et al.*, this volume
Microphis manadensis (Bleeker, 1856) – Keith *et al.*, this volume
Microphis retzii (Bleeker, 1856) – Keith *et al.*, this volume
Microphis sp. (cf. *spinachoides*) – Keith *et al.*, this volume
Syngnathoides biaculeatus (Bloch, 1785) – USNM material
Trachyrhamphus bicoarctatus (Bleeker, 1857) – BPBM material

••• Scorpaenidae
Subfamily Pteroinae
Dendrochirus zebra (Cuvier *in* Cuvier & Valenciennes, 1829) – CAS material
Ebosia bleekeri (Döderlein *in* Steindachner & Döderlein, 1884) – MNHN material; photograph 2006
†*Pterois antennata* (Bloch, 1787) – CAS material; video sequence 2006; underwater photograph 2006
†*Pterois volitans* (Linnaeus, 1758) – CAS, MNHN and USNM material; video sequence 2006; underwater photograph 2006

Subfamily Scorpaeninae
Neomerinthe procurva (Chen, 1981) – MNHN material; photograph 2006
Parascorpaena mcadamsi (Fowler, 1938) – MNHN material; photograph 2006
Phenacoscorpius sp. – MNHN material
Pontinus sp. – MNHN material
†*Scorpaenodes guamensis* (Quoy & Gaimard, 1824) – FMNH and SU material
Scorpaenopsis oxycephala (Bleeker, 1849) – Visual census, R. Pyle 2006
Scorpaenopsis papuensis (Cuvier *in* Cuvier & Valenciennes, 1829) – First record by Randall & Eschmeyer (2001: 48-49); CAS and MNHN material; photographs 2006
Scorpaenopsis possi Randall & Eschmeyer, 2001 – Underwater photograph 2006
Sebastapistes cyanostigma (Bleeker, 1856) – USNM material
Sebastapistes fowleri (Pietschmann, 1934) – First record by Randall & Poss (2002: 58-63); USNM material
Taenionotus triacanthus Lacepède, 1802 – Video sequence 2006

••• Tetrarogidae
Tetraroge niger (Cuvier in Cuvier & Valenciennes, 1829) – Keith *et al.*, this volume

••• Synanceiidae
Erosa erosa (Cuvier [ex Langsdorf] *in* Cuvier & Valenciennes, 1829) – MNHN material; photograph 2006
Inimicus didactylus (Pallas, 1769) – CAS and MNHN material; photograph, 2006; visual census, R. Pyle, 2006

••• Plectrogeniidae
**Plectrogenium nanum* Gilbert, 1905 – MNHN material

••• Setarchidae
†*Setarches guentheri* Johnson, 1862 – Photograph 2006
**Setarches longimanus* Alcock, 1894 – MNHN material; photograph 2006

••• Caracanthidae
**Caracanthus maculatus* (Gray, 1831) – USNM material

••• Peristediidae
Peristedion pothumaluvae Deraniyagala, 1936 – First record by Rivaton & Bourret (1999: 312)
†*Satyrichthys quadratorostratus* (Fourmanoir & Rivaton, 1979) – Photograph 2006
**Satyrichthys rugosum* Fowler, 1943 – MNHN material

••• Triglidae
**Lepidotrigla* sp. – MNHN material
**Pterygotrigla ryukyuensis* Matsubara & Hiyama, 1932 – MNHN material

••• Hoplichthyidae
**Hoplichthys gilberti* Jordan & Richardson, 1908 – MNHN material; photograph 2006

••• Bembridae
**Bembradium roseum* Gilbert, 1905 – MNHN material; photograph 2006
**Bembras* sp. – MNHN material; photographs 2006

••• Platycephalidae
**Onigocia pedimacula* (Regan, 1908) – MNHN material; photograph 2006
**Rogadius pristiger* (Cuvier *in* Cuvier & Valenciennes, 1829) – MNHN material; photographs 2006

••• Psychrolutidae
**Psychrolutes* sp. (Fricke, MS) – MNHN material; photograph 2006

••• Ambassidae
Ambassis interruptus Bleeker, 1852 – First record by Allen & Burgess (1990: 165-167); USNM material
Ambassis miops Günther, 1872 – Keith *et al.*, this volume
†*Ambassis urotaenia* Bleeker, 1852 – BMNH and USNM material

••• Serranidae
Subfamily Anthiinae
**Plectranthias* sp. 1 – Video sequence 2006; photograph 2006
**Plectranthias* sp. 2 – MNHN material; photograph 2006
**Plectranthias nanus* Randall, 1980 – MNHN material; photograph 2006
**Pseudanthias bicolor* (Randall, 1979) – MNHN material; photograph 2006
**Pseudanthias carlsoni* Randall & Pyle, 2001 – Video sequence 2006 – (Fig. 441)
†*Pseudanthias dispar* (Herre, 1955) – Video sequence 2006
**Pseudanthias fasciatus* (Kamohara, 1954) – Video sequence 2006
**Pseudanthias flavicauda* Randall & Pyle, 2001 – MNHN material; photograph 2006; video sequence 2006 – (Fig. 442)
**Pseudanthias hutomoi* Allen & Burhanuddin, 1976 – Video sequence 2006
**Pseudanthias hypselosoma* Bleeker, 1878 – Video sequence 2006
**Pseudanthias lori* Lubbock & Randall *in* Fourmanoir & Laboute, 1976 – Video sequence 2006
**Pseudanthias pascalus* (Jordan & Tanaka, 1927) – MNHN material; photograph 2006
**Pseudanthias pleurotaenia* (Bleeker, 1857) – MNHN material; photograph 2006; video sequence 2006;
    underwater photograph 2006

Figure 441: *Pseudanthias carlsoni*. Carlson's Anthias is a new record for Vanuatu, previously known from PNG, Fiji, and the Loyalty Islands. We found small groups of this species in 30-40 m, usually one male, as seen in this image, with several females nearby.

Figure 442: *Pseudanthias flavicauda*. The Yellowtail Anthias is a deep water species named by Randall and Pyle in 2001 from Fiji, and observed by us in 100 m..

Figure 443: *Cephalopholis boenak*. The small grouper pictured here appears to be one of this widely distributed Western Pacific species (though a new record for Vanuatu). It is usually found in shallow silty habitats. However, the only specimens we saw in Vanuatu were in rocky habitat at 100 m or deeper. This is a mystery, and one wishes specimens were collected for close examination!

Figure 444: *Hoplolatilus starcki* is one of three species of this genus observed in Vanuatu. All are new records for Vanuatu and one appears to be not yet scientifically described. This species was usually seen in pairs at around 40 m, and would dive into a hole at the approach of a diver.

Pseudanthias squamipinnis (Peters, 1855) – Video sequence 2006
†*Pseudanthias tuka* Herre & Montalban *in* Herre, 1927 – Video sequence 2006
Selenanthias sp. – MNHN material; photograph 2006
Serranocirrhitus latus Watanabe, 1949 – Video sequence 2006

Subfamily Serraninae
Aethaloperca rogaa (Forsskål *in* Niebuhr, 1775) – Video sequence 2006
†*Belonoperca chabanaudi* Fowler & Bean, 1930 – Video sequence 2006
†*Cephalopholis argus* Bloch & Schneider, 1801 – FMNH and USNM material; video sequence 2006
Cephalopholis boenak (Bloch, 1790) – Video sequence 2006 – (Fig. 443)
Cephalopholis leopardus (Lacepède, 1801) – Video sequence 2006
†*Cephalopholis microprion* (Bleeker, 1852) – USNM material; video sequence 2006
†*Cephalopholis miniata* (Forsskål *in* Niebuhr, 1775) – Video sequence 2006
Cephalopholis sexmaculata (Rüppell, 1830) – Video sequence 2006
Cephalopholis spiloparaea (Valenciennes *in* Cuvier & Valenciennes, 1828) – Video sequence 2006
†*Cephalopholis urodeta* (Bloch & Schneider [ex Forster], 1801) – CAS, FMNH, MNHN, SU and USNM
    material; photograph 2006; video sequence 2006
Chelidoperca lecromi Fourmanoir, 1982 – MNHN material; photograph 2006
Chelidoperca pleurospilus (Günther, 1877) – MNHN material
†*Diploprion bifasciatum* Cuvier [ex Kuhl & Hasselt] *in* Cuvier & Valenciennes, 1828 – Video sequence 2006

Epinephelus bleekeri (Vaillant, 1878) – MNHN material; photograph 2006
†*Epinephelus coeruleopunctatus* (Bloch, 1790) – Video sequence 2006
Epinephelus coioides (Hamilton, 1822) – Underwater photograph 2006
†*Epinephelus corallicola* (Valenciennes [ex Kuhl & Hasselt] *in* Cuvier & Valenciennes, 1828) – Video sequence 2006
†*Epinephelus fuscoguttatus* (Forsskål *in* Niebuhr, 1775) – Video sequence 2006
Epinephelus macrospilos (Bleeker, 1855) – Video sequence 2006
†*Epinephelus merra* Bloch, 1793 – SU material; video sequence 2006
†*Epinephelus ongus* (Bloch, 1790) – MNHN and USNM material; video sequence 2006
Epinephelus polyphekadion (Bleeker, 1849) – Video sequence 2006
Gracila albomarginata (Fowler & Bean, 1930) – Video sequence 2006
Grammistes sexlineatus (Thunberg, 1792) – First record by Fowler (1934: 411); ANSP, CAS and USNM material; underwater photograph 2006
Plectropomus areolatus Rüppell, 1830 – Video sequence 2006
†*Plectropomus laevis* (Lacepède [ex Commerson], 1801) – Video sequence 2006
†*Plectropomus leopardus* (Lacepède, 1802) – Video sequence 2006
Pogonoperca punctata (Valenciennes *in* Cuvier & Valenciennes, 1830) – Video sequence 2006
Pseudogramma polyacantha (Bleeker, 1856) – First record by Randall & Baldwin (1997: 39-45); USNM material
†*Suttonia lineata* Gosline, 1960 – Video sequence 2006
Variola albimarginata Baissac, 1953 – Video sequence 2006
†*Variola louti* (Forsskål *in* Niebuhr, 1775) – USNM material; video sequence 2006

••• Pseudochromidae
†*Cypho purpurascens* (DeVis, 1884) – Video sequence 2006
Pictichromis porphyrea (Lubbock & Goldman, 1974) – First record by Lubbock & Goldman (1974: 107-110); video sequence 2006
†*Pseudochromis fuscus* Müller & Troschel, 1849 – CAS and USNM material; video sequence 2006
Pseudochromis marshallensis Schultz *in* Schultz, Herald, Lachner, Welander & Woods, 1953 – USNM material

••• Plesiopidae
†*Plesiops coeruleolineatus* Rüppell, 1835 – FMNH, SU and USNM material
†*Plesiops corallicola* Bleeker, 1853 – SU material
Plesiops verecundus Mooi, 1995 – First record by Mooi (1995: 57-60); FMNH and USNM material

••• Opistognathidae
Opistognathus n. sp. 1 – MNHN material; photograph 2006; video sequence 2006
Opistognathus sp. 2 – MNHN material; photograph 2006

Figure 445: *Cookeolus japonicus*. The Bullseye is a spectacular deep water species, circumtropical in distribution. This specimen seen at 100 m is the first record from Vanuatu

••• **Priacanthidae**
Cookeolus japonicus (Cuvier *in* Cuvier & Valenciennes, 1829) – Video sequence 2006 – (Fig. 445)
Heteropriacanthus cruentatus (Lacepède, 1801) – MNHN material; photograph 2006
†*Priacanthus hamrur* (Forsskål *in* Niebuhr, 1775) – MNHN material; photograph 2006; video sequence 2006
Pristigenys niphonia (Cuvier *in* Cuvier & Valenciennes, 1829) – MNHN material; photographs 2006

••• **Teraponidae**
†*Mesopristes argenteus* (Cuvier *in* Cuvier & Valenciennes, 1829) – BMNH material
Terapon jarbua (Forsskål *in* Niebuhr, 1775) – First record by Fowler (1934: 416); USNM material; video sequence 2006

••• **Kuhliidae**
†*Kuhlia marginata* (Cuvier *in* Cuvier & Valenciennes, 1829) – BMNH and USNM material
†*Kuhlia mugil* (Forster *in* Bloch & Schneider, 1801) – ANSP material
†*Kuhlia munda* (DeVis, 1885) – BMNH material
†*Kuhlia rupestris* Lacepède, 1802 – BMNH material

••• **Acropomatidae**
†*Acropoma japonicum* Günther, 1859 – MNHN material
Synagrops analis (Katayama, 1957) – MNHN material

••• **Dactylopteridae**
Dactyloptena papilio Ogilby, 1910 – MNHN material; photograph 2006
Dactyloptena peterseni (Nyström, 1887) – MNHN material; photographs 2006

••• **Apogonidae**
†*Apogon amboinensis* Bleeker, 1853 – USNM material
Apogon crassiceps Garman, 1903 – USNM material
Apogon posterofasciatus Allen & Randall, 2002 – Video sequence 2006
Apogon rhodopterus Bleeker, 1852 – USNM material
Apogonichthys ocellatus (Weber, 1913) – First record by Whitley (1959: 314, as *Apogonichthys ahimsa santoensis*); USNM material
Archamia fucata (Cantor, 1850) – First record by Gon & Randall (2003: 28-29); USNM material; video sequence 2006
Archamia leai Waite, 1916 – Underwater photograph 2006
Archamia macroptera (Cuvier *in* Cuvier & Valenciennes, 1828) – First record by Gon & Randall (2003: 36-39); USNM material; video sequence 2006
Archamia zosterophora (Bleeker, 1856) – First record by Gon & Randall (2003: 43-45); CAS material; video sequence 2006
Cheilodipterus alleni Gon, 1993 – Video sequence 2006
Cheilodipterus artus Smith, 1961 – Video sequence 2006; CAS material
Cheilodipterus isostigmus (Schultz, 1940) – CAS material
†*Cheilodipterus macrodon* (Lacepède, 1802) – Video sequence 2006
†*Cheilodipterus quinquelineatus* Cuvier *in* Cuvier & Valenciennes, 1828 – CAS and USNM material; video sequence 2006; underwater photograph 2006
Fowleria isostigma (Jordan & Seale, 1906) – USNM material
Fowleria punctulata (Rüppell, 1838) – USNM material
Fowleria vaiulae (Jordan & Seale, 1906) – MNHN material; photohraph 2006
†*Fowleria variegata* (Valenciennes, 1832) – USNM material
†*Nectamia fusca* (Quoy & Gaimard, 1825) – FMNH and SU material
Nectamia viria Fraser, 2008 – First record by Fraser (2008: 41-44, figs 3B, 6C, 7, 16, 18, tabs 2-3, 10); USNM material
†*Ostorhinchus angustatus* (Smith & Radcliffe *in* Radcliffe, 1911) – Video sequence 2006
Ostorhinchus compressus (Smith & Radcliffe *in* Radcliffe, 1911) – Video sequence 2006; underwater photograph 2006
Ostorhinchus cyanosoma (Bleeker, 1853) – CAS and USNM material; video sequence 2006
†*Ostorhinchus lateralis* (Valenciennes, 1832) – USNM material
†*Ostorhinchus nigrofasciatus* (Lachner *in* Schultz, Herald, Lachner, Welander & Woods, 1953) – FMNH and USNM material; video sequence 2006
Ostorhinchus poecilopterus (Cuvier [ex Kuhl & Hasselt] *in* Cuvier & Valenciennes, 1828) – MNHN material; photograph 2006
Ostorhinchus septemstriatus (Günther, 1880) – MNHN material; photograph 2006

Figure 446: *Caranx papuensis*. The Brassy Trevally is another new record for Vanuatu, which was surprising to us. Not only is this fish one that would be expected to turn up in local fishermen's catches, this species also would approach a diver so closely that one might consider collecting a specimen with a dinner fork. Perhaps this fish, distinct in small details like the black and white spot at the upper gill opening and the white border of the lower tail, has been confused with similar-looking and equally delicious species.

Ostorhinchus sp. (Fricke, MS) – MNHN material; photograph 2006

†*Ostorhinchus taeniophorus* (Regan, 1905) – SU material

Pristiapogon exostigma (Jordan & Starks, 1906) – USNM material; video sequence 2006

†*Pristiapogon fraenatus* (Valenciennes, 1832) – Video sequence 2006

Pristiapogon kallopterus (Bleeker, 1856) – CAS, SU and USNM material; video sequence 2006

Pristicon trimaculatus (Cuvier *in* Cuvier & Valenciennes, 1828) – First record by Randall & Fraser (1999: 619-624); FMNH and USNM material; video sequence 2006; underwater photograph 2006

Rhabdamia cypselura Weber, 1909 – CAS material; video sequence 2006; photograph 2006

Siphamia versicolor (Smith & Radcliffe, 1911) – MNHN material; photograph 2006

Zoramia fragilis (Smith, 1961) – Video sequence 2006

†*Zoramia leptacantha* (Bleeker, 1856) – CAS material

••• Epigonidae

Sphyraenops bairdianus Poey, 1861 – First record by Fourmanoir (1970: 27-28) and Bauchot & Desoutter (1989: 26) as *Scombrosphyraena oceanica*)

••• Malacanthidae

Hoplolatilus sp. Pyle, MS – Video sequence 2006

Hoplolatilus cuniculus Randall & Dooley, 1974 – Video sequence 2006

Hoplolatilus starcki Randall & Dooley, 1974 – Video sequence 2006 – (Fig. 444)

Malacanthus brevirostris Guichenot, 1848 – Video sequence 2006

†*Malacanthus latovittatus* (Lacepède [ex Commerson], 1801) – Video sequence 2006

••• Sillaginidae

Sillago macrolepis Bleeker, 1858 – USNM material

••• Echeneidae

†*Echeneis naucrates* Linnaeus, 1758 – BMNH material; video sequence 2006

••• Carangidae

Carangoides ferdau (Forsskål *in* Niebuhr, 1775) – Video sequence 2006; underwater photograph 2006

Carangoides oblongus (Cuvier *in* Cuvier & Valenciennes, 1833) – Video sequence 2006

†*Carangoides orthogrammus* Jordan & Gilbert, 1881 – USNM material; video sequence 2006

Carangoides plagiotaenia Bleeker, 1857 – Video sequence 2006

†*Caranx ignobilis* (Forsskål *in* Niebuhr, 1775) – Video sequence 2006

†*Caranx melampygus* Cuvier *in* Cuvier & Valenciennes, 1833 – CAS material; video sequence 2006

Caranx papuensis (Alleyne & Macleay, 1877) – Video sequence 2006 – (Fig. 446)

†*Caranx sexfasciatus* (Quoy & Gaimard, 1825) – USNM material

†*Elagatis bipinnulata* (Quoy & Gaimard, 1825) – USNM material; video sequence 2006

†*Gnathanodon speciosus* (Forsskål *in* Niebuhr, 1775) – Video sequence 2006

Selar crumenophthalmus (Bloch, 1793) – MNHN material; photograph 2006

Seriola rivoliana Valenciennes *in* Cuvier & Valenciennes, 1833 – Video sequence 2006

†*Trachinotus blochii* (Lacepède, 1801) – USNM material; video sequence 2006

Figure 447: *Paracaesio sordida*. The Smallscale Snapper is a new record for Vanuatu, observed in small schools or as individuals along steep slopes in 70-100 m. *P. xanthura* was observed in the same deep habitat.

••• Leiognathidae
Gazza rhombea Kimura, Yamashita & Iwatsuki, 2000 – First record by Kimura *et al.* (2000: 2-7); USNM material

••• Lutjanidae
†*Aphareus furca* (Lacepède [ex Commerson], 1801) – Video sequence 2006
†*Aprion virescens* Valenciennes *in* Cuvier & Valenciennes, 1830 – Video sequence 2006
**Lutjanus adetii* (Castelnau, 1873) – USNM material
†*Lutjanus argentimaculatus* (Forsskål *in* Niebuhr, 1775) – BMNH and USNM material
**Lutjanus bohar* (Forsskål *in* Niebuhr, 1775) – Video sequence 2006
**Lutjanus fulvus* (Bloch & Schneider, 1801) – SU and USNM material; video sequence 2006
**Lutjanus fuscescens* (Valenciennes *in* Cuvier & Valenciennes, 1830) – USNM material
†*Lutjanus gibbus* (Forsskål *in* Niebuhr, 1775) – USNM material; video sequence 2006
†*Lutjanus kasmira* (Forsskål *in* Niebuhr, 1775) – USNM material; video sequence 2006; underwater photographs 2006
**Lutjanus lemniscatus* (Valenciennes *in* Cuvier & Valenciennes, 1828) – USNM material
†*Lutjanus monostigma* (Cuvier *in* Cuvier & Valenciennes, 1828) – Video sequence 2006
†*Lutjanus quinquelineatus* Bloch, 1790 – USNM material
**Lutjanus rivulatus* (Cuvier *in* Cuvier & Valenciennes, 1828) – Video sequence 2006
†*Lutjanus rufolineatus* (Cuvier *in* Cuvier & Valenciennes, 1830) – Video sequence 2006
†*Lutjanus semicinctus* Quoy & Gaimard, 1824 – FMNH material; video sequence 2006
**Macolor macularis* Fowler, 1931 – Video sequence 2006
†*Macolor niger* (Forsskål *in* Niebuhr, 1775) – Video sequence 2006; underwater photographs 2006
**Paracaesio sordida* Abe & Shinohara, 1962 – Video sequence 2006 – (Fig. 447)
**Paracaesio xanthura* (Bleeker, 1869) – Video sequence 2006

••• Symphysanodontidae
**Symphysanodon* sp. – Video sequence 2006

••• Caesionidae
†*Caesio caerulaurea* (Lacepède [ex Commerson], 1801) – Video sequence 2006
†*Caesio cuning* (Bloch, 1791) – Video sequence 2006
**Caesio lunaris* Cuvier [ex Ehrenberg] *in* Cuvier & Valenciennes, 1830 – Video sequence 2006
†*Caesio teres* Seale, 1906 – Video sequence 2006
†*Pterocaesio digramma* (Bleeker, 1864) – Video sequence 2006
**Pterocaesio marri* (Schultz *in* Schultz, Herald, Lachner, Welander & Woods, 1953) – Video sequence 2006
†*Pterocaesio pisang* (Bleeker, 1853) – Video sequence 2006
†*Pterocaesio tessellata* Carpenter, 1987 – Video sequence 2006
**Pterocaesio trilineata* Carpenter, 1987 – Video sequence 2006

••• Gerreidae
†*Gerres filamentosus* Cuvier, 1829 – BMNH material
†*Gerres oyena* (Forsskål *in* Niebuhr, 1775) – USNM material

••• Haemulidae
Diagramma pictum (Thunberg, 1792) – First record by Johnson *et al.* (2001: 663-664); video sequence 2006
†*Plectorhinchus albovittatus* (Rüppell, 1838) – Video sequence 2006
**Plectorhinchus chaetodonoides* Lacepède, 1801 – Video sequence 2006
Plectorhinchus lessonii (Cuvier *in* Cuvier & Valenciennes, 1830) – First record by Randall & Johnson (2000: 479-480); AMS material; video sequence 2006
**Plectorhinchus lineatus* (Linnaeus, 1758) – Video sequence 2006
**Plectorhinchus picus* (Cuvier *in* Cuvier & Valenciennes, 1830) – Video sequence 2006
†*Plectorhinchus vittatus* (Linnaeus, 1758) – Video sequence 2006; underwater photograph 2006

••• Lethrinidae
†*Gnathodentex aureolineatus* (Lacepède [ex Commerson], 1802) – Video sequence 2006
**Gymnocranius audleyi* Ogilby, 1916– MNHN material; photograph 2006
**Gymnocranius grandoculis* (Valenciennes *in* Cuvier & Valenciennes, 1830) – Underwater photograph 2006
**Lethrinus erythracanthus* Valenciennes *in* Cuvier & Valenciennes, 1830 – Video sequence 2006
**Lethrinus erythropterus* Valenciennes *in* Cuvier & Valenciennes, 1830 – Video sequence 2006
†*Lethrinus harak* (Forsskål *in* Niebuhr, 1775) – FMNH and SU material; video sequence 2006; underwater photograph 2006
†*Lethrinus obsoletus* (Forsskål *in* Niebuhr, 1775) – Video sequence 2006
**Lethrinus olivaceus* Valenciennes *in* Cuvier & Valenciennes, 1830 – MNHN material
†*Lethrinus xanthochilus* Klunzinger, 1870 – USNM material; video sequence 2006
†*Monotaxis grandoculis* (Forsskål *in* Niebuhr, 1775) – Video sequence 2006

••• Nemipteridae
**Nemipterus* sp. – MNHN material
†*Pentapodus caninus* (Cuvier *in* Cuvier & Valenciennes, 1830) – Video sequence 2006
†*Scolopsis bilineata* (Bloch, 1793) – CAS, FMNH and USNM material; video sequence 2006
†*Scolopsis lineata* Quoy & Gaimard, 1824 – Video sequence 2006
†*Scolopsis margaritifera* (Cuvier *in* Cuvier & Valenciennes, 1830) – Video sequence 2006

••• Polynemidae
Polydactylus longipes Motomura, Okamoto & Iwatsuki, 2001 – Record by H. Motomura (personal communication, Sep. 2010); MNHN material
Polydactylus plebeius (Broussonet, 1782) – Record by Motomura, Iwatsuki & Kimura (2001: 87-88) and Motomura, Iwatsuki & Yoshino (2001: 118-122); USNM material

••• Mullidae
†*Mulloidichthys flavolineatus* (Lacepède, 1801) – BMNH and FMNH material; video sequence 2006
**Mulloidichthys vanicolensis* (Valenciennes *in* Cuvier & Valenciennes, 1831) – Video sequence 2006
**Parupeneus barberinoides* (Bleeker, 1852) – Video sequence 2006
Parupeneus barberinus (Lacepède, 1801) – First record by Randall (2004: 12-14); USNM material; video sequence 2006
†*Parupeneus crassilabris* (Valenciennes *in* Cuvier & Valenciennes, 1831) – Video sequence 2006
†*Parupeneus cyclostomus* (Lacepède [ex Commerson], 1801) – MNHN material; photograph 2006, video sequence 2006
†*Parupeneus indicus* (Shaw, 1803) – USNM material
†*Parupeneus multifasciatus* (Quoy & Gaimard, 1825) – FMNH, SU and USNM material; video sequence 2006
†*Parupeneus pleurostigma* (Bennett, 1831) – Video sequence 2006
**Upeneus tragula* Richardson, 1846 – Video sequence 2006
**Upeneus* sp. (undescribed, Fricke, MS) – MNHN material; photographs 2006

••• Pempheridae
**Parapriacanthus ransonneti* Steindachner, 1870 – Video sequence 2006
**Pempheris molucca* Cuvier, 1829 – MNHN material; photograph 2006
**Pempheris oualensis* Cuvier *in* Cuvier & Valenciennes, 1831 – Video sequence 2006

••• Kyphosidae
†*Kyphosus cinerascens* (Forsskål *in* Niebuhr, 1775) – USNM material; video sequence 2006

••• Scatophagidae
†*Scatophagus argus* (Linné, 1766) – BMNH and USNM material

Figure 448: *Forcipiger longirostris* and *F. flavissimus* are here reported as new records for Vanuatu. The pair on the right are *F. longirostris*, with a single *F. flavissimus* to the left. Both species were frequently seen on shallow coral reefs from 5-20 m.

Figure 449: *Centropyge colini* is a small anglefish found throughout the Western Pacific in deep water. Though this is a new record for Vanuatu, we were not surprised to find this species around caves and ledges in 100 m. Perhaps of greater interest is the grouper below the blue and yellow angelfish. It is likely the deep water species *Cephalopholis aurantia*, which is not on the Vanuatu fish list, but, alas, this image is not of sufficient quality to identify this species with full confidence. The diver is embarassed to admit that he was so intent on photographing the angelfish he did not even notice the much more interesting grouper.

Figure 450: *Amblyglyphidodon orbicularis*. The orbicular Damselfish is a new record for Vanuatu, previously known from Samoa, Fiji and New Caledonia. We found this species common among staghorn coral in 15-30 m.

••• **Ephippidae**
**Platax orbicularis* (Forsskål *in* Niebuhr, 1775) – Video sequence 2006
†*Platax pinnatus* (Linnaeus, 1758) – Video sequence 2006

••• **Chaetodontidae**
†*Chaetodon auriga* Forsskål *in* Niebuhr, 1775 – Video sequence 2006
Chaetodon baronessa Cuvier, 1829 – First record by Fowler (1944: 254, as *Gonochaetodon baronessa*); FMNH and MNHN material; photograph 2006; video sequence 2006
**Chaetodon bennetti* Cuvier *in* Cuvier & Valenciennes, 1831 – Video sequence 2006
**Chaetodon burgessi* Allen & Starck, 1973 – Video sequence 2006
Chaetodon citrinellus Cuvier *in* Cuvier & Valenciennes, 1831 – First record by Fowler (1934: 424); CAS material; video sequence 2006
†*Chaetodon ephippium* Cuvier *in* Cuvier & Valenciennes, 1831 – Video sequence 2006
**Chaetodon kleinii* Bloch, 1790 – MNHN material; video sequence 2006; photograph 2006; underwater photograph 2006
†*Chaetodon lineolatus* Cuvier [ex Quoy & Gaimard] *in* Cuvier & Valenciennes, 1831 – Video sequence 2006
†*Chaetodon lunula* (Lacepède [ex Commerson], 1802) – Video sequence 2006
†*Chaetodon lunulatus* Quoy & Gaimard, 1825 – CAS, FMNH, MNHN and SU material; video sequence 2006; photographs 2006; underwater photograph 2006

Chaetodon melannotus Bloch & Schneider, 1801 – Video sequence 2006
†*Chaetodon mertensii* Cuvier *in* Cuvier & Valenciennes, 1831 – Video sequence 2006
Chaetodon meyeri Bloch & Schneider, 1801 – MNHN material; photograph 2006
Chaetodon ornatissimus Cuvier [ex Solander] *in* Cuvier & Valenciennes, 1831 – Video sequence 2006
Chaetodon oxycephalus Bleeker, 1853 – Video sequence 2006
†*Chaetodon pelewensis* Kner, 1868 – MNHN material; photograph 2006; video sequence 2006
†*Chaetodon rafflesii* Anonymus [Bennett], 1830 – Video sequence 2006
Chaetodon reticulatus Cuvier *in* Cuvier & Valenciennes, 1831 – Video sequence 2006
†*Chaetodon semeion* Bleeker, 1855 – Video sequence 2006
Chaetodon speculum Cuvier [ex Kuhl & Hasselt] *in* Cuvier & Valenciennes, 1831 – Video sequence 2006
†*Chaetodon trifascialis* Quoy & Gaimard, 1825 – Video sequence 2006
Chaetodon ulietensis Cuvier *in* Cuvier & Valenciennes, 1831 – Video sequence 2006
†*Chaetodon unimaculatus* Bloch, 1787 – FMNH material; video sequence 2006
Chaetodon vagabundus Linnaeus, 1758 – First record by Fowler (1934: 424); MNHN and USNM material; photograph 2006; video sequence 2006
Forcipiger flavissimus Jordan & McGregor *in* Jordan & Evermann, 1898 – MNHN material; photograph 2006; video sequence 2006 – (Fig. 448)
Forcipiger longirostris (Broussonet, 1782) – MNHN material; photograph 2006; video sequence 2006 – (Fig. 448)
Heniochus acuminatus (Linnaeus, 1758) – First record by Plessis & Fourmanoir (1966: 131); MNHN material; video sequence 2006; underwater photograph 2006
Heniochus chrysostomus Cuvier *in* Cuvier & Valenciennes, 1831 – SU material; video sequence 2006
Heniochus monoceros Cuvier *in* Cuvier & Valenciennes, 1831 – Video sequence 2006
†*Heniochus singularius* Smith & Radcliffe, 1911 – USNM material; video sequence 2006
†*Heniochus varius* (Cuvier, 1829) – Video sequence 2006

••• Pomacanthidae
Apolemichthys trimaculatus (Cuvier [ex Lacepède] *in* Cuvier & Valenciennes, 1831) – Video sequence 2006
Centropyge aurantia Randall & Wass, 1974 – Video sequence 2006
†*Centropyge bicolor* (Bloch, 1787) – MNHN and USNM material; photograph 2006; video sequence 2006; underwater photograph 2006
†*Centropyge bispinosa* (Günther, 1860) – MNHN material; photograph 2006; video sequence 2006
Centropyge colini Smith-Vaniz & Randall, 1974 – Video sequence 2006 – (Fig. 449)
Centropyge fisheri (Snyder, 1904) – Video sequence 2006
†*Centropyge flavissima* (Cuvier *in* Cuvier & Valenciennes, 1831) – CAS, MNHN and USNM material; photograph 2006; video sequence 2006; underwater photograph 2006
Centropyge multifasciata (Smith & Radcliffe, 1911) – Video sequence 2006
Centropyge nox (Bleeker, 1853) – MNHN material; photograph 2006; video sequence 2006
†*Centropyge vrolikii* (Bleeker, 1853) – USNM material; video sequence 2006
Genicanthus bellus Randall, 1975 – Video sequence 2006
Genicanthus melanospilos (Bleeker, 1857) – MNHN material; photograph 2006; video sequence 2006
Genicanthus watanabei (Yasuda & Tominaga, 1970) – Video sequence 2006
Pomacanthus imperator (Bloch, 1787) – First record by Fowler (1934: 425); video sequence 2006
†*Pomacanthus semicirculatus* (Cuvier *in* Cuvier & Valenciennes, 1831) – USNM material; video sequence 2006; underwater photograph 2006
Pomacanthus sexstriatus (Cuvier [ex Kuhl & Hasselt] *in* Cuvier & Valenciennes, 1831) – Video sequence 2006
Pomacanthus xanthometopon (Bleeker, 1853) – Video sequence 2006
Pygoplites diacanthus (Boddaert, 1772) – MNHN and USNM material; photograph 2006; video sequence 2006

••• Cirrhitidae
Cirrhitichthys falco Randall, 1963 – Video sequence 2006
Cyprinocirrhites polyactis (Bleeker, 1874) – Video sequence 2006
Oxycirrhites typus Bleeker, 1857 – Video sequence 2006; underwater photograph 2006
Paracirrhites arcatus (Cuvier *in* Cuvier & Valenciennes, 1829) – Video sequence 2006
Paracirrhites forsteri (Schneider *in* Bloch & Schneider, 1801) – Video sequence 2006; underwater photograph 2006

••• Cepolidae
Acanthocepola sp. – MNHN material; photographs 2006

••• Cichlidae
Sarotherodon occidentalis (Daget, 1962) – Introduced; Keith *et al.*, this volume

Figure 451: *Chromis brevirostris*. The shortsnout chromis was common in 60 m depth in Vanuatu. It was scientifically described in 2008 by Pyle, Earle and Green, and a specimen collected in Vanuatu during the Santo expedition is now one of the paratypes at the MNHN in Paris.

Figure 452: *Chromis earina*. The holotype of the Spring Chromis, now in the *Museum National d'Histoire Naturelle*, Paris, was collected in Vanuatu during the Santo Expedition in 63 m on the dive when this image was taken (In fact this may be a picture of the holotype before collection). This species was named by Pyle, Earle and Greene in 2008 along with four other deep water *Chromis* damselfish, including *C. brevirostris*, also collected in Vanuatu.

••• **Pomacentridae**

†*Abudefduf sordidus* (Forsskål *in* Niebuhr, 1775) – ANSP and CAS material

†*Abudefduf vaigiensis* (Quoy & Gaimard, 1825) – USNM material; video sequence 2006; underwater photograph 2006

†*Acanthochromis polyacanthus* (Bleeker, 1855) – CAS and USNM material; video sequence 2006

**Amblyglyphidodon aureus* (Cuvier [ex Kuhl & Hasselt] *in* Cuvier & Valenciennes, 1830) – Video sequence 2006

†*Amblyglyphidodon curacao* (Bloch, 1787) – Video sequence 2006

**Amblyglyphidodon leucogaster* (Bleeker, 1847) – Video sequence 2006

**Amblyglyphidodon orbicularis* Hombron & Jacquinot *in* Jacquinot & Guichenot, 1853 – Video sequence 2006; underwater photograph 2006 – (Fig. 450)

†*Amphiprion chrysopterus* Cuvier *in* Cuvier & Valenciennes, 1830 – MNHN and USNM material; photograph 2006; video sequence 2006

**Amphiprion clarkii* (Bennett, 1830) – First record by Plessis & Fourmanoir [1966: 133, as *Amphiprion sebae* (non Bleeker, 1853)]; USNM material; video sequence 2006

**Amphiprion melanopus* Bleeker, 1852 – CAS and USNM material; video sequence 2006; underwater photograph 2006

**Amphiprion perideraion* Bleeker, 1855 – USNM material; video sequence 2006; underwater photograph 2006

Cheiloprion labiatus (Day, 1877) – First record by Herre (1936: 273); FMNH and SU material 2006

Chromis alpha Randall, 1988 – First record by Randall (1988: 74-76); video sequence 2006

**Chromis amboinensis* (Bleeker, 1871) – MNHN material; video sequence 2006; photograph 2006; underwater photograph 2006

**Chromis atripectoralis* Welander & Schultz, 1951 – Video sequence 2006

**Chromis atripes* Fowler & Bean, 1928 – MNHN material; photograph 2006; video sequence 2006; underwater photograph 2006

Chromis brevirostris Pyle, Earle & Greene, 2008 – (Fig. 451)

**Chromis chrysura* (Bliss, 1883) – Video sequence 2006

**Chromis delta* Randall, 1988 – Video sequence 2006

Chromis earina Pyle, Earle & Greene, 2008 – Described from Espiritu Santo Group by Pyle, Earle & Greene (2008: 21-25); BPBM material; video sequence 2006 – (Fig. 452)

**Chromis elerae* Fowler & Bean, 1928 – Video sequence 2006

**Chromis iomelas* Jordan & Seale, 1906 – FMNH and SU material; video sequence 2006

**Chromis lepidolepis* Bleeker, 1877 – Video sequence 2006

**Chromis lineata* Fowler & Bean, 1928 – Video sequence 2006

**Chromis margaritifer* Fowler, 1946 – CAS and USNM material; video sequence 2006; underwater photograph 2006

**Chromis retrofasciata* Weber, 1913 – Video sequence 2006

Chromis ternatensis (Bleeker, 1856) – First record by Herre (1936: 272); CAS, FMNH, SU and USNM material; video sequence 2006; underwater photograph 2006

†*Chromis viridis* (Cuvier [ex Ehrenberg] *in* Cuvier & Valenciennes, 1830) – USNM material; video sequence 2006; underwater photograph 2006

**Chromis weberi* Fowler & Bean, 1928 – Video sequence 2006

**Chromis xanthochira* (Bleeker, 1851) – USNM material

**Chromis xanthura* (Bleeker, 1854) – Video sequence 2006

†*Chrysiptera biocellata* (Quoy & Gaimard, 1825) – ANSP and SU material

Figure 453: *Neoglyphidodon carlsoni*. Carlson's damselfish, previously known only fron the type locality of Fiji and Tonga, is a new record for Vanuatu. This is a large juvenile found in a small colony of this species around caves in 7 m.

Chrysiptera brownriggii (Bennett, 1828) – First record by Fowler (1934: 436, as *Abudefduf xanthozona*); ANSP and FMNH material; video sequence 2006

**Chrysiptera caeruleolineata* (Allen, 1973) – Video sequence 2006

†*Chrysiptera cyanea* (Quoy & Gaimard, 1825) – ANSP material

Chrysiptera glauca (Cuvier *in* Cuvier & Valenciennes, 1830) – First record by Fowler (1934: 435, *Abudefduf glaucus*); ANSP material

**Chrysiptera rex* (Snyder, 1909) – USNM material; video sequence 2006

**Chrysiptera rollandi* (Whitley, 1961) – Video sequence 2006; underwater photograph 2006

**Chrysiptera talboti* (Allen, 1975) – Video sequence 2006

†*Chrysiptera taupou* (Jordan & Seale, 1906) – Video sequence 2006

**Chrysiptera tricincta* (Allen & Randall, 1974) – Video sequence 2006

†*Dascyllus aruanus* (Linnaeus, 1758) – CAS and USNM material; video sequence 2006

†*Dascyllus reticulatus* (Richardson, 1846) – Video sequence 2006

Dascyllus trimaculatus (Rüppell, 1828) – First record by Plessis & Fourmanoir (1966: 135); USNM material; video sequence 2006; underwater photograph 2006

**Dischistodus melanotus* (Bleeker, 1858) – SU and USNM material; video sequence 2006

**Dischistodus perspicillatus* (Cuvier *in* Cuvier & Valenciennes, 1830) – Video sequence 2006

**Dischistodus prosopotaenia* (Bleeker, 1852) – Video sequence 2006

**Neoglyphidodon carlsoni* (Allen, 1975) – Video sequence 2006 – (Fig. 453)

†*Neoglyphidodon melas* (Cuvier [ex Kuhl & Hasselt] *in* Cuvier & Valenciennes, 1830) – Video sequence 2006

**Neoglyphidodon nigroris* (Cuvier *in* Cuvier & Valenciennes, 1830) – USNM; material; video sequence 2006

**Neopomacentrus azysron* (Bleeker, 1877) – USNM material

†*Neopomacentrus nemurus* (Bleeker, 1857) – CAS and USNM material

Neopomacentrus violascens (Bleeker, 1848) – First record by Herre (1936: 281); FMNH, MNHN and SU material; photograph 2006

**Plectroglyphidodon dickii* (Liénard, 1839) – Video sequence 2006

**Plectroglyphidodon johnstonianus* Fowler & Ball, 1924 – Video sequence 2006

†*Plectroglyphidodon lacrymatus* (Quoy & Gaimard, 1825) – Video sequence 2006

**Pomacentrus adelus* Allen, 1991 – USNM material

†*Pomacentrus amboinensis* Bleeker, 1868 – CAS and USNM material; video sequence 2006; underwater photograph 2006

Pomacentrus arenarius Allen, 1987 – First record by Allen (1987: 9-11)

**Pomacentrus bankanensis* Bleeker, 1853 – ANSP material; video sequence 2006

**Pomacentrus brachialis* Cuvier *in* Cuvier & Valenciennes, 1830 – USNM material; video sequence 2006

**Pomacentrus burroughi* Fowler, 1918 – Video sequence 2006

Pomacentrus chrysurus (Cuvier [ex Broussonet] *in* Cuvier & Valenciennes, 1830) – First record by Fowler (1934: 434, *Pomacentrus rhodonotus*); video sequence 2006

**Pomacentrus coelestis* Jordan & Starks, 1901 – CAS material; video sequence 2006

**Pomacentrus imitator* (Whitley, 1964) – CAS material; video sequence 2006

**Pomacentrus lepidogenys* Fowler & Bean, 1928 – Video sequence 2006

†*Pomacentrus littoralis* Cuvier *in* Cuvier & Valenciennes, 1830 – SU material

**Pomacentrus moluccensis* Bleeker, 1853 – CAS, SU and USNM material; video sequence 2006

**Pomacentrus nagasakiensis* Tanaka, 1917 – MNHN material; video sequence 2006; photograph 2006

**Pomacentrus nigromanus* Weber, 1913 – CAS material; video sequence 2006

Figure 454: *Bodianus bimaculatus*. The Twinspot Hogfish, has been reported from deep water from Madagascar to New Caledonia, and now is reported as a new record for Vanuatu. We found this small colorful wrasse to be common at 85-110 m depth.

Figure 455: *Bodianus paraleucosticticus*. This new record from Vanuatu is also known from New Caledonia in the South Pacific. An inquisitive species, the Fivestripe Hogfish, was occasionally observed, usually observing the diver, at around 100 m.

Pomacentrus nigromarginatus Allen, 1973 – MNHN material; photograph 2006; video sequence 2006
Pomacentrus philippinus Evermann & Seale, 1907 – Video sequence 2006
Pomacentrus reidi Fowler & Bean, 1928 – Video sequence 2006
Pomacentrus spilotoceps Randall, 2002 – Video sequence 2006
Pomacentrus vaiuli Jordan & Seale, 1906 – Video sequence 2006
Pomachromis richardsoni (Snyder, 1909) – Video sequence 2006
Stegastes albifasciatus (Schlegel & Müller, 1839) – First record by Fowler (1934: 434, *Pomacentrus albofasciatus*); video sequence 2006
Stegastes aureus (Fowler, 1927) – Underwater photograph 2006
Stegastes fasciolatus (Ogilby, 1889) – Video sequence 2006
†*Stegastes nigricans* (Lacepède, 1803) – FMNH, SU and USNM material; video sequence 2006
†*Stegastes punctatus* (Quoy & Gaimard, 1825) – FMNH and SU material; video sequence 2006

••• **Labridae**
Anampses melanurus (Quoy & Gaimard, 1825) – Video sequence 2006
Anampses meleagrides (Quoy & Gaimard, 1825) – Video sequence 2006
Anampses neoguinaicus Bleeker, 1878 – Video sequence 2006
Anampses twistii Bleeker, 1856 – Video sequence 2006
†*Bodianus anthioides* (Bennett, 1832) – Video sequence 2006
†*Bodianus axillaris* (Bennett, 1832) – Video sequence 2006
Bodianus bimaculatus Allen, 1973 – Video sequence 2006 – (Fig. 454)
Bodianus dictynna Gomon, 2006 – Video sequence 2006
Bodianus loxozonus (Snyder, 1908) – Video sequence 2006
Bodianus mesothorax (Bloch & Schneider, 2001) – Video sequence 2006
Bodianus paraleucosticticus Gomon, 2006 – Video sequence 2006 – (Fig. 455)
†*Cheilinus chlorourus* (Bloch, 1791) – FMNH and USNM material
†*Cheilinus fasciatus* (Bloch, 1791) – Video sequence 2006
Cheilinus oxycephalus Bleeker, 1853 – USNM material; video sequence 2006
†*Cheilinus trilobatus* (Lacepède [ex Commerson], 1801) – Video sequence 2006
†*Cheilinus undulatus* Rüppell, 1835 – Video sequence 2006
Choerodon jordani (Snyder, 1908) – Video sequence 2006
Cirrhilabrus bathyphilus Randall & Nagareda, 2002 – Video sequence 2006
Cirrhilabrus condei Allen & Randall, 1996 – Video sequence 2006
Cirrhilabrus exquisitus Smith, 1957 – Video sequence 2006
Cirrhilabrus punctatus Randall & Kuiter, 1989 – Video sequence 2006 – (Fig. 456)
Cirrhilabrus pylei Allen & Randall, 1996 – Video sequence 2006 – (Fig. 457)
Cirrhilabrus roseafascia Randall & Lubbock, 1982 – Video sequence 2006 – (Fig. 458)
Cirrhilabrus rubrimarginatus Randall, 1992 – First record by Randall (1992: 114-118); video sequence 2006
†*Cirrhilabrus* n. sp. – MNHN material; photograph 2006
Coris batuensis (Bleeker, 1856-1857) – Video sequence 2006
†*Coris gaimard* (Quoy & Gaimard, 1824) – Video sequence 2006

Figure 456: *Cirrhilabrus punctatus*. The Dotted Wrasse is a highly variable species throughout its range. Males we observed in Vanuatu have a yellow streak on the upper body, and differ in other ways from this species in Fiji and the Great Barrier Reef. We are provisionally calling this species *C. punctatus*, but specimens should be collected for study.

Figure 457: *Cirrhilabrus pylei*. Pyle's Fairy Wrasse is a new record from Vanuatu. This image is of a male from 40 m depth, where we occasionally saw small groups of this species, named after Santo Expedition rebreather diver Richard Pyle.

Figure 458: *Cirrhilabrus roseafascia*. This wrasse is a new record for Vanuatu, known from the South Pacific in Samoa, Fiji and New Caledonia. Small groups of this species, usually a male and his harem, were observed at 90-100 m in rubble areas.

Diproctacanthus xanthurus (Bleeker, 1856) – Video sequence 2006

†*Epibulus insidiator* (Pallas, 1770) – USNM material; video sequence 2006

†*Gomphosus varius* Lacepède, 1801 – USNM material; video sequence 2006

†*Halichoeres argus* (Bloch & Schneider, 1801) – Video sequence 2006

Halichoeres biocellatus Schultz *in* Schultz, Chapman, Lachner & Woods, 1960 – Video sequence 2006

Halichoeres chrysus Randall, 1981 – Video sequence 2006

†*Halichoeres hortulanus* (Lacepède, 1801) – USNM material; video sequence 2006

†*Halichoeres margaritaceus* (Valenciennes *in* Cuvier & Valenciennes, 1839) – Video sequence 2006

†*Halichoeres marginatus* (Rüppell, 1835) – Video sequence 2006

†*Halichoeres melanurus* (Bleeker, 1851) – ANSP and USNM material; video sequence 2006; underwater photograph 2006

†*Halichoeres miniatus* (Valenciennes [ex Kuhl & Hasselt] *in* Cuvier & Valenciennes, 1839) – USNM material

Halichoeres ornatissimus (Garrett, 1863) – Video sequence 2006

Halichoeres prosopeion (Bleeker, 1853) – Video sequence 2006

Halichoeres richmondi (Fowler & Bean, 1928) – Video sequence 2006

†*Halichoeres trimaculatus* (Quoy & Gaimard, 1834) – USNM material; video sequence 2006

Halichoeres zeylonicus (Bennett, 1833) – Video sequence 2006

†*Hemigymnus fasciatus* (Bloch, 1792) – Video sequence 2006

†*Hemigymnus melapterus* (Bloch, 1791) – Video sequence 2006

†*Hologymnosus annulatus* (Lacepède, 1801) – Video sequence 2006

Hologymnosus doliatus (Lacepède [ex Commerson], 1801) – Video sequence 2006

Labrichthys unilineatus (Guichenot, 1847) – Video sequence 2006

Labroides bicolor Fowler & Bean, 1928 – Video sequence 2006

†*Labroides dimidiatus* (Valenciennes *in* Cuvier & Valenciennes, 1839) – CAS and USNM material; video sequence 2006

Labroides pectoralis Randall & Springer, 1975 – Video sequence 2006; underwater photograph 2006

Labropsis alleni Randall, 1981 – Video sequence 2006

†*Labropsis australis* Randall, 1981 – Video sequence 2006

Labropsis xanthonota Randall, 1981 – Video sequence 2006

Macropharyngodon meleagris (Valenciennes *in* Cuvier & Valenciennes, 1839) – Video sequence 2006

Macropharyngodon negrosensis Herre, 1932 – Video sequence 2006

†*Novaculichthys taeniourus* (Lacepède, 1802) – USNM material; video sequence 2006

Figure 459: *Oxycheilinus nigromarginatus*. The Blackmargin Wrasse is a new record from Vanuatu. This species, scientifically described in 2003, is known from the Coral Sea, New Caledonia and Tonga, so its presence in Vanuatu was not unexpected. We observed occasional individuals in around 35 m.

**Oxycheilinus arenatus* (Valenciennes *in* Cuvier & Valenciennes, 1840) – Video sequence 2006
**Oxycheilinus bimaculatus* (Valenciennes *in* Cuvier & Valenciennes, 1840) – Video sequence 2006
**Oxycheilinus celebicus* (Bleeker, 1853) – Video sequence 2006
Oxycheilinus digramma (Lacepède, 1801) – First record by Herre (1936: 327, *Cheilinus diagrammus*); SU material; video sequence 2006
**Oxycheilinus nigromarginatus* Randall, Westneat & Gomon, 2003 – MNHN material; photograph 2006; video sequence 2006 – (Fig. 459)
**Paracheilinus rubricaudalis* Randall & Allen, 2003 – Video sequence 2006
†*Pseudocheilinus evanidus* Jordan & Evermann, 1903 – Video sequence 2006
†*Pseudocheilinus hexataenia* (Bleeker, 1857) – USNM material; video sequence 2006
†*Pseudocheilinus octotaenia* Jenkins, 1901 – Video sequence 2006
**Pseudocoris yamashiroi* (Schmidt, 1931) – Video sequence 2006
**Pseudodax moluccanus* (Valenciennes *in* Cuvier & Valenciennes, 1840) – Video sequence 2006
**Pseudojuloides cerasinus* (Snyder, 1904) – Video sequence 2006
**Pseudojuloides severnsi* Bellwood & Randall, 2000 – Video sequence 2006
**Pteragogus cryptus* Randall, 1981 – Video sequence 2006
†*Stethojulis bandanensis* (Bleeker, 1851) – CAS and USNM material; video sequence 2006
Stethojulis trilineata (Bloch & Schneider, 1801) – First record by Randall (2000: 36-38); USNM material; video sequence 2006
**Thalassoma amblycephalum* (Bleeker, 1856) – USNM material; video sequence 2006
†*Thalassoma hardwicke* (Bennett, 1830) – BMNH and SU material; video sequence 2006; underwater photograph 2006
†*Thalassoma lunare* (Linnaeus, 1758) – Video sequence 2006
†*Thalassoma lutescens* (Lay & Bennett [ex Solander], 1839) – Video sequence 2006
**Thalassoma nigrofasciatum* Randall, 2003 – Video sequence 2006
**Thalassoma quinquevittatum* (Lay & Bennett, 1839) – MNHN material; photograph 2006; video sequence 2006
†*Thalassoma trilobatum* (Lacepède [ex Commerson], 1801) – Video sequence 2006
†*Wetmorella nigropinnata* (Seale, 1901) – Video sequence 2006

••• **Scaridae**
†*Bolbometopon muricatum* (Valenciennes *in* Cuvier & Valenciennes, 1840) – Video sequence 2006
**Calotomus carolinus* (Valenciennes *in* Cuvier & Valenciennes, 1840) – Video sequence 2006
**Cetoscarus ocellatus* (Valenciennes *in* Cuvier & Valenciennes, 1840) – Video sequence 2006
†*Chlorurus bleekeri* (Beaufort *in* Weber & Beaufort, 1940) – Video sequence 2006
†*Chlorurus japanensis* (Bloch, 1789) – Video sequence 2006
**Chlorurus microrhinos* (Bleeker, 1854) – Video sequence 2006
†*Chlorurus sordidus* (Forsskål *in* Niebuhr, 1775) – SU material; video sequence 2006
†*Scarus altipinnis* (Steindachner, 1879) – Video sequence 2006
**Scarus chameleon* Choat & Randall, 1986 – Video sequence 2006
**Scarus dimidiatus* Bleeker, 1859 – Video sequence 2006
**Scarus flavipectoralis* Schultz, 1958 – Video sequence 2006

Scarus forsteni (Bleeker, 1861) – Video sequence 2006
Scarus frenatus Lacepède, 1802 – Video sequence 2006
Scarus fuscocaudalis Randall & Myers, 2000 – Video sequence 2006
Scarus globiceps Valenciennes *in* Cuvier & Valenciennes, 1840 – Video sequence 2006
†*Scarus niger* Forsskål *in* Niebuhr, 1775 – Video sequence 2006
†*Scarus oviceps* Valenciennes *in* Cuvier & Valenciennes, 1840 – Video sequence 2006; underwater
    photograph 2006
Scarus quoyi Valenciennes *in* Cuvier & Valenciennes, 1840 – Video sequence 2006
Scarus rubroviolaceus Bleeker, 1847 – Video sequence 2006
†*Scarus schlegeli* (Bleeker, 1861) – Video sequence 2006
Scarus spinus (Kner, 1868) – Video sequence 2006

••• **Chiasmodontidae**
Dysalotus alcocki Gilchrist, 1905 – First record by Fourmanoir (1970: 26, fig. 4); photograph 2006

••• **Champsodontidae**
Champsodon snyderi Franz, 1910 – MNHN material

••• **Percophidae**
Bembrops caudimacula Steindachner, 1876 – MNHN material; photograph 2006
Bembrops filifera Gilbert, 1905 – MNHN material
Bembrops philippinus Fowler, 1939 – MNHN material; photographs 2006
Bembrops platyrhynchus (Alcock, 1894) – MNHN material
Pteropsaron sp. (undescribed, Fricke, MS – MNHN material; photographs 2006

••• **Pinguipedidae**
Parapercis clathrata Ogilby, 1910 – CAS and USNM material; video sequence 2006
†*Parapercis hexophtalma* (Cuvier [ex Ehrenberg] *in* Cuvier & Valenciennes, 1829) – Video sequence 2006
Parapercis multiplicata Randall, 1984 – MNHN material; photograph 2006
Parapercis schauinslandi (Steindachner, 1900) – MNHN material; photograph 2006
Parapercis snyderi Jordan & Starks, 1905 – MNHN material; photograph 2006
Parapercis sp. (undescribed, Fricke, MS) – MNHN material; photograph 2006

••• **Trichonotidae**
Trichonotus setiger Bloch & Schneider, 1801 – MNHN material; photograph 2006

••• **Uranoscopidae**
Uranoscopus fuscomaculatus Kner, 1868 – MNHN material; photograph 2006
Uranoscopus sp. – MNHN material; photograph 2006

••• **Tripterygiidae**
Enneapterygius niger Fricke, 1994 – First record by Fricke (1997: 248-254); USNM material
Enneapterygius philippinus (Peters, 1869) – First record by Herre (1935: 432-433, as *Enneapterygius punctulatus*);
    SU material
Helcogramma sp. Williams, MS – First record by Fricke (1997: 408, *Helcogramma* sp. 7); USNM material
Helcogramma hudsoni (Jordan & Seale, 1906) – First record by Fricke (1997: 445).
Helcogramma springeri Hansen, 1986 – First record by Hansen (1986: 345); USNM material
Helcogramma striata Hansen, 1986 – Underwater photograph 2006

••• **Blenniidae**
Blenniella caudolineata (Günther, 1877) – First record by Fowler (1934: 446, *Salarias caudolineatus*);
    BMNH, FMNH and SU material
Blenniella cyanostigma (Bleeker, 1849) – First record by Rofen (1958: 196-197)
Cirripectes stigmaticus Strasburg & Schultz, 1953 – CAS and USNM material; video sequence 2006
Crossosalarias macrospilus Smith-Vaniz & Springer, 1971 – USNM material
†*Ecsenius bicolor* (Day, 1888) – CAS and USNM material; video sequence 2006
Ecsenius isos McKinney & Springer, 1976 – First record by McKinney & Springer (1976: 5); USNM material
†*Ecsenius yaeyamaensis* (Aoyagi, 1954) – Video sequence 2006
†*Entomacrodus decussatus* (Bleeker, 1858) – ANSP material
†*Entomacrodus striatus* (Quoy & Gaimard, 1836) – BMNH material
Glyptoparus delicatulus Smith, 1959 – USNM material

Figure 460: *Amblyeleotris fontanesii*. The Giant Shrimpgoby, a new record from Vanuatu, is the largest of the shrimp gobies, and lives in association with *Alpheus fenneri*, the largest of the symbiotic snapping shrimp, which maintains their mutual burrow. Note that the myopic shrimp is keeping one of its antennae in contact with the shrimp goby. The keen-eyed goby acts as a sentinel and signals the shrimp to return to the burrow at the approach of a predator by flicking its fins. Encountered near the wreck of the Calvin Coolidge in a typical soft bottom habitat.

Figure 461: *Amblyeleotris guttata*. The yellowspotted shrimp goby, which lives in symbiotic association with an alpheid snapping shrimp (likely *Alpheus ochrostriatus*), is a new record for Vanuatu. We found it not uncommon in mixed sand and coral rubble habitat.

Istiblennius bellus (Günther, 1861) – ANSP material

Istiblennius edentulus (Forster *in* Bloch & Schneider, 1801) – First record by Fowler (1934: 446); BMNH, CAS, FMNH, SU and USNM material

†*Istiblennius lineatus* (Valenciennes *in* Cuvier & Valenciennes, 1836) – ANSP, CAS and SU material

Istiblennius meleagris (Valenciennes *in* Cuvier & Valenciennes, 1836) – First record by Fowler (1934: 446, *Salarias meleagris*); BMNH material

Meiacanthus anema (Bleeker, 1852) – USNM material

†*Meiacanthus atrodorsalis* (Günther, 1877) – USNM material

Petroscirtes mitratus Rüppell, 1830 – SU material

Petroscirtes xestus Jordan & Seale, 1906 – CAS and USNM material

†*Plagiotremus rhinorhynchos* (Bleeker, 1852) – USNM material; video sequence 2006; underwater photograph 2006

†*Plagiotremus tapeinosoma* (Bleeker, 1857) – Video sequence 2006

Praealticus bilineatus (Peters, 1868) – CAS and SU material

Praealticus striolatus Bath, 1992 – ANSP material

Salarias fasciatus (Bloch, 1786) – First record by Fowler (1934: 445, as *Salarias nitidus*)

†*Salarias guttatus* Valenciennes *in* Cuvier & Valenciennes, 1836 – USNM material

Salarias segmentatus Bath & Randall, 1981 – USNM material

••• Gobiesocidae

Lepadichthys bolini Briggs, 1962 – First record by Briggs (1962: 424); SU material

••• Callionymidae

Callionymus kanakorum Fricke, 2006 – MNHN material; photograph 2006

Callionymus keeleyi Fowler, 1941 – MNHN material; photograph 2006

Callionymus rivatoni Fricke, 1993 – MNHN material; photographs 2006

Callionymus sp. (undescribed, aff. *spiniceps*; Fricke, MS) – MNHN material; photograph 2006

Callionymus tethys Fricke, 1993 – MNHN material; photograph 2006

Eleutherochir opercularis (Valenciennes *in* Cuvier & Valenciennes, 1837) – USNM material

Synchiropus sp. 1 Fricke, MS – MNHN material; photograph 2006

Synchiropus sp. 2 Fricke, MS – MNHN material; photograph 2006

Synchiropus sp. 3 Fricke, MS – MNHN material; photograph 2006

••• Rhyacichthyidae

Rhyacichthys sp. – Keith *et al.*, this volume

••• Eleotridae

Belobranchus belobranchus (Valenciennes *in* Cuvier & Valenciennes, 1837) – BMNH material

Bunaka gyrinoides (Bleeker, 1853) – Keith *et al.*, this volume

Figure 462: *Signigobius biocellatus*. The Twinspot Goby, was occasionally seen in 20 m, and is also listed as a new record for Vanuatu. The two dorsal fin spots resemble eyes, and this species moves in a back and forth crab-like scuttle motion. While it has been suggested by some that it is thus a crab mimic, one might also note that crabs are sought by many predators so mimicry here may be self-defeating.

Figure 463: *Nemateleotris decora*. The Elegant Firefish is a new record for Vanuatu, occasionally observed in 40 m or deeper. A *Tryssogobius colini* is seen below and to the left of the firefish. This is another new record for Vanuatu.

Butis amboinensis (Bleeker, 1853) – Keith *et al.*, this volume
Eleotris acanthopoma Bleeker, 1853 – Keith *et al.*, this volume
Eleotris fusca (Bloch & Schneider, 1801) – First record by Fowler (1934: 441)
Hypseleotris guentheri (Bleeker, 1875) – Keith *et al.*, this volume
†*Ophieleotris aporos* (Bleeker, 1854) – BMNH material
Ophieleotris sp. – Keith *et al.*, this volume
**Ophiocara porocephala* (Valenciennes *in* Cuvier & Valenciennes, 1837) – BMNH material

••• Kraemeriidae
**Gobitrichinotus radiocularis* Fowler, 1943 – Keith *et al.*, this volume

••• Gobiidae
**Amblyeleotris fasciata* (Herre, 1953) – Video sequence 2006
**Amblyeleotris fontanesii* (Bleeker, 1852) – Video sequence 2006; underwater photographs 2006 – (Fig. 460)
**Amblyeleotris guttata* (Fowler, 1938) – Video sequence 2006; underwater photographs 2006 – (Fig. 461)
**Amblyeleotris novaecaledoniae* Goren, 1981 – Photograph 2006
**Amblyeleotris ogasawarensis* Yanagisawa, 1978 – Photograph 2006
**Amblyeleotris randalli* Hoese & Steene, 1978 – Video sequence 2006
**Amblyeleotris steinitzi* (Klausewitz, 1974) – CAS material; photograph 2006
†*Amblygobius decussatus* (Bleeker, 1855) – USNM material; video sequence 2006; underwater photograph 2006
**Amblygobius hectori* (Smith, 1956) – USNM material
†*Amblygobius nocturnus* (Herre, 1945) – Photograph 2006
†*Amblygobius phalaena* (Valenciennes *in* Cuvier & Valenciennes, 1837) – CAS and USNM material
†*Asterropterix semipunctata* Rüppell, 1830 – CAS, FMNH, SU and USNM material
**Asterropterix striata* Allen & Munday, 1995 – Video sequence 2006
†*Awaous ocellaris* (Broussonet, 1782) – BMNH material
**Bathygobius albopunctatus* (Valenciennes *in* Cuvier & Valenciennes, 1837) – BMNH material
Bathygobius fuscus (Rüppell, 1830) – First record by Fowler (1934: 442); ANSP material
†*Bryaninops yongei* (Davis & Cohen, 1969) – Underwater photograph 2006
Callogobius okinawae (Snyder, 1908) – First record by Herre (1935: 416, as *Macgregorella santa*); FMNH and SU material
†*Callogobius sclateri* (Steindachner, 1880) – CAS and SU material
**Cryptocentrus leptocephalus* Bleeker, 1876 – Underwater photograph 2006
**Ctenogobiops pomastictus* Lubbock & Polunin, 1977 – Underwater photograph 2006
**Ctenotrypauchen microcephalus* (Bleeker, 1860) – Photographs 2006
**Eviota albolineata* Jewett & Lachner, 1983 – CAS material
**Eviota queenslandica* Whitley, 1932 – CAS material
**Eviota sebreei* Jordan & Seale, 1906 – Video sequence 2006
**Eviota smaragdus* Jordan & Seale, 1906 – ANSP material
†*Exyrias puntang* (Bleeker, 1851) – MNHN and USNM material; photograph 2006

Fusigobius signipinnis Hoese & Obika, 1988 – Video sequence 2006
†*Gnatholepis anjerensis* (Bleeker, 1851) – Video sequence 2006
†*Gnatholepis cauerensis* (Bleeker, 1853) – CAS material; video sequence 2006
Gobiodon quinquestrigatus (Valenciennes *in* Cuvier & Valenciennes, 1837) – SU material
Istigobius decoratus (Herre, 1927) – USNM material; video sequence 2006
†*Istigobius ornatus* (Rüppell, 1830) – FMNH and SU material
†*Kelloggella cardinalis* Jordan & Seale, 1906 – ANSP material; photograph 2006
Koumansetta rainfordi (Whitley, 1940) – USNM material; video sequence 2006; photograph 2006
Lentipes kaaea Watson, Keith & Marquet, 2002 – Keith *et al.*, this volume
Lotilia graciliosa Klausewitz, 1960 – Video sequence 2006
Mahidolia mystacina (Valenciennes *in* Cuvier & Valenciennes, 1837) – Photographs 2006
Oplopomops diacanthus (Schultz, 1943) – USNM material
Oplopomus oplopomus (Valenciennes *in* Cuvier & Valenciennes, 1837) – Photograph 2006
Paragobiodon echinocephalus (Rüppell, 1828) – First record by Herre (1936: 363); CAS and FMNH material
†*Periophthalmus argentilineatus* Valenciennes *in* Cuvier & Valenciennes, 1837 – USNM material; photograph 2006
Priolepis cincta (Regan, 1908) – USNM material
Priolepis inhaca (Smith, 1949) – USNM material
†*Priolepis semidoliatus* (Valenciennes *in* Cuvier & Valenciennes, 1837) – ANSP, FMNH and SU material
Priolepis sp. – MNHN material; photographs 2006
Psammogobius biocellatus (Valenciennes *in* Cuvier & Valenciennes, 1837) – USNM material
Redigobius bikolanus (Herre, 1927) – Keith *et al.*, this volume
Schismatogobius vanuatuensis Keith, Marquet & Watson, 2004 – First record by Keith, Marquet & Watson (2004: 238); MNHN material
Sicyopterus aiensis Keith, Watson & Marquet, 2004– Keith *et al.*, this volume
Sicyopterus lagocephalus (Pallas, 1770) – BMNH material; Keith *et al.*, this volume
Sicyopus chloe Watson, Keith & Marquet, 2001 – Keith *et al.*, this volume
Sicyopus zosterophorus (Bleeker, 1857) – Keith *et al.*, this volume
†*Signigobius biocellatus* Hoese & Allen, 1977 – Video sequence 2006 – (Fig. 462)
Stenogobius yateiensis Keith, Watson & Marquet, 2002 – Keith *et al.*, this volume
Stiphodon astilbos Ryan, 1986 – First record by Ryan (1986: 656-660); AMS material
Stiphodon atratus Watson, 1996 – Keith *et al.*, this volume
Stiphodon kalfatak Keith, Marquet & Watson, 2007 – First record by Keith, Marquet & Watson (2007: 34-37); MNHN material
Stiphodon mele Keith, Marquet & Pouilly 2009 – First record by Keith, Marquet & Pouilly (2009: 473); MNHN paratypes.
Stiphodon rutilaureus Watson, 1996 – Keith *et al.*, this volume
Stiphodon sapphirinus Watson, Keith & Marquet, 2005 – Keith *et al.*, this volume
Trimma caesiura Jordan & Seale, 1906 – USNM material
Trimma okinawae (Aoyagi, 1949) – USNM material
Trimma striata (Herre, 1945) – USNM material
Trimma tevegae Cohen & Davis, 1969 – Video sequence 2006
Tryssogobius colini Larson & Hoese, 2001 – Photograph 2006; video sequence 2006
†*Valenciennea puellaris* (Tomiyama, 1955) – USNM material; video sequence 2006
Valenciennea randalli Hoese & Larson, 1994 – Photographs 2006
Valenciennea sexguttata (Valenciennes *in* Cuvier & Valenciennes, 1837) – USNM material; video sequence 2006
†*Valenciennea strigata* (Broussonet, 1782) – Video sequence 2006
Yongeichthys criniger (Valenciennes *in* Cuvier & Valenciennes, 1837) – MNHN material; photograph 2006

••• Microdesmidae

Gunnellichthys curiosus Dawson, 1968 – MNHN material; photograph 2006; video sequence 2006
Nemateleotris decora Randall & Allen, 1973 – Video sequence 2006 – (Fig. 463)
Nemateleotris helfrichi Randall & Allen, 1973 – Video sequence 2006
†*Nemateleotris magnifica* Fowler, 1938 – Video sequence 2006
†*Ptereleotris evides* (Jordan & Hubbs, 1925) – CAS, FMNH, SU and USNM material; video sequence 2006
Ptereleotris heteroptera (Bleeker, 1855) – MNHN material; photograph 2006
Ptereleotris microlepis (Bleeker, 1856) – USNM material
Ptereleotris monoptera Randall & Hoese, 1985 – First record by Randall & Hoese (1985: 24); USNM material

••• Xenisthmidae

Xenisthmus chapmani (Schultz *in* Schultz, Woods & Lachner, 1966) – First record by Schultz, Woods & Lachner (1966: 8-9); USNM material

••• Zanclidae
†*Zanclus cornutus* (Linnaeus, 1758) – Video sequence 2006

••• Acanthuridae
**Acanthurus blochii* Valenciennes *in* Cuvier & Valenciennes, 1835 – Video sequence 2006
**Acanthurus dussumieri* Valenciennes *in* Cuvier & Valenciennes, 1835 – Video sequence 2006
†*Acanthurus lineatus* (Linnaeus, 1758) – FMNH, SU and USNM material; video sequence 2006
†*Acanthurus mata* (Cuvier, 1829) – Video sequence 2006
†*Acanthurus nigricans* (Linnaeus, 1758) – Video sequence 2006
**Acanthurus nigricauda* Duncker & Mohr, 1929 – SU material
**Acanthurus nigrofuscus* Forsskål *in* Niebuhr, 1775 – USNM material; video sequence 2006
†*Acanthurus olivaceus* Forster *in* Bloch & Schneider, 1801 – Video sequence 2006
†*Acanthurus pyroferus* Kittlitz, 1834 – MNHN material; photograph 2006; video sequence 2006
**Acanthurus thompsoni* (Fowler, 1923) – Video sequence 2006
†*Acanthurus triostegus* (Linnaeus, 1758) – ANSP, BMNH, FMNH and USNM material
†*Ctenochaetus binotatus* Randall, 1955 – USNM material; video sequence 2006
†*Ctenochaetus cyanocheilus* Randall & Clements, 2001 – FMNH material
†*Ctenochaetus striatus* (Quoy & Gaimard, 1825) – SU material; video sequence 2006
†*Ctenochaetus tominiensis* Randall, 1955 – Video sequence 2006
†*Naso annulatus* (Quoy & Gaimard, 1825) – Video sequence 2006
†*Naso brachycentron* (Valenciennes *in* Cuvier & Valenciennes, 1835) – Video sequence 2006
†*Naso brevirostris* (Cuvier, 1829) – Video sequence 2006
†*Naso hexacanthus* (Bleeker, 1855) – Video sequence 2006
†*Naso lituratus* (Bloch & Schneider [ex Forster], 1801) – Video sequence 2006
**Naso lopezi* Herre, 1927 – Video sequence 2006
**Naso macdadei* Johnson, 2002 – Video sequence 2006
**Naso minor* (Smith, 1966) – Video sequence 2006
**Naso thynnoides* (Cuvier, 1829) – Video sequence 2006 – (Fig. 464)
†*Naso unicornis* (Forsskål *in* Niebuhr, 1775) – Video sequence 2006
†*Naso vlamingii* (Valenciennes *in* Cuvier & Valenciennes, 1835) – Video sequence 2006
†*Zebrasoma scopas* (Cuvier, 1829) – FMNH, MNHN and USNM material; photograph 2006; video sequence 2006
†*Zebrasoma veliferum* (Bloch, 1797) – SMNS and USNM material; video sequence 2006

Figure 464: In the foreground of this image from 10 m one may see schooling *Naso thynnoides*, or Singlespine Unicornfish, whose common name may be a misnomer because this species is without a horn. This is a new record for Vanuatu, as is the *Variola albimarginata*, or Lyretail Grouper, seen below the Naso school. In the right background one may also note the clownfish *Amphiprion clarkii*, yet another Vanuatu record, while in the left background the dark surgeonfish *Acanthurus pyroferus* is "merely" a new Espititu Santo record.

••• **Siganidae**
†*Siganus argenteus* (Quoy & Gaimard, 1825) – Video sequence 2006; underwater photograph 2006
†*Siganus corallinus* (Valenciennes *in* Cuvier & Valenciennes, 1835) – Video sequence 2006
†*Siganus puellus* (Schlegel, 1852) – Video sequence 2006
**Siganus punctatissimus* Fowler & Bean, 1929 – Video sequence 2006
Siganus spinus (Linnaeus, 1758) – First record by Fowler (1934: 429)
†*Siganus vulpinus* (Schlegel & Müller, 1845) – Video sequence 2006

••• **Nomeidae**
†*Cubiceps* sp. – MNHN material; photograph 2006

••• **Scombridae**
**Gymnosarda unicolor* (Rüppell, 1836) – Video sequence 2006
†*Rastrelliger kanagurta* (Cuvier, 1816) – Video sequence 2006

••• **Sphyraenidae**
†*Sphyraena forsteri* Cuvier *in* Cuvier & Valenciennes, 1829 – Video sequence 2006
†*Sphyraena qenie* Klunzinger, 1870 – Video sequence 2006

••• **Gempylidae**
**Rexea* sp. – MNHN material

••• **Caproidae**
**Antigonia malayana* Weber, 1913 – MNHN material; photographs 2006
**Antigonia rubescens* (Günther, 1860) – MNHN material

••• **Paralichthyidae**
**Psammodiscus ocellatus* Günther, 1882 – MNHN material; photograph 2006
**Pseudorhombus dupliciocellatus* Regan, 1905 – MNHN material
**Pseudorhombus elevatus* Ogilby, 1912 – MNHN material; photograph 2006
**Pseudorhombus spinosus* McCulloch, 1914 – MNHN material; photograph 2006

••• **Bothidae**
**Arnoglossus japonicus* Hubbs, 1915 – MNHN material
**Arnoglossus macrolophus* Alcock, 1889 – MNHN material
†*Bothus mancus* (Broussonet, 1782) – MNHN material; photograph 2006
†*Bothus pantherinus* (Rüppell, 1830) – Photograph 2006
**Engyprosopon bellonaensis* Amaoka, Mihara & Rivaton, 1993 – MNHN material
**Engyprosopon longipterum* Amaoka, Mihara & Rivaton, 1993 – MNHN material
Engyprosopon vanuatuensis Amaoka & Séret, 2005 – First record by Amaoka & Séret (2005: 15); MNHN material
**Kamoharaia megastoma* (Kamohara, 1936) – MNHN material
**Neolaeops microphthalmus* (Bonde, 1922) – MNHN material; photograph 2006
**Parabothus filipes* Amaoka, Mihara & Rivaton, 1997 – MNHN material
**Parabothus kiensis* (Tanaka, 1918) – MNHN material
**Psettina variegata* (Fowler, 1934) – MNHN material

••• **Pleuronectidae**
**Nematops grandisquamus* Weber & Baufort, 1929 – MNHN material
**Nematops macrochirus* Norman, 1931 – MNHN material
**Poecilopsetta plinthus* (Jordan & Starks, 1904) – MNHN material

••• **Citharidae**
**Brachypleura novaezeelandiae* Günther, 1862 – MNHN material

••• **Soleidae**
†*Pardachirus pavoninus* (Lacepède, 1802) – BMNH material

••• **Cynoglossidae**
**Cynoglossus* sp. (Fricke MS) – MNHN material; photograph 2006
**Symphurus* sp. – MNHN material

••• **Samaridae**
Samariscus longimanus Norman, 1927 – Photograph 2006

••• **Triacanthodidae**
Atrophacanthus japonicus (Kamohara, 1941) – MNHN material; photograph 2006
Halimochirurgus sp. – MNHN material
Macrorhamphosodes uradoi (Kamohara, 1933) – MNHN material; photograph 2006
Triacanthodes ethiops Alcock, 1894 – MNHN material; photograph 2006
Tydemania navigatoris Weber, 1913 – MNHN material

••• **Balistidae**
†*Balistapus undulatus* (Park, 1797) – CAS, FMNH and SU material; video sequence 2006
†*Balistoides conspicillum* (Bloch & Schneider, 1801) – Video sequence 2006; underwater photograph 2006
†*Balistoides viridescens* (Bloch & Schneider, 1801) – Video sequence 2006; underwater photograph 2006
†*Melichthys vidua* (Richardson [ex Solander], 1845) – Video sequence 2006
†*Odonus niger* (Rüppell, 1836) – Video sequence 2006
†*Pseudobalistes fuscus* (Bloch & Schneider, 1801) – USNM material
†*Rhinecanthus aculeatus* (Linnaeus, 1758) – Underwater photograph 2006
†*Sufflamen bursa* (Bloch & Schneider, 1801) – MNHN material; photograph 2006, video sequence 2006
†*Sufflamen chrysopterum* (Bloch & Schneider, 1801) – Video sequence 2006
†*Sufflamen fraenatum* (Latreille, 1804) – Video sequence 2006

••• **Monacanthidae**
†*Amanses scopas* (Cuvier, 1829) – Video sequence 2006
†*Cantherhines pardalis* (Rüppell, 1837) – Video sequence 2006
†*Oxymonacanthus longirostris* (Bloch & Schneider, 1801) – Video sequence 2006
†*Paraluteres prionurus* (Bleeker, 1851) – Video sequence 2006
Paramonacanthus curtorhynchus (Bleeker, 1855) – MNHN material; photograph 2006
†*Pervagor melanocephalus* (Bleeker, 1853) – Video sequence 2006

••• **Ostraciidae**
†*Lactoria cornuta* (Linnaeus, 1758) – MNHN material; photograph 2006
†*Ostracion cubicus* Linnaeus, 1758 – Video sequence 2006
Ostracion meleagris Shaw in Shaw & Nodder, 1796 – Video sequence 2006
Ostracion solorensis Bleeker, 1853 – Video sequence 2006

••• **Triodontidae**
Triodon macropterus Lesson, 1831 – MNHN material

••• **Tetraodontidae**
†*Arothron hispidus* (Linnaeus, 1758) – MNHN material; photographs 2006
Arothron manilensis (Marion de Procé, 1822) – MNHN material; photograph 2006
†*Arothron nigropunctatus* (Bloch & Schneider, 1801) – CAS and USNM material; video sequence 2006
Canthigaster axiologus Whitley, 1931 – Video sequence 2006; taxonomic decision of Randall, Williams & Rocha (2008: 6-7, as *C. axiologa*)
Canthigaster bennetti (Bleeker, 1854) – First record by Fowler (1934: 449); USNM material
†*Canthigaster compressa* (Marion de Procé, 1822) – MNHN and USNM material; photograph 2006
Canthigaster epilampra (Jenkins, 1903) – MNHN material; video sequence 2006; photographs 2006
Canthigaster janthinoptera (Bleeker, 1855) – First record by Fowler (1934: 449); USNM material
Canthigaster papua (Bleeker, 1848) – Video sequence 2006
Canthigaster rivulata (Temminck & Schlegel, 1850) – MNHN material; photograph 2006
Canthigaster solandri (Richardson [ex Solander], 1845) – Video sequence 2006
†*Canthigaster valentini* (Bleeker, 1853) – CAS, MNHN and USNM material; video sequence 2006; photographs 2006; underwater photograph 2006
Tylerius spinosissimus (Regan, 1908) – MNHN material

••• **Diodontidae**
Diodon holocanthus Linnaeus, 1758 – MNHN material; photograph 2006
†*Diodon hystrix* Linnaeus, 1758 – Video sequence 2006

UNUSUAL AND SPECTACULAR CRUSTACEANS

Tin-Yam Chan, Masako Mitsuhashi, Charles H.J.M. Fransen, Régis Cleva, Swee Hee Tan,
Jose Christopher Mendoza, Marivene Manuel-Santos & Peter K. L. Ng

In Santo 2006, the crustaceans mainly from the Order Decapoda (ie. shrimps and crabs) were surveyed. It was estimated about 1000 species were collected during the expedition. This number suggests that a high diversity of crustaceans could be found around the island. This is especially considering that only 671 species of decapod crustaceans have been reported from the entire Mariana Islands while 2262 species of shrimps and crabs are know from the whole of Australia. On the other hand, the diversity found in Santo is about two-thirds that collected by similar expeditions in the Philippines, which is believed to be in the centre of marine biodiversity. Although the diversity of decapod crustacean from Santo 2006 is lower than in the Philippines, it was estimated that about one-third of the species collected were different from those found in the Philippines. Moreover, the roughly 500 species of macrurans (i.e. shrimps and lobsters) and anomurans (i.e. hermit crabs and alike) collected belong to 45 families — that constituted about 60% of the world marine families (76) known for these groups. Preliminary sorting revealed that at least two new genera and 32 new species were present in the Santo 2006 collection.

Most species collected from the Santo 2006 were relatively small in size. Large commercial species (such as *Penaeus* and spiny lobsters) are not well represented but this was probably due to the sampling methods used. For example, one of the most renowned crustaceans in Vanuatu, the coconut crab *Birgus latro*, was not collected although the villagers know it well. In shallow marine habitats, the caridean shrimp from the families Palaemonidae (mainly the subfamily Pontoniinae) and Alpheidae were abundant as in most parts of the world. However, the Santo 2006 expedition also collected numerous specimens from the supposed rare caridean shrimp family Ogyrididae, which is unique in have very long eye stalks. In deeper waters, species belonging to the genera *Metapenaeopsis* and *Penaeopsis*, the families Aristeidae and Solenoceridae (Penaeoidea), Pandalidae (Caridea), and Galatheidae (Anomura) were dominant in the catches, a situation similar in other deep sea sampling efforts. However, clawed lobsters (Astacidea) were poorly represented in the Santo 2006 deep-water samples. Other than the decapods, only the crustaceans from the order Stomatopoda (mantis shrimps) were collected and with preliminary investigations suggest a rather high diversity could be expected. Similar to the situation seen in the decapods, the stomatopod collection was also mainly represented by small-sized species. Other crustaceans collected at Santo 2006 were Cirripedia (barnacles), Mysida, Amphipoda, Isopoda and Euphausiacea, but they were present in lesser numbers probably due to the collecting gears used.

••• Penaeoidea (Fig. 465)

This group includes the most commercial important shrimp genus *Penaeus* s.l. but adult specimens were collected for only one species, *P. marginatus*, in Santo. On the other hand, the genus *Metapenaeopsis*

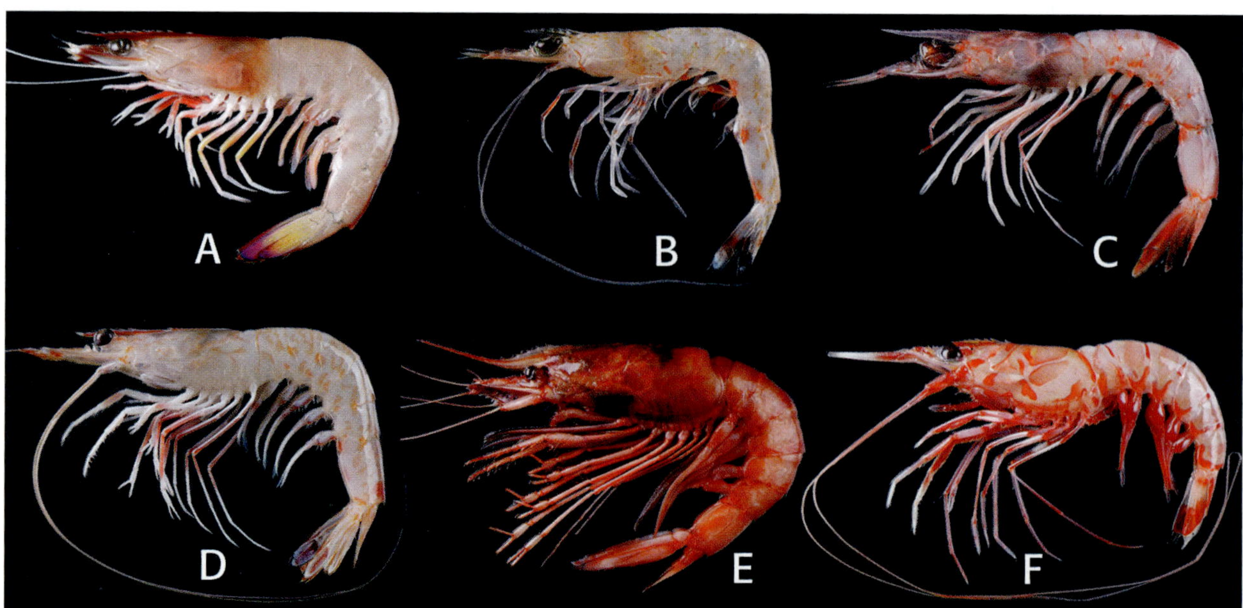

Figure 465: Penaeoidea (id. by T.Y. Chan). **A**: *Penaeus (Melicertus) marginatus* (Penaeidae). **B**: *Metapenaeopsis* sp. (Penaeidae). **C**: *Penaeopsis* sp. (Penaeidae). **D**: *Parapenaeus* sp. (Penaeidae). **E**: *Aristaemorpha foliacea* (Aristeidae). **F**: *Solenocera rathbuni* (Solenoceridae).

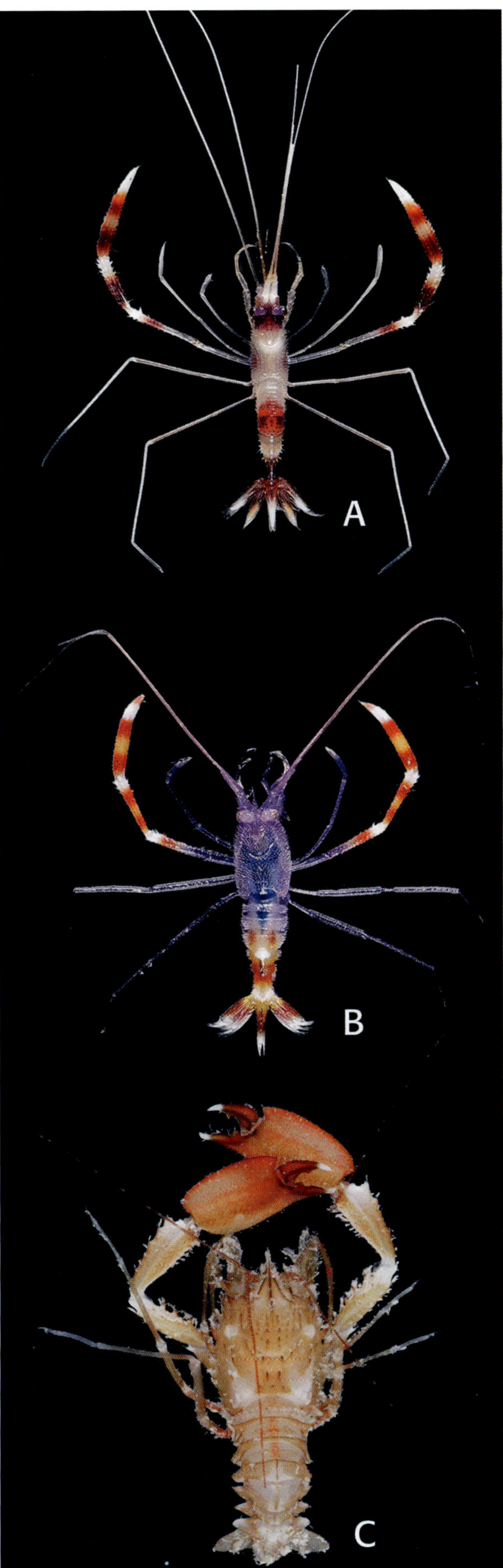

Figure 466: Stenopodidea (id. by T.Y. Chan). **A**: *Stenopus hispidus* (Stenopodidae). **B**: *Stenopus tenuirostris* (Stenopodidae). **C**: *Microprosthema* sp. (Spongiocolidae).

is very common in both shallow and deep waters. In deep water, species of the genus *Penaeopsis* and sometimes *Parapenaeus* are also abundant in the catches. All these genera belong to the family Penaeidae. Other penaeoids that sometimes can be found in large numbers belong to the families Aristeidae and Solenoceridae.

Stenopodidea (Fig. 466)

Of the three families in this infraorder, two were found in Santo. The well-known cleaner shrimp genus *Stenopus* has two species collected, the common *S. hispidus* and the less common *S. tenuirostris*. In the Philippines, many stenopodids were found to live inside sponges but this association is uncommon in Santo. Therefore, the diversity of Stenopodidae is not particularly high in Santo.

Caridea

The highest diversity for shrimps is in the Caridea. As in most of other parts of the world, the main diversity of carideans in Santo are in the families Palaemonidae (largely the subfamily Pontoniinae, Fig. 467) and Alpheidae (Fig. 468) for shallow waters, and Pandalidae (Fig. 469) for deep water. Most of the Pontoniinae and Alpheidae are symbiotic with other invertebrates (e.g. corals, sea anemones, sea stars, sea urchins, crinoids, sponges, etc.) and with brilliant coloration and/or bizarre body forms. Many of them even with body coloration are able to change with the host color and thus making a perfect camouflage. Beside these two families, shallow water members of the family Hippolytidae (Fig. 470), which have similar symbiotic life styles, are also well represented in the samples but with a much lower diversity. Of particular in the shallow water samples of Santo are many specimens belonging to the family Ogyrididae, which is generally considered to be a rare family. Most *Ogyrides* were found in samples from sandy bottoms. Such a habitat probably has some relationships with the characteristic long eyes in these shrimps. In deep water, the caridean family Pandalidae is predominant. Those belonging to the genus *Heterocarpus* are moderate in size and of commercial potential. The others mainly belong to the genus *Plesionika* which often have distinctive color markings on the body. The classical deep-water caridean family Glyphocrangonidae is also well represented in the Santo samples. The Santo expedition also collected some uncommon or rare families (Fig. 471), such as Disciadidae from shallow water, and Psalidopodidae and Thalassocarididae from deep water. At least one caridean species new to science has been discovered and it is believed that more new species of this group will be found when the Santo material of this group is studied in detail.

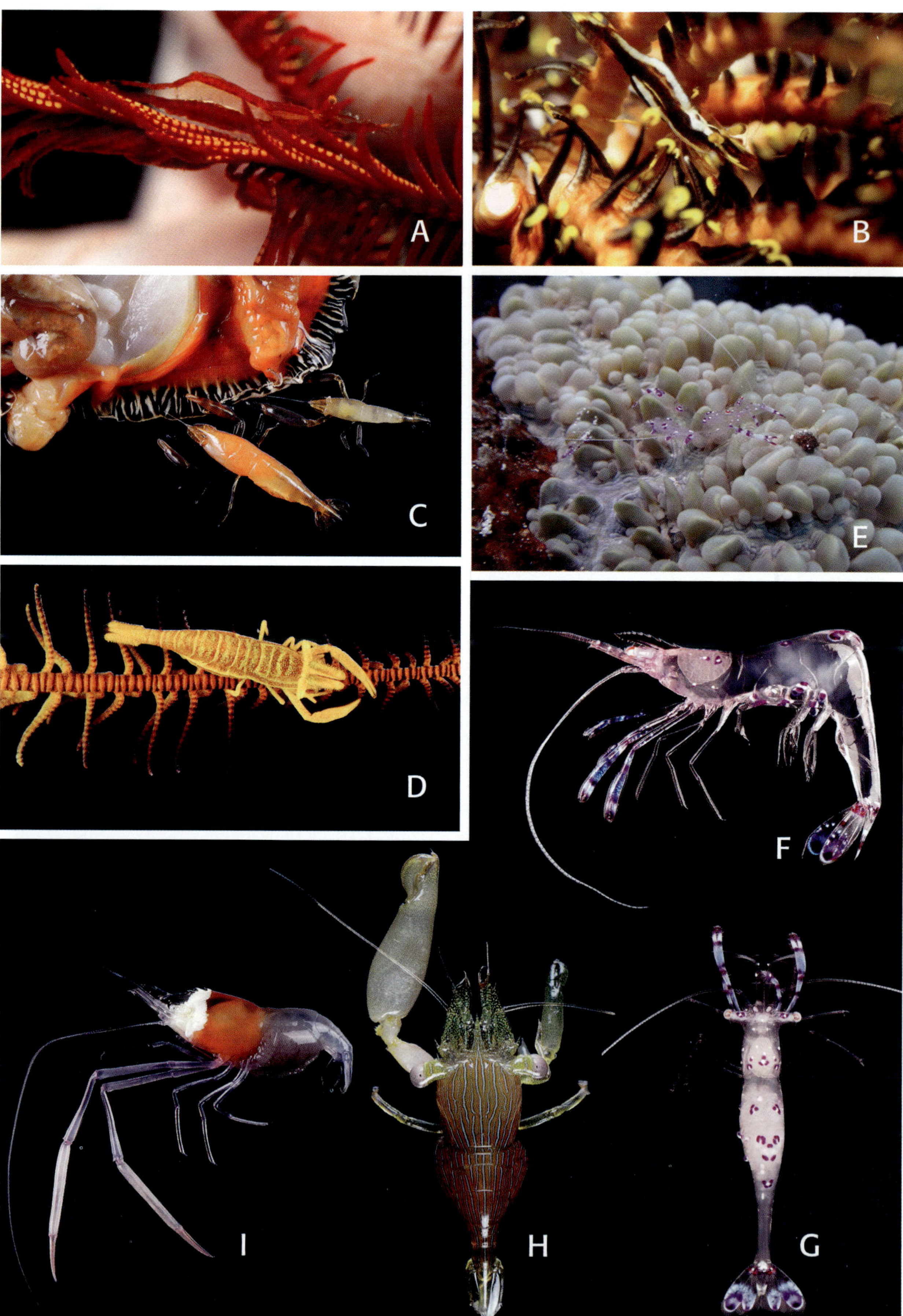

Figure 467: Palaemonidae (Pontoniinae) (id. by M. Mitsuhashi & C.H.J.M. Fransen). **A**: *Brucecaris tenuis*, in association with red crinoids. **B**: *Brucecaris tenuis*, in association with black and yellow crinoids. **C**: *Anchistus custoides*, live inside clams. **D**: *Laomenes amboinensis*, in association with crinoids. **E**: *Ancylomenes sarasvati* in association with the bubble coral *Physogyra lichtensteini* (photo S. Schiaparelli). **F**: *Ancylomenes sarasvati*, side view. **G**: *Ancylomenes sarasvati*, upper view. **H**: *Coralliocaris graminea* found in association with scleractinian corals. **I**: *Cuapetes kororensis* found in association with the coral *Heliofungia actiniformis*.

Figure 468: Alpheidae (id. by A. Anker). **A**: *Alpheus* sp. **B**: *Alpheus lottini*. **C**: *Alpheopsis* aff. *yaldwyni*. **D**: *Aretopsis* aff. *amabilis*. **E**: *Synalpheus stimpsoni* living on the oral disc of crinoids.

Figure 469: Pandalidae (id. by T.Y. Chan). **A**: *Heterocarpus hayashii*. **B**: *Heterocarpus corona* (recently described new species based on Santo specimens). **C**: *Plesionika* aff. *binoculus*.

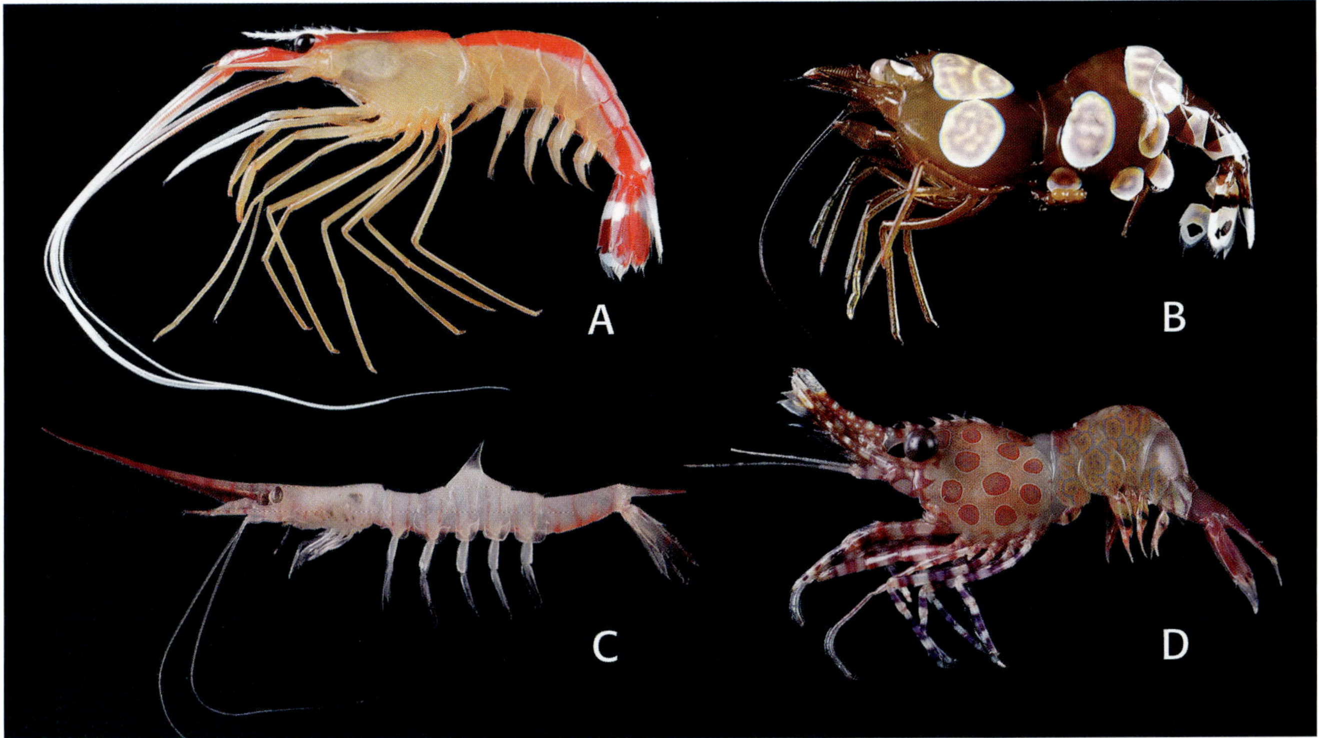

Figure 470: Hippolytidae (id. by T.Y. Chan & T. Komai). **A**: *Lysmata amboinensis*, one of the cleaner shrimps. **B**: *Thor amboinensis*. **C**: *Tozeuma* sp. **D**: *Saron* sp. (probably a new species).

Figure 471: Rare and uncommon shrimps collected (id. by T.Y. Chan & T. Komai). **A**: *Ogyrides* sp. (Ogyrididae). **B**: *Discias* sp. (Disciadidae). **C**: *Chlorotocoides spinicauda* (Thalassocarididae). **D**: *Phyllognathia ceratophthalma* (Hymenoceridae). **E**: *Pycnocaris chagoae* (Gnathophyllidae). **F**: *Psalidopus huxleyi* (Psalidopodidae). **G**: *Glyphocrangon similior* (Glyphocrangonidae).

••• Lobsters

Lobsters (Fig. 472) are not well represented in the Santo samples. Only two species of clawed lobsters (Astacidea) and a few species of spiny lobsters (Palinuridae) were collected. It seems that the low diversity of lobsters in the Santo expedition was a result of the sampling gears used. Nevertheless, at least two new species of lobsters were found during the expedition.

••• Anomurans

As in the other parts of the world, anomurans are well represented in Santo by the hermit crabs (Paguroidea, Fig. 473), squat lobsters (Galatheidae and Chirostylidae, Fig. 474), procellanid crabs (Porcellanidae, Fig. 475), the mud shrimps (Thalassinidae, Fig. 476) and the king crabs (Lithodidae, Fig. 477). The porcellanid crabs are largely from shallow water, but the three other groups have high diversity in both shallow and deep water. The king crabs are all from the deep sea. For the hermit crabs, a genus new to science has already been discovered and at least two new species of squat lobsters have also been found.

Figure 472: Lobsters (id. by T.Y. Chan & S. Ahyong). **A**: *Panulirus femoristriga* (Palinuridae). **B**: *Justitia vericeli* (Palinuridae). **C**: *Puerulus* new species (Palinuridae). **D**: *Petrarctus holthuisi* (Scyllaridae, recently described new species based on Santo specimens). **E**: *Ibacus novemdentatus* (Scyllaridae). **F**: *Stereomastis* aff. *surda* (Polychelidae). **G**: *Nephropsis* aff. *sulcata* (Nephropidae).

Figure 473: Hermit crabs (id. by T. Komai, D.L. Rahayu & P.A. McLaughlin). **A**: *Xylopagurus caledonicus*, lives in hollow wood tubes (Pylochelidae). **B**: *Paguritta* sp., lives in burrows in coral (Paguridae). **C**: *Coenobita rugosus*, land hermit crab (Coenobitidae). **D**: *Pumilopagurus tuberculomanus* (Paguridae, recently described new genus based on Santo specimens). **E**: *Calcinus pulcher* (Diogenidae). **F**: *Ciliopagurus strigatus* (Diogenidae). **G**: *Pseudopaguristes* aff. *kuekenthali* (Diogenidae). **H**: *Pseudopaguristes bollandi* (Diogenidae). **I**: *Dardanus megistos* (Diogenidae).

Figure 474: Squat lobsters (id. by E. Macpherson, K. Baba & C.W. Lin). **A**: *Allogalathea elegans* (Galatheidae). **B**: *Galathea* aff. *tropis*, associated with sunken woods (Galatheidae). **C**: *Raymunida vittata* (Galatheidae). **D**: *Galatha* aff. *maculiabdominalis*, associated with hard corals (Galatheidae). **E**: *Lauriea* new species, associated with sponges (Galatheidae). **F**: *Uroptychus joloensis*, assocaited with black corals (Chirostylidae). **G**: *Uroptychus* new species (Chirostylidae). **H**: *Uroptychus* new species (Chirostylidae).

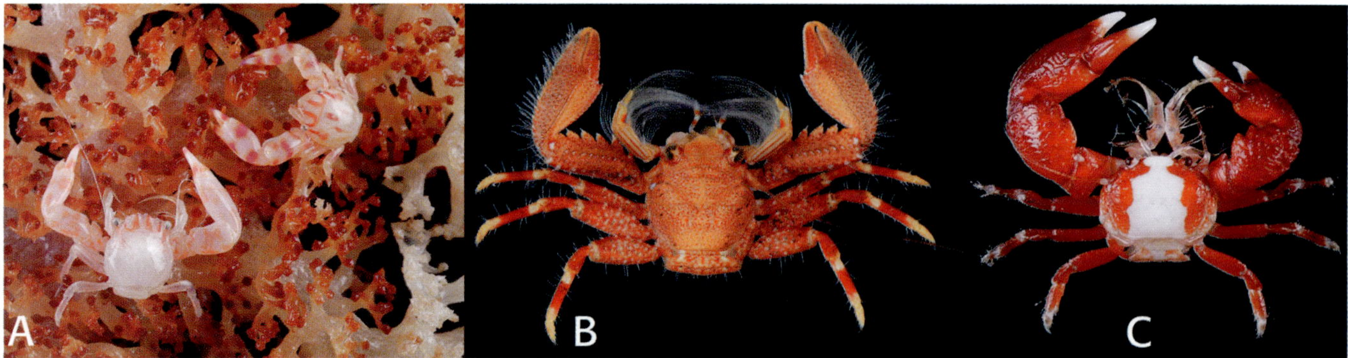

Figure 475: Porcellanid crabs (id. by M. Osawa). **A**: *Lissoporcellana nakasonei*, associated with soft corals (Porcellanidae). **B**: *Petrolisthes militaris* (Porcellanidae). **C**: *Pachycheles sculptus* (Porcellanidae).

Figure 476: Mud shrimps (id. by P.C. Dworschak). **A**: *Eiconaxius* sp., associated with ?encrusting sponges (Axiidae). **B**: *Paraxiopsis* sp. (Axiidae). **C**: *Callianidea typa* (Callianideidae). **D**: *Upogebia* sp. (Upogebiidae).

••• Crabs

For crabs (Brachyura) (Fig. 478), it was estimated that some 500 species were collected and an estimated 10 species are undescribed. Investigations on the family Calappidae revealed 11 species in the genus *Calappa*, and one species each from the genera *Izanami* and *Mursia*. In the family Leucosiidae, the following number of species were recorded from the following genera: four for *Arcania*; two each for *Iphiculus*, *Pariphiculus* and *Raylila* and *Toru*; one each for *Nursilia*, *Oreotlos*, *Alox*, *Heterolithadia*, *Soceulia*, *Tokoyo*, *Urashima*. One beautiful new species of deep water leucosiid of the genus *Euclosiana*, *E. guinotae*, was recently described. The largest family represented in the entire crab collection was from the family Xanthidae with some 125 species. One of them is

Figure 477: King crabs (id. by E. Macpherson). *Neolithodes* aff. *brodiei* (Lithodidae).

a new species, *Liagore pulchella*, which is only the third known species in the genus and is unique in that it lack spots on the carapace, unlike its congeners.

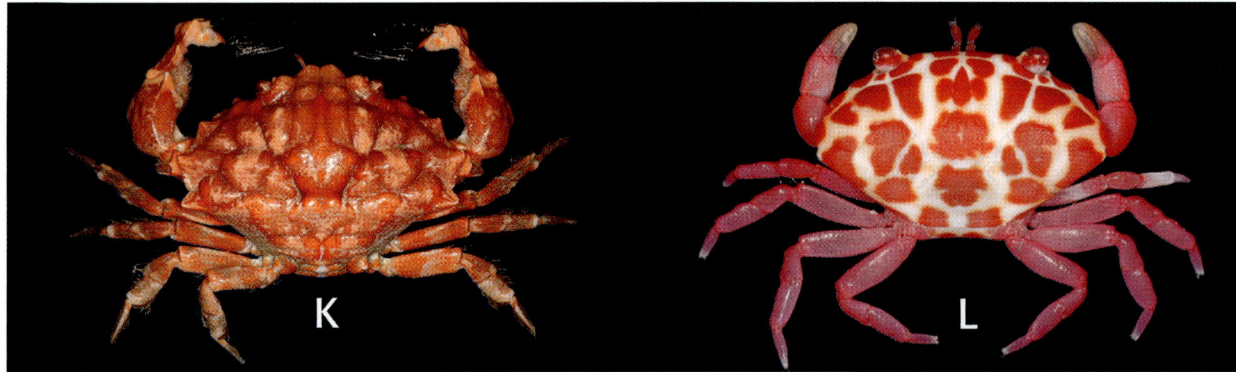

Figure 478: Crabs (id. by S.H. Tan, J.C. Mendoza, M. Manuel-Santos & P.K.L. Ng). **A**: *Homolomannia sibogae* (Homolidae). **B**: *Calappa japonica* (Calappidae). **C**: *Pleistacantha cervicornis* (Inachidae). **D**: *Xenocarcinus depressus* (Epialtidae). **E**: *Platypilumnus* cf. *gracilipes* (Mathildellidae). **F**: *Orcovita mcneiceae* (Varunidae). **G**: *Benthochascon hemingi* (Portunidae). **H**: *Trapezia rufopunctata* (Trapeziidae). **I**: *Liagore pulchella* (Xanthidae, recently described new species based on Santo specimens). **J**: *Demania splendida* (Xanthidae). **K**: *Euxanthus ruali* (Xanthidae). **L**: *Liomera* sp. (Xanthidae).

Figure 479: Stomatopods (id. by S. Ahyong). **A**: *Neoanchisquilla tuberculata*. **B**: *Odontodactylus scyllarus*. **C**: *Pullosquilla* new species. **D**: *Chorisquilla hystrix*.

••• Stomatopods

Stomatopods (Fig. 479), although commonly called mantis shrimps, are far distant from the true shrimps and crabs (i.e. they belong to different orders), but they are often grouped with the latter as large crustaceans. In Santo, the diversity of stomatopods is rather high with at least 36 species collected and five new.

THE MARINE MOLLUSCS OF SANTO

Philippe Bouchet, Virginie Héros, Pierre Lozouet, Philippe Maestrati & Rudo von Cosel

With about 60 000 valid species currently known, molluscs are the most diversified phylum in the sea. They exhibit a fantastic range of size and body form — from micromolluscs adult at sizes < 1 mm to giant clams — and their biology and ecology also display many different modes and patterns — filter-feeders, grazers, predators, or parasites. Many lay-persons believe they know "molluscs" because they know "seashells", however this is a very misleading approximation (Fig. 480). In fact, most species of molluscs are small to minute, with adult sizes in the range of 2-10 mm. Many have reduced shells, or even no shell at all. And many live in close relationship with other invertebrates, as commensals, associates and parasites. This means that a scientific inventory of marine molluscs resembles only remotely shell collecting, as it is practiced leisurely by many hobbyists, divers and vacationers.

The Santo 2006 expedition was the fifth organized by our research group to document the biodiversity of marine molluscs in complex tropical waters in the West and South Pacific: Touho and Koumac (New Caledonia) 1993, Lifou (Loyalty Islands) 2000, Rapa (Austral Is, French Polynesia) 2002, Panglao (Philippines) 2004, were all forerunners to the marine part of Santo 2006. During each expedition, observations were made, conclusions were drawn and lessons were learnt.

... Different forms of biodiversity surveys

The world of (marine) biodiversity research and monitoring is organized in schools that reflect the intrinsic motivations of its participants. These schools publish in different journals, meet in different congress, and are supported by different programmes and institutions.

The oldest of these approaches to documenting marine biodiversity is undoubtedly the taxonomic school. Taxonomists travel the world to discover species, document where they live, name them and establish their classification. To a taxonomist, "every species count". After two and a half century of such exploration, taxonomists have successfully documented around a quarter million marine species, of which 60 000 are molluscs. Based on species inventories, by the end of the 19th century scientists already had a clear vision of the major biogeographical provinces of the world, and by the middle of the 20th century they had recognized South-East Asia and the West Pacific as a center of high species concentration.

The second approach to documenting biodiversity is that of (quantitative) ecology. Ecologists are interested in documenting and understanding such descriptors as biomass, productivity, recruitment efficiency, or (un)evenness in species composition, and this is achieved by counting individuals,

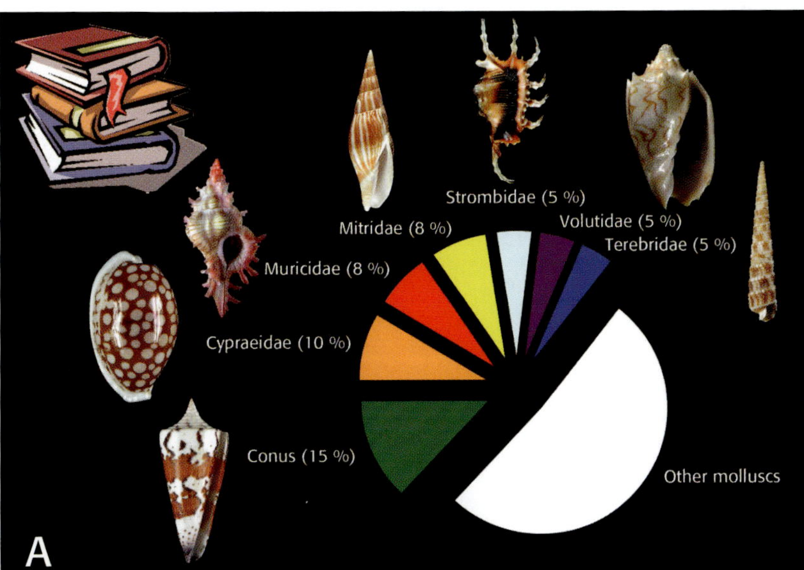

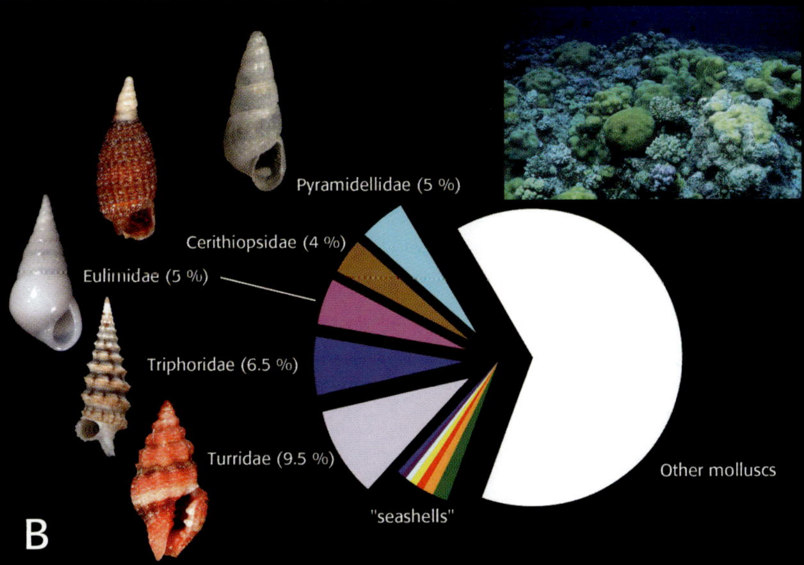

Figure 480: Tropical molluscan biodiversity as represented in "shell books" (**A**) as opposed to their "real" diversity in complex tropical coastal environments (**B**). Percentages for the different seashell families are averaged from a blend of recent popular guides; percentages for the families of micromolluscs are based on their weight in the fauna of Koumac, New Caledonia.

measuring and weighting them, measuring their respiratory rate, etc. In a nutshell, the approach of ecologists is all about "processes", when the approach of taxonomists is all about "patterns". It is essential for ecologists that sampling follows a repeatable protocol and is done quantitatively per surface or volume, i.e. per square meter or hectare or cubic meter. To an ecologist, species *per se* do not matter, except perhaps as a measure of how complex the ecosystem is, i.e. how many species there are that make it work. In the case of molluscs, an ecologist is, as a rule, blind to empty shells, because they do not carry any information on ecosystem function.

A third approach to biodiversity surveys is biodiversity assessments. Conservationists need "immediate" science-based facts to inspire decisions on management and use of land and sea areas that are measured in 1 000s or 10 000s of hectares, i.e. the spatial scale of landscapes/seascapes. But conservationists are daunted by the magnitude of the biodiversity they want to highlight, promote, and conserve. As a consequence, biodiversity assessments focus on a few selected taxa for which there is the work force to identify them on the spot: fishes, reef corals, sea grasses and mangroves, and a handful of charismatic megavertebrates such as turtles and dugongs. The Rapid Assessment approach has been successful in highlighting areas of conservation interest, in raising and disseminating environmental awareness, and in bringing together the worlds of public agencies (the World Bank, USAID, etc.), private funding (corporate and foundations) and the public opinion.

••• Background: The New Caledonia and Panglao case studies

Three schools, three irreconcilable schools, it would seem. Back in the 1990s, partly in reaction to the strangely small numbers of species reported in quantitative ecology papers from the Australian Great Barrier Reef or Fiji, we organized the Montrouzier expedition in New Caledonia in 1993. But we wanted both specimens and data. We wanted simultaneously to champion the taxonomist's soft spot for "every species" and to return home with data that would have ecological significance beyond species lists. Already then, the Montrouzier expedition used innovative collecting devices (the time-approved vacuum cleaner of ecologists and a newly developed "brushing basket" that did

marvels), deployed large-scale manpower for processing and sorting samples, and ambitioned to saturate the sampling at the seascape scale. Two sites, one on the west coast (Koumac), one on the east coast of New Caledonia (Touho), were sampled each during a period of four weeks, each with a 400 day-persons collecting and field sorting effort. It was the first time marine molluscs were documented at that scale, and it took almost 10 years to analyze the data. The results of the Koumac site comforted our initial intuitions: over an area of less than 30 000 hectares, we had sampled 127 652 specimens of molluscs representing 2 738 species. Of these, despite 42 discrete collecting stations had been sampled, 32% were collected only once, and — despite the intensity of the collecting effort — 20% of the species were represented by just singletons. The parallel so often drawn between coral reefs and rainforests was justified, and the molluscs were truly the beetles of the sea!

After New Caledonia, like any marine biologist, we were itching to investigate the Coral Triangle, the cradle of marine biodiversity richness. The Panglao 2004 Marine Biodiversity Project was a joint venture between Muséum national d'Histoire Naturelle

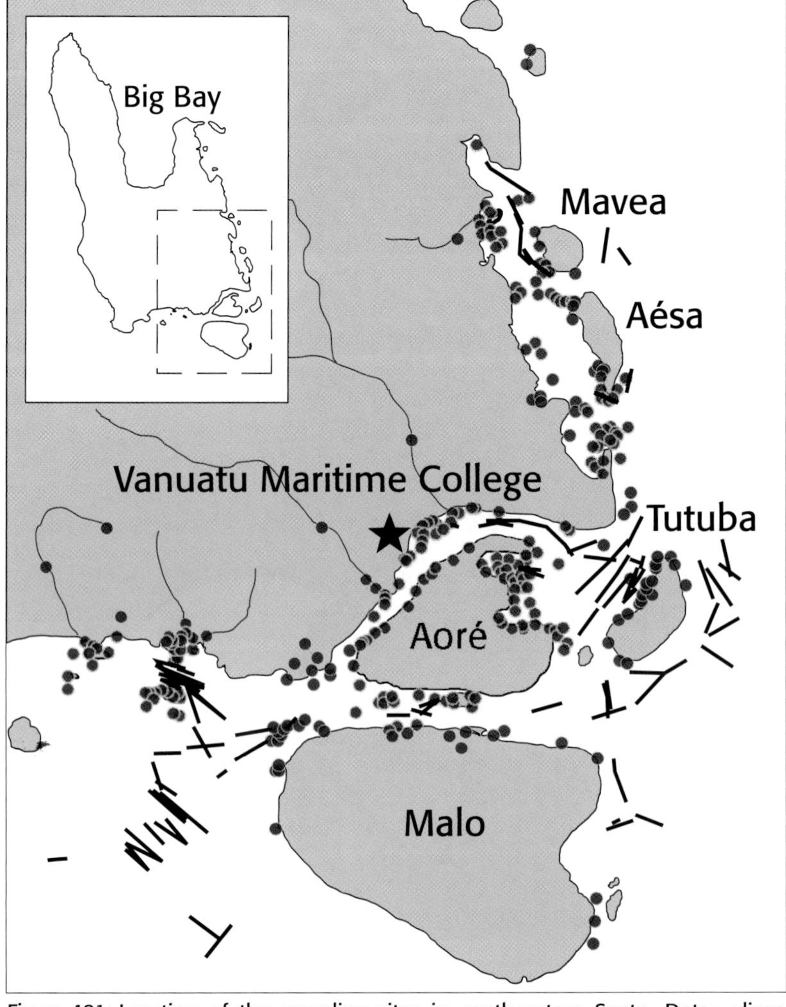

Figure 481: Location of the sampling sites in southeastern Santo. Dots = dives, intertyidal collects, triangular dredge, moored nets; segments = dredging and trawling from *Alis*.

and the University of San Carlos, based in Cebu City. Beside molluscs, the expedition also involved a strong crustacean component. From our base on the island of Panglao, in the province of Bohol, in the Visayas region of the central Philippines, a total of 59 sites were sampled intertidally; 78 general samples, 53 suction samples and 42 brushings were obtained by SCUBA; 42 trawl and 14 dredge samples were taken between 60-130 m; in addition, traps, tangle nets and lumun lumun, a speciality of the Visayas, were deployed on reefs drop-offs. Admittedly, Panglao was certainly not an "untouched" place. The human population pressure was obvious, with people everywhere at low tide on the narrow intertidal platform, collecting anything that can be eaten or sold at the market. Likewise, many echinoderms are collected subtidally, and large holothurians have become rare there as in many other parts of the world. However, we were surprised that scarcity affected not only the large edible specimens, but also the micromolluscs, as if the excessive taking of many large specimens had a cascading effect in the ecosystem on all size classes of benthic molluscs.

Compared to New Caledonia, the Philippines experience had been invaluable in confirming the validity of our diversified sampling techniques and in upscaling and expanding the organization of the field party. The Panglao expedition was also the first one immediately following Paul Hebert's "big bang" barcode paper of 2003, and was thus the first expedition where a row of participants was specifically dedicated to tissue clipping for molecular sequencing.

••• Back to the South Pacific

The Panglao expedition had undoubtedly refined the art of taxonomic inventory, but it was sending the gospel of tropical species richness to the same, already converted, community of taxonomists. With the Santo 2006 expedition, our ambition was to actually confront on a single site the different approaches to biodiversity surveys: we would have an "all-species" taxonomic survey, of course, but we would also attempt to conduct a rapid assessment and we would do quantitative sampling. This required to extend the roster of participants beyond the traditional core of our previous expeditions. The rapid assessment was conducted by Fred Wells, then at the Western Australian Museum, and "the" malacologist on many Conservation International's Rapid Assessments. The quantitative sampling was conducted by John Gray, benthic ecologist at the University of Oslo, and a giant in statistical approaches to measuring species richness and species diversity. For the "all-species" survey, we also innovated by bringing on board Richard Pyle, of the Bernice P. Bishop Museum in Honolulu, who is pioneering deep diving in the "twilight zone" with trimix and rebreathers.

Vanuatu, and more specifically the island of Espiritu Santo, was selected for a variety of reasons. Some of the reasons had to do with the overall project architecture, as Santo had a lot of different habitats to offer to botanists, entomologists, biospeleologists, and freshwater biologists, beside marine scientists. The other reasons were environmental (low population pressure, good overall environmental quality), and logistical (facilities offered by the *Vanuatu Maritime College*, and IRD's Nouméa center and facilities in the background, so to say).

The Rapid Assessment was conducted in August 2006, and its results are presented in a separate chapter. Regrettably, although 22 grab samples were taken from R.V. *Alis*, the quantitative survey did not work as expected. John Gray and his team (Camilla Friseid, Gorild Hoel, Karen Webb, Annelise Fleddum) quickly found out that nearshore soft bottoms near Luganville were too coarse to allow for the grab to operate smoothly, and they had to shift their bathymetric target to 60-80 m. However, as these depths were too great to be sampled by SCUBA by "every-species-counts" as well as rapid assessment methodologies, the comparison between the three approaches to measuring biodiversity was, in the end, not possible. Nevertheless, John Gray was inspired by what he saw in Santo and we had lengthy discussions on the different strategic approaches to measuring and documenting marine biodiversity. It is sad that John Gray passed away on 21 October 2007, at the age of 66, before we could implement the ideas and plans resulting from these discussions. (Incidentally, the editorial to issue 56 of *Marine Pollution Bulletin* published after his death carries a picture showing him at work in Santo).

••• Organisation of field operations

••• Intertidal work

The maximum tidal amplitude in Santo is in the order of 1.4 m. As the coastline profile is generally rather steep, the intertidal zone is usually

Figure 482: Jean-Claude Plaziat and Samson Vilvil-Fare sampling the high intertidal in a mangrove area.

quite narrow, often in the range of 20-50 m. New moon and full moon periods are naturally the best, but intertidal samples were actually taken nearly every day, as some habitats (high intertidal, estuarine, river/mangrove transition) do not require spring tides to be properly sampled. The rivers of SW Santo were sampled upstream until freshwater (Fig. 482). Intertidal samples typically consisted of specimens of large/medium-sized species picked with the naked eye on the shore or in hand dredge (in sand); examination of residues of rock/algal wash; and echinoderms hosts specially examined for eulimid gastropods. A total of 84 sites were sampled intertidally. Some sites were sampled twice, thrice, or even four times. Participants specially tasked to intertidal work were: Takuma Haga, Yasunori Kano, Pierre Lozouet, Timea Neusser, Jean-Claude Plaziat, Stephen Vutilolo.

••• Diving

Subtidal sampling was central to the Santo project and at least three small boats were assigned to sampling by SCUBA at any one time. The Maritime College's *Emm-nao* was dedicated, first to the Rapid Assessment (Fred Wells, Dominique Lamy), then to the nudibranch group (Yolanda Camacho, Mike Miller, Marta Pola, José Templado, Angel Valdés, assisted by Steve Vira), while the whaleboat from the Department of Fisheries and the *Evolan* or *Aldric* would carry divers with special equipment (brushing baskets and suction samplers) and divers focussing on hand-picking specimens (especially molluscs living in association with other invertebrates: echinoderms, octocorals and corals, etc.). Because the processing of the residues takes a long time, dives that generated such samples (by brushing or suction) were, as far as possible,

Figure 483: The air-operated suction sampler in operation by SCUBA divers.

Figure 484: The team of trimix/rebreather divers on the pier and in operation, with some of the bizarre molluscs collected by electric suction sampling: a eulimid with wing-like growth varices (right), an arcid bivalve (left), and a gastropod of unknown family position.

taken in the morning. A few dives were made at night, mostly in the Segond Channel in the vicinity of our shore base. A total of 204 general samples, 47 suction samples and 66 brushings were made (Fig. 483). Participants specially tasked to the SCUBA team were: Laurent Albenga, Jacques Dumas, Eric Folcher, Adriaan Gittenberger, Marco Oliverio, Jacques Pelorce, Patrice Petit de Voize, Stefano Schiaparelli (Fig. 484).

As mentioned above, beside regular SCUBA, the Santo expedition also made use of helium diving (with helium tanks specially freighted to Santo from Australia!). The small team (John Earle, Brian Greene, Richard Pyle, assisted by Jean-François Barazer) of trimix divers, operating from IRD's *Aldric*, were fish specialists but Richard took nine suction samples down to 120 m using a small electric vacuum cleaner built for the expedition.

••• Trawling and dredging

R.V. *Alis*, the 27 m research trawler of *Institut de Recherche pour le Développement* (IRD), was used for taking quantitative grap samples in 40-80 m, and for dredging and trawling to greater depths, from about 50 m to 850 m. Although the ship has autonomy for cruises of up to three weeks, it was essentially deployed on a day-to-day basis with catches and residues brought back alive every afternoon to the lab for photography and tissue clipping. However, a cluster of five consecutive days was set aside to sample sunken wood in Big Bay, as part of the SANTOBOA cruise. In more shallow water, a small triangular dredge was deployed by hand from Maritime College's *Evolan* (or other small boats) to sample soft bottom patches in 5-30 m. A total of 135 dredge/trawl were taken from *Alis*, and 37 triangular dredge samples were taken subtidally. Participants specially tasked to dredging and trawling were Magalie Castelin, Christophe Chevillon, Rudo von Cosel, Olivier Gros, Bertrand Richer de Forges, Sarah Samadi.

••• Tangle nets and lumun lumun

Lumun lumun is a cebuano word designating tangle net material surrounded by a stronger net with large

Figure 485: The sturdy Warén Dredge used to sample mixed and hard bottoms.

Figure 486: Jo Arbasto (right) and Steve Vira (left) deploying tangle nets off Aore.

mesh size, that is usually left on the sea floor for about a month or more. Tangle nets and lumun lumun are routinely used in the Philippines to catch commercial seashells from deep (30-200 m) and steep hard bottoms (Fig. 486). These collecting methods have been responsible for the discovery of dozens of new species, many of which have never been collected by another technique. It was thus felt desirable to use them also in Santo. Jo Arbasto, a local fisherman from Panglao, the Philippines, who had been working for us during the Panglao 2004 expedition, agreed to try his skills in Vanuatu, a completely new environment for him. Initially, we had planned to deploy lumun lumuns ahead of the main party, so that they would be retrieved at the beginning of the expedition and then redeployed. However, the shipment of equipment from Manila was delayed and the lumun lumun were deployed only days before the main party arrived. Additionally, it took time for Jo Arbasto to become familiar with spots and currents in the area, and it can be said without exaggeration that by the time he had become acquainted with it, it was regrettably time for him to return to the Philippines. Altogether, 30 units of lumun lumun/tangle nets were deployed.

••• Laboratory work

••• Sieving and Sorting

Bulk samples and residues were sieved fresh in seawater and fractioned through a set of sieves from 0.5 to 10 mm, by which the light and heavy fractions were separated (Fig. 488). The coarse fractions were sorted with the naked eye, fractions below 3 mm were sorted with dissecting microscopes (Fig. 487). Living molluscs were screened for "interesting" species/specimens that would be channelled to the photography and/or barcoding units. The remaining live-taken specimens, as well as the empty shells, were preserved in bulk in ethanol for further sorting. Participants more specially allocated to sieving and on-site sorting were Ritchie Franck, Virginie Heros, Tania

Figure 487: The "sorting factory" during the marine leg of the Santo 2006 expedition. At rush hour, up to 20 persons were sorting, photographing, labelling and tissue clipping molluscs.

Kantor, Julien Lorion, Philippe Maestrati, Kerstin Rigneus, Marilyn Schotte, Charles Tari, Emily Tasale, Anders Warén, Maria Yorley. Coarse sediment samples were also treated separately for the extraction of meiofauna (see "Marine interstitial" by Timea Neusser).

In France, the bulk mollusc samples were drained, rinsed in freshwater, and dried. The second-tier sorting took place in November 2007 at the Station Hydrobiologique in Besse and, it took another three weeks and 12 persons (Philippe Bouchet, Mauricette Bourgeois, Rudo von Cosel, Virginie Heros, Mandë Holford, Anna Holmes, Philippe Maestrati, Robert Moolenbeek, Graham Oliver, Stefano Palazzi, Jacques Pelorce, Anselmo Peñas, Danièle Plaçais) to sort all the material to family level.

••• Photography

Books on molluscs mostly deal only with their shells and museum collections mostly consist of empty shells. Even spirit collections of live-collected specimens consist of discoloured animals retracted deep inside their shells. All the characters of the living animals are lost in the process. In Santo, we documented as many species as possible of living, crawling molluscs, mostly taken in the laboratory, but some taken *in situ*. All digitals photographs have an identifier that connects them to the exact specimen that was photographed. Participants more specially allocated to mollusc photography were Delphine Brabant and Annelise Fleddum.

••• Barcoding and Fixations for anatomical work

The capacity to identify all living organisms from a specific sequence of their gene coding for cytochrome oxidase 1 (CO1) is known as the "Barcode of Life" Initiative. The Santo 2006 expedition was a unique

Figure 488: Graham Taridia (in the back) and a local helper sieving residues at our base at Vanuatu Maritime College.

Figure 489: An assortment of free-living, commensal and endoparasitic galeommatoid bivalves from the Santo study area. Most species are in the 3-10 mm range. Many are probably new to science.

occasion to preserve a vast collection of tropical marine gastropods in high-grade ethanol. A special difficulty for snails is that for a proper fixation, the animal must not be retracted deep inside the shell, especially if it closes with an operculum; species-level taxonomy requires examination of the intact shell. In Santo, we used a combination of approaches to ensure proper fixation and preservation of shell characters. This required either breaking the shell of one specimen and conserving it side by side with an intact specimen of the same species from the same sample; or relaxing and extending the animal with magnesium chloride. A total of 5019 barcode mollusc samples have been preserved for this exercise, each sample consisting of one to 3-5 (occasionally more) hypothetically conspecific specimens. A subset of these (3019 specimens) were tissue-clipped already in Santo.

Taking into account redundancy (either accidental or intentional) between samples, we believe that this represents a set of 2000-2200 species of molluscs. This is the largest such systematic collection ever made. In parallel, further specimens were relaxed and fixed for anatomical or microscopy work, in different fixatives for different purposes (glutaraldehyde for electron microscopy; Bouin's, formalin or alcohol for dissecting), totalling 739 such samples. Participants more specially allocated to molluscan barcoding and fixing were Jason Biggs, Yuri Kantor, Nicolas Puillandre and Ellen Strong.

••• Station coding and labelling

With such a large amount of samples and specimens taken, and their associated photographs and documentation, we built a database with three different files:
- sample identifier;
- specimen identifier;
- photo identifier.

Sample identifiers have a two-letters code where the first letter corresponds to the boat/vehicle used and the second letter corresponds to the type of gear/collect, followed by a sequential number for that boat/vehicle. For example, LB26 [26th operation on *Aldric*, a brushing], AT115 [115th operation on *Alis*], etc. The database contains location (latitude, longitude and depth), brief description of site and information such as the boat used.

Specimen/photo identifiers also consist of a two-letter code corresponding to the initials of the observer/operator, followed by a sequential number. For example, LB26_BC2456 [brushing sample 26 on *Aldric*, barcode sample 2456], AT115_DB247 [haul 115 on *Alis*, Delphine Brabant photo 247]. Didier Molin was specially allocated to the tracking of station data and associated information.

Letters used to represent the method of collection are:
- B – SCUBA, brushing;
- D – triangular dredge;
- L – lumum lumun;
- M – intertidal collection;
- R – SCUBA, hand-collecting;
- S – SCUBA, suction;
- T – beam trawl and Warén dredge on *Alis*.

••• Networking with specialists

No single person and no single institution has the skills to study, identify and describe such a rich and complex tropical fauna as that of Santo. In a group as diverse as molluscs, species-level taxonomic expertise is developed at the level of a family, superfamily or any other discrete operational group, and throughout the years we have nurtured a network of specialists that collaborate with us on the material generated by our expeditions. The Santo material may be worked up by them as a separate publication, or it may be included in a broader dataset that includes material from other South and West Pacific localities, or it may even be studied as part of a worldwide revision of a genus. However, many taxonomical groups are orphaned, with not a single specialist anywhere in the world having the skills to work up material from that group. Ironically, this especially true for the mega-diverse families of micromolluscs ("Turridae", Eulimidae, Triphoridae, Pyramidellidae, Cerithiopsidae, Galeommatoidea).

••• **Results**

••• Species composition and species richness

The islands of Vanuatu are typically separated from other lands (New Caledonia, Fiji, the Solomon Islands) by depths greater than 2000-4000 m. Because these great depths are inhospitable to coastal molluscs, it means that all coastal mollusc species have reached Vanuatu as swimming larvae or by rafting of some sort. This is of course also true for New Caledonia, or Hawaii, or Easter Island, to take just a few other examples. With surface currents of, e.g. 0.5 knots (i.e. 0.5 nautical mile, or 0.925 kilometer, per hour), a planktonic larva can be passively transported over a distance of 350 km in just two weeks. This means that, when it comes to its marine biota, and in contrast with terrestrial biota, Vanuatu is not an isolated world, but shares its species with other South Pacific island groups. In a recent review of Indo-Pacific reef fishes for hotspots of diversity and endemism, ichtyologist Gerald Allen, found that the "megadiverse" countries include Indonesia, Australia, Philippines, Papua New Guinea, Malaysia, Japan, Taiwan, Solomon Islands, Palau, Vanuatu, Fiji, New Caledonia, and the Federated States of Micronesia. By contrast, the top-ranked areas based on percentage of endemism are Easter Island, Baja California, the Hawaiian Islands, the Galapagos Islands, the Red Sea, Clipperton, the Marquesas, Isla del Coco, the Mascarenes and Oman, i.e. regions that are peripheral to the vast Indo-West Pacific province. Endemism and richness thus carry different messages. So, even if no in-depth survey of the marine mollusc fauna of Santo had ever been carried,

Figure 490: An assortment of the micromolluscs (adult sizes 2-8 mm) that form the bulk of the diversity of molluscan tropical faunas.

discovery of new species there is thus not likely to result from its geographic isolation. As a matter of fact, all the species of Conidae and Cypraeidae found in Santo are known to also occur elsewhere.

The expedition brought together a remarkable array of field skills and experienced collectors and, as a result, the molluscan inventory of Santo is exceptionally complete. It is still impossible to evaluate how many species were collected, but one can guess it was in the order of 4000 species. This is a guess based on comparing species numbers in Santo with Koumac (New Caledonia) for two families of gastropods, the Terebridae and the Triphoridae, that have been segregated to species, respectively by Yves Terryn and Paolo Albano. The respective numbers were 38 and 174 (Koumac), vs 53 and 259 (Santo), i.e. there were 39-49% more species in Santo. Extrapolating from the total number of species censused in Koumac (2738), this would suggest that we sampled 3818-4075 species in Santo. Whatever the exact number, it is obvious that significantly more species were collected in Santo than in Koumac. This may be because Santo is intrisically biologically richer, and/or because the deeper habitats were better sampled here, but it may also be because the sampling was qualitatively and quantitatively more intensive in Santo. For instance, 23238 specimens of Triphoridae were collected in Santo vs "only" 2839 in Koumac, i.e. an 8x more intensive effort in Santo. This suggests that the sampling was rather considerably more saturated in Santo.

Santo is thus undoubtedly a species-rich place, yet the participants in the expedition were not overwhelmed by the abundance of life in its waters. In fact:
- habitats were extremely patchy and most species were sampled only once or twice;
- as elsewhere in complex tropical ecosystems, most species were rare or very rare.

For instance, out of a total of 588 sampling events, only 187 (32%) yielded triphorids, and 28% of the triphorid species were collected at just one (16%) or two (12%) stations. Twenty-one percent of all triphorids are represented by just one or two specimens, i.e. 0.3% of the catch (77 specimens) account for 21% of the species. How these results apply to the rest of the mollusc fauna is debatable, but there is no reason to suspect that other species-rich specialist families of microgastropods should have different patterns of abundance and distribution.

••• Unusual catches, remarkable animals
Separate chapters report on molluscs in sulfide rich environments —with the special case of sunken wood— molluscs in mangroves, molluscs (and crustaceans) living as commensals, associates and parasites, and finally meiofaunal molluscs and nudibranchs. However, unusual catches and strange animals were by far not confined to these habitats and/or mollusc groups. Among the bivalves,

for instance, Santo had a remarkable diversity of free-living galeommatoids (Fig. 489), living under stones at low tide with their colourful mantle covering their super-thin shells, and crawling on their foot like gastropods. Among microgastropods, the diversity of Rissoininae, Iravadiidae, Triphoridae and Cerithiopsidae was noticeable, reflecting for the latter two families the diversity of sponges off Santo (Figs 490 & 491).

Discovering new species was one of the declared goals of the Santo 2006 expedition. New species can be discovered as a result of exploring new or little known geographic locations, or as a result of sampling previously ignored or inaccessible habitats, but new species are also discovered in groups that are difficult to study (e.g. because species are very small and/or have few evident morphological characters) and/or have been left aside by other specialists (e.g. because the literature or type material is difficult to access). The slug *Pseudunela espiritusanta* Neusser & Schrödl, 2009, is a good example of such habitat-driven exploration: it was discovered on the underside of intertidal rocks deeply embedded into coarse sand, the interstices of which are filled with a mixture of fresh subsoil and seawater. The cockle *Frigidocardium helios* ter Poorten & Poutiers, 2009, was described from Santo based on specimens dredged offshore, but it is also known from the Philippines from specimens taken by tangle nets, and *Iotyrris devoizei* and *I. musivum* Kantor, Puillandre, Olivera & Bouchet, 2008, both described based on holotypes from Santo, are also known from New Caledonia. The auger shell *Myurella lineaperlata* Holford & Terryn, 2008, is so far known only from Santo, but can be expected to be found elsewhere in the South and West Pacific.

••• Molecular systematics
In this age of intense re-evaluation of many branches of the Tree of Life, access to a wide range of families, genera and species is critical to many research projects. With more than 5000 samples, the molluscan molecular collection put together during the expedition is attractive to scientists worldwide. Indeed, the Santo molluscan molecular collection has been chosen by the Sloan Foundation as one of the flagship collections to be part of their MarBoL (*Marine Barcoding of Life*) initiative. As tissues are sequenced and species identified, the data are uploaded on the freely accessible Barcode of Life Database (BOLD, http://www.boldsystems.org/), thus enhancing the visibility of Vanuatu's marine biota to the world.

As examples of projects that have relied heavily on the molecular samples taken during Santo 2006, one can mention Mandë Holford's molecular phylogeny of the gastropod family Terebridae, based on sequences obtained from 154 specimens of 64 species, of which respectively 97 and 39 were obtained during the Santo 2006 expedition. Likewise, Sarah

Samadi's work on the mussels living on sunken organic substrates includes a significant proportion of specimens originating from Vanuatu. Such works have broad evolutionary and ecological significance beyond the Santo case study.

••• Epilogue

A 60-persons marine molluscs expedition is a big machine to set in motion, and it never runs entirely smoothly. We had our share of failing engines, power outages, and bad weather, but fortunately we had no injuries or serious accident. Rufino Pineda, Danielle Plaçais, Noel Saguil, Dave Valles and Samson Vilvil-Fare spared no time and no energy to make sure that everyone and everything would be at the right place at the right time.

It takes on average nine years between the moment a new species (of plant, insect, fish, mollusc, whatever...) is collected in the field and the moment its description is published in the scientific literature. Despite four years have passed since the Santo 2006 expedition, we are still far from the peak of academic studies on the specimens that were sampled then. The Convention on Biological Diversity has termed *Taxonomic Impediment* "the knowledge gaps in our taxonomic system, the shortage of trained taxonomists and curators, and the impact these deficiencies have on our ability to conserve, use and share the benefits of our biological diversity". Given the current chaos in triphorid systematics, it might take a decade, or more, to actually bridge the gap between the 259 triphorid species collected during the expedition and named entities. However names are essential to communicate about properties and attributes of species, and there is justifiable concern for the gap between discovering and documenting the diversity of the world and backing this exercise with sound nomenclature.

Figure 491: The animal of cystiscid and marginellid microgastropods have stunning colour patterns, but their shells are glassy and colourless. The species on this figure are all 1.5-2 mm as adults.

A RAPID ASSESSMENT OF THE MARINE MOLLUSCS OF SOUTHEASTERN SANTO

Fred E. Wells

Since October 1997, the Washington, D.C. based nongovernmental organization Conservation International has conducted a series of Marine Rapid Assessment (Marine RAP) surveys of the fauna of coral reefs in the Indo-West Pacific. The surveys have been centred in the "coral triangle" of the western Pacific Ocean. The goal of the expeditions has been to develop information on the biodiversity of three key animal groups: corals, fish, and molluscs, for use in assessing the importance of the reefs for conservation purposes. Surveys have occurred in the following areas: Milne Bay, Papua New Guinea (two surveys); Calamian Islands, Philippines; Togean and Banggai Islands, Sulawesi, Indonesia; and Raja Ampat Islands, West Papua, Indonesia. Results on molluscs are reported by Wells in a series of RAP reports (Milne Bay: 1998, *RAP Working Papers Number* 11: 35-38, and 2003, *RAP Bulletin of Biological Assessment* 29; Calamianes Islands, Philippines: 2001, *RAP Bulletin of Biological Assessment* 17: 27-30; 81-94; Gulf of Tomini: 2002: *RAP Bulletin of Biological Assessment* 20; Raja Ampat: 2002, *RAP Bulletin of Biological Assessment* 22; NW Madagascar: 2005, *RAP Bulletin of Biological Assessment* 31). In turn, the Conservation International trips are based on a parallel sampling of coral reefs in northwestern Australia and adjoining areas undertaken by the Western Australian Museum (See Table 47 for a list of sites surveyed). In 2005, a similar survey was undertaken for the *National Geographic Magazine* by the New England Aquarium to assess the effects of the 26 December 2004 tsunami in the area near Phuket, Thailand. While the Phuket region was not surveyed for molluscs in this manner before the tsunami, data collected during the Conservation International and Western Australian Museum surveys provided a solid basis for assessing the effects of the tsunami.

As the basic building blocks of the reef, corals are an obvious group to include in surveys of this type.

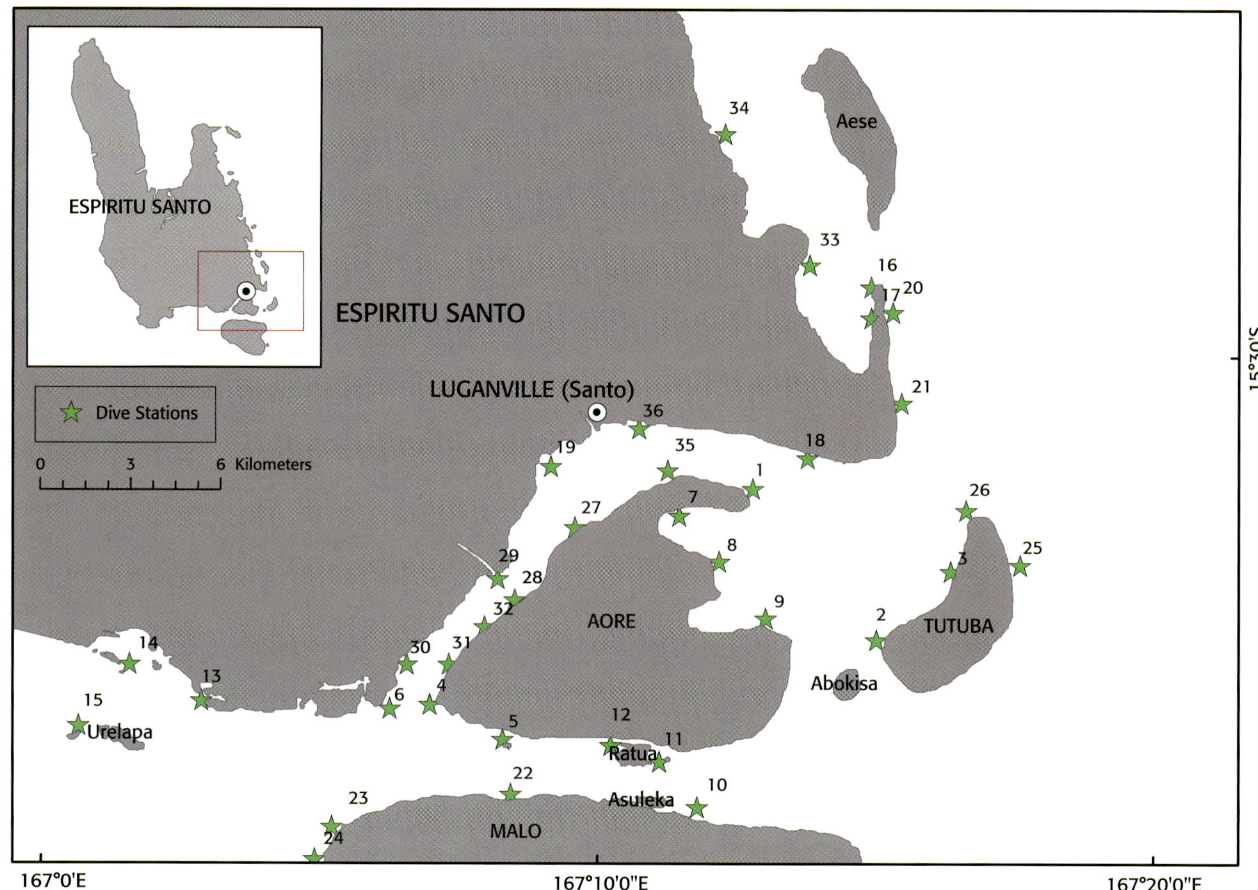

Figure 492: Map of southern Espiritu Santo, Vanuatu, showing localities sampled during the mollusc rapid assessment survey.

Fish are also an important component because of their ecological and economic importance in the system. Molluscs are included for several reasons. They dominate diversity in many marine systems, including coral reefs. Molluscs can be used as a surrogate for measuring the diversity of invertebrates other than corals in the reefs. With their high diversity, often high density, and variety of lifestyles, molluscs are ecologically important to trophic flows within the system. Many groups, such as giant clams, spider shells, trochus, pearl oysters, cephalopods and others are economically important. For all of these reasons, molluscs have been included in the Western Australian Museum and Conservation International surveys.

••• Methods

The survey was conducted from August 7-19, 2006, with a total of 36 sites being examined (Fig. 492). Twenty-three sites were surveyed by scuba diving. Collecting times ranged from 44 to 127 minutes; eighteen of the 23 dives were between 70 and 105 minutes. Each site was examined by starting at depths of 10-25 m and working up the reef slope. Most of the time was spent in shallow (<6 m) water, as the greatest diversity of molluscs occurs in this region; the shallow depth also maximized diving time. To obtain as many species as possible, all habitats encountered at each site were examined for molluscs, including living coral, the upper and lower surfaces of dead coral, shallow and deep sandy habi-

tats, and, where possible, intertidal habitats. For the same reason, no differentiation was made between species collected alive or as dead shells, as the dead shells would have been living at the site. Thirteen of the sites were sampled by intertidal collecting.

While this collecting approach allows the rapid assessment of a variety of mollusc species, it is by no means complete. For example, no attempt was made to break open the corals to search for boring species, such as *Lithophaga* spp. Similarly, arcid bivalves burrowing into the corals were not thoroughly examined. Furthermore, micro molluscs were not sampled. However, the sampling method is the same used for all six of the Conservation International Marine RAP trips and surveys by the Western Australian Museum. Therefore, the sampling results provide a good indication of diversity relative to previous surveys.

Several standard shell books and field guides were available for reference during the expedition and were used for identification. All of the gastropods, chitons, and cephalopods were sorted to species and provisionally identified to genus, or in most cases, species. All bivalves were also sorted to species, but not all were identified.

Detailed analyses were conducted on 199 species that were collected at four or more sites. The presence/absence of each mollusc species recorded at each site was used to construct a Bray-Curtis similarity matrix which was subjected to hierarchical agglomerative clustering (CLUSTER), similarity profiles (the SIMPROF test) and also to non-metric multidimensional scaling (MDS) ordination. The construction of the SIMPROF test employs the multivariate structure of a species data matrix to test for the presence of significant group structure in *a priori* unstructured sets of samples. This SIMPROF procedure is carried out hierarchically, proceeding downwards through the nodes of the dendrogram. Groups of samples that cannot be distinguished (i.e. appear to have a common community structure) are identified in the dendrogram by dashed lines. In the current context, the SIMPROF test was used to identify groups of sites with common community structure, which were then used as an *a priori* factor in the labelling of the resultant MDS ordination and ANOSIM and SIMPER analyses. The PRIMER 6.1 statistical package was used for cluster analysis, ordination and associated tests.

••• Results

A total of 572 species of molluscs were identified at least to genus: 402 gastropods, 167 bivalves, one cephalopod, one scaphopod and one chiton (Table 41). The most abundant species at each site were generally arcid bivalves, *Lithophaga* spp. and *Pedum spondyloidaeum*, which live in the coral and the coral dwelling gastropod *Coralliophila neritoides*. Common species in the intertidal included the nerite, *Nerita costata* on rocks and the littorinid *Littoraria scabra* on mangroves.

The number of species collected per day declined rapidly after the first day, when all species collected were collected for the first time (Table 42, Fig. 493). On the first day, 148 species were collected. This declined to 60 species on the second day, but increased to 77 species on the third day. The number of new species collected progressively declined during the middle of the trop. During the last three days 23, 17 and 9 new records were made, with only two stations being sampled on the final day. This indicates that even at the end of the trip new records were being made, and a longer trip would have produced more species. However, the number would not have been great.

There were 2267 species records made at the 36 stations, giving a mean of 63.0 ± 3.1 species per site; the range was from 26 to 106 species (Table 43). Sites with both the largest and smallest recorded mollusc diversity varied geographically and with whether they were intertidal or subtidal stations (Tables 44 & 45). The 13 intertidal stations had a mean of 71.7 ± 4.5 species per station, compared to the 23 dive stations, which had a mean of 58.0 ± 3.9 species per station. One of the richest areas included the intertidal sites on Palikulo Bay and the mainland of Espiritu Santo Island to the northwest (stations

Table 41: Taxonomic composition of mollusc species identified from sites at Santo, Vanuatu.

Class	Families	Genera	Species
Polyplacophora	1	1	1
Gastropoda	56	116	402
Bivalvia	29	70	167
Scaphopoda	1	1	1
Cephalopoda	1	1	1
Totals	88	189	572

Table 42: New records collected every day during the 13 days of collecting at Vanuatu.

Day	New records
1	148
2	60
3	77
4	52
5	27
6	51
7	23
8	27
9	17
10	41
11	23
12	17
13	9
Total	572

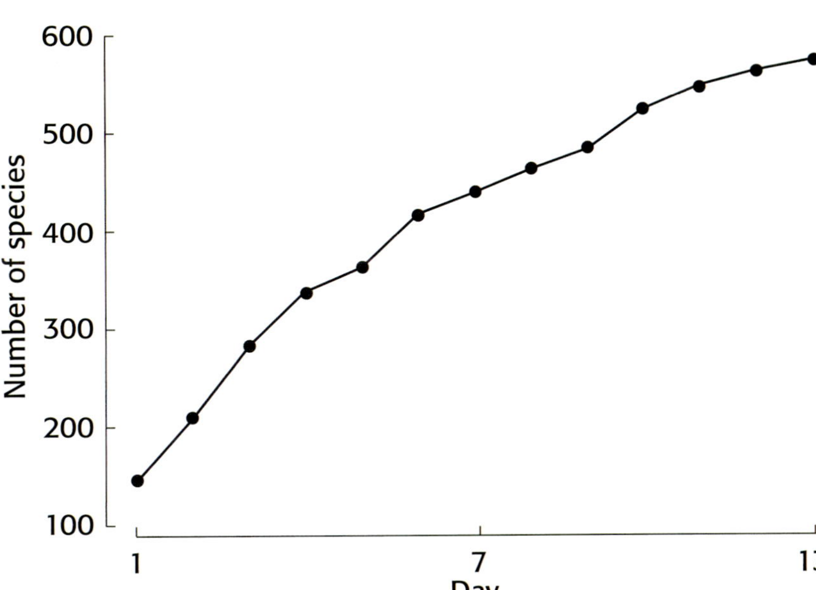

Figure 493: Number of new species collected per day during the mollusc rapid assessment survey conducted in Vanuatu in August 2006.

Table 43: Total number of mollusc species collected at each site at Santo, Vanuatu.

Site	Type	Nb. species	Site	Type	Nb. species	Site	Type	Nb. species
1	Intertidal	70	13	Intertidal	51	25	Dive	44
2	Dive	67	14	Intertidal	67	26	Dive	78
3	Dive	54	15	Dive	57	27	Dive	62
4	Intertidal	73	16	Dive	65	28	Dive	76
5	Dive	56	17	Intertidal	82	29	Dive	42
6	Dive	41	18	Dive	26	30	Dive	65
7	Intertidal	85	19	Dive	71	31	Dive	58
8	Intertidal	85	20	Intertidal	50	32	Dive	55
9	Dive	106	21	Intertidal	47	33	Intertidal	90
10	Intertidal	61	22	Dive	31	34	Intertidal	99
11	Intertidal	72	23	Dive	81	35	Dive	30
12	Dive	67	24	Dive	42	36	Dive	61

Table 44: Ten sites observed to have the richest mollusc diversity among the 36 sites surveyed at Santo, Vanuatu.

Site	Location	Number of species
9	North side Ambei Bay, Aore I. Intertidal reef flat at low tide and dive to 8 m.	106
34	South of Vunaora Point, Espiritu Santo I. Intertidal reef flat adjacent to club house of golf course.	99
33	Intertidal reef flat between shore of Espiritu Santo I. and Maltinérava I.	90
7	North side Ambei Bay, Aore I. Intertidal reef flat at low tide.	85
8	Southern entrance to Ambei Bay, Aore I. Intertidal reef flat at low tide.	85
17	Abandoned Japanese fishing company south of Naoréuré Point, Espiritu Santo I.	82
23	Near Wombwabavua Point, Malo Island. Dive on mixture of coral and sand to about 6 m.	81
26	Northwest corner of Tutuba Island. Dive on mixture of coral and sand to about 13 m.	78
28	West side of Aore I. Dive to 24 m on steep slope. Intertidal sand but slope has some areas of coral.	76
11	Mélevatu I., south of Aore I. Intertidal reef flat at low tide. Rocky upper intertidal, lower is reef flat with sand and coral with few dead corals to turn over.	72

Table 45: Ten sites observed to have the lowest mollusc diversity among the 36 sites surveyed at Santo, Vanuatu.

Site	Location	Number of species
18	Wreck of *President Coolidge*, Espiritu Santo I. Dive to 25 m on upper surface of wreck.	26
35	Northern side of Aore I. Dive to 25 m on steep slope, sand all the way down with very limited areas of coral.	30
22	Mouth of Anduélé River, Malo Island. Fine sand with heavy detrital content. Subtidal flat at about 3 m gives way to steeply sloped sand to 28 m.	31
29	Mouth of Wambu River, Espiritu Santo I. Dive to 21 m. Shallows fine silt over gravel where river mouth. At slope down into channel gravel takes over with some larger rocks.	42
24	Near Avunambulko, Malo Island. Dive on mixture of coral and sand to about 6 m.	42
25	Eastern side of Tutuba Island. Dive from 6-22 m on exposed reef to the east of the island.	44
21	3 km south of Palikulo, Espiritu Santo I. Exposed intertidal reef flat.	47
6	Brigatoche Point, Espiritu Santo I. Dive off reef to 4 m in sand and coral.	41
20	Palikulo, Espiritu Santo I. Exposed intertidal reef flat.	50
13	Tsingoniaru Point, Espiritu Santo I. Dive to 24 m. Lower portion is fine sand, upper flat about 5 m is coral bombies and coral rocks.	51

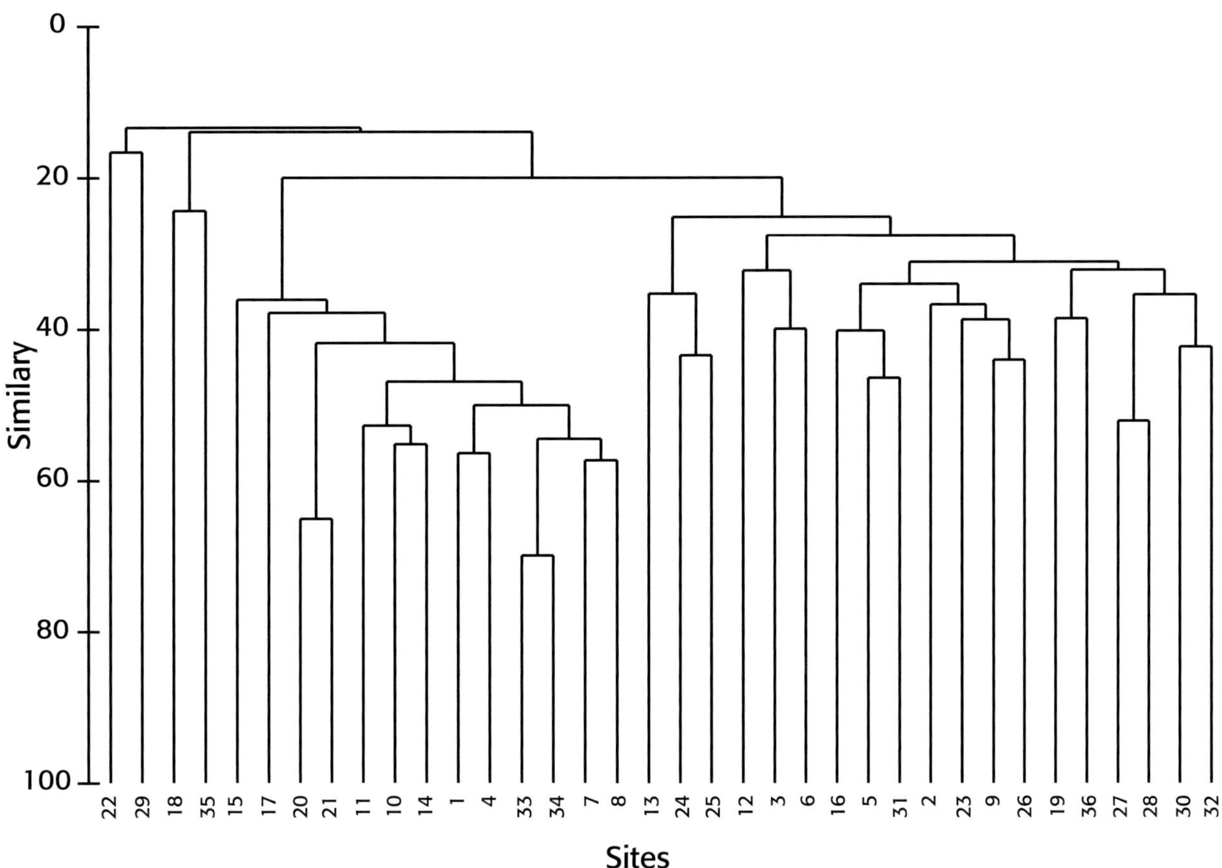

Figure 494: Dendrogram of species overlap for the 36 stations examined during the mollusc rapid assessment survey conducted in Vanuatu in August 2006.

16, 17, 33, 34). These four stations had a mean of 84.0 ± 7.2 species per station. In contrast, the two stations (20 and 21) on the broad platform to the south of Naoréuré Point and station 25 off the platform on the eastern side of Tutuba Island averaged only 47.0 ± 1.8 species per station. The tides were not particularly good when stations 20 and 21 were sampled, but the area appeared to be both restricted in terms of available habitats and there were many people on the shoreline collecting materials; it appears these sites have been thoroughly picked over. The richest area was the four stations (1, 7, 8, 9) on the eastern side of Aore Island, which had a mean of 86.5 ± 7.4 species per station. All of these stations were sampled intertidally. Station 9 was sampled briefly in the intertidal and was then dived; since most of the time spent collecting was during the dive this is listed as a dive station. The intertidal rocks were different, making this an area for further work. It was the station with the greatest recorded diversity (106 species). The five stations (5, 10, 11, 12, 22) in the Bruat Channel had a moderate diversity of 57.4 ± 7.1 species per station. Station 22, at the mouth of the Anduélé River was poor (31 species). This was an area of sand with considerable organic material. The low diversity was a real feature, but diversity was also lowered because local people objected to the dive and it was terminated. Five sites to the south of Espiritu Santo Island (13, 14 and 15) and to the west of Malo Island (23 and 24) had, in general, a disappointing diversity,

with a mean of 59.2 ± 6.6 species per station. The exception was the reef at station 23, which had 81 species. Twelve stations (4, 6, 18, 19, 27, 28, 29, 30, 31, 32, 35, 36) had a mean diversity of 55.0 ± 4.8 species per station. This area was variable. Stations in the southwest corner of the channel and most of the stations along Aore Island (4, 6, 27, 28, 29, 30, 31, 32) tended to have at least small reefs in the upper subtidal and a higher diversity. These stations averaged 61.4 ± 4.2 species per station. Station 35 lacked the reef and had only 30 species. It, and the dive on the Coolidge (18), were the stations with the lowest diversity. Station 19, adjacent to the Vanuatu Maritime College, had both a moderate diversity (71 species) and the most species of opisthobranchs (10) collected at any station.

The dendrogram (Fig 494) indicates that there were three major groupings of species. The first group (stations 22, 29, 18, and 35) were environmentally similar. Stations 22 and 29 were sandy areas at the mouth of rivers. Station 35 was not at the mouth of a river, but was a sandy beach with a steep sandy slope. Station 18 was the wreck of the *President Coolidge*. These sites were characterized by relatively low diversity of molluscs, dominated by bivalves which attached to hard surfaces. The MDS plot (Fig. 495) shows that these stations were relatively loosely related. The middle grouping on the dendrogram (Fig. 494) is essentially comprised of intertidal stations. All of the stations in this

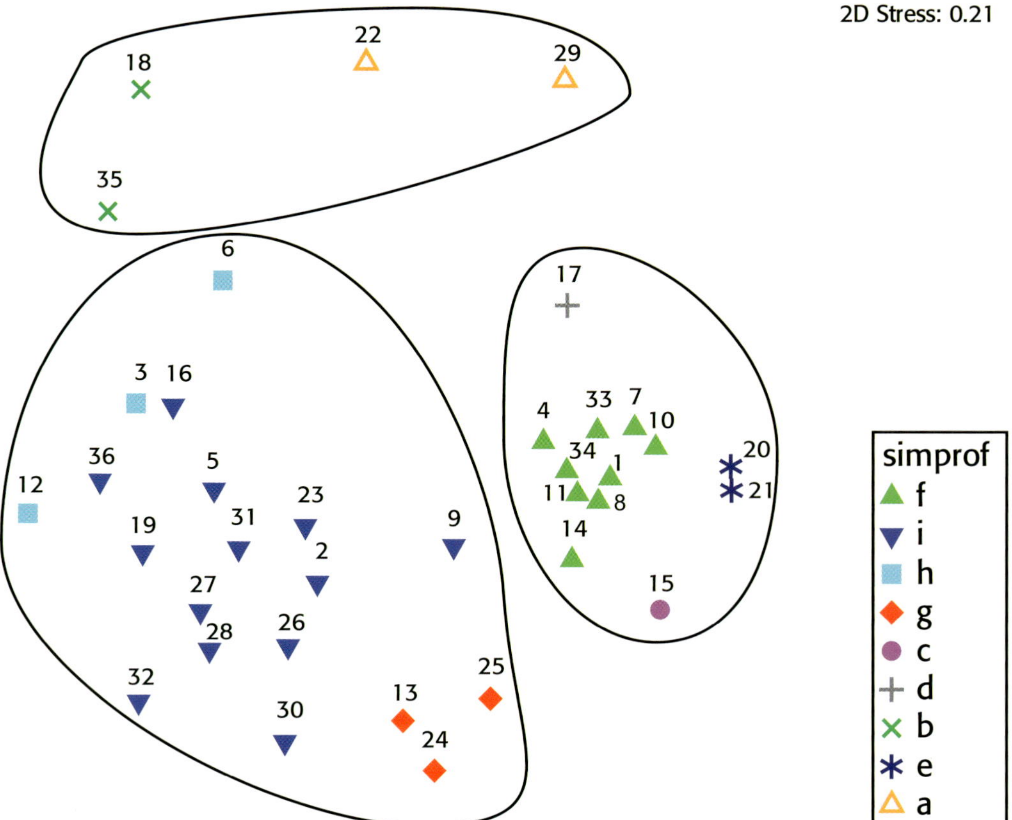

Figure 495: MDS plot of the 36 stations examined during the mollusc rapid assessment survey conducted in Vanuatu in August 2006.

group were intertidal except station 15. Conversely, all of the intertidal stations are in the group except Station 13. The MDS plot shows that these stations are grouped well together. The remaining stations are, with the exception of Station 13, the subtidal stations, which are relatively loosely grouped.

Not enough is known about the distributions of many mollusc species collected in Vanuatu area to place them within a geographical context. Table 46 shows the distributions of 202 species, as recorded in the literature. The vast majority (84%) of the species are widespread Indo-West Pacific forms; 21 species (10.5%) are regarded as being characteristic of the Western Pacific; and 5.5% occur in the eastern Indian Ocean and western Pacific Ocean.

A number of commercially important edible mollusc species occurred widely at the sites surveyed, including spider shells (*Lambis*), conchs (*Strombus luhuanus*), pearl oysters (*Pinctada margaritifera*), and giant clams (*Tridacna* spp.). However, commercial quantities were never found of any species. The most commonly found species was *Tridacna squamosa*.

••• Discussion

A total of 572 species of molluscs were collected at 36 stations during the 13 days of collecting in southern Espiritu Santo, Vanuatu. The species collected were generally widespread Indo-West Pacific species (84%), though there was a small component (11.5%) restricted to the Western Pacific and an even smaller (5.5%) proportion of species which occur in the eastern Indian Ocean and western Pacific. Three mollusc assemblages were detected in the Vanuatu stations: a grouping of intertidal stations, one of subtidal stations; and a second subtidal group which included stations of low diversity. Three of the four stations in this group were sandy bottoms with small coral bombies. The attached molluscs on the bombies had a similar composition to species occurring on steel plates of the *President Coolidge*.

Table 46: Geographical distributions of selected species recorded at Santo, Vanuatu.

Geographical Range	Number of species	Percentage of species
Indo-West Pacific	170	84.0
Western Pacific Ocean	21	10.5
Eastern Indian Ocean-Western Pacific	11	5.5
Totals	202	100.0

The 572 species collected in 13 days indicates the area has a moderate diversity of molluscs (Table 47). Diversity at Vanuatu was higher than Marine RAP surveys conducted at Phuket, Thailand and Nosy Be, Madagascar but lower than similar work undertaken in two surveys of Milne Bay, Papua New Guinea, and at the Raja Ampat Islands, Indonesia, and Calamian Islands, Philippines. Diversity in Vanuatu was similar to that recorded in the Togean and Banggai Islands, Indonesia. It was greater than some of the collecting areas in Western Australia, but less than in others. The key factor restricting mollusc diversity in Vanuatu was relatively low habitat diversity in the study area. The study region was smaller than most of the other areas surveyed, where there was a broader range of habitats available. In addition, local people collect shells for sale, which meant that no large cassids, tonnids, etc were collected.

The present study demonstrated that a considerable amount of information could be collected cost effectively by a single person working on mollusks in a 13-day period. However, the present survey differed from all previous such expeditions in that there no other groups being examined simultaneously. In all previous surveys fish and corals were also examined. Some of the previous surveys also examined groups such as echinoderms, crustaceans, etc. The additional groups can be surveyed at little additional cost, but contribute substantially to the results obtained, and provide a much broader picture of overall diversity in the study area. Most of the previous studies also include examinations of reef structures, human fishing pressure, and other human utilization.

Table 47: Numbers of mollusc species collected during previous Marine RAP surveys undertaken by Conservation International and similar surveys by the Western Australian Museum.

Location	Survey days	Mollusc species	Reference
Espiritu Santo, Vanuatu	13	572	Present survey
New England Aquarium Survey			
Phuket, Thailand	12	380 (estimate)	Wells 2005b
CI Marine RAP Surveys			
Northern Madagascar	16	525	Wells 2005a
Raja Ampat Islands	15	665	Wells 2002b
Togean-Banggai Islands, Indonesia	11	541	Wells 2002a
Calamian Group, Philippines	16	651	Wells 2001
Milne Bay, Papua New Guinea	19	638	Wells 1998
Milne Bay, Papua New Guinea	11	643	Wells & Kinch 2003
Western Australian Museum Surveys			
Cocos (Keeling) Islands	20	380 on survey; total known fauna of 610 species	Abbott 1950; Maes 1967; Wells 1994
Christmas Island (Indian Ocean)	12 plus accumulated data	313 on survey; approx. 520 total	Iredale 1917; Wells *et al.* 1990; Wells & Slack-Smith 2000
Ashmore Reef	12	433	Wells 1993; Willan 1993
Cartier Island	7	381	Wells 1993
Hibernia Reef	6	294	Willan 1993
Scott/Seringapatam Reef	8	279	Wilson 1985; Wells & Slack-Smith 1986
Rowley Shoals	7	260	Wells & Slack-Smith 1986
Montebello Islands	19	633	Preston 1914; Wells *et al.* 2000
Muiron Islands and Exmouth Gulf	12	655	Slack-Smith & Bryce 1995
Bernier and Dorre Islands, Shark Bay	12	425	Slack-Smith & Bryce 1996
Abrolhos Islands	Accumulated data	492	Wells & Bryce 1997
Other surveys			
Chagos Islands	Accumulated data	384	Shepherd 1984

MOLLUSCS ON BIOGENIC SUBSTRATES

Anders Warén

Pieces of sunken driftwood, sulphide rich emissions from geologic activities or reduced water circulation like in caves or oxygen minimum zones all constitute available, chemically bound energy. Energy is usually short in marine environments, especially in the deep-sea. This chemically bound energy is not evenly distributed in the sea but highly scattered and often present only during a short time, from a few weeks up to a few years. This puts special requirement on the organisms that want to use it. At first to find it. Then to utilise it before anybody else. And finally to disperse to a new place. These are probably the driving forces behind the evolution of an array of different organisms found on whale carcasses, driftwood, cold seeps and hot vents and many other such temporary concentrations of energy. Usually the chemical energy is first utilised by bacteria, which in turn are consumed by grazers or filter feeders, or less commonly, the bacteria are cultivated inside an organism which directly ingests the bacteria. In other cases the energy is contained in tissues that can be used directly, like scavengers feeding on a whale carcass, or pectinodontid limpets chewing up a log.

There is thus a continuum from scavenging sharks that will localise many preys during their life, to gastropods whose larvae settle on a cephalopod beak that will have to last long enough for the snail to grow up and reproduce and spread to next beak, or in some cases to pass through more than one generation on the same log, and bacteria that will go through thousands of generation also on the same log.

The word "biogenic" is used for the various substrates produced by living organisms. When including also hot vents and cold seeps, anoxic water bodies and similar biotopes, the term "chemosynthetic environments" is often used, referring to the fact that bacteria utilizing reduced compounds are involved. It should however be noticed that most animal life in chemosynthetic environments do not harbour symbiotic bacteria.

• • • Santo 2006 collects yielding biogenic substrates

- During the intertidal and shallow subtidal work, coconuts and pieces of wood were found and searched for molluscs.
- In the shallow water dredgings, biogenic substrates like plant debris, were plentiful enough to be washed separately and examined for biogenic fauna. Also occasional coconuts, and on one occasion tubes of the polychaete *Hyalinoecia* were found with associated gastropods.
- Deep water trawling brought up pieces of wood, leaves, coconuts and other debris which was carefully searched for inhabitants.

- Traps baited with pieces of whale skeleton, different kinds of wood, turtle carapaces, and coconuts were set SW of Malo Island in 400-700 m depth.
- Tangle nets do not only trap beautiful shells, but also driftwood. Fishes and gastropods die after being trapped and attract molluscs with affinity for chemosynthetic biota.

• • • Survey of Mollusca in chemosynthetic environments

This is a list of all mollusc families known from biogenic substrates and other chemosynthetic environments, with comments on those we found, to set the Santo fauna into its context.

• • • Polyplacophora
Chitons (Fig. 496A) were quite common in deep water, especially on sturdy and large pieces of wood but the species have not yet been identified.

• • • Patellogastropoda (True limpets)
- **Family Acmaeidae.** Species of *Pectinodonta* (Fig. 496E) were common on large (typically >10 cm diameter) logs of hardwood only. They have the ability to digest cellulose and excavate distinct depressions, possibly getting better protection as a side result.
- **Family Neolepetopsidae.** This true limpet family is well known from cold seeps and hot vents, but has also been found on whale skeletons, off southern California. No specimens found in Vanuatu.

• • • Cocculiniformia
- **Superfamily Cocculinoidea** includes a single family, **Cocculinidae** (Figs 498C & 499E), with more than 50 described species. Almost all live on sunken driftwood, but a few species have been found on whale bone and cephalopod beaks, also species that normally live on wood. More than 10 species were found on sunken driftwood, coconuts and other plant debris. Cocculiniformia is here used in a restricted sense. Earlier, Lepetelloidea, Pseudococculinidae and a few other families were included; these are now classified in the Vetigastropoda (see below).

• • • Vetigastropoda
- **Superfamily Lepetelloidea**, family **Lepetellidae**. Subfamily **Lepetellinae**. A couple of described genera and a few species. Only known from tubes of the polychaete genus *Hyalinoecia* (Fig. 496B). The species often occur in large numbers, up to hundreds of specimens on a single tube. Surprisingly, a few specimens were found in a tube dredged in ca 10 m depth, by far the most shallow record of the family. Subfamily **Choristellinae**. Species of Choristellinae differ from other lepetellids, in being coiled (Fig. 500B). They live in elasmobranch egg cases, empty or more rarely with a developing

Figure 496: Mixed molluscs. **A**: A chiton of the family Leptochitonidae, 10 mm long. **B**: Limpets of the family Lepetellidae, genus *Lepetella* (diam. 0.5-2 mm), on an old tube of the worm *Hyalinoecia*. **C**: Shipworm, Teredinidae sp., removed from its calcareous tube. **D**: Bathysciadiid limpets on an old cephalopod beak. Diameter of shells 2-3 mm. **E**: Limpets of the genus *Pectinodonta*, diameter 15-20 mm, on a piece of wood. **F**: A xylophagaid bivalve, partly removed from its burrow (not coated by calcareous material). **G**: A mussel of the genus *Idas*, 8 mm long.

embryo, but seem not to harm the embryo. No species were found during Santo 2006.

• **Family Addisoniidae**, Subfamily **Addisoniinae** The type genus consists of limpets (Figs 498B & 499B) in the size range around 10 mm and a slightly asymmetrical shell. They live in Elasmobranch egg cases but no species was found during the Santo project. Subfamily **Helicopeltinae**. Species of *Helicopelta* are known from cephalopod beaks. They are highly apomorphic, small (~2 mm) coiled shells with a Choristellid type radula. No species were found during the Santo project.

• **Family Bathysciadiidae**. The familiy contains species normally living on old cephalopod beaks (Figs 496D & 497A-B). They can be recognised by the evenly conical shell, usually with concentrically arranged periostracal tufts. No species were found during Santo 2006; they are more common in less eutrophic offshore areas.

• **Family Bathyphytophilidae** with a single genus, *Bathyphytophilus*, contains two named species that have been found on sea-grass deposited in deep water. Some species have a remarkably reduced radula. The shell is small and slender. A single very small cocculiniform (Fig. 499D) that may belong here, occurred in hundreds on certain types of leathery leaves which evidently break down very slowly, in 50-200 m. Its radula also resembles that of the Pyropeltidae and is a good examples of the classification problems in the cocculiniforms.

• **Family Caymanabyssiidae**. At least one species of the type genus *Caymanabyssia* was found (Fig. 498F). Our records are much more shallow than the distribution recognised in the literature, from ca 120 m; previously deeper than 900 m.

• **Family Cocculinellidae**. A few specimens of what seems to be a species of Cocculinellidae (Figs 497E-F) were found in residues from tangle-nets, depth

Figure 497: Some diagnostic limpet shells. **A-B**: *Bathysciadium* sp. (diameter 2.5 mm) live on old cephalopod beaks, lieing on the seafloor after the squid has died. Living specimens of the limpet have radiating rows of tufts of periostracum. **C**: Neritimorph limpets of the genus *Phenacolepas* (length 3-5 mm) may live in burrows in pieces of wood. **D**: Slit limpets of the family Fissurellidae have a strongly sculptured shell with a slit or a hole for the exhalant water. In this case a young specimen of *Puncturella*, (diam. 2.5 mm attains 20 mm) with a slit that later closes to a round hole. **E-F**: Species of *Cocculinella* (length of shell 1.5-2.5 mm) live on old fish bones. (Photos A. Warén).

10-50 m. It differs from previously known species in having a more normal cocculinid-type radula and protoconch while the shell and external body morphology come very close to *Cocculinella*.

• **Family Osteopeltidae**. A very few species known from whale bone. No protoconchs are known, but the relations seem to be to the Pseudococculinoidea, not to the Cocculinidae. No species at Santo.

• **Family Pseudococculinidae**. Probably 10 species representing four genera were found. *Notocrater* with what seems to be three species found during Santo 2006, can be taken as an example with typical external morphology of the soft parts, protoconch and shell (Figs 499A-C).

• **Family Pyropeltidae**. This monotypic family is assumed to be restricted to cold seeps, hot vents and whale bone, but the radula is suspiciously similar to the suspected Bathyphytophilid mentioned above. The genus *Pyropelta* was not found during Santo 2006.

• **Superfamily Seguenzioidea**. A few "skeneimorph" genera of seguenzioids live on wood and in hot vents (*Xyloskenea*, *Trenchia*, *Ventsia*) but the number is likely to increase when more precise information becomes available. Many specimens of

Xyloskenea (Fig. 500I) were found in depths below 500 m. Subfamily **Cataegiinae**. The genus *Cataegis* lives in cold seeps and on sunken driftwood, and *C. leucogranulatus* (Fu & Sun, 2006) was common on driftwood during Santo 2006 (Fig. 502D). Subfamily **Chilodontinae**. Several specimens of a single species of *Euchelus* (Fig. 500D) were found on wood at two occasions.

• **Superfamily Fissurelloidea**, family **Fissurellidae**. Species of *Puncturella* (Fig. 497D) occur regularly on wood and usually only a single species in an area. The group is also known from seeps and vents, but there seems to be no direct relations between the species and they probably represent local adaptation. A single species was found regularly on wood during Santo 2006.

• **Superfamily Scissurelloidea**. Species of *Anatoma* and *Scissurella* are regularly found on sunken driftwood as well as in seeps and vents, but only scattered specimens. Only a couple of specimens of *Anatoma*, *Scissurella* (Fig. 496C) and one *Larochaea* were found during Santo 2006.

• **Superfamily Lepetodriloidea**. Family **Lepetodrilidae**. This family occurrs almost exclusively

Figure 498: Some not diagnostic limpet shells. **A**: Species of *Pyropelta* (diameter 2-4 mm) usually live in cold seeps but also on whale skeletons and *Provanna* shells. They are, except when very young, extremely corroded with a terrace like corrosion pattern. **B**: Species of *Addisonia* (shell diameter 10-20 mm) live in skate- and shark egg capsules. **C**: Species of Cocculinidae, here *Cocculina* sp. (diameter 5 mm), seem to be restricted to sunken driftwood. **D**: *Osteopelta* spp. are known from pieces of whale bone only. **E**: The genus *Notocrater* (diameter of shell 2-4 mm) belongs to Pseudococculinidae and are very similar to 19. **F**: *Caymanabyssia* sp. (4 mm; Caymanabyssiidae). Both live on Sunken driftwood. (Photos A. Warén).

in vents, more rarely in seeps, but a few specimens have been found on sunken driftwood along the American West Coast. Not found during Santo 2006.

• **Superfamily Neomphaloidea**. The Neomphaloidea or Neomphalines are best known as a large radiation in hot vents. A smaller group of somewhat uncertain position comprises the genera *Leptogyra*, *Xyleptogyra*, and *Leptogyropsis*. They are mainly known from sunken driftwood but *Leptogyra* occurs also in seeps and vents. At least two species of *Leptogyra* (Fig. 500F) were found during Santo 2006, deeper than 500 m.

• **Superfamily Trochoidea**. family **Trochidae**. The genus *Pseudotalopia* occurrs on sunken driftwood from Japan to Vanuatu in depths between *c.* 250 and 750 m. Two species were found during Santo 2006 (Fig. 502A).

• **Superfamily Turbinoidea**. Family **Turbinidae**, subfamily **Colloniinae**: Several species of *Cantrainea*

(> 400 m) and *Homalopoma* (> 50 m, Fig. 502B) have been found on sunken driftwood in intermediate depths from 50 to 1000 m, Four species during Santo 2006.

• **Family Skeneidae**. The family Skeneidae is used both as a family for species related to the genus Skenea and as a storage bin for species of an undefined skenea-like appearance. Species of Skeneidae in the former sense live in all kinds of chemosynthetic environments, from interstitially in sand, and under rocks with reduced circulation to cold seeps and hydrothermal vents. In tropical areas and deep water the family is represented by the genus *Dillwynella* (Fig. 500A) on sunken driftwood. At least three species of this genus in Vanuatu.

••• **Neritimorpha**

• **Superfamily Neritoidea**, family **Phenacolepadiidae**. Species of *Phenacolepas* live in all kinds of reducing environments, from sewage out-

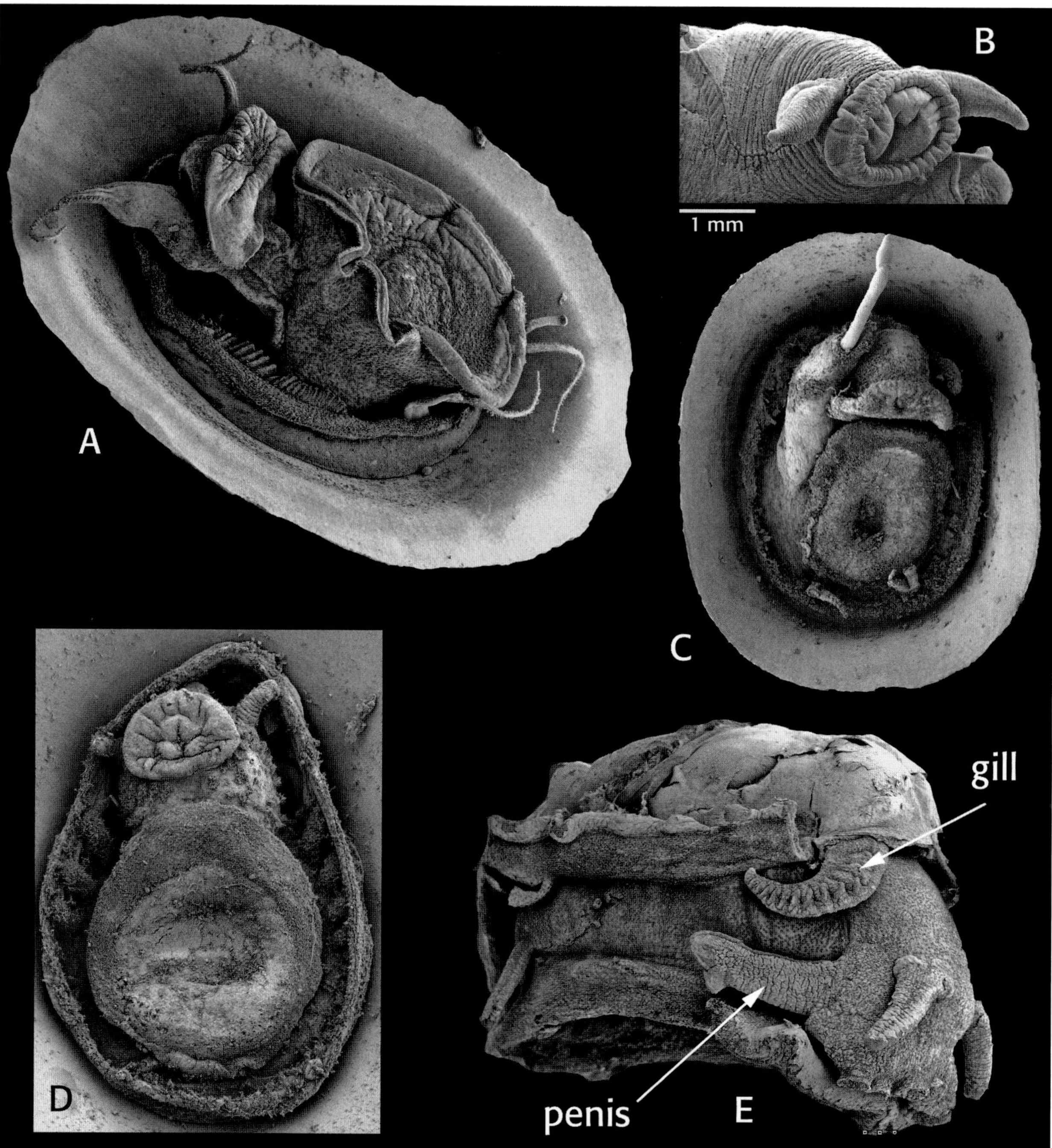

Figure 499: Limpet bodies. **A**: Pseudococculinidae, uncertain genus, diameter 7 mm, from wood. Right cephalic tentacle much larger than left one. **B**: Head of *Addisonia* sp., with an open sperm gutter behind the right cephalic tentacle. **C**: *Notocrater* sp. (shell 2.5 mm), from wood, in the process of expelling a spermatophore through the right, expanded, cephalic tentacle. **D**: Bathyphytophilidae, uncertain genus, from submerged leaves, soft parts inside shell. Diam. of soft body 1.7 mm. Notice paired gills in the pallial furrow. **E**: *Cocculina* sp. body removed from shell (diameter 5 mm). Notice the penis (drawn out from the right side of the snout) and the gill which in this species is bipectinate. (Photos A. Warén).

lets and subterranean drainage canals in tidal coral platforms, to sunken driftwood down to several hundred meters depth. Related genera live in hot vents. 3-4 species were found in burrows of xylophagaid bivalves in a few hundred meters depth (Fig. 497C).

••• Caenogastropoda

• **Family Provannidae.** This family has its main occurrence in hot vents and cold seeps, but some species occur also on whale carcasses and sunken driftwood. One species of *Provanna* found during Santo 2006, in 800 m depth (Figs 501D-E).

• **Superfamily Rissooidea.** The subfamily **Rissoininae** has its main distribution under rocks and in sediment intertidally to subtidally. Specimens are frequently found on wood in shallow water, also during Santo 2006, but they never occur in large numbers and the occurrence is probably accidental and can be blamed on the similarity between a rock and a log. The subfamily Rissoinae has a small radiation in seeps, vents and on wood in the North Atlantic (Fig. 501F), but no species are known from such environments elsewhere.

Figure 500: "Skeneimorph" gastropods. **A**: *Dillwynella* sp., a real Skeneidae living on sunken driftwood, diameter 1.5 mm. **B**: *Choristella* sp. (fam. Choristellidae), diameter of shell 5 mm, from skate egg capsules. **C**: *Ponderinella* sp. (fam. Vitrinellidae), common on sunken driftwood in Vanuatu with an unidentified tube attached. Diameter of shell 2 mm. **D**: *Euchelus* sp. (fam. Chilodontidae), from driftwood, diameter 6 mm. **E**: Species of *Hyalogyrina* (fam. Hyalogyrinidae) often live among sulphide bacteria in low oxygen areas around organic debris and cold seeps. This species from whalebone is 2.5 mm high. **F**: *Leptogyra* sp. (provisionally in Melanodrymiidae) from driftwood, diameter 2 mm. **H**: *Scissurella* sp. (fam. Scissurellidae), diameter 2.3 mm. **I**: *Lurifax* sp. (fam. Orbitestellidae), diameter 2.5 mm. **J**: *Xyloskenea* sp, (uncertain family in superfamily Seguenzioidea), from sunken wood, diameter 1.8 mm. **J-K**: *Xylodiscula* (fam. Xylodiscidae, diam. 2 mm), from sunken driftwood. (Photos A. Warén).

• **Family Iravadiidae**. Species of this family are usually associated with chemosynthetic environments, from decapod burrows and under rocks, to cold seeps and sunken wood (Fig. 501C), intertidally to several thousand meters depth. A few species were commonly associated with sunken wood from 50 m, down to several hundred meters depth.

• **Family Caecidae** is normally found interstitially in sand but two species were regularly found on sunken driftwood in 50-150 m, during the Santo workshop (Fig. 501B).

• **Family Elachisinidae**. A very little known family. A very strange, undescribed species (Fig. 501G) that may belong here was found during Santo 2006, as well as a couple of less deviating species (*Laeviphitus*, Fig. 501D).

• **Family Tornidae** (~Vitrinellidae). The genus *Ponderinella* (Fig. 500C) is definitively associated with sunken driftwood and occurred commonly in 200-800 m depth. Several other species of Tornidae or Vitrinellidae were found, but only scattered specimens or empty shells. Although the biology of tornids is poorly known it seems that they frequently live under rocks or in burrows and sulphides or bacteria from the wood may attract larvae outside their normal dispersal range.

• **Superfamily Cerithioidea**. This very large group must be noted for its scarcity in chemosynthetic environments. Scattered specimens of *Bittium*-like gastropods were found in 50-150 m depth on wood (Fig. 501A) during the workshop, but they may have been using the wood as any other hard bottom. Species of *Lirobittium* have been found in seep environments off California.

• **Superfamily Vanikoroidea**, family **Vanikoridae**. Specimens are regularly found on sunken driftwood, but like most tornids, they are never present in high numbers and the occurrence may be fortuitous, perhaps aided by a need to hide in cracks and crevices as species of *Vanikoro* do. Scattered specimens during the workshop (Fig. 501K).

• **Superfamily Buccinoidea**, family **Buccinidae**. Species of the genera *Eosipho* and *Manaria* are often found on sunken driftwood, especially on large tree trunks. The reasons for this habitat preference are not clear; they certainly do not eat wood or bacteria; probably they are general predators attracted by the large concentrations of potential prey. The genus *Eosipho* includes species from hydrothermal vents. A few species were found during the Santo workshop, especially the large *E. smithi* (Schepmann, 1911) (Fig. 502H).

• **Family Columbellidae**. Columbellids are regularly found on shallow wood-falls, also at Santo (Fig. 502E), and they are known from mass occurrences in Californian cold seeps, but it is uncertain if they can be considered part of the wood fauna.

• **Family Fusinidae** (Fig. 502F) and **Nassariidae** (Fig. 501I). These two families are sometimes found on sunken driftwood, also in fair numbers, but also here it may be attraction to the rich animal life, perhaps combined with deposition of egg capsules on the wood. (Santo trap 06-10-15, AT96).

• **Superfamily Muricoidea**. Some tropical species of Muricidae are sometimes found on sunken driftwood, like *Orania gaskelli* (Melvill, 1891) (Subfamily Ergalataxinae), which occurs on wood falls in 50-150 m (Fig. 502G). No other muricids occurred in more than one or two specimens.

• **Family Costellariidae**. Several specimens of a single species were caught in the traps (Fig. 502C).

• **Superfamily Conoidea**. No species seem to be dependent on sunken wood or other biogenic substrates, but several species are common in and seem to be endemic to hot vents and cold seeps (especially the genus *Phymorhynchus*).

••• **Heterobranchia**

• **Superfamily Pyramidelloidea**, family **Pyramidellidae**. A few species have been found on wood and in cold seeps. They may very well be unspecific bivalve or polychaete parasites, but at least a couple of them were found in several specimens and so far only on wood (Fig. 501J). Probably parasitic on some wood inhabitant.

• **Superfamily uncertain, Family Cimidae** (Fig. 501H). A few records are known from cold seeps and sunken driftwood, but the family is extremely poorly known with a few species from more normal subtidal environments. Not found during Santo 2006.

• **Family Xylodisculidae**. This family is known from seeps and vents but is more common on sunken driftwood, from deposits of *Posidonia* (sea grass) fibers and one species from sulphide rich bottoms under a commercial fish farm. Two species were found at Santo (Figs 500J-K).

• **Family Orbitestellidae**. The family was originally described from shallow water and intertidal environments but now a few more species are known also from deeper bottoms. No species were found at Santo but I have encountered a species of *Orbistestella* on sunken driftwood at the Solomon Islands and the genus *Lurifax* (Fig. 500H) occurs in hot vents and cold seeps.

• **Family Hyalogyrinidae** (Fig. 500E). This family, with two genera was originally described from sunken driftwood, but is now known also from rotting coco nuts, hot vents, cold seeps and caves. Found inside decaying coco nuts during Santo 2006.

••• **Bivalvia**

The number of bivalve groups found during Santo 2006 is small, but personal experience from other areas do not lead to expectations of much more. Some other typical hard bottom taxa like pectinoids, dimyids, and anomids are regularly found, but those species also live on any other solid substrate in the same area.

• **Family Mytilidae**, subfamily **Bathymodiolinae**. Species of *Idas* (= *Idasola* = *Adipicola*) live byssally attached on driftwood and whale bone, especially along and in cracks. The related genus *Bathymodiolus* lives at cold seeps and hot vents.

Figure 501: Mixed gastropods. **A**: *Bittium* sp. (fam. Cerithiidae, shell 3 mm), occurrence on wood accidental? **B**: *Caecum* sp. (fam. Caecidae, 2 mm), common on sunken driftwood in 40-80 m. **C**: *Nozeba* sp. (fam. Iravadiidae, 2.3 mm), vey common on driftwood in 50-500 m. **D**: *Laeviphitus* sp. (fam. Elachisinidae, 1.5 mm), a genus from biogenic substrates and hydrothermal vents. **E**: *Provanna* sp. (fam. Provannidae, 4 mm), a genus regular but rare on sunken wood in deep water. **F**: *Alvania* sp. (Rissoidae, 2 mm), member of a radiation of wood associated species in the North Atlantic. **G**: A possible member of the Elachisinidae (height 3 mm), not rare on wood in Vanuatu. **H**: *Cima* sp. (fam. Cimidae, 1.3 mm), a group of poorly known species mainly occurring in reducing environments. **I**: *Nassarius* sp. (fam. Nassariidae, 6 mm), a species of uncertain affinity to, but common on sunken wood at Santo in 50-150 m. **J**: A species of fam. Pyramidellidae (1.5 mm) frequently found on sunken wood, but probably parasitic on bivalves or polychaetes living there. **K**: A species of Vanikoridae, shell 2 mm), a family regularly found on sunken driftwood, but probably not dependent on wood. (Photos A. Warén).

Figure 502: Gastropods. **A**: *Pseudotalopia* sp. (Trochidae, 12 mm), common on wood in traps off Santo, in 500 m depth. **B**: *Homalopoma* aff. *laevigatum* (Sowerby, 1914), (Colloniinae, 5 mm), common on wood in 50-500 m depth. **C**: *Costellaria* sp. (Costellariidae, 12 mm). **D**: *Cataegis leucogranulatus* (fam. Cataegidae), 12 mm), occurs on wood, in cold seeps and hot vents. **E**: *Mitrella* sp. (fam. Columbellidae, 7 mm). **F**: *Pseudolatirus* sp., (fam. Fasciolariidae, 25 mm). **G**: *Orania gaskelli* (Melvill, 1891) (fam. Muricidae, 20 mm). The species in figs 49 and 51-53 are common on wood off Santo, in 50-500 m, but are probably more attracted to the local enrichment of animals than to the wood. **H**: *Eosipho smithii* (Schepman, 1911) (Buccinidae, 58 mm), is probably one of the largest gastropod species regularly found on driftwood. (Photos A. Warén).

Seven species of *Idas* (Fig. 496G) were found in depths between 1 and 1 200 m during Santo 2006.

• **Family Lucinidae**. Species of Lucinidae live as infauna and some species live mainly in reducing sediments, and have symbiotic bacteria. At least one species, *Bretskya scapula* Glover & Taylor, 2007 (Figs 503C-D) seems to be attracted to oxygen depleted sediments in direct contact with sunken wood. The family is known also from cold seeps (incl the World's largest lucinid) and hot vents. The gills of the Santo species are white from deposits of elementary sulphur.

• **Family Thyasiridae**. Species of Thyasiridae live as infauna and are often found in reducing sediments. Many species have symbiotic bacteria. No species seem to be especially attracted to oxygen depleted sediments in direct contact with sunken wood, but some have a more general liking for reducing environments. A couple of species were found in blackish mud adjacent to wood during Santo 2006 (Figs 503E-F).

• **Family Vesicomyidae** (or **Kelliellidae**). The family Vesicomyidae includes the "giant clams" from hot vents and cold seeps; the poorly known minute species usually referred to as "Kelliellidae" are more typical for normal deep-sea environment. Interestingly enough, one species similar to *Kelliella* (Figs 503 A-B), but with a hinge approaching *Vesicomya* was regularly obtained from washes of highly degraded driftwood in fairly shallow water. The precise biotope was not recognised. The very similar *Kelliella japonica* was described from submarine caves in the Ryukyu Islands.

• **Family Mesodesmatidae**. One species of *Rochefortina* (Figs 503G-H) was regularly obtained from washes of wood, but the precise habitat could not be determined. The species was only found when sediment was present, so it may live in sediment accumulations in the wood. One species is known from submarine caves of this otherwise poorly known genus.

• **Superfamily Pholadoidea**. Family **Pholadidae**: One species, *Martesia* sp., in shallow water wood and coconuts. Family **Teredinidae** (Fig. 496C): More than 15 species. Family **Xylophagaidae** (Figs 496F & 503I-J): At least 10 species, all seemingly undescribed. A small oddity: Some species of Xylophagaidae have a mesoplax, a small shell element in connection with the hinge, that may be mistaken for a strange limpet (Figs 503I-J).

Concluding remarks

New taxa. It is still too early to try to estimate species numbers and percentage of undescribed taxa, especially among the cocculiniforms which are poor in characters for recognising species and higher taxa.

Diversity. The diversity on sunken driftwood is usually not very high; usually up to maximum of some 10 species was found on a large log. Higher number have been seen in the Solomon Islands and the Philippines (*c.* 30 species, unpublished). However, the numbers of individuals may be very high, especially on logs with external and internal galleries of burrows and cracks and crevices in the wood. Also the bark may form a slightly different environment since it seems that it becomes more rapidly colonised than nearby areas of naked wood.

Taxa absent from wood. The most conspicuous absentee among the gastropods is the Cephalaspidea, a group which is very common on and in soft bottoms. No such species has been noticed. They are, however, usually soft bottom infauna, and since they are active predators, they may be discouraged by low oxygen sediments. No species of the Conoidea, the largest of all mollusc superfamilies, is even suspected to be associated with sunken driftwood, although several seem to be endemic to hot vents and cold seeps. Could it be that wood and other biogenic substrates are too ephemeral for them?

Size and type of substrate. The gastropods associated with chemosynthetic environments are usually not very specific in their choice. It seems that they can inhabit most species of wood, as long as the wood not is especially soft, porous or rotten. And the wood associated species are never found on leaves. Wood associated species are not found on the husk of coco nuts but occur regularly on the hard, internal shell.

Some gastropod species are mainly found on large logs, for example the costellariids and buccinids. A possible explanation is that they need to pass through one reproductive cycle to get population densities high enough to make the species recognizable as a wood associate. At least the species of *Manaria* and *Eosipho* have lecithotrophic development with the young specimens hatching in a crawling stage. *Pectinodonta*, on the contrary, has a planktonic dispersal phase, but is also mainly found on large logs; here it may be that these species really eat the wood, not a salad of wood and bacteria.

Faunal composition. One interesting observation is that the there is a distinct bathymetric zonation of the families. During Santo 2006 the caenogastropods were common on wood in 0-500 m but have no representatives deeper than ca 700 m, except a single species of *Ponderinella* (Tornidae), a couple of buccinids which occurred down to 1 000 m, and *Provanna*. *Ponderinella* was described from ca 400 m depth in New South Wales, Australia. I have observed it down to similar depth in the Philippines and the Solomon Island. Also some of the buccinids, e.g. *Eosipho* and *Manaria* there occurred down to slightly more than 1 000 m.

The Caenogastropods associated with sunken wood also have latitudinal gradient. Except the Provannidae, the southernmost record is

Ponderinella from New South Wales, at 35°S. The northernmost records are from southern Japan, at about 30-35°N where some species of *Manaria* and *Eosipho* are likely members of this fauna, as well as *Nozeba* (Iravadiidae) and *Ponderinella* (Tornidae).

The provannids on the contrary, occur from the Aleutians (54°N) south to at least 47°S in New Zealand, and in depths between ca 500 and 7 000 m. This may indicate that the provannids with suspected affinity to some palaeozoic taxa have been present on biogenic substrates and chemosynthetic environ-ments for a much longer time than the other caeno-gastropods, which are of more Caenozoic origin.

In addition, this distributional pattern may be interpreted as a result of a general spreading from shallow water to deeper in the maximum biodi-versity area, as suggested by David Jablonski and coauthors for the deep water fauna in general. My interpretation of the absence of most caenogas-tropods on biogenic substrates in temperate and boreal areas as well as the deep sea is that they have simply not had the time to spread there yet.

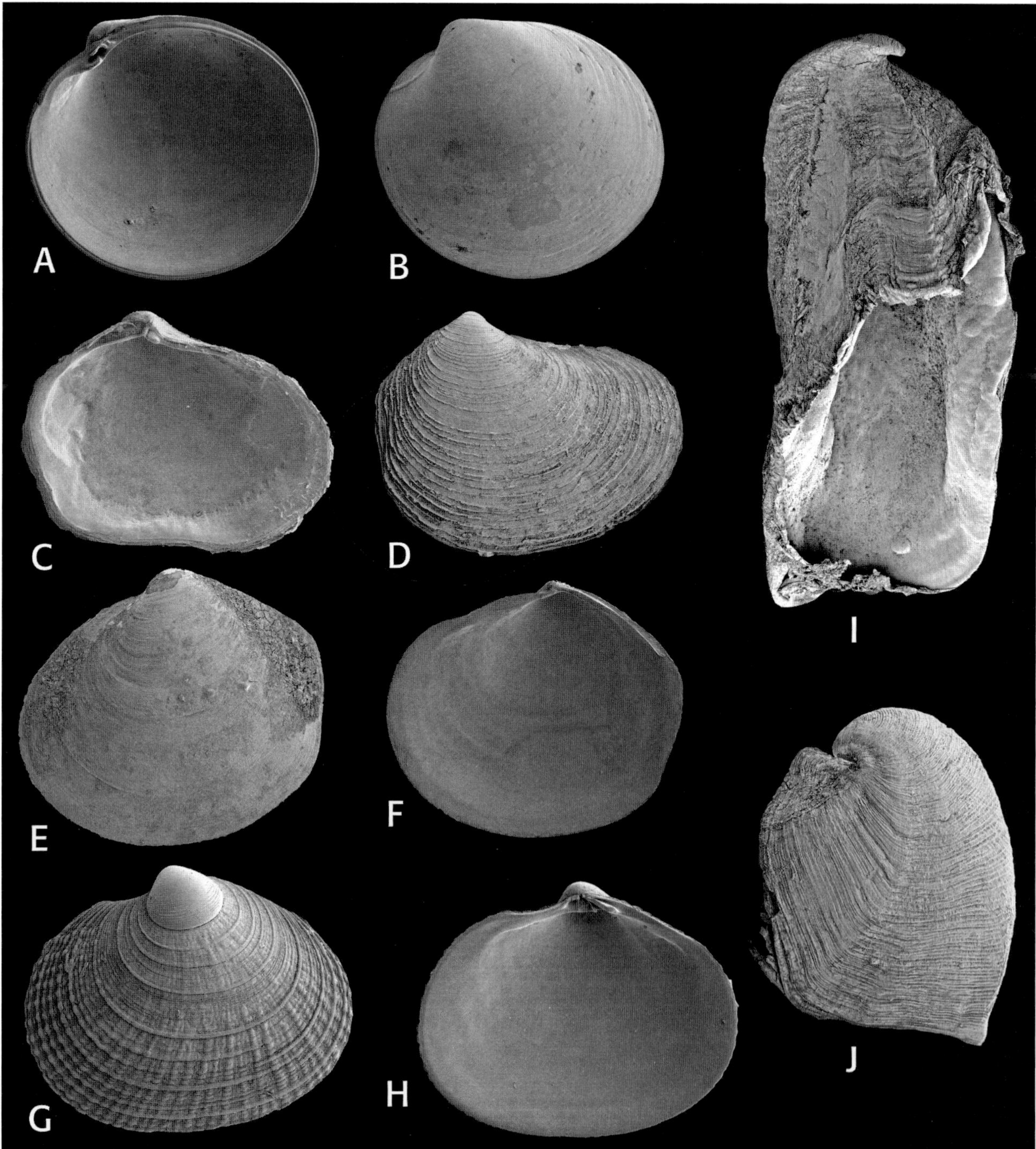

Figure 503: **A-B**: *Kelliella* sp. (Kelliellidae, 2 mm), common on driftwood at Santo. **C-D**: *Bretskya scapula* Glover & Taylor, 2007 (Lucinidae, 7 mm), common in sediment on drifwood at Santo. **E-F**: Species of Thyasiridae occur in sediment on and in driftwood. The deposits on the shell are often bright brown or black and consist of iron compounds precipitated by bacterial activity. **G-H**: *Rochefortina* sp. (Mesodesmatidae, 3 mm) was regularly found on driftwood off Santo. **I-J**: *Xylophaga* sp. (fam. Xylophagaidae), mesoplax, an additional shell element, with a striking similarity to a mm-sized limpet. (Photos A. Warén).

MARINE PARTNERSHIPS IN SANTO'S REEF ENVIRONMENTS: PARASITES, COMMENSALS AND OTHER ORGANISMS THAT LIVE IN CLOSE ASSOCIATION

Stefano Schiaparelli, Charles Fransen & Marco Oliverio

The Santo 2006 expedition provided an exceptional opportunity to scientists for exploring and gathering field data on one of the most important aspects of marine ecology: the animal associations. In Nature, living organisms interact with the environment and with other organisms since the beginning of life. The relationships between organisms of the same species are, with very few exceptions, not harmful and are largely dominated by events linked to reproduction, as the choice of a partner, or cures of the offspring in species with parental care or brooding.

Instead, if we look at the possible interaction between organisms belonging to different species, rarely we can define these as not harmful. What we immediately see is that they are all intimately linked to each other within complex trophic webs, where a species may play the role of prey, of predator, or both roles at the same time.

These relationships have been refined through millions of years of evolution, both on a biochemical base (e.g. producing compounds used as deterrents to escape predation or "venoms" used to ease predation) and on a morpho-functional one (e.g. increase of strength of skeletal structures in preys and, in parallel, of the capability to destroy these by predators), leading to a sort of "arms race", where the two partners, prey and predator, keeps refining their reciprocal capabilities and techniques to escape predation or catch the prey, respectively.

However, besides this kind of direct relationships, there are others, commonly defined "symbioses", in which organisms establish partnerships (Fig. 504) that do not necessarily determine the death of either organisms, but that allow to gain food or protection and, in turn, to feed and survive.

These interactions have received comparatively less attention from an ecological point of view, as predation is commonly recognised as a major factor in shaping community structure, both in qualitative and quantitative terms. "Symbioses", in fact, have generally been considered oddities or curiosities, certainly not important from a quantitative point of view. Only very recently, the ecological role of "symbiotic" interactions has been re-evaluated and it has been evidenced as, in many cases, they may affect the community structure to an extent that is as important as predation or physical disturbance.

If we think to an example of symbioses, our mind immediately goes to coral reef environments, which are (correctly) considered the cradle of these kinds of

Figure 504: Among the decapods, many species of shrimps, crabs and anomurans form associations with a variety of other organisms. Shrimps and crabs can be found on cnidarians like stony corals, soft corals and sea anemones.
The beautifully coloured shrimps of the genus *Vir* live among the tentacles of bubble corals. *Vir philippinensis* and *Vir colemani* (here depicted on *Plerogyra sinuosa*), among the several species known, are particularly well known by underwater photographers. During the Santo 2006 expedition most of these have been collected, as well as an undescribed species.

Figure 505: A classic example of association in the Santo reefs: the clown fish *Amphiprion* sp. and its anemone (*Entacmaea* sp.).

Figure 506: The gastropod genus *Colubraria* includes two dozens of mostly shallow-water species which inhabit rocky-coral environments in tropical, subtropical and temperate seas. They use their long proboscis to feed on the blood of parrotfishes (Scaridae) and occasionally on other fishes, and live in crevices, coves, caves, where fishes are expected to go to sleep. At least three species have been observed in their parasitic habits during the Santo 2006 expedition: *C. muricata* (**A-C**: here a specimen 98 mm long, feeding on a resting *Siganus* sp.), *C. nitidula*, and *C. obscura*. **D**: Parasitism on fish has been reported for species of two other neogastropod families: Cancellariidae and Marginellidae. This small (4 mm long) marginellid *Hydroginella* sp. has been observed on parrotfishes during their "siesta" at night in the reefs of Santo. (Photos A-C, S. Schiaparelli; Photo D, D. Brabant).

associations. In fact, from a numerical point of view, most examples of "symbioses" can be found only in these environments, spanning from the association between stony corals and their symbiotic unicellular algae (the zooxanthellae of the genus *Symbiodinium*), which is the base of the whole coral reef ecosystem, to the well know one between clown fishes (*Amphiprion* spp.) and their anemones (Fig. 505).

The term "symbiosis" has a variety of interpretations. It is commonly used to define a beneficial association between an organism, the symbiont (usually smaller), and its animal or plant host (usually larger). However, in Biology, the use of this term has changed with time, shifting from its original 19th century definition by de Bary, indicating a close physical (and/or metabolic) association between two organisms, regardless of any ranking of relative benefits, to a modern one which restricts its meaning only to mutualistic interactions, being defined as the "union of two organisms whereby they mutually benefit". From an ecological point of view, studying 'symbioses' means dealing with interactions between organisms where the evaluation of costs/benefits is crucial to the definition itself of the particular partnership involved. In many instances —as in the case of the hematophagous gastropods (Fig. 506)—the kind of relationship is clear.

Figure 507: Coralliophilines (Muricidae, Coralliophilinae) include approximately 200-250 species, distributed world-wide in warm temperate and tropical oceans, from intertidal habitats down to depths of over 1 000 m, always associated with anthozoans (Alcyonacea, Gorgonacea, Antipatharia, Actiniaria, Corallimorpharia, Scleractinia, Zoantharia). Nearly 30 species have been found at Vanuatu. **A**: These specimens of *Coralliophila violacea*, from Station ZR12, act as parasites of the colony, yet they are predators of each single polyp. **B-C**: Animal associations are important also in deep waters. The majority of coralliophilines (over 75 % on a global scale) live deeper than 100-150 m, the lower limit for the hermatypic scleractinians, where the latter are out competed for space by Alcyonaria, Stylasterids and Porifera. In this case (**B**) *Coralliophila solutistoma* (31.4 mm long) has been collected fixed on its host, a deep water gorgonian at Vanuatu in 210 m (MUSORSTOM 8 stn CP 1102). A smaller (12.6-13 mm long) undescribed coralliophiline species of deep water collected associated to a 20 mm wide solitary *Flabellum* coral (**C**), has been named *Mipus boucheti* by Oliverio (2008). (Photo A, S. Schiaparelli; Photos B-C, M. Oliverio).

Figure 508: **A**: *Lysmata amboinensis* are day-active hippolytid shrimps perching on prominent points in coral reefs, swaying their bodies, and lashing their long, white antennal flagella to attract fishes. Fishes respond by changing color and showing ritualized, slow movements. The shrimps hop onto the fish and start picking ectoparasites even on the gills or inside the mouth. **B**: The Barber-pole shrimp or *Stenopus hispidus* (here depicted), of the infraorder Stenopodidea, can be observed in the neighbourhood of large predatory fishes like Moray eels. It removes dead tissue, algae and parasites even from larger fishes.

Unluckily for our need to reduce the astonishing diversity of living forms and interactions into an understandable framework, these associations frequently shift from positive, to neutral, to negative, depending on the environmental and community context. Moreover, these partnerships may be temporary, unilateral (when adaptation by the "symbiont" side does not implies a counter-adaptation in the host), or easily shift from not harmful to parasitic ones, making their definition a hard (and sometimes impossible) task. This holds particularly true for those associations occurring in deeper waters (Fig. 507).

The demonstration that the "symbiont" gains a benefit by interacting with another species is not always easy to demonstrate. It relies on the empirical prediction that the former should have a higher fitness in doing this, relative to the opposite situations, and this kind of studies necessitates a rigorous statistical approach. "Classics" of this type are the cases of cleaner fishes and of cleaner shrimps (Fig. 508), among the best examples of mutualistic interactions in reef environments.

However, there is a growing body of evidence that host-mediated adaptations in the sea may drive

Figure 509: A remarkable radiation of wentletrap snails (Epitoniidae) associated with mushroom corals (Fungiidae) has been recently discovered in the Indo-West Pacific. Nine species of this parasitic group have been found during the Santo 2006 expedition. One of these was *Epifungium twilae*, here pictured with clusters of eggs underneath the fungiid coral *Herpolita limax*.

the behaviour of even single populations of small invertebrates that interact with their larger plant or animal hosts, suggesting as these local interactions may play a fundamental role in differentiation, perhaps driving speciation. In some cases (e.g. the coralliophiline genus *Leptoconchus* and the wentletrap snails Epitoniidae studied by A. Gittenberger

Figure 510: The Eulimidae are a large family of gastropods that parasitize echinoderms, with genera specifically adapted to feed on Asteroidea, Ophiuroidea, Echinoidea, Holoturoidea and Crinoidea. Feeding is performed by penetrating the host skin with a proboscis and sucking body fluids from internal cavities. In this case, the eulimid *Parvioris* sp. is clearly visible on the top of the sea star arm on the left of the picture (arrow, highlighted in **B**). This external (epizoic) parasite has the proboscis inserted in the echinoderm body and is feeding on its body fluids. As the sea star is a burrowing species, this small parasite has probably to spend part of the day below the sediment surface, according to the sea star feeding behaviour. (Photos S. Schiaparelli).

of Naturalis, Leiden), unexpected radiations have been discovered in groups involved in associations (Fig. 509).

Another way to define partnerships between two species, decoupled from the quantification of relative costs and benefits, is the topological description of the position of the "symbiont" relative to the host. In this case, a series of descriptive terms such as epizoism, endoecism, inquilinism and phoresy can be adopted to indicate — respectively — those associations where one organism gains protection by living on (Fig. 510), in (Figs 511 & 512), together with (Fig. 513), or being transported by another one. All these terms are sometimes encompassed by the word "aegism". Instead, if the "symbiont" lives within the cytoplasm of its multicellular partner, as in the case of lichens or the stony coral with zooxanthellae, it is used the word "endosymbiosis".

In multiple associations (Figs 514 & 515), partnerships involve more that two species. In fact, cases of a host harbouring more than one symbiont are not rare, especially if the host is large and has a three dimensional shape (as gorgonians, corals, anemones and sponges) which provides shelter and refuge.

One of the most striking phenomena implied in many associations is the use of particular colouration by the participants. The need for camouflage requires that the guest species mimics the colour (and often the shape) of the host organism, sometimes with spectacularly cryptic results (Figs 516-519).

Figure 511: **A**: In other cases, as this one, the mollusc shell becomes completely embedded in the host skin and hardly visible from the outside (arrow). **B**: An 'unusual' hole in the skin of the holothuroid *Bohadschia argus* reveals the presence of the endoecious parasite: the eulimid *Prostilifer subpellucida*. **C**: Once part of the skin has been dissected, it is possible to observe a couple of eulimids, a male (the small one on the right) and a female. (Photos S. Schiaparelli).

In some cases, analogously to what is better known in butterflies, or in nudibranchs, the coloration is quite probably aposematic, bearing a message of toxicity or disgusting taste for predators (Figs 520-521).

Scientists interested in animal associations have derived from the participation in the Santo 2006 expedition and from the first study of the collected data and samples, a reinforced conviction that in the next years, the accumulation of data on these marine partnerships, will shed new light on the functioning of such highly diverse ecosystems.

Figure 512: Within muricids only a few coralliophiline groups have exploited endoecious life styles. Among these, the genus *Rapa* makes galls in the tissues of the soft corals *Sarcophyton* and *Lobophyton*. **A-C**: As all Coralliophilinae, *Rapa rapa* (here a group infesting a colony of *Lobophyton*) is a protandric hermaphrodite. The small white snail pointed by the arrow (**C**) is the dwarf male that lives in the same gall with the large female. (Photos S. Schiaparelli).

Figure 513: This galatheid crab, *Allogalathea elegans,* is perfectly camouflaged and live on vagile crinoid echinoderms: a typical case of phoresy.

Figure 514: Ovulidae, or egg-cowries, is a family of specialised browsing carnivorous molluscs that feed on polyps and tissues of Anthozoa, especially soft corals and sea fans. This species, *Phenacovolva rosea*, is here part of a multiple association centred around a large host, a gorgonian of the family Plexauridae. This octocoral is fully covered by minute brittle stars, whose arms also embrace the mollusc. The small ophiuroids are "kleptocommensals" (they steal food particles that become entrapped by the octocoral mucous) of the gorgonian. Small shrimps (not visible at this magnification) have also been collected from this gorgonian.

Figure 515: Hermit crabs use empty gastropod shells for protection. Often they actively attach sea anemones on the shell to enhance protection, benefiting from their presence and probably supporting (by food residuals escaping manipulation) the anemone by food. Deep sea dredging at Vanuatu recovered specimens of the carrier shell *Xenophora pallidula*, with hormathiid sea anemones attached on the shell and several coralliophiline gastropods (16-17 mm long) parasitizing the sea anemones in a curious and unusual "chain of associations", Vanuatu [MUSORSTOM 8: stn CP 1087]. The undescribed coralliophiline species has been named *Coralliophila xenophila*, a combination of the words *Xenophora* and *Coralliophila*; also, the Latin adjective *xenophilus, -a, -um*, means friend of the stranger. Hence, "living on something strange, foreign" refers to the strange association between the coralliophiline, the carrier shell and the hormathiid sea-anemone. (Photos M. Oliverio).

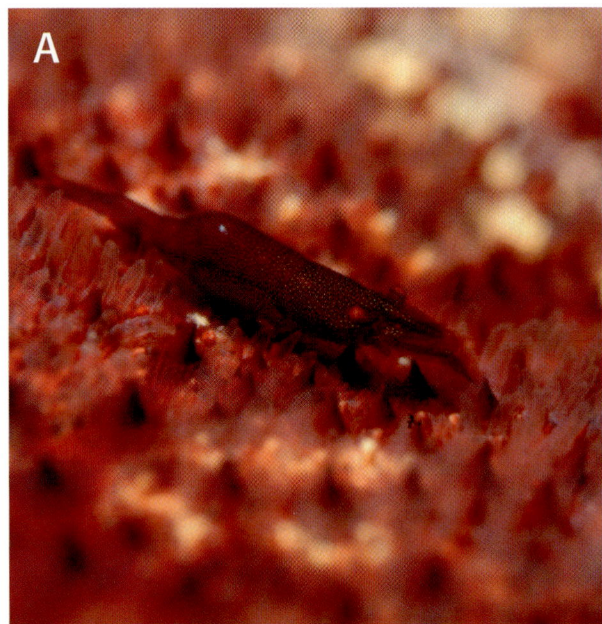

Figure 516: In this case, the ovulid *Prosimnia piriei* can be hardly detected at the centre of the picture (**A**, arrow) within the gorgonian (*Euplexaura* sp.) branch network. The shape and colour pattern of both shell (**B**) and mantle, perfectly match those of the host, probably due to phenomenon of alimentary homochromy. Scale bar is 1 cm. (Photos S. Schiaparelli).

Figure 517 : On asteroids one can often find small shrimps perfectly camouflaged with the colour of their host. These often belong to the species *Periclimenes soror*. During the Santo expedition the species has been found on various sea stars like the cushion star *Culcita novaeguineae* (**A**), *Choriaster granulatus* (**B**), *Linckia laevigata* and the Crown-of-Thorns *Acanthaster planci*. (Photos C. Fransen).

Figure 518 : In the hippolytid shrimps, few species are known to associate with cnidarians. Here *Hippolyte commensalis* is photographed on a on soft coral (*Xenia* sp.).

Figure 519: *Calpurnus verrucosus* is an ovulid that feed on soft corals of the genera *Sarcophyton* (as depicted here) and *Lobophyton*. In **A** the mollusc is visible at the centre of the picture (arrow), crawling and feeding at the soft coral edge. The colour pattern of *Calpurnus verrucosus* is unique (**B**): differently to other ovulid species whose camouflage pattern resembles the extended polyps of the host, in this case it is the other way round, and the texture of *Sarcophyton* with retracted polyps has been successfully imitated. The shell (**C**) has no pigmentation. Scale bar is 1 cm. (Photos S. Schiaparelli).

Figure 520: Few species of ovulids, as *Crenavolva tigris* (**A**) (here depicted on a colony of *Euplexaura*), show an aposematic coloration which should alert potential predators about the unpalatability of the mollusc. The shell of *Crenavolva tigris* is monochromatic (**B**), as in most ovulids, the aposematic colouration being entirely due to the mantle. Scale bar is 1 cm. (Photos S. Schiaparelli).

Figure 521 : *Periclimenes imperator* is as spectacularly coloured as peculiar is its host choice: it is known to live on various nudibranchs as well as on sea cucumbers.

SEASLUGS: THE UNDERWATER JEWELS OF SANTO

Yolanda E. Camacho & Marta Pola

"Opisthobranchs are to the mollusks what orchids are to the angiosperms or butterflies to the arthropods" – Thompson, 1976.

Balancing on the bow of the USS *President Coolidge*, a shipwreck 38 m below the surface of the sea, our minds swim with images —not of sharks or big intimidating dugongs, but of the treasure we seek. We search the wreck for just a few minutes when the first jewel of the evening lights up in the beam of our flashlights: *Hypselodoris maritima*, a yellow-gold sea slug twice the size of a precious pearl (Fig. 522). Called "Kala Blong Solwata" in Bislama, the primary language of Vanuatu, sea slugs are some of the most beautiful gems in the world's oceans.

Figure 522: *Hypselodoris maritima*, one of the many colorful opisthobranchs found in Santo.

••• What is a sea slug?

Sea slugs such as *Hypselodoris maritima* are fascinating soft-bodied molluscs, commonly known by scientists as "opisthobranchs". Related to snails, opisthobranchs are not well known among shell collectors because they have evolved a reduced, internalized, or completely absent shell (Fig. 523). Opisthobranchs without any trace of a shell are called nudibranchs, or "naked molluscs", and they constitute the vast majority of sea slugs.

Opisthobranchs have received much attention, however, from amateur naturalists and underwater photographers due to their vivid colours and dazzling array of shapes and sizes. The vast majority of opisthobranchs have a specialized feeding structure called "radula". The tongue-like organ is studded with hard chitinous teeth, which the slugs use to rasp and scrape the substrate on which they feed. However, not all opisthobranchs have a radula. Such species suck up their prey with a muscular tube-like extension of the digestive system. This method of feeding is used by *Philinopsis* spp. to rapidly engulf other opisthobranchs and

by *Phyllidia* spp. and *Doriopsilla* spp. to forage on the tissues of sponges. Most opisthobranchs are fairly specialized in their choice of food.

Opisthobranchs have several club-like feathery structures that are thought to serve as sensory organs. Called rhinophores when located on the dorsal part of the head, or oral tentacles when present on both sides of the mouth, these structures detect chemical molecules in the water and compounds given off by preferred foods, helping the sea slugs locate their food resources. Other appendages that may or may not be present in different species serve as breathing (gills) and digestive (cerata) organs. Gills are feathery and play a

Figure 523: **A:** *Micromelo undatus*. **B:** *Haminoea cymbalum*. **C:** *Durvelodoris lemniscata*, showing different stages of shell reduction or complete absence of the shell.

Figure 524: The aeolid nudibranch *Flabellina rubrolineata*, showing the cephalic tentacles and cerata.

role in respiration. In some groups, the gills may also contain branches of the digestive gland. Cerata are club-shaped or finger-like structures arranged along each side of the dorsal surface. They are present in the suborders Aeolidacea, Dendronotacea, and some species of Arminacea. In some aeolids (Fig. 524), the tips of the cerata contain tiny capsules, or cnidosacs, that house dart-like stinging structures called nematocysts. The nematocysts are in fact produced by anemones and hydroids, upon which some nudibranchs feed. The sea slugs digest the prey, but the nematocysts are shuttled to the sea slug's cnidosacs for use in their own defence.

Most opisthobranchs are simultaneously hermaphroditic. That is, individual animals have both male and female reproductive organs, and partners interchange sperm during copulation. In Santo, it is very common to see mating pairs of *Chromodoris elisabethina* and *Risbecia tryoni* (Fig. 525).

Opisthobranchs are important study animals in different fields of biology. For example, several species of the genus *Aplysia* have been used in neurobiological studies due to the large size of their individual nerve cells. In addition, defensive chemicals of opisthobranchs are being surveyed for potential pharmaceutical use. They are also good indicators of healthy marine environments.

Figure 525: The courtship rituals. **A**: *Chromodoris elisabethina*. **B**: *Risbecia tryoni*. **C**: *Dendrodoris tuberculosa*.

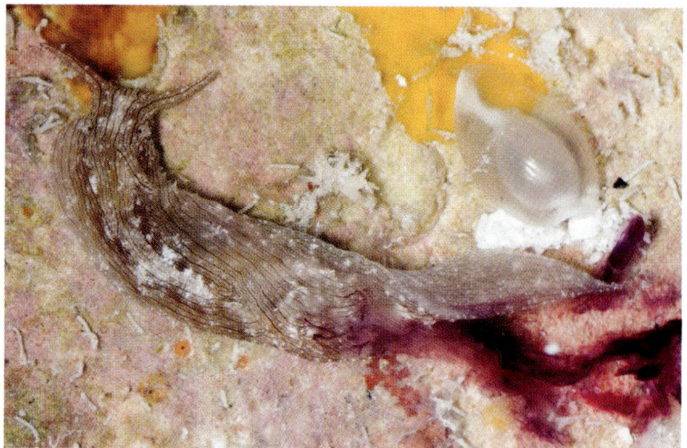

Figure 526: When disturbed, *Stylocheilus striatus* releases ink as a chemical defense.

••• How do sea slugs defend themselves?

Most opisthobranchs shine like jewels and are a striking combination of bright colours, stripes, and eyespots. This striking appearance advertises the sea slug's toxicity to predators, a defensive strategy called aposematic coloration. Without a protective shell or the ability to escape quickly, opisthobranchs have evolved chemical defences (Fig. 526).

Many species, however, are not toxic, but mimic the warning coloration (Fig. 527) of truly toxic species to discourage predators from eating them, a strategy called Batesian mimicry. Sometimes Batesian mimicry may involve other organisms as well, such as flatworms, sea cucumbers, crustaceans, or sea snails. In Santo, for example, both the flatworm

Figure 527: Many opisthobranchs have flamboyant colors and patterns, which may signal their toxicity to potential predators. The eye-catching reds and yellows of **A**: *Hypselodoris krakatoa* and **B**: *Hypselodoris kaname* are examples of such warning coloration..

Pseudoceros imitatus and the sea slug *Chromodoris geometrica* mimic the nudibranch *Phillidiella pustulosa*, which secretes toxins to keep away predators (Fig. 528). The purpose of this kind of mimicry is to fool predators into believing that the mimic has all the toxic chemical defences of the model. Another type of mimicry also exists among opisthobranchs, where a group of similarly coloured species are all toxic. The similar coloration reinforces the message to predators.

Sea slugs have other defence strategies as well. If attacked, they may release a specific body part, such as cerata or part of their mantle, that can readily be regenerated (autotomy), the same way many lizards release their tails. To warn foes that they will fight under threat, they may act aggressively.

Figure 528: Examples of aposematic mimicry. The flatworm **A**: *Pseudoceros imitatus* (left) bears a striking resemblance to its model *Phyllidiella pustulosa* (right). **B**: *Chromodoris geometrica* also appears to be a mimic of *P. pustulosa*.

Figure 529: The sap-sucking slug *Placida* sp., showing a network of ducts that branch throughout the body and store chloroplasts derived from their diet. The chloroplasts, which remain viable, synthesize food for the slug.

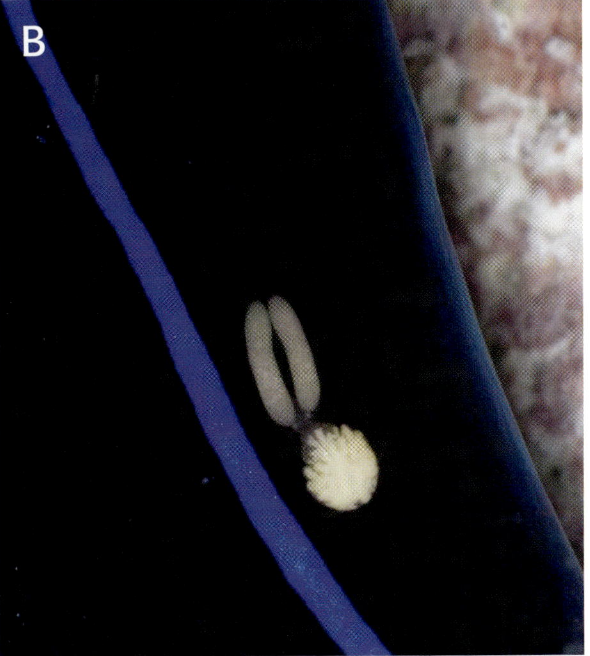

Figure 530: Examples of parasitic copepods on nudibranchs. Egg sacs on the gills of **A**: *Chromodoris lochi* and a copepod on the surface of **B**: *Chelidonura varians*.

They may swim quickly and thrash their bodies from side to side to flash their bright colours or puff up their cerata perhaps to look bigger than they really are. Some sea slugs also emit light (bioluminescence) from organs located on the dorsal surface when distressed, which is thought to also be a defence strategy. Clearly these soft-bodied creatures are anything but fragile and defenceless.

Opisthobranchs often form associations with other species of marine life. For example, some opisthobranchs have evolved symbiotic relationships with microscopic algae called zooxanthellae. The algae are protected from predators by living within the tissues of the opisthobranchs and in return offer their hosts important sugars and nutrients that they produce via photosynthesis (Fig. 529). Small crustaceans called copepods have been recorded living on the surface or inside (Fig. 530) several species of nudibranchs.

• • • Where to find sea slugs

Are sea slugs easy to find in Santo? Well, yes and no! You need to know what you are looking for, and where to look. Many opisthobranchs are large enough to be seen easily by even inexperienced eyes, while others are very difficult to find because they may be only a few millimetres long and spend most of their lives underneath rocks. For example, you won't miss the giant forms of the Spanish dancer (*Hexabranchus sanguineus*) or *Asteronotus hepatica* (Fig. 531), both of which can be more than 40 cm long. Nevertheless, you will need more patience and practice to find the tiny *Eubranchus* or *Mexichromis* species (Fig. 532), which are about the size of a grain of rice. Another reason why even an experienced diver could miss some species is because they are nocturnal, only coming out of their crevices to look for food at night. And while most opisthobranchs are brightly coloured, some species are exceedingly difficult to see because their body colours (and sometimes textures) are camouflaged to resemble the substrate upon which they are normally found (a phenomenon called

Figure 531: Giant form of *Asteronotus hepatica* found during a night dive in Santo.

Figure 532: Examples of tiny nudibranchs found in Santo: the aeolid **A**: *Eubranchus* sp. feeding on hydroids, and the chromodorid **B**: *Mexichromis mariei*.

cryptic coloration). Examples of cryptic species in Santo include *Phyllaplysia* sp. and *Atagema* sp. (Fig. 533).

However, if you take the time to look carefully for opisthobranchs in the wild, you will be rewarded with some of the most beautiful biological jewels of the world's oceans. Sea slugs can be found in almost every available habitat in the marine environment, from high up in the intertidal zone to more than 3 000 m deep. A wide variety of these major habitats, as well as microhabitats, are represented in Santo. Good places to look for opisthobranchs are under rocks and pieces of dead coral. In fact, any submerged object may provide sufficient shelter from the sunlight and predators for a large number of species. The most exciting way to find species is to look inside the holes, on, in, among, and under coral rubble. Also, sea slugs lay their eggs in delicate, almost artful ribbons or egg masses. If you see either of these, look closer, as where there are eggs, there are adults nearby. Something important to remember is to return any rocks or coral rubble you move to their original positions once you've finished your search so as to minimize disturbance and

maintain the protective environment for all plants and animals in that microhabitat.

Calm sandy areas can also be good places to look for some opisthobranch species, such as those in the order Cephalaspidea, called "head shield slugs" (Fig. 534). These sea slugs are named after their shield-like heads, which they use to plough beneath the surface of the sand. The vast majority of head shield slugs also possess a shell, though it may be reduced and internal. Cephalaspidean slugs are usually predatory on other molluscs, including opisthobranchs, which are on the surface of the sand. A great number of these species, such as *Chelinodura inornata*, *C. hirundinina*, *C. sandrana*, *C. varians*, *Philinopsis gardineri*, *P. pilsbryi*, and many others can be found in sandy areas of Santo, between reefs or in lagoons.

Other good places to find sea slugs are in sea grass meadows and on rocky reefs covered in algae, tunicates and sponges. These substrates, including vertical walls, provide both food and shelter for a great variety of opisthobranchs and are therefore a good bet for finding sea slugs —especially considering that, as small animals with limited mobility, they tend to remain as close as possible to their food source. But there is a catch: opisthobranchs are as diverse in their food preferences as they are in their colours, shapes, and patterns. Depending on the species, they may feed on algae, sea grass, cnidarians, ascidians, sponges, bryozoans, worms, other molluscs, fish eggs, or even small fishes. But despite such a range of possible prey species, each opisthobranch species is typically very selective in its food choice. The more knowledge you have about the specific

Figure 533: The cryptic species. **A**: *Phyllaplysia* sp. and **B**: *Atagema* sp. resemble the substrate on which they are usually found.

Figure 534: In calm sandy areas of Santo, the headshield slug *Chelinodura varians* lays its eggs in the form of a tube by twisting its body inside the egg mass as it releases the eggs.

food preferences of certain species, the greater the chances you will find them.

The waters of Santo includes massive, healthy coral reefs that are beautiful to investigate. Interestingly, however, these reefs do not have a great diversity of opisthobranch species, because there is in fact relatively little diversity of food resources. Nevertheless, some opisthobranch species are reef specialists, several of which are commonly found in Santo. A large number of species in the family Phyllidiidae prefer live coral reefs. Species of this family are instantly recognizable by their warty mantles and lack of dorsal gills. The genera *Phyllidia*, *Phyllidiella*, *Phyllidiopsis*, and *Reticulidia* are also very common on reefs. Among all opisthobranchs, *Phyllidia elegans* and *Chromodoris lochi* are some of the most common species in Santo (Fig. 535).

Although diving in a harbour is not typically one's preference in a diving paradise such as Santo, harbours can be very productive habitats for opisthobranchs. Besides the murky waters, you will find a great variety of artificial habitats, such as glass jars and plastic Coke bottles — all kinds of human trash. Sea slugs, as well as the sponges, tunicates, algae, and bryozoans they eat, take advantage of any and all such microhabitats. Some of the species

Figure 535: Two of the most common opisthobranch species found in Santo waters. **A**: *Phyllidia elegans*. **B**: *Chromodoris lochi*.

Figure 536: *Tambja victoriae* is usually found in the harbours of Santo.

usually found in Santo's harbours include *Tambja victoriae* and *Tambja tentaculata* (Fig. 536).

Finally, in Santo, don't forget to look for sea slugs among those other artificial reefs deep below the surface of the sea —those amazing wrecks like *The Coolidge*!

••• Biodiversity

With more than 3 000 documented species worldwide, and another 3 000 yet to be named and described, the highest diversity of opisthobranch molluscs occurs in the tropical regions, especially in the triangle formed by Papua New Guinea, the Philippines, and Indonesia. Of the 3 000 described species, at least 1 500 are thought to be present in the Indo-Pacific region.

On our expedition, we found more than 450 opisthobranch species in the waters of Santo. This number represents 30 % of the total sea slug biodiversity of the Indo-Pacific. We believe that in Santo there could still be hundreds of species waiting to be discovered. This belief is based on our diving experiences there: we found those 450 species during six short weeks of diving in just a small part of the island of Santo. Moreover, 62 of those species were new to science (Fig. 537). During some dives, we surfaced with 20-30 species (and during other dives, none!).

So if you find yourself searching for opisthobranchs in the waters of Santo, don't be disappointed if you come up empty handed, so to speak. The unpredictable nature of the quest, and abundance or rarity of your findings, makes every dive a unique adventure: The fun is in the finding. And these underwater treasures are worth the journey.

Figure 537: *Plakobranchus* sp. is one of 62 new species to science found during the Santo 2006 expedition.

Please never collect these animals, unless you are part of a scientific team. Opisthobranchs are nearly impossible to keep in aquariums and will die immediately without the proper food and environmental conditions. Admire them and take a picture, but please leave them undisturbed.

Man and Nature

coordinated by Michel Pascal

Times

Pre-European

VERTEBRATE PRE-HUMAN FAUNA OF SANTO: WHAT CAN WE EXPECT TO FIND?

Joseph Antoni Alcover

Because no relevant fossil deposits have been discovered on Espiritu Santo Island to date, the vertebrate fauna found by the earliest human settlers remains poorly known. Aiming to discover such fossils, the Santo 2006 expedition explored *c.* 25 caves and rock-shelters. No pre-human deposits were discovered during this expedition. Such deposits, of which the most relevant are pre-Holocene, are extremely elusive in archaeological assemblages of all tropical Pacific islands. Nevertheless, early humans frequently acted as bio-accumulator agents of autochthonous vertebrate bone remains. Thus, early archaeological sites can contain notable evidence of the autochthonous fauna. During September 2006, several archaeological sites of unknown age were discovered and explored, providing only a few bones of pigs, chicken and fish. The local scarcity of relevant sites partially derives from karstic processes driven by a high tectonic uplift rate and a high erosion rate. Only one site documenting the early human history of Santo has so far been discovered, near Luganville. The rich faunal material dating to the first centuries BC has yet to be analysed.

Thus, the scarce information that can be drawn for Santo pre-human fauna can be derived only from a few remains collected by palaeontologists, and from the extant and extinct fauna of other Vanuatu islands, or from neighbouring archipelagoes like New Caledonia, Santa Cruz, and Fiji.

Species extinction patterns following human arrival in Pacific islands have often been described. These consist of the quick disappearance of several native taxa, perhaps taking only one or two human generations. Vertebrate species that humans consumed (seabirds, large size reptiles, and terrestrial birds) often included vanished taxa. The range of human-related species extinction and disappearance of populations has been huge and has been detected only through the fossil record. The impacts of human arrival on the native fauna were not only caused by species consumption, but also by substantial modification of natural ecosystems through agriculture and the introduction of alien species. Among the introduced species, rats such as the kiore (*Rattus exulans*) emerge as key transformers of natural ecosystems, with a direct influence on the decline and extinction of native species.

The human colonization of Vanuatu was a part of the spread of the Lapita culture. Recent archaeological research has gradually increased the antiquity of human arrival in this archipelago. Currently, it is generally accepted that this occurred between 3 300 and 3 200 BP. This date comes from analyses of charcoal provided by one recent excavation of Teouma on Efate Island. As dates determined from charcoal may be subject to the "old wood" effect, the cited date may be slightly modified in the future, although this effect has been extremely rare on tropical oceanic islands. To date, Teouma is the earliest collective burial site known on islands affected by the Lapita spread. A similarly old site has been excavated on the island of Aore. This coastal camp site has not produced much fauna.

••• Prehuman fossil veretebrates from Santo

••• Amphibians – *Platymantis*
Amphibians are usually considered unable to colonise islands. Nevertheless, the two extant (*Platymantis vitianus* and *P. vitiensis*) and one extinct Fijian species (*P. megabotoniviti*) show that some overseas colonisation occurred in the past. As Vanuatu is midway between the Solomon Islands (the most probable source of the Fijian *Platymantis*) and Fiji, the presence of extinct *Platymantis* species can be predicted for the Vanuatuan fossil record. If the occurrence of *Platymantis* species in the pre-human Santo fauna is demonstrated, its current absence may be related to rat introduction.

••• Reptiles – Iguanidae and *Mekosuchus*
Two extant tree iguanas are known from Fiji, *Brachylophus fasciatus*, which was introduced to Vanuatu and Tonga, and *Brachylopus vitiensis*. A fossil of a giant iguana (*Lapitiguana impensa*) was described from Fiji, and a large iguana (*Brachylophus gibbonsi*) was described from Tonga. Although

Vanuatu and Fiji share some taxa (e.g. *Platymantis*), it seems highly improbable, but not impossible, that iguanas lived on Vanuatu before human arrival. The presence of iguanas on Fiji and Tonga results from a dispersal event from South America, founding a single clade. The distance between Fiji and Vanuatu is among the greatest between tropical Pacific archipelagoes, and colonisation of Fiji and Tonga by American iguanas is considered a biogeographic oddity. For these reasons, iguana presence among Vanuatuan fossils seems highly improbable.

An archaic crocodilian clade that includes relatively small species of several islands of Remote Oceania fauna has been described during the last twenty years. *Mekosuchus inexpectatus* from New Caledonia, *Mekosuchus kalpokasoi* from Efate (Vanuatu) and *Volia athollandersoni* from Viti Levu (Fiji) were described from fossil remains and were all included within the Mekosuchinae, a subfamily with unclear relationships. The discovery of *Mekosuchus kalpokasoi* on Efate and the evidence that Mekosuchinae colonized different islands through over-water dispersal suggest that Mekosuchinae may have been present in the Santo pre-human fauna. It is likely that a *Mekosuchus* species, probably *M. kalpokasoi*, will be discovered in the Santo fossil deposits. If this is indeed the case, this *Mekosuchus* species probably was the top predator on Santo, with birds, small reptiles and fish as regular components of its diet.

••• Birds

Bones of six bird species identified at the species level and a bone belonging to an unidentified Passeriform have been collected in the prehistoric deposit of Malososaba, a small shelter near Hokua on the northwestern point of Santo dated to around 1000 BP). All the identified species belong to extant birds on the island. These findings provide no relevant data on the pre-human bird fauna but indicate that the recorded species have long been present.

For the purposes of the present paper, it is relevant to consider the fossil bird fauna collected on the neighbouring island of Malakula (Table 48). During the last glacial era, Malakula, Santo, Aore and Malo islands formed a single large island of more than 8000 km². The fauna should have been widely distributed through this island, with only small regional differences. Nine prehistoric sites have been discovered on Malakula Island, from which remains of 31 bird species have been collected. Among these, are a still undescribed rail species (a large *Porzana*) and an *Eclectus* parrot, the two fragmented bones having be tentatively attributed to *Eclectus infectus* by Steadman in 2006. *Eclectus infectus*, the second described species of the genus, was recently described from fossils from one Tonga archaeological deposit and may have belonged to the Santo pre-human fauna. The distribution of the unique extant species

Table 48: Birds obtained at archaeological deposits from Santo (site: Malososaba) and Malekula (site: Woplamplam, Yalu, Navaprah, Navapule, Waprap, Woapraf, Ndavru, Wambraf, Malua Bay Cave) after Bedford (2006) and Steadman (2006a, b).

Island Species	Santo	Malekula
Puffinus cf. *gavia*		*
Megapodius layardi		*
Gallus gallus	*	*
Gallirallus philippensis		*
Porzana tabuensis		*
Porzana undescribed sp.		*
Porphyrio porphyrio		*
Heteroscelus incanus	*	
Columba vitiensis		*
Macropygia mackinlayi		*
Ptilinopus cf. *tannensis*	*	*
Ptilinopus greyii		*
Ducula pacifica		*
Ducula sp.		*
Calcophaps indica	*	*
Columbidae sp.		*
Trichoglossus haematodus		*
Charmosyna palmarum		
Eclectus infectus		*
Cacomantis flabelliformis	*	
Chrysococcyx lucidus		*
Eudynamys taitensis		*
Tyto alba		*
Collocalia esculenta		*
Collocalia sp.	*	*
Halcyon chloris		*
Turdus poliocephalus		*
Zosterops lateralis		*
Zosterops flavifrons		*
Zosterops cf. *lateralis*		*
Aplonis sp.		*
Erythrura sp.		*
Coracina caledonica		*
Lalage sp.		*
Gerygone flavolateralis		*
Pachycephala pectoralis		*
Rhipidura sp.		*
Lichmera incana		*
Myzomela cardinalis		*
Phylidonyris notabilis		*
Passeriformes undet.	*	*

of this genus, *Eclectus roratus*, encompasses the Moluccas, New Guinea, the northernmost part of Australia, and islands from the Bismarck and Solomon archipelagos. This highly peculiar parrot species exhibits a reversed colour sexual dimorphism; females are bright red, while males are green. The extinct *E. infectus* had slightly smaller wings than *E. roratus*.

The discovery of a flightless *Porzana* on Malakula suggests that the same species or some close related taxon was also present on Santo. To date, at least 25 *Porzana* species have been recorded from Remote Oceania Islands. Currently, only *Porzana tabuensis* is present in the region and may be described from future Santo fossil records.

The distribution and species combination patterns emerging from studies of tropical Pacific island fossil assemblages suggest that more than one megapode species would have been present in the Santo pre-human fauna. Currently, only *Megapodius layardi* is present on the island, and future fossil records will probably include one to three more species that probably disappeared after human settlement. One can suspect that extinct species were larger than *M. layardi*. It is also reasonable to predict that future fossil record will include two or three species of flightless rails belonging to the genera *Gallirallus* and *Porzana*, and extinct pigeon and parrot species or vanished populations.

••• Mammals

Except for bats, the pre-human mammalian fauna of Santo lacked any terrestrial species. The pre-human bat fauna probably included more species than the current one, and several species of monkey-faced bat may be described as fossils are found. Six species of monkey-faced bats, all in decline, are currently distributed over old-growth forests of the Solomon Islands (genus *Pteralopex*, five species) and Fiji (genus *Mirimiri*, one species). Some monkey-faced bat are supposed to inhabit unexplored mountain forests of Vanuatu.

••• Vertebrates consumed by prehistoric inhabitants of Santo

The zooarchaeological records: pigs may have been introduced in the earlier phase of the human settlement, as well as the chicken, kiore, and probably *Rattus praetor* (present at several sites of Malekula Island). Moreover, during the earliest settlement phase humans regularly consumed fishes, turtles, shellfish, fruit bats and, to a lesser extent, wild birds, together with introduced pigs and chicken. According to the limited zooarchaeological record, consumption of fruit bats, important during the initial settlement phases, declined later, although it continues. Fruit bats, formerly a major food item, are now a complementary resource. Noteworthy is the total lack of dogs in the zooarchaeological record from Vanuatu.

The vertebrate fauna from Santo and its use by the earliest human settlers remains largely unknown owing mainly to the difficulty of finding suitable sites. The few available data, together with the general patterns of species distribution and extinction in other Remote Oceania Islands, allow only the prediction of future discoveries of new taxa. It is hoped that the very recent discovery of an early site near Luganville will help confirm these predictions.

THE PREHISTORY OF SANTO

Jean-Christophe Galipaud

••• The first Oceanians

At present, Archaeology considers that discovery of the Pacific region by Man was performed through two main distinct processes. The first one took place during the Pleistocene and had concerned the Australian continent, New Guinea and the close archipelagos joined up by the past in a single continental landmass called Sahul. The second one began around the fourth millennium BP and ended with the discovery of the Pacific Islands followed by human settlement. The first sea crossing from South East Asia to Sahul took place 80 000 years ago, maybe earlier. Human presence is attested 40 000 years ago in New Guinea, and 10 000 years later in the Bismarck Archipelago islands, east of New Guinea. This first settlement, which developed slowly, has allowed man to reach the Solomon Islands up to Guadalcanal, thank to past existing natural bridges between islands and thank to seafaring techniques which allowed crossing at sight. There is a strong correlation between the distribution of the Papuan languages and this very ancient colonisation.

The marine gap between the Santa Cruz Islands and the smaller archipelagos of Vanuatu and further east to the Fiji Islands was only crossed about 3 200 years ago, when nautical knowledge enabled deep-sea navigation. This natural border in the Santa Cruz Islands divides Near Oceania, inhabited for at least 30 000 years from Remote Oceania (Fig. 538).

The discovery of Remote Oceania archipelagos, as far as Samoa and Tonga, was performed by very small mobile groups ultimately from Southeast Asia who appeared in the Bismarck Archipelago islands about 3 500 years ago and quickly spread

Figure 538: The second step of human diaspora into the Pacific: the discovery of Remote Oceania. Dates shown are mean accepted values for the settlement of each archipelago. (Carte J.-C. Galipaud - IRD).

towards the South and the East reaching Samoa and Tonga 3 000 years ago.

Their presence is noticed in the Santa Cruz Islands and in the North of Vanuatu 3 200 years ago and we believe that this area has been an advanced base for their later movements. The distinctive pottery they left, called Lapita, marks out their peregrination. The Austronesian languages are the only spoken languages in Remote Oceania and are logically associated with the Lapita diaspora (Fig. 539).

••• The discovery of Santo

The settlement of the Vanuatu islands follows the settlement process of other Remote Oceania islands. In Vanuatu, Lapita sailors favoured the small offshore coral islands of South Santo and North Malakula as well as Efate and Erromango (Fig. 540). They locally produced a very characteristic and richly decorated pottery, the Lapita pottery. The North Vanuatu islands offered many resources to these discoverers and were the melting pot where they became acclimatised to their new world.

These sailors, being used to marine environment, first preferred coral island's beaches, which offered marine resources and havens for canoes. They exploited local resources, but quickly introduced plants and animals for a more lasting installation. Among transported plants were probably many trees with edible nuts, fruit trees including banana, and

some variety of taro or yam. The animals imported were the Pacific rat (*Rattus exulans*), the chicken (the *Gallus gallus* domestic form) and maybe the dog (the domestic form of *Canis lupus*); the pig (the domestic form of *Sus scrofa*) is absent from the very early settlements associated with decorated Lapita pottery, but becomes, after some centuries, an essential part of the fauna in archaeological sites. The large spiny rat (*Rattus praetor*) is very rare in Lapita settlements beyond the Santa Cruz.

Nevertheless, these first seafarers did not neglect the resources of large islands like Santo where they settled after few centuries. They preferred coastal environments such as the Shokraon site in Luganville, one of the few well-preserved sites of this period in Santo. At the beginning of our era they left the coast to move inside the island.

From Man arrival, 3 200 years ago, until the 350 BP first contacts with the European world, these populations had modeled Santo, adapting their society to the specificities of a diversified and changing environment.

••• Sailors of the New World

Many Lapita sites revealing the beginning of human colonization of the islands are found in Vanuatu and they are especially numerous in Malo, Aore, and in the South of Santo. Dozens of sites have been located along the north and east coast of Malo and around

Figure 539: Lapita pottery fragment from the Makué archaeological site in Aore Island (North Vanuatu).

Aore. All these sites are now buried under approximately one meter of sediment on uplifted terraces that are several meters above today's seashore. Such situation is due to uprising episodes that affected these islands during the last three millennia.

The site of Makué, North of Aore, in front of Luganville is a well-kept testimony of the initial populating period. Discovered then excavated between 2002 and 2006, it entailed several successive settlements, which took place at the very beginning of the human installation in the island. The oldest layers give evidence of a seasonal camp from sailors from the Bismarck Archipelago, 2 000 km further north. Many obsidian flakes, which have been traced successfully to the region of Talasea in New-Britain attest of the origin of these first discoverers of the Vanuatu shores. The large amount of marine turtle and shells remains, of large size, give evidence of the marine economy of these first Vanuatu inhabitants. The successive levels of occupation (Fig. 541) suggest more a landing place that was used during a period not exceeding a few hundreds years, rather than a coastal village occupied for many centuries. The varying decorative patterns of the Lapita pottery in the layers also suggest multiple origins for these first inhabitants (Fig. 541).

After few centuries, the Lapita pottery disappears while these sailors remain and settle down in most of the islands. During this founding period, they remain close to the sea, a doubtless indication of a widely marine economy and of a high mobility. At the beginning of the Christian Era, most of the shore sites are deserted.

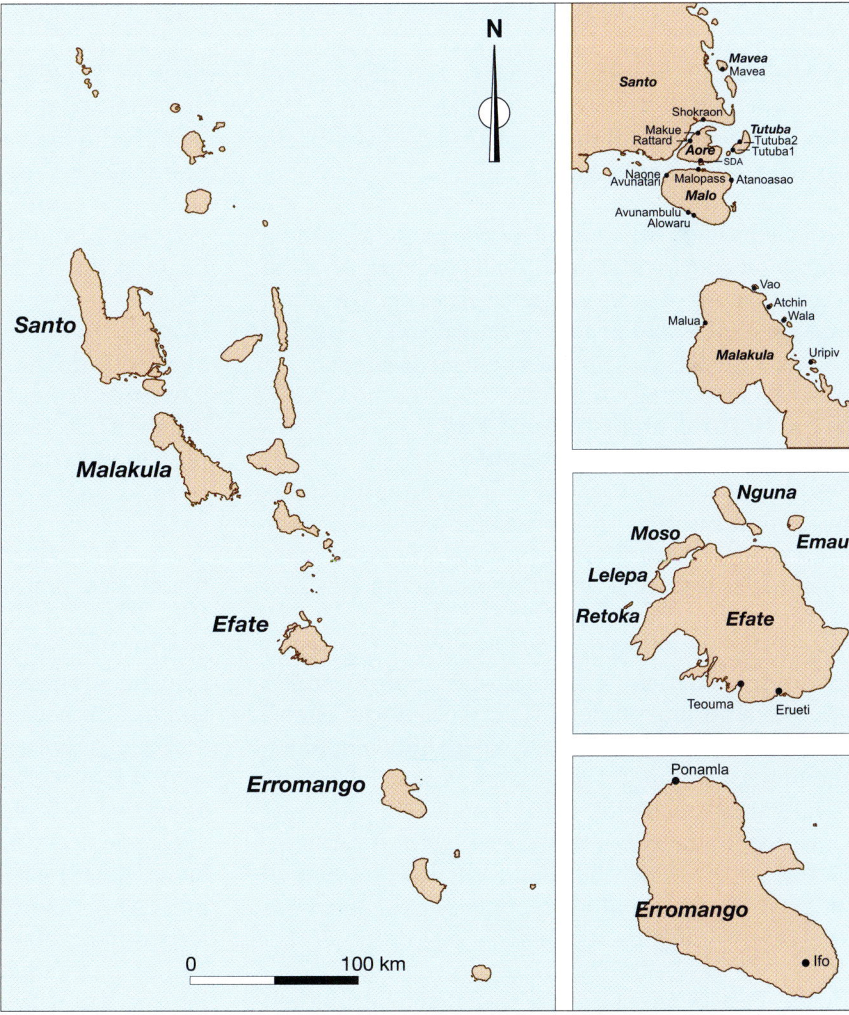

Figure 540: Location of the main Lapita sites in Vanuatu. (Carte L. Billaut - IRD).

Figure 541: The three Lapita layers of the Makué site from Aore Island (zones one to three).

Figure 542: The Shokraon late Lapita site from Santo Island.

Few evidences of this founding settlement were discovered on the big island of Santo whereas all the smaller offshore islands close to the south and east Santo coast were peopled 3 000 years ago. This large island must have been nevertheless visited and its resources used. For example, we find in the Lapita site of Makué, in the North of Aore, rocks originating from Santo.

The site of Shokraon (Fig. 542), in the suburbs of Luganville, is the only site known to have some traces of the Lapita period in Santo. The Lapita pottery in the site is however very rare and the main archaeological finding at Shokraon is a simple shape undecorated pottery, which is associated with numerous pig remains. These remains characterise the period, following the island discovery by Lapita, a period during which the colonists begin to settle down on the island.

The sailors and potters of the Lapita period had some impact on the marine and land resources of these islands in spite of their occasional presence. The introduction of new species such as rats and pigs, certainly had disastrous effects on this virgin environment.

••• Natural environment and anthropisation in Santo

Santo is the largest island of the Vanuatu Archipelago and one of the most ancient at the geological scale with Mallicolo. It is composed of a volcanic substratum on the West and an elevated reef formation in the East. In the North a big bay, where the river Jordan flows, separates these two geological entities. This singular geology influences island morphology. The West of the island is a high mountainous range where Tabwemasana, the highest summit of the island, peaks at 1 700 m. Strong and steep slopes limit access to the west coast where the coastal plain is often non-existent. In the northwest of the island, the lower and well irrigated Cumberland Plain offers places more convenient for human settlement.

In the East of the island, a succession of uplifted reef formations shows the role that tectonic forces play in the island build up. Off this east coast and off the south one, many coral islands stretch.

The centre of the island and the foothills of the volcanic massifs block clouds brought by trade winds and are well watered supporting a tropical rain forest while the much drier, west coast, shelters a vegetation scattered by savannahs excepted in the alluvial valleys.

The two largest rivers, Apuna and Ora (or the Jordan) spring up in the wet centre of the island and run into Big Bay. During flood season, their rate of flow and stream speed are so high that a dense system of alluviums and plant waste reaches a depth of more than 800 m in the bay. All along the centuries these rivers built up a wide alluvial plain, which is very fertile and convenient for agriculture but may be flooded both by fresh and sea-water during tropical gales.

Man settled first and foremost in the plain of Big Bay and along rivers. On the West part of the island, he preferred wetter and cooler high altitude places and apparently did not occupy the west coast formerly (Fig. 543).

••• The origin of island societies

About two thousand years ago, these first settlers have deserted the coastal environments of Santo and its offshore islands, probably forced by tectonic adjustments, which have remodelled the coastal zone, and in turn have destroyed archaeological evidences. Traces of human occupation dated from *c.* 1450 BP were discovered on the East calcareous terraces and along the West of the Cumberland Peninsula. A large exchange network covering the North of Vanuatu as far as the Santa Cruz islands (Tikopia) took place during the first millennium of our era and is characterised by a painted pottery decorated with incisions named "Sinapupu pottery" (Fig. 544). In Santo, this Sinapupu pottery marks the beginning of the human settlement inside the island.

Until the beginning of the second millennium of our era, remains of human activity are restricted to this pottery. Man has settled the mountainous foothills and the calcareous terraces where fertile soil and heavy rainfall favoured agriculture. The mobility probably remained important, the demography was quite low and the extended exchange networks are an indication of the cultural and linguistic homogeneity of these early societies.

Just before the year 1000, many innovations mark the beginning of a new cycle, announcing modern societies. New types of pottery emerge on Vanuatu islands where clay sources are numerous and abundant (Malakula, Santo) and exchange networks seem to shrink. Small offshore Islands are widely populated and regulate the movement of people and goods between the main islands, especially in the North where a new social order develops: the graded societies.

In Santo, several pottery manufacturing centres appear on the West coast, where clay deposits are abundant (Figs 545 & 546). The diversity in styles is an indication of the diversification of groups and the beginning of regionalism. This evolution is linked with increasing evidences of settlements and the emergence of big villages inside islands and on hillsides in front of the sea. This evolution culminates just prior to the European contact with an obvious increase in demography. Small offshore islands are the residence of powerful leaders who have the control over inter-island exchanges.

At the beginning of the second millennium of our era, irrigated taro gardening was introduced on the well-watered islands and in less sloping places. Such intensive culture allows for an uninterrupted production and generates surplus available for exchanges. It is probably also a necessary innovation linked with a drier climate. Irrigated garden are still in use on the West and North-West coast (Fig. 547).

Figure 544: Sinapupu pottery style (northwest Santo).

Figure 545: traditional pottery from the Wusi Village located on the west of Santo.

Figure 543: The Wusi village located on the West coast of Santo was funded in historical time by people inhabiting the highlands of the Tabwemasana area.

Figure 546: Traditional pottery from the Olpoï Village located on the north-west of Santo.

Figure 547: The Hokua irrigated taro gardens in the North-West of Santo.

During the 19th century, Europeans seriously disturbed this social balance by introducing into these networks new prestigious objects, as well as religion and the imported diseases which led quickly to the decline of the traditional systems.

••• Rank societies and alliance networks

During the first millennium of our era, increasing demography, probably associated with competition for land, induced an evolution of the social order and a new hierarchy based on the merit of chiefs emerged in the northern part of Vanuatu. This "Big Men" society joins in its essence that of New Guinea Plateau. As this system is not at first hereditary, it is the merit of the leader that allows him to stay at the head of the group. These Big Men, or rather High Men as they are called in Vanuatu, are in the centre of a complex social network allowing the entire society to live in peace and prosperity. The leader's merit as well as his capacity as strategist and judge influenced the trust of the group and so his future as a leader. The High Men recognize and support one another in the view to rise in the system during solemn ceremonies, where ritual slaughters of pigs and presentation and exchange of goods allow the transition to a higher rank. It is in fact the ability of these High Men to generate, produce or share with their allies that enables them to rise. Their status within the regional group reflects on the society they control, and encourages the wealth and renown of the whole group.

The study of anthropological texts and the distribution of some archaeological structures suggest that at least two big systems co-existed in Santo during the period we are interested in (Fig. 548). The mountainous area of Cape Cumberland and part of the West coast participated in the graded society of the Suqe system; this system coming from the Banks islands was characterized, among others, by the importance of shell currencies in the exchanges. On the East coast, in the South

of Santo as in the offshore islands, the Sumbwe system was mostly dominant entailing the breeding and the exchange of pigs with overdeveloped teeth. The inhabitants of Malo Island and of some Southwestern villages of Santo facilitated the selection of small-sized bisexual pigs, called Narave, which were very valuable in these exchanges.

In ancient sites of Santo, the Suque system is symbolized by the presence of stone-tables (Fig. 549) on which the High Men climb during ceremonies. The geographical distribution of these tables provides an indication of the extent of this system: they are unknown on the West coast South of Wusi village and they do not either appear in the alluvial plain of Big Bay but are present in ancient villages of the centre of the island.

In the Sumbwe system, large-sized coral platforms with sometimes several steps, mark the celebration places of the grade-taking ceremonies (Fig. 550). These platforms occur specially along the ancient seaside villages on the East and South coast of Santo, and are often seen in offshore islands. The situation in the North East of the island, around Cape Quiros and in Sakao Island is not very clear. This part of Santo is linguistically different from the Southern coast, and although the rank system is practised there as elsewhere, it is difficult to compare it with the Sumbwe system further South due to the lack of precise information.

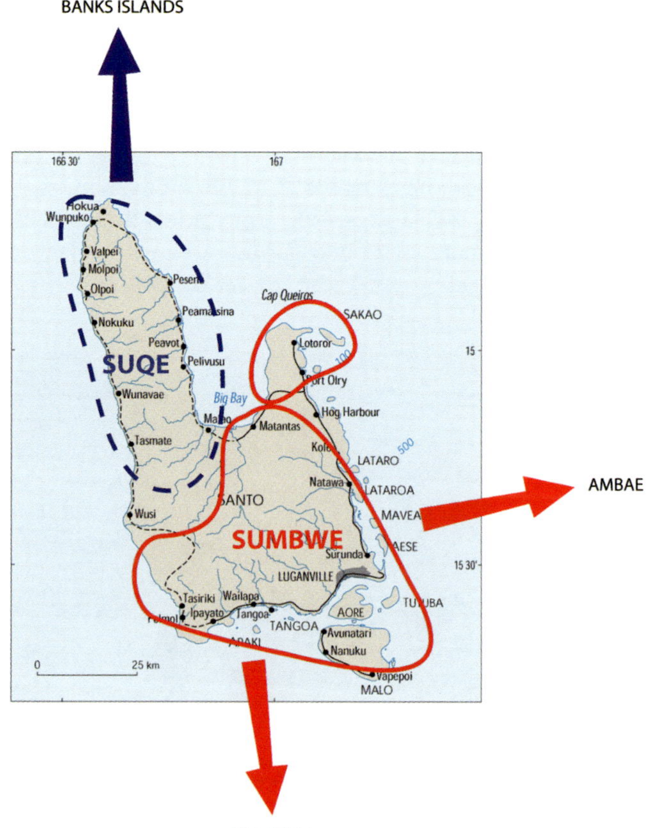

Figure 548: The graded societies in Santo and the extend of the network system during the last millennium. (Carte J.-C. Galipaud - IRD).

Figure 549: Stone table of the Suqe system. Salaea Village located in North-West Santo.

Figure 550: coral platforms of the Sumbwe system. Mavea Island.

In order to keep up and develop, these systems had to maintain large networks of relations to facilitate the movement of goods, but also to allow marital exchanges and therefore insure the perpetuity of the groups. The inhabitants of the Cumberland region had close relations with the inhabitants of the Banks islands and their network extended probably as far as the Santa Cruz Islands at certain times.

East Santo's inhabitants kept relation with people of far East islands. It is the case of Aoba, and Malo people who exchanged pigs and pottery with the inhabitants of the North of Malakula Island.

... Santo chronology in a wider context

The chronology of the human evolution in Santo closely matches the chronology of the evolution in the remaining islands of the Vanuatu archipelago. The initial phase of discovery and settlement, confined to the coastal fringe of the islands is characterised by Lapita pottery. A period of integration follows, characterised by plain ware pottery of different styles and an increase in settlements, which cover all islands. Starting in the second half of the first millennium of our era, during the third phase of this evolution, the roots of pre-European societies emerge.

The cultural representations differ from one island to the other especially towards the last millennium when an increase in demography triggers a heavier reliance to land, which in turn provokes a diversification of cultural markers. This tendency, which is Melanesian wide and not only limited to Vanuatu, is expressed in the following figure 551.

... Conclusion

This rapid overview of the human evolution in Santo and its offshore islands shows that the gradual rise of population led to differing social strategies. These strategies developed in response to cultural and natural necessities but never were the result of isolation. From the Lapita sea-nomads to the potters of the late West Santo societies, movements of men and goods helped shape the people to the islands. Santo, with its natural and cultural diversity, is a good example of this process.

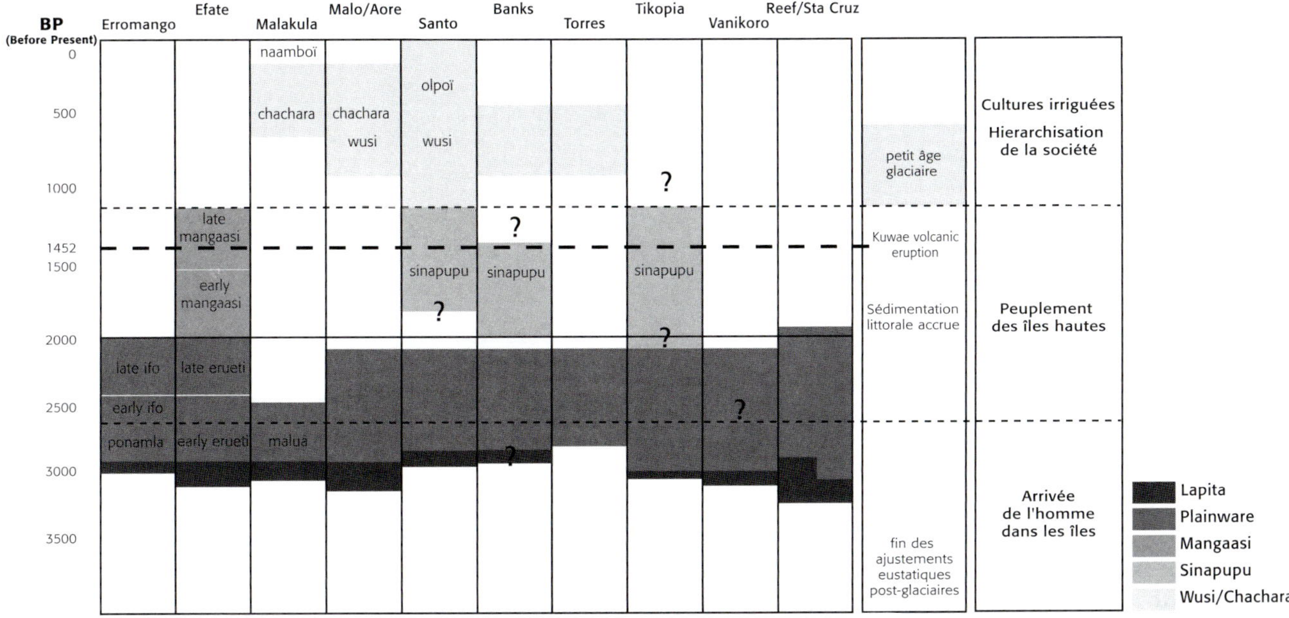

Figure 551: comparative table of the cultural evolution in Vanuatu and adjacent islands. (J.-C. Galipaud).

Biota

OVERVIEW: INTRODUCED SPECIES, THE "GOOD", THE "WORRISOME" AND THE "BAD"

Michel Pascal, Olivier Lorvelec, Nicolas Barré,
Michel de Garine-Wichatitsky & Marc Pignal

It may appear strange to find a theme specifically devoted to common and widely distributed alien species included in an expedition devoted to biodiversity inventories mainly focused on finding new taxa. But, before summarising major reasons for this decision, we must define several essential words used in the following text.

In this text, a species induces a biological invasion when it enlarges its initial distribution area and builds up self-sustaining population(s) in invaded area(s), with or without human assistance. "Allochthonous" or "alien" are adjectives often applied to populations that are established in invaded areas. "Invasive" is an adjective used for alien populations that are introduced by humans and have detrimental effects on biodiversity and ecosystem functions.

● ● ● Biological invasion: a very old process

Since life appeared some three billion years BP, biological invasions have occurred. Charles Darwin in 1859 mentioned this phenomenon as one of the processes contributing to evolution. Evidence of natural invasion is apparent when one considers the native flora and fauna constituting the biodiversity of "true" islands (or oceanic islands) which have never been in contact with continents during geological times.

New Caledonia, an island that is geographically close to Vanuatu, offers a remarkable and clear example of an extraordinary biodiversity produced by biological invasions. Part of the continent of Gondwana that drifted from Australia, New Caledonia has a geological history that was especially tumultuous after the Cretaceous period. Total submersion during the Paleocene was followed in the Eocene by the slippage of the small tectonic plate carrying New Caledonia beneath the sea floor. Rocks of the seafloor covered the plate, parts of which rose above the sea again during Pliocene mountain-building. All terrestrial life disappeared during submersion, and the island has been totally isolated from continents since these events. The extraordinary diversity today results from radiations of species from Australia, New Zealand and the Melanesian archipelagos that have invaded the island since it emerged.

● ● ● Biological invasion: a process very recently under human control

Although biological invasions may be a natural process, humans have significantly interfered with this phenomenon at least since the Neolithic period and have recently generated a dramatic increase worldwide in the number of biological invasions.

One documented example among several is the increase of the number of vertebrate biological invasions in France during the Holocene period that began 11 200 BP in Western Europe (Fig. 552). From less than one species per century between 11 200 BP and 400 BP, the vertebrate invasion rate has reached 135 species per century between 1945 and 2005 (Fig. 552). Similar trends have been reported for Italy during the Holocene period by Scalera in 2001 and for all Europe between 500 BP and the present according to the recent European project DAISIE (Delivering Alien Invasive Species Inventories for Europe).

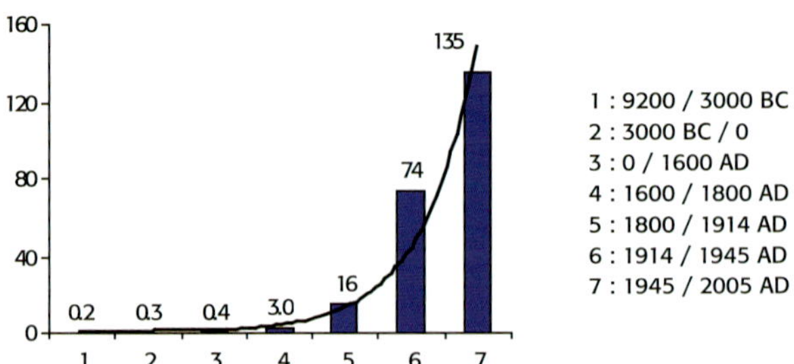

Figure 552: Vertebrate invasion rates in France during the Holocene period (Pascal *et al.* 2006).

If we consider New Caledonia, as an example of a Pacific island closer to Vanuatu than the Old World is, the recent change in the invasion rate is similar to that observed in Europe. The oldest New Caledonian Lapita sites suggest that humans reached the island 4 000 BP. Since then, humans have occupied New Caledonia without interruption until the European arrival with Cook in 1774. During these last four millennia, 42 vertebrate species have invaded the island, four autonomously, 38 as the result of human activities. Only one was introduced by Melanesians, corresponding to a rate of 0.003 introductions per century, and 37 were introduced during the historical period, corresponding to a rate of 14.8 introductions per century.

Vertebrates are large animals with a mineralised skeleton that may form long-lasting fossils. Moreover, many of them are of special interest for man. For these reasons, the level of knowledge of vertebrates is much higher than for any other animal or plant group. Thus, we can assume that invasion rates cited above are the best we can find that will allow us to perceive the role humans played in biological invasion. Nevertheless, we must bear in mind that vertebrates constitute a very limited part of the biological diversity in the world, and we can imagine the total number of biological invasions if the invasion rate estimated for vertebrates resembles that for other taxa.

Moreover, we must realise that, for at least the last six thousands years, throughout the world humans have introduced many plant and animal species that are presently cultivated or bred. Moreover, recent interest in pets and ornament plants has increased the introduction rate, while the shortened travel duration has increased the survival rate of stowaways in planes or in ballasts cargoes and hulls of ships. What are the consequences of these introductions?

• • • Introducing species: the "good"

In western Eurasia, the Neolithic Revolution began 11 000 years ago in the Near East. Recent genetic investigations have confirmed what zooarchaeological data suggested for several domesticated mammals. For example, goats, sheep, pigs, oxen and cats were first domesticated in the Fertile Crescent (Middle East). They were later introduced from this area to Europe by human westward migration. This process yielded the majority of the most important domesticated animal and plant species currently in Europe.

However, there may be multiple initial areas of domestication for some species, as Larson and coauthorshave shown for the pig. These authors suggested that western Islands of southeast Asia may be an independent centre of pig domestication because specimens from New Guinea, Hawaii, Vanuatu and Halmahera belonged to the same distinct cluster in their genetic analysis.

In any event, for a period of several millennia, all these introductions led to the Neolithic Revolution that drove human societies from the stage of hunter-gathers to that of breeders and farmers.

Though the introduction rate for domesticated species in new areas was low until 1500, it increased afterwards with the European voyages around the world. The majority of European cultivated plants seem unable to colonise natural habitats; the suggested reason is that long-term domestication has entailed genetic modifications preventing successful establishment. However, many domesticated animals have proven able to do so. Individuals that escaped accidentally or were deliberately released founded feral populations in many areas. If they are exploited appropriately, these feral populations may be maintained at a level that prevents detrimental effects to the invaded ecosystems. But when the exploitation ceases, or the harvest rate is too low to prevent uncontrolled outbreaks of the populations, negative environmental effects occur. For example, from the 16th to the 19th centuries, sailors introduced rabbits to Australia and more than 800 oceanic islands, aiming to escape starvation if wrecked. As almost all of these populations were never controlled, they produced dramatic and irreversible perturbations in ecosystem functioning. There are also many cases of feral populations of goats (e.g. Galapagos Islands) or pigs (e.g. Clipperton Island) established by human settlers on Pacific Islands that have caused dramatic environmental disasters.

• • • Introducing species: the "worrisome" and the "bad"

Despite the low rate that prevailed before 1 500, introductions by Palaeolithic and Neolithic populations were consequential for biodiversity. 35 000 years ago, humans colonised the Bismarck Archipelago in Near Oceania, and after 19 000 BP they introduced seven mammal species there. Among these aliens were two rodents, the Pacific rat (*Rattus exulans*) and *R. praetor*, the dog, and the pig. Since these introductions, 11 bird species at least that were identified by palaeontologists have disappeared and six among these 11 are currently considered extinct. The first extinction occurred 14 000 years ago, suggesting a direct relationship between alien mammal introductions and extinctions. Nevertheless, there is not enough evidence to say that these introductions were the only or major cause of extinction. More recently, 3 500 years ago, Human began to colonise Remote Oceania islands, and introduced the Pacific rat, in many places, probably as a deliberate action. This rodent together with the pig and the dog, which were also introduced, contributed to 127 land bird species recorded extinction soon after human arrival. Moreover, in its recent and exhaustive review Steadman wrote in 2006: "With more data, I now would estimate, conservatively but still crudely, that 559 to 1 696 species of non passerine landbirds and 82 species of passerines have been lost since human arrival in Oceania, excluding the Hawaiian Islands and New Zealand."

Other parts of the world did not escape this phenomenon. Except for bats, two endemic shrews and one endemic mouse, all mammals currently present on large Mediterranean islands were introduced, and all the other native mammals are presently extinct. Similarly, except for bats, all the terrestrial mammals currently present in the French West Indies were introduced and all the native ones are now extinct.

Of course, detrimental effects of biological invasions dramatically increased with the spectacular increase of introduction rate that started one and a half centuries ago. Moreover, recent introductions and their consequences are better documented than older ones. For example, the Nile perch (*Lates niloticus*), introduced between 1954 and 1957 to Lake Victoria in Africa, caused the extinction by 1970 of 200 of its 300 cichlid fish species, 99 % of which were endemic to this lake. Such disappearances drastically simplified the food chain and modified the functioning of the entire lacustrine ecosystem via cascading effects, with major impacts on human communities.

The previous chapter describing the consequences of Neolithic introductions mentions only the local disappearance or extinction of native species. This is mainly due to the fact that it is difficult for palaeontologists to demonstrate subtler consequences. Recent introductions give us the opportunity to do so.

Some biological invasions have had strong economic impacts. For example, the introduction of the potato blight and phylloxera during the nineteenth century from North America produced the last European peacetime famine in Ireland and a 75 % drop in wine production in France respectively. More recently, the economic impact of biological invasions in the USA was estimated over 137 billion US $ per year. Unfortunately, similar assessments have been conducted for few other countries.

Some biological invasions have had strong impacts on human health. One well-known case is the European Middle Age black plague outbreaks, which were consequences of the *Yersinia pestis* bacillus introduction. This bacterium was spread by the rat flea vector (*Xenopsylla cheopis*) and ship rat reservoir (*Rattus rattus*), both introduced previously to Europe. More recently, the tse-tse fly that hosted the parasite of trypanosomiasis was introduced to Principe Island at the end of the nineteenth century. In less than twenty years, the disease threatened the extinction of the human

population of the island. Even more recently, during the 1990s, the tiger mosquito (*Aedes albopictus*) was introduced from eastern Asia to the USA, and then to many parts of the world by trade in used tires. This mosquito was the vector of the viruses responsible for dengue and chikungunya fever. The 2005-2006 chikungunya outbreak in the Indian Ocean affected more than half of the population of Réunion Island. The number of examples could be amplified, as emergent diseases and their vectors are currently considered as one of the major global threats to human health.

Examples mentioned above are simple because they involved only introduction of pathogens, vectors or reservoirs. More sophisticated are processes that involve co-evolution. A mammal or a bird species may be introduced totally or partially deprived of its specific pathogenic parasites, bacteria or viruses. Native parasites, bacteria or viruses of the invaded area may infect the introduced species. The new pairs of host and pathogens will evolve, possibly increasing or decreasing the impacts of the disease agents on wildlife, human or livestock species.

The above examples enlighten the potential impact of biological invasions on biodiversity through the disappearance or extinction of native or endemic species. Sometimes, the impact is more drastic, disrupting ecosystem functioning. The example of Lake Victoria ecosystem disruption following the Nile perch introduction began with the extinction of many native species. But species extinction is not necessary to cause huge changes in ecosystems. On Christmas Island for instance, the introduction of the crazy ant (*Anoplolepis gracilipes*) followed by the introduction of a sucking insect led to a large decrease in the number of native terrestrial crabs. As these crabs play a major role in recycling dead material produced by the tropical rain forest that covered the island, the entire insular ecosystem was modified.

••• Why are insular ecosystems more sensitive to biological invasions than continental ones?

Most of the above examples took place on islands. Indeed, island ecosystems appear to be more sensitive to biological invasions than continental ones, although the nature of this vulnerability is still debated. According to MacArthur and Wilson in 1967, the number of animal and plant species present on a "true" island is lower than in a patch with the same area on the nearest continent. Several taxa may be totally absent from oceanic islands. For example, many islands lack terrestrial mammals. Because of changed selective pressures and lack of gene flow, the endemicity ratio is higher on islands than in continents. Insular endemic species that evolved in the absence of continental predators, competitors and pathogens could be particularly sensitive to the effects of introduced continental species.

Does biological invasion always end in disaster?

Advocates of alien species often use the "concept"of vacant niche, saying that certain alien species have no impact on the invaded ecosystem because they occupy an "empty" place. This argument is fallacious because the empty place concept appears to be rigorously empty itself. Nevertheless, and fortunately, not all biological invasions are catastrophic. Paradoxically, such a felicitous situation has an uncomfortable aspect because science is presently unable to predict accurately which species are likely to become invasive; nor can they yet predict precisely the consequences of introductions except when the species has been introduced previously in a similar ecosystem. This is the main reason why we must be cautious when planning to introduce a species.

Why a theme specifically devoted to aliens in the Santo 2006 expedition?

Previous chapters briefly summarized what biological invasions are, their dramatic and recent increase, and their consequences, showing that alien species may produce major perturbations in ecosystem functioning and biodiversity. In 1989 Diamond named the phenomenon as one of the major processes causing global biodiversity loss. More recently in 2000, Mooney & Hobbs ranked it at the same level as climate change as a cause of global change. As biodiversity was the major goal of the Santo 2006 expedition, it seems appropriate to address a major process interfering with it.

Since the WWII, Santo has remained outside the major trade routes that have introduced many species elsewhere. Moreover, Santo is of special interest with two very different parts. The southeastern portion of the island, calcareous and more or less flat, is subjected to varying degrees of pressure from European settlers, while the western, volcanic and mountainous part hosts small traditional Melanesian villages. This situation offered the opportunity to combine alien species inventories with an assessment of alien invasion along human perturbation gradients.

A brief summary of results produced by the "Fallows & Aliens" theme of the Santo 2006 expedition.

The staff of the "Fallows & Aliens" theme was the smallest staff among those for the four themes of the Santo 2006 expedition. Among the 14 members, the areas of expertise of the 11 scientists were restricted to vertebrates, insects (Hymenoptera and Coleoptera), plants, and sociology and ethnology. The following chapters will specifically address these taxa and subjects, but some general conclusions are possible.

Among the 16 alien vertebrate species recorded in the wild (six mammals, four birds, two reptiles, one amphibian and three fish), six belong to the list of the 100 worst invasive alien species in the world published by the Invasive Species Specialist Group of the World Conservation Union. Among those species are four mammals (the cat [*Felis silvestris*], the pig [*Sus scrofa*], the house mouse [*Mus musculus*] and the ship rat [*Rattus rattus*]), one bird (the Indian myna [*Acridotheres tristis*]), and one freshwater fish (the western mosquito fish [*Gambusia affinis*]). Among species introduced to Espiritu Santo belonging to other taxa that are found on the IUCN list are two ant species (the crazy ant [*Anoplolepis gracilipes*] and the little fire ant [*Wasmannia auropunctata*]), two terrestrial molluscs (the giant African snail [*Achatina fulica*] and the rosy wolf snail [*Euglandina rosea*]), and four terrestrial plants (the mile-a-minute weed [*Mikania micrantha*], the coffee bush [*Leucaena leucocephala*], the lantana [*Lantana camara*] and the blady grass [*Imperata cylindrica*]). One other species on this list that we did not find on Santo but that is mentioned elsewhere in Vanuatu is Brazilian pepper [*Schinus terebinthifolius*]. In any event, we must bear in mind the following warning from editors of the IUCN list: "*Species in the list were selected to illustrate important issues for biological invasions. Absence from the list does not imply that a species poses a lesser threat.*"

We also must bear in mind that several human activities that severely perturb native ecosystems increase invasions of the perturbed areas, whether or not the invaders are native. The specific case of the liana *Merremia peltata*, which is considered a highly invasive species by PIER (Pacific Island Ecosystems at Risk), will be discussed in a subsequent chapter.

Although Espiritu Santo Island has remained outside major trade routes since the last World War, at least 14 species that are known to be highly detrimental to biodiversity when introduced are currently present in the island's ecosystems.

Preventing introduction remains less expensive and more effective than controlling established alien populations. Thus, prevention must be promoted in a global strategy to counter biological invasions. But, in this respect, Vanuatu faces a specific challenge shared with other countries: its territory encompasses several islands, so control of domestic inter-island trade must be added to the classical control of trade with foreign countries. Control of domestic trade is of major importance in preventing introduction of alien species previously introduced to other islands of the country, but also to prevent introduction of species native to one island to other islands where this species does not occur.

FOCUS ON SYNANTHROPIC MAMMALS

Olivier Lorvelec & Michel Pascal

"Synanthropic" refers to animal or plant species associated with human habitats. Several rodent species belonging to the Muridae have been synanthropic since the Neolithic Revolution that drove human societies from the stage of hunter-gatherers to that of livestock breeders and farmers. This document deals with synanthropic Muridae of Espiritu Santo Island.

Vanuatu lacks native terrestrial mammals except for bats, as is true for all archipelagoes of Remote Oceania. Therefore, all terrestrial mammals now present on Santo, except for bats, were introduced by humans. On islands of Remote Oceania, some mammals were introduced long ago when humans colonised these islands, and others were recently introduced by Europeans. Among the Muridae, the Pacific rat, *Rattus exulans* (Fig. 553), was introduced during the first period, while the ship rat, *Rattus rattus*, the Norway rat, *Rattus norvegicus* (Fig. 554), and the domestic mouse, *Mus musculus domesticus*, a subspecies of the house mouse, were introduced during the second one.

Few data are reported dealing with rodents on Santo. In 1928, Baker reported only *R. rattus* from Vanuatu, without specifying which islands. In 1975, Medway and Marshall recorded three rodent species from Santo: *R. exulans*, *R. rattus* and *R. norvegicus*. Having recorded *M. musculus* on Effate and Aneityum Islands, these authors did not record the species on Santo.

Biogeographical and archaeological data suggest that other rodent species may be present or have been present on Santo. The large spiny rat, *Rattus praetor*, a New Guinea species, was probably introduced and is currently present in the Bismarck Archipelago and in the Solomon Islands, two archipelagoes in Near Oceania. Several remains from Remote Oceania archaeological sites have been referred to this species, which is presently absent from Remote Oceania. In Vanuatu, *R. praetor* was found in three archaeological sites of Malakula Island and one of Efate Island, but was absent from Erromango Island sites. Several questions remain about the status of rodent species on Santo. Is the *R. rattus* recorded on Santo the true ship rat, or is it the Oriental house rat, *Rattus tanezumi*, a closely related species? Are both species now present? Is the subspecies of *M. musculus* recorded on Santo *Mus musculus domesticus* or *Mus musculus castaneus*, the Asian house mouse? Once again, are both subspecies present today?

••• Methods for trapping

Trap lines were set (Fig. 555) by the "Fallows and Aliens" team during October 2006 at four sites:
- Luganville town and its harbours, where the main commerical activity occurs;
- The "Vanuatu Agricultural Research and Technical Center" (VARTC near Saraoutou village), where several alien species of interest for agronomy and breeding with their associated alien flora and fauna were introduced more than 50 years ago;
- The "Vatthe Conservation Area", which is close to the Matantas Melanesian village;
- The Butmas Melanesian village, which was first connected by a road to Luganville few years ago.

Each trap line included 20 to 30 trapping stations depending on the sampled habitat; distance between two consecutive stations was *c.* 10 meters. Each station included two live-traps: a Manufrance© trap targeting large rodents such as *R. rattus* and *R. norvegicus*, and an INRA© trap aimed at smaller species like *M. musculus*. *Rattus exulans* was caught by both types of live-traps. Traps were baited with a mixture

Figure 553: *Rattus exulans*.

Figure 554: *Rattus norvegicus*.

Table 49: Sampling effort (**T.d.**: number of trapping devices; **T.n.**: number of nights; **T.e.**: T.d. x T.n. x 2), and number of caught rodents of each species (**R.e.**: *Rattus exulans*; **R.r.**: *Rattus rattus*; **R.n.**: *Rattus norvegicus*; **M.m.**: *Mus musculus*) according to field identification, sites and habitats (**italic**: escaped; *****: hand-caught).

Site	Habitat	T.d.	T.n.	T.e.	R.e.	R.r.	R.n.	M.m.
Luganville	Building	-	-	-	-	-	-	*1
	Harbour 1	20	2	80	0	1	0	0
	Harbour 2	10	1	20	0	1	0	*1+2*
	Total				0	2	0	*1*+3
VARTC	Building	5	4	40	1	0	0	*1
	Coffee	20	4	160	4	2	0	0
	Coconut	29	3	174	1	0	0	0
	Cacao	20	4	160	0	0	0	0
	Melanesian garden	20	4	160	1	0	0	0
	Pasture	20	3	120	4	0	0	1
	Ecotone	30	4	240	1	1	0	4
	Secondary forest	20	4	160	*1*+4	0	0	0
	Total				*1*+16	3	0	6
Vatthe	Melanesian garden	20	3	160	*1*+3	6	8	0
	Ecotone	20	3	160	*1*+2	1	10	2
	Secondary forest	20	3	160	2	1	0	0
	Total				*2*+7	8	18	2
Butmas	Village huts	3	1	6	0	0	0	3
	Melanesian garden	20	3	160	0	0	0	1
	Ecotone	20	3	160	*1*+6	0	1	7
	Secondary forest	20	3	160	1	1	1	0
	Total				*1*+7	1	2	11
Total		-	-	2280	**4+30**	**14**	**20**	**1+22**

Figure 555: Setting a trap line in a Butmas Melanesian garden.

of peanut butter, oat flakes and sardine oil and were checked daily. The number of trap-nights for each habitat of each sampled site is listed in Table 49. Trapped mammals were autopsied (Fig. 556), and information about morphology (weight, corps, foot, and tail length), reproduction (number of fœtuses, placental scars, suckled teats, lactation, length of seminal vesicles and larger diameter of testis) and ectoparasites was collected. Crania and tissue samples were preserved in 90° ethyl alcohol for morphometry and DNA analysis.

••• The recorded species of Muridae

Among the 91 sampled rodents, two were caught by hand and 89 were trapped, but five (the shape and behaviour suggesting the species *R. exulans* and *M. musculus*) escaped when extracted from

Figure 556: Autopsies in the field laboratory in Vatthe Conservation Area.

the traps, and one was represented only by its head, all that its congeners left. Eighty-six specimens were preserved. According to field identification, which must be confirmed by cranial and DNA analysis, the species distribution of these 86 specimens was as follow: 30 *R. exulans*, 14 *R. rattus*, 20 *R. norvegicus*, and 22 *M. musculus* (Table 49).

Our sample contains one species, *M. musculus*, not recorded on Santo. Mice represented 26% of our total rodent sample. Moreover, although present in human settlements as it usually is, the house

Figure 557: *Rattus rattus*.

mouse was numerous in ecotones along paths, and at Butmas, our more isolated and less anthropogenic site.

As is usual on Pacific Islands, *R. exulans* was the most common species (35%) and was present at all the investigated sites and in the majority of habitats.

Rattus rattus (16%, Fig. 557) was found at all sites and was especially abundant in the Vatthe Melanesian garden.

More surprising was the absence of *R. norvegicus* at the Luganville harbours despite the trapping effort in this habitat, which was considered to be favourable for this species. Moreover, we recorded *R. norvegicus* (23%) only at our more isolated sites (Vatthe and Butmas), and its highest abundance was observed in the Vatthe Melanesian gardens and ecotones.

Autopsies led to the conclusion that all trapped animals were in good health, with no evident pathology. Fleas that were recorded on seven rodents (*R. norvegicus* and *R. exulans*) belong to the species *Xenopsylla vexabilis* (species identification by Jean-Claude Beaucournu), which was introduced by humans to almost all the Pacific islands. This is the first mention of this species for Vanuatu. Two tick nymphs belonging to the species *Amblyomma cyprinus* (species identification by Nicolas Barré) were collected on one young *R. exulans*. Future analysis of preserved digestive tracts will offer the opportunity to assemble a list of internal parasites of the Santo rodent community.

••• Conclusions

No unexpected new rodent species were found during our survey. Nevertheless, field examination did not allow us to separate *R. rattus* from *R. tanezumi* and *M. m. domesticus* from *M. m. castaneus*. Therefore, future examinations of the collected samples may change this conclusion. Except for *R. norvegicus*, the distribution of the other species is the classical one reported for other Pacific Islands.

Our stay on Santo was too short to allow us to identify and assess the impacts of the alien rodents on island biodiversity. Nevertheless, according to the IUCN, *M. musculus* and *R. rattus* belong to the list of the World's 100 worst invasive alien species. Moreover, the roles that *R. rattus*, *R. norvegicus* and *M. musculus* play as agricultural pests and reservoirs and vectors of human diseases in tropical areas are well documented. As an example, those species play a major role in leptospirosis, a bacterial disease (*Leptospira interrogans*) of humans and cattle. The prevalence of this human disease in Grande Terre of New Caledonia, an island close to Vanuatu, is 200 times higher than in France.

FOCUS ON FERAL MAMMALS

Michel de Garine-Wichatitsky & Anthony Harry

Feral mammals have seldom been the focus of reports or papers published following scientific expeditions to tropical islands. Apart from the bats, the diversity of mammal faunas of Pacific islands is little studied compared to other taxa such as birds or invertebrates. The attention of mammalogists has often been restricted to a few rat species plus feral populations of domestic ungulates or carnivores. It is only recently that documented studies on the ecology and impacts of feral mammals have been carried out, mostly to improve control or eradication of these "alien invasive species".

Thus, most records of terrestrial mammals made by naturalists during scientific expeditions consisted of occasional sightings of individuals (or their droppings or tracks), located at more or less remote places. They also sometimes referred to reports by local villagers that "wild" dogs (or cattle or cats or goats) occur in places, but acknowledged the fact that the extent to which these populations are genuinely self-sustaining is not clear. This is a crucial point, as populations of domestic animals are considered feral if they do not depend on domestic populations for their survival and if they reproduce independently. This is an additional difficulty when one deals with feral mammals, because it is often impossible to assess the frequency and the relative importance of contacts and exchanges with domestic populations, including supplemental feeding provided by humans and exchanges of mates.

••• Positive and negative impacts of feral mammals in the Pacific

The paucity of detailed information collected on feral mammals during past scientific expeditions in the Pacific is unfortunate, for a least three reasons:
- The feral mammals are widespread;
- They are usually very important (culturally and nutritionally) to the local populations;
- They have major impacts on island ecosystems.

The introduction of domestic ungulates and carnivores by man onto Pacific Islands was initiated centuries ago. Goats, for instance have become established on countless islands worldwide, generally having been introduced by humans to serve as food. James Cook, during his remarkable journey in the Pacific, was an efficient disseminator of pigs and dogs throughout the Pacific. In New Caledonia, the famous navigator is believed to be responsible for the introduction of the first dogs, given as a gift to the local chief Tihabouna, although the "success" of "failure" of this first introduction remains undocumented to date. There are also many examples on Pacific Islands of domestic populations of ungulates (including cattle, sheep [e.g. sheep on Eiao Islands in the Marquesas Archipelago])

that were released and became feral after the failure of a livestock production venture.

The establishment of feral mammal populations on Pacific Islands was not exclusively the consequence of European settlers. Previously, Lapita people introduced at least the pig, the dog, the red jungle fowl and the Pacific rat, *Rattus exulans* (see "Vertebrate pre-human fauna what can we expect to find?" by J.A. Alcover, and "The prehistiry of Santo" by J.C. Galipaud in this section of this book) on the islands they colonised, and many feral populations descend from domestic stocks introduced by the early settlers. On Espiritu Santo, feral pigs are widespread, and it is believed that pig introduction occurred with the first permanent human settlements, prior to the beginning of the first millennium BP.

Domestic mammals, introduced for food, draft-power, leather, etc., have considerable economic and cultural values for the human populations that introduced them, whether native or European. For instance, the pig is the most important domestic animal in the Pacific region, and the cultural and nutritional values of this animal for the people of Esperitu Santo is immense (Fig. 558). We are not going to comment extensively on these aspects during this chapter dedicated to feral mammals, although we will briefly report on some information we collected on hunting practises and on the perception of alien species by Ni-Vanuatu people.

Figure 558: Espiritu Santo domestic pig.

Some feral populations have acquired some positive cultural, economic or nutritional values, often through hunting. This is the case for feral pigs, praised as a prime game species by Melanesian and Polynesian populations throughout the Pacific (Fig. 559). This is also the case for "wild" species of ungulates that have been introduced for hunting, such as deer species (*Cervus* spp.) in New Zealand, Australia and New Caledonia. Rusa deer, *Cervus timorensis*, introduced to the main island of New Caledonia less than 150 years ago, has acquired major cultural, economic and nutritional importance for both Melanesian and European populations. From a management perspective, the human dimension is thus of prime importance when one deals with feral populations.

Feral animals can be among the most aggressive and damaging alien species to the natural environment, especially on islands. In the French Pacific Territories, for instance, invasive alien species are recognised as a major threat to endemic species, ranked second just after habitat degradation. Alien mammals have been demonstrated to have major direct and indirect effects on the native floras and faunas of insular ecosystems. Introduced goats, for instance, are responsible for wholesale impacts on island floras, including altering the structure and composition of plant communities, causing extinction, and accelerating soil erosion. The pig is another devastating species for both plant and animal native species: eating or uprooting tree seedlings, ferns and orchids, breaking open tree-fern trunks in searching for starch, preying on invertebrates and birds, and possibly spreading seeds of several alien invasive plants. However, these impacts have seldom been quantified, even for pigs, and few detailed studies on Pacific Islands have been published.

• • • Observations about feral mammals on Santo

A careful observer walking for 1 km along the streets of Luganville (we sampled six transects for a total length of 6.5 km in October 2006) will on average encounter 5.4 dogs, 4.0 chickens, 0.2 cats and 0.1 sheep. Obviously, the animals encountered would be domestic and do not meet the criteria listed previously to qualify as feral. What is more, they would not, at first glance, match the picture of "the most aggressive and damaging alien species to the natural environment" described above, especially in an urban environment. One should remember, however, that the "gentle pussy cat" turns out to be a very efficient and devastating killer of birds and lizards. Whether domestic or feral, "feline delinquency" has a strong impact on vertebrate biodiversity throughout the world, and there is no reason to think that it might be otherwise on Santo.

If the mammal observer moves out of town to the CETRAV Research station at Saraoutou, he or she is likely to encounter the following animals along a

Figure 559: Feral pig hunting party in the nearby rain-forest of Butmas village.

similar 1 km long transect (we did 16 transects for a total of 8.8 km in October 2006): 52.0 cattles, 0.1 dog, 0.1 rat and 0.1 unidentified flying fox. A single look at the beautiful Charolais cows will convince the observer that this livestock herd is in no way feral and that its impacts are confined to the man-made, and well maintained, pastures of the CETRAV station. The dog is a bit more worrisome from a conservation perspective, as an investigation with the workers of the station led to the conclusion that a couple of "wild" dogs, which had no known owner, were roaming the area. The animal we had seen was apparently in poor condition, and it is thought that these "wild" dogs were in part feeding on the carrion left in a dumping area after the occasional slaughtering of cattle. Although these animals are not feral *sensu stricto*, as they probably maintain contact with domestic dogs and indirectly receive some food from human activities, they are likely to have a significant impact on native biodiversity in forest patches surrounding the cattle station, including native birds such as the megapode *Megapodius freycinet layadri*.

The situation in the Matantas village and the adjacent Vatthe Conservation Area is quite different than that in the Luganville-Saraoutou urbanized area. The mammal observer is likely to encounter the following species of interest (to him) along 1 km long transects, located inside the village and in the forested and grassland areas of the Vatthe Conservation

Figure 560: Domestic goat with movements restrained by a leash, suburbs of Luganville.

Area (we did 27 transects for a total of 30.6 km in October 2006): 1.7 cattle, 1.3 dogs, 0.2 goat, 0.1 unidentified flying fox, 0.1 cat, 0.1 horse and 0.03 pig. All the goats and horses seen along these transects were located inside or in the vicinity of the village, their movements efficiently restrained by either a leash or a well maintained fence (Fig. 560). Two-thirds of the cattle seen along the transects were also domestic stock, restrained in paddocks or leashed by the side of the main roads or trails, and attended by villagers on a daily basis. But the remaining one-third of the observed cattle (two herds, a total 17 individuals), were spotted in a grassland area situated several kilometres away from the village and surrounded by primary forest. Their elusive and wary behaviour clearly indicated that contacts with humans were not welcomed, and local guides indicated that they were feral animals, occasionally hunted to provide meat (see below). The only pig seen during our transects (abundant indirect signs were noted in some areas of the primary forest) was a feral pig, roaming in the middle of the primary forest, who ran away as soon as it detected our presence [11]. This is a peculiar situation for Vanuatu, where the pig is culturally the most important animal, but we did not see any domestic pig in the village of Matantas. This absence of pigs is apparently the consequence of the religious beliefs of a proportion of the population of the village converted to a religion that precludes the consumption of pork. As for domestic carnivores, both dogs and cats were seen inside and in the vicinity of the village. However, we also spotted several dogs and at least one cat (several spoors and droppings were also recorded) at a great distance from the village, in the middle of the primary forest.

11 - Pigs have poor eyesight, but very efficient olfaction and hearing: be silent and pay attention to the direction of the wind if you want to approach feral pigs at close range, something you should not do if you are not accompanied by an experienced guide, as they can be dangerous.

Figure 561: During the West to East crossing of the Cumberland peninsula, a feral cattle was killed.

The dogs were often seen accompanying hunters, and their familiar behaviour, even when we encountered them alone, indicates that they were domestic animals. This is not the case for cats, which were almost certainly feral.

Finally, the mammal observer will find transects in the Butmas region very interesting. Along a standardized 1 km long transect (we did 11 transects for a total distance of 16.0 km), the observer will on average encounter the following mammals: 0.6 cattle, 0.3 dog, 0.25 pig and 0.3 unidentified flying fox. Given the ecological conditions in Butmas (high precipitation, almost continuous forest cover with few open and flat areas), the presence of cattle, whether domestic or feral, was unexpected altogether. It is even more remarkable to notice that all the cattle seen were feral, ranging in remote primary forest areas (spoors and dung were seen in several areas, including on top of very steep tracks). Local hunters indicate that they occasionally hunt these

Table 50: Feral mammals of Espiritu Santo Island, Vanuatu. Latin names (ICZN 2003/Usual synonym), English name, acknowledged feral population (exact location in Santo whenever possible). Sources: 1 Medway & Marshall, 1975; 2 Atkinson & Atkinson, 2000; ** this expedition, direct observation; * this expedition, indirect observations.

Latin name	English name	Feral populations	Remarks
Bos primigenius/B. taurus	Cattle	Matantas * Butmas * Santo [1]	
Canis lupus/C. familiaris	Dog		No direct or indirect evidence of established feral population during this expedition
Capra aegagrus/C. hircus	Goat		No direct or indirect evidence of established feral population during this expedition
Equus ferus/E. caballus	Horse		No direct or indirect evidence of established feral population during this expedition
Felis silvestris/F. catus	Cat	Matantas * Butmas ** Santo [1]	
Ovis orientalis/O. aries	Sheep		No direct or indirect evidence of established feral population during this expedition
Sus scrofa	Pig	Matantas * Butmas * Santo [1, 2]	Feral pig populations appear to be widespread and locally abundant

"wild buluk", and we actually witnessed during our stay the gutting/skinning/butchering/transport of a cow… and tasted the meat (Fig. 561). As for domestic carnivores, the same comments as for Matantas apply for dogs in Butmas. Several dogs were seen during our transects, but they were most often accompanying hunters or their behaviour indicated that they were not feral animals. No direct observation of a cat was made during our transects in the Butmas region. However, we saw recent cat tracks and droppings in the middle of remote primary forest, on top of a mountain, located several kilometres from the village. These indirect signs are most likely attributable to a feral population of cats.

Table 50 summarises the current status of feral mammal populations on Santo Island. It is a compilation of published records of feral species and observations made during this expedition.

• • • Focus on the feral cattle of Matantas

In the middle of the forested area of Matantas, designated as the Vatthe Conservation Area, visitors end up in an open grassland zone, the "White Grass", after an hour-long walk in pristine primary forests. With a bit of luck, a good local guide, and an early morning departure, you might encounter individuals of the feral herd of cattle that roam in the area. And if you don't have the chance to spot the animals directly, there is much indirect evidence of their presence (Fig. 562): tracks, signs of grazing and browsing and cattle dung. Indeed, the density of cattle dung in the western hedge of the open grassland area, which we estimated using line-transects, was nearly as high as in the paddocks of the CETRAV station, where intensive cattle production occurs.

The Chief of Matantas reported to us the history and management of this feral herd of cattle (Fig. 563)

Figure 562: Skull of a feral cattle found in Vatthe Conservation Area.

Figure 563: Domestic cattle grazing in a paddock, Eastern coast of Espiritu Santo.

Figure 564: On site interview with the Chief of Matantas about the feral cattle herds of Vatthe.

during our stay in Vatthe (October 2006). This herd was apparently founded decades ago, following the accidental escape/release of domestic stock from farms held by Australian farmers (near the villages of Tavunapui, Tchuriviu and Hanura). After a short period of defiance towards these frightening animals, the villagers of Matantas began exploiting them through hunting. As opposed to feral pigs, which can be hunted without limits and regulation, hunting of cattle situated on the territory controlled by the Chief (including the "White Grass" area) requires permission granted by the Chief. Feral cattle are usually slaughtered and consumed during the celebrations of the Independence Day of Vanuatu (30th July) and Christmas/New Year's day. Adult bulls are usually targeted, although the Chief might give permission to hunt adult females when the herd is thought to be "too large". The animals are shot by hunters with a gun (when available with appropriate ammunitions), or they are chased to exhaustion by hunting dogs and eventually killed with a spear by hunters.

The Chief, as well as most of the hunters with whom we have been discussing this issue, is perfectly aware of the impacts of the cattle on the vegetation (Fig. 564). Mention is made of several species of grass and sedges consumed by cattle (e.g. *Mariscus* sp., *Pycreus* sp.), including a "water grass" ("*pore pala pala*", sp. N.id) also collected for food by villagers and the introduced big leaf rope, *Merremia peltata*. Trampling by cattle is also mentioned as both a problem, as hoof marks left by cattle when the soil is wet allows the collection of water which in turn favours the development of mosquito larvae, but also as an efficient way of controlling the spread of the "American rope", *Mikania micrantha*. The direct and indirect effects of cattle on the vegetation are thus used as management indicators: hunting is allowed when "the wind passes through the forest" —i.e. excessive browsing of the vegetation at the hedges of the grassland area— but hunting is restricted when the grasses and sedges in the open grassland are overgrown, especially "*pore pala pala*" in the depressions of the ground filled with water.

The feral cattle of Matantas are illustrative of the dilemma faced by conservationists regarding control of introduced species. This population of feral mammals has been introduced in the native forest of the Vatthe Conservation Area, and there is no doubt that this alien species has an impact on the native vegetation, although the extent of these impacts has not been documented. However, the cattle could also have a positive effect by contributing to the control of two major invasive introduced vines (*M. peltata* and *M. micrantha*), but detailed information on the efficiency of this control is again lacking. Moreover, this population of feral mammals has acquired a nutritional and cultural importance, being consumed during major social or religious events. The local population has established some management practices through hunting, which could control this feral population efficiently. Should conservationists bother about these feral aliens?

••• Perception of feral mammals by local people of Santo

During our stay in Santo (October 2006), we also conducted a questionnaire survey on the perception of "introduced invasive animals". We asked 44 adult men from Luganville/Saraoutou, Matantas village and Butmas village, to list the animals they considered to be "introduced invasive animals" (the definition given by IUCN/Invasive Species Specialist Group was translated into bishlama). The results of this "free-listing" exercise are compiled in Table 51. The two species that were cited most frequently and with the smallest rank (e.g. frequently on top of the list) were conspicuous birds, the common myna, *Acridotheres tristis*, and the black-headed munia, *Lonchura malacca*, that were frequently seen at the three sites surveyed. The third one is the giant African snail, *Achatina fulica*, also an abundant species native from Africa and known to have been introduced in many Pacific islands. The first feral mammal mentioned in the list, the pig, is only fourth, despite the fact that it is a widespread and abundant species, acknowledged by local people as having strong impacts on crops as well as on the natural vegetation. The fact that nearly 63 % of the solicited persons did not mention feral pigs in their lists probably reflects the cultural importance of pigs to Ni-Vanuatu people. Most of them consider the pig to have been in Santo for as long as humans, and thus to them this species does not match the definition given for an introduced species. Feral cattle were also mentioned by 23 % of the respondents, whereas only 9 % (four persons) mentioned feral dogs.

As emphasized previously, the results of this questionnaire survey illustrate the crucial importance of the human dimension when dealing with the management of invasive species, particularly with feral mammals. Despite their negative impacts on the native biodiversity, some species have acquired positive cultural, economic or nutritional values in the local population. Species targeted as priorities for control or eradication should be selected according to their perception by local populations, as well as the expected conservation benefits from control.

Table 51: Species (or group of species) cited as introduced invasive species by Ni-Vanuatu adult men of Santo. Results of a "free-listing" questionnaire survey conducted in October 2006 (n = 44).

Latin name	Bislama Name	Frequency	Average Rank
Acridotheres tristis	Pidjin blong buluk, Sako	38	1.816
Lonchura malacca	Bengali	19	2.158
Achatina fulica	Snail, African snail	17	3.412
Sus scrofa	Wael pig	12	2.833
N.I. Insect	Bebet	11	2.727
Bos primigenius/B. taurus	Wael Buluk	10	3.200
Polistes olivaceus	Honet, Guepe	7	2.857
Rattus sp.	Rat	6	4.500
Porphyrio prophyrio	Black pidjin red hed, V'Ndrai	4	2.500
Canis lupus/C. familiaris	Wael dog	4	4.000

FOCUS ON ALIEN BIRDS

Nicolas Barré

On the same sites and habitats where mammalogists studied alien ungulates, carnivores and rodents, we conducted a survey in order to find alien birds species and to assess their abundance and potential interactions with the native birds.

The field study was implemented in October 2006. Five days were devoted to the bird survey in Luganville town and its harbours, and nine days to the Saraoutou village area. This lowland area encompasses the CETRAV agronomic station, where the landscape is dominated by pastures and coconut plantations with patches of remnant dry and half-humid forest. Aside these localities heavily modified by human activities, a four-day survey took place in two less heavily modified areas: the vicinities of Matantas village and the primary forest of Vatthe Conservation Area located at the far end of Big Bay, and the Butmas hilly village in the interior (300-600 m). At each site, we performed bird surveys in the morning and in the evening, respectively, using the point-count method, each count lasting 15 minutes. A total of 128 point counts were performed: 10 in Luganville, 42 in the Saraoutou area, 37 in the Matantas area and 39 in the Butmas area. Distance between consecutive points was at least 300 m. In each locality, checkpoints were distributed as far as possible in the following habitats: Melanesian gardens, cultures, pastures, secondary and primary forests.

All bird species were recorded, but we paid special attention to the alien ones. In this note, we focus our comments on aliens. The status and distribution of native species as well as native bird community composition are depicted in "Terrestrial bird communities" in the part "Terrestrial fauna".

● ● ● List of the birds introduced to Santo

Six alien birds species from Esperitu Santo Island have been reported: a phasianid, the red jungle fowl (*Gallus gallus*); two estrildids, the chestnut-breasted and the black-headed munia (*Lonchura castaneothorax* and *L. malacca*); a ploceide, the house sparrow (*Passer domesticus*); and a sturnid, the common myna (*Acridotheres tristis*) (Fig. 565). Among them and as confirmed by this study and by other surveys, only the fowl, the black-headed munia and the common myna are established on Santo to date. In addition, we recorded the feral rock pigeon, *Columba livia*, in Luganville (a group of 10 sighted in the market) and its suburbs.

It should be emphasised that the number of bird species introduced to and established on Santo (four), and even on Vanuatu (six), is very low compared to the 14 introduced species established on New Caledonia or the 12 in Fiji, but not so low if compared to the at most three in the entire Salomon Archipelago.

● ● ● Origin and abundance of the three dominant alien species

The red jungle fowl is native to southeast Asia. It was introduced for food to the south-western Pacific Islands by the earliest settlers several thousands years ago and this is why this species is now perceived by Santo people as native (see "Focus on feral mammals"). This is the only alien bird that inhabits mature rainforests in Vanuatu; it is also present in secondary forests, scrubs, clearings and boundaries of cultivated patches. It is said to occur up to mid-level elevations (*c.* 500 m) but may be locally common up to 900 m. We recorded the red jungle fowl in the four inventoried sites; those heard in the Luganville area belonged to the domestic form. This favoured game species was abundant in the Matantas area, especially in the Vatthe Conservation Area forests, where its shrilly repeated crowing was detected deep in the primary forest (83 % of the 18 checkpoints). It was also detected at the woody borders of pastures (71 %) and in gardens around the village (42 %), where it may be confused with domestic fowl that are more or less feral. It was also quite common in the primary forests of Saraoutou (31 % of the 13 checkpoints in this habitat) but absent from other cultivated, pastoral or disturbed forest habitats. Probably owing to intense hunting pressure, the bird was rare in the Butmas secondary forests located at the elevation of the village (4 % of the 23 checkpoints) and absent from primary forests at higher elevations (16 checkpoints). We speculate that in most places, the bird does not interbreed with feral domestic chickens and remains genetically close to the Asian ancestor. This omnivorous ground feeding bird may compete with the native and endemic Vanuatu megapode, *Megapodius layardi*. However, both species were recorded simultaneously in most of the forest habitats of the three studied sites with similar frequencies at Saraoutou and Butmas (and a fourfold lower frequency of the megapode at Matantas), suggesting that the impact of this competition —which may have occurred for hundreds/thousands of years— is questionable.

Europeans introduced the common myna (Fig. 565), also called Indian myna, from southern Asia to control insect pests of crops and cattle (ticks). It is one of the most successful introduced and naturalised birds in the world, especially in tropical and subtropical islands. It was collected on Tanna Island by Macmillan in 1935 but is supposed to have been introduced to this island in the 1880s, when a ship carrying caged birds to

Figure 565: the common myna (*Acridotheres tristis*) is the most common introduced bird. It is restricted to human-modified lowlands.

Fiji was wrecked at Lenakel. Introduction to other Vanuatu islands is not precisely dated, but whereas Scott did not observe the species on Santo in 1944, Medway and Marshall recorded it in 1971 in the Segond Channel area, including the Aore and Malo islands, where it was widely dispersed and did not see the bird in the northeastern and northern settled strips of Malo Island. During our visit, the common myna was observed in the four investigated sites. The frequency of this bird decreased from Luganville (100 % of the checkpoints), to Saraoutou (8 % in forest edges, 68 % in secondary growth, 100 % in pastures), Matantas (none in forests, 33 % in Melanesian gardens, 57 % in pasture) and Butmas (only two birds seen at one point located in a Melanesian garden). According to data from all four sites, the distribution of the common myna among habitats is as follow: 100 % of the urban checkpoints, 82 % of the pasture and meadow checkpoints, 33 % of cultivated and secondary open vegetation checkpoints, and 2 % of forest edge checkpoints. It was also seen at low elevations near Tasmate and Penaoru villages, located on the Cumberland Northwest coast. The common myna is typically a synanthropic species closely associated with human habitats, buildings and agro-pastoral habitats. This behaviour pattern combined with its quite recent spread explains why Santo people view the common myna as in the first rank of the introduced species (see "Focus on feral mammals"). Its highest densities —sometimes with flocks and roosts of hundreds birds, as was recorded in the Saraoutou area— are observed at Luganville and around the livestock and coconut plantations all over the southeast Santo coast.

As soon as there are cattle or horses, the common myna will be found in the vicinity. It is called "*pijin blong buluk*" in Bislama, which means "the bird that belongs to the beef". Recorded at an altitude of 300 m in the Butmas area, where it is very rare, this bird prefers lowlands. It is one of the three bird species on the list of the 100 worst invasive alien species in the world. The common myna is frequently accused of excluding native forest birds. In fact, restricted forest occupation by most native birds must be related to their inability to adapt to anthropogenic habitats. The common myna may be in competition —for example for nesting holes— with native species of open habitats. But those native species also take advantage of human activities, and the common myna, absent from primary forests, has no interactions with native forest species in this habitat, even in small patches, whatever the elevation.

The black-headed munia is a cage bird native to Asia, from India to Indonesia and the Philippines. Among the two Estrildidae noted by Bregulla in 1992, it is the only species that "Fallow and Alien" team members observed during the Santo 2006 expedition. In Vanuatu, it occurs only on Santo, where it is said to have escaped from an aviary in the 1960s. From there it spread on its own in 1965 to Auta Plantation on Aore Island, the only place where these authors saw the bird in 1971. This small granivorous and gregarious bird occupies very similar habitats to those of the common myna, but it is more specialised to short grassy areas, shrublands and boundaries of mature pastures. It thrives everywhere from urban areas (present in 70 % of the Luganville checkpoints) to agricultural lands, herbaceous clearings and trail and roadsides. At the Saraoutou site, it was recorded in 60 % of the pasture checkpoints and 5 % of the secondary vegetation and clearings checkpoints. At the Matantas (large flocks) and Butmas (two birds) sites, it was seen feeding on Graminae (*Panicum* sp.) along trails and recorded with a frequency of 17 % and 4 %, respectively. The black-headed munia was also observed at low elevations along the Penaoru River and near Tasmate village located in the northwest of the Cumberland Peninsula. At all sites combined, the black-headed munia was recorded in 70 % of the urban habitat checkpoints, 35 % of the pasture habitat checkpoints, and 7 % of the secondary-open lowland habitat checkpoints. It is absent from forests. We can assume that this bird, which took advantage of the opening of the native habitat and the introduction of fodder Graminae and various weeds, has no impact on the native avifauna, which lacks granivores.

• • • Conclusions

Except for the red jungle fowl, whose favoured habitats are primary and secondary forests, the other introduced birds are adapted to suburban and

Figure 566: The white-breasted woodswallow (*Artamus leucorhynchus*), an insectivore, may compete with the common myna for food and nesting holes. However, this native bird also took advantage of the clearings made by humans in the original landscape.

cultivated open lowlands. The red jungle fowl may compete with the endemic megapode for food and space. However, they have lived in sympatry for a very long time and both species remain quite common, suggesting that at least some important components of their ecological niche do not overlap. So long as the rain forest is preserved, there is no reason to fear that this equilibrium will be altered.

As in New Caledonia and Fiji, the introduced omnivorous common myna and the small seedeating black-headed munia are adapted and restricted to anthropogenic habitats. These species, which avoid any type of forest, were recorded on Santo along a clearly declining gradient of frequency from urban areas, coastal lands, cultivated lowlands and especially livestock grazed pastures, to shrubs, Melanesian gardens and, to a lesser extent, forest clearings. These anthropogenic habitats are distributed at low elevations on Santo, and only a few native bird species favour these habitats. The black-headed munia has a very specific weed/graminae niche and may not compete with any native seedeater. The distribution of the rare endemic royal parrotfinch, *Erythrura cyaneovirens*, is restricted to mountain forest where it has no opportunity to encounter the black-headed munia.

The impact of the common myna may be more questionable. It has a bad reputation globally, but it is not clear how severe its impact is on native species. Its habitat is entirely artificial and the bird feeds mainly on insects and fruits, themselves introduced, or by taking advantage of the anthropogenic environment it favours. However, native birds like the white-breasted woodswallow (Fig. 566), the collared kingfisher, *Todiramphus chloris*, etc., use such anthropogenic habitats. The hypothesis that the common myna competes for nest holes with the rare rusty-winged starling, *Aplonis zelandicus*, a mountain forests species, is unfounded. The species do not share the same habitat. In conclusion, we believe that in the particular case of Santo, opprobrium should be cast on these exotic species with caution, or at least without further evidence.

FOCUS ON INTRODUCED AMPHIBIANS AND REPTILES

Olivier Lorvelec & Michel Pascal

Espiritu Santo Island currently hosts populations of one species of amphibian and eighteen species of terrestrial reptiles.

The one amphibian on Santo is a bell frog (Hylidae: *Litoria aurea*). The reptile fauna encompasses six geckos (Gekkonidae: *Gehyra oceanica*, *Gehyra vorax*, *Hemidactylus frenatus*, *Lepidodactylus lugubris*, *Lepidodactylus vanuatuensis*, and *Nactus multicarinatus*, which we consider distinct from *Nactus pelagicus*), ten skinks (Scincidae: *Cryptoblepharus novohebridicus*, which we consider distinct from *Cryptoblepharus poecilopleurus*, *Emoia atrocostata*, *Emoia caeruleocauda*, *Emoia cyanogaster*, *Emoia cyanura*, *Emoia impar*, which we consider distinct from *E. cyanura*, *Emoia nigra*, *Emoia nigromarginata*, *Emoia sanfordi*, and *Lipinia noctua*), a blind snake (Typhlopidae: *Ramphotyphlops braminus*), and a boa (Boidae: *Candoia bibroni*).

None of these nineteen species is strictly endemic to Santo, but four are endemic to Vanuatu (*L. vanuatuensis*, *C. novohebridicus*, *E. nigromarginata*, and *E. sanfordi*). Four species with relatively restricted insular distributions can be regarded as native in Vanuatu. These are *N. multicarinatus* (the Solomon Islands and Vanuatu), *E. cyanogaster* (from the Bismarck Archipelago to Vanuatu), *E. nigra* (from the Bismarck Archipelago to Tonga), and *C. bibroni* (from the Solomon Islands to Samoa). The distributions of eight other species are larger than the previous ones. Among these eight species, two that are represented in Vanuatu by infra-specific taxa with a restricted distribution can be regarded as native: *E. atrocostata freycineti* (the Solomon Islands and Vanuatu) and the local morphotype of *G. vorax* (Vanuatu and the Loyalty Islands). The six remaining species of this group (*G. oceanica*, *L. lugubris*, *E. caeruleocauda*, *E. cyanura*, *E. impar*, and *L. noctua*) can be regarded as cryptogenic, because the present lack of palaeontological data from Vanuatu precludes determining whether they are native (natural pre-human dispersal) or introduced (human-mediated dispersal). However, in 1999, Austin indicated no genetic isolation of *L. noctua* in Vanuatu, suggesting its introduction. At last, three species (*L. aurea*, *H. frenatus*, and *R. braminus*) were recently introduced to Santo.

Three other gecko species (Gekkonidae), widely distributed in the Pacific, were occasionally seen in Vanuatu in the past. *Gehyra mutilata* is known only in Efate by an isolated specimen collected in 1924 or 1925 that suggests an introduction without naturalization. *Hemidactylus garnotii* and *Hemiphyllodactylus typus* were reported from unspecified islands of Vanuatu by Whitaker & Whitaker on the basis of 1993 personal communications that suggest recent introductions. To conclude this inventory, we should add the 1985 record of *Rhinella marina* (formerly *Bufo marinus*, Bufonidae) from Santo, which suggests an introduction without naturalization.

••• Material and methods

Nocturnal and diurnal visual investigations were performed, allowing captures and pictures of specimens. The few explorations of limited parts of the shoreline yielded no marine species, so the following text refers only to terrestrial ones (including *E. atrocostata*, a foreshore skink) and freshwater species. We limited our collections to forty specimens for alien, cryptogenic or abundant species, and five for others. Those numbers were seldom achieved.

To study alien species along a gradient of decreasing anthropogenic impact, four sites were investigated:
- Luganville town, its harbours, and the "Maritime College";
- The "Vanuatu Agricultural Research and Technical Center" (VARTC near Saraoutou village), where many species of agricultural interest were introduced beginning more than fifty years ago;
- Matantas village and the "Vatthe Conservation Area";
- Butmas village and its surrounding forest, which was first connected by a road to Luganville few years ago. Poor weather conditions prevented our searching in the last site as intensively as the other ones.

••• Results and discussion

Table 52 includes survey data (number of preserved specimens and number of identified species) for each sampled habitat of each site.

We recorded and preserved specimens of seventeen species among the nineteen currently reported (total number: 162). We did not detect *E. nigromarginata* and *G. vorax*. Although a large *Gehyra* was seen on a Luganville lodge wall, we did not assign it to *G. vorax* because of the habitat.

We summarized the number of recorded species in the four sites declined along the gradient of anthropogenic impact, i.e. Luganville, VARTC, Matantas, and Butmas. The figures were 8, 12, 12 and 5, respectively. One should bear in mind that surveys on Butmas were cursory.

The three introduced species previously reported from Santo were recorded during the present expedition. Published data allow us to infer introduction date and potential ecological impact of each of these.

Solem in 1959, using material previously collected by E. Kuntz, seems to be the first author to quote the green and golden bell frog (*Litoria aurea*) in Vanuatu, more precisely on Santo, and Whitaker & Whitaker in 1994 suggested 1944 as the first record for this island (cf. material collected by E. Kuntz). Although Tyler in 1979 cited Fischthal & Kuntz in 1967 as the first to record the species in Vanuatu, we found, as Whitaker & Whitaker in 1994, that this publication did not mention the species, although it is based on material collected by E. Kuntz. Next, the species was collected in 1970 and 1971 on several Vanuatu islands (Aore, Malakula, Norsup, Efate), including Santo

Table 52: Recorded species and preserved specimen numbers for different habitats and sites. **To.**: total. **Luganville**: the capital. **Ha.**: harbour and warehouses. **Tw.**: town and maritime college. **VARTC**: agricultural centre. **Qu.**: living quarters and sheds. **Pa.**: agricultural parcels, Melanesian gardens, and woods. **Co.**: coastal forest. **Un.**: origin unclear. **Matantas**: Matantas village and conservation area. **Fo.**: Melanesian gardens and secondary forest. **Vi.**: village and coastal rocks. **But.**: Butmas village, Melanesian gardens, and secondary forest. **Man.**: mangrove swamp, South-East Santo (S. Samadi). **PS**: number of preserved specimens. **s**: seen (O. Lorvelec). **]s**: seen but species unclear (O. Lorvelec). **h**: heard (P. Bouchet).

Species	Luganville			Luganville					Matantas			But.	Man.	PS
	Ha.	Tw.	To.	Qu.	Pa.	Co.	Un.	To.	Fo.	Vi.	To.			
Litoria aurea		h	h							1	1	s		1
Gehyra oceanica		]s	]s			1		1		3	3			4
Gehyra vorax														0
Hemidactylus frenatus	2	1	3	17	1			18		1	1			22
Lepidodactylus lugubris	4	1	5	4	1	1		6		2	2			13
Lepidodactylus cf. *vanuatuensis*				2				2		1	1			3
Nactus multicarinatus	1	2	3	9	11	15	2	37		1	1	1	2	44
Cryptoblepharus novohebridicus	3		3											3
Emoia atrocostata										2	2			2
Emoia caeruleocauda					1			1	1	2	3	4		8
Emoia cyanogaster					2			2				1		3
Emoia cf. *cyanura*				2	4	9		15	1		1			16
Emoia cf. *impar*		2	2		7	6		13	3	1	4			19
Emoia nigra									s	4	4			4
Emoia nigromarginata														0
Emoia sanfordi		1	1	1	s			1	s		s			2
Lipinia noctua					1			1						1
Ramphotyphlops braminus				15	1			16						16
Candoia bibroni													1	1
Recorded species	4	7	8	7	9	6	1	12	5	10	12	5	1	17
Preserved specimens	10	7	17	50	28	33	2	113	5	18	23	7	2	162

(Big Bay area) in 1971, and recorded in 1993 on Santo (Matantas area). According to Medway & Marshall in 1975, the species was introduced to control mosquitoes, and Tyler in 1979 suggested its 1967-68 introduction to Efate by planters on the basis of testimony. We preserved one adult from Matantas (Table 52 & Fig. 567) and saw larvae in Butmas. Also, the species was heard at Luganville "Maritime College".

L. aurea is native to southeast Australia, where it is considered endangered. *L. aurea* was introduced during the late 19th century to "Grande Terre" (New Caledonia), where it was widely distributed by the early 20th century. However, archaeological clues suggest a previous introduction before historical period. This hypothesis must be considered cautiously. On Grande Terre, *L. aurea* is found in many habitats, especially gardens, ditches, and secondary forests, and tadpoles grow in ponds and slow rivers.

Photo M. Pascal, O. Lorvelec

Figure 567: *Litoria aurea* (Matantas village, Santo, Vanuatu, Oct. 2006, M. Pascal & O. Lorvelec).

The species apparently selects ephemeral habitats lacking predators. In New Caledonia, *L. aurea* is also currently present in the Loyalty Islands, despite the absence of a hydrographic system, and on the Isle of Pines. In addition to New Caledonia and Vanuatu, the species was introduced to New Zealand during the late 19th century and perhaps to Wallis and Futuna. *L. aurea* is carnivorous with a large prey spectrum that includes vertebrates like Australian Elapidae in its native distribution area and skinks (*Caledoniscincus austrocaledonicus*) in Grande Terre, and it is occasionally cannibalistic. In 2002, we saw specimen catching a gecko on a house wall on Grande Terre.

Absent from the specimen reported by Medway & Marshall in 1975, the Pacific house gecko (*Hemidactylus frenatus*) was abundant in Santo and Efate towns twenty years later. Introduction of this conspicuous, edificarian gecko to Vanuatu is recent and probably occurred after 1971. We preserved specimens from Luganville (Table 52 & Fig. 568), VARTC, and Matantas and saw the species on Efate (Port Vila). It was very abundant in VARTC buildings.

Native to south and southeast Asia and Australia, *H. frenatus* was introduced to Grande Terre during the WWII, then colonized the Isle of Pines and Lifou in the Loyalty Islands. On Grande Terre, *H. frenatus* lives in houses and cultivated area and is usually absent from primary and secondary forests. However, it can live in bushes in the northwest part of the island, and in some coastal habitats. Wherever *H. frenatus* has been introduced to Pacific Islands, it has displaced other edificarian geckos (*Gehyra mutilata*, *Gehyra oceanica*, *Hemidactylus garnotii*, and *Lepidodactylus lugubris*) in human dwellings. There is no information about the potential interaction between *H. frenatus* and the endemic Vanuatu *Lepidodactylus vanuatuensis*. In human dwellings of Grande Terre, where *H. frenatus* is common, the males develop aggressive behaviour against other geckos and exclude parthenogenetic *H. garnotii* and *L. lugubris*. Moreover, adults of *H. frenatus* consume juveniles of *L. lugubris*. However, no competition for food resources occurs between *H. frenatus* and *H. garnotii*, and no aggressive behaviour of *H. frenatus* males towards *H. garnotii*, but the males of *H. frenatus* can mate with parthenogenetic *H. garnotii*, without producing offspring. Furthermore, introduction of *H. frenatus* in the Mascarene Islands led to a drastic decline in endemic gecko populations belonging to the Indo-Pacific genus *Nactus* by virtue of competition for habitat refuges. Taxa undergoing the decline were *N. coindemirensis*, *N. serpensinsula durrelli*, and probably *N. serpensinsula serpensinsula*.

The braminy blind snake (*Ramphotyphlops braminus*) was first recorded in Vanuatu in 1971 at Port Vila

Figure 568: *Hemidactylus frenatus* (Luganville, Santo, Vanuatu, Sept. 2006, M. Pascal & O. Lorvelec).

(Efate). Later, the species was reported from Santo (Luganville and Palikulo village) including dark brown and blue specimen. The habitat distribution of the specimens collected in 1971 (Port Vila gardens and plantations) suggests a recent introduction to Efate by importation of agricultural and horticultural equipment. We preserved specimens from VARTC (Table 52 & Fig. 569), where the population was abundant under coconut palm seedlings and planks located near a coconut furnace. But one specimen was collected outside this site, in one of the old coconut palm plantations at VARTC. Several VARTC specimens exhibited the blue pattern (Fig. 569). This blue colour is linked to the period during which the snake's skin is shedding. Apart from VARTC, one unconfirmed record was from Matantas.

Perhaps native in south and southeast Asia, this parthenogenetic, fossorial and insectivorous species was introduced to many tropical and subtropical areas of the world. It has been introduced to various Pacific Islands, where it lives usually in close association with humans, as do several geckos. Its success may be due to the lack of fossorial

Figure 569: *Ramphotyphlops braminus* (VARTC [agricultural centre], Santo, Vanuatu, Oct. 2006, M. Pascal & O. Lorvelec).

mammals on these islands, at least until recently. The first capture on Grande Terre occurred in 1974. In New Caledonia, *R. braminus* was also introduced to Maré in the Loyalty Islands. The impact of *R. braminus* on invaded ecosystems is not yet documented. However, competition with endemic Typhlopidae like *Ramphotyphlops willeyi* (Loyalty Islands), or *Ramphotyphlops exocoeti* (Christmas Island, Australian territory) has been suggested. As no native Typhlopidae have been found in Vanuatu, there is no reason to address the possibility of such an interaction.

••• Conclusion

No new species was discovered on Santo by the "Fallows and Aliens" Team of the Santo 2006 expedition. In addition, our stay was too short for us to attempt to assess alien species impact on invaded ecosystems. We only have confirmed that these alien species can reach high density in anthropogenic habitats.

However, the impact of *Litoria aurea* and *Hemidactylus frenatus* has begun to be documented on other invaded Pacific Islands, and these data must be taken into account when planning future investigations on Santo.

FOCUS ON INTRODUCED FISH

Philippe Keith, Clara Lord, Donna Kalfatak & Philippe Gerbeaux

Figure 570: *Gambusia affinis*.

Figure 571: *Poecilia reticulata*.

The different survey campaigns led by the MNHN since 2000 have recorded three alien fish species in fresh and brackish waters from Esperitu Santo:
- The tilapia *Sarotherodon occidentalis* (Daget, 1962);
- The gambusia *Gambusia affinis* (Baird & Girard, 1853);
- The guppy *Poecilia reticulata* Peters, 1859.

All were introduced during the 20th century.

••• Introduction histories

The South Pacific Commission (SPC) has led most of the Pacific fish introductions. Indeed, it used to be thought that the development of food self-sufficiency had to be based on aquaculture. Other introductions are due to the army, during the WWII. Other species have even been introduced by fishermen, or by ornamental fish-keepers.

Owing to the tilapia's ability to act as a biological control (on mosquito larvae or on aquatic vegetation), and because it is easy to breed, several tilapia species have been introduced on many Pacific islands since 1950. The Mozambique tilapia (*Oreochromis mossambicus*, (Peters, 1852)), the first to have been introduced, is naturally found in eastern African rivers discharging in the Indian Ocean. Its distribution stretches from lower Zambezi to Algoa bay in South Africa. It first appeared in Java. A few individuals probably escaped or were released from aquariums, just before the WWII. Then, owing to deliberate or accidental introductions throughout the world, this species became cosmopolitan. Much advertising, first by the FAO, then by the SPC, has led to the establishment of this species in this region.

In October 1955, 40 tilapia coming from Manila (presumably *O. mossambicus*) were introduced to New Caledonia by the SPC. Presumably, this species then colonised streams and rivers. It was probably at this time that another tilapia, *Sarotherodon occidentalis*, arrived. However, the fact that the latter species is present is a conundrum. Indeed, its original geographical area is very restricted in Africa. Moreover, this species has never been used for commercial rearing. It was probably introduced with *O. mossambicus*, individuals from both species might have been mixed. *Sarotherodon occidentalis* was then rapidly established throughout New Caledonia's running waters, and from New Caledonia, was then introduced to Efate and Tanna Vanuatu Islands. Humans moved this species to Santo, where it is acclimatised. The date of arrival of the species on Santo is unknown.

The guppy, *P. reticulata*, is native to South America. It can bear temperatures under 15° C, and can live in brackish waters and even in poorly oxygenated waters. These abilities to adapt to a range of environments, as well as rapid sexual maturity and high fecundity, have allowed this species to colonise the entire world. *Poecilia reticulata* has widely been introduced on all continents. Indeed, not only was it used to control mosquito larvae, it was also very popular as an ornamental, because males bare bright colours. Americans are thought to have introduced this species during WWII to Vanuatu for mosquito control. This guppy is now well acclimatised on Santo, especially on the east coast and in the swamps of the centre of the island.

The gambusia, *G. affinis*, is native to the southern United States. This species was probably introduced with the guppy by the American army, also to control mosquitoes. It is now found on Santo's east coast and in the northern parts of the island. *G. affinis* belongs to the list of the World's 100 worst invasive alien species published by the World Conservation Union.

Figure 572: *Sarotherodon occidentalis*.

FOCUS ON ALIEN LAND SNAILS

Olivier Gargominy, Benoît Fontaine & Vincent Prié

If one decides to collect snails on Santo and particularly if one searches close to one's house or hotel, most of what will be found could have been collected in Hawaii, Tahiti or Jamaica. Indeed, the land snail fauna of Santo is totally dominated by introduced species found on most tropical islands. Moreover, in some heavily disturbed areas such as the secondary forest close to Luganville, no native species remain, whereas introduced ones can be very abundant (Fig. 573).

Methods

At each collecting site, we searched at ground level and in the vegetation (leaves, bark) for live snails; then leaf-litter and a few millimeters of topsoil were collected. We processed this sample on-site with a Winkler sieve (1 cm mesh), checking the coarse material by eye for snails, then discarding it. The remaining material was bagged and sun-dried as soon as possible. The molluscs collected alive were drowned overnight and fixed in 95 % ethanol. Once dried, the leaf-litter was passed through 5 mm, 2 mm and 0.6 mm sieves. The two larger fractions were thoroughly searched by eye; the third one will be sorted later under a dissecting microscope.

Results and discussion

This introduced fauna counts only 13 species, all recently introduced (Table 53, Figs 574 & 575). This figure fits with the general situation in the Pacific Ocean: most islands host between one to 20 alien terrestrial mollusc species.

Figure 573: In degraded forests around Luganville, the ground is often littered with shells of introduced species, here *Achatina fulica*.

Subulinidae, a family of cosmopolitan (sub)tropical snails, represent the bulk of introduced species on Santo. *Subulina octona*, *Paropeas achatinaceum* and *Allopeas gracile* are among the most common snails found on the island. The floor seldom lacks one of those species at any site and often supports

Figure 574: Introduced land snails of Santo. **A**: *Subulina octona* alive, with two eggs visible inside the penultimate whorl of the shell, Hiu, Torres Islands. **B**: *Opeas hannense* alive, Matantas, Santo. **C**: *Allopeas* cf. *kyotoense* (= *A. clavulinum* Auct.), Matantas, Santo, h = 6.1 mm. **D**: *Allopeas gracile*, Matantas, Santo, h = 7.2 mm. **E**: *Allopeas* cf. *oparanum*, Port Olry, Santo, h = 6.5 mm. **F**: *Opeas hannense*, Port Olry, Santo, h = 4.5 mm. **G**: *Paropeas achatinaceum*, Matantas, Santo, h = 11.2mm. **H**: *Subulina octona*, Matantas, Santo, h = 14.2 mm. **I**: *Euglandina rosea*, Peavot, Santo, h = 58.5 mm. **J**: *Huttonella bicolor*, Aore, h = 5.4 mm. Scale bar (shells only): 5 mm, x8. (Photos O. Gargominy).

Figure 575: Introduced land snails of Santo. **A**: *Subulina octona* (same as Fig. 574H). **B**: *Opeas hannense* (same as Fig. 574F). **C**: *Achatina fulica*, W Luganville, Santo, h = 96.9 mm. **D**: *Achatina fulica* alive, same specimen as **C**. **E**: *Euglandina rosea*, Peavot, Santo, h = 58.5 mm. **F**: *Euglandina rosea* alive devouring a *Partula* species, Tahiti, French Polynesia. **G**: *Bradybaena similaris*, Butmas, Santo, d = 14.5 mm. **H**: *Bradybaena similaris* alive, Butmas, Santo. **I**: Veronicellidae, Penaoru, Santo. Scale bar (shells and slug only): 20 mm, x1. (Photos O. Gargominy).

hundreds of shells, particularly in disturbed areas —far more than any indigenous species (Fig. 576). Subulinidae are the most common snails on Santo and can even be found in relatively undisturbed areas, such as the inner Cumberland Range.

Despite their abundance and large distribution, taxonomy of the Subulinidae family remains confused, with no recent revision. Moreover, introduction of several species of this family throughout the tropics has led to many taxonomic and nomenclatural problems, because a species collected in different areas of the world could have received different names, or the same name could have been given to two different taxa collected in distant areas. For instance, *Allopeas oparanum* was originally described from Rapa, one of the most remote islands in the world, where Subulinidae are definitely aliens, and mentions of *Lamellaxis clavulinum* from the Pacific are likely to refer to *Allopeas kyotoense*. Thus, the names we use in this publication may be provisional.

Table 53: Terrestrial alien snail species recorded on Espiritu Santo Island (Vanuatu). **Date**: date of first specimen in collection or date of first publication for Santo Island. Between brackets are dates of first record for Vanuatu. **IUCN worst list**: species that are listed in the IUCN 100 worst alien species. **1**: usually recorded as *Opeas pumilum* in the Pacific. **2**: usually recorded as *Allopeas clavulinum* in the Pacific. **3**: described from Santo as *Semperula solemi* by Forcart, 1969.

Family	Species	Date	IUCN worst list
Achatinidae	*Achatina fulica* Bowdich, 1822	1968-1969? (1967)	*
Bradybaenidae	*Bradybaena similaris* (Férussac, 1821)	1944	
Oleacinidae	*Euglandina rosea* (Férussac, 1821)	1982-1983? (1973-1974)	*
Streptaxidae	*Huttonella bicolor* (Hutton, 1834)	2006 (2006)	
Subulinidae	*Allopeas gracile* (Hutton, 1834)	1943 (1895)	
Subulinidae	*Allopeas oparanum* (L. Pfeiffer, 1846)	2006 (1903)	
Subulinidae	*Paropeas achatinaceum* (L. Pfeiffer, 1846)	2006 (2006)	
Subulinidae	*Opeas hannense* (Rang, 1831) [1]	1944 (1944)	
Subulinidae	*Subulina octona* (Bruguière, 1792)	1943 (1903)	
Subulinidae	*Allopeas kyotoense* Pilsbry, 1904 [2]	2006 (2006)	
Veronicellidae	*Angustipes plebeius* (Fisher, 1868)	1925 (1925)	
Veronicellidae	*Laevicaulis alte* (Férussac, 1822)	1958 (1958)	
Veronicellidae	*Semperula wallacei* (Issel, 1874) [3]	1958 (1958)	

Beside Subulinidae, the Santo alien mollusc fauna includes two famous species listed in the IUCN 100 worst alien species: the rosy wolf snail, *Euglandina rosea*, and the giant African snail, *Achatina fulica*. Indeed, the introduction of the giant African snail and the subsequent introduction of the predatory rosy wolf snail are a famous example of a biological control program that has led to the decline and extinction of dozens of endemic land snail species in Pacific Islands. *Achatina fulica*, native to eastern Africa, is a large species that can reach 200 mm in length. Since the beginning of the 19th century, it has been introduced to most islands located between 30° N and 30° S, and even to continental areas (Fig. 577). It reached the most remote islands during and after the WWII due to bulldozers, development and maritime trade increase. When established, the giant African snail populations increased rapidly due to the lack of predators and pathogens and became a major agricultural pest. A well-intentioned but ill-conceived biological control effort was planned to control these populations. It consisted of introducing snail predators such as *Gonaxis spp.* and, more often, *E. rosea*. This latter species, native to southern North America, was introduced to Hawaii in 1955. Hawaii then Guam played the role of stepping-stones, and, during the 1960s and 1970s, the species was introduced to the most isolated Pacific Islands, wherever people complained about damage caused by *A. fulica* (Fig. 578).

Wherever it was introduced, *E. rosea* spread out of cultivated areas and reached natural forests. Because *E. rosea* is not a specific predator of *A. fulica*, it fed on native snails, leading all endemic *Achatinella* from Hawaii to the verge of extinction, or even eliminating them. In the course of a few years, the island of Moorea in French Polynesia was totally invaded and seven endemic *Partula* species disappeared. Nowadays, almost all Polynesian

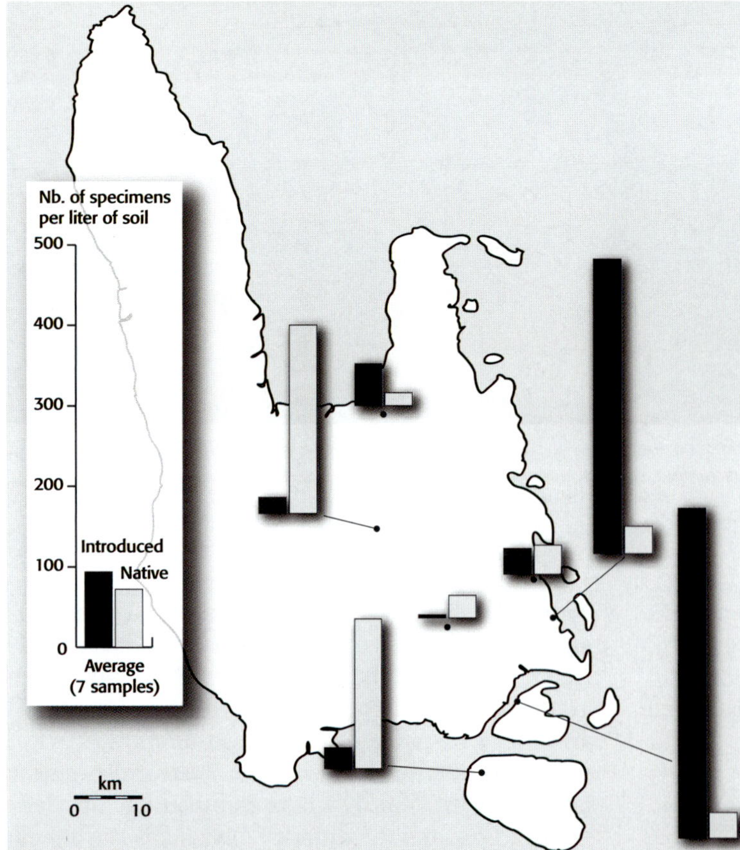

Figure 576: Introduced vs native species: introduced species dominate the fauna in human impacted areas (mainly near the coast) where they seem to replace native ones almost.

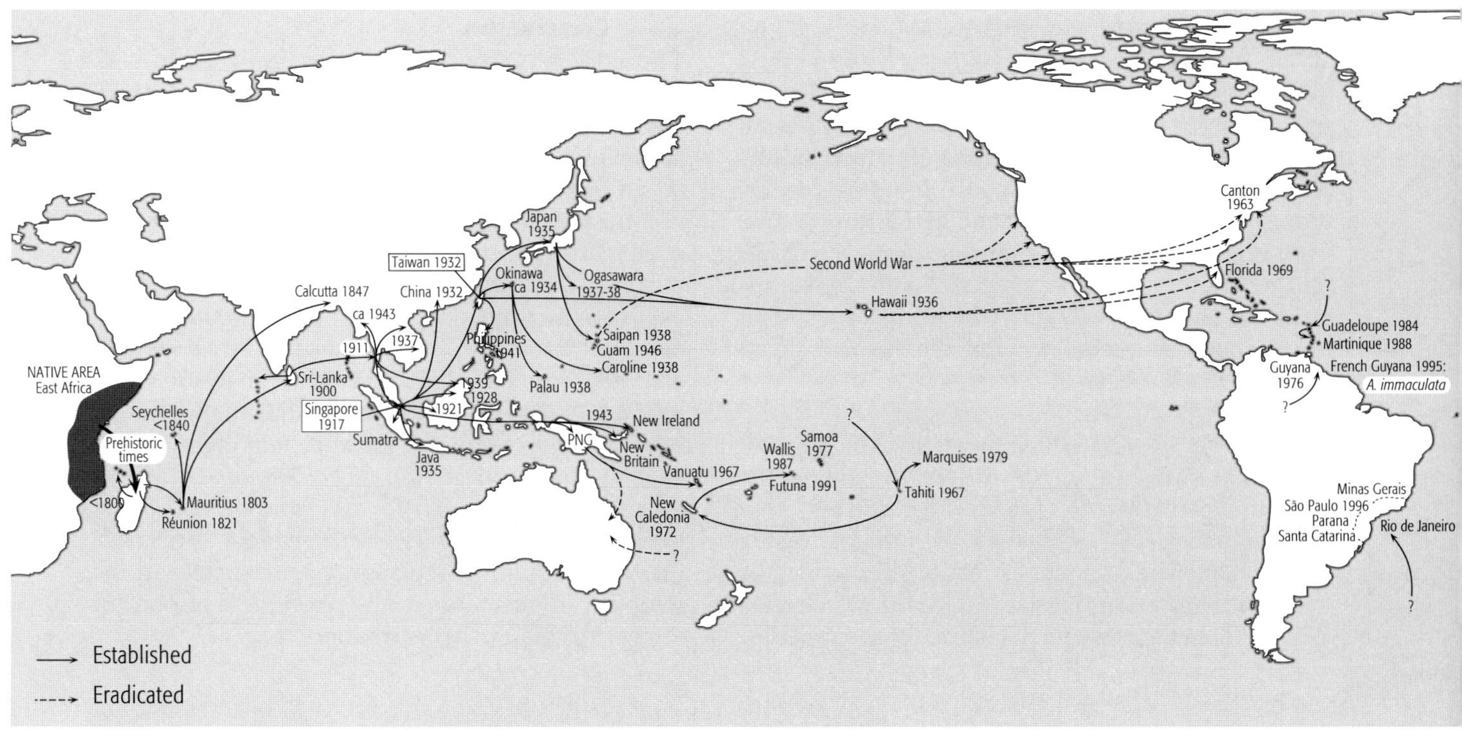

Figure 577: The spread of *Achatina fulica* around the world. Grey area: native range.

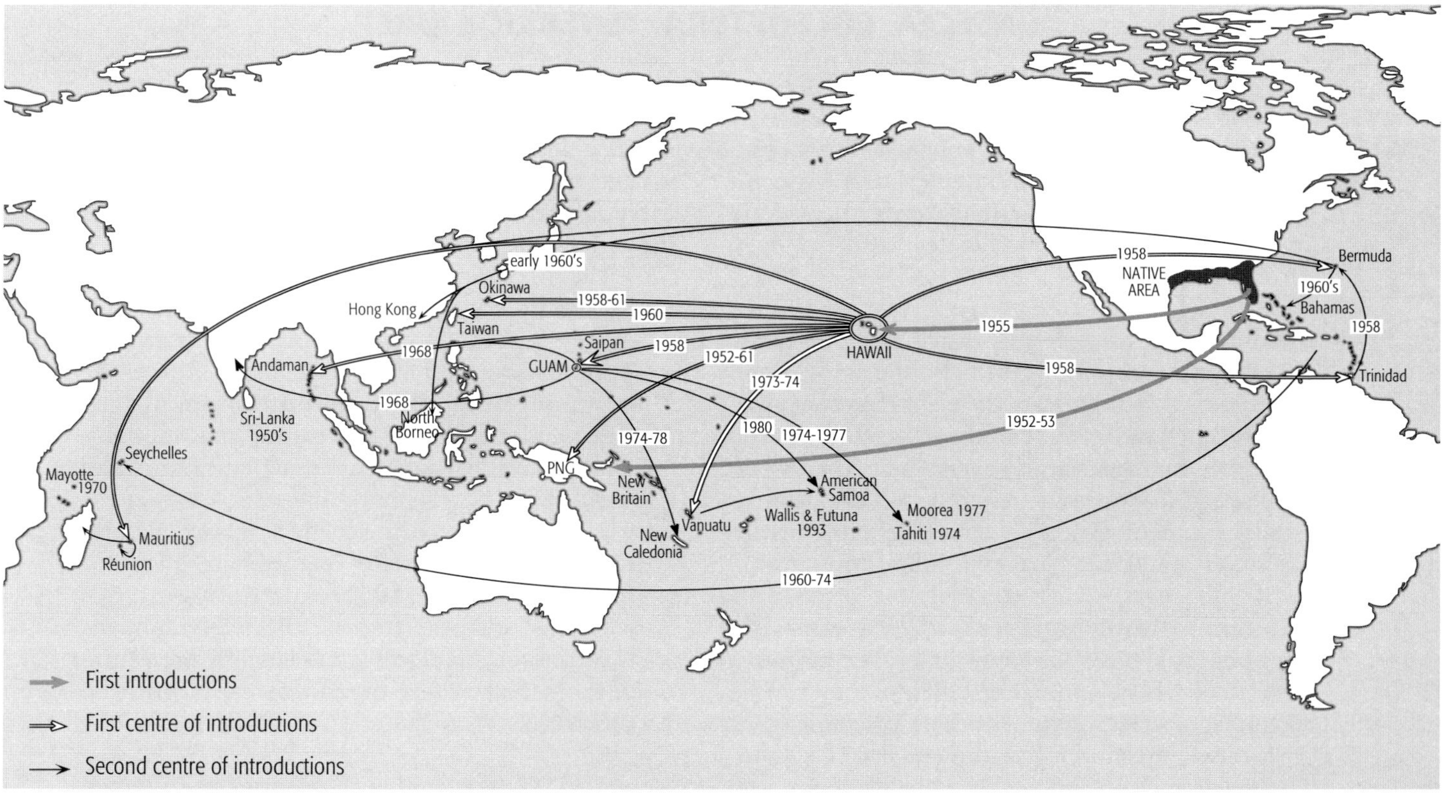

Figure 578: The spread of *Euglandina rosea* around the world. Grey area: native range.

islands are invaded, and 55 endemic *Partula* species disappeared in only 20 years, while 10 others are following the same path. While those extinctions happened, *A. fulica* remained abundant, *E. rosea* having no impact on its demography. To date, the number of extinct snail species (302 according to the 2007 IUCN Red List) is greater than the number of terrestrial vertebrate extinctions (271), mainly because of the disastrous impact of *E. rosea* on the unique snail fauna of Pacific Islands. Moreover, impact on minute species that are the bulk of the native fauna, but are overlooked by most naturalists, is completely unknown.

E. rosea was introduced in 1982 or 1983 to Santo by a Bob Wheeler, quarantine officer, who was asked to find a way to eradicate *A. fulica*. Except for this species, most mollusc introductions probably were accidental on Santo: eggs or even fully-grown adults are easily transported with soil, moss or vegetation spread by horticultural and agricultural trade. However, Santo was an important American base during the WWII and transport of military equipment (in particular vehicles, which can have soil on their wheels or tracks) has probably helped spread several alien species.

After habitat loss, introductions are the second most significant cause of species extinction throughout the word and particularly on islands. Alien species either prey on native ones or compete with them, causing their decline and eventual extinction. These mechanisms are enhanced by habitat degradation, which often favours the settlement and dominance of alien species.

••• Conclusion

There is virtually no way to remove a successfully established alien snail species, except in very restricted areas such as islets or exclosures. In this context, the only action that could (and should) be taken is preventing further introductions through raising public awareness and rigorous control in ports and airports. In Hawaii, one new alien species is introduced every year. It is important to keep excluding new introductions, especially since introductions are not inevitable: some Pacific island such as Fiji have succeeded in avoiding invasion by *A. fulica*. Faunas and floras throughout the world are becoming homogenized: some alien mollusc species such as *Lamellidea pusilla*, *Achatina fulica*, *Allopeas gracile* or *Subulina octona*, all found on Santo, are present on nearly every Pacific island. With rats, flies and subulinids replacing native species in the remotest islands all over the world, it is important to try to avoid further introductions.

ENDEMIC, NATIVE, ALIEN OR CRYPTOGENIC? THE CONTROVERSY OF SANTO DARKLING BEETLES (INSECTA: COLEOPTERA: TENEBRIONIDAE)

Laurent Soldati

Among Coleoptera (Greek, *koleos* = sheath, and *pteron* = wing), the most species-rich order of insects with more than 300 000 known species, are Tenebrionidae or darkling beetles, with *c.* 19 000 described species.

Darkling beetles are globally distributed, but many of them are xerophilous and abound in arid and desert areas, where they are the main representatives of the beetles and even of all the insects. In tropical areas, they can occur nearly everywhere, even when local species richness is very low, but they can be locally abundant, especially where decaying matter like rotten wood, dead leaves or litter is present.

The tenebrionid representatives are mainly saprophagous, but poor environmental conditions often make them opportunist. Thus, they can be mycophagous (*Platydema*, *Toxicum*), psichaphagous (*Tribolium*), rarely necrophagous and even occasionally predators. Some cases of cannibalism have been observed in the larvae. Several species are synanthropic, and some are stored-products pests, mainly of cereals (barley, corn, oat...) and their derivatives (flour) (*Tribolium* spp.). In those cases, they can be of economic importance. Most tenebrionid beetles are nocturnal or crepuscular.

••• Material and methods

Members of the "Fallow & Aliens" team of the Santo 2006 expedition, focused mainly on the balance between native and introduced species in several habitats of Santo Island. For this purpose, three different areas showing an increasing human influence were studied:
- The first area is located in the surroundings of Luganville (South-East of Santo) on the lands of the CETRAV (*Centre Technique de Recherches Agronomiques du Vanuatu*). Here, since the beginning of the 20th century, European agriculture and breeding has turned natural habitats into fields, huge coconut plantations and meadows;
- The second area studied is the Vatthe Conservation Area (Big Bay), where there is a mixture of natural and secondary forest;
- The last area and the least affected by humans is Butmas, a small village located in the centre of the island, where the rain forest is still well preserved.

About 25 Tenebrionidae species were collected during the one month the "Fallow & Aliens" team spent in Santo. This material clearly shows a main (and foreseeable) result: the species richness obviously decreases from the first inventoried area to the third and last one. This decrease can be explained by the fact that human activities and settlements favour several animal species, especially insects. The introduction of many plants and animals strongly modified native habitats (agriculture, breeding, plantation and ornamental plants...).

Commercial or trade routes not only introduced foods, but also their pests and parasites. Moreover, several parts of the island, especially in the south-east, were greatly modified during WWII, when US Army and Air Force logistic and operational bases were established. Recent tourist activities also have an impact on the island's biodiversity. By contrast, traditional Melanesian places like Butmas, which are less accessible were far less perturbed. But this does not necessary mean that all tenebrionid species collected there are native.

••• Difficulties in interpreting the biogeographical origin of the species

The major question connected to the "Fallow & Aliens" project regarding Tenebionidae is: among the 25 species collected by the team, which ones are native and perhaps endemic species, and which ones are aliens? Answering this question requires the availability of paleontological, archaeological or historical and/or biogeographical data. For Vanuatu and more precisely Santo Island, there are neither paleontological, archaeological nor historical data devoted to Tenebrionidae. With respect to the bio-geographical aspect, there are no monographs or other contributions on the Vanuatu Tenebrionidae, except for two original descriptions written by German authors. The first one dates from the early 1950s and is the description of *Thesilea mallicolensis* from Mallicolo Island by Kulzer in 1951, and the second one is the description of *Corticeus levis* by Bremer in 1993. Finally, in the introduction to his fauna of New Caledonia, Kaszab summarized in 1982 the tenebrionid component of the entomological fauna of Vanuatu and mentioned a few species such as *Brachyidium irroratum*, without further details.

I have chosen four species, among the 25 collected in Santo, in order to illustrate different issues encountered when I attempted to establish their status as native or alien to Santo.

••• *Tribolium castaneum* (Herbst, 1797) (Fig. 579)
Tribolium castaneum is a cosmopolitan or sub-cosmopolitan species and one of the most important food grain pests worldwide. It is known to infest rice, corn, bran, maize, barley, sorghum, millet, manioc, tapioca, yam, dried fruits, peanut, cacao, nutmegs, pepper, ginger and different kind of flours. This species can also greatly damage entomological collections!

A single specimen was found in Luganville, CETRAV in Saraoutou, the locality most heavily disturbed by humans among those inventoried by the "Fallow & Aliens" team.

The species of *Tribolium* that have become pests of stored products have been widely distributed by man for a very long time. It is believed that *T. castaneum* originated in the Oriental (Indian?) region, while

Figure 579: *Tribolium castaneum*. (Photo L. Soldati).

Figure 580: *Brachyidium irroratum*. (Photo L. Soldati).

other *Tribolium* belonging to the same species-group (e.g. *T. madens* (Charpentier, 1825), *T. waterhousei* Hinton, 1948...) seem to be of an Indo-Australian origin. In fact, at present, nothing can be clearly established for such insects, which are introduced and often established worldwide.

••• *Brachyidium irroratum* **(Fauvel, 1867) (Fig. 580)**
Brachyidium irroratum is known from New Caledonia, the Loyalty Islands and New Hebrides (Vanuatu). Several specimens were recently collected (November 2003) on Vanikoro Island (Santa Cruz group of the Solomon Islands) by my friend and colleague H.-P. Aberlenc (CIRAD).

On Santo, *B. irroratum* was found only at the isolated Butmas site, in the centre of the island (Butmas, rain forest, 21.X.2006, light trap, L. Soldati leg). As this small Melanesian village is very far from the island main track network, and because this small insect does not feed on cultivated vegetables or fruits, and consequently is absent from plot crops, there is no chance that human activities have favoured its dispersal. Moreover, even if it is able to fly (but perhaps not for very long distances), *B. irroratum* lives mainly on the ground and probably feeds on leaf litter, humus or other decaying plant matter. Though establishing the entire distribution range of this species is difficult because it is not easy to collect, the conditions in which specimens were captured in Santo strongly suggest that it is native to the island.

••• *Corticeus (Cnemophloeus) cephalotes* Gebien, 1913 **(Fig. 581)**
This insect has an unusually wide distribution. It is known from the major part of South and Southeast Asia (Indonesia, Philippines, Vietnam, Thailand and India), from New Guinea and probably the whole Melanesian Arc, and from Northern Australia and the Comoro Islands. This distribution explains why it was described several times under different names, even by the same author! The lectotype of *C. cephalotes* was collected in Taihorin (Formosa) and the following synonyms are presently known:
- *Hypophloeus cornutus* Pic, 1914, from Malacca;
- *Hypophloeus cornutus* var. *subcastaneus* Pic, 1914, from Andaman Island;
- *Hypophloeus palawanus* Pic, 1945, from Palawan;
- *Hypophloeus andaiensis* Pic, 1946, from New Guinea;

C. cephalotes has also been collected by my friend and colleague H.-P. Aberlenc during the 2003 Vanikoro expedition (Santa Cruz group of the Solomon Islands).

On Espiritu Santo, *C. cephalotes* was found under bark in an ecotone between secondary rainforest and the coffee tree plantation of Saraoutou.

Figure 581: *Corticeus (Cnemophloeus) cephalotes.* (Photo L. Soldati)

Figure 582: *Corticeus (Cnemophloeus) levis.* (Photo L. Soldati).

Many *Corticeus* species have huge distributions, especially in the Indo-Pacific region. As these insects live under the bark of dead or decaying trees, it is improbable that human activities play an important role in their dispersal. Tropical storms and floating wood are probably major dispersal agents. Dr. H. J. Bremer, a world specialist of the genus *Corticeus*, emphasized that specimens could be transported over long distances by typhoons as aerial plankton. Because the huge range of this species is probably due to natural agents, and because Vanuatu belongs to the Melanesian Arc, which is highly likely to be part of the native range of the species, it seems reasonable to consider *C. cephalotes* as native to Santo.

••• *Corticeus* (*Cnemophloeus*) *levis* Bremer, 1993 (Fig. 582)

Species described from a single specimen, sex unknown, collected in "Malao village in the Big Bay Area, Santo, Vanuatu, 6-16 Sept. 71", (Holotype in South Australian Museum, Adelaide).

This species has never been mentioned since the original description, until the present expedition. Many specimens were collected in the same area as the holotype, under the bark of a large dead trunk lying on the ground in the Vatthe Conservation Area (Matantas, Big Bay). This rich sample allowed me to confirm that *C. levis* is a valid species and that Bremer was right to assign this species to the subgenus *Cnemophloeus* Bremer, 1998. Because of the restricted known range of the species, *C. levis* can be considered endemic to Santo… until evidence shows otherwise!

••• Conclusion

These few examples illustrate the difficulties we face trying to answer the question of alien or native for taxa living on islands or in archipelagos. It is even more difficult when the systematics of some insect groups, such as Tenebrionidae, is still unclear, as there are sometimes great difficulties in identifying the different taxa, even for specialists. For Vanuatu, the nearly total lack of references makes the study even more difficult. Comparison with data from adjoining areas like New Guinea, Fiji or New Caledonia proved to be very helpful, but not sufficiently in several cases. It is highly probable that many Tenebrionidae have a large distribution over the Melanesian Arc and could therefore be considered as native to many islands of this region. The main problem in documenting this hypothesis is the collection of specimens. Performing such exhaustive collection is quite impossible because the lack of knowledge of the biology and behaviour of many taxa impedes the sampling effort. Moreover, Vanuatu has been insufficiently inventoried from an entomological point of view. Finally, several species are really rare.

One of the major purposes of the Santo 2006 expedition was to realize the most complete inventory of the animals and plants of Santo, and the results will contribute to the general understanding of the origin and history of the island flora and fauna. But, for many species, establishing their local status as endemic, native or alien will probably long remain a question without a definitive answer. This situation gives a major role to the cryptogenic uncomfortable status.

THE CASE OF TWO INVASIVE SPECIES: *MIKANIA MICRANTHA* AND *MERREMIA PELTATA*

Marc Pignal

The impact of exotic species on biodiversity, especially in the case of island ecosystems, was taken into account in the Convention on Biological Diversity (CBD). Biological invasion is regarded as the third most important cause of biological diversity loss, after habitat destruction and excessive harvesting. The CBG therefore notes that any voluntary or involuntary introduction could impact entire ecosystems and the survival of indigenous flora and fauna. In addition to biological considerations, those who drafted the Convention were also aware of the significant socio-economic implications of introductions involving allochthonous species.

Research on exotic species that pose a risk is important, as is monitoring in natural environments, but increasing public awareness is critical. The most striking example is surely the water hyacinth,

Eichhornia crassipes (Mart.) Solms, which is a member of the family Pontederiaceae and native to Brazil. The beauty of its colorful inflorescences led to its introduction in water bodies throughout the tropics. But these introductions turn out to have been most unfortunate because the vigor of this species and its rapid reproduction rate have made this plant the primary scourge of rivers everywhere in warm parts of the globe.

Pantropical invasive plants are the subject of monitoring in most country, and even within native ecosystems. Within the Pacific region, PIER (Pacific Plant Ecosystems at Risk) is in charge of compilation and dissemination of information on invasive exotic species. Several organizations and programs devote their efforts to this question and produce documents aimed at the general public. The Invasive Species Specialist Group (ISSG), and in

Figure 583: *Mikania micrantha*, flowers.

Figure 584: *Merremia peltata*, flowers.

particular the Pacific Invasives Initiative (PII) program, will be cited frequently in this article.

Insular floras, with high proportions of endemic species, are particularly sensitive to invasive exotics. The problem of invasive species in Vanuatu can be illustrated by two examples, *Mikania micrantha* and *Merremia peltata*, whose impact may at first appear similar (co-occurrence with both autochthonous and cultivated species), but their origin and the dynamics of their invasion are radically differe.

Paradoxically, the locals use the same local name, *lianes américaines*, for both. However, farmers and people living in forested areas make no confusion and are able to distinguish both species morphologically and semantically.

••• *Mikania micrantha* Kunth ex H.B.K.

The Asteraceae family, to which *M. micrantha* belongs, is certainly the largest one, with ca. 1 100 genera and 25 000 species. It is cosmopolitan and occupies all strata within various vegetation types, from herbs and shrubs to lianas and trees. Numerous Asteraceae species present problems for agriculture. In temperate areas, this is the case for dandelion (*Taraxacum officinale* L.), and in the tropics, among many examples, one could mention *Wedelia trilobata* (L.) Hitchc. and especially *M. micrantha*.

Mikania micrantha has rudimentary capitula comprising four white florets (Fig. 583). It has opposite leaves that are soft and heart-shaped, and lack any type of indument. The flowers, each 3-5 mm long, are arranged in terminal or axillary corymbs. Individual florets are white to greenish-white. The seed is black, linear-oblong, five-angled and about 2 mm long. The *M. micrantha* stem is flexible, breaks easily, and can produce adventitious roots.

••• Environmental impacts

This species invades natural ecosystems and is also a serious pest in various kinds of plantation, including cocoa, coconut, and oil palm. Cock and coauthors cited "*In 1981, about 5% of the total weed control costs in rubber and oil palm plantations were directly attributable to control of* M. micrantha; *this amounted to about 10-12 million Malaysian Ringgit (approx. US$ 3 million at the 2002 exchange rate)*". In Queensland, Australia, this species is listed in Class 1 in the "Rural Lands Protection Act", and it has also been included in the ISSG's list of the 100 most invasive species in the world.

In Vanuatu, while the available data are sketchy, we encountered *M. micrantha* in all of the human-modified area visited, including CRTAV's cultivated plots, gardens, clearings, forest openings and secondary forests at Butmas, and old secondary forest at Vatthe.

••• Modes of dispersal

• Dispersal by seeds

As in many genera of this family, *Mikania* is anemochorous, in other words, it is wind dispersed. Its seeds, or more precisely, its small, dry fruits known as achenes, are crowned by a plume and can travel several kilometers and pass through areas that are impenetrable to man. As a consequence, this plant can occur in inaccessible rocky areas and in openings in the deepest parts of the forest.

Botanists who prepare herbarium specimens know very well this another biological characteristic of *M. micrantha*: when collected in flower, this species taps its last reserves to develop fruits even after being cut. This feature has an impact on mechanical control efforts, making it necessary to kill the plants as soon as they are cut and piled up in order to ensure that fruits do not develop and subsequently disperse.

• Vegetative dispersal

The ability of *M. micrantha* to produce adventitious roots, combined with the fragility of its stems, are two biological characteristics which increase greatly its dispersal capability as even the smallest piece

Figure 585: CETRAV, agricultural field invaded by *Merremia peltata*.

with a single node is capable of developing into a full plant. Clearing can thus actually increase the risk of dispersal unless it is done with great care.

••• Methods of control

• Uprooting
Hand clearing, mentioned earlier, is only practical in small areas, and even in the case of gardens, *M. micrantha* is practically impossible to eradicate, for the reasons outlined above. Mechanical uprooting which is performed in agricultural situations is the most widely used and easily employed method because the stems of *Mikania* are flexible, but many stems remain on the ground.

• Chemical control
This species is sensitive to numerous pesticides, in particular 2,4 D. However, the PIER suspects that *M. micrantha* is capable of re-sprouting from the base after chemical treatment.

• Biological control
Several prospects are being developed, from the parasitic fungus *Puccinia spegazzinii* De Toni, native to South America (several projects are ongoing) to the use of Cuscuta. While these methods should be used with caution as they involve introduction of another living organism, with all the risks that entails, they nevertheless remain promising.

••• Origin
Apparently *M. micrantha* has been introduced in Santo during the WWII. Discussions with villagers confirmed that this species appeared at that time, probably via introduction with organic matter. This

Figure 586: Vatthe, tree of second groth forest covered by *Merremia peltata*.

Figure 587: CETRAV, old agricultural parcel covered by *Merremia peltata*.

explains its local name, *liane américaine* or American vine, as well as the legend that it was introduced as camouflage for military installations. However, despite the efficiency of military planning and the rapid growth potential of *M. micrantha*, it efficacy for protecting such installations nevertheless seems doubtful.

• • • *Merremia peltata* (L.) Merr.

The family Convolvulaceae comprises 50 genera and 1 500 species, many of which are creeping adventive lianas in gardens and agricultural fields. In ornamental gardens, members of this family can often be difficult to control. In Vanuatu, several species are found in human-modified areas and can have an impact on secondary zones. However, only one species, *Merremia peltata*, impacts both the natural environment and crops.

••• Brief botanical description

Merremia peltata is a robust liana with alternate, simple, cordate to orbicular, peltate leaves with an obtuse to short-acuminate apex and very evident venation. The flexible stems exude white latex. Inflorescences can have more than 10 funnel-shaped flowers that are white or rarely yellow (Fig. 584). The fruit is a capsule about 1.5 cm in diameter that opens at maturity to form lanceolate valves that release brown, pilose seeds.

••• Mode of dispersal

In 2006, Paynter *and coauthors* in their report on *M. peltata*, mentioned that it is regarded as thalassochorous, i.e. that it's seeds are transported by the sea.

••• Effect on ecosystems

Individuals of *M. peltata* can reach 20 m in length and can occupy large areas (Figs 585 & 587). In understory of human-modified areas, it can invade all strata, and its stems and roots can carpet the ground. Field observations reveal up to 70 rhizomes per square meter. The thick, twining stems spread toward forest openings produced by natural tree-fall, cyclones and human activity. Ultimately, this liana covers the canopy (Fig. 586), covering the surrounding trees and crushing them under the

Figure 588: Butmas, lean-to covered by *Merremia peltata*.

weight of its branches and leaves. Once established, it appears impossible to get rid of this plant, and all the elements seem to be in place for expanded colonization. However, reference is made that pioneer species such as *Pometia pinnata* J.R.Forst. & G.Forst. (Sapindaceae) are able to resist the *M. peltata* invasion (see the entry for this species in the ISSG database at: http://www.issg.org/database/species/ecology.asp?si=163&fr=1&sts).

Merremia peltata can be found in all heavily human-impacted habitats (Fig. 588), but it is infrequent in better conserved forest areas. It takes advantages of all types of environmental modifications, including some that are not of human origin. Schmid, in his study of western Melanesian vegetation, indicated in 1987: "*The destructive effects of frequent cyclones have an important influence on the evolution of vegetation*". At Butmas, a site was observed where *Merremia* had become established following a cyclone five years ago and has persisted since.

••• Methods of control

The mechanical and chemical control methods mentioned above for *Mikania* are also recommended for *Merremia*. Intensive pasturing is also used to eliminate this species. Preventive control that aims to avoid the introduction of *Merremia* and the dispersal of fragments, as well as the disturbance of natural habitats, is also recommended. Research has also been conducted in Samoa on the use of biological control involving the introduction of species resistant to canopy suffocation, such as Ylang-ylang and *Cananga odorata* (Lam.) Hook.f. & Thomson. The use of a pathogenic fungus, *Glomerella cingulata* (Stoneman) Spauld. & H. Schrenk, an anthracnose present in the Pacific, could also be an effective means of control.

••• Origin and native area of distribution

Is *M. peltata* native or not to Vanuatu? The question is under debate. Its distribution area is large, encompassing the entire inter-topical zone from the western Indian Ocean (Madagascar and the Mascarene Islands) through Sumatra, Java, Timor, and on to Australia. *Merremia* is present in West Pacific Islands, reaching its eastern limit in Samoa. PIER provides a detailed summary of its occurrence in the region based on the published literature (see: http://www.hear.org/pier/species/merremia_peltata.htm).

Depending on the author and the island, *M. peltata* is considered to be indigenous or introduced, sometimes long ago by the local inhabitants. In reality, it is difficult to date the arrival of this species when it is invasive, and it statue as native or not must be decided with a great deal of caution.

Strictly speaking, no element of the flora or fauna of an oceanic island can be regarded as truly indigenous. What matters is when a species arrives and the perturbations it causes on the ecosystems present at that time. In the case of *M. micrantha*, whose date of arrival and the fact that it was introduced by humans leave no doubt, it is clearly not indigenous. An allochthonous origin for *M. peltata*, on the other hand, is much more doubtful.

Has *Merremia* been present in Vanuatu for a long time? To date the answer is not clear. Data collected on local names in the Butmas area indicate that recently arrived plants do not have a name in the language of this village (*ati* or *Butmas-tur*), and local residents principally use the name in Bislama. In the case of *M. peltata*, the name used locally is *Aïdraveri*, which differs from the Bislama name (*big lif*).

The published literature and label data on herbarium specimens provide no additional information on this species, and it is difficult to evaluate the impact it has had or the surface area it has occupied on Santo in the past. However, Levat collected it in 1883 at Port Vila, indicating that it occupies large areas: "*Croît sur les coteaux partout sauf dans les vallées sablonneuses du bord de la mer. Atteint dans les alluvions de l'intérieur des proportions gigantesques. Envahi des plaines entières et couvre les plus grands arbres*".

Merremia peltata is not among the first collections made on Santo by Campbell in 1872 and 1873, and later by Quaife in 1902 and Aubert de la Rüe in 1936. While these botanists made only about 30 collections each, they were mostly of species typical of human-modified habitats, such as *Sida rhombifolia*, *Ocimum basilicum*, *Euphorbia hirta*, *Boehmeria platyphylla*, *Cenchrus calyculatus*, etc. If *Merremia* had been present at the time, it seems hard to imagine that they would have missed it.

The first time this species is mentioned was in Hog Harbour in the 1934 material collected by Mr. and Ms. Baker. Today, *Merremia peltata* is present throughout this part of the island. In 1948, it was also cited on the islands of Epi and Efate. Guillaumin does not report the presence of *Merremia peltata* in other nearby island and archipelagos, including New Caledonia, but does indicate that it occurs in Tahiti and Malaysia. Just because a species is not cited as being present somewhere does not mean that it is absent from the flora, but in the case of *Merremia*, which is so abundant, it can hardly go unnoticed.

More detailed studies especially in sediments will be required to assess when *M. peltata* reached Santo. It is not out of the question that this took place prior to the first human colonization of the archipelago, in which case *M. peltata* would be a model illustrating the aggressiveness of a non-exotic species as a consequence of perturbation of its native habitat rather than as a result of its introduction.

Man Santo

FOOD-GARDEN BIODIVERSITY IN VANUATU

Sara Muller, Vincent Lebot & Annie Walter

• • • A brief typology of Vanuatu gardens

• • • Rain-fed gardens

The Melanesian rain-fed garden uses a small plot (500 to 2 500 m²) opened by clearing and burning the forest cover (Fig. 589). Long rotation cycles include a lengthy fallow period of seven to 30 years, or more, depending on local access to the land. It is therefore necessary each year to clear a new garden within high, regrown trees and then to fell and burn the woody species. Roots or vegetable remains are eliminated by burning. Generally no organic matter and no trees excepting fruit trees are left on the plot. However, minerals resulting from combustion seep into the ground in rainwater. The plot, roughly delimited by curtains of trees, is not plowed, but holes are dug for every plant, using a stick or a clearing knife. Within-garden organization varies from one island to another. Generally, taro or yam comes first in the cropping cycle. The former are generally planted in wet grounds of the interior forests (Fig. 590) while the latter are planted on drier coastal slopes. The ground is loosened before planting the propagules (taro head-sets or small yam tubers) into holes with great care. The soil is then buttressed nicely around the tuber and here again, the mounds vary in shape and size from one island to another and from a local group to another. Constructing the mounds crushes the ground aggregates and disturbs the galleries of insects and worms or remaining roots. The vegetal remains or debris are removed from the garden in order to avoid the formation of rodents' nests and bacterial or fungal contamination of the crops.

A few months later, the secondary species are planted: sugarcane (*Saccharum officinarum*), naviso (*Saccharum edule*), island cabbage (*Abelmoschus manihot*), corn (*Zea mays*), onion (*Allium cepa* var. *cepa*), kava (*Piper methysticum*) and recently introduced vegetable species. Various species of fruit trees are planted around the plot in addition to those that were preserved from clearing during the initiation of the garden. Such trees include banana, papaw (*Carica papaya*), breadfruit (*Artocarpus altilis*), cut-nuts (*Barringtonia* spp.), and many *Citrus*. The

Figure 589: Typical landscape of the interior valleys of south Santo (here in the valley of Navaka River). In the background, a new taro rain garden has just been slashed and burned.

organization of the species within the plot is primarily based on agronomic goals. The noble species (taro and yams) are placed in the centre, i.e. most distant from the surrounding forest curtains, while the tallest species —island cabbage, corn, manioc (*Manihot esculenta*)— are planted around them as windbreaks. But organization is also based on cultural criteria, as is shown by the intercropping of kava between yam tubers in some gardens of Pentecost. Finally, coloured species such as cordyline or croton are added for decorative or magic purposes.

Formerly, each plot in rain-fed cultivation was used approximately three years before being left fallow.

Figure 590: Taro rain garden at the village of Lolosori on the island of Ambae.

However, the garden continued to be exploited: kava, bananas, residual tubers, fruits and nuts, which continue to be produced naturally, are then harvested in the fallows. The re-use of the plot is evaluated according to the growth of fast-growing woody species. This very powerful system requires the annual opening of a new garden to provide the soil fertility, which the demanding yams require. However, the fallow period is generally much shorter nowadays, mainly because of increasing pressure on lands.

••• Irrigated gardens

When the environment is not appropriate for the major cultivated species, as is the case in the dry zones for social groups having taro as main crop, the horticulturists radically transform their environment in order to adapt it to their favorite crop. Large irrigated complexes are established (Fig. 591) on the western coast of Santo but also in the Banks Islands, Maewo, Ambae and Pentecost and formerly on Aneytium in southern Vanuatu. In pondfield systems, the water is diverted from rivers upstream through long channels dug into the ground or using bamboo pipes, and thus kept constantly flowing. The ground released by slashing and burning is arranged in successive terraces that follow the slope line. They are delimited by low walls and communicate with each other by small outlets through the walls. The main inlet is located upstream on the highest terrace. The water then flows out downstream gradually from one pond to another. Along with taro, secondary plants — including many fruit trees (taken from wild or domestic stock) — are often grown on the low walls.

The irrigated pondfields are cultivated for many years. Actually, new terraces are built downstream each year while others are left fallow upstream. The hydraulic system then must be rebalanced. The reasons a new pondfield is dug are generally related to a decrease in yields or to a specific requirement for prestigious taro, in anticipation of a ceremony (an easy way to restore fertility after a short fallow period is to remove the mud from the pondfield and spread it over the walls).

••• Diversity and main purposes of the gardens

As is well known, social, cultural and biological factors always overlap to constitute a coherent system. In addition to the ecosystem variation directing the choice of the main crop, taro or yam, cultivation must be considered in the context of a complex exchange network that exploits differences in agricultural production. Thus, taro is the main crop on Santo while the small neighboring island of Malo, much dry, specialized in cultivating yams. On Santo itself, the agricultural landscape presents some variation. On the west coast, sheltered from the easterly wind and thus drier, yam cultivation, although subordinate to that of taro, has a more important place than elsewhere,

Figure 591: Irrigated taro pondfields in Hokua village (N-W Santo).

in gardens, in diets as well as in cultural and ritual contexts. This ritual function is not found in villages located on the slopes exposed to trade winds. In southern Santo valleys, taro is cultivated in rainfed gardens and the landscape contrasts greatly with that of the irrigated gardens.

The gardens (rain-fed or irrigated) were, at least until recently, designed and managed according to knowledge and practices elaborated over centuries. As pointed out by Morelli in 2003, yields obtained in Vanuatu gardens, varying from 20 to 30 tons per hectare per year, are much higher than what is usually observed in other intensive tropical farming systems. Vertical stratification —from tree tops, to tall plants like bananas and cassava, down to intermediary plants like island cabbage, beans, corns and finally down to the herbaceous stratum— allows use of the whole vertical and horizontal space. The system is very well balanced, so it is difficult to intensify cultivation further, and farmers have no other choice than multiplying the number of the cultivated plots in order to increase production.

Scattering those plots allows risk management (particularly in regards to climatic hazards), permits benefit from ecological variations, and minimizes contamination risks by pests and diseases in a given cropping cycle. The environment is also respected, as re-growth is easier on small plots than in larger fields. Last but not least, the dispersion of the cultivated units is a way to secure food resources in regard to social risks such as the destruction of a garden by enemies, a break-up of a former alliance resulting in the loss of plots, etc. Land strategies must be taken into account as well: plot cultivation in remote ancestral lands, sometimes as far as several walking hours away from the village, is also a way to claim the land's ownership as well as, perhaps, to maintain its memory.

The Melanesian garden thus appears as the optimum answer given by farmers to the conditions of their natural and social environment. From a social point of view, the gardens also constitute a place of great distinction, because their beauty bespeaks the farmers' ability to acquire traditional knowledge, magic and gardening skills.

Such a relation involves, in addition, specific practices and gestures that alone testify to a deep respect (and ethics) of the farmer for his cultivated clones. In Pesena, during the harvest of the so-called *uha* taro, the leaves of the head-sets to be replanted in the garden are folded with great care along the stems instead of being cut off (Fig. 592).

••• Biodiversity production

Vanuatu farming systems thus rely on vegeculture, and one can be astonished to see so many varieties within vegetatively propagated species. The farmers knew indeed how to benefit from the species brought by their ancestors as well as those imported later on, such as sweet potato and cassava, introduced in the second half of the 19th century.

First, the personalized mode of culture, where each clone is cultivated individually and receives distinct care, allows the farmer to know perfectly the material he is handling. The least variation in color, texture, shape or taste is indeed immediately identified in the field, in the pan or the prepared dish. Farmers select different varieties and thereby diversify their stocks.

••• Selection process

The production of various morphotypes or chemotypes operates mainly in the natural environment, from wild stocks that have escaped from cultivation or residual crops in the fallows. Farmers working within a long revolution system, in places where wild and cultivated forms belonging to the same species coexist, sometimes manage (unconsciously) to produce interactions between both forms. Thus, some taro produces seeds while intercrossing and produce morphotypes or a particular chemotype that the farmers can observe during a walk in the forest. For other species, like island cabbage, cassava or sweet potato, wild forms do not exist. But the plants can remain in the ground many years in the old fallows. They produce seeds resulting from sexual recombination and thus the production of diversity. All these wild morphotypes are recognized by the farmer who will collect and transplant them in his gardens in the view of vegetative multiplication. First, the wild stock collected has to be multiplied in order to observe the aspect, the texture and the behavior of the resulting clones. Next, the gardener evaluates the role it could play among the range of varieties already cultivated. This is a slow and complex process (as the subterranean useful part of the plant is not

Figure 592: Harvest and replanting of *Uha* taro in Pesena: once the taro corm has been harvested, the leaves are folded carefully along he stems, the head-set is then replanted.

seen) that can only be done with a few specimens. Indeed, the plant must be grown to maturity for the gardener to be able evaluate the quality of the edible root or tuber. However, this process allows the improvement of the planting material, as it always involves selection of the best clones.

Finally, if the new cultivar is satisfying, the first need is to name it. It is generally named after the man who discovered it, after the place where it was discovered, or sometimes in reference to a known cultivar similar to it and whose proper name is supplemented by an adjective referring to its color or shape. The named cultivar is then propagated and integrated definitively into the existing stock.

Selection of fruit- or nut-bearing species is a different process, since reproduction is sexual. Fruits, seedlings or germinated seeds are collected under trees —growing wild in the forest or in the domestic space— whose fruits seem to show new characteristics. They are then transplanted inside the village, along pathways leading to the gardens, or in orchards in the village's surroundings. Their growth is then observed. This process is simpler than the former, but also much slower. The nomenclature applied is thus looser and the intraspecific variability is less important. There is, however, an important exception: that is the breadfruit tree (*Artocarpus altilis*), which is propagated by both sexual and vegetative means in Vanuatu. Indeed, the cultivation of this tree is somewhat closer to the cultivation of root and tuber crops, as it produces suckers (young shoots emerging from the tree's roots) that can be taken and transplanted, thus allowing the preservation of selected characteristics. One single island, Malo, thus shelters more than one hundred breadfruit varieties, each with its own name and uses.

Although it is not possible to embrace at a glance the whole range of one's cultivated plants, the gardener has perfect knowledge of his stocks. He often knows about the stocks cultivated by the other villagers as well, and he often knows about those of the neighboring villages too: farmers are then able to name, locate, and even describe hundreds of forms. This precise knowledge of the agrobiodiversity maintained in a broad geographical area (or rather within a vast network of alliances) combined with the individual care given to each clone enables farmers to locate and select every single variation. However, everyone does not select new forms or at least select them at the same rate and intensity. The efficiency of the selection depends very much on the farmer's curiosity as well as on his observational skills and his taste for experimentation. Each of those qualities is indeed necessary and selection is a process that is continually implemented throughout a lifetime. Those who are doing particularly well in that field therefore gain a reputation as "specialists".

The introduction of a new morphotype into the village's stocks is often followed by a double check of its originality, especially when the discoverer is poorly experienced or if he is not inclined to selection: a young farmer would then consult family and friends, a specialist would consult another specialist in a distant village etc. Checking is based on pre-existing social networks, which are partially maintained by this activity. This great knowledge and interest in selection are unfortunately being lost nowadays as a consequence of the degradation of the social prestige formerly associated with this activity.

••• Management practices

• Stocks handling

Inhabitants of Vanuatu always say that each plant species has its own use. If someone does not know the use of one plant, then someone else does. And if nobody knows, someone will surely discover it one day. Thus every plant is respected and named. Any species can be phenotypically diversified, as an amenity, to satisfy a need, or in the search for plants to use as food or medicine, for example. Knowledge of the plants and their diversity is garnered by observation, from relatives, and from those who know. It is a voluntary personal quest, during which the youngster secures himself within the group, increases his alliance network and finds his place in it. Knowledge and improved practice go hand in hand with the establishment of the social network in which the individual evolves. Conversely, the maintenance and the regular activation of this social network contribute to the dynamics of the selection processes and of biodiversity management.

The village of Pesena (300 inhabitants) on the eastern point of the horn of Santo constitutes one of the largest holders of taro cultivars diversity, at the island level or perhaps within the whole archipelago. The heritage of cultivated varieties is, indeed, so vast that this site merits detailed attention. This site exemplifies the probable limits of human faculties of discernment, with nearly 150 varieties of taro, each with its own name. As we will see below, the culture "plays" (this expression is borrowed from Panoff in 1972) with all this phenotypic variation, to define a wide range of uses, in various technological, cultural or aesthetics' contexts.

In Pesena, the diversity of varieties cultivated on irrigated surfaces is arrayed in the fields in a "Matrushka doll" pattern. The gathering of a few clones inside a small pond defines the basic unit of management or *pal-pal*. Several *pal-pal*, separated by stone lines, then form a pondfield *woura* (garden). These are of different types: small gardens (*woura parav*) are less than 10 m²; long gardens (*woura wasé*), are narrow pondfields, from 10 to 50 m long and two or three meters across. Finally, half-moon shaped canoe gardens (*woro owo*) are the widest. The gathering of several *woura* separated by low walls (*tin-tin*) defines the irrigated taro garden. And finally, there are several irrigated gardens within a village territory.

Diversity is favoured at every level. Even a small *pal-pal* containing no more than ten taro plants would count five or six different varieties. According to gardeners, taro plants grow better when mixed together, even on small surfaces. Indeed, this practice does constitute a sort of "filter" limiting the dissemination of pathogens.

The constant rotation of the clones from one pondfield to another is related to the mode of cultivation: as the varieties collected daily are immediately replanted, so as to secure a continuous production throughout the year, each garden shelters taro at various stages of maturity. Thus, the only way for a household to gather enough planting material when a new pond is opened is to pick seedlings in each of its pondfields. What is more, as the farmers visit various gardens during the week, the taro eaten in the evening is often replanted the following day in a different garden.

The losses caused by the fallow are not permanent, since lost cultivars will be recovered during a visit to a relative or friend. The fragmentation of the cultivated units, indeed, is counterbalanced by an intense circulation of varieties within the whole territory and between households. Sharing varieties is particularly involved in every social event in the village. Every ceremony (whatever the context, traditional or modern) involves tubers (as food or as an item of ritual exchange) and implies a certain redistribution of varieties between households. Marriage, because it creates bonds between families and clans, is a major vector of cultivated varieties. Indeed, if the young couple inherits the gardens and the taro of the husband, they will be able later on to draw from the heritage of the wife's family to complete their stock.

Table 54: Biodiversity of root and tuber species in Pesena (Santo) and Ipota (Erromango).

Botanic group	Bislama	English	Linnaean	number of land races	
				Pesena	**Ipota**
Dioscoreaceae	Wael yam	Wild yam	*Dioscorea. nummularia*	1	1
		Round yam	*Dioscorea bulbifera*	3	3
		Five-leaved yam	*Dioscorea pentaphylla*	1	3
	Sopsop yam	Greater or water yam	*Dioscorea alata*	14	20
	Swit yam or Wovilé	Lesser yam	*Dioscorea esculenta*	3	3
	Houaïlou	Guinea yam	*Dioscorea cayensis* *Dioscorea rotundata*		2
	Africa	Kus kush	*Dioscorea trifida*		1
			Dioscorea spp		1
Araceae	*Taro*	Taro	*Colocasia esculenta*	156	9
	Navia	Giant taro	*Alocasia macrorrhyza*	4	3
	Taro fidji	Cocoyam	*Xanthosoma sagittifolium*	3	4
Euphorbiaceae	*Maniok*	Cassava	*Manihot esculenta*	15	16
			Manihot glaziovii		1
Convolvulaceae	*Kumala*	Sweet potato	*Ipomoea batatas*	8	7
TOTAL				**208**	**82**

Taro is by far the dominant crop in Pesena as well as the most culturally valued. However (and surprisingly), the villagers also cultivate a wide range of other roots. Indeed, the rain-fed gardens located on the slopes of the many hills of Pesena shelter a diversity of yams (five different species, with a total of 22 varieties), cassava (*Manihot esculenta*), sweet potato (*Ipomoea batatas*), cocoyam (*Xanthosoma sagittifolium*) and giant taro (*Alocasia maccrorhyza*). All in all, the villagers in Pesena maintain a diversity of root crops totaling more than 200 varieties (Table 54). To our knowledge, there is no other example in the world of such variety within vegetatively propagated crops.

Some other villages of Vanuatu, although less diversified at the intraspecific level, grow an even larger number of root species. The distribution of varieties is then more evenly shared among all the different species. This is particularly the case of villages with drier agro-systems that have yams as the main crop. Yams are seasonal, so villagers cannot rely on this crop all year long, and production must be diversified. This is demonstrated by the example of Ipota, a "yam-based" village in Erromango (southern Vanuatu) that harbors 82 varieties of root crops, shared among at least 15 different species (Table 54). In some localities however, yam remains the main component of the agrosystem and the noble species, the Greater Yam (*Dioscorea alata*) is extremely diversified. That is the

Figure 593: Yam diversity in Brenwe, North-West Malekula (newly harvested yams).

case for exemple in North-West Malekula where the Greater Yam (which has more than 40 cultivars) is still an important emblem of identity (Fig. 593).

• A management oriented towards use

As stated above, agrobiodiversity management in Vanuatu is closely linked to a wide diversity of uses belonging to different spheres of cultural life, from cooking to traditional medicine, rituals and even aesthetic considerations.

Thus, every group differentiates in its own stocks of cultivars those that are to be used for daily

Figure 594: One of Pesena's many pondfields: practical and esthetic considerations are associated, producing rich, beautiful gardens.

consumption only and others whose role is to serve in rituals. These latter cultivars always belong to the traditionally culturally valued species: yam (*D. alata*) and taro (*C. esculenta*). These are also characterized by fine organoleptical properties (palatability, a firm consistency, and often a pleasant smell). Those varieties are cultivated according to the process of ceremonial horticulture allowing production of massive tubers. They are then given away during ritual exchange ceremonies or used for the preparation of ritual dishes such as *nalot* on Santo. Cultivars used for daily food consumption are of lower quality and are cultivated according to less intensive horticultural practices. They are many of them, and they are always divided into different categories fulfilling different cooking needs: some varieties are to be roasted, others boiled (using a cooking pot), cooked in hot stone ovens, or grated for *laplap* .

Beyond those main uses, esthetic considerations are also taken into account: some taro varieties are appreciated because of the colour and design of their leaves, while a specific cassava variety named *flowa manioc* (literally, flower cassava) is found today in different villages of Vanuatu (when growing, it takes the shape of an umbrella providing shelter from the sun). In Pesena, sweet potatoes and cassava chosen for the colour and shapes of their leaves are sometimes planted to decorate the low walls (*tintin*) between pondfields. The low walls also shelter a diversity of more purely ornamental plants, among which *nangaria* (*Cordyline* sp.) is the most widely used in Vanuatu (this plant commonly delimits gardens of different users).

Many of these plants also have magic or therapeutic properties, improving fertility of gardens or protecting them against pathogenic agents or herbivorous animals. In Pesena, elderly people remember an ancient practice in which the juice of particular leaves poured into water draining to the pondfields is used to improve taro yield. Conversely, some other leaves were used in the past to poison gardens of enemies. A number of colorful creeping plants planted throughout the gardens are also used to treat some human diseases, while others have contraceptive and abortive properties.

The gardens of Vanuatu thus express an "ethical" relationship between human beings and their environment, i.e. a relationship filled with meaning (a world view) in which magic and knowledge are associated to produce a beautiful garden bearing quality products (Fig. 594).

••• What does the future hold for the gardens of Vanuatu?

Again, the case of Pesena is illustrative. Far from being an untouched haven of tradition, the western horn of Santo went through many crises during the twentieth century. Contacts with the Europeans (mostly traders) resulted, as in other places in Vanuatu, in massive depopulation. The survivors of the interior valleys found safety by joining the religious missions established on the seashores, but at the price of having to adopt a new model of social and economic organization. For Pesena, this means that the farmers succeeded in re-creating, in spite of the crisis, consistent social structures favorable to the preservation of a rich agrobiodiversity. The new "roads" opened by the Catholic mission there (connecting to other missions of Vanuatu and to the emerging urban centres) were themselves followed, so as to increase the number of cultivated varieties (as shown by the number of varieties of root crops introduced via those channels). One of Pesena's assets here is probably to have kept as well, along with the new connections, close relations with other groups in the region according to traditional social networks (namely in the villages of Olpoï, Hokua, and Wunpuko on the west coast).

In addition, cultivation of taro in irrigated systems is itself a great asset, because it readily maintains a great diversity of cultivars. Indeed, while rain gardens require the clearing of new gardens every year, the pondfields produce continuously, and maintenance is restricted to weeding. Farmers themselves admit that, because of their pondfields, it is easy for them to keep taro as their main food and still have enough time to focus on cash crops (mainly kava and copra nowadays) and community works.

Nevertheless, there is now a gap between youngsters who were brought up according to the mission's standards and elders, who are the keepers

Figure 595: Kava at Hokua.

Figure 596: Recently harvested kava roots, Pesena.

of traditional knowledge. In the last decade, trading of kava (Figs 595 & 596)has allowed a flow of cash into the village. This has resulted in the establishment of small retail stores offering a range of imported foods (mainly rice and canned goods) that are already affecting diets. Today the villagers dream of material wealth, and they tend to shift their focus from the taro pondfields to the kava gardens. The system may already be reaching its limits, as is shown by the general decrease in taro yields, caused by mud and algae accumulating in the pondfields owing to lack of maintenance.

Another source of worry is that farmers today tend to favor a few high-performance varieties that produce acceptable yields even when grown in old gardens. As a consequence, much of agrobiodiversity relies today on the activities of elders, either because they still live in accordance with a traditional system of values in which the richness of the gardens accounts for the value of a man, or because they are aware of the value of the heritage that is slowly disappearing and they voluntarily take on the role of preserving it. Genetic erosion is even faster when farmers deal with vegetatively propagated crops, because the varieties that are not planted any more are bound to disappear. Pesena is probably very representative of Vanuatu itself: a place rich in agrobiodiversity, but a very vulnerable one.

As for food security, the drastic increase in prices of imported foodstuffs is now a major problem because these have gradually replaced a consistent major part of the gardens' food production. How, then, is the rapidly growing population of Vanuatu going to meet its increasing needs? As the country has only limited scope for increasing export income in order to sustain food imports, it is likely that the people of Vanuatu will have to rely increasingly on their own crops. It is thus urgent to increase food production: this will have to come about by intensifying use of the existing agricultural land and by fostering the existing high cultivar diversity of food crops.

A way to achieve that goal is to promote conservation and breeding strategies allowing the conservation of genetic resources of traditional root crops before they are lost, as well as broadening the existing genetic base, if those crops are to be able to respond to rapid environmental changes. Conventional strategies encounter many difficulties when applied to minor root crop species (low financial support, difficulties in distributing genetically improved material to the growers, genotypes/environment (G x E) interactions, etc.) Thus, an alternative strategy based on the geographical distribution of allelic diversity rather than on localized *ex situ* preservation of genotypes has been launched in Vanuatu. It aims to increase farmers' long-term access to useful genes by providing genotypes representing the useful genetic diversity of the species. Farmers then select varieties adapted to local conditions.

There are of course other possible solutions. The economic potential of these crops on the fast-growing urban market should be assessed. This implies finding a way to reduce transportation costs from the scattered production sites to the market so that these food resources could constitute a real alternative to imported foodstuffs, and thus fill in the "missing link" between urban and rural areas. Other economic outlets should also be considered, namely in the food industry, which requires compounds existing in traditional root crop varieties (e.g. for the processing of natural coloring or hypoallergenic food products).

The future of Vanuatu's rich agrobiodiversity in the end rests in the hands of both political leaders and aid donors. Unfortunately, traditional crops do not attract the attention they deserve. Rather, they tend to be disregarded as the remains of primary, underdeveloped agricultural practices. The main task might be to assign their real value to these plants, because, in Vanuatu as elsewhere, the future can only be built on the heritage of the past.

AT THE JUNCTION OF BIOLOGICAL CYCLES AND CUSTOM: THE NIGHT OF THE PALOLO

Laurent Palka

Twice a year, the shore of Santo becomes the theatre of a strange invasion. During the seventh night after the full moon in October and November, thousands of headless worms called palolo (Figs 597 & 598) swarm at the surface of the water on the East coast. The phenomenon lasts only a few hours but gives rise to a traditional celebration by people that predict the date of emergence and gather to fish the worms. The night of the palolo is a natural and social event known since the mid-19th century.

• • • Description of worms

Palolo is the Samoan name of the marine worm *Palola viridis*, formerly *Eunice viridis*. Swarming specimens, collected in the Samoan Islands (Polynesian region) by the Reverend J.B. Stair, were first taxonomically identified by Gray of the British Museum in 1847. Since then, twelve species of Palola have been reported, mostly in tropical zones, although some species do inhabit temperate latitudes. All the species have shown the same swarming behaviour as *P. viridis*, but do not necessarily swarm at the same time of year.

The genus *Palola* belongs to the phylum of annelids that comprises all segmented worms, and to the class of polychaetes where each segment bears two lateral appendages (parapods) terminating with cirri. *P. viridis* belongs to the family of Eunicidae and has a slender body (soma) made of many short and homologous segments containing all the organs and muscles. The anterior region has a short head (prostomium) with cylindrical or tapering antennae and palps, and two simple eyes. The mouth possesses a movable jaw and a typically scoop-shaped mandible. *P. viridis* is also characterized by the absence of hooks and the presence of branchiae as single filaments from the mid-body region. Males and females are morphologically different, brown-reddish and green-bluish, respectively (Fig. 598).

• • • The effect of the moon

During the year, the worms are benthonic and inhabit tunnels that they dig in coral reefs at several meters depth. They apparently feed mainly on algae, but also appear interested in protists, small crustaceans and, sometimes, their own young. As the reproductive time approaches, males and females undergo a metamorphosis called epitoky, well known among polychaetes.

Figure 597: The palolo worms wriggling on the surface of the water.

Figure 598: The palolo worms, once fished.

Figure 599: Santo villagers fishing for palolo.

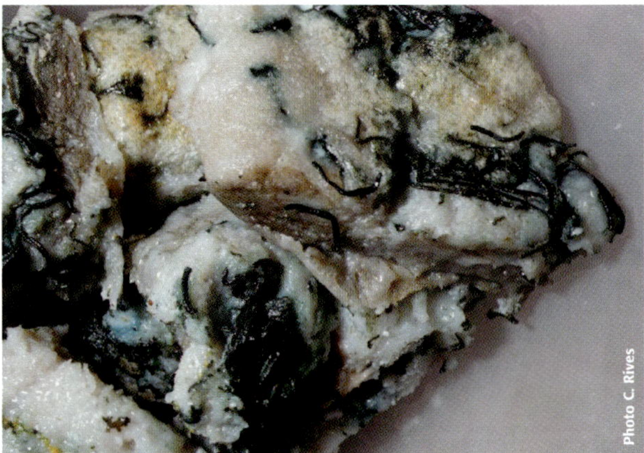

Figure 600: A traditional dish filled with worms: the lap-lap.

The posterior segments begin to degenerate by histolysis and phagocytosis. The reproductive organs increase in size and produce gametes while two photosensitive pigment spots develop at the ventral face of each segment. The development of supplementary eyes allows the very low illumination produced by the moon in its last quarter, the light signal that activates the swarming, to be detected.

Then the epitokous portions filled with eggs or spermatozoids detach as a mature sexual satellite (stolon) and wriggle to the surface (Fig. 597), adopting a pelagic lifestyle for some hours. Fertilization takes place when the epitokes disintegrate releasing eggs and sperm into the water column. Such swarming can cause a spectacular change of the sea's appearance over wide areas because of the milky and opaque aspect of the mucus. The anterior end (atokous) containing the head survives and remains in the substratum where it regenerates new segments in order to reproduce again the next year. Epitoky improves the chance of sexual partners encountering each other, for instance in bringing individuals to the surface of the water, i.e. in 2-dimensional space.

••• Celebration by the locals

The fishing is greatly anticipated by the islanders, who spend the entire day preparing for the big moment. The families, who take turns from year to year, gather in front of the campfire to sing and dance, with the exception of any young or "unpredictable" children. This is not because the

feast will finish late, but because it is forbidden to pronounce the name of the palolo. In fact, local mythology holds that the worms are the precious jewels of the spirit of the ocean's wife. The legend says, if the spirit hears that people are preparing to catch the jewels, he will warn his wife who will keep them and nothing will take place.

The critical moment comes towards midnight (Fig. 599). People arm themselves with lamps or torches to attract the worms and wade into the sea on pirogue or by foot, depending on the location of the fishing site. Only the elderly seem to know exactly where the worms will appear, that is, not necessarily in the same place from year to year. It depends, for example, on the intensity of the moon light. Experienced people also detect the presence of worms because of a strange smell, the smell of palolo. Then the fishing begins. The wriggling worms get stuck on everything. Formerly, people plunged their arms into the water, then made them swirl to catch them. Today, people use nets and sieves. After two hours of fishing, people thank the gods and head back home.

The following day, the women prepare the traditional recipe (the lap-lap) in which the worms are deposited on pasta made of plantain banana and manioc, a bit like a "pizza" (Fig. 600) that is subsequently wrapped into a wild banana tree leaf and baked in an oven made of volcanic rocks. When the fishing has been good, it's frequent to see the lap-lap sold on the market of Luganville.

NI-VANUATU PERCEPTIONS AND ATTITUDES VIS-À-VIS BIODIVERSITY

Florence Brunois & Marine Robillard

Understanding the level of perception that the Ni-Vanuatu populations have of Santo biodiversity was one of the objectives of the Santo 2006 expedition, in parallel with the inventories of that same biodiversity by non Ni-Vanuatu academic

scientists. The scope of the study is obviously immense, and one only needs to emphasise the social and environmental diversity encountered on this island to be convinced of that! With an area of $3\,959\,km^2$, no less than forty languages have been catalogued, that is forty distinct societies with a fabulous diversity of ecological environments and socio-biological histories as numerous as they are varied. This present study does not therefore aspire to completeness, nor to making generalisations on a Santo-wide scale. With more humble ambitions, it deals with the comparative ethnological research we conducted for two months across five communities evolving in very different environments: the island of Malo, the tropical forest of the foothills (Butmas), the coastal forest of Vatthe (Matantas) and Okwa (West Coast), and around urban Luganville.

During our investigations in Bislama, we systematically combined both ethnoscientific and social anthropological approaches. This methodological choice was fundamental in order to translate as accurately as possible what biodiversity represented for each community. Indeed, our desire was not to impose a naturalistic view of life, nor to make an inventory of living beings and their practices that would illustrate local populations. Our intention was rather to seize their unique perception of the environment by questioning their own modes of learning and of identifying living beings and the practices and relationships that are associated with them. The results obtained during just two months of investigation have enabled us to highlight the tremendous complexity governing the Ni-Vanuatu awareness of biodiversity by means of certain constants that we will now develop.

••• Diversity of life as part of a complex and territorialised world

The first lesson the Ni-Vanuatu teach us is how to look at what biologists call "biodiversity" in a much wider context while at the same time remaining strongly territorialised. Indeed, animals and plants are not the only "living" beings to participate fully in their world. The creators, the ancestors, the spirits of the clan, the spirits of certain spaces or species (plant and animal), and even evil spirits are an integral part of life, and are created in the image of humans. Biodiversity is therefore defined as part of an extremely complex universe: to its phenomenal dimension has to be added another dimension, albeit invisible, but equally vivid. Origin myths are at the centre of this unique concept, which does not dissociate the world of people from that of non-humans (Fig. 601). Thus, in Okwa, the various clans comprising the community originated from a plant, or from a bird, or from a fish, or from a source of fresh water, whilst in Vatthe they would derive from the taro plant. In Butmas, although "*Atarr*" is well recognised as the creator of the earth, it is the activities of human subsistence that give rise to the morphogenesis of the caves whose guardian is "*Sosole*", the Monarch of the banks. In both places, mythology eulogises this metamorphosis. Cosmology, the rules governing the functioning of the world, adds amazingly to this permeability of boundaries. For all the communities interviewed, each of the spiritual entities present in their world holds the power to intervene appropriately in the visible world in the form of living beings (birds, wild pigs, snakes or fish) or speaks through natural phenomena such as rain, hurricanes or thunder. In addition, spirits can also affect the lives of humans (through illness or death), by attacking either the visible person or their double in the invisible world of dreams. This cosmological context therefore encourages the human populations to maintain a certain relational harmony by practicing rituals, sacrifices and offerings as well as observing the numerous taboos, which relate to the spirits. Compliance with this ethical behaviour is vital for the survival of humanity, but also to prevent the spirits from leaving the territory, and with them the living beings and the spaces of which they are custodians.

The world of the communities studied is complex; it is also very territorialised. As Joel Bonnemaison stressed in 1985, "*Identity here is a "geographical identity" which is derived from the memory and values attached to places. [...] Each local group is thus a sort of "geographically-based society" which is defined by an area or even a "territorial society" which derives its identity not only from ownership of a common territory, but from its identification with it*". This close identification with territory is indeed remarkable: it becomes formative in transforming the island of Santo (and understanding its biodiversity) into a multitude of islands with clearly defined contours, linked only by the paths of their alliances and the wanderings of their ancestors. Does this phenomenon explain that the myths of origin of each community encountered extol immanence and that the identity of each community is built according to its experience with the beings and singular landscapes which make up its territory?

Figure 601: Willy, a Butmas chief, explains to Florence Brunois (anthropologist) how some plants are usefull not only for humans or animals but also for the spirits which inhabit the forest.

To understand what constitutes biodiversity among Santo populations therefore requires an assimilation of this context, which is as complex as it is geographically sharply defined. It is in the light of this that we can seize the modes of knowledge and acknowledgement of beings who inhabit their environment. And as we shall see, this context, as original as it is, does not at all produce a confused vision of reality and its events. Rather, it imposes an observation as demanding of others as a global and integral vision of relations linking people to local biodiversity.

••• Comprehending the diversity of life

The singular world of which biodiversity is a part exerts a definite influence on how living beings comprehend things. First and most logically, the territorial boundaries of the everyday universe delimits the species and the spaces to be comprehended. The diversity of lives is therefore experienced locally, and the scale on which this is measured is primarily that of the territory and local alliances, and rarely extends beyond these borders, except for those who have left to work elsewhere, in town or in forestry or agricultural farms. To this geographical demarcation is added an historical delimitation [12]. Genealogical memory extends only to the fifth generation, and beyond this historical reach, memory evokes the journey of the ancestors as the immanence of beings and of the world. To sum up, the presence of an invisible dimension to each territory, and hence each geographical community, makes it more complex to understand the phenomenal dimension. It requires the Ni-Vanuatu to be strictly attentive to the context of their interaction with the world. The same holds true for any ethnological study. Indeed, only such observation *in situ* is likely to give meaning to the different ways the Ni-Vanuatu have to engage with other beings. This is true, for example, of the attitude of Willy (chief of Butmas village) while we were walking together in the woods. Seeing me (the first author) struggling, he told me to be quiet, and to stop questioning him about the trees that I saw. His look, like the tone of his voice, was authoritarian. Finally arriving at the top of the hill, he explained his attitude. We had just crossed the territory assigned to "*Taïfali*", the spirit of wild pigs, who causes disease to any intruder who dares to speak the language of humans, and even worse, it may refuse to provide pigs for the hunters. Different attitudes towards the shark from the "*man blo bush*" who lives on the coast and from the "*man blo salt wota*", in Malo are equally eloquent. While the first will fish for the shark with all the care needed for this large predator, the second will do everything to avoid meeting him because this fisherman knows he may face an "*Ambalewo*", a spirit or a man who has borrowed the form of a shark. Also, unlike the first man, the "*man blo salt wota*" will simply not fish at night along the reef or in areas known for their abundance of shark. This difference of comprehension (and the practices that flow from it) reflects the distinct impact of their engagement with the living being, the sharks. For the fisherman of Malo, the impact of this act actually extends beyond the only framework he knows for the gathering of a resource. As part of his territory, where reefs, sea, and land are intrinsically linked to the spirits and his ancestors, the gathering of resources requires him to take on this relationship with animal-spirits and its implications, which may extend into his community.

Vigilance is required. It certainly explains the concern of the Ni-Vanuatu that they are in the presence of a living being when they demonstrate their knowledge. The morphological criteria, such as they are described in naturalist books, are not sufficient even if known and named with great precision by the authors. The recognition of a living being demands the collection of additional information that the Ni-Vanuatu have themselves observed during their daily routine activities. Thus, on the basis of data collected, which cover 111 marine fishes and shellfishes, 31 freshwater fishes and freshwater shrimps, all birds and mammalian wildlife, 10 or so types of ants and 250 plants, the recognition of all living beings first requires recognition of their "habitat", which may be confined to a rock, a tree, an area of freshwater, a sandbar, or it may extend to a reef, a cave, even an animal species such as cows for the common Myna bird, or a forest ecosystem defined by its altitude, a clearing, a path, the household, and so on. However, this

12 - Another delimitation of a physiological order is relevant here: that of vision, an essential sense according to the Ni-Vanuatu, used to access knowledge of the living world. Someone who came to visit the marine laboratory in the Maritime College freely agreed: "*We did not know the name of that animal, but that is because we do not have the equipment. But we have known for a long time that there are tiny things that you can not see, you see when you look at the sand that it moves, only it's so small you can not look and know its name. If we had the device (the microscope), we would know for sure*". Let us also note that the sacredness of certain karstic sites can slow down the knowledge of its inhabitants.

Figure 602: The more famous and precious mammal of Vanuatu: the pig. This animal occupies a fundamental place in the history and everyday life of islanders. It arrived with them during their initial migration; it's body is a substitute for that of a human, and sometimes is considered as the lifestock of the spirits. Its complex taxonomy reflects the crucial role it plays in these societies.

typology, which gives names to the places that are occupied, does not prevent the Ni-Vanuatu from recognising that the plants and animals have the ability to migrate, to introduce themselves into new territory or conversely to disappear altogether. Such is the case with the unexplained disappearance of the Scarlet Robin from the territory of Butmas or conversely the arrival of cows that ran away from Matantas and many plants which "came with trucks from loggers" or with seeds scattered by the wind and cyclones. A living being is first and foremost an acting being. It can act on its own, it can also act on the initiative of its appointed spirit as is the case with the wild pigs of Butmas, who are treated as the livestock of the spirits, who depending on their goodwill, make them fatten, distribute them to respectful hunters, and make them migrate within the territory (Fig. 602).

According to the Ni-Vanuatu interviewed, a living being is an inherently relational being: it certainly exhibits specific behaviour but also inter-specific behaviour. Thus, according to the constancy of the being, it's a matter of recognising a type of behaviour, a relationship to an animal, a plant or even a human. Whilst the sea anemone is primarily known for being the "House of fish", the blood loach is regarded as a close friend of the small black loach, and stays close on the rocks in deep waters. As for the Pacific pigeon, the people of Vatthe recognise it in terms of its mocking behaviour (the female systematically recovers the shell from her eggs and takes it away from the nest to deceive predators), while the yellow-fronted zosterops is characterised by its sad song, which announces the forthcoming death of one of their own. Walking in the woods, Ni-Vanuatu canot fail to recognise the "*Hawuraï*" tree (unidentified), which emits a very strong smell or to hear (see) the rustle of leaves of the "*Natiritiri*" tree (unidentified) because the "wind sings in this tree". In Butmas and likewise in Okwa, it is well known that the purple swamphen is a thief, and the sudden appearance of the glossy swiftlet announces a hurricane or a storm. Inter-specific relationships can be so narrow that the observation of one of the actors is enough to call to mind the other, as is the case with the barn owl, which is supposed to be the materialisation of a magic spirit in Butmas, or in a completely different way, the case of the tree *Pterocarpus indicus*, whose vegetative cycle serves as a time marker for the reproductive cycle of fish in Malo, or the case of the "*Naho atena malao*" tree (unidentified) in Vatthe at the foot of which the Vanuatu scrub-fowl (called "*Malao*" locally) builds its nest.

••• Naming the diversity of life

Recognition of living beings is therefore governed by the assimilation of several criteria: morphological, ethological, ecological and cosmological. The same is true of their naming and their classification (Fig. 603). Without claiming to identify the linguistic logic governing local nomenclatures, lacking of course a mastery of their language, we were able, on the basis of all the names collected, to isolate a few relevant traits in local nomenclatures. The most widespread is undoubtedly that which designates a living being by a monolexeme —a name composed of a single segment— since this rule concerns the vast majority of flora or fauna identified. A being thus named could be regarded as the basic local unit. However, it is impossible to apply an absolute value to this phenomenon. Indeed, a monolexeme could equally be applied to many animals which, while being recognised as different, are nonetheless called the same thing. This is true of small freshwater fishes. So, whilst the people of Butmas distinguish 11 species of loaches and clingfish, all are designated by a single name, "*Famendro*". Another widespread scenario is to designate two species of dimorphique shrimp by the same term, considering that one is male and one female. In other circumstances, a monolexeme can become a generic term encompassing a whole genus, family or order. Such is the case, for example, in Vatthe where they replace the individual names of different ants with the general term for an ant, "*Natsitsiu*". We also found this phenomenon with plants. Depending on the context of the conversation, rather than specifically naming the specimen, the Ni-Vanuatu define it by its category, named also by a monolexeme: vine, tree, grass, etc. This lack of systematic correspondence between nomenclature and classification was confirmed with the study of beings whose name is composed of two segments. Here again, this naming system may cover several hierarchical realities. It can correspond to basic binomial nomenclature, gender + species. Thus, to the base term will be added an identifier which specifies either a morphological trait (referring to a part of the animal, a colour…) such as in the

Figure 603: Some Ni-Vanuatu people are asked to identify and name river fishes by looking at photos in a scientific book. This kind of investigation is often unsuccessfull as people have a contextual rather than literary knowledge of living beings.

case of loaches, "*Avae*" with "*Avae Mboe*", a loach with black spots, "*Avae One*", a loach with blue spots, or specific or interspecific behaviour such as with loaches, "*Raue*" with "*Raue Até Até*", those who remain at the bottom without moving, "*Raue Bouaha*", deep water loache or "*Raue Revevure*", the red loach, friend of small black loaches. These consolidations in nomenclature are given further meaning in these terms: "This is his uncle", "his brother", "his friend".

Moreover a taxon composed of two segments can be attributed to a being without their being considered as forming a subcategory. The banded sea krait exemplifies this while revealing another singular phenomenon of local taxonomy: a living being may be from two classifications and therefore receive two separate names. Thus, in everyday life in Malo, the snake will be named "*Vine Vine*"; however in the context where it houses the spirit "*Tanoume*", it is called "*Lororua*". And finally, the formation of a binomial may refer to another type of hierarchy, as is the case with pigs in Okwa. Like many Melanesian societies, wild pigs are the subject of an elaborate classification based on the colour of their bristles, reiterating the crucial role that this animal plays in these societies. Thus, under the global generic name of "*Boy*", we find "*Boy Tawuka*": white pig; "*Boy Talolo*": black pig; "*Boy Tavetwet*": coloured pig; "*Boy Tawukmuti*": black and white pig; "*Boy Tahahara*": red pig, "*Boy Tachich*": pig with a corkscrew tail; "*Boy Tamarack*": pig with curved tusks, and finally "*Boy Wot*": the castrated pig. In Butmas, the naming of a pig depends on its state: wild, semi-domesticated, "marronneur"…

The study of local Santo nomenclatures and classifications is therefore of great complexity. The analysis of names given to non-native plants confirms this. Indeed, of the 16 plant species recognised as non-native by the Butmas community, ten were introduced towards the end of the 1990s and are simply not (yet?) named, even though their individuality and their ecology have been identified, causing this humorous reaction from Willy when asked to name them: "*You have to ask the white people what the name is!*"; five were introduced with exogenous names; and a single plant, *Mikania micrantha*, which was introduced in the 1940s, has received a local name. These data are exciting because they have helped highlight a phenomenon that deserves more general study. In Vatthe, not only have we found the same exogenous names for the same non-native plants but also the same use for the *M. micrantha*, that of treating gaping wounds. This diffusion of the name, associated with the use of a plant, may explain how some animals such as the Vanatu scrubfowl or a species of small bats, retain the same name from "*Okwa to Vatthe*" via Butmas.

Figure 604: Marine Robillard (Anthropology student) is engaged in a deep discussion in pidgin with Ni-Vanuatu women to understand clearly their knowledge regarding crabs and how they relate to them.

One thing is certain: understanding, recognising and naming a living plant or animal is not conditioned by what use it can be to humanity (Fig. 604). As useless as they may seem, people from Malo call seaslugs "*Nemetamate*" and coral mushrooms "*Stagas mauru*". And yet, examples of language use that the Ni-Vanuatu have borrowed from the living world are manifold. Those attributed to plants are obviously more varied. By way of illustration, for 142 plants identified by the Butmas, and excepting the six plants deemed unnecessary — although they nevertheless have names— we have listed 153 uses. They break down as follows: 3.9% are useful for wildlife exclusively, 18.3% for wildlife, spirits and humans, then for the use of humans only: 7.7% for medicinal purposes, 12.3% for construction, 15.5% for subsistence techniques; 10.4% for tools; 9.7% for consumption; 5.2% for firewood; 4.5% for rituals; 5.8% for body decoration; 1.3% for crafts; and finally 5.9% for various uses such as, suicide, indicating the season, building a musical instrument, domesticating beetles, treating semi-domesticated pigs or driving away evil spirits. The multiplicity of uses is a measure of the diversity of flora in this territory. It is none the less clear that the usefulness of the plant world is not exclusive to humans: wildlife and the spirits share this privilege.

••• An integrated vision of the diversity of life

Although the discontinuities of the living world are recognised, this knowledge does not produce a discontinuous vision of the world among the Ni-Vanuatu. Human society is not alienated from the diversity of other lives. From an early age, individuals are asked to identify themselves closely with plants and animals, with the landscapes of their territory, by means of rituals and narratives recounting the wanderings of their ancestors, by means of subsistence activities, or simply by the discovery of their lands as though they were an invisible part of them. Whilst each Okwa family recognises a particular affiliation with a living being, they will also learn that it is the cardinal honeyeater bird who taught them hunting, war and peace, that the forest, the sea or the gardens are under the respective supervision of a spirit called "*Tamate*", that plants and animals, like human families, have a "*Sar*", i.e. a head, and that the wind, the rain or a hurricane are produced by "*Sarya Nahoy*"… In Butmas, at the birth of a child, a banana cutting is planted near the household. If after ten days, the child is still alive, they are given a first name as is the banana, which is then transplanted into the taro garden. Then, when the taro is ready, a ceremony is held to honour and to compensate the maternal parents of the child. The good roots of the plant, and its fertility, will confirm that the child is well-rooted in the world. In the forest, the child is taught to see the remains of dead spirits in the banyan, to place an offering in order to enter a cave, not to fish eels in such and such a stream, which could be lethal, or to notice if this tree is in fruit so that a trap can be laid at its base, or there again to know not to touch nor feel such and such a plant at the risk of vomiting. The senses and emotions are the only tools to comprehend the diversity of life: they are the first things to be developed. In this way, as a result of these experiences of growing up, the individual nourishes its identity with the relations that bind beings to their environment. That certainly explains how in each community, the diversity of life is cultivated in their gardens, around the household. By the way, after any trip, it is usual to take back if possible a new cutting to plant immediately in the garden. Additionally, any new plants arriving by road or by air do not cause a strong *a priori* negative reaction. Judgement will only be made after observing how the plant fits into the world of everyday life. However, even if some plants are found after observation to be "unpleasant", like the thorny bigo *Solanum torvum*, which forces the gardener to cut it down, or even *Bidens pilosa*, whose fruit sticks the wings of chicks together, it does not qualify as "harmful" as defined by biologists [13]. *Mikania micrantha* is certainly considered a "tree thief", but it is a property shared by many other endogenous climbers such as the famous *Meremia*. The difference lies more in the fact that its invasive propensity is not controlled (integrated) like that of the *Meremia* in the traditional horticultural know-how: gardeners always take care not to leave any space for it when starting a new plot. This tolerance in integrating the plant "intruders" is not shared by farmers around Luganville, whose horticulture practices have long been decontextualised from the forest environment and from the know-how associated with it (Fig. 605). A similar divergence in attitudes is reflected with respect to non-native animals, which are said by some to be invasive, such as ants. In rural areas, where work in the garden predominates, these ants are not harmful, rather they are allies of people because they are predators of

13 - With the exception of the cannibals spirits introduced by white settlers who allegedly killed many people in Butmas.

species which devastate crops. They maintain the integrity of the garden, which leads to good harvests, but also act as decomposers of organic waste left after garden maintenance. Whilst some of the ants will venture into the house to gorge on sugar as well as food, they also protect the inhabitants of the household from millipedes, a species perceived as dangerous. However, in suburban areas, these ants are viewed somewhat differently. The changes induced by the lifestyle there reduce the opportunities to enter the garden and measure the usefulness of these insects. Not yet considered as harmful, with the exception of *Wasmania auropunctata*, ants, however, are perceived by urban dwellers as causing some inconvenience.

Decontextualised from their territory, city dwellers are also decontextualised from its sense of time. In Malo as in Butmas or Okwa, people are still very attentive to the growth of each form of life and their ecological dynamic: they herald and harmonise the cycle of human activities. Thus, "*Rarambut*", "the time to plant yams" in Malo comes when the bats are fat, when the seas are calm, when tides are often low and when the "*Vuvilay*" (*Pterocarpus indicus*) starts flowering. The concurrence of all these phenomena forewarns men that the earth is ready to be worked but also that it is the period of fertility for fish and that in a few months, when *P. indicus* loses its flowers, fish will be had in large quantities on the beach. The start of fishing for "*Ely*" (a wandering polychaete), which is so much anticipated, is announced by observation of the procession that the "*Naoura*" shrimp will form, then the "*Nawuita*" octopus (*Octopus* sp.), followed finally by the father of "*Ely*", the banded sea krait [14], "*Vine Vine*". Here, like there, people's sense of time adjusts to the sense of time of their biodiversity.

14 - The banded sea krait (*Laticauda colubrina*) is also named: colubrine sea krait or yellow-lipped sea krait.

This ecological awareness, in which the voice of man participates fully, explains how the Ni-Vanuatu are sensitive to the vulnerability of their ecological system. Whilst they recognise its fundamental

Figure 605: For the people of Luganville, the vine *Meremia* is regarded as non-native, whereas the people of Butmas, Vatthe and Okwa who regard it as indigenous

dynamics, they also know from experience — hurricanes, epidemics, logging — that living beings like humans or spirits are not around in infinite numbers. The balance is precarious and is the rigour of relational ethics, which imposes local cosmologies on people. In this traditional context, man cannot behave as master and owner of other living beings. We have seen this in Butmas, where pigs are under the guidance of the spirit "*Taïfali*", who makes the decision whether to deliver up the pig to the hunters after judging its good behaviour; or in Okwa, where the spirit of the seas will not fail to pound the soul of the fisherman who has not been respectful, like "*Tamate Nahavela*", the spirit of gardens who requires the practice of certain rituals to clear a plot or even to plant taro, whilst the "*Tamate*" of the forest causes the disrespectful hunter to get fever and tremors. In Malo, non-respect of taboos related to maritime activities can have devastating effects on land-based activities.

In this way, there is the collective fishing of the epitoke form of "*Ely*", the wandering polychaete, which come to reproduce en masse on the coasts for a period of three days in the year (a period which correlates to the full moon). Access to this resource is abundant, but it is also prohibited to menstruating women and bereaved individuals. In addition, it is strictly prohibited in this period of marine fertility to eat the "island cabbage". Transgressing these bans inexorably leads to rotting in the fields and causes the "*Ely*" to scatter so it can no longer be fished. Finally, even a breach of the rules governing social relations can encourage

the start of an ecological disaster: torrential rain in Okwa, the decline of coral reefs in Malo.

••• Conclusion

Men, living beings and spirits are all interlinked in a single relational system as complex as it is fragile (Fig. 606). Breaking these ecological ethics and the traditional knowledge which maintains it is to break the balance of their world. In Malo, schoolchildren are aware of this when they say "*before there were lots of animals and plants in Malo, but the older generation have not passed them on to us. Now there are a lot fewer*". We can wager that new generations will recover the formidable ecological know-how of their elders and enrich this with the naturalist knowledge that the Santo scientific expedition has brought to re-invent their world of such great riches.

Figure 606: After collecting plants in the Butmas forest with the chief Willy, Marc Pignal (botanist) and Florence Brunois (anthropologist) record the botanical knowledge of children to assess to level to which traditional folk-knowledge is transmitted or lost.

The Santo

2006 Expedition

Expedition from an Ethnologist's Point of View

Elsa Faugère

As an ethnologist, taking the Santo 2006 expedition as a subject of study was an excellent opportunity to penetrate the core of some of the major contemporary scientific, political and economic challenges linked to the exploration and conservation of biological diversity. It also gave me the chance to examine several new research ideas from an ethnological standpoint. The conservation of biodiversity and its scientific exploration were new for me and I had never performed any field studies in this research area. At the time, my vision of this theme was narrow and limited and grounded only on what I had read. My position was that of an ethnologist landing on uncharted shores, albeit Parisian and in spite of the fact that I travelled by high speed train and underground network rather than in a dugout canoe.

In order to study the 2006 Santo expedition at the invitation of Philippe Bouchet back in March 2005, I made several methodological choices that I was never entirely satisfied with. However, the subject of my study did not lend itself easily to ethnography owing to the way it spread out in space and time. The Santo expedition involved networks of individuals, institutions and objects, scattered across many different countries and progressively woven around a joint objective: to establish a biodiversity inventory for the island of Santo. It took close to two years to set up the expedition. The preparation work entailed discussions, meetings, lunches, e-mail exchanges, the writing of various texts (statements of intent, preliminary project description, project description, etc.), telephone calls, trips, ground-breaking missions to Vanuatu, etc. The question facing me was how to apply ethnography to such a multitude of individual events. This was obviously impossible. I therefore decided to follow, record and transcribe the following events:
- The ten preparatory meetings taking place between April 2005 and June 2006;
- Two post-expedition meetings;
- Seven interviews with the expedition's organisers and staff in charge of Corporate Affairs Development at the French Natural History Museum in 2005;
- Informal discussions with the members of the expedition's steering committee;
- An eight-day ground-breaking mission carried out by Philippe Bouchet and Hervé Le Guyader in Vanuatu in October/November 2005;
- Two of the expedition's modules over a four-week period: the Marine module, and a small group of three entomologists and a botanist from the Forests Mountains and Rivers module.

Finally, I was allowed access to all of Philippe Bouchet's e-mail exchanges in 2005 concerning the setting up of the expedition. All of this provided me with some extremely rich, although disparate, material. Furthermore, the decision to systematically record and transcribe the meetings proved to be very time-consuming. However, proceeding somewhat like an entomologist collecting whatever insects may come across her path, being unsure which group or family to specialise in, I collected all kinds of ethnographic data before I could actually decide upon a specific research topic. I first had to build up a general culture and understanding of this naturalist expedition, along with its challenges, objectives, operating conditions and various other dimensions (scientific, legal, political, financial, logistical, etc.).

The main difficulty facing ethnologists when they commence work in a new research field is that of defining the actual topic, in other words picking the "right questions" to explore. As Olivier Schwartz points out in his paper "L'empirisme irréductible" (Postface to Nels Anderson's *Le Hobo*; Nathan, 1993):

"But what do you look out for when there is too much material, or it seems too "common" at first glance to be able to get anything from it? As we all know, the official epistemological tradition has a ready-made answer to this question: you can't have good observation if you don't first define the subject, if you don't establish a corpus of hypotheses all converging towards a problem to be explored. Of course, we can disagree and say that such a model does not tie in with a real ethnographic situation. Whoever decides to embark on a study of this kind does not do so from scratch, but it's quite possible for the initial questioning to be extremely confused, and even remain so for quite some time. The first objective of the study is not to answer questions but to discover the questions to be explored and this simple discovery operation requires time: time to understand where, in the

universe of those being studied, the problems and challenges lie, and to achieve a solid enough perception of their life to be able to pick out what is worth being studied."

I thus explored several research ideas at the same time and have briefly summed these up below:
- The links between science and money and more especially between scientific exploration of biodiversity and corporate foundation sponsoring;
- The links between science and politics, with a particular focus on the profession of biologist in the international context of the post Convention on Biological Diversity;
- The building of scientific knowledge on biological diversity.

The Santo 2006 expedition questions the links between sciences and societies, and does so in an international context where access to biodiversity and the sharing of the benefits stemming from its use were profoundly modified by the Convention on Biological Diversity signed in 1992. Owing to its size, the expedition went some way to laying bare certain social dynamics and some of the imbrications that are ever present in scientific research but less visible in small-scale field work.

The first, but by no means the least, of these imbrications are those between science and money. The fact that the expedition was mostly financed by private funds is interesting in itself for several reasons. The privatisation of financing for such a scientific expedition does not point to a lack of means in French public research but more to the priorities of this research. And, clearly, in spite of the discourse held by policy-makers, the financial priorities do not lie in the exploration and knowledge of the diversity of living species.

The fact that corporate foundations were willing to sponsor an expedition like the Santo 2006 expedition reflects several trends underlying the contemporary capitalist economy:
- On the one hand, a renewed interest in philanthropy and corporate sponsorship since the early 1980s, a process that some sociologists link to the financiering of the capitalist economy since that period;
- And, on the other hand, the appearance and development of socially and environmentally responsible companies, or companies which at least claim to be so.

Lastly, the private funding of this scientific expedition did not generate any controversies over, or criticism of the scientific quality, integrity or independence of the research being carried out. In a country like France, where the link between the private and public sectors, notably in the field of research, leads to much controversy, the tranquil atmosphere surrounding the Santo expedition might be interpreted as the sign of a significant change in our relationship to money, private corporations, science and the capitalist economy. For an ethnologist, these links between private funding, scientific exploration and conservation of biodiversity represent an especially new and interesting avenue to be explored.

The second imbrication uncovered through the Santo expedition is that of science and politics. The expedition was organised by western biologists, i.e. French, in a country that gained independence in 1980 after being a French-British condominium. Furthermore, Vanuatu is considered one of the poorest on the planet (according to UN indicators, Vanuatu is listed as one of the twenty-seven "Least Advanced Countries"). These factors lend themselves to a macro-political interpretation in terms of the North's domination over the South. This is something that it is difficult to ignore especially given that the current international context, in terms of access to biodiversity and the sharing of the benefits stemming from the use of biological resources, is tending to instil a particularly pernicious climate with respect to bioprospecting and biopiracy.

This geopolitical context means that, at worst, western biologists are suspected of coming to southern countries to pillage their biodiversity in order to draw illegal financial profit from it and, at best, of coming to study this biodiversity in order to fulfil their scientific objectives without providing the host countries and their local populations with sufficient compensation. The Santo expedition did not escape either of these suspicions.

The first criticism that emerged focused on the expedition's ethnological module and, in particular, its ethno-pharmacological and ethno-botanical objectives to study Santo inhabitants' uses and representations of biodiversity. As the expedition was being set up, these aspects sowed confusion and put a strain on the relations between the organisers and some members of the Vanuatu government. Worried of being suspected of bioprospecting and not wishing this to jeopardise the whole naturalist expedition, the organisers decided to completely abandon the ethno-pharmacological aspects and other research questions relating to the use of biodiversity by the local populations.

Yet, this part of the expedition might have generated economic value from the island's biodiversity, which is something that would have certainly interested Vanuatu. It would also have made it possible to go further in the famous "sharing of the benefits" stemming from the use of its genetic resources, as outlined in article 15 of the CBD. All of this would have been possible without discrediting the integrity of Santo's scientists. However, although the Santo expedition organisers wanted to abide by the CBD down to the smallest detail, especially in terms of sharing the benefits with the host country and its local population, the monetary and non-monetary compensations offered by the expedition were judged differently by different observers.

What were these compensations?
First, there was the renovation of a ship, the *Euphrosyne*, belonging to the Vanuatu Maritime College. The restoration work accounted for 10% of the expedition's total operating budget (i.e. 105 000 €). The inhabitants of Santo and the northern islands of Vanuatu (especially Torres) should draw long-term benefits from the ship as it transports persons and goods, but also benefits in the form of training. The Maritime College promised to send annual reports to the Santo organisers to keep them informed of the *Euphrosyne*'s activity.

Then there was the fairly considerable redistribution of money in the different parts of the island where the members of the expedition worked. This money was spent in exchange for guides, porters, cooks and accommodation, but also for local purchase of various products, especially food.

Finally, there was the training side of the expedition. This consisted in associating Vanuatu students with the different groups of scientists on the expedition. In the marine module, they were mainly involved in sorting activities. And with the group of entomologists and the botanist that I followed, two local students participated in the sampling work, which was mainly botany.

Some considered that the trade-offs were insufficient and that the Santo expedition should also have generated local economic development or collaborated with development professionals. For these people it would therefore seem that scientific research cannot (or can no longer) find legitimacy in the pursuance of academic scientific objectives alone. For it to be legitimate, it should also be useful socially, economically, medically, etc. For others, including myself, scientific research such as that implemented through the Santo expedition is legitimate in itself: producing knowledge about nature and/or knowledge about human culture are legitimate goals in themselves.

Finally, the third type of dynamics I wish to bring up here is specific to biological sciences, and to the extreme diversity, compartmenting and specialisation of the disciplines and knowledge produced about biological diversity. The primary objective of the Santo expedition was to produce naturalist scientific knowledge about the biodiversity compartments of this island. This objective belongs to a specific professional universe, that of biologists, even though to be fulfilled it required a plurality of commitments and social inscriptions going beyond the purely scientific sphere. To a certain extent, the Santo expedition constituted a "total social fact", in other words a complex system with multiple dimensions: political, legal, cultural, social, but also scientific, of course.

For these scientists, who are professional researchers, the living world and its diversity form the core of their work. Although the Santo expedition involved a fair number of enlightened amateurs, its objectives were embedded in professional logic specific to the scientists present. This of course entails significant differences compared with the rest of the population, differences in terms of the way biodiversity is addressed and understood, in terms of what it means and represents, and in terms of a personal commitment to a professional objective.

These professional scientists are indeed governed by logic specific to their work as taxonomists, classifiers, molecular biologists, invertebrate experts, entomologists, ecologists, etc. These disciplines follow their own operating rules and logic (such as the publication of results in specific journals with review boards), and have their own internal hierarchies, controversies, quarrels, etc. To properly describe and understand the scientific *modus operandi* of this expedition, it is thus necessary to plunge into the diversity of these biological disciplines and explore their logic.

Here again, it is the size of the expedition which, by bringing together in the same space-time over 150 researchers, so strongly and colourfully highlighted the internal splits, controversies and quarrels in biology. Indeed, for an ethnologist, the differences permeating the expedition were striking, revealing the extreme intra-specific diversity of *Homo sapiens biologicus*: invertebrate versus vertebrate experts, taxonomists versus molecular biologists; malacologists versus carcinologists; Coleoptera versus Orthoptera specialists; amateurs versus professionals; trappers versus hunters; divers versus dredge /trawl operators; fundamental researchers versus applied scientists; collective methods versus individualistic approaches, etc.

The most powerful opposition was undoubtedly between "old-fashioned taxonomists" and "trendy molecular biologists", in other words between Ancients and Moderns! The former were basically accused of being simple "stamp collectors", who were only looking to boost their figures, so to speak, in other words to collect and accumulate as many specimens and "new species for science" as possible. As for the latter, they were accused of being "young cretins", entirely incapable of recognising any species from a morphological point of view, only interested in collecting "vulgar bits of tissues" to sequence and lacking any genuine interest in or curiosity about nature. Furthermore, their inability to recognise species made them entirely dependent on the "real taxonomists", whom they liked to criticise so much. In spite of this reciprocal criticism, the Santo expedition actually managed to get them to work together...

These internal quarrels in biology in particular raise some interesting questions about the use of sensorial perception in the building of naturalist scientific knowledge (in the case of alpha-taxonomy) and the delegation of sophisticated technical instruments (in the case of molecular biology).

But they also point to what is considered by researchers as a major problem: the famous "Taxonomic Impediment": "taxonomic expertise is first founded on long personal experience and accumulated knowledge". Can we do without this first visual step in the process of building knowledge and understanding the living world? This indeed puts us before a choice of society or rather a choice of relations between science and society. Some criticise taxonomy for its social uselessness: given the alarming erosion of biodiversity, they claim that

it is better to focus means and financial efforts on protecting specific spaces and specific species (strategies adopted by the main conservation NGOs, such as *Conservation International* with its hotspot approach, or *WWF* with its ecoregion approach). Others consider that this biodiversity needs to be documented before it disappears from the surface of the Earth, and hence vast collection and inventory campaigns need to be set up, like the Santo operation. For these people, it is a question of building up scientific knowledge about nature and understanding the evolution of the living world. These differences in opinion arise notably (but not only) from different conceptions of the role of science in our contemporary societies, as mentioned above; between a more utilitarian conception and a more encyclopaedic conception.

While the Santo expedition was above all designed to explore the biodiversity of the island of Santo, from an ethnological point of view it also made it possible to explore a certain number of social logics specific to our contemporary society and of which I have only briefly touched on here. *In fine*, the ethnography of the Santo expedition invites us to adopt an anthropological approach to thinking about the relations between nature and culture.

Santo 2006

Philippe Bouchet, Hervé Le Guyader & Olivier Pascal

The discovery of the faunas and floras of the Pacific did not begin with Captain James Cook. The early Lapita explorers some 3 500 years ago undoubtedly discovered, in their way, the plants and animals of the new territories they colonized. Yet even though he was not the first European to explore the Pacific Ocean, it was James Cook (1728-1779) who began the scientific description of biodiversity in the sense that we use this word today. Eleven scientists accompanied him during his first trip onboard *HMS Endeavour* (1768-1771). Although the goals of the trip were primarily astronomy (to observe the transit of Venus on the 9th June, 1769) and geography (to search for the mythical continent of Terra Australis), Cook accepted to allow three botanists on board, Daniel Solander, Herman Spöring and Joseph Banks (1743-1820). The participation of the latter proved to be particularly important through his scientific skills, but also because he became in 1778 the President of the Royal Society. During the course of his first two trips — the second on *HMS Resolution* and *HMS Adventure* from 1772 to 1775 — Cook charted much of the modern map of the South Pacific and the naturalists discovered the extraordinary biodiversity of the islands. At the same time, France equipped the expedition of Louis Antoine de Bougainville (1729-1811) who, on *La Boudeuse* and *L'Etoile*, reached Vanuatu in 1768. The botanist on board, Philibert Commerson (1727-1773), described in particular the Bougainvillea. These voyages have been the inspiration for numerous expeditions that have brought naturalists to the Pacific Islands during the past two centuries.

The type of people who have sampled the biota of the Pacific Islands changed significantly during the 19th century. At the start of that time, most collectors were missionaries/priests, administrators or naval officers, but rarely scientists. They tried to collect "a bit of everything" wherever they went. The result of these collections was a superficial assortment of the most visible organisms, that is to say the biggest or most common, usually mammals, birds or fish, but also some amphibians and reptiles.

From the 1870s, the collectors started to specialize in a particular group (plants, butterflies, snails, birds, fish…), and provided specimens to institution-based zoologists and botanists who essentially never make it to the field. This situation changed gradually in the first half of the 20th century with the general natural history expeditions that were mounted by the major scientific institutions of the Pacific region (Bernice P. Bishop Museum in Honolulu, Australian Museum in Sydney), as well as by Europeans or North Americans coming to the South Pacific. However, despite the voyage to Santo of the "Whitney South Sea Expedition" (1926) of the American Museum of Natural History, the "Oxford University Expedition to the New Hebrides" (1933), "The Royal Society-Percy Sladen Expedition to the New Hebrides" (1971), the islands of the "New Hebrides" remained secondary scientific goals compared to New Caledonia or Fiji. From this point of view, the Santo 2006 expedition can be considered to represent both a continuation of earlier expeditions and one with a more targeted interest in the island.

Islands have always fascinated naturalists, and continue to attract the interest of contemporary biologists. Island ecosystems are simplified biological communities, containing a smaller number of species than continental ecosystems: they exhibit simultaneously the abundance of species of tropical ecosystems and the biological impoverishment of island environments. Their geographical and ecological isolation are drivers of evolution and speciation, but also causes of their vulnerability: islands are particularly rich reservoirs of endemic species, as well as of microcosms that are threatened by the introduction of invasive species. To date, 72% of the extinctions listed by the International Union for Conservation of Nature (IUCN) in five animal groups (mammals, birds, amphibians, reptiles and molluscs) are for island species.

Naturalists at the beginning of the 21th century therefore face a paradox. On the one hand they have, over the last twenty years, realised the enormous magnitude of biodiversity and increased their estimates from:

"We know 1.6 million species and there remain perhaps as many to discover."

to:

"We know 1.9 million species, but the actual number is probably between 8 and 30 million."

The species already characterized and named thus

represent between just 5 and 20% of the actual number of species present on the planet, despite climate change and biodiversity loss being at the forefront of society's concerns about the environment. We do not know if "a quarter or half" of all species could disappear "by the middle or end of the century", but the biodiversity crisis is a concept that is no longer disputed and, worse, this course seems to be inevitable.

The paradox is that despite the high stakes, the pace of exploration and description of the planet's biodiversity is ridiculously low. At the current rate of progress, it will take 250 to 1 000 years to complete the biodiversity inventory demanded by policy makers, scientists and managers. That's what the Convention on Biological Diversity (CBD) calls the "Taxonomic Impediment" (http://www.cbd.int/gti/).

Another impediment obscures the staggering figures behind this reality. This is rarely expressed or carefully hidden as it hinders the strategies —mostly based on "flagship species"— of those who "make money" from "nature": most of the unexplored components of biodiversity concern animals that are considered to be unattractive— i.e. most invertebrates (crustaceans, insects, mites, molluscs) that contain the highest concentrations of unknown species in biodiversity. But it is difficult to create empathy for a weevil, a worm or bacteria and basic research on these neglected components of biodiversity traditionally suffers a deficit of attention from the public. This indirectly leads to unwanted difficulties in obtaining the necessary resources to study these components.

Accelerating the exploration of biodiversity while overcoming these impediments is thus the great challenge for modern naturalists. We believe that the answers to these challenges lie in a change of the scale of biological surveys as a whole and the working methods of experts from all disciplines involved in them. That is what the Paris Museum of Natural History (MNHN) and Pro-Natura International (PNI) had been doing separately over the past ten years and decided to do together and in partnership with Institut de Recherche pour le Développement (IRD) for the Santo 2006 expedition.

WHY VANUATU AND WHY SANTO?

Santo (or Espiritu Santo) is the largest island of Vanuatu: 3 959 km², three times the size of Tahiti or half the size of Corsica, with only 30 000 inhabitants and... some forty languages! Santo is also the island that reaches the highest elevation in the archipelago, crossed by a mountain range, with four peaks over 1 700 m and peaking at 1 879 m at Mt Tabwemasana. The natural wealth of the island is summarized by a passing comment in the *Lonely Planet* guide:
> *"Sparkling blue holes, unlogged rainforests & the world's largest accessible shipwreck."*

Santo has "a bit of everything" to offer for curious naturalists to discover: coral reefs, caves, mountains, forests, rivers, but also gardens, plantations, and increasing pressures on coastal land. In short, the island is a compendium of the natural and the manmade habitats that can be encountered on the large high islands of Melanesia.

Given its size, topography and geological age, Santo was clearly under-explored. Thus, one of the most recent botanical collections undertaken on Santo in 1988 revealed six new species of orchid. For terrestrial invertebrates, endemism is estimated at 30-50%, but sometimes reaches 80% (snails). Vanuatu is recognized by Bird Life International as an Endemic Bird Area and, together with the Solomon Islands and the large islands northeast of Papua New Guinea, forms part of Conservation International's East Melanesian Islands Hotspot.

Beyond the diversity of environments to explore and taxa to discover, the scientific objectives of the expedition's various modules were underpinned by the same major questions: what is the real extent of biodiversity in terms of diversity and richness? What fraction of the species composition are rare species? What is the spatial dimension of biodiversity or, in other words, how representative are the sites at eco-region level? What influence do invasive species have on local ecosystems?

Monitoring the impact of climate change on biodiversity is a goal for many scientists. But how can you measure future changes if you do not have a baseline, that is to say a credible knowledge of an initial state? We have to recognise that many existing collections in museums —including the Paris National Museum of Natural History— are of little use for providing this information, because the samples that have been deposited are not sufficiently accurately referenced, in particular with regard to geography. Many specimens are simply listed as being "from the New Hebrides" and at best indicate the island of the archipelago where they were taken, but without further precision on locality etc.

It was therefore important that the inventory compiled in 2006 would serve as a baseline for monitoring the medium-and long-term evolution of the faunas and floras of these islands, during the 21th century and beyond. These future changes will at a minimum see a depletion and extinction of native species, as well as further introductions and the establishment of more alien species. We therefore wanted to label in the inventory those species considered native and those considered as exotic species, whether they are invasive or not. It therefore seemed right to include a module called "Fallows and aliens" (aliens referring to non-native species of plants and animals) in this early twenty-first century natural history expedition.

Santo remained relatively isolated from the outside world until the first half of the twentieth century. But during the WWII it became a rear base for U.S. troops engaged in the Pacific War, housing up to 250 000 U.S. soldiers and their infrastructure, and the island thereby experienced heavy human disturbance. Thus, biological invasions are known to have taken place over several historical periods, which can be summarized as follows: the arrival of the first Melanesians over 3 000 years ago; the arrival of the first Europeans during the 18th and 19th centuries; the WWII; and the last half century which has seen a major increase in trade within the archipelago and with the rest of the Pacific.

Thus, beyond a "classic" natural history inventory, the objectives of the Santo 2006 expedition have also included an inventory of alien species in certain taxonomic groups, comparing the relative weight of non-native species present in both natural and manmade habitats, highlighting the evidence for an increasing fraction of introduced species since the arrival of man, and finally a study of the perception of non-native species by local people.

PREPARATIONS FOR THE EXPEDITION

Once the outline of the scientific program had been decided upon and even before its budget had been completed, an initial visit of three persons was conducted to Vanuatu in March 2005 to present the project to the political and administrative authorities of the country and to the French Embassy, and to do a first check of the logistical facilities on Santo itself. During that first visit, Russell Nari —then of the Environment Unit of Vanuatu—, and Bernard Sexe —then of the French Embassy— demonstrated their strong support for the project, which continued throughout the expedition. An exchange of letters with the Minister of Lands marked the first formal commitment of the government of Vanuatu to the expedition.

From there, further events unfolded. Rufino Pineda was appointed as the representative of the project in Luganville, the capital of Santo. His long experience, together with his mastery of local languages and his activities in the field of rural development (more than 25 years in Melanesia, including 20 years in Vanuatu, plus a stint in 2003-2005 as head of the reforestation program LEARN funded by the European Union) made him the obvious choice as the local coordinator. From July to November 2005, other visits were made to refine the objectives of the "Karst" and "Forests, mountains and rivers" modules. It was planned that the latter module would set up camp near a forest that could be considered as "primary" as possible (and probably far from populated areas), with three main goals:
- Serve as a transit camp for small groups of itinerant scientists collecting in the wider area;
- Permit the study of changes in plant and animal biodiversity along an altitudinal gradient (program IBISCA);
- Serve as experimental and operational base

for a new powered balloon for studying the high forest canopy: hereafter known as the Canopy-Glider.

An aerial reconnaissance survey revealed potential areas of interest for the expedition on the west coast of Cape Cumberland, where three potential sites were visited in November 2005: Tasmate, Olpoi and Penaoru. In each local community, we presented the objectives of the project, indicating that we were looking for a site to study the vegetation and local fauna. We concluded that Penaoru would be the ideal site given that the Kaori forest found above the village seemed the best area to study. At our second meeting with Chief Wortut, the Penaoru village leader, he informed us that the whole village where he was the spokesman would welcome the Santo 2006 expedition. We then laid the groundwork for the agreement that would bind the community of Penaoru and the project. That agreement included the construction of a base camp an hour's walk uphill from the village of Penaoru, beside the river of the same name, to include eight bamboo huts on stilts, a refectory, a kitchen hut, two laboratory huts, a storage hut, two toilets, a landing pad for the Canopy Glider and a bridge to span the river. A quote was made for each building, which were to be predominantly constructed out of local building materials (wood, bamboo, Natangora leaves), with the expedition purchasing materials that were not available locally, such as nails and plastic sheeting. The cost of constructing the camp was estimated by mutual agreement between both parties, with a contractual commitment plus a potential bonus for "successful completion".

Meanwhile in Luganville, the existing infrastructure of the town made it easier to provide places to live and work for the scientists of the expedition who would be based there. The directors of the Vanuatu Maritime College (VMC) and the Vanuatu Agricultural Research and Technical Centre (VARTC) agreed to install the "Marine Biodiversity" module (VMC), and the "Karst" and "Fallows and aliens" modules (VARTC) at their premises. In France, the three lead agencies — the National Museum of Natural History, the Institute of Research for Development, and Pro-Natura International — were bound by a tripartite agreement. A steering committee consisting of the representatives of the three signatories of this agreement was formed, as well as a scientific committee composed of the leaders of the main modules.

A follow up visit in November 2005 resulted in a Memorandum of Understanding between the project and the Vanuatu authorities. In the weeks that followed, this memorandum of understanding was circulated to government services in Port Vila but reservations arose on some of the objectives of the expedition in the field of ethnobotany — especially ethnopharmacology. Given that this was a very minor part of our project (consisting of one researcher), and to avoid being suspected of biopiracy, we decided to remove this component. The agreement was ultimately signed on the 24th March 2006 by the then Minister of Lands, Maxime Carlot, and the then director of the Paris Museum, Bertrand-Pierre Galey, who represented the Santo 2006 expedition. This agreement then paved the way for a research permit issued collectively to all members of the expedition, signed on the 2nd June, 2006 by Ernest Bani, director of Vanuatu's Environment Unit.

During that time, the various foundations, corporations and public institutions supporting the expedition pledged 1.1 million euros of public and private funds. Foremost among these institutions were the Stavros Niarchos Foundation, the Total Foundation and the Pacific Fund that deserve a special mention (Fig. 607). Our salaries and services in kind from our three organizations are not included in the budget of the expedition. The

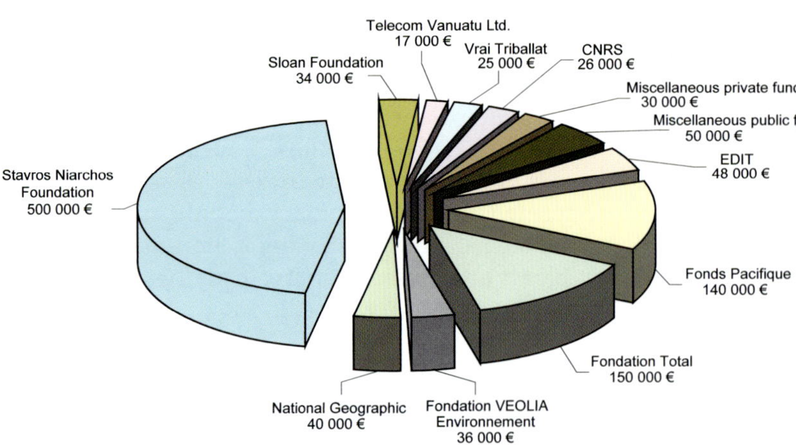

Figure 607: Sources of funding for the operational budget of the Santo 2006 expedition. The total of 1.1 million € does not take into account the "goods in kind" (research vessel, vehicles, equipment, personnel) or the salaries of the participants.

deployment of IRD-funded research vessel *Alis* alone amounted to some 250 000 € but was not formally accounted in the cost of the expedition. Similarly, the camp at Penaoru benefitted from the establishment of a satellite dish provided courtesy of Telecom Vanuatu.

At the local level, we knew that permission from the Minister in Port Vila or the Luganville Provincial Secretary were essential but not enough. In Santo, as in the rest of Vanuatu, the real power to accept visitors comes from the communities themselves, at the level of customary chiefs and villages. We therefore had to contact all members of these communities and inform them about our projects with the support of the traditional chiefs, as well as the field workers of Vanuatu Kaljarol Senta (VKS) and the member of parliament for West Santo, Sela Molissa, who personally took part in informing the people of "his" constituency. A large meeting was held in Luganville at the nakamal (meeting area) of the chiefs of Sanma Province, where the objectives of the expedition were the subject of a presentation in English and Bislama, before dozens of participants who came expressly from South Santo, Kerepoa, Malo, Aore, Tutuba, Port Olry, etc. The Rotary Club of Santo, the Tourist Board... no one was forgotten. In August 2006, everything was ready for the start of the expedition.

PROGRESS OF THE EXPEDITION

The first scientists arrived in small numbers in August 2006 (Fig. 608). Despite the reconnaissance missions in 2005 and 2006, we had overestimated the adequacy of the local facilities to meet our needs and underestimated all kinds of tropical delays. All cargo shipments arrived three to eight weeks late and we therefore had to modify the timing of the first arrivals. The first month would be a start-up month with a smaller team: it managed to fulfil this role and the first participants all arrived in August (Claude Payri's seaweed group from Noumea, and the small Rapid Assessment Team with Fred Wells from Perth) who faced the first difficulties: Torrential rains (August is statistically the driest month of the year!), the incomplete transformation of a boathouse into a marine "laboratory", engine troubles with the small motor boats of the VMC, miscommunication with some tourism operators... Despite all these setbacks, the expedition got underway. In late August, Philippe Bouchet was

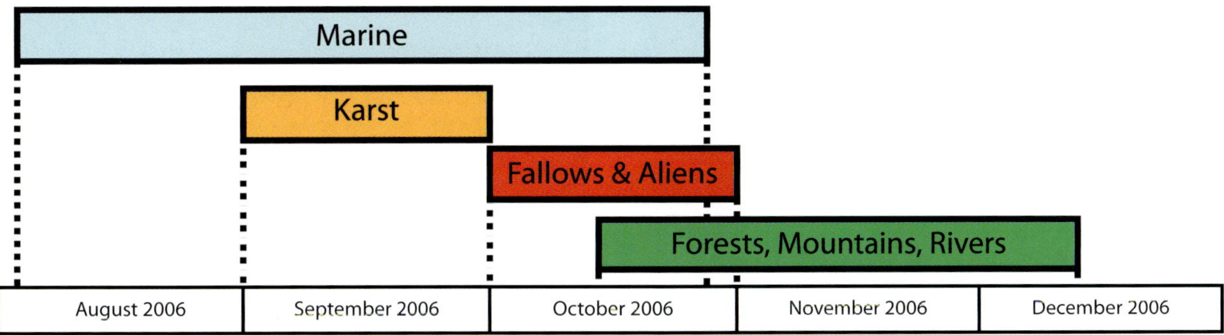

Figure 608: Timetable of the presence of the four modules of the Santo 2006 expedition in the field.

Figure 609: The marine group towards mid October. A number of participants had already left, many others were in the field. From left to right. Front row (kneeling): Jo Arbasto, Dave Valles, Samson Vilvil-Fare, Mike Miller, Yolanda Camacho, Marco Oliverio, Jason Biggs, Marta Pola, Timea Neusser, Christopher Mendoza, Mitsuhashi Masako, Steven Vitulolo. Back row (standing): Regis Cleva, Claire Goiran, Damien Hinsinger, Rudo von Cosel, Delphine Brabant, Noel Saguil, Mark Erdmann, Magalie Castelin, Yuri Kantor, Danielle Plaçais, Marivene Manuel Santos, Virginie Héros, Laurent Albenga, Tanya Kantor, Philippe Maestrati, Jacques Dumas, Philippe Bouchet, Jacques Dumas, Sandrine Perey, Brian Greene, Richard Pyle, Nicolas Puillandre, Julien Lorion, John Earle, Roger Swainston, Anders Warén, Arjan Gittenberger, Ellen Strong, Emmanuel Vincent.

joined by a small group of Filipinos with whom he had mounted the "Panglao 2004" expedition together with a volunteer from the capital. They did a great job in finalising the organization of the laboratory and the "outdoor services" (sieving areas, compressor site, sorting tables) and ensured that domestic services were ready for the rapid influx of September.

The marine side of the expedition (Fig. 609) was not spared of its own incidents, the latest being a strike for reasons related to a change in status of part of the crew of the *Alis*, just as it was to sail from Nouméa on the 6th September. With the mass arrival of sixty people on the 8th September, it required that Plan B be put into action, which involved a ship from the French Navy. Saturday the 9th therefore turned into a day of organising chaos, and sampling could only begin on the 10th. The *Alis* (Fig. 610) finally arrived late in the day on Tuesday the 12th.

Field trips started at 7 am, with different boats taking out groups of divers, each with a specific objective (general collection, targeted collection of nudibranches, brushing stones, laying nets...). In the laboratory the day started a bit later and ended at around 10 or 11 pm, since we had to treat the fresh collections of the day (sorting specimens, processing

Figure 610: The Nouméa-based Research Vessel *Alis* was deployed during 40 days to sample the deep water benthos (to 800 m deep) and also, occasionally, to transport expedition members to more distant locations on Malo.

Figure 611: IRD's *Aldric* was used daily to take samples by diving. Jean-François Barazer (pilot), Arjan Gittenberger (holding the tube of the "vacuum cleaner") and Patrice Petit-Devoize (getting ready). Laurent Albenga (left) and Jacques Dumas are already in the water.

Figure 612: A sieving area was installed on the pontoon of the Vanuatu Maritime College, to receive bulk samples from diving, dredging and intertidal collects. Very large fractions were inspected on the spot, while smaller fractions were taken indoors in the lab. From left to right: Philippe Maestrati, Maria Yorley (seated), Steven Vitulolo, Frank Ritchie.

Figure 613: Dave Valles (forefront) and Charles Tari operating the hand winch to retrieve tangle nets deployed in 60-150 m.

digital camera images, fixing material for genetic sequencing) (Figs 611 & 612). The working language was English, but here and there other languages were spoken among smaller groups. In particular, in the boathouse, now the Marine Laboratory, there was the potential to hear (assuming that at least two people could speak that language!): Chinese, Japanese, Tagalog, Cebuano, Swedish, Norwegian, Italian, Spanish, Dutch, Russian, German and, of course, French and Bislama (Fig. 613). In total, the expedition involved participants from twenty-five countries.

The team for the Karst module was complete, with the some twenty people operational from the 1st September. They split in small groups of three to seven people working in parallel on multiple scattered sites over eastern and central Santo and nearby islands: some concentrating on dry rock shelters, others in guano filled caves, while yet others explored underground aquifers. A small group explored the tip of Cape Cumberland, a very remote area (five days walk from the nearest "road"), which they reached with the *Etelis*, a small boat from the Vanuatu Maritime College. The Karst group was hosted at CTRAV about 15 km from Luganville

Figure 614: Down into the Fapon doline.

Figure 615: Tarius cave entrance.

Figure 616: A well decorated gallery in the Fapon cave.

and our paths rarely crossed. We exchanged our thoughts, viwes and stories when they visited the marine team to get fuel, formaldehyde and ethanol. We therefore organized two social evenings at the site of the Vanuatu Maritime College, which was a

Figure 617: Florence Brunois in discussion with the Ni-Vanuatu.

Figure 618: Examining traps in the Loren sump.

Cumberland team discovered some interesting sites in terms of archaeology, but returned empty-handed with respect to finding Quaternary vertebrate remains. The week was also marked by the arrival of the first photographers, filmmakers and journalists sent from France and New Caledonia by the ATOM and Gamma agencies. Given the exciting progress of the Karst module, they naturally began to collaborate with this group. Franck Bréhier and Nadir Lasson were their guides in the tunnels of the Loren Cave, reaching almost to the end of Cape Quiros (Fig. 618). On the margins of the caves, Louis Deharveng did not hold back with superlatives to describe the soil fauna of Santo: the paradise of Collembola!

In the marine field, the work was a little disturbed by the trade winds that blew much of the week (18 to 22 September 2006) and we were forced to tackle less exposed coasts. The divers were struck by the fact that many reefs were very damaged, though we could not attribute the cause of this mortality to over-fishing (there are very few fishermen and no dynamite fishing in Vanuatu, unlike in the Philippines) or any other identified human activity. Perhaps the combination of cyclones, earthquakes (a nice little tremor in early August and several other insignificant ones thereafter) and increasing seawater temperatures may be responsible for bleaching the corals? Despite the complex topography of the coastal zone, the habitats showed little diversity, and we did not find sea grass beds and only limited mangrove stands. That week was also marked by the start of so-called night dives (night falls at 18 hours and night dives were held between 19:30 and 22:00). Sunday 24th September marked the first rest day when we could enjoy the season of citrus and mango trees in bloom.

The Karst cavers handed over to biologists of the "Fallow and aliens" module, while the number of marine biologists remained the same. Just over half of the participants of Santo 2006 had already left and only the big "Forests, mountains, rivers" component had not started, but preparations for it at the Penaoru camp progressed normally.

Fatigue and climate began to take their toll on people and equipment. Small sores or fever struck several participants and kept them out of the water or away from the lab benches. A moray eel bite required five stitches to the recipient. Another researcher became the victim of a centipede bite. On the equipment side, there was a fracture of a vehicle axle, and the propeller shaft on a boat was put out of service until the end of the mission. Nothing to really worry about.

For the Karst module, Murphy's law struck again. The previous week had been marked by the discovery of the largest cave network on the island; a new record was achieved this week (25 to 30 September 2006), with the discovery of a new

good opportunity for the expedition to invite the staff of the organizations that hosted us, as well as schoolteachers and other Luganville's VIPs.

The expedition continued to experience small incidents and the risks remained, but fortunately without causing any injuries, and without the expedition going over budget or any confrontations between participants. In the karst research group the first week was marked by the discovery of the largest cave network on the island (Figs 614-616), with three kilometres of tunnels identified and mapped in central Santo, and the exploration of a large sinkhole of 50 m depth on Malo, the largest of the "small islands" near Luganville. These caves were partly known to the villagers, but were not immediately revealed to us. As our team started to explore the field, the tongues of the inhabitants of Santo gradually loosened. Eventually, curiosity prevailed and traditional custom owners began to offer to show us new caves (Fig. 617). The Cape

network and huge galleries, discovered a few days before the team was due to leave, which meant it was not possible to explore them fully. These new networks seem to have the condition necessary for harbouring typical cave biota, while the caves explored until then were mainly inhabited by the same fauna as that of the surface soils (Figs 619 & 620). Biologically, the underground waters on Santo have generally been very poor, which reflects the recent geological age of this part of the island, and also the torrential flows of water that go through them. This poverty applies also to Quaternary vertebrate remains. However, and to his great satisfaction, Louis Deharveng left Santo with about a hundred new species of Collembola (Fig. 621).

In the marine field, our sampling sites were this week at Pallicolo, Surunda, Aesa, Mavea, Turtle Bay and Oyster Island (Figs 622 & 623). We based our small boats about 30 km from the lab, which helped to save transit time for the divers but required a large number of transhipments by road for the various tanks, bins, baskets and appliances (Fig. 624). Some beautiful reefs and reef walls are present on the east coast and the collection of species did not diminish there. It rained less since the Saturday and at last the wind died down and we could dedicate the following week to the most exposed areas of the coast, off the localities of Tutuba, Malo and Urelapa.

Figure 619: Mapping Nanda Blue Hole.

Figure 620: Pootering arthropods in the Vobananadi Aven on Malo.

Figure 621: The extraction room for cave and soil fauna at the karst basecamp of CETRAV.

Figure 622: The cod end of the beam trawl is brought on board *Alis*. From left to right, Christian Fitialeata, Ferrand N'Ganyane, Jean Lamata and Felise Liufau work in coordination like musicians in an orchestra.

Figure 623: Everyone is busy on the deck of *R.V. Alis*: Bertrand Richer de Forges (with hat) and Anders Warén (seated) are examining samples, while Ferrand N'Ganyane (yellow helmet) and Jean Lamata (background) are assembling a beam trawl.

Figure 624: Vehicles were used to ferry people and equipment to the east coast of Santo, and thus avoid the long, windy, choppy and unpleasant transit by small boats. Steven Vitulolo and Jean-François Barazer (in the background) are tying the equipment for the bumpy road.

Thursday 5th October was a national holiday in Vanuatu and we held an Open Day on the 6th at our base at the Vanuatu Maritime College. We invited school principals, science teachers, officials of the Province and tribal chiefs to visit our facilities, where they could come and ask questions, and look at animals under a dissecting microscope...

Figure 625: One of the classrooms of Vanuatu Maritime College had been transformed into an office and "media" center, with the benefit of one of the very first high-speed internet connections in Luganville. Seated on the left: Damien Hinsinger, Stefano Schiaparelli, Yuri Kantor. Facing each other: Didier Molin (left) and Noel Saguil (right).

Figure 626: Training and capacity building was an integral part of the Santo 2006 expedition. Charles Tari (right), from University of the South Pacific, gets a personal tutoring from algae expert Claude Payri.

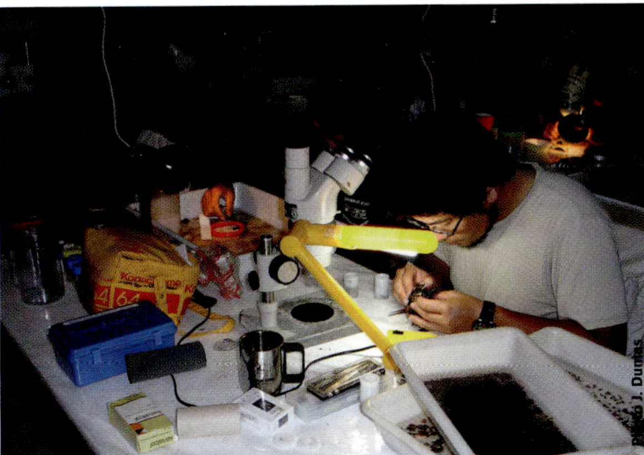

Figure 627: Takuma Haga (in front) and Yasunori Kano screening the catch for minute gastropods. In the background, Marilyn Schotte at the microscope behind a mosquito coil.

Figure 628: Trimix divers Brian Greene (left) and Richard Pyle (right) at the end of a dive. The cylinder of the electric vacuum cleaner (center) contains precious bottom material from otherwise inaccessible overhangs and crevices at depths between 80 and 120 m.

(Figs 625 & 626). After the few small mishaps during our start-up in August our relationship with the authorities and population had become excellent, and it was apparent that there was no suspicion of us being there to loot the "genetic resources" of Vanuatu. The *Daily Post*, the only daily newspaper in the country, regularly devoted a full page to Santo 2006. In short, the progress of our expedition was now entering the regular news on Santo.

The moon was full on the 7th October and the week was thus marked by good tides around Santo. We took advantage of these tides to sift through a seemingly mundane reef, beside which we drove most days just outside of the city, right next to the fuel depot! A sensational discovery was waiting for us. Our Japanese colleagues Yasunori Kano and Takuma Haba (Fig. 627), who had the curiosity to dig under some large blocks in fresh water seeping in the sea at high tide, got their hands on an entirely new fauna of small gastropods.

While the first week of October was marked by high tides, the second would be by very deep dives. Our dream team of divers who were used to using trimix (Fig. 628) arrived from Honolulu, and sorted out all the problems of fittings and valves. Richard Pyle's team could then take on the deep reefs of Tutuba up to 120 m deep. With decompression times amounting to hours, the patience of Jean-François Barazer, who ensured safety at the surface, was tested, basically through reading thick books and doing lots of sudoku.

We continued to experience very rough weather. We pitied in advance the botanists and entomologists who were going to work in the dripping wet forests and then have to spend their nights in tents. With the arrival of Laurent Soldati last week, together with Hervé Jourdan, and Laure Desutter's group arriving in Santo in a few days, entomological research would now be strengthened on the Santo 2006 expedition. Similarly, the complement of the botanical team having been so far limited to only Marc Pignal, was also about to be built up.

The "Fallows and aliens" module had by now already completed half of its program. The purpose of this module was to sample sites modified by man, which are obviously less natural than the sites studied by the karst module, and comprised the port area of Luganville, coconut plantations, coffee and cocoa farms, and Melanesian vegetable gardens. An alarming preliminary result emerged, that the electric ant *Wasmannia auropunctata* had already arrived on Santo, apparently from the exchange of plants with the Banks Islands, which have been ant-infested for several years. Wherever it has been introduced in the Pacific Islands, this American ant causes major environmental damage and creates considerable discomfort to humans and domesticated animals. For the moment, it appears that the infestation is limited to less than one hectare on Santo, which makes it possible to curb. Other introduced species of ants are present on Santo, but Hervé Jourdan was surprised to note that unlike on New Caledonia, they have managed to coexist well with local ants.

With the exception of bats, all wild mammal species on Santo have been introduced. The Pacific rat (*Rattus exulans*), the pig and the dog can be attributed to the first Melanesian people; the black rat (*Rattus rattus*) and the Norwegian brown rat (*Rattus norvegicus*), the grey house mouse, the cow and the cat to the Europeans. Remains of large spiny rats from New Guinea (*Rattus praetor*) have been found in archaeological sites.

Two invasive plant species, *Merremia peltata* (Convolvulaceae) and *Mikania micrantha* (Asteraceae) were the subject of detailed observations by Marc Pignal. *Merremia* was reported to be highly invasive, yet there is a name for it in the language of Butmas, a village in central Santo, where it is found in its natural habitat. It is likely that its aggressive behaviour (once a patch of light appears in the canopy, it emits vertical tendrils that get into trees and eventually kill them) is a consequence of uncontrolled logging. *Mikania* is a native of tropical America, hence its name the "American liana". There is a story that attributes its local introduction to the U.S. military, who allegedly used it during the WWII to camouflage their facilities. Even though this story is probably untrue, it nonetheless suggests that the plant was introduced at about that time.

By now, the first "Forest" module roaming group had been in deep forest for several days. Beyond about 20 miles around Luganville, they could only be reached by satellite phone. So there was radio silence with that group.

All marine biology textbooks tell the story of palolo worms that gather once a year to release their gametes and reproduce. Yet few authors of these manuals have ever themselves seen the gathering of these marine worms. In any case none of the scientists who took part in Santo 2006 had seen the gathering. The phenomenon occurs with astronomical regularity. In Santo, six days after the October full moon, we had the chance to experience this event in the village of Tongoa during the night of October 12 to 13. The fishing starts on the falling tide, around 22 pm and ends around 1 am at moonrise. And the next day, we ate a delicious palolo laplap, a dish based on cassava topped with green worms.

Richard Pyle's group from the Bishop Museum in Honolulu dived this week to 140 m depth and successfully tested a new method of sucking samples from up to 120 m depth. Several remarkable discoveries have already been made and reward those efforts and the trouble we had taken to bring the helium from Noumea and 300 kg of freight from Honolulu.

VIP visits punctuated the expedition from time to time. This week it was Ham Lini Vanuaroroa, the Prime Minister of Vanuatu, who spent just under an hour with us.

The sounds of activities of the "Marine" module had begun to die down. The botanists and entomologists were now taking over. The group of Laure Desutter and Odile Poncy have already been prospecting for over a week, while the team of Adeline Soulier-Perkins, Ronan Kirsch and Gregory Plunkett would embark the next day to take on Mt Tabwemasana, the highest point on Santo and Bruno Corbara, Jérome Munzinger and Marika Tuiwawa selected and marked out IBISCA sites to describe the effects of the altitudal gradient on flora and insects in the kaori forest above Penaoru (Fig. 629).

Two earthquakes this week on Santo. The first reached 6.2 on the Richter scale. Its epicentre was

Figure 629: The village of Penaoru, situated beside the sea at the mouth of the river of the same name.

Figure 630: The *Euphrosyne* leaving Luganville with the first contingent of scientists on board. The journey time to Penaoru on the west coast took 10 to 12 hours according to sea conditions. The boat did a dozen round trips and spent about 30 days at sea to service the project needs.

located 50 km from Luganville. It lasted a good 30 seconds (just enough to intimidate us yet not quite enough to scare us). The second was felt at the base camp located in the valley of Penaoru. In this third week of October, the *Euphrosyne* brought by sea the last wave of logisticians and the first wave of researchers to the Penaoru base camp (Fig. 630). A satellite antenna, generators, computers, the first phone call and the first internet connection were established from the camp, which until now had been lit by kerosene lamps at night. The excitement and enthusiasm felt by Olivier Pascal was evident when he called Luganville to describe all the villagers of Penaoru glued in front of the computer, seeing their country, their island, and their corner of the valley on Google Earth. A technologically small achievement, but one that meant a lot to the participants of the "Forests" module who had to sail 10 hours through rough seas with the *Euphrosyne* to reach Penaoru: the remaining pristine forests on Santo can now only be reached by sea.

To echo the earthquakes, two hurricanes came to frame the beginning and end of work in the remnants of natural forest on the island and complicate land operations. No physical damage was sustained, but morale and nerves were seriously tested. The first, "Hurricane Xavier", was stuck for several

days about 200 nautical miles northwest of the archipelago, preventing all navigation around the outskirts of Santo and forcing the ship carrying our container from New Caledonia to make a U-turn. This was particularly irritating, after having battled with the shippers to deliver our equipment that was stuck in Singapore and to find an alternative sea route (and not least to find a ship, as there are few that service Vanuatu) to ensure that the equipment would arrive in due time on Santo. At midnight we would have benefitted from the weekly visit of the barge *Brisk*, the only vessel capable of delivering our equipment directly on the beach without pontoons. The consequence of all this for the Penaoru camp was a much more disorganised start than expected... For two days, instead of finishing the setting up of the camp, the villagers were occupied with dismantling roofs and felling the most dangerous trees. The main body of the IBISCA project landed with the *Euphrosyne* before the arrival of the barge containing the equipment, and so came to find an empty lab, with all equipment still in the container... Luckily everything was soon back to normal in no time.

Access to the west coast of the island (Fig. 631) was far from easy. This is a straight coastline without a barrier reef and virtually free of mooring sites to

Figure 631: The mountainous west coast of Santo at Cape Cumberland.

Figure 632: The barge *Brisk* arriving on the west coast to deliver equipment for the "Forest, Mountains and Rivers" module of the expedition.

Figure 633: There was chaos on the beach after disembarking the equipment.

protect from the threat of random "squalls from the west". Landing the teams, after 10 hours at sea from Luganville, on the pebble beach at the mouth of the Penaoru River during a heavy swell, was no simple task. Unloading 10 tons of equipment, including the Canopy-Glider and the gas tanks needed for its operation (propane and helium) was even less easy. This was also a matter of luck, for if the swell is too strong, the barge cannot wash up on shore. The sea rippled slightly on this morning of the 11th November, signifying our truce with the elements and with our shippers. Helped by the villagers of Penaoru and Peavot, a human chain was organized to clear the barge Brisk as quickly as possible. The scene on the beach looked like a shipwreck, where crates and containers of all kinds were strewn all over the place (Figs 632 & 633). It took three days to carry all the expedition's equipment over the four

kilometres from the shore to Penaoru camp, located deep in the valley of the same name.

The second cyclone came late, about three weeks after "Xavier", at the end of November. Strong winds forced us to evacuate "Camp 2" situated on a ridge at 900 m elevation: this fly camp proved to be too exposed; it had served as a relay point for the teams working in the Saratsi range between 1 000 and 1 500 m elevation, the highest point in the area, and was temporarily abandoned for the main camp, jammed in the valley overlooking the Penaoru River, but more sheltered than the first camp. The base camp hosted up to 50 people per day during its "high season" (Figs 634-636). Built

Figures 634 & 635: For some weeks the base camp became an object of curiosity and a site to visit for the surrounding villages.

Figure 636: The kitchen. Hundreds of meals a day were served during the high season of the expedition at the Penaoru camp.

Figure 637: The women of Penaoru cooking a stuffed pig in an oven made of hot stones.

on a former taro plantation, that offered a flat open area, it overlooked the Penaoru River that provided the water needed for a camp of this size. Even though comfort was basic, the "restaurant" was gargantuan. Trained by the village women, our bakers became masters at cooking pigs through contact with hot stones (Fig. 637), a difficult recipe to export, but the memories of which still cause our mouths to salivate long after our return.

Lots of good food —that is the secret of a successful expedition. The statistics recorded by Dan Molczadzki, our camp manager, are telling. Five oxen, 14 pigs, four goats, some 80 chickens and about four tons of fruit and vegetables came under the knife of our cooks, Guillaume Chipy and Luc Zelmat. All of these consumables were an essential resource for the scientists, and were "ordered" six months in advance with the villages in the region to avoid causing shortages in these isolated communities that are still largely self-sufficient.

The last intact forests of Penaoru have not survived by chance. The teams climbed a combined altitude of tens of thousands of meters to work every day in the forests that cling to the steep terrain. Below 600 m the forest is degraded with an impoverished species composition. The "good" forest begins at about 800-900 m, and two hours of brisk walking are needed to reach the target sites each morning. Kaori (*Agathis macrophylla*), a conifer, is at this altitude the most abundant tree. Kaori forests are known only from two other islands in Vanuatu, from Anatom and Erromango, where they are particularly threatened and overexploited, which means that those on Santo are unique in the archipelago.

At higher altitudes, above 1 200 m, there is a different type of forest dominated by trees belonging to the genus *Metrosideros*. Here there is a "wet" moss garden that was difficult to extricate Australian Elizabeth Brown from, who is a specialist in these plants. Even though getting to Camp 2 was a relief for those who worked up to the 1 500 m summit ridge, there were many who went there and back in a day.

Between the two hurricanes there were no mishaps, no accidents or major incidents. The only notable incident, and material for the chronicles of adventure, was the breaking of the propeller of the Canopy-Glider. It was not the breaking of a blade that caused the problems. Breaking a propeller on an engine is not serious in itself (the Canopy-Glider made its first flights on Santo and, like all prototypes, needed some improvements) but finding another one for this canopy-harvesting machine, and getting it unscathed to the west coast of Santo, could be a real headache (Figs 638-640).

Ordering by phone (Fig. 641) (from the only "telephone booth" in the area, installed by the project

Figure 638: The Canopy Glider provides a similar role to a boat for the marine module, and was used for the first time during the Santo 2006 expedition.

Figure 639: The Canopy Glider can take two people on board in addition to the pilot, and permits the direct collection of specimens from the top of the canopy.

Figure 640: Professional tree climbers provide indispensable assistance to botanists collecting specimens from the tree canopy.

Figure 641: Penaoru camp had the only "telephone booth" on the west coast of Santo. Gilles Ebersolt, architect and inventor of the inflatable canopy raft, is an expert in mounting and organising simple but fully functional bush camps in the field.

as mentioned above) a custom made propeller from the workshops of Duc-hélices in France, took alone a few hours. The propeller was moulded according to original drawings, sent to Paris, then travelled by various aircrafts to Port Vila, to be finally sent as cargo to Luganville. Here it was received by Rufino Pineda, placed in a small plane specially hired for the occasion to its destination at the airfield of Lajmoli (the only one on the west coast of Santo, near the village of Olpoy, 15 km as the crow flies from Penaoru). Here it was picked up by Faustin (the "second in command" to Dan Molczadzki, our camp manager), who left Penaoru at dawn by horse to reach Lajmoli. Returning by

the same route along the coast, but at high tide, the horse did not like the idea of having to cross a stretch of the sea, and threw both propeller and rider over its head. Faustin finished the journey on foot, taking the horse by the bridle and the "surviving" and "miraculously unscathed" propeller under his arm. This adventure took a total of one week.

Expeditions do not, in most cases, turn out to involve quite the extreme hardship and adventure that we might imagine, although it does sometime happen. We found that investigating the leeward side of the mountains of the Cumberland Peninsula to not be enough to give an overview of the flora and fauna of forests in the west and north of the island, so a traverse from west to east of the cape was deemed necessary to collect specimens on the windward slopes (Figs 642 & 643). For the 15 villagers, 10 "volunteer" scientists, our doctor and film crew who took part, the joy of the first 24 hours soon receded: for six whole days the rain fell in buckets without stopping, turning a mountain trek into a wet nightmare marathon.

It was hard, if not impossible, work for meagre results: the anticipated variations were not observed. And apart from significant differences in the relative abundance of different species, the floristics appeared about the same on one side of the main ridge as the other. A "boulouk" (feral cow, especially abundant in these forests even at the highest altitudes) paid the price for this trip. An improvised hunt thus brought a few kilos of meat for the team who were running low on food (Fig. 644). Fairly exhausted but overjoyed, the little band reached the beach of Piamatsina village, on the east coast of the Cumberland Peninsula where the *Euphrosyne* was waiting to take them back to Luganville, its penultimate trip with the scientists of the Santo expedition.

Figure 642: The team embarking on the traverse of the Cumberland Peninsula, still smiling prior to the rain that would accompany them for six days.

The ship finally arrived to collect the last of the voluntary castaways in the Penaoru valley on the 5th December. The *Euphrosyne* and her crew had spent nearly 30 days at sea in total, transporting the waves of researchers who made it to Penaoru to work on terra firma.

Figure 643: Difficult progress in the valleys hemmed in by the central mountain chain of the Cumberland Peninsula.

Figure 644: The *boulouk* hunted that morning provided a feast for the *c.* 20 people who took part on the traverse.

RESULTS

Compared to conventional natural history missions, that usually mobilise just one to three collectors on a specific topic, Santo 2006 was innovative in the diversity of the sampling means used and the size of the teams deployed on site during the mission, as well as in laboratories after the mission. With the exception of birds, we expected to discover new species in all groups of animals and plants. However, even when you're the first to explore a cave, reef or the canopy of a large kaori on Santo, it is rare for a scientist to be in the situation where they can immediately shout eureka! at the time of collection. (Figs 645 & 646).

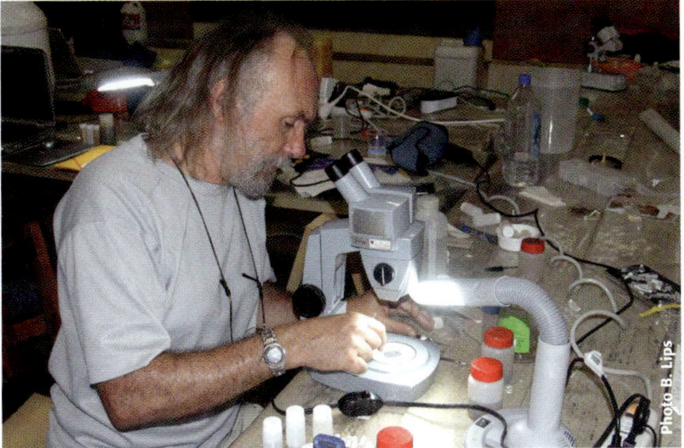

Figure 645: Denis Wirrmann in the karst team lab of CETRAV.

Figure 646: Elizabeth Brown in the « field laboratory » of the Penaoru camp.

In most groups of invertebrates, the number of species already known is simply too high for even a specialist to remember just ten per cent of the species in "his" group. Most of the time, it is not known at the time of collection if what is being collected is new or already known. And even if a certain crab or beetle is unknown, it does not mean that it is a new species. An expedition therefore ends with containers filled with bags and boxes filled with alcohol or formaldehyde, piles of dried plants pressed between sheets of newspapers, images and sounds recorded on computers, annotated maps, lists of collection localities, GPS points... The containers left Santo in late December 2006 and arrived at the Paris Museum in March 2007. The chain of operations continued then into specialist hands. Before that, all samples must be sorted into families; each specimen must be recorded by a "collection event" or a location, with its latitude, longitude and altitude or depth. It goes without saying that it is not the same people who are able to sort flies caught in a Malaise trap or snails in a dredged sample. Sorting is the bottleneck that constricts the outcome of many inventory projects, because while the field phase is fun, meticulously sorting specimens can quickly become monotonous. Finally, when all the material in a given group is fully sorted into families, the real work of systematic study can begin. This relies on an international network of specialists. Systematics is a branch of biology where there is little competition among researchers: there is plenty to study and we are few in number! Taxonomists thus form an "International" team where everyone has his/her micro-specialty. Thus, Roland Gerstmeier of the University of Munich is a specialist in beetles of the family Cleridae, and so naturally it is he who has studied the Cleridae of Santo, while Yves Terryn of Ghent, Belgium, is a specialist in marine gastropods of the family Terebridae, and so naturally it is he who has studied the Terebridae of Santo. It is these specialists, and their ilk, who, two, three or even ten years after the collection of a sample, are able to say "this species is new." The eureka! moment may be deferred even longer, as there are families where there are absolutely no specialists in the world. Of course, it would be unscientific to ignore or discard the samples of such families: they were costly to collect, sort and label, and it is almost certain that we will not be able to return and do a collection when there is an opportunity to study them. It is precisely one of the roles of museums to archive such temporarily orphaned collections.

At the time of writing these lines, the expedition had ended four years previously, the specialized sorting has been completed for the marine fauna and flora, and is on track to complete this for the insects and soil fauna. It is therefore possible to take stock of the collections and observations and to report specific results, but it is too early to make a summary review of all the results. In total, we believe we have identified so far about 650 species of plants (higher plants, ferns, mosses and liverworts), 350 species of fungi and well over 1 700 species of terrestrial animals — mostly invertebrates, of course — but there are still a lot to study. In the marine realm we have recognized 1 100 species of decapod crustaceans, 4 000 species of molluscs, and 650 species of fish. To put these numbers into perspective, note that in all the seas around Europe, from Spitsbergen to the Canaries and from Iceland to the Eastern Mediterranean, there are 672 known species of decapod crustaceans! Hundreds of new species have without doubt been collected

—we predict, cautiously, 1 000-2 000— and these are in almost all groups.

But behind these numbers lies a more complex biological reality. Paradoxical as it may seem, the scientists on the Santo 2006 expedition have not found the fauna and flora of Santo to be all that rich. Throughout the planet, cloud forests above 1 200 m, despite the visual appearance of exuberant jungle, turn out to be particularly poor in species of insects and snails. Whether at sea or on land, more than half the species have been seen only once or twice, and in low numbers at every encounter. Species diversity is especially high out at sea, but with extremely low numbers and restricted to one or a few sites of occurrence; it is therefore necessary to work hard to discover these species one by one, one after another, far from the image of plenty that coral reefs convey.

Outwardly the Santo 2006 expedition differs little from the natural history expeditions conducted 50 or 100 years ago, but, in reality, the nature of the data obtained from the samples has changed a lot. We now know that in all ecosystems most species are naturally rare and small, and accordingly we deployed relevant methods for collecting and sorting: Malaise traps for small flying insects, for example, Berlese funnels for soil fauna, and brushing baskets for flushing out sessile invertebrates off coral rubble, sieving residues down to a 0.5 mm mesh, routinely sorting samples with a binocular microscope. Although we returned from Santo with many new species, it was primarily because we collected animals in a generally very poorly sampled size range. Another difference with the natural history expeditions of previous generations was the georeferencing of samples and the collection of images and associated sounds. At least 3 000 species of Santo were photographed alive and the calls of dozens (grasshoppers, bats) have been recorded, providing taxonomic characters that are obviously not available from a preserved specimen. Finally, the massive everyday presence of molecular techniques in systematics requires —or permits according to one's point of view— the collection of fresh tissues for gene sequencing. In this regard, the Santo harvest is exceptional because our working conditions —whether in camp in Penaoru, at the Maritime College, or CTRAV— benefitted from a "scientific comfort" that most natural history expeditions would have envied.

With 16 km of subterranean tunnels newly explored and charted, Santo 2006 has practically multiplied by tenfold the known underground network on Santo. New archaeological sites were discovered but, unfortunately, none of these sites contained Quaternary vertebrate remains as have been found on Efate. Pollen cores from swampy areas will allow for paleoclimatic analysis.

WHAT DIFFICULTIES DID WE MEET?

The taxonomic impediment that we mentioned in the introduction to this article is exacerbated by the new modalities of access to biodiversity. Whether for basic or applied research, the collection and export of scientific samples is only possible under strict laws that have their origin in the Convention on Biological Diversity (CBD). Thinking that they will get the ear of decision makers and opinion leaders, and therefore funding for their research, some scientists have made themselves the heralds of a "useful" biodiversity, for which one would only have to stretch out one's hands to discover new molecules and genes. The monetary valuation of ecosystem services arises from that same attitude of mind.

This commercialisation of biodiversity has created a sense of suspicion and even hostility towards scientists, who are now routinely suspected of biopiracy by lawyers and diplomats from developing countries, and who therefore do not want to take any political or economic risk by authorizing the exploration of the biodiversity of their countries. Outside the taxonomists camp, for whom "each species counts", biodiversity tends to be strongly considered as a sum of "resources".

The CBD also does not speak of "species" but of "genetic resources", and gave birth in the South to hopes and expectations of income, patents and royalties that we do not hesitate to describe as unreasonable. Scientists from the North —like ourselves— motivated by a selfless interest in research, are under suspicion, at best, of doing irrelevant research for the host country, and at worst, of lying and of being biopirates in disguise. It is alarming to note that while everyone pretends to be concerned about the future of biodiversity and species extinctions, the work of exploring and describing biodiversity is severely hampered by numerous regulatory barriers and attitudes.

On the 28th June, 2006, the *Daily Post* of Port Vila published an article entitled "*Scientific research gets first butting from Government institution*" in which Ralph Regenvanu, Director of the VKS (Vanuatu Kaljarol Senta) was concerned about the inability of Vanuatu to supervise and monitor the research on its biodiversity, making particular reference to Santo 2006, and noting that Vanuatu is "*simply relying on the goodwill of our French colleagues to ensure misappropriation of national heritage does not take place*". During the expedition, Mr. Regenvanu suggested a moratorium on all research conducted by foreigners as long as the country does not have a Scientific Research Council.

We are here at the heart of the misunderstanding about the meaning of biodiversity. For taxonomists, biodiversity means millions of species of plants, fungi, animals —mostly insects and parasites— among which a tiny portion (less than 1%) is cultivated, reared or used by man. For ethnobiologists, lawyers and economists, biodiversity consists of the hundreds of species consumed, used or bred by man, plus a few others for which it has not yet been found what they could be used for but are considered to be reservoirs of genes and molecules awaiting patents.

In response to the fears of Ralph Regenvanu, we sent on the 8th July 2006 a letter to Russell Nari, who in the meantime had become Director of the Ministry of Lands, to describe the measures we had taken to ensure that there would be no "looting" of the natural resources of Vanuatu. We reminded him that we had taken note of the reservations made by the authorities in Port Vila regarding research in ethnobotany and ethnopharmacology, and we emphasized that the scientific program of the expedition no longer contained any topic that could be equated with bioprospecting for genes or molecules of commercial interest. We also stressed that to prevent any conflicts of interest and ambiguity, we had refrained from seeking funding from foundations or private corporations whose business is related to living organisms or their derivatives; so we had deliberately refrained from seeking support from companies in the seed industry, cosmetics or pharmaceuticals. We emphasised again that the agreement signed between the Government of Vanuatu and the MNHN anticipated the concerns of local authorities as it clarified that the expedition "commits to collect information and specimens for academic and management purposes only. Under « academic and management purposes », the parties include any purpose other than commercial and industrial". Reflecting the concern of the VKS Director vis-à-vis bioprospectors, the memorandum even states that "the Cultural Center will give its prior consent to any publication that might contain elements of indigenous or traditional knowledge gathered by the project."

Finally, despite the absence of a Scientific Research Council, we have ourselves actively sought and solicited participation of local scientists and technicians whenever possible on the expedition, and in cooperation with the Ministry of Education, have sought to involve Ni-Vanuatu students. In total, despite the obvious lack of local academic institutions in the field of biodiversity, ten Ni-Vanuatu (Sam Chanel, Tarere Garae, Anthony Harry, Donna Kalfatak, George Matariki, Vatumaraga Molisa, Charles Tari, Graham Taridia, Emily Tasale, Samson Vilvil-Fare) participated on the expedition within the scientific team. Following the expedition, Samson Vilvil-Fare received a grant from the Territory of New Caledonia to do a Masters degree at the Pierre and Marie Curie University in Paris. This will provide him with the foundation needed to initiate, conduct or evaluate the management of biodiversity when he returns to his country.

In response to the concerns raised in Port Vila, we required each participant prior to their arrival in the field to sign the "General Conditions" which state that "a research and collecting permit has been issued by the government of Vanuatu collectively for the Santo 2006 expedition. The Participant/Visitor declares that he/she has been informed of the agreement signed between the co-organizers and the Government of Vanuatu and undertakes to comply with the rules as stipulated in that agreement"

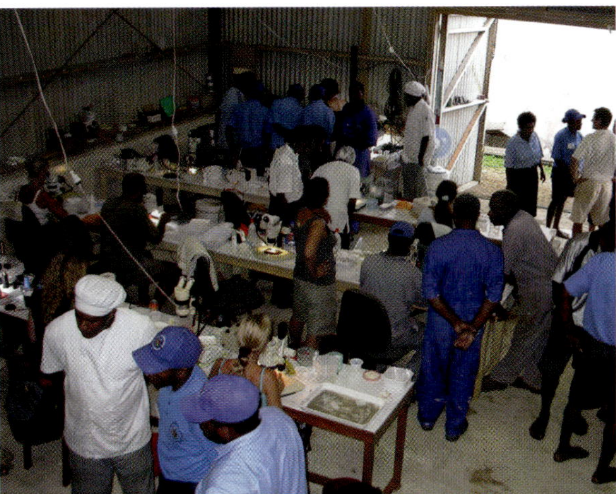

Figure 648: Several open days were organized to reach out to schoolchildren, VMC staff and other interested parties, who could visit the laboratory and interact with the scientists.

Figure 647: The expedition "headquarters" at Vanuatu Maritime College received the visit of Ham Lini Vanuaroroa, the Prime Minister of Vanuatu.

WHAT DO WE LEAVE BEHIND?

The marine and "Fallows and aliens" modules operated mostly in the "developed" parts of Santo, while the Karst, but especially the "Forests, Mountains, Rivers" modules worked in the poorly developed regions without infrastructure. It is difficult to answer the question "what have you , ethically and economically, left behind?". As organizers of the project we have always been careful, whoever our interlocutors were in Vanuatu, to clarify our intentions and goals and stress that a scientific inventory of biodiversity is not a development project. We have, however, wherever possible and in accordance with our commitments vis-à-vis all partners, including our financial backers, sought to act in the longer term interests of Vanuatu and the island of Espiritu Santo in particular.

The splendid isolation of the village of Penaoru made it difficult to transport people and equipment from Luganville to the west of Santo. We considered leasing a boat from an Australian operator in Port Vila, but we finally chose to rehabilitate, within the budget of the expedition, the *Euphrosyne*, an ancient boat of the former British High Commissioner at the time of the condominium and today the property of the Vanuatu Maritime College (VMC). This choice, although much more expensive from a simple accounting point of view (105 000 €, or 10 % of total operational budget of the expedition), was done so that the island of Santo and the communities of the far north of Vanuatu (in particular the Torres Islands) could benefit in the longer-term from the expedition. The activity reports of *Euphrosyne* sent to us by the VMC confirm *a posteriori* our choice, since all the communities of the island now potentially benefit from the services of this vessel (training, transportation, medical assistance), and thus indirectly from the investment made by the project.

Even though the presence of the scientific team has not transformed the lives of Penaoru or the west coast of Santo, it has helped put focus on one of the villages (Penaoru), which has benefited from this situation. By entrusting the leadership of the camp to Dan Molczadzki, the young French agronomist who speaks Bislama, we could be confident that his experience gained during two years with the rural communities of Santo (POPACA project assistance to coconut farmers in copra, coffee, cocoa) would result in the implementation of a number of impeccable ethical/social practices: a fixed salary of 1 000 VT/day per person according to the tariff defined by the VKS for its fieldworkers, and applied to all workers (porters, guides, assistant cooks); never more than six working days per person per week, and on average three days/week/person; a rotation system to offer jobs to as many people as possible (especially for the camp assistants); a rotation system to supply fresh produce to guarantee an income to a maximum number of people. A village management committee was formed with regular meetings (almost daily) between Dan and Ruben Boe, the secretary of the committee, which helped bring attention to the comments/claims of the villagers. The biggest tension was in fact "internal" within our hosts, with workers from nearby villages fearing that the wages due to them would not be paid by the management committee, which was composed solely of residents from Penaoru. In total, 96 residents of the west coast of Santo "participated" in the project, 44 were from Penaoru (the village has 103 inhabitants, including women and children over the age of 3), the 52 others were mostly from the neighbouring villages of Nokuku, Petawat, Sules, and Wunon. The expedition directly injected (through camp construction, portering, camp assistants, supply fees, guides...) approximately 2.4 million VT (approximately 18 000 euros) into the economy of the village and surrounding villages. In addition, of course, there are the buildings of the camp itself, and the equipment that was left behind at the end of the expedition, under the responsibility of the village management committee. After the expedition, Rufino Pineda, with the support of the Rotary Club of Santo came back to install a large 1 100 litre water tank and a rainwater catchment system. The idea of transforming the camp structure to accommodate ecotourists has unfortunately fizzled out, as the traditional owner of the site where the camp was located did not accept to share the management and any profits derived from such a venture with the community.

But our most important legacy, from our point of view, is the present work, which finally brings together the key discoveries to date of the expedition, and is a sort of settlement of our debt towards Vanuatu. It is the collective legacy of the participants in one of the greatest natural history expeditions ever organized, and it embodies the commitment of its organizers to return the acquired knowledge in a form appropriate to the authorities of the Republic of Vanuatu. The amount of information collected here on the nature of the island of Santo will doubtless be a reference to biologists and conservation practitioners for the island of Santo for a long time.

Taxonomists publish their results in the form of species descriptions that appear in articles scattered in highly specialized journals read only by specialists, and the publication of the results of a large natural history expedition is never over. New findings still emerge from the "Mangarevan Expedition" led by the Bishop Museum in Honolulu in 1934... Santo 2006 will be no exception: it is likely that descriptions of species collected during the expedition will continue to appear in 20 or 50 years. However, it is not necessary to have identified every snail and every beetle to publish the overall results that are awaited by the biological community. *The Natural History of Santo* fulfils this function. In a little over 500 pages, the geography of the island, the main habitats, and the uniqueness of the fauna and flora, are all presented in a scientific language that is also free, as far as possible, of specialist terminology. This is the book we would all have liked to have had with us when we arrived in Santo, and we hope that this will be a book that will inspire other explorers in the South Pacific, and will be useful in Vanuatu for the management of its natural heritage.

EPILOGUE

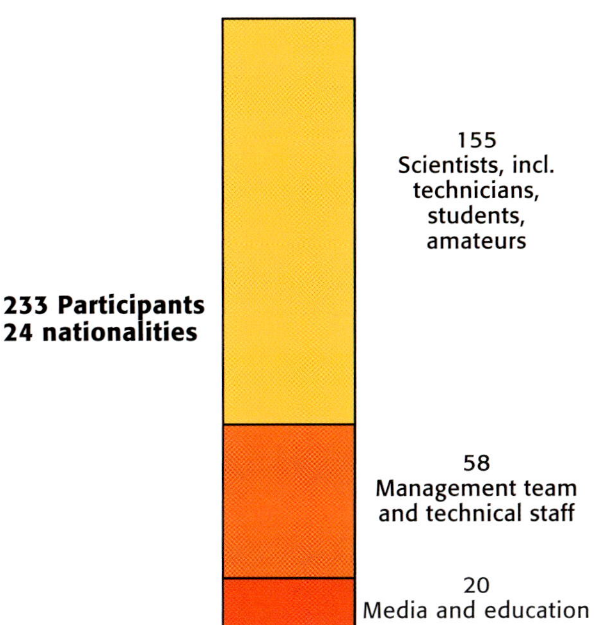

Figure 649: Participants on the Santo 2006 expedition: listed by function.

The Santo 2006 expedition showed that ambitious inventories are possible by deploying the appropriate logistical and human resources (Fig. 649). On Santo, the inventory of the fauna and flora has reached a high degree of completeness and the initial objective to sample "from the breaking of the waves to the top of the mountains" has been achieved. The expedition was not the first mounted by the three project partners, but it was undoubtedly the first to involve so many scientists in the field, to cause as much impact in public media and, somehow, to restore the prestige of natural history exploration. While it was predictable that its researchers would return from a Pacific Island with new species of fish, shrimp or crickets, we could not have expected that the French business newspaper *Les Echos* would devote half a page to Santo 2006 with the title "The return of the great natural history expeditions?", or that the book by Vincent Tardieu and Lise Barnéoud (*Santo, les explorateurs de l'île planète*) published by Belin would be so well received, or that the documentary movie entitled "*Santo, l'île planète*" broadcasted on France 3 in connection with the program Thalassa, later rebroadcasted in two parts on TV channel France 5 and internationally on TV5-Monde would reach over 4 millions viewers. A hoped for but not expected result of Santo 2006 was that it would demonstrate that it is possible to get the public interested in all areas of biodiversity.

What were, in retrospect, the reasons for this success? To an enthusiastic public, who are fully saturated —even jaded— by images, the wealth of digital photographs collected during the expedition was a real revelation of the world of the tiny/cryptic components of biodiversity that are so poorly known. Then, when we set up the expedition, we tried to put the inventory of biodiversity in its human context and rooted in its history: paleoclimatologists, archaeologists, dendrochronologists have turned their attention on the past millennia; ethnologists, lawyers and biodiversity

Figure 650: Peter Lowry, a botanist from the Missouri Botanical Garden, linked to the world by Wi-fi in the mess hut. High technology reached Penaoru before Luganville, thanks to Telecom Vanuatu Ltd.

Figure 651: Nature artist Roger Swainston at the Penaoru camp. Large expeditions like Santo 2006 bring on board other professions than just scientists.

economists have been concerned with the perceptions of this biodiversity and, finally, journalists, photographers, illustrators, filmmakers and teachers have shared the life of the expedition with great numbers of people, in the host country as in France (Figs 650-652).

This change of scale in biological surveying, with an operation involving 233 participants from various levels and nationalities in the field work, has facilitated the mobilization of legal resources and diplomatic support to respond to changes in the way of accessing biodiversity. Beyond Santo, we hope to have shown how the pursuit of legitimate academic research and the growth of research collections of fauna and flora must take place through the development of good institutional practices, including input from the human sciences, to negotiate transparent conditions for access to biodiversity, to give guarantees on the resulting research (academic vs genes / molecules) that meet negotiated commitments, and to learn how to implement the means of restitution and share the "benefits" with the countries of the South.

Figure 652: Preparation for diving at Amarur cave.

This chapter is a translation - slightly edited - of:
BOUCHET P., LE GUYADER H. & PASCAL O. 2008. — Des voyages de Cook à l'expédition Santo 2006: un renouveau des explorations naturalistes des îles du Pacifique. *Journal de la Société des Océanistes* 126-127: 167-185.

2006 Expedition in the Classroom

Sophie Pons & Alain Pothet

An expedition like Santo2006 is a good vehicle to increase the awareness of schoolchildren on the diversity of life, and to introduce them to 'science in action'. While the expedition was in Santo, Emmanuel Vincent presented the expedition in classrooms in several of the schools of Luganville, and organized teachers visits of our marine

laboratory, at the Vanuatu Maritime College. But Vanuatu is an insular country, and we had the desire to reach beyond Luganville and Santo. Thanks to the European Commission office in Vanuatu and the French Embassy in Port Vila, a series of 10 posters was designed by expedition scientists and printed locally in Port Vila. With a print run of 500 copies, it was possible to disseminate them through the Vanuatu Ministry of Education to every school in the country. This educational material addresses a variety of biodiversity themes in the three main languages of Vanuatu (Bislama, English and French):

- Nudibranchs/Kala blong solwota
- Stopem faea Anis blong kam long Vanuatu! [Prevent the arrival of the fire ant in Vanuatu]
- Mimicry/Animol we i kiaman luk olsem nara wan
- Endemism
- Bats of Vanuatu
- Life hidden underground/Laef blong andanit graon
- Santo 2006: Expedesen blong stadi deaversiti long laef [An expedition to study biodiversity]
- Terrestrial mammals of Santo
- When vines steal our gardens
- Looking for insects.

After the expedition, additional educational and support material was produced in France through a project launched by Gérard Bonhoure, General Inspector of Education, implemented by the ACCES team of the Institut National de Recherche Pédagogique (Françoise Morel, Alain Pothet, Naoum Salamé and Gérard Vidal,) and the MNHN Education Department

(Sophie Pons), and involving a large group of educators and teachers (Alain Boissières, Claudie Bonnet, Gilles Camus, Julien Chamboredon, Christophe Charillon, Monique Dupuis, Anne Florimond, Michelle Folco, Gilles Furelaud, Françoise Jauzein, Samuel Jean, Philippe Lejamble, Marie-Laure Le Louarn Bonnet, Sophie Mouge, Elisabeth Mouilleau, Jean-François Rodes, Anne Viguier). A specially developed website made a variety of resources available to classes: summary articles, interviews of researchers, educational strategies, discussion forums, etc. Throughout the 2006-2007 academic year, twelve pilot classes (teachers: Benoit Benard, Claire Calmet, Christine Chevade, Martine Depacheterre, Valérie Fages, Sabine Lavorel, Jean Marc Lepage, Françoise Poujol, Jérôme Rechagneux, Khadija Tehhoune, Delphine Tessier and Thierry Wolniak) developed projects about or around the expedition. While some classes chose to discover the biodiversity of Santo or the customs of Vanuatu, others deployed the techniques of the researchers to organize their own "expedition" near their school. The work of these classes was presented by the children themselves at the occasion of a conference held at MNHN on June 5, 2007, through presentations, posters and "publication" of their work.

2006 Expedition Participants List

Total number of paticipants: 233

Including
Scientists: 155
Management team, technical staff & field support: 58
Media & outreach: 20

(Institutional affiliations as of 2006)

••• Australia

Elizabeth BROWN, Royal Botanic Gardens, Sydney,
Stefan EBERHARD, Department of Conservation and Land Management, Wanneroo,
Roger KITCHING, Griffith University, Brisbane,
Roger SWAINSTON, wildlife artist, Perth,
Fred WELLS, Dept of Fisheries, Perth,

••• Belgium

Thibaut DELSINNE, Institut royal des Sciences naturelles de Belgique,
Maurice LEPONCE, Institut royal des Sciences naturelles de Belgique,
Yves ROISIN, Université Libre de Bruxelles,

••• Brazil

Milton BARBOSA da SILVA, Federal University of Ouro Preto,

••• Brunei

David LANE, Universiti Brunei Darussalam,

••• Fiji

Philippe GERBEAUX, IUCN Regional Office,
Marika TUIWAWA, South Pacific Regional Herbarium, University of the South Pacific,

••• France

Laurent ALBENGA, MNHN,
Marc ATTIE, Université de la Réunion,
Jean-François BARAZER, Captain,
Lise BARNEOUD, science journalist,
Anne BEDOS, MNHN,
Florence BELLIVIER, Université Paris 10 Nanterre,
Denis BERTRAND, camera operator, film crew,
Grégory BEUNEUX, Groupe Chiroptères Corse,
Lucie BITTNER, MNHN,
Emmanuel BOITIER, Université de Limoges,
Jean-Michel BORET, camera operator IRD,
Philippe BOUCHET, expedition co-director, MNHN,
Frank BOUZIDI, Genavir, R/V Alis shipboard party,
Delphine BRABANT, MNHN,
Franck BREHIER, free-lance speleo diver,
Florence BRUNOIS, MNHN,
Bart BUYCK, MNHN,
Magalie CASTELIN, MNHN,
Vincent CHEVALIER, Genavir, R/V Alis shipboard party,
Guillaume CHIPY, Universal-Sodexho,
Regis CLEVA, MNHN,

Dany CLEYET-MARREL, pilot,
Alain COMPOST, camera operator and special effects, film crew,
Bruno CORBARA, Université Blaise Pascal, Clermont-Ferrand,
Louis DEHARVENG, MNHN,
Xavier DESMIER, photographer,
Laure DESUTTER, MNHN,
Jean DROUAULT, mechanical engineer (Canopy-Glider),
Jacques DUMAS, dive master,
Joelle DUPONT, MNHN,
Frédéric DURAND, Soc. d'Histoire Naturelle Alcide d'Orbigny,
Gilles EBERSOLT, logistics,
Franck ESCUDIER, camera operator, film crew
Elsa FAUGERE, INRA Avignon,
Christian FLEURY, logistics,
Benoît FONTAINE, MNHN,
Olivier GARGOMINY, MNHN,
Jean-Baptiste GOASGLAS, tree climber & logistics,
Olivier GROS, Université Antilles-Guyane,
Bernard GUERRINI, film director, film crew
Thomas HAEVERMANS, MNHN,
Virginie HEROS, MNHN,
René HEUZEY, underwater cameraman, film crew,
Damien HINSINGER, MNHN,
Frédéric HONTSCHOOTE, web journalist,
Sylvain HUGEL, CNRS,
Ivan INEICH, MNHN,
Philippe KEITH, MNHN,
Ronan KIRSCH, associate MNHN,
Jean-Noël LABAT, MNHN,
Dominique LAMY, associate MNHN,
Nadir LASSON, free lance speleo diver,
Hervé LE GUYADER, expedition co-director, Université Paris 6,
Hervé Le HOUARNO, Genavir, R/V Alis shipboard party,
Philippe LEMARCHAND, web journalist,
Véronique LESNARD, assistant director, film crew,
Bernard LIPS, Fédération Française de Spéléologie,
Josiane LIPS, free lance speleologist,
Clara LORD, MNHN,
Julien LORION, MNHN,

Olivier LORVELEC, INRA,
Selim LOUAFI, Institut du Développement Durable et des Relations Internationales,
Pierre LOZOUET, MNHN,
Michel MARIN, camera operator, RFO,
Philippe MAESTRATI, MNHN,
Frederic MATHIAS, tree climber / arborist,
Didier MOLIN, MNHN,
Dan MOLCZADZKI, Penaoru camp manager,
Jerôme ORIVEL, Université Paul Sabatier,
Michel PASCAL, INRA,
Olivier PASCAL, Pro-Natura international, expedition co-director,
Jacques PELORCE, associate MNHN,
Patrice PETIT-DEVOIZE, dive master,
Marc PIGNAL, MNHN,
Danielle PLACAIS, associate MNHN,
Jean-Claude PLAZIAT, Université Paris 11,
Odile PONCY, MNHN,
Marc POUILLY, IRD,
Vincent PRIÉ, attaché MNHN,
Pierrick PRIGENT, Genavir, R/V Alis shipboard party,
Nicolas PUILLANDRE, student MNHN,
Laurent PYOT, logistics (Canopy-Glider),
Eric QUEINNEC, Université Paris 6,
Frédérick RANNOU, Genavir, R/V Alis shipboard party,
Marc REBUTINI, camera operator, film crew,
Claude RIVES, photographer,
Marine ROBILLARD, MNHN,
Tony ROBILLARD, MNHN,
Christine ROLLARD, MNHN,
Germinal ROUHAN, MNHN,
Sophia SALABASCHEW, production sound mixer, film crew,
Sarah SAMADI, IRD,
Mathias SCHMITT, journalist, scenarist, film crew,
Sylvestre SEGAUD, technical staff,
Anne-Marie SEMAH, IRD Bondy,
Jean-Yves SEREIN, tree climber / arborist,
Jerémy SIMONNOT, assistant camera, film crew,
Laurent SOLDATI, INRA,
Adeline SOULIER-PERKINS, MNHN,
Vincent TARDIEU, science journalist,
Jean-Marc THIBAUD, MNHN,
Xavier TRAVEL, Genavir, R/V Alis shipboard party,
Renaud TUFFREAU, Genavir, R/V Alis shipboard party,
Fabienne TZERIKIANTZ, Centre de Recherche et de Documentation sur l'Océanie,
Claire VILLEMANT, MNHN,
Emmanuel VINCENT, Ecole Normale Supérieure,
Rudo VON COSEL, MNHN,
Denis WIRRMANN, IRD,
Magali ZBINDEN, Université Paris 6 / CNRS,
Robert ZAMORA, Genavir, R/V Alis shipboard party,
Luc ZELMAT, Universal-Sodexho,

Germany
Timea NEUSSER, Zoologisches Staatsammlung, Munich,
Milan PALLMANN, Museum für Naturkunde, Stuttgart,
Jürgen SCHMIDL, University of Erlangen,
Arnold STANICZEK, Museum für Naturkunde, Stuttgart,

Indonesia
Mark ERDMANN, Conservation International, Raja Ampat Program,
Cahyo RAHMADI, Museum Zoologicum Bogoriense,

Italy
Marco OLIVERIO, University of Roma,
Stefano SCHIAPARELLI, University of Genoa,

Japan
Yasunori KANO, University of Miyazaki,
Takuma HAGA, student, University of Tokyo,

Netherlands
Charles FRANSEN, Naturalis, Leiden,
Adriaan GITTENBERGER, Naturalis, Leiden,
Bert HOEKSEMA, Naturalis, Leiden,
Willem RENEMA, Naturalis, Leiden,

New Caledonia
Nicolas BARRÉ, Institut Agronomique Néo-Calédonien (IAC) / CIRAD,
Jean-Michel BORÉ, IRD,
Cyril CHEVALIER, medical doctor,
Christophe CHEVILLON, IRD,
Valérie DEMORY, medical doctor,
Christian FITIALEATA, Genavir, R/V Alis shipboard party,
Eric FOLCHER, IRD, professional diver,
Jean-Christophe GALIPAUD, IRD,
Michel de GARINE-WICHATITSKY, IAC/CIRAD,
Bruno GATIMEL, IRD,
Cathy GEOFFRAY, IRD, professional diver,
Claire GOIRAN, IRD,
Nicolas JOSSE, Genavir, R/V Alis shipboard party,
Hervé JOURDAN, IRD,
Jean LAMATA, Genavir, R/V Alis shipboard party,
Gregory LASNE, IRD,
Felise LIUFAU, Genavir, R/V Alis shipboard party,
Lydiane MATTIO, IRD,
Jean-Louis MENOU, IRD, professional diver,
Elisabeth MERLIN, medical doctor,
Tristan MICHINEAU, medical doctor,
Jérôme MUNZINGER, IRD,
Ferrand N'GANYANE Genavir, R/V Alis shipboard party,
Claude PAYRI, IRD,
Sandrine PEREY, medical doctor,
Yohan PILLON, IRD,
Jacques QENENOJ, Genavir, R/V Alis shipboard party,
Gwénaelle QUERMELIN, nurse,
Bertrand RICHER DE FORGES, IRD,
Frédéric RIGAULT, IRD,
Marcel SAGEL, Genavir, R/V Alis shipboard party,

New Zealand
Alan BEU, GNS Science, Lower Hutt,
Jonathan PALMER, Gondwana Tree-ring Laboratory,

Norway
Anne Lise FLEDDUM, University of Oslo,
John GRAY, University of Oslo,
Camilla FRISEID, University of Oslo,
Gorild HOEL, University of Oslo,
Karen WEBB, University of Oslo,

••• Philippines

Jo ARBASTO, fisherman, Panglao Island, Bohol,

Marivene MANUEL SANTOS, National Museum of the Philippines,

Noel SAGUIL, University of San Carlos,

Dave VALLES, University San Carlos,

••• Russia

Yuri KANTOR, Russian Academy of Sciences, Moscow,

Tanya STEYKER, Russian Academy of Sciences, Moscow

••• Singapore

Jose Christopher E. MENDOZA, National University, Singapore,

Peter NG, National University, Singapore,

TAN Swee Hee, National University, Singapore,

TAN Heok Hui, National University, Singapore,

••• Spain

Josep Antoni ALCOVER TOMAS, Institut Mediterrani d'Estudis Avançats, Mallorca,

Damia JAUME, Institut Mediterrani d'Estudis Avançats, Mallorca,

Marta POLA, Universidad de Cadiz,

Jose TEMPLADO, Museo Nacional de Ciencias Naturales, Madrid,

••• Sweden

Anders WAREN, Swedish Museum of Natural History, Stockholm,

Kerstin RIGNEUS, Swedish Museum of Natural History, Stockholm,

••• Switzerland

Valérie HOFSTETTER, Duke University,

••• Taiwan

Tin-Yam CHAN, National Taiwan Ocean University, Keelung,

Masako MITSUHASHI, National Taiwan Ocean University,

••• United Kingdom

Geoff BOXSHALL, The Natural History Museum, London,

Paul CLARK, The Natural History Museum, London,

••• USA

Jason BIGGS, University of Utah, Salt Lake City,

Yolanda CAMACHO, California Academy of Sciences, San Francisco,

John EARLE, Associate Bernice P. Bishop Museum, Honolulu,

Brian GREENE, Associate Bernice P. Bishop Museum, Honolulu,

Porter P. LOWRY II, Missouri Botanical Garden, St Louis,

Gordon McPHERSON, Missouri Botanical Garden, St Louis,

Mike MILLER, California Academy of Sciences, San Francisco,

Gregory PLUNKETT, Virginia Commonwealth University,

Richard PYLE, Bernice P. Bishop Museum, Honolulu,

Marilyn SCHOTTE, Smithsonian Institution, Washington DC,

Ellen STRONG, Smithsonian Institution, Washington DC,

Alexey TISHECHKIN, Louisiana Sate University,

Angel VALDES, Los Angeles County Museum of Natural History,

••• Vanuatu

Joseph BOE, guide, Penaoru,

Ruben BOE, guide, Penaoru,

Sam CHANEL, Forest Department, Port-Vila,

André FRANCK, logistical assistant,

Faustin FRANCK, logistical assistant,

Joséphine GRIMAUD, cook, Santo,

Anthony HARRY, trainee IAC,

Charley JOHNSON, logistical assistant,

Malvina JOKONMAL, cook, Penaoru,

Tarere GARAE, Forest Department, Santo,

Donna KALFATAK, Environment Unit,

Rupen MARAÏ, assistant to the Penaoru Camp Manager,

George MATARIKI, student, Port Vila,

Vatumaraga MOLISA, student USP, Wounpuko,

Rufino PINEDA, local co-ordinator,

Charles TARI, student USP, Port Vila,

Graham TARIDIA, student USP, Port Vila,

Leimas Emily TASALE, student USP, Port Vila,

Samson VILVIL-FARE, student, Port Vila,

Maria YORLEY, Santo,

Stephen VUTILOLO, Santo,

Steve VIRA, boat pilot, Santo,

Ritchie FRANCK, Santo,

Rony TOM, guide Penaoru

An expedition of the dimension of Santo 2006 was only possible through the support and understanding of numerous organisations, institutions and individuals. First and foremost, special thanks go to Russell Nari and Maxime Carlot, then of the Ministry of Lands of Vanuatu; Joel Path, Secretary General, and Daniel Lekele, Assistant Secretary General, in Sanma Province; and Bernard Sexe at the French Embassy in Port Vila, for their trust in the project, before, during and after the expedition, and to the traditional chiefs and the communities of Aore, Butmas, Finmelé, Funafus, Hokua, Kerepua, Malo, Natsara, Natawa, Peavot, Penaoru, Tasmate, Tangoa, Vatthe, Wunpoku, and many other villages who welcomed the scientists with a mixture of patience, unselfishness and bemused curiosity. They should know that they all have left their mark in our memories.

The path from project to expedition was a long one, but one that benefitted from much interest, good will and support, in-house and outside. In our home institutions, Muséum National d'Histoire Naturelle's (MNHN) Director General Bertrand-Pierre Galey enthused for the combination of discovery, adventure and international linkage contained in the project, and he and Marie-Louise Seitz, MNHN Director of Financial Services, personally saw that it would not —at times— sink in the quagmire of timeless institutional bureaucracy. At Institut de Recherche pour le Développement (IRD), Marie-Noëlle Favier, Director of Communication, Fabrice Colin, Director of the institute's branch in Nouméa, and Patrice Cayré established the necessary institutional linkage and authorized the deployment of vehicles, boats and, last but not least, the research vessel *Alis*. Pro-Natura International's (PNI) President Guy Reinaud agreed to take the risk of accepting much of the project management under the umbrella of *Pro-Natura*, and Franck de Rouville and Pierre-Yves Simond saw that the expedition would be accountable despite some of its unorthodox expenditures.

The Board of Directors of the *Stavros Niarchos Foundation* (http://www.SNF.org) generously provided major support for the backbone of the operational costs of the expedition, notably the terrestrial component. In this respect, the commitment of the Foundation's Co-President, Andreas Dracopoulos, is gratefully acknowledged. It is also necessary to acknowledge that the Santo 2006 expedition was realized thanks to the help and advise of Leonidas Cambanis. The *Total Foundation* (http://fondation.total.com) specifically funded the marine component of the expedition, a continuation of its sponsorship of MNHN marine expeditions since 2000, and the support of President Bernard Tramier, Gina Sardella-Sadiki and Laure Fournier is gratefully acknowledged. We thank the Permanent Secretariat for the Pacific and its French permanent representative with the Pacific Community, Ambassador Patrick Roussel, which funded the Santo 2006 project within the regional cooperation framework. Pascale Joannot, MNHN overseas delegate, played a key role in approaching the Permanent Secretariat. European Commission funding came through two channels, the *European Distributed Institute of Taxonomy* (EDIT) and the office of the European Commission in Port Vila, and we thank Simon Tillier and Nicolás Berlanga, respectively, for their contributions. Three of the themes of the expedition (Marine Biodiversity, Botany, and Karst) received grants from the *National Geographic Society's* Committee for Research and Exploration, and the *Sloan Foundation* specifically funded the exploration of deep reefs as part of the *Census of Marine Life* CReefs component, and the advice of Peter Raven and Jesse Ausubel, respectively, was instrumental in the success of our applications. Danielle Plaçais and Gwenaëlle Le Garrec arranged a last-minute grant from *Triballat*; Michel Dupuis organized in-kind support from *Telecom Vanuatu* for our communication needs; *Solvin/Solvay* made it possible to bring the Canopy Glider to Santo; and *Sodexo* provided personnel for catering services at our Penaoru base camp.

In Santo, Captains Kenneth Barnett and John Lloyd, and Caroline Nalo arranged for the facilities of the *Vanuatu Maritime College* (VMC) —from boats and classrooms to refectories, dormitories and fuel tank— to be used by the Marine Biodiversity party and others. VMC was the expedition base for more than 3 months and this placed high pressure on its personnel, at sea as well as on land. Additionally, Alsen Obed, from the Department of Fisheries, kindly agreed to our using their large boat shed on the VMC campus as our marine laboratory. Likewise, the *Vanuatu Agricultural Research Technical Centre* (VARTC/CTRAV), outside town, was the base of the Karst and Aliens teams, and we thank its director Marie Malterus and personnel for having responded to our needs despite the disruption this brought.

We had a programme, we had the permits, we had the funding. After Christian Fleury had battled hard with the shipping company to get the expedition equipment to Santo in time, we had the logistics in place, and we had a scientific party. At the risk of omitting anyone —to whom we apologize in advance— the expedition leaders and all the participating scientists are grateful to all those who contributed their skills to the success of the field work: pilots and seamen (Jo Arbasto, Jean François Barazer, Frank Bouzidi, Vincent Chevalier, Napoleon Colombani, Ferrand N'Ganyane, Christian Fitialeata, Nicolas Josse, Jean Lamata, Hervé Le Houarno, Felise Liufau, Pierrick Prigent, Jacques Qenenoj, Frédérick Rannou, Marcel Sagel, Xavier Travel, Renaud Tuffreaud, Steve Vira, Robert Zamora; captain and crew of VMC's *Euphrosyne*), divers (Laurent Albenga, Jacques Dumas, Eric Folcher, Cathy Geoffray, Jean-Louis Menou, Jacques Pelorce, Patrice Petit-Devoize), climbers (Frederic Mathias, Jean-Yves Serein, Jean-Baptiste Goasglas), cooks (Guillaume Chipy, Luc Zelmat), vehicle drivers / guides (Franck Faustin, Charley Johnson, Stephen Vutilolo), helpers (Ritchie Franck, Maria Yorley, Roman), photographers, illustrators and cameramen (Jean-Michel Boré, Delphine Brabant, Xavier Desmier, Emmanuel Boitier, Claude Rives, Roger Swainston), IT technicians (Francis Gallois, Didier Molin), and students and trainees (Lucie Bittner, Magalie Castelin, Camilla Friseid, Anthony Harry, Damien Hinsinger, Gorild Hoel, Gregory Lasne, Julien Lorion, Lydiane Mattio, Nicolas Puillandre, Charles Tari, Graham Taridia, Emily Tasale, Samson Vilvil-Fare, Karen Webb). In Kerepua, custom owner Olboy allowed free access to Mt Tabwemasana, and Matt Drury, the resident U.S. Peace Corps volunteer, facilitated the work of the expedition party. Special thanks go to the Vanuatu Cultural Center fieldworkers of the Santo-Malo area and particularly Lenki Bilal from Wusi, Ezekiel Aldi from Valpei, Joseph Vira from Avunatari in Malo, and Takau Mwele from Aore. In Penaoru, Chief Wortut and local Ni-Vanuatu guides Rony Tom, Joseph and Ruben Boe shared with the scientists their intimate knowledge of the forest.
We wish to highlight the role of Rufino Pineda, our project representative in Santo, who during months oversaw the local logistics from custom brokers and awareness meetings with village elders, to carpentry, renovation, hiring personnel and more; his and Dan Molczadzki's, our camp manager in Penaoru, local experience with attitudes and people eased many difficulties and avoided a number of faux pas. We also wish to highlight Philippe Maestrati, Noel Saguil and his buddy Dave Valles, who coordinated the logistics, equipment needs for, and unfolding of, the large marine theme; and Chantal Parent, Virginie Héros, Charlotte Leroy and Samson Vilvil-Fare, who arranged the complex paper work, international travel arrangements and smooth local transits for close to 200 persons. Finally the expedition organizers would like to thank the participants themselves, for their dedication, camaraderie, and spirit of discovery. All came on the time of their institutions, and over 40 institutions in 25 countries thus provided in-kind support through personnel and their equipment.

Through Lise Barnéoud, Etienne Collomb and Marie Lescroart, representing on site various French press outlets, Santo 2006 was the focus of considerable attention —during and after the expedition. Julia Bigot, at MNHN, and Emilie Dumont, then at *Atom Production*, coordinated press conferences and the associated documentation. Thanks to Jack Guichard, its Director, and Marie Canard, several of these meetings with the press took place at the *Palais de la Découverte*. Roger Genet, advisor to the minister, was instrumental in having this Santo 2006 press point inaugurated by François Goulard, then Minister of Research and Higher Education. Arjan Gittenberger Frédéric Hontschoote, Philippe Lemarchand and Danielle Plaçais fed our website almost daily remotely from Santo. Emmanuel Vincent and Claire Goiran brought the expedition to the Luganville classrooms, and brought the Luganville schoolchildren to visit our facilities. The post-expedition education component, coordinated by Gérard Bonhoure, Sophie Pons and Naoum Salamé, involved a large group of teachers and educators, cited by name in the *Santo 2006 in the classroom* chapter. Stéphane Frey oversaw, from cradle to market, Vincent Tardieu and Lise Barnéoud's Santo *Les explorateurs de l'île planète*, published by Belin. And, last but not least, *Atom Production's* (now Docside Production) Director Yves Bourgeois took the financial risk to produce the full-length film *Les aventuriers de l'île planète*, authored by Mathias Schmitt and realized by Bernard Guerrini.

To a large extent, *The Natural History of Santo* reflects observations and research conducted in Santo during the 2006 expedition, but is not restricted to this. Several researchers who had been Santo "long timers", but did not take part in the expedition, have contributed to this volume, and we are grateful in particular to Benoit Antheaume, Guy Cabioch, Christophe Maes, Patricia Simeoni, James Terry, Sara Muller and their co-authors who kindly accepted their "assignments" to make this book more comprehensive. The production of the volume has also been a long process... It is not easy to herd a flock of over 100 cats! We would not have made it without Laurent Palka, who scanned the many contributions for redundancies and polished numerous internal inconsistencies, and Gwenaëlle Chavassieu, who juggled with hundreds of files, laid out the volume, and handled feedback from the authors. Thank you, Laurent and Gwenaëlle! Phil Clarke, Mark Judson and Daniel Simberloff, kindly revised the language and made many helpful suggestions that have improved the contents of many chapters. Gilberto Marani reformated and redesigned graphs and maps throughout the volume.

The authors themselves also wish to direct specific thanks to colleagues who helped sort specimens and/or analyze data and contributed expertise in many ways. James Terry (*The Climate of Santo and Drainage, Hydrology and Fluvial Geomorphology*) wishes to thank Swastika Devi of the Climate Services Division, Fiji Meteorological Services, Nadi, and Philip Malsale of Vanuatu Meteorological Services, Port Vila, for kindly providing cyclone reports and data on Vanuatu rainfall. Ketty Newpatt and Alfred Maoh kindly provided organisational assistance and The University of the South Pacific generously funded three of his field surveys. Pierre Lozouet and his co-authors (*The Holocene and Pleistocene Marine Faunas Reconsidered*) thank Céline Chaignon for her help with species separation, Jean-Claude Plaziat for very useful suggestions, and Luc Dolin, Georges Richard and Jean-Maurice Poutiers for identification of species. Radiometric analyses were performed by Laboratoire des Sciences du Climat et de l'Environnement (CNRS-CEA, Gif-sur-Yvette). Warren Blow (Smithsonian Institution, Washington DC), a veteran of the 1973 paleontological expedition to Santo, sent all the information he could. It was a major disappointment for Warren not to be able to join the expedition; unfortunately, he passed away just after the spring 2006 trip to Santo. Jean-Christophe Galipaud (*The Prehistory of Santo*) wishes to thank the former vice Prime Minister of Vanuatu, the Honourable Seti Regenvanu, and the former Director of the Vanuatu Cultural Center, Mr Kirk Hoffman, for their support during the initial phase of his research, which started in Santo and adjacent islands back in 1992. Ralph Regenvanu, director of the Vanuatu Cultural Center and his successor Marcellin Abong played an important role in the developments of his archaeology research. Rufino Pineda (*Deforestation on Santo and Logging Operations*) thanks Phyllis Kamasteia (Mapping/GIS Officer, Vanuatu Dept of Forests) for the map of logging concessions. Laurent Palka and Rufino Pineda (*The Impact of WWII on Infrastructure and Landscape*; *The Night of the Palolo*) thanks Matt G. Bentley (Newcastle University, UK) for helpful comments on the biology of palolo, Samson Vilvil-Fare (University Paris 6, France) for his Ni-Vanuatu narration of the palolo story, and Meghann Cassidy (Paris Sorbonne University) for linguistic assistance with the English manuscript. Ivan Ineich (*Amphibians and Reptiles*) addresses special thanks to A.H. Whitaker for numerous exchanges and bibliographic information, and to George R. Zug for providing data, continued support and friendship and his comments on a first draft of that paper. Useful information was contributed by Edouard-Raoul Brygoo (MNHN), John Thorbjarnarson (Wildlife Conservation Society), Andreas Schmitz (Geneva), the late Eugen Kramer (Basel), Colin McCarthy (NHM, London), and Ross Sadlier (Australian Museum, Sydney, Australia). Claire Vilemant (*Focus on Bees and Wasps*) is much indebted to colleagues who carried out identifications and provided information about nomenclature and distribution of species in the Hymenoptera families of their speciality: Denis Brothers (University of Kwazulu-Natal, Pietermaritzburg), James Carpenter (AMNH, New York), Gérard Delvare (CIRAD, Montpellier, France); Jacques Dubois (AMNH, New York); Mattias Forshage (Uppsala University, Sweden); Norman Johnson & Luciana Musetti (Ohio State University); Massimo Olmi (Università della Tuscia, Viterbo); Alain Pauly (IRSNB, Bruxelles) and Raymond Wahis (Faculté Universitaire des Sciences agromomiques, Gembloux). She is also very grateful to Frédéric Durand (Société d'Histoire Naturelle Alcide d'Orbigny) who collected by hand and sorted a large quantity of specimens, to Patrick Burguet who mounted them and to Agnièle Touret-Alby for photography. Jürgen Schmidl (*Beetles in Saratsi Range, Santo*) is grateful to Belinda Fleming and Camille Dejonquères for extensive help in sorting, mounting and analyzing numerous beetle samples. Bruno Corbara (*Myrmecophily*) likewise extends his thanks to Patrick Burguet (Société d'Histoire Naturelle Alcide d'Orbigny, SHNAO, Clermont-Ferrand) who mounted the ant specimens and to Benjamin Calmont (SHNAO) who photographed them. Nicolas Barré and his co-authors (*Terrestrial Bird Communities in Santo*) thank Anthony Harry who translated and spelled the names of birds in bislama and they are particularly grateful to Dick Watling (NatureFiji-MareqetiViti) for his constructive revision and comments on the manuscript. Denis Wirrmann and Anne-Marie Sémah. (*Caves as archives*) wish to thank Erickson Sammy, former Acting Director of the Vanuatu Department of Geology Mines and Water Resources, Toney Tevi, Commissioner of Mine, Esline Garaebiti, Manager of Geophysics, and their staff for their very helpful administrative participation during their previous work and field surveys the year before Santo 2006. Louis Deharveng and his co-authors (*Karst habitats of Santo* and *Karst biota of Santo*) acknowledge the advice of Richard Harris (Royal Adelaide Hospital, Adelaide), pioneer diver and caver in Santo, and the assistance of Arnaud Faille (MNHN), Christopher Glasby (Museum and Art Gallery, Darwin,

Northern Territory, Australia), Sergei Golovatch (Institute for Problems of Ecology and Evolution, Russian Academy of Sciences, Moscow), Peter Jaeger (Senckenberg Museum, Frankfurt), Mark Judson (MNHN), Ivan Löbl (Muséum d'Histoire Naturelle, Genève), Wilson Lourenço (MNHN), Quentin Rome (MNHN), Stefano Taiti (Istituto per lo Studio degli Ecosistemi, CNR, Firenze), Wanda Weiner (Institute of Systematics and Evolution of Animals, Polish Academy of Sciences, Krakow), Sun Xin and Feng Zhang (both Nanjing University, Nanjing), and other expedition participants, for the identification of specimens. Claude Payri (*Benthic Algal and Seagrass Communities*) acknowledges a grant received from the CRISP [Coral Reefs Initiative for the South Pacific] programme that permitted Antoine de N'Yeurt to process the algae collections in Nouméa. She also thanks Hannah Stewart for improving the English of her manuscript. The results presented by Philippe Bouchet and his co-authors (*The Marine Molluscs of Santo*) are the collective results of a large group of people, cited by name in that chapter. Fred Wells (*A Rapid Assessment of the Marine Molluscs of Southeastern Santo*) acknowledges the partnership of his dive buddy Dominique Lamy, and the assistance of Mike Travers and Gareth Parry, of the Western Australian Fisheries and Marine Research Laboratories, with analyses and development of the figures in his paper. Olivier Lorvellec (*Focus on Introduced Amphibians and Reptiles*) acknowledges the assistance of Thierry Frétey for his help in documentation investigation and analysis, and Samson Vilvil-Fare for his participation in documentation search. Florence Brunois and Marine Robillard (*Ni-Vanuatu Perceptions and Attitudes vis-à-vis Biodiversity*) thank Paul Sally for translating their text from the French original. The data presented by Sara Muller and co-authors (*Food-garden Biodiversity in Vanuatu*) were collected during a FFEM- [French Fund for World Environment] and CIRAD- [Centre International de Recherche Agronomique pour le Développement] funded project to document and strengthen agrobiodiversity of root crops in Vanuatu.

The present bibliography includes references of scientific works dealing specifically with Espiritu Santo, or Vanuatu in general but including Santo.

It does not include general works that may include marginal reference to Vanuatu, nor taxonomic papers that contain only a few records on Santo or Vanuatu.

An asterisk* marks papers resulting from the Santo 2006 expedition.

The references are arranged thematically in eight sections:
- Geography and climate
- Human settlement
- Man and nature
- Quaternary geology and paleontology
- History of biodiversity exploration
- Vegetation and flora
- Terrestrial fauna
- Marine ecosystems

GEOGRAPHY AND CLIMATE

ANONYMOUS [ORSTOM] 1985. – *Iles d'Efaté et de Santo. Évaluation des ressources en eau. Résultats des campagnes 1981 à 1984*. Centre ORSTOM, Nouméa, 149 p.

BONNEMAISON J. 1986. – *Vanuatu*. Les Editions du Pacifique, Papeete, 128 p.

ANTHEAUME B. & BONNEMAISON J. 1988. – *Atlas des îles et Etats du Pacifique-Sud*. Publisud-Reclus, Paris-Montpellier, 126 p.

ANTHEAUME B., BONNEMAISON J., BRUNEAU M. & TAILLARD C. (Eds) 1995. – *Asie du Sud-Est, Océanie*. Belin-Reclus, Paris-Montpellier, 480 p. (Géographie universelle; 7).

CHURCH J.A., WOODWORTH P.L., AARUP T. & WILSON W.S. 2010. – *Understanding Sea-Level rise and variability*. Wiley-Blackwell, Chichester, 428 p.

COLLINS M. *et al.* 2010. – The impact of global warming on the tropical Pacific Ocean and El Niño. *Nature Geoscience* 3: 391-397, doi: 10.1038/ngeo868.

FMS 2003. – *List of tropical cyclones in the south west Pacific 1969/70 - Present*. Fiji Meteorological Service, Nadi Airport, Fiji. (Information Sheet; 121).

GANACHAUD A. *et al.* 2011. – Observed and expected changes to the Pacific Ocean, *in* BELL J.D., JOHNSON J.E. & HOBDAY A.J. (Eds), *Vulnerability of Fisheries and Aquaculture in the Pacific to Climate Change*. Secretariat of the Pacific Community, Noumea, New Caledonia, in press.

GOURDEAU L., KESSLER W.S., DAVIS R.E., SHERMAN J., MAES C. & KESTENARE E. 2008. – Zonal jets entering the Coral Sea. *Journal of Physical Oceanography* 38: 715-725.

GOUYET R.C. 1984. – *Etudes hydrologiques en République de Vanuatu*. Office de la Recherche Scientifique et Technique Outre-Mer (ORSTOM), Centre de Nouméa, Nouvelle-Calédonie.

HAREWOOD J. 2009. – *Vanuatu & New Caledonia travel guide*. 6th ed. Lonely Planet Publications, Melbourne, 224 p.

LONGWORTH W.M. 1991. – *Tropical cyclones in Vanuatu: 1847-1991*. Vanuatu Meteorological Service, Port Vila, 24 p.

MAES C. 2006. – Océanographes des mers du Sud, 50 ans d'océanographie physique dans l'océan Pacifique sud-ouest. *MetMar* 213: 4-11.

MITTERMEIER R., ROBLES GIL P., HOFFMAN M., PILGRIM J., BROOKS T., GOETTSCH-MITTERMEIER C., LAMOREUX J. & DA FONSECA G.A.B. (Eds) 2005. – *Hotspots revisited. Earth's biologically richest and most endangered terrestrial ecoregions*. Conservation International, Washington; Cemex; University of Chicago Press, Chicago, 391 p.

O'BYRNE D. & HARCOMBE D. 1999. – *Vanuatu*. 3rd ed. Lonely Planet Publications, Melbourne, 304 p.

QU T. & LINDSTROM E.J. 2002. – A climatological interpretation of the circulation in the western South Pacific. *Journal of Physical Oceanography* 32: 2492–2508.

SCHILLER A., RIDGWAY K.R., STEINBERG C.R. & OKE P.R. 2009. – Dynamics of three anomalous SST events in the Coral Sea. *Geophysical Research Letters* 36: L06606, doi:10.1029/2008GL036997.

SIMEONI P. 2009. – *Atlas du Vanouatou*. Éditions Géo-Consulte, Port Vila, Vanuatu, 392 p.

STATTERSFIELD A.J., CROSBY M.J., LONG A.J. & WEGE D.C. 1998. – *Endemic bird areas of the world: Priorities for biodiversity conservation*. Birdlife International, University of California Press, 846 p.

TERRY J.P. 2007. – *Tropical cyclones: climatology and impacts in the South Pacific*. Springer, New York, 210 p.

VMS 2009. – *The Climate of Vanuatu*. Vanuatu Meteorological Services, accessed February 2009 from <http://www.meteo.gov.vu/>.

HUMAN SETTLEMENT

ANDERSON A. 2002. – Faunal collapse, landscape change and settlement history in remote Oceania. *World Archaeology* 33: 375-90.

BAKER J.R. 1928. – Notes on the New Hebridean customs with special references to the intersex pig. *Man* 80-81: 113-18.

BEDFORD S. 2006. – Pieces of the Vanuatu puzzle: Archaeology of the North, South and Centre. *Terra Australis* 23: pages ??

BEDFORD S., SPRIGGS M. & REGENVANU R. 2006. – The Teouma Lapita site and the early human settlement of the Pacific Islands. *Antiquity* 80: 812-828.

BONNEMAISON J. 1985. – Territorial control and mobility within Vanuatu societies, *in* PROTHERO K.M & CHAPMAN M. (Eds), *Circulation and population movement : substance and concepts from the Melanesian case*. Routledge & Keegan, Londres: 57-79

BONNEMAISON J. 1985. – Les lieux de l'identité : vision du passé et identité culturelles dans les îles du sud et du centre du Vanuatu (Mélanésie). *Cahiers de l'ORSTOM*, série sciences humaines, 21(1): 151-170.

BONNEMAISON J. 1996. – *Gens de pirogue et gens de la terre. Les fondements géographiques d'une identité (L'archipel du Vanuatu)*. Livre I. Orstom Editions, Paris, 460 p.

BONNEMAISON J. 1996. – Le tissu de Nexus, *in* BONNEMAISON J., HUFFMAN K., KAUFFMANN C. & TRYON D. (Eds), *Vanuatu, Océanie: Arts des îles de cendre et de corail*. Réunion des Musées Nationaux; ORSTOM, Paris: 176-183.

BONNEMAISON J. 1997. – *Les gens des lieux. Histoire et géosymboles d'une société enracinée : Tanna. Les fondements géographiques d'une identité (L'archipel du Vanuatu)*. Livre II. Orstom Editions, Paris, 562 p.

GALIPAUD J.C. 1996. – Le Rouge et Le Noir: la poterie de Mangaasi et le peuplement des iles de Mélanesie, *in* ORLIAC M., ORLIAC C. & JULIEN M. (Eds), *Mémoire de Pierre, mémoire d'homme: tradition et archéologie en Océanie*. Publications de la Sorbonne, Paris: 115-31.

GALIPAUD J.C. 1996. – Poteries et potiers de Vanuatu, *in* BONNEMAISON J., HUFFMAN K., KAUFFMANN C. & TRYON D. (Eds),*Vanuatu, Océanie: Arts des îles de cendre et de corail*. Réunion des Musées Nationaux; ORSTOM, Paris: 365.

GALIPAUD J.C. 1997. – *Fouille de l'abri sous roche de Malsosoba, Hokua, côte nord-ouest de Santo. Rapport de terrain*. Orstom, Port-Vila.

GALIPAUD J.C. 2000. – The Lapita site of Atanoasao, Malo, Vanuatu. *World Archaeological Bulletin* 12: 41-55.

GALIPAUD J.C. 2002. – *Ancient cooking strategies and development of irrigated taro systems in North-West Santo*. Archaeology in Oceania special Publications. University of Sydney.

GALIPAUD J.C. 2004. – Settlement history and landscape use in Santo, Vanuatu. *Records of the Australian Museum*, Supplement 29: 59–64.

GALIPAUD J.C. 2007. – Les sociétés du Vanuatu au temps de Quiros, *in* ANGLEVIEL F. (Ed.), *Pedro Fernandez de Quiros et le Vanuatu. Découverte Mutuelle et Historiographie d'un acte fondateur, 1606*. Editions du GRHOC, Nouméa: 17-30.

GALIPAUD J.C. & LILLEY I. (Eds) 1999. – *Le Pacifique de 5000 à 2000 avant le présent : suppléments à l'histoire d'une colonisation / The Pacific from 5000 to 2000 BP colonisation and transformations. [Actes de la troisième conférence Lapita, Vanuatu, 31 juillet-6 août 1996].* IRD, Paris, 619 p.

GALIPAUD J.C. & DI PIAZZA A. 2006. – Taro pondfields and demography: the example of the Hokua water gardens in Santo, Vanuatu. *Asia-Pacific Forum*, Taiwan 31: 114-129.

GALIPAUD J.C. & SWETE KELLY M.C. 2007. – Makué (Aore island, Santo, Vanuatu): A new Lapita site in the ambit o.f New Britain obsidian distribution. *Terra Australis*, Canberra 18:1-12. .

GALIPAUD J.C. & SWETE KELLY M.C. 2007. – New evidence relating to the transport of obsidian from New Britain to Vanuatu. *Antiquity* 81(312) http://www.antiquity.ac.uk/ProjGall/swete/index.html

GUIART J. 1958. – *Espiritu Santo (Nouvelles Hébrides)*. Plon, Paris, 236 p. (L'Homme; 2).

HARRISSON T. 1936. – Living in Espiritu Santo. *Journal of the Royal Geographical Society* 88: 243-61.

KIRCH P.V. 1997. – *The Lapita Peoples. Ancestors of the Oceanic World (The Peoples of South-East Asia and the Pacific).* Blackwell, Malden, 353 p.

KIRCH P.V. & HUNT T. (Eds) 1997. – *Historical ecology in the Pacific Islands. Prehistoric environmental and landscape change.* Yale University Press, New Haven, 331 p.

MEAD J.I., STEADMAN D.W., BEDFORD S.H., BELL C.J. & SPRIGGS M. 2002. – New extinct mekosuchine crocodile from Vanuatu, South Pacific. *Copeia* 3: 632–641.

NOURY A. 2005. – *Le reflet de l'âme Lapita*. Noury éditions, Versailles, 120 p.

PINEDA R. & GALIPAUD J.C. 1998. – Évidences archéologiques d'une surrection différentielle de Île de Malo (archipel du Vanuatu) autour de l'Holocène récent. *Comptes Rendues de l'Académie des Sciences, Sciences de la Terre et des Planètes* 327: 777-779.

QUIROS P.F. de 2001. – *Histoire de la découverte des régions australes (Îles Salomon, Marquises, Santa Cruz, Tuamotu, Cook du nord, Vanuatu).* L'Harmattan, Paris, 345 p.

SPEISER F. 1990. – *Ethnology of Vanuatu, an early twentieth century study.* transl. by STEPHENSON D.Q. Crawford House publ., Bathurst 1996, 643 p.

STRECKER M.R., BLOOM A.L. & LECOLLE J. 1987. – Time span for karst development on Quaternary coral limestones: Santo Island, Vanuatu, *in* GODARD A. & RAPP A. (Eds), *Processus et Mesure de l'érosion/ Processes and Measurement of Erosion. 25ème Congrès International de Géographie (UGI).* CNRS, Paris, 1984: 369-386.

MAN AND NATURE

ANONYMOUS 1999. – *Plan d'action et stratégie pour la biodiversité nationale. Stratégie pour la conservation de la Diversité Biologique du Vanuatu.* Service de l'Environnement, Port Vila, 87 p. http://www.cbd.int/doc/world/vu/vu-nbsap-01-fr.pdf / *Vanuatu National Biodiversity Strategy and Action Plan Project. National Biodiversity Conservation Strategy.* Environment Unit, Port Vila, 84 p. http://www.cbd.int/doc/world/vu/vu-nbsap-01-en.pdf

ANONYMOUS 2006. – *Third national report to the Conference of Parties of the Convention on Biodiversity.* Environment Unit, Port Vila, 207 p. http://www.cbd.int/doc/world/vu/vu-nr-03-en.pdf

ATKINSON A.E. & ATKINSON T.J. 2000. – Land vertebrates as invasive species on islands served by the South Pacific Regional Environment Programme, *in* SHERLEY G. (Ed.), *Invasive species in the Pacific: a technical review and draft regional strategy.* South Pacific Regional Environment Programme, Apia, Samoa: 19-84.

BAKER J.R. 1929. – *Man and animals in the New Hebrides.* Georges Routledge & sons Ldt, London, 200 p.

BARRAU J. 1956. – L'agriculture indigène vivrière aux Nouvelles Hébrides. *Journal de la Société des Océanistes* 12(12): 185-215.

BOURDY G., CABALION P., WALTER A. & DJIAN-CAPORALINO C. 1995. – Plantes magiques, plantes protectrices : quelques techniques d'horticulture traditionnelle à Vanuatu. *JATBA* 37(2): 51-78.

CARVAJAL A. & ADLER G.H. 2005. – Biogeography of mammals on tropical Pacific Islands. *Journal of Biogeography* 32: 1561-1569.

ELLISON C.A. & MURPHY S.T. 2000. – *Dossier on Puccinia spegazzini de Toni (Basidiomycetes: Uredinales): a potential biological control agent for Mikania micrantha H.B.K. (Asteraceae) in India.* CABI Bioscience UK Centre, Ascot, UK, 49 p.

GALIPAUD J.C. 2002. – Under the volcano: Ni-Vanuatu and their environment, *in* TORRENCE R. & GRATTAN J. (Eds), *Natural disasters and cultural changes.* Routledge, Londres: 162-172.

LEBOT V. 2002. – La domestication des plantes en Océanie et les contraintes de la voie asexuée. *Journal de la Société des Océanistes* 114-115: 45-61.

LEBOT V., IVANCIC A. & ABRAHAM K. 2005. – The geographical distribution of allelic diversity, a practical means of preserving and using minor root crop genetic resources. *Experimental Agriculture* 41: 475-489.

LEBOT V., WALTER A. & SAM C. 2007. – The Domestication of Fruit and Nut Tree Species in Vanuatu, Oceania, *in* AKINNIFESI F.K. *et al.* (Eds), *Indigenous Fruit Trees in the Tropics: Domestication, Utilization and Commercialization.* CAB International: 120-136.

LOWE S., BROWNE M. & BOUDJELAS S. 2000. – *100 of the world's worst invasive alien species. A selection from the global invasive species database.* Invasive Species Specialist Group (ISSG), Species Survival Commission (SSC),The World Conservation Union (IUCN), ISSG, Auckland, New Zealand.

LUM J.K., MCINTYRE J.K., GREGER D.L., HUFFMAN K.W. & VILAR M.G. 2006. – Recent Southeast Asian domestication and Lapita dispersal of sacred male pseudohermaphroditic "tuskers" and hairless pigs of Vanuatu. *Proceedings of National Academy of Sciences USA* 103: 17190-17195.

MATISOO-SMITH E. & ALLEN J.S. 2001. – Name that rat: molecular and morphological identification of Pacific rodent remains. *International Journal of Osteoarchaeology* 11: 34-42.

MATISOO-SMITH E. & ROBINS J.H. 2004. – Origins and dispersals of Pacific peoples: evidence from mtDNA phylogenies of the Pacific rat. *Proceedings of the National Academy of Sciences* 101: 9167-9172.

MATISOO-SMITH E., ALLEN J.S., ROBERTS R.M., IRWIN G.J. & LAMBERT D.M. 1999. – Rodents of the sunrise: mitochondrial DNA phylogenies of Polynesian Rattus exulans and the settlement of Polynesia, *in* GALIPAUD J.-C. & LILLEY I. (Eds), *Le Pacifique de 5000 à 2000 avant le présent. Suppléments à l'histoire d'une colonisation.* Institut de Recherche pour le Développement (IRD), Paris: 259-276.

MATISOO-SMITH E., ROBERTS R.M., IRWIN G.J., ALLEN J.S., PENNY D. & LAMBERT D.M. 1998. – Patterns of prehistoric human mobility in Polynesia indicated by mtDNA from the Pacific rat. *Proceedings of the National Academy of Sciences* 95: 15145-15150.

MORELLI C. 2003. – *Évaluation des performances agronomiques des jardins au Vanuatu. Estimation de leur durabilité agroécologique et proposition d'intensification par association aux cocoteraies.* Mémoire DAA, ENSAR, Rennes, 99 p.

MORITZ C., CASE T.J., BOLGER D.T. & DONNELLAN S. 1993. – Genetic diversity and the history of Pacific island house geckos (*Hemidactylus* and *Lepidodactylus*). *Biological Journal of the Linnean Society* 48: 113-133.

MULLER S. 2010. – *Le jardin mélanésien et la mondialisation : quand l'agrobiodiversité révèle la dynamique des espaces au Vanuatu.* Thèse Université de Strasbourg, Strasbourg, France, 426 p.

PANOFF F. 1972. – Maenge Taro and Cordyline: Elements of a Melanesian Key. *The Journal of the Polynesian Society* 81: 375-390.

PAYNTER Q., HARMAN H. & WAIPARA N. 2006. – *Prospects for biological control of Merremia peltata.* Landcare Research Contract Report: LC0506/177. Conservation International & The Pacific Invasives Initiative (PII), 34 p. (http://www.cepf.net/ImageCache/cepf/content/pdfs/landcare_2eresearch_2ereport_2epdf/v1/landcare.research.report.pdf).

*ROBILLARD M. 2008. – Perception plurielle de la biodiversité de Santo: scientifiques et Ni-van, un double regard. *Journal de la Société des Océanistes* 126-127: 221-229.

SIMBERLOFF D. 1995. – Why do introduced species appear to devastate islands more than mainland areas? *Pacific Science* 49: 87-97.

SPRIGGS M. 1981. – *Vegetable kingdoms. Taro irrigation and Pacific history.* Doctoral thesis in archaeology, The Australian National University, Canberra, Australia, 203 p.

SWARBRICK J.T. 1997. – *Weeds of the Pacific Islands.* South Pacific Commission, Noumea, New Caledonia, 124 p. (Technical paper; 209).

TEOH C.H., CHUNG G.F., LIAU S.S., GHANI I., TAN A.M., LEE S.A. & MOHAMMED M. 1985. – Prospects for biological control of *Mikania micrantha* HBK in Malaysia. *Planter* 61: 515-530

TYLER M.J. 1979. – The introduction and current distribution in the New Hebrides of the Australian hylid frog Litoria aurea. *Copeia* 1979(2): 355-356.

VEITCH C.R. & CLOUT M.N. (Eds) 2002. – *Turning the tide: the eradication of invasive species.* IUCN, Gland, Switzerland, 414 p.

WALTER A. & LEBOT V. 2003. – *Jardins d'Océanie, les Plantes Alimentaires du Vanuatu.* Presses de l'IRD, Montpellier, 320 p. (Collection Didactiques).

WALTER A. & SAM C. 1999. – *Fruits d'Océanie.* Presses de l'IRD, Montpellier, 310 p. (Collection Didactiques).

WARDS G. 1980. – Rural-urban connexions: the missing link in Melanesia. *The Malaysian Journal of Tropical Geography* 1: 57-63.

WEIGHTMAN B. 1989. – *Agriculture in Vanuatu. A historical review.* British Friends of Vanuatu, Surrey, UK, 320 p.

WHITE J.P., CLARK G. & BEDFORD S. 2000. – Distribution, present and past, of *Rattus praetor* in the Pacific and its implications. *Pacific Science* 54: 105-117.

YOSIDA T.H., UDAGAWA T., ISHIBASHI M., MORIWAKI K., YABE T. & HAMADA T. 1985. – Studies on the karyotypes of the black rats distributed in the Pacific and South Pacific islands, with special regard to the border line of the Asian and Oceanian type black rats on the Pacific Ocean. *Proceedings of the Japan Academy* 61: 71-74.

QUATERNARY GEOLOGY AND PALEONTOLOGY

ANONYMOUS (New Hebrides Geological Survery) 1977. – *South Santo*. 1:100 000 New Hebrides Geological Survery Sheet 4. British Ministry of Overseas Development (Directorate of Overseas Surveys).

BLOOM A.L. & YONEKURA N. 1985. – Coastal terraces generated by sea-level change and tectonic uplift, *in* WOLDENBERG M.J. (Ed.), *Models in Geomorphology*. Allen & Unwin, Winchester, Mass.: 139-154.

BURR G.S., BECK W., TAYLOR F.W., RECY J., LAWRENCE EDWARDS R., CABIOCH G., CORRÈGE T., DONAHUE D.J. & O'MALLEY J.M. 1998. – A High-Resolution Radiocarbon Calibration between 11.7 and 12.4 kyr BP Derived from ^{230}Th ages of Corals from Espiritu Santo Island, Vanuatu. *Radiocarbon (INTCAL98)* 40(3): 1093-1106.

CABIOCH G. 2003. – Postglacial reef development in the South-West Pacific: case studies from New Caledonia and Vanuatu. *Sedimentary Geology* 159(1-2): 43-59.

CABIOCH G., BANKS-CUTLER K., BECK W.J., BURR G.S., CORRÈGE T., EDWARDS R.L. & TAYLOR F.W. 2003. – Continuous reef growth during the last 23 ka in a tectonically active zone (Vanuatu, SouthWest Pacific). *Quaternary Science Reviews* 22: 1771–1786.

CABIOCH G., TAYLOR F.W., CORRÈGE T., RÉCY J., EDWARDS L.R., BURR G.S., LE CORNEC F. & BANKS K.A. 1999. – Occurrence and significance of microbialites in the uplifted Tasmaloum reef (SW Espiritu Santo, SW Pacific). *Sedimentary Geology* 126: 305-316.

CABIOCH G., TAYLOR F.W., RÉCY J., EDWARDS R.L., GRAY S.C., FAURE G., BURR G. & CORRÈGE T. 1998. – Environmental and tectonic influences on growth and internal structure of a fringing reef at Tasmaloum (SW Espiritu Santo, New Hebrides Island Arc, SW Pacific), *in* CAMOIN G. & DAVIES P.J. (Eds), Reefs and carbonate platforms in the Pacific and Indian Oceans. *IAS special publication* 25: 261-277.

COLLOT J.Y., DANIEL J. & BURNE R.V. 1985. – Recent tectonics associated with the subduction collision of the D'Entrecasteaux Zone in the Central New-Hebrides. *Tectonophysics* 112: 325-356.

CUTLER K.B., GRAY S.C., BURR G.S., EDWARDS R.L., TAYLOR F.W., CABIOCH G., BECK J.W., CHENG H. & MOORE J. 2004. – Radiocarbon calibration and comparison to 50 kyr BP with paired 14C and 230Th Dating of corals from Vanuatu and Papua New Guinea. *Radiocarbon* 46(3): 1127-1160.

PINEDA R. & GALIPAUD J-C. 1998. – Evidences archéologiques d'une surrection différentielle de l'ile de Malo (Archipel du Vanuatu) au cours de l'holocène récent. *Comptes-Rendus de l'Académie des Sciences* 327: 777-79.

GILPIN L. 1982. – *Tectonic geomorphology of Santo Island, Vanuatu*. M.S. Thesis, Cornell Univ., Ithaca, N.Y.

GREENE H.G. & WONG F.L. (Eds) 1988. – *Geology and Offshore Resources of Pacific Islands Arcs-Vanuatu Region*. Circum-Pacific Council for Energy and Mineral Resources, 442 p. (Earth Science Series; 8).

JOUANNIC C., TAYLOR F.W. & BLOOM A.L. 1982. – Sur la surrection et la déformation d'un arc jeune : l'arc des Nouvelles-Hébrides. *Travaux et Documents ORSTOM* 147: 223-246.

JOUANNIC C., TAYLOR F.W., BLOOM A.L. & BERNAT M. 1980. – Late Quaternary uplift history from emerged reef terraces on Santo and Malekula, Central New Hebrides island arc. Symposium on Petroleum potential in island arcs, small basins, submerged margins and related areas, Suva, Fidji, 1979, UN/ESCAP, CCOP/SOPAC, tech. Bull. 3: 91-108.

LADD H.S. 1982. – Cenozoic fossil mollusks from western Pacific Islands; Gastropods (Eulimidae and Volutidae through Terebridae). *United States Geological Survey Professional Paper* 1171: 1-100, 41 pls.

LADD H.S. 1977. – Cenozoic fossil mollusks from western Pacific Islands; Gastropods (Eratoidae through Harpidae). *United States Geological Survey Professional Paper* 533: 1-84, 23 pls.

LADD H.S. 1976. – New Pleistocene Neogastropoda from the New Hebrides. *The Nautilus* 90(4): 127-138.

LADD H.S. 1975. – Two Pleistocene Volutes from the New Hebrides (Mollusca: Gastropoda). *The Veliger* 18(2): 134-138.

MALLICK D.I.J. 1975. – Development of the New Hebrides Archipelago. *Philosophical Transactions of the Royal Society of London* B 272: 277-285.

MALLICK D.I.J. & GREENBAUM D. 1975. – The Navaka fossiliferous sands and the Kere Shell Bed. *Annual Report of the Geological Survey of New Hebrides for 1973*: 8-12.

MALLICK D.I.J. & GREENBAUM D. 1977. – *Geology of Southern Santo*. Regional Report, New Hebrides Condominium Geological Survey, 84 p.

MAWSON D. 1905. – The geology of the New Hebrides. *Proceedings of the Linnean Society of New South Wales* 30: 400-485.

MEFFRE S. & CRAWFORD A.J. 2001. – Collision tectonics in the New Hebrides arc (Vanuatu). *The Island Arc* 10: 33-50.

NEEF G. & VEEH H.H. 1977. – Uranium series ages and late Quaternary uplift in the New Hebrides. *Nature* 269: 682-683.

OBELLIANNE J.M. 1958. – *Contribution à la connaissance géologique de l'Archipel des Nouvelles-Hébrides : Iles Vaté, Pentecôte, Maewo, Santo*. Thèse de Doctorat, Université de Nancy, Nancy, France, 234 p.

QUINN T.M., CROWLEY T.J. & TAYLOR F.W. 1996. – New stable isotope results from a 173-year coral from Espiritu Santo, Vanuatu. *Geophysical Research Letters* 23 : 3413-3416.

STRECKER M.R, BLOOM A.L., GILPIN L.M. & TAYLOR F.W. 1986. – Morphology of uplifted Quaternary coral limestone terraces - Santo-island, Vanuatu. *Zeitschrift fur Geomorphologie* 30(4): 387-405.

TAYLOR F.W. 1992. – Quaternary vertical movements of the central New Hebrides island arc, *in* COLLOT J.-Y., GREENE H.G. & STOKKING L.B. (Eds), Proceed. O.D.P. Init. Rep. 134: 33-42.

TAYLOR F.W., FROHLICH C., LECOLLE J. & STRECKER M. 1987. – Analysis of partially emerged corals and reef terraces in the central Vanuatu arc: comparison of contemporary coseismic and nonseismic with Quaternary vertical movements. *Journal of Geophysical Research* 92: 4905-4933.

TAYLOR F.W., ISACKS B.L., JOUANNIC C., BLOOM A.L. & DUBOIS J. 1980. – Coseismic and Quaternary vertical tectonic movements, Santo and Malekula islands, New Hebrides island arc. *Journal of Geophysical Research* 85: 5367-5381.

TAYLOR F.W., JOUANNIC C. & BLOOM A.L. 1985. – Quaternary uplift of the Torres islands, northern New Hebrides frontal arc: comparison with Santo and Malekula islands, central New Hebrides frontal arc. *Journal of Geology* 93: 419-438.

TAYLOR F.W., MANN P., BEVIS M.G., EDWARDS R.L., CHENG, H., CUTLER K.B., GRAY S.C., BURR G.S., BECK J.W., PHILLIPS D.A., CABIOCH G. & RECY J. 2005. – Rapid forearc uplift and subsidence caused by impinging bathymetric features: Examples from the New Hebrides and Solomon arcs. *Tectonics* 24(6), TC6005 (15 November 2005).

URMOS J.P. 1985. – *Oxygen isotopes, sea levels, and uplift of reef terraces, Araki Island, Vanuatu*. M.S. Thesis, Cornell Univ., Ithaca, NY, 122 p.

HISTORY OF BIODIVERSITY EXPLORATION

BAERT A. 1999. – *Le paradis terrestre, un mythe espagnol en Océanie. Les voyages de Mendana et de Quiros 1567-1606*. L'Harmattan, Paris, 352 p.

BAKER J.R. & GOODENOUGH W. 1935. – Espiritu Santo, New Hebrides: Discussion. *The Geographical Journal* 85: 230-233.

BAKER J.R. 1929. – The Northern New Hebrides. *The Geographical Journal* 73: 305-325.

BAKER J.R. 1949. – Surveying in the New Hebrides. *The Geographical Journal* 145: 114-115.

BAKER J.R., BIRD T.F., HARRISSON T.H., BAKER S.J. & CAMPBELL SMITH W. 1935. – Espiritu Santo, New Hebrides. *The Geographical Journal* 85: 209-229.

BOUCHET P., LE GUYADER H. & PASCAL O. 2008. – Des voyages de Cook à l'expédition Santo 2006 : un renouveau des explorations naturalistes des iles du Pacifique. *Journal de la Société des Océanistes* 126-127: 167-185.

BOUCHET P., LE GUYADER H. & PASCAL O. 2009. – The SANTO 2006 Global Biodiversity Survey: an attempt to reconcile the pace of taxonomy and conservation. *Zoosystema* 31(3): 401-406.

CHAPMAN F.M. 1935. – The Whitney South Sea Expedition. *Science* 81: 95-97.

CORNER E.J.H. & LEE K.E. (Eds) 1975. – A discussion on the results of the 1971 Royal Society/Percy Sladen Expedition to the New Hebrides. *Philosophical Transactions of the Royal Society of London*, Series B 272: 267-486.

ESTENSEN M. 2006. – *Terra Australis Incognita: The Spanish quest for the mysterious Great South Land*. Allen & Unwin, Crows Nest, 274 pp.

*FAUGERE E. 2008. – L'exploration contemporaine de la biodiversité. Approche anthropologique de l'expédition Santo 2006. *Journal de la Société des Océanistes* 126-127: 195-206.

IWASHINA T. 1998. – Botanical expedition to Vanuatu and adjacent countries in 1996 and 1997 (Contributions to the Flora of Vanuatu). *Annals of the Tsukuba Botanical Garden* 17: iii-vii.

IWASHINA T. 2002. – Botanical expedition to Vanuatu in 2000 and 2001 (Contributions to the Flora of Vanuatu, Vol. 2). *Annals of the Tsukuba Botanical Garden* 21: iii-vii.

LEE K.E. 1974. – *Royal Society and Percy Sladen Expedition to the New Hebrides, 1971*. The Royal Society, 77 p. (Collection data).

MURPHY R.C. 1922. – The Whitney South Sea Expedition of the American Museum of Natural History. *Science* 56: 701-704.

*PASCAL M., LORVELEC O. & SOLDATI L. 2008. – "Friches et Aliens" : un module de l'expédition Santo 2006. *Le Courrier de l'Environnement de l'INRA* 55: 27-36.

RICHER de FORGES B., FALIEX E. & MENOU J.L. 1996. – La campagne MUSORSTOM 8 dans l'archipel de Vanuatu. Compte rendu et liste des stations, *in* CROSNIER A. (Ed.), Résultats des Campagnes MUSORSTOM, Volume 15. *Mémoires du Muséum national d'Histoire naturelle* 168: 9-32.

TARDIEU V. & BARNEOUD L. 2007. – *Santo. Les explorateurs de l'île planète*. Belin, Paris, 288 p.

VEGETATION AND FLORA

BEVERIDGE A.E. 1975. – Kauri forests in the New Hebrides. *Philosophical Transactions of the Royal Society of London* B 272: 369-383.

BRAITHWAITE A.F. 1975. – The phytogeographical relationships and origin of the New Hebrides fern flora. *Philosophical Transactions of the Royal Society of London* B 272: 293-313.

CHAMBERS T.C., JERMY A.C. & CRABBE J.A. 1971. – A collection of ferns from Espiritu Santo, New Hebrides. *British Fern Gazette* 10(4): 175-182.

CHEW W.L. 1975. – The phanerogamic flora of the New Hebrides and its relationships. *Philosophical Transactions of the Royal Society of London* B 272: 315-328.

DOWE J.L. & CABALION P. 1996. – A taxonomic account of Arecaceae in Vanuatu, with descriptions of three new species. *Australian Systematic Botany* 9(1):1-60.

EBIHARA A., MATSUMOTO S., IWASHINA T., SUGIMURA K. & IWATSUKI K. 2002. – Hymenophyllaceae (Pteridophytes) of Vanuatu. *Annals of the Tsukuba Botanical Garden* 21: 61-72.

ELLIS L. 2002. – Studies on the bryophyte flora of Vanuatu. 2. Calymperaceae (Musci). *Annals of the Tsukuba Botanical Garden* 21: 79-77.

FURUKI T. 2002. – Studies on the bryophyte flora of Vanuatu. 5. Metzgeriales and Marchantiales. *Annals of the Tsukuba Botanical Garden* 21: 95-101.

GILLISON A.N. 1975. – Phytogeographical relatioship of the northern islands of the New Hebrides. *Philosophical Transactions of the Royal Society of London* B 272: 385-390.

GUILLAUMIN A. 1931. – Contribution to the flora of the New Hybrides, plants collected by S. F. Kajewski in 1928 and 1929. *Journal of the Arnold Arboretum* 12: 221-317.

GUILLAUMIN A. 1937. – Contribution à la flore des Nouvelles-Hébrides. Plantes recueillies par M. et Mme Aubert de la Rüe dans leur deuxième voyage (1935-1936). *Bulletin du Muséum national d'Histoire naturelle*, Séries 2, Paris 9: 283-306.

GUILLAUMIN A. 1938. – A florula of the island of Espiritu Santo, one of the New Hebrides. *The Linnean Society's Journal - Botany* 51(340): 547-566.

GUILLAUMIN A. 1948. – Compendium de la flore phanérogamique des Nouvelles-Hébrides. *Annales du Musée colonial de Marseille*: 1-56.

HASEGAWA 2002. – Studies on the bryophyte flora of Vanuatu. 6. Anthocerotae. *Annals of the Tsukuba Botanical Garden* 21: 103-107.

HIGUCHI M. 2005. – Studies on the bryophyte flora of Vanuatu. 8. Field studies in 2000 and 2001 and Haplomitriaceae and Treubiaceae (Hepaticae). *Bulletin of the National Science Museum, Series B, Botany* 31: 11-17.

HIGUCHI M. 2003. – Studies on the bryophyte flora of Vanuatu. 7. Stereophyllaceae (Musci). *Bulletin of the National Science Museum, Series B, Botany* 29: 123-125.

HIGUCHI M. 2002. – Studies on the bryophyte flora of Vanuatu. 1. Introduction and Mniaceae. *Annals of the Tsukuba Botanical Garden* 21: 73-77.

HIGUCHI M. & NISHIMURA N. 2006. – Studies on the bryophyte flora of Vanuatu. 10. Additions to the Hypnaceae (Musci). *Bulletin of the National Science Museum, Series B, Botany* 32: 175-179.

HIGUCHI M. & NISHIMURA N. 2002. – Studies on the bryophyte flora of Vanuatu. 4. Hypnaceae (Musci). *Annals of the Tsukuba Botanical Garden* 21: 91-94.

IWASHINA T., HASHIMOTO T. & BANI E. (Eds) 2002. – *Contributions to the Flora of Vanuatu*. Vol. 2. Tsukuba Botanical Garden, National Science Museum, Japan, 133 p.

IWATSUKI Z. & SUZUKI T. 1995. – Fissidens (Musci, Fissidentaceae) in Vanuatu (New Hebrides) collected by Dr. M.Higuchi. *Fragmenta Floristica et Geobotanica* 40: 153-158.

JOVET-AST S. 1951. – Hépatiques des Nouvelles-Hébrides. Récoltes de E. Aubert de la Rüe 1934. *Revue bryologique et lichenologique* 20: 96-98.

KONISHI T., HASHIMOTO T., YUKAWA T., MATSUMOTO S., HIRAYAMA R., CHANEL S. & IWASHINA T. 1998. – A list of live specimens from Vanuatu, collected in 1996 and 1997. *Annals of the Tsukuba Botanical Garden* 17: 23-50.

LEWIS B. & CRIBB P. 1989. – *Orchids of Vanuatu*. Royal Botanic Gardens, Kew, London, 171 p.

LOWRY P.P.II 1989. – A revision of Araliaceae from Vanuatu. *Bulletin du Muséum national d'Histoire naturelle*, Série 4, B, Adansonia, Paris 11: 117-155.

MATSUMOTO S., IWASHINA T., SUGIMURA K. & TANAKA N. 2002. – A list of Pteridophytes herbarium specimens (without Hymenophyllaceae) from Vanuatu, collected in 2000 and 2001. *Annals of the Tsukuba Botanical Garden* 21: 37-60.

MATSUMOTO S., IWASHINA T., SUGIMURA K., HASHIMOTO T. & NAKAMURA T. 1998. – A list of Pteridophytes herbarium specimens from Vanuatu, collected in 1996 and 1997. *Annals of the Tsukuba Botanical Garden* 17: 75-100.

MILLER H.A., WHITTIER H.O. & WHITTIER B.A. 1978. – *Prodromus florae muscorum Polynesiae*. J Cramer, Vaduz.

MILLER H.A., WHITTIER H.O. & WHITTIER B.A. 1983. – *Prodromus florae hepaticarum Polynesiae*. J Cramer, Vaduz.

MUELLER-DOMBOIS D. & FOSBERG R. 1998. – *Vegetation of the tropical Pacific Islands*. Springer, New York, 733 pp.

RAMSAY H.P., SEUR J., WILSON P.G. & GOODWIN T. 1990. – Register of type specimens of mosses in Australian herbaria. General introduction and part I. Special collections at NSW: Lord Howe Island, Vanuatu (New Hebrides). *Telopea* 3(4): 571-592

SCHMID M. 1987. – Conditions d'évolution et caractéristiques du peuplement végétal insulaire en Mélanésie occidentale : Nouvelle-Calédonie, Vanuatu. *Bulletin de la Société Zoologique de France* 112(1-2):233-254.

SCHMID M. 1973. – *Flore des Nouvelles Hébrides, Pteridophytes*. Orstom.

St. JOHN H. 1989. – *Revision of the genus* Pandanus *Stickman, Part 60. Pandanus of the New Hebrides*. Published privately, Honolulu.

STREIMANN H. & REESE W.D. 2001. – Vanuatu moss records. *Journal of the Hattori Botanical Laboratory* 91: 295-300.

SUZUKI & IWATSUKI 2002. – Studies on the bryophyte flora of Vanuatu. 3. Fissidentaceae. *Annals of the Tsukuba Botanical Garden* 21: 87-90.

THÉRIOT I. 1938. – Sur une collection de mousses des Nouvelles-Hébrides. (Mission E. Aubert de la Rüe, 1934). *Revue bryologique et lichenologique* 10: 128-135.

TIXIER P. 1975. – Exotic bryophytes of the New Hebrides South Pacific expedition of the Royal Society. *Bulletin du Muséum National d'Histoire Naturelle* Botanique 16: 33-46.

TIXIER P. 1973. – Exotic bryophytes. *Bulletin du Muséum National d'Histoire Naturelle* Botanique 10: 73-86.

TIXIER P. 1972. – Exotic bryophytes. *Bulletin du Muséum National d'Histoire Naturelle* Botanique 4: 89-98.

WHEATLEY J.I. 1992. – *A guide to the common trees of Vanuatu with lists of their traditional uses & ni-Vanuatu names*. Department of Forestry, Port Vila, 308 p.

YAMADA K. & HAYASHI M. 2003. – Studies on the bryophyte flora of Vanuatu. 8. Heteroscyphus (Geocalycaceae, Hepaticae). *Bulletin of the National Science Museum*, Series B, Botany 29: 149-152.

TERRESTRIAL FAUNA

ANGEL F. 1935. – Liste des reptiles récoltés par la Mission Aubert de la Rüe aux Nouvelles Hébrides ou dans les îles voisines. *Bulletin du Muséum national d'Histoire naturelle* Série 2, 7(1): 54-56.

ANONYMOUS 1991. – An unusual sea snake. *Naika* 35: 3-4.

ANONYMOUS 1985a. – Leatherback turtle at South West Bay, Malakula. *Naika* 17: 11.

ANONYMOUS 1985b. – Does Vanuatu have native frogs? *Naika* 18: 11.

ANONYMOUS 1981. – A new snake from Efate. *Naika* 2: 13.

BAKER J.R. 1928. – The non-marine vertebrate fauna of the New Hebrides. *The Annals and Magazine of Natural History*, serie 10, 2: 294-302.

BALFOUR-BROWNE J. 1939. – On the aquatic Coleoptera of the New Hebrides and Banks Islands. Dytiscidae, Gyrinidae and Palpicornia. *Annals and Magazine of Natural History Society* 11(3): 459-479.

BAUER A.M. 1988. – Hypothesis: A geological basis for some herpetological disjunctions in the southwest Pacific, with special reference to Vanuatu. *Herpetological Journal* 1: 259-263.

BERLAND 1938. – Araignées des Nouvelles-Hébrides. *Annales de la Société Entomologique de France* CVII: 121-190.

BICKEL D.J. 2005. – The *Plagiozopelma flavipodex* species group (Diptera: Dolichopodidae:Sciapodinae) from Fiji, Vanuatu, and the Solomon Islands. *Bishop Museum Occasional Papers* 82: 47–61.

BIRCHENOUGH A.C., DOUGLAS G.W. & EVANS S.M. 2003. – Assessing the distribution of estrilid finches on Vanuatu using local knowledge. Bird Conservation International, 13: 29-44.

BOWEN J. 1997. – The status of the avifauna of Loru Protected Area, Santo, Vanuatu. *Bird Conservation International* 7: 331-344.

BREGULLA H.L. 1992. – *Birds of Vanuatu*. A. Nelson, Oswestry: 1-294.

BURT C.E. & Burt M.D. 1932. – Herpetological results of The Whitney South Sea Expedition. VI. *Bulletin of the American Museum of Natural History* 63, article v: 461-597.

BUTLER A.G. 1875. – On a collection of butterflies from the New Hebrides and Loyalty Islands with descriptions of new species. *Proceedings of the Zoological Society of London* 1875: 610-619.

CHALLACOMBE J. 1986. – Frogs of Vanuatu. *Naika* 23: 13-15.

CHEESMAN L.E. 1948. – XXVI. Bees of New Guinea and the New Hebrides. *The Annals and Magazine of Natural History* 12: 318-335.

CHEESMAN L.E. 1937. – Sphecoidea of the New Hebrides, Banks Islands, and New Caledonia. *The Annals and Magazine of Natural History* 10(20): 203-208.

CHEESMAN L.E. 1936. – Hymenoptera of the New Hebrides and Banks Islands. *The Transactions of the Royal Entomological Society of London* 85(7): 169-191.

CHEESMAN L.E. & PERKINS M.A. 1939. – Halictine bees from the New Hebrides and Banks Islands (Hymen.). *The Transactions of the Royal Entomological Society of London* 88(6): 161-172.

COCKERELL T.D.A. 1916. – Some bees from Australia, Tasmania and the New Hebrides. *Proceedings of the Academy of natural Sciences of Philadelphia* 68: 360-375.

COGGER H. & HEATWOLE H. 2006. – *Laticauda frontalis* (de Vis, 1905) and *Laticauda saintgironsi* n.sp. from Vanuatu and New Caledonia (Serpentes: Elapidae: Laticaudinae) – a new lineage of sea kraits? *Records of the Australian Museum* 58: 245-256.

CRANBROOK, THE EARL OF [FORMERLY: MEDWAY, LORD] & PICKERING R. 1981. – A checklist and preliminary key to the lizards of Vanuatu. *Naika* 4: 3-6.

CRANBROOK, THE EARL OF [FORMERLY: MEDWAY, LORD] 1985a. – The herpetofauna of Vanuata [sic] (Pacific Ocean). *British Herpetological Society Bulletin* 11: 13.

CRANBROOK, THE EARL OF [FORMERLY: MEDWAY, LORD] 1985b. – The lizards of Vanuatu. *Association for the Study of Reptilia and Amphibia Journal* 2(4): 47-70.

DE POUS P.E. 2008. – Geographic distribution – *Hemidactylus frenatus* (Common House Gecko). Vanuatu: Malekula: Nabelchel Bungalows, Norsup. *Herpetological Review* 39(4): 482-483.

*DESUTTER-GRANDCOLAS L. 2009. – New and little known crickets from Espiritu Santo Island, Vanuatu (Insecta, Orthoptera, Grylloidea, *Pseudotrigonidium* Chopard, 1915, Phaloriinae and Nemobiinae p.p.). *Zoosystema* 31(3): 619-659.

DIAMOND J.M. & MARSHALL A.G., 1977. – Distributional ecology of New Hebridean birds: a species kaleidoscope. *Journal of Animal Ecology* 46: 703-727.

DICKENSON D. 1981. – Marine crocodiles (*Crocodylus porosus*) in Vanuatu. *Naika* 3: 5-6.

*EITSCHBERGER U. & SCHMIDL J. 2007. – *P. vanuatui*, eine neue *Psilogramma* Rothschild & Jordan, 1903 -Art von Vanuatu (Lepidoptera, Sphingidae). *Neue Entomologische Nachrichten* 60: 171-174.

FISCHTHAL J.H. & KUNTZ R.E. 1967. – Digenetic trematodes of amphibians and reptiles from Fiji, New Hebrides and British Solomon Islands. *Proceedings of the Heminthological Society Washington* 32: 244-251.

*GATTOLLIAT J.L. & STANICZEK A.H. 2011. – New larvae of Baetidae (Insecta: Ephemeroptera) from Espiritu Santo, Vanuatu. *Stuttgarter Beiträge zur Naturkunde* A, Neue Serie, 4: 75-82.

*GERSTMEIER, R. & J. SCHMIDL, 2007. – *Omadius santo* sp. nov. from Espiritu Santo, Vanuatu (Coleoptera, Cleridae, Clerinae). *Entomologische Zeitschrift*, Stuttgart 117(2): 85-87.

GIORDANI SOIKA A. 1981 (1980). – Notulae vespidologicae XLIII. Nuovi Polistes delle Nuove Ebridi. *Bollettino del Museo Civico di Storia Naturale di Venezia* 31: 117-119.

*GOLOVATCH S., GEOFFROY J.J., MAURIES J.P. & VANDENSPIEGEL D. 2008. – The first, new species of the millipede family Pyrgodesmidae to be recorded in Vanuatu, Melanesia, southwestern Pacific (Diplopoda: Polydesmida). *Arthropoda Selecta* 17(3-4): 145-151.

*GOMY Y., LIPS J. & SOLDATI L. 2009. – Contribution à la connaissance des Histeridae de l'archipel du Vanuatu. *Bulletin Mensuel de la Société Linnéenne de Lyon* 78(9-10): 217-228.

GROSS G.F. 1975. – The land invertebrates of the New Hebrides and their relationships. *Philosophical Transactions of the Royal Society (B)*, London 272: 391-421.

*HAASE M., FONTAINE B. & GARGOMINY O. 2010. – Rissooidean freshwater gastropods from the Vanuatu archipelago. *Hydrobiologia* 637(1): 53-71.

HAMILTON A.M., HARTMAN J.H. & AUSTIN C.C. 2009. – Island area and species diversity in the southwest Pacific Ocean: is the lizard fauna of Vanuatu depauperate? *Ecography* 32(2): 247-258.

HAMILTON A.M., KLEIN E.R. & AUSTIN C.C. 2010. – Biogeographic breaks in Vanuatu, a nascent oceanic archipelago. *Pacific Science* 64:149-159.

HORROCKS M. 1989a. – Pacific Boa identification. *Naika* 31: 30-32.

HORROCKS M. 1989b. – More about Pacific Boas. *Naika* 32: 31.

*HUGEL S. 2009. – Gryllacrididae and Tettigoniidae (Insecta, Orthoptera, Ensifera) from Espiritu Santo, Vanuatu. *Zoosystema* 31(3): 525-576.

*INEICH I. 2009. – The terrestrial herpetofauna of Torres and Banks Groups (northern Vanuatu), with report of a new species for Vanuatu. *Zootaxa* 2198: 1-15.

INEICH I. 2008. – A new arboreal *Lepidodactylus* (Reptilia: Gekkonidae) from Espiritu Santo Island, Vanuatu: from egg to holotype. *Zootaxa* 1918: 26-38.

*JAUME D. & QUEINNEC E. 2007. – A new species of freshwater isopod (Sphaeromatidea: Sphaeromatidae) from an inland karstic stream on Espiritu Santo Island, Vanuatu, southwestern Pacific. *Zootaxa* 1653: 41-55.

*JAUME D., SKET B. & BOXSHALL G.A. 2009. – New subterranean Sebidae (Amphipoda: Gammaridea) from Vietnam and the SW Pacific. *Zoosystema* 31(2): 249-277.

JOURDAN H., BONNET DE LARBOGNE L. & CHAZAU J. 2002. – The recent introduction of the Neotropical tramp ant *Wasmannia auropunctata* (Hymenoptera: Formicidae) into Vanuatu Archipelago (Southwest Pacific). *Sociobiology* 40: 483-509.

KEITH P., GALEWSKI T., CATTANEO-BERREBI G., HOAREAU T., & BERREBI P. 2005. – Ubiquity of *Sicyopterus lagocephalus* (Teleostei: Gobioidei) and phylogeography of the genus *Sicyopterus* in the Indo-Pacific area inferred from mitochondrial cytochrome b gene. *Molecular Phylogenetics and Evolution* 37(2005): 721-732.

*KEITH P., MARQUET G. & POUILLY M. 2009. – *Stiphodon mele* n. sp., a new species of freshwater goby from Vanuatu and New Caledonia (Teleostei, Gobiidae, Sicydiinae), and comments about amphidromy and regional dispersion. *Zoosystema* 31(3): 471-483.

KEITH P., MARQUET G. & WATSON R.E. 2007. – *Stiphodon kalfatak*, a new species of freshwater goby from Vanuatu (Teleostei: Gobioidei: Sicydiinae). *Cybium* 3(1): 33-37.

KEITH P., MARQUET G. & WATSON R.E. 2004. – *Schismatogobius vanuatuensis*, a new species of freshwater goby form Vanuatu, South Pacific. *Cybium* 28(3): 237-241.

KEITH P., WATSON R.E. & MARQUET G. 2004b. – *Sicyopterus aiensis*, a new species of freshwater goby from Vanuatu (Teleostei: Gobioidei). *Cybium* 28(2): 111-118.

KEITH P., WATSON R. & MARQUET G. 2000. – Découverte d'*Awaous ocellaris* (Broussonet, 1782) (Perciformes, Gobiidae) en Nouvelle-Calédonie et au Vanuatu et conséquences biogéographiques. *Cybium* 24(4): 350-400.

KELLEY R.W. 1989. – New species of micro-caddisflies (Trichoptera: Hydroptilidae) from New Caledonia, Vanuatu and Fiji. *Proceedings of the Entomological Society of Washington* 91(2): 190-202.

KIMMINS D.E. 1958. – Miss L.E. Cheeseman's expedition to New Hebrides, 1955. Orders Odonata, Neuroptera and Trichoptera. *Bulletin of the British Museum (Natural History) Entomology* 6(9): 239-250.

KIMMINS D.E. 1936. – Odonata, Ephemeroptera, and Neuroptera of the New Hebrides and Banks Islands. *Annals and Magazine of Natural History*, serie 10 18: 68-88.

KOWALSI J. 1917. – Un ennemi du cocotier aux Nouvelles-Hébrides. Le *Promecotheca opacicollis* Gestro. *Annales du Service des Epiphyties* 4: 286-327.

KRATTER A.W., KIRCHMAN J.J. & STEADMAN D.W. 2006. – Upland bird communities on Santo, Vanuatu, Southwest Pacific. *The Wilson Journal of Ornithology* 118(3): 295-308.

KURAHASHI H. 1982. – Blow flies from Vanuatu (New Hebrides), with descriptions of three new species of the genus *Onesia* (Diptera: Calliphoridae). *Pacific Insects* 24(3-4): 235-249.

KUSCHEL G. 1998. – The subfamily Anthribinae in New Caledonia and Vanuatu (Coleoptera: Anthribidae). *New Zealand Journal of Zoology* 25: 335-408.

LAYARD E.L. & LAYARD E.L.C. 1878. – Notes on some birds collected or observed by Mr Leopold C. Layard in the New Hebrides. *Ibis* 4(2): 267-280.

LEE K.E. 1981. – Earthworms (Annelida : Oligochaeta) of Vanua Tu (New Hebrides Islands). *Australian Journal of Zoology* 29: 535-572.

LEVER C. 2003. – *Naturalized reptiles and amphibians of the world.* Oxford University Press, Oxford: i-xx + 1-318.

*LOURENÇO W.R. 2009. – Scorpions collected in the island of Espiritu Santo (Vanuatu) and description of a new species of *Lychas* C.L. Koch, 1845 (Arachnida, Scorpiones, Buthidae). *Zoosystema* 31(3): 731-740.

MAFFI M. 1977. – Contribution to the knowledge of the mosquito fauna of the New Hebrides island group proper (Diptera : Culicidae). *Rivista di Parassitologia* 38(2-3): 194-214.

*MALZACHER P. & STANICZEK A.H. 2007. – *Caenis vanuatensis*, a new species of mayflies (Ephemeroptera: Caenidae) from Vanuatu. *Aquatic Insects* 29(4): 285-295.

MARQUET G., TAIKI N., CHADDERTON L. & GERBEAUX P. 2002. – Biodiversity and biogeography of freshwater crustaceans (Decapoda: Natantia) from Vanuatu, a comparison with Fiji and New Caledonia. *Bulletin français de la pêche et de la pisciculture* 364: 217-232.

MARQUET G., KEITH P. & KALFATAK D. 2009. – *Caridina gueryi*, a new species of freshwater shrimp (Decapoda, Atyidae) from Santo Island, Vanuatu. *Crustaceana* 82(2): 159-166.

MAYR E. 1945. – *Birds of the Southwest Pacific. A field guide to the birds of the area between Samoa, New Caledonia, and Micronesia.* Macmillan, New York, 316 p.

McEVEY S.F. & POLAK M. 2005. – *Mycodrosophila* (Diptera: Drosophilidae) of Fiji and Vanuatu, with description of nine new species. *Bishop Museum Occasional Papers* 84: 35–67.

MEAD J.I., STEADMAN D.W., BEDFORD S.H., BELL C.J. & SPRIGGS M. 2002. – New extinct mekosuchine crocodile from Vanuatu, South Pacific. *Copeia* 2002(3): 632-641.

MEDWAY Lord 1974. – A new skink (Reptilia: Scincidae: Genus *Emoia*) from the New Hebrides, with comments on the status of *Emoia samoensis loyaltiensis* (Roux). *Bulletin of the British Museum of natural History (Zoology)* 27(2): 53-57.

MEDWAY L. & MARSHALL A.G. 1975. – Terrestrial vertebrates of the New Hebrides: origin and distribution. *Philosophical Transactions of the Royal Society of London*, B 272: 423-165.

MESSEL H. (Ed.) 1994. – *Review of the status of crocodilians in the Pacific Island nations of Palau, Solomon Islands, and Vanuatu, in Crocodiles. Proceedings of the 2nd Regional (Eastern Asia, Oceania, Australasia) meeting of the Crocodile Specialist Group.* IUCN-The World Conservation Union, Gland, Suisse.

MESSEL H. & KING F.W. 1993. – Survey and plan for recovery of the crocodile population of Vanuatu and a project for the sustainable use of wildlife resources based at Port Patteson on Vanua Lava, Banks Islands. *Naika* 40: 4-11.

MESSEL H. & KING F.W. 1992. – *Survey and plan for recovery of the crocodile population of the Republic of Vanuatu, southwestern Pacific Ocean and a project for the sustainable use of wildlife resources based at Port Patteson on Vanua Lava, Banks Islands, in the Banks-Torres Conservation Region. A report to the Government of the Republic of Vanuatu, Port Vila, Vanuatu, in Crocodile Conservation Action. A special Publication of the Crocodile Specialist Group of the Species Survival Commission of the IUCN.* The World Conservation Union, Gland, Suisse: 102-108

MOUND L.A. & WALKER A.K. 1987. – Thysanoptera as tropical tramps: New records from New Zealand and the Pacific. *New Zealand Entomologist* 9: 70-85.

*OBER S.V. & STANICZEK A.H. 2009. – A new genus and species of coenagrionid damselflies (Insecta, Odonata, Zygoptera, Coenagrionidae) from Vanuatu. *Zoosystema*, 31(3): 485-497.

*OLMI M. & VILLEMANT C. 2009. – Les Dryinidae (Insecta, Hymenoptera, Chrysidoidea) du Vanuatu et des îles du Pacifique. *Zoosystema* 31(3): 691-705.

OTA H., FISHER R.N., INEICH I., CASE T.J., RADTKEY R.R. & ZUG G.R. 1998. – A new *Lepidodactylus* (Squamata: Gekkonidae) from Vanuatu. *Herpetologica* 54(3): 325-332.

*PASCAL M., LORVELEC O., BARRE N. & de GARINE-WICHATITSKY M. 2008. – Espèces allochtones d'Espiritu Santo. Premiers résultats de l'expédition Santo 2006. *Journal de la Société des Océanistes* 126-127: xxx

*PAULY A. & VILLEMANT C. 2009. – Hyménoptères Apoidea (Insecta) de l'archipel du Vanuatu. *Zoosystema* 31(3): 719-730.

*PLANT A.R. & DAUGERON C. 2009. – A new species of *Phyllodromia* Zetterstedt, 1837 (Insecta, Diptera, Empididae, Hemerodromiinae) from Vanuatu. *Zoosystema* 31(3): 519-524.

*ROBILLARD T. 2009. – Eneopterinae crickets (Insecta, Orthoptera, Grylloidea) from Vanuatu. *Zoosystema* 31(3): 577-618.

ROBINSON G.S. 1976. – Biogeography of the New Hebrides Macrolepidoptera. *Journal of the Entomological Society of Australia* 9: 47-53.

ROUX J. 1913. – Note sur quelques reptiles des Nouvelles Hébrides, des îles Banks et Santa Cruz, in SARASIN & ROUX, *Nova Caledonia*, A, Zoologie I: 153-160.

RYAN P.A. 1986. – A new species of *Stiphodon* (Gobiidae: Sicydiaphiinae) from Vanuatu, in UYENO T., ARAI R., TANIUCHI T. & MATSUURA K. (Eds), *Indo-Pacific Fish Biology. Proceedings of the Second International Conference on Indo-Pacific Fishes.* Ichthyological Society of Japan, Tokyo: 655-662

SCOTT W.E. 1946. – Birds observed on Espiritu Santo, New Hebrides. *Auk* 63: 362-368.

SOLEM A. (1959). – Systematics of the land and fresh-water Mollusca of the New Hebrides. *Fieldiana: Zoology* 43(1): 1-238, 34 pl.

SOLEM A. (1959). – Zoogeography of the land and fresh-water Mollusca of the New Hebrides. *Fieldiana: Zoology* 43(2): 241-359.

SOLEM A. (1962). – Notes on, and descriptions of New Hebridean land snails. *Bulletin of the British Museum (Natural History)* 9(5): 217-247. Pls. 1-2.

STEADMAN D.W. 2006a. – *Extinction and biogeography of tropical Pacific birds.* Chicago University Press, Chicago, 594 p.

STEADMAN D.W. 2006b. – New species of extinct parrot (Psittacidae: *Eclectus*) from Tonga and Vanuatu, South Pacific. *Pacific Science* 60(1):137–145.

SUNDE R.G., MADDISON P.A. & TOCKER M.F. 1987. – Notes on some aphids (Homoptera) from Vanuatu. *New Zealand Entomologist* 9: 86-88.

TENNENT W.J. 2009. – *A field guide to the butterflies of Vanuatu.* Storm Entomological Publishing, Dereham, U.K.

*THIBAUD J.M. 2009. – Les collemboles (Collembola) interstitiels des sables littoraux de l'île d'Espiritu Santo (Vanuatu). *Zoosystema* 31(3): 499-505.

THOMPSON F.G. & HUCK E.L. 1985. – The land snail family Hydrocenidae in Vanuatu (New Hebrides Islands), and comments on other Pacific island species. *Nautilus* 99: 81–84.

*TISHECHKIN A.K. 2009. – Discovery of Chlamydopsinae (Insecta, Coleoptera, Histeridae) in Vanuatu with the description of eight new species from Espiritu Santo Island. *Zoosystema* 31(3): 661-690.

TYLER M.J. 1979b. – The introduction and current distribution in the New Hebrides of the Australian hylid frog Litoria aurea. *Copeia* 1979(2): 355-356.

VIETTE P.E.L. 1950a. – The Noctuidae Catocalinae from New Caledonia and the New Hebrides. *Pacific Science* 4: 139-157.

VIETTE P.E.L. 1950b. – Les Lithosiidae de Nouvelle-Calédonie et des Nouvelles Hébrides. *Annales de la Société Entomologique de France* 119: 81-96.

VIETTE P.E.L. 1951. – Les Noctuidae Noctuinae de Nouvelle-Calédonie et des Nouvelles Hébrides. *Annales de la Société Entomologique de France* 118: 29-50.

*WAHIS R., DURAND F. & VILLEMANT C. 2009. – Pompiles de l'île d'Espiritu Santo, Vanuatu (Insecta, Hymenoptera, Pompilidae). *Zoosystema* 31(3): 707-718.

WATSON R.E., KEITH P. & MARQUET G. 2007. – *Akihito vanuatu*, a new genus and new species of freshwater goby from the South Pacific (Teleostei: Gobioidei: Sicydiinae). *Cybium* 31(3): 341-349.

*WEINER W.M., BEDOS A.& DEHARVENG L. 2009. – Species of the genus *Friesea* (Collembola, Neanuridae) from New Caledonia and Vanuatu. *Zoosystema* 31(3): 507-518.

WHITAKER A.H. & WHITAKER V.A. 1994. – *A survey of the herpetofauna near Matantas, Espiritu Santo; with notes on reptiles elsewhere on Espiritu Santo and Efaté, Vanuatu.* Unpublished report to Environment Unit. Ministry of Natural Resources, Port Vila, : i-iv + 1-47.

WILLIAMS D.J. & BUTCHER C.F. 1987. – Scale insects (Hemiptera: Coccoidea) of Vanuatu. *New Zealand Entomologist* 9: 88-99.

WOMERSLEY H. 1928. – Apterygota from the New Hebrides. *Annals and Magazine of Natural History* 10: 55-61.

WOMERSLEY H. 1937. – On some Apterygota from New Guinea and the New Hebrides. *Proceedings of the Royal Entomological Society of London* 6: 204-210.

YOSHII R. 1995. – Notes on Collembola of Vanuatu. *AZAO* 3: 43-50.

ZUG G.R. & MOON B.R. 1995. – Systematics of the Pacific slender-toed geckos, *Nactus pelagicus* complex: Oceania, Vanuatu, and Solomon Islands populations. *Herpetologica* 51(1): 77-90.

MARINE ECOSYSTEMS

AMAOKA K. & SÉRET B. 2005. – *Engyprosopon vanuatuensis*, a new species of bothid flounder (Pleuronectiformes: Bothidae) from off Vanuatu, South West Pacific. *Ichthyological Research* 52(1): 15-19.

*BAMBER R.N. 2009. – Two new species of shell-inhabiting tanaidaceans (Crustacea, Peracarida, Tanaidacea, Pagurapseudidae, Pagurapseudinae) from the shallow sublittoral off Vanuatu. *Zoosystema* 31(3): 407-418.

BENTLEY M.G., OLIVE P.J. & LAST K. 2001. – Sexual satellites, Moonlight and the Nuptial Dances of Worms: the influence of the Moon on the Reproduction of Marine Animals. *Earth, Moon and Planets* 85-86: 67-84.

BRIGGS J.C. 1962. – A new clingfish of the genus Lepadichthys from the New Hebrides. *Copeia* 1962(2): 424-425.

*CABEZAS P., MACPHERSON E. & MACHORDOM A. 2010. – Taxonomic revision of the genus *Paramunida* Baba, 1988 (Crustacea: Decapoda: Galatheidae): a morphological and molecular approach. *Zootaxa* 2712: 1-60.

DONE T.J. & NAVIN K.F. 1990 (Eds). – *Vanuatu marine resources: Report of a biological survey*. Australian Institute of Marine Science, Townsville: 272 p.
Includes:
Done TJ. Introduction: 1-9.
Done TJ and Navin KF. Shallow water benthic communities on coral reefs: 10-37.
Williams DMcB. Shallow-water reef fishes: 66-76.
McKinnon AD. Zooplankton: 82-85.
Chambers MR, Nguyen F and Navin KF. Seagrass communities: 92-103.
Zann LP, Ayling AM and Done TJ. Crown of thorns starfish: 104-113.
Benzie JAH. Genetic relationships of crown-of-thorns starfish: 114-118.
Ayling AM, Andrews GJ, Navin KF and Benzie JAH. Quantitative surveys around Malakula Island: 119-135.
Williams DMcB and Ayling AM. Checklist of shallow-water reef fishes: 224-229.

FAUCHALD K. 1992. – Review of the types of *Palola* (Eunicidae: Polychaeta). *Journal of Natural History* 26: 1177-1225.

FOURMANOIR P. 1970. – Notes ichtyologiques (I). *Cahiers de l'O.R.S.T.O.M.* Série Océanographie 8(2): 19-33.

FOURMANOIR P. & LABOUTE P. 1976. – *Poissons des mers tropicales. Nouvelle Calédonie. Nouvelles Hébrides*. Papeete (Éditions du Pacifique): 1-376.

FOURMANOIR P. & RIVATON J. 1979. – Poissons de la pente récifale externe de Nouvelle-Calédonie et des Nouvelles-Hébrides. *Cahiers d'Indo-Pacifique* 1(4): 405-443.

FOWLER H.W. 1944. – Fishes obtained in the New Hebrides by Dr. Edward L. Jackson. *Proceedings of the Academy of Natural Sciences of Philadelphia* 96: 155-199.

*GALIL B.S. & NG P.K.L. 2010. – On a collection of calappoid and leucosioid crabs (Decapoda, Brachyura) from Vanuatu, with description of a new species of Leucosiidae, *in* CASTRO P., DAVIE P.J.F, NG P.K.L. & RICHER DE FORGES B. (Eds), Studies on Brachyura. *Crustaceana Monographs* 11: 139-152.

GUILCHER A. 1974. – Coral reefs of the New Hebrides, Melanesia, with particular reference to open-sea, not fringing, reefs, *in Proceedings of the Second International Coral Reef Symposium, Vol. 2*. I.S.R.S., Brisbane: 523-535

HASHIMOTO T., HATTA H., YUKAWA T., SUGIMURA K., CHANEL S. & IWASHINA T. 1998. – A list of herbarium specimens from Vanuatu, collected in 1966 and 1997, excluding the Orchidaceae. *Annals of the Tsubuka Botanical Garden* 17: 1-22. [Marine phanerogams included]

HOEKSEMA B.W. 2007. – Delineation of the Indo-Malayan Centre of Maximum Marine Biodiversity: The Coral Triangle, *in* RENEMA W. (Ed.), *Biogeography, Time and Place: Distributions, Barriers and Islands*. Springer, Dordrecht: 117-178

*KANTOR Y., PUILLANDRE N., OLIVERA B. & BOUCHET P. 2008. – Morphological proxies for taxonomic decision in turrids (Mollusca, Neogastropoda): a test of the value of shell and radula characters using molecular data. *Zoological Science* 25: 1156-1170.

*LANE D.J.W. & ROWE F.W.E. 2009. – A new species of *Asterodiscides* (Echinodermata, Asteroidea, Asterodiscididae) from the tropical southwest Pacific, and the biogeography of the genus revisited. *Zoosystema* 31(3): 419-429.

*MACPHERSON E. 2009. – New species of squat lobsters of the genera *Munida* and *Raymunida* (Crustacea, Decapoda, Galatheidae) from Vanuatu and New Caledonia. *Zoosystema* 31(3): 431-451.

*MATTIO L., PAYRI C. & VERLAQUE M. 2009. – Taxonomic revision and geographic distribution of the subgenus *Sargassum* (Fucales, Phaeophyceae) in the western and central Pacific Islands based on morphological and molecular analyses. *Journal of Phycology* 45(5): 1213-1227.

*McLAUGHLIN P.A. & RAHAYU D.L. 2008. – A new genus and species of hermit crab of the family Paguridae (Crustacea: Anomura: Paguroidea) from the Vanuatu Archipelago. *Proceedings of the Biological Society of Washington* 121(3): 365-373.

*NARUSE T., CASTRO P. & NG P.K.L. 2009. – A new genus and new species of Ethusidae (Decapoda, Brachyura) from Vanuatu, Western Pacific. *Crustaceana* 82(7): 931-938.

*NEUSSER T.P. & SCHRÖDL M. 2009. – Between Vanuatu tides: 3D anatomical reconstruction of a new brackish water acochlidian gastropod from Espiritu Santo. *Zoosystema* 31(3): 453-469.

*NG P.K.L. & MANUEL-SANTOS M.R. 2007. – Establishment of the Vultocinidae, a new family for an unusual new genus and new species of Indo-West Pacific crab (Crustacea: Decapoda: Brachyura: Goneplacoidea), with comments on the taxonomy of the Goneplacidae. *Zootaxa* 1558: 39-68.

*NG P.K.L. & NARUSE T. 2007. – *Liagore pulchella*, a new species of xanthid crab (Crustacea: Decapoda: Brachyura) from Vanuatu. *Zootaxa* 1665: 53-60.

*N'YEURT A. & PAYRI C.E. 2007. – *Etude de la collection d'algues marines de l'île de Santo, Vanuatu*. www.crisponline.net/Portals/1/PDF/ENG-Algae-Santo-Vanuatu.pdf

*N'YEURT A. & PAYRI C. 2008. – *Sebdenia cerebriformis* sp. nov. (Sebdeniaceae, Sebdeniales) from the south and western Pacific Ocean. *Phycological Research* 56(1): 13-20.

*N'YEURT A. & PAYRI C. 2009. – Four new species of Rhodophyceae from Fiji, Polynesia and Vanuatu, South Pacific. *Phycological Research* 57(1): 12-24.

*PAILLERET M., HAGA T., PETIT P., PRIVE-GILL C., SAEDLOU N., GAILL F. & ZBINDEN M. 2007. – Sunken wood from the Vanuatu Islands: identification of wood substrates and preliminary description of associated fauna. *Marine Ecology* 28: 233-241.

PAYRI C. & VERBRUGGEN H. 2009. – *Pseudocodium mucronatum*, a new species from New Caledonia, and an analysis of the evolution of climatic preferences in the genus (Bryopsidales, Chlorophyta). *Journal of Phycology* 45(4): 953-961.

*PYLE R.L., EARLE J.L. & GREENE B.D. 2008. – Five new species of the damselfish genus *Chromis* (Perciformes: Labroidei: Pomacentridae) from deep coral reefs in the tropical western Pacific. *Zootaxa* 1671: 3-31.

RICHER DE FORGES B., FALIEX E. & MENOU J.L. 1996. – La campagne MUSORSTOM 8 dans l'archipel de Vanuatu. Compte rendu et liste des stations, *in* CROSNIER A. (Ed.), Résultats des campagnes MUSORSTOM, volume 15. *Mémoires du Muséum National d'Histoire Naturelle* 168: 9-32.

SCHULZE A. 2006. – Phylogeny and genetic diversity of Palolo worms (*Palola*, Eunicidae) from the tropical North Pacific and the Caribbean. *The Biological Bulletin* 210: 25-37.

TANAKA N. 2002. – Notes on the seagrasses of Vanuatu. *Annals of the Tsubuka Botanical garden* 21: 31-35.

TAYLOR F.J. 1978. – Seagrasses in the New Hebrides. *Aquatic Botany* 4: 373-375.

*TERRYN Y. & HOLFORD M. 2008. – The Terebridae of Vanuatu with a revision of the genus *Granuliterebra*, Oyama 1961. *Visaya* Supplement 3: 1-96.

TILBROOK K.J., HAYWARD P.J. & GORDON D.P. 2001. – Cheilostomatous Bryozoa from Vanuatu. *Zoological Journal of the Linnean Society*, 131: 35–109.

VERON J.E.N. 1990. – Checklist of the hermatypic corals of Vanuatu. *Pacific Science* 44: 51-70.

WEST J.A., ZUCCARELLO G.C., SCOTT J.L., WEST K.A. & LOISEAUX DE GOER S. 2007. – *Pulvinus veneticus* gen. et sp. nov. (Compsopogonales, Rhodophyta) from Vanuatu. *Phycologia* 46(3): 237-246.

*YANG C.H., CHEN I.S. & CHAN T.Y. 2008. – A new slipper lobster of the genus *Petrarctus* (Crustacea: Decapoda: Scyllaridae) from the West Pacific. *Raffles Bulletin of Zoology* Supplement 19: 71-81.

*YANG C.W., CHAN T.Y. & CHU K.H. 2010. – Two new species of the "*Heterocarpus gibbosus* Bate, 1888" species group (Crustacea: Decapoda: Pandalidae) from the western Pacific and north-western Australia. *Zootaxa* 2372: 206-220.

Josep Antoni Alcover Tomas
Instituto Mediterráneo de Estudios Avanzados
(IMEDEA), C/ Miquel Marquès 21, 07190-Esporles
Islas Baleares - Spain.
josepantoni.alcover@uib.es

Nicolas Barré
Institut Agronomique Néo-Calédonien (IAC) [retired]
Present address:
403 Route de Bel Air Desrozières
97170 Petit Bourg - Guadeloupe - FWI
belairbarre@hotmail.fr

Anne Bedos
Muséum national d'Histoire naturelle
Département Systématique et Evolution, CP 50
45 rue Buffon
75005 Paris - France.
bedosanne@yahoo.fr

Alain Beu
GNS Science, 1 Fairway Drive, Avalon
PO Box 30-368, Lower Hutt - New Zealand.
A.Beu@gns.cri.nz

Philippe Bouchet
Muséum national d'Histoire naturelle
Département Systématique et Evolution, CP 51
55 rue Buffon
75005 Paris - France.
pbouchet@mnhn.fr

Geoff Boxshall
Natural History Museum
Cromwell Road
London SW7 5BD - United Kingdom.
gab@nhm.ac.uk

Franck Bréhier
09800 Alas - France
brehier-franck@orange.fr

Florence Brunois
Muséum national d'Histoire naturelle
Laboratoire Eco-Anthropologie et Ethnobiologie,
CP 135, 57 rue Cuvier
75231 Paris Cedex 05 - France.
Present address:
Laboratoire d'Anthropologie sociale
Collège de France
11, place Marcelin Berthelot
75231 Paris Cedex 05 - France.
florence.brunois@college-de-france.fr

Elizabeth Brown
The Royal Botanic Gardens Sydney
Mrs Macquaries Road
Sydney, NSW 2000 - Australia
elizabeth.brown@rbgsyd.nsw.gov.au

Bart Buyck
Muséum national d'Histoire naturelle
Département Systématique et Evolution, CP 39B
12-16 rue Buffon, 75005 Paris - France.
buyck@mnhn.fr

Guy Cabioch
Laboratoire d'Océanographie et du Climat:
Expérimentations et Approches Numériques
LOCEAN
Institut Pierre-Simon Laplace
Centre IRD France Nord, Unité 182
32 Avenue Henri-Varagnat
93143 Bondy cedex - France.
guy.cabioch@ird.fr

Yolanda E. Camacho García
Department of Invertebrate Zoology and Geology
California Academy of Sciences
875 Howard Street
San Francisco, CA. 94103 - USA
Present address:
Centro de Investigaciones Marinas y Limnológicas
(CIMAR), Universidad de Costa Rica
Apdo. 2060. San Pedro, San José - Costa Rica.
ycamacho_99@yahoo.com

Tin-Yam Chan
Institute of Marine Biology
National Taiwan Ocean University
2 Pei-Ning Rd., Keelung 20224 - Taiwan R.O.C.
tychan@ntou.edu.tw

Regis Cleva
Muséum national d'Histoire naturelle
Direction des collections, CP 63
61 rue Buffon, 75005 Paris - France. [retired]

Bruno Corbara
LMGE/CNRS, UMR 6023, Université Blaise Pascal
34 Avenue Carnot
63037 Clermont-Ferrand Cedex - France.
Bruno.Corbara@univ-bpclermont.fr

Rudo Von Cosel
Muséum national d'Histoire naturelle
Département Systématique et Evolution, CP 51
55 rue Buffon, 75005 Paris - France.
cosel@mnhn.fr

Louis Deharveng
Muséum national d'Histoire naturelle
Département Systématique et Evolution
UMR 7205 "Origine, Structure et Evolution de la
Biodiversité", CP 50
45 rue Buffon, 75005 Paris - France.
deharven@mnhn.fr

Thibaut Delsinne
Royal Belgian Institute of Natural Sciences
Section of Biological Evaluation
Rue Vautier, 29, B-1000 Brussels - Belgium.
Thibaut.Delsinne@sciencesnaturelles.be

Laure Desutter-Grandcolas
Muséum national d'Histoire naturelle, Département
Systématique et Evolution, CP 50, 45 rue Buffon
75005 Paris - France.
desutter@mnhn.fr

Jean-Michel Dupuyoo
Jardin d'Oiseaux Tropicaux, Conservatoire
Biologique Tropical
83250 La Londe-les-Maures - France.
jmdupuyoo@yahoo.fr

John Earle
1221 Victoria Street, Honolulu, HI 96814 - U.S.A.

Stefan Eberhard
Subterranean Ecology Pty Ltd,
95A Flora Tce, North Beach WA 6020 - Australia
stefan@subterraneanecology.com.au

Elsa Faugère
Institut National de la Recherche Agronomique
INRA PACA, Domaine Saint Paul
Site Agroparc, 84914 Avignon cedex 9 - France.
faugere@avignon.inra.fr

Benoît Fontaine
Muséum national d'Histoire naturelle, UMR 7204
Conservation des espèces, suivi et restauration des
populations
Département Ecologie et Gestion de la Biodiversité
CP 51, 55 rue Buffon, 75005 Paris - France.
fontaine@mnhn.fr

Charles Fransen
National Museum of Natural History Naturalis
PO Box 9517, NL-2300 RA Leiden - The Netherlands.
fransen@naturalis.nl

Ronald Fricke
Ichthyology, Staatliches Museum für Naturkunde
Rosenstein 1, 70191 Stuttgart - Germany.
fricke.smns@naturkundemuseum-bw.de

Jean-Christophe Galipaud
Centre IRD de Nouméa
BP A5, 98848 Nouméa cedex - New Caledonia.
galipaud@arkeologie.net

Olivier Gargominy
Muséum national d'Histoire naturelle
Service du Patrimoine Naturel (USM 0308), CP41
36, rue Geoffroy saint-Hilaire, 75005 Paris - France.
gargo@mnhn.fr

Michel de Garine-Wichatitsky
Institut Agronomique Néo-Calédonien (IAC)
Port Laguerre, BP 73 98890 Paita - New Caledonia.
Present address:
CIRAD-French Embassy, 37 Arcturus Road
PO BOX 1378, 99 Harare Zimbabwe.
michel.de_garine-wichatitsky@cirad.fr

Philippe Gerbeaux
Department of Conservation
Torrens House Level 4/195 Hereford Street
Christchurch 8011 - New Zealand.
pgerbeaux@doc.govt.nz

Adriaan Gittenberger
National Museum of Natural History Naturalis
PO Box 9517, NL-2300 RA Leiden - The Netherlands.
Gittenberger@yahoo.com

Matthias Glaubrecht
Museum für Naturkunde Berlin
Leibniz-Institut für Evolutions- und Biodiversitäts-
forschung an der Humboldt-Universität zu Berlin
Invalidenstrasse 43, D-10115 Berlin - Germany.
Matthias.Glaubrecht@mfn-berlin.de

Thomas Haevermans
Muséum national d'Histoire naturelle
Département Systématique et Evolution
CP 39A, 12-16 rue Buffon 75005 Paris - France.
haever@mnhn.fr

Takuma Haga
Department of Geology and Paleontology
National Museum of Nature and Science (NMNS)
3-23-1 Hyakunin-chô, Shinjuku-ku
Tôkyô 169-0073 - Japan.
haga@kahaku.go.jp

Anthony Harry
Institut Agronomique Néo-Calédonien (IAC)
Port Laguerre, BP 73 98890 Paita - New Caledonia.
lenamangki@yahoo.fr

Virginie Héros
Muséum national d'Histoire naturelle
Département Systématique et Evolution
CP 51, 55 rue Buffon, 75005 Paris - France.
malaco@mnhn.fr

Bert Hoeksema
National Museum of Natural History Naturalis
PO Box 9517, NL-2300 RA Leiden - The Netherlands
bert.hoeksema@ncbnaturalis.nl

Sylvain Hugel
Institut de Neurosciences Cellulaires et Intégratives
LC2, UMR 7168 CNRS, 21, rue Descartes
67083 cedex Strasbourg - France.
hugel@neurochem.u-strasbg.fr

Ivan Ineich
Muséum national d'Histoire naturelle
Département Systématique et Evolution, CP 30
25 rue Cuvier 75005 Paris - France.
ineich@mnhn.fr

Damia Jaume
Instituto Mediterráneo de Estudios Avanzados
(IMEDEA), C/ Miquel Marquès 21
07190-Esporles (Mallorca, Illes Balears) - Spain.
damiajaume@imedea.uib-csic.es

Donna Kalfatak
Department of Environment and Conservation
Private mail bag 063, Port Vila - Vanuatu.
dkalfatak@vanuatu.gov.vu

Yasunori Kano
Department of Marine Ecosystems Dynamics,
Atmosphere and Ocean Research Institute
The University of Tokyo, 5-1-5 Kashiwanoha
Kashiwa, Chiba 277-8564 - Japan.
kano@aori.u-tokyo.ac.jp

Philippe Keith
Muséum national d'Histoire naturelle
Département des milieux et peuplements aquatiques
CP 26, 43 rue Cuvier 75005 Paris - France.
keith@mnhn.fr

Roger Kitching
Centre for Innovative Conservation Strategies
The Griffith School of Environment
Griffith University, Nathan, QLD 4111 - Australia.
r.kitching@griffith.edu.au

Jean-Noël Labat †
Muséum national d'Histoire naturelle
Département Systématique et Evolution, CP 39A
12-16 rue Buffon, 75005 Paris - France.

Nadir Lasson
Le Bourg 46090 Cours - France.
nadir-lasson@netcourrier.com

Vincent Lebot
Centre de coopération internationale en recherche
agronomique pour le développement (CIRAD)
PO Box 946 Port-Vila - Vanuatu.
lebot@vanuatu.com.vu

Hervé Le Guyader
UMR 7138 - Systématique, Adaptation, Evolution
UPMC, Bâtiment A, 4 ème étage, Case 05
7, Quai Saint Bernard, 75252 Paris Cedex 05 - France.
herve.le_guyader@upmc.fr

Maurice Leponce
Biological Evaluation Section. Royal Belgian
Institute of Natural Sciences
29 rue Vautier, 1000 Brussels - Belgium.
Maurice.Leponce@naturalsciences.be

Bernard Lips
4 avenue Allende, 69100 Villeurbanne - France
bernard.lips@free.fr

Josiane Lips
4 avenue Allende, 69100 Villeurbanne - France
josiane.lips@free.fr

Clara Lord
Muséum national d'Histoire naturelle
Département des milieux et peuplements aquatiques
CP26, 43 rue Cuvier, 75005 Paris - France.
Present address:
ORI, Tokyo University,
5-1-5 Kashiwanoha, Kashiwa, Chiba 277-8564
Tokyo - Japan.
claralord@aori.u-tokyo.ac.jp

Olivier Lorvelec
Institut National de la Recherche Agronomique
UMR (INRA/Agrocampus Rennes) Écologie et
Santé des Écosystèmes
Équipe Écologie des Invasions Biologiques
Avenue du Général Leclerc - Campus de Beaulieu
Bâtiment 16A, 35042 Rennes Cedex - France.
Olivier.Lorvelec@rennes.inra.fr

Porter Lowry
Missouri Botanical Garden
P.O. Box 299, St. Louis, MO 63166-0299 - U.S.A.
&
Muséum national d'Histoire naturelle
Département Systématique et Evolution
CP 39, 57 rue Cuvier, 75231 Paris Cedex 05 - France.
Pete.Lowry@mobot.org

Pierre Lozouet
Muséum national d'Histoire naturelle
Département, Direction des Collections
CP 51, 55 rue Buffon, 75005 Paris - France.
lozouet@mnhn.fr

Christophe Maes
Laboratoire d'Etudes en Géophysique et
Océanographie Spatiales (LEGOS)
Institut de Recherche pour le Développement
(IRD), Centre IRD de Nouméa
BP A5, 98848 Nouméa cedex - New Caledonia.
Christophe.Maes@noumea.ird.nc

Philippe Maestrati
Muséum national d'Histoire naturelle
Département Systématique et Evolution
CP 51, 55 rue Buffon, 75005 Paris - France.
maestrat@mnhn.fr

Masako Mitsuhashi
Laboratory of Biology, Osaka Institute of Technology
Ohmiya, Asahi-ku, Osaka 535-8585 - Japan
mitsuhashi@ge.oit.ac.jp

Gordon McPherson
Missouri Botanical Garden
P.O. Box 299, St. Louis, MO 63166-0299 - U.S.A.
gordon.mcpherson@mobot.org

Marivene Manuel-Santos
National Museum of the Philippines
P. Burgos St., Manila - Philippines
vinrmanuel@yahoo.com

Jose Christopher E. Mendoza
National University of Singapore
Systematics and Ecology Lab - DBS, Science Drive 4
Singapore 117543 - Singapore.
jcmendoza@nus.edu.sg

Sara Muller
Vanuatu Agricultural Research Center
Santo - Republic of Vanuatu.
Present address:
University of Strasbourg
Faculty of Geography and Planning
3, rue de l'Argonne, 67000 Strasbourg - France.
sara.muller@unistra.fr

Jérôme Munzinger
Centre IRD de Nouméa
B.P. A5, 98 848 Nouméa cedex - New Caledonia.
jerome.munzinger@ird.fr

Timea Neusser
Zoologische Staatssammlung München
Münchhausenstr. 21
D-81247 München - Germany.
timea-neusser@gmx.de

Peter Ng
Department of Biological Science
National University of Singapore, 14 Science Drive 4
Singapore 117543 - Singapore.
dbsngkl@nus.edu.sg

Marco Oliverio
Università di Roma "La Sapienza"
Dipartimento di Biologia Animale e dell'Uomo
Viale dell'Università 32, I-00185 Rome - Italy.
marco.oliverio@uniroma1.it

Jerôme Orivel
Université Paul Sabatier
Laboratoire Evolution et Diversité Biologique
118 route de Narbonne
31062 Toulouse Cedex 9 - France.
orivel@cict.fr

Laurent Palka
Muséum national d'Histoire naturelle
Département d'Ecologie et Gestion de la Biodiversité
61, rue Buffon, 75005 Paris - France.
palka@mnhn.fr

Jonathan Palmer
Gondwana Tree-Ring Laboratory, P.O. Box 14
Little River, Canterbury, 7546 - New Zealand.
jonathan@dendro.co.nz

Michel Pascal
Institut National de la Recherche Agronomique
UMR (INRA/Agrocampus Rennes) Écologie et
Santé des Écosystèmes
Équipe Écologie des Invasions Biologiques
Avenue du Général Leclerc, Campus de Beaulieu
Bâtiment 16A, 35042 Rennes Cedex - France.
Michel.Pascal@rennes.inra.fr

Olivier Pascal
Pro-Natura International
15, avenue de Ségur, 75007 Paris - France.
oli.pascal@gmail.com

Claude Payri
Centre IRD de Nouméa, 101 Promenade Roger Laroque
Anse Vata, BP A5, 98848 Noumea - New Caledonia.
claude.payri@ird.fr

Marc Pignal
Muséum national d'Histoire naturelle
Département Systématique et Evolution, CP 39A
12-16 rue Buffon, 75005 Paris - France.
pignal@mnhn.fr

Yohan Pillon
Laboratoire de Botanique, Institut de Recherche
pour le Développement
BP A5, 98848 Nouméa Cedex - New Caledonia.
Present address:
Department of Biology, University of Hawai'i at Hilo
200 West Kawili Street, Hilo, HI 96720 - U.S.A.
pillon@hawaii.edu

Rufino Pineda
PMB 004, Luganville - Republic of Vanuatu.
learn@vanuatu.com.vu

Jean-Claude Plaziat
Formerly University Paris 11
Present address:
7, rue du Panorama, 91400 Orsay - France [retired].
jc.plaziat@dbmail.com

Gregory Plunkett
Department of Biology, Virginia Commonwealth University
P.O. Box 842012, Richmond, VA 23284-2012 - U.S.A.
gmplunke@saturn.vcu.edu

Marta Pola Perez
California Academy of Sciences
55 Music Concourse Dr., Golden Gate Park,
San Francisco, CA 94118-4503 - U.S.A.
Present address:
Laboratorio de Biología Marina
Departamento de Biología, Edificio de Biología
C/ Darwin, 2, Universidad Autónoma de Madrid
28049 Madrid - Spain.
mpolaperez@gmail.com

Sophie Pons
Muséum national d'Histoire naturelle
Direction de l'enseignement de la pédagogie et des formations, CP 27
57 rue Cuvier, 75005 Paris - France.
spons@mnhn.fr

Alain Pothet
Rectorat de l'académie de Créteil
 4, rue Georges Enesco
94010 Créteil Cedex - France.
alain.pothet@ac-creteil.fr

Marc Pouilly
Muséum national d'Histoire naturelle
Département des milieux et peuplements aquatiques
UMR CNRS 7208 BOREA, CP 53
61 rue Buffon, 75005 PARIS - France
marc.pouilly@ird.fr

Vincent Prié
Biotope, 22, Boulevard du Maréchal Foch
F-34140 Mèze - France
vprie@biotope.fr

Richard Pyle
Bernice P. Bishop Museum
1525 Bernice Street, Honolulu, HI 96817 - U.S.A.
deepreef@bishopmuseum.org

Eric Queinnec
UMR 7138, Systématique, Adaptation & Evolution
Université Pierre& Marie Curie
7 quai Saint-Bernard, 75005 Paris - France.
eric.queinnec@upmc.fr

Cahyo Rahmadi
Museum Zoologicum Bogoriense, Research Center for Biology, Indonesian Institute of Sciences
Jl. Raya Jakarta-Bogor Km. 46 Cibinong 16911 Indonesia.
cahyo.rahmadi@gmail.com

Jean-Louis Reyss
Laboratoire des Sciences du Climat et de l'Environnement, Domaine du CNRS
Avenue de la Terrasse
91198 Gif-sur-Yvette cedex - France.
Jean-Louis.reyss@lsce.cnrs-gif.fr

Bertrand Richer De Forges
Centre IRD de Nouméa, BP A5
98848 Nouméa cedex - New Caledonia. [retired]
b.richerdeforges@gmail.com

Marine Robillard
Muséum national d'Histoire naturelle
Département Homme, Nature et Sociétés
43 rue Cuvier, 75231 Paris cedex 05 - France
robillard@mnhn.fr

Tony Robillard
Muséum national d'Histoire naturelle
Département Systématique et Evolution, CP 50
45 rue Buffon, 75005 Paris - France.
robillar@mnhn.fr

Yves Roisin
Université Libre de Bruxelles
Eco-éthologie évolutive, CP 160/12
Avenue F. D. Roosevelt, 50
B - 1050 Bruxelles - Belgium.
yroisin@ulb.ac.be

Christine Rollard
Muséum national d'Histoire naturelle
Département Systématique et Evolution, CP 53
61 rue Buffon, 75005 Paris - France.
chroll@mnhn.fr

Germinal Rouhan
Muséum national d'Histoire naturelle
Département Systématique et Evolution, CP 39
16 rue Buffon, 75005 Paris - France.
rouhan@mnhn.fr

Stefano Schiaparelli
Dipartmento per lo Studio del Territorio e delle sue Risorse (Dip. Te. Res.)
Viale Benedetto XV, 5, I-16123 Genova - Italy
stefano.schiaparelli@unige.it

Jürgen Schmidl
Ecology, landscape & nature conservation group
Department for biology/Developmental biology unit
University of Erlangen-Nuremberg
Staudtstr. 5, D-91058 Erlangen - Germany.
jschmidl@biologie.uni-erlangen.de

Anne-Marie Sémah
IRD, UR Paléotropique, 32 Av. Henri Varagnat,
93143 Bondy Cedex - France.
Anne-Marie.Semah@bondy.ird.fr

Bernard Séret
Muséum national d'Histoire naturelle
Département Systématique et Evolution
CP 51, 55 rue Buffon, 75005 Paris - France.
seret@mnhn.fr

Patricia Siméoni
Boîte Postale 946, Port-Vila - Vanuatu
patricia@vanuatu.com.vu

Laurent Soldati
CBGP, Campus International de Baillarguet
CS 30016
34988 Montferrier-sur-Lez cedex - France.
soldati@supagro.inra.fr

Arnold H. Staniczek
Staatliches Museum für Naturkunde
Abt. Entomologie
Rosenstein 1, D-70191 Stuttgart - Germany.
Staniczek.smns@naturkundemuseum-bw.de

Ellen Strong
Smithsonian Institution, PO Box 37012, MRC 163
Washington, DC 20013-7012 - U.S.A.
StrongE@si.edu

Swee Hee Tan
Raffles Museum of Biodiversity Research (RMBR)
National University of Singapore
Department of Biological Sciences
6 Science Drive 2, #03-01, Singapore 117546
Singapore.
dbstansh@nus.edu.sg

Frederick Taylor
University of Texas Institute for Geophysics
John A. and Katherine G. Jackson School of Geosciences
J.J. Pickle Research Campus, Bldg. 196 (ROC)
10100 Burnet Road (R2200)
Austin, TX 78758-4445 - U.S.A.
fred@ig.utexas.edu

James P. Terry
Department of Geography
National University of Singapore
AS2, 1 Arts Link, Kent Ridge, 117570 Singapore
geojpt@nus.edu.sg

Alexey Tishechkin
Louisiana State Arthropod Museum
Department of Entomology
404 Life Science Building
Louisiana State University
Baton Rouge, LA 70803-1710 - U.S.A.
atishe1@lsu.edu

David Varillon
US191 IMAGO, Instrumentation, Moyens
Analytiques, Observations en Géophysique et
Océanographie - (IRD)
Centre IRD de Nouméa, BP A5
98848 Nouméa cedex - New Caledonia.
David.Varillon@noumea.ird.nc

Claire Villemant
Muséum national d'Histoire naturelle
Département Systématique et Evolution, CP 50
45 rue Buffon, 75005 Paris - France.
villeman@mnhn.fr

Samson Vilvil-Fare
81 A Canberra Avenue
Griffith, ACT, 2603 - Australia
samsonvfare@yahoo.com

Annie Walter
IRD, CP9214, La Paz - Bolivia.
anniwalter@hotmail.com

Anders Warén
Swedish Museum of Natural History
Department of Invertebrate Zoology
Box 50007, SE-104 05 Stockholm - Sweden.
anders.waren@nrm.se

Fred Wells
Western Australian Department of Fisheries
Level 3, 168 St. Georges Terrace
Perth, Western Australia 6000, Australia
Present address:
Enzer Marine Environmental Consulting
P.O. Box 4176, Wembley, WA 6014 - Australia
molluscau@yahoo.com.au

Denis Wirrmann
Centre IRD de Nouméa, BP A5
98848 Nouméa cedex - New Caledonia.
Present address:
Laboratoire d'Océanographie et du Climat:
Expérimentations et Approches Numériques
LOCEAN
Institut Pierre-Simon Laplace
Centre IRD France Nord, Unité 182
32 Avenue Henri-Varagnat
93143 Bondy cedex - France.
denis.wirrmann@ird.fr

Muséum national d'Histoire naturelle
Publications Scientifiques
Diffusion :
57 rue Cuvier - CP 41 - F 75231 Paris Cedex 05
Tel. : [33] 01 40 79 48 05
Fax : [33] 01 40 79 38 40
e-mail : diff.pub@mnhn.fr
http://www.mnhn.fr/pubsci

IRD
Diffusion :
32 avenue Henri-Varagnat - F 93143 Bondy Cedex
Tél. : [33] 01 48 02 56 49
Fax : [33] 01 48 02 79 09
e-mail : diffusion@ird.fr

PHOTOGRAPHS (alphabetic order)

Tonyo Alcover, Australian Bureau of Meteorology, M. Barbosa da Silva, Nicolas Barré, L. Billaut, Emmanuel Boîtier, J.M. Boré, Boulenger, Geoff Boxshall, D. Brabant, Franck Bréhier, Elizabeth A. Brown, Florence Brunois, Frédéric Busson, Bart Buyck, B. Calmont, Yolanda Camacho, P. Camus, Tim-Yam Chan, Chesher, Régis Cleva, Bruno Corbara, S. Cronin, Louis Deharveng, Xavier Desmier, J. Dumas, Jean-Michel Dupuyoo, J.L. Earle, Stefan Eberhard, Fiji Meteorological Service, Benoît Fontaine, Charles Fransen, M. Fuji, Jean-Christophe Galipaud, Olivier Gargominy, Michel de Garine-Wichatitsky, Géo-Consulte, GeoMapApp, F. Girard, GIS Unit, Adrian Gittenberger, C. Grignon, Anthony Harry, Sylvain Hugel, Ivan Ineich, Japan Meteorological Agency, Damià Jaume, Yasunori Kano, Philippe Keith, R. Kostaschuk, Jean-Noël Labat †, D. Lane, Nadir Lasson, M. Leponce, Bernard Lips, Josiane Lips, Clara Lord, Olivier Lorvelec, Porter P. Lowry, Pierre Lozouet, Marivene Manuel-Santos, Jose Christopher Mendoza, Jean-Louis Menou, M. Miller, Masako Mitsuhashi, Sara Muller, Jérôme Munzinger, Timea Neusser, Peter K. L. Ng, NIWA, Marco Oliverio, J. Orivel, M. Pallmann, Jonathan Palmer, Michel Pascal, Olivier Pascal, H. Pierce (NASA-GSFC), Marc Pignal, Yoan Pillon, Jean-Claude Plaziat, G.M. Plunkett, Marc Pouilly, Vincent Prié, Laurent Pyot, Cahyo Rahmadi, J. Reibnitz, Bertrand Richer de Forges, Claude Rives, Marine Robillard, Tony Robillard, Yves Roisin, Nick Rollings, Germinal Rouhan, Stefano Schiaparelli, Jürgen Schmidl, M. Schotte, Anne-Marie Sémah, Laurent Soldati, Arnold Staniczek, Roger Swainston, Swee Hee Tan, F. Taylor, James P. Terry, B. Thaman, Alexey Tishechkin, P. Torres, A. Touret-Alby, University of the South Pacific, UTIG, A. Valdés, Vanuatu Lands Department, Vanuatu Meteorological Services, Erick Vigneux, Claire Villemant, Anders Warén, Denis Wirrmann.

Date de distribution : le 14 avril 2011
Achevé d'imprimer en avril 2011
par Bialec à Nancy
Dépôt légal : n° 74862 - avril 2011